OLIGOMER TECHNOLOGY AND APPLICATIONS

PLASTICS ENGINEERING

Founding Editor

Donald E. Hudgin

Professor
Clemson University
Clemson, South Carolina

1. Plastics Waste: Recovery of Economic Value, *Jacob Leidner*
2. Polyester Molding Compounds, *Robert Burns*
3. Carbon Black-Polymer Composites: The Physics of Electrically Conducting Composites, *edited by Enid Keil Sichel*
4. The Strength and Stiffness of Polymers, *edited by Anagnostis E. Zachariades and Roger S. Porter*
5. Selecting Thermoplastics for Engineering Applications, *Charles P. MacDermott*
6. Engineering with Rigid PVC: Processability and Applications, *edited by I. Luis Gomez*
7. Computer-Aided Design of Polymers and Composites, *D. H. Kaelble*
8. Engineering Thermoplastics: Properties and Applications, *edited by James M. Margolis*
9. Structural Foam: A Purchasing and Design Guide, *Bruce C. Wendle*
10. Plastics in Architecture: A Guide to Acrylic and Polycarbonate, *Ralph Montella*
11. Metal-Filled Polymers: Properties and Applications, *edited by Swapan K. Bhattacharya*
12. Plastics Technology Handbook, *Manas Chanda and Salil K. Roy*
13. Reaction Injection Molding Machinery and Processes, *F. Melvin Sweeney*
14. Practical Thermoforming: Principles and Applications, *John Florian*
15. Injection and Compression Molding Fundamentals, *edited by Avraam I. Isayev*
16. Polymer Mixing and Extrusion Technology, *Nicholas P. Cheremisinoff*
17. High Modulus Polymers: Approaches to Design and Development, *edited by Anagnostis E. Zachariades and Roger S. Porter*
18. Corrosion-Resistant Plastic Composites in Chemical Plant Design, *John H. Mallinson*
19. Handbook of Elastomers: New Developments and Technology, *edited by Anil K. Bhowmick and Howard L. Stephens*
20. Rubber Compounding: Principles, Materials, and Techniques, *Fred W. Barlow*
21. Thermoplastic Polymer Additives: Theory and Practice, *edited by John T. Lutz, Jr.*
22. Emulsion Polymer Technology, *Robert D. Athey, Jr.*
23. Mixing in Polymer Processing, *edited by Chris Rauwendaal*
24. Handbook of Polymer Synthesis, Parts A and B, *edited by Hans R. Kricheldorf*
25. Computational Modeling of Polymers, *edited by Jozef Bicerano*
26. Plastics Technology Handbook: Second Edition, Revised and Expanded, *Manas Chanda and Salil K. Roy*
27. Prediction of Polymer Properties, *Jozef Bicerano*
28. Ferroelectric Polymers: Chemistry, Physics, and Applications, *edited by Hari Singh Nalwa*

29. Degradable Polymers, Recycling, and Plastics Waste Management, *edited by Ann-Christine Albertsson and Samuel J. Huang*
30. Polymer Toughening, *edited by Charles B. Arends*
31. Handbook of Applied Polymer Processing Technology, *edited by Nicholas P. Cheremisinoff and Paul N. Cheremisinoff*
32. Diffusion in Polymers, *edited by P. Neogi*
33. Polymer Devolatilization, *edited by Ramon J. Albalak*
34. Anionic Polymerization: Principles and Practical Applications, *Henry L. Hsieh and Roderic P. Quirk*
35. Cationic Polymerizations: Mechanisms, Synthesis, and Applications, *edited by Krzysztof Matyjaszewski*
36. Polyimides: Fundamentals and Applications, *edited by Malay K. Ghosh and K. L. Mittal*
37. Thermoplastic Melt Rheology and Processing, *A. V. Shenoy and D. R. Saini*
38. Prediction of Polymer Properties: Second Edition, Revised and Expanded, *Jozef Bicerano*
39. Practical Thermoforming: Principles and Applications, Second Edition, Revised and Expanded, *John Florian*
40. Macromolecular Design of Polymeric Materials, *edited by Koichi Hatada, Tatsuki Kitayama, and Otto Vogl*
41. Handbook of Thermoplastics, *edited by Olagoke Olabisi*
42. Selecting Thermoplastics for Engineering Applications: Second Edition, Revised and Expanded, *Charles P. MacDermott and Aroon V. Shenoy*
43. Metallized Plastics, *edited by K. L. Mittal*
44. Oligomer Technology and Applications, *Constantin V. Uglea*

Additional Volumes in Preparation

Electrical and Optical Polymer Systems: Fundamentals, Methods, and Applications, *edited by Donald L. Wise, Gary E. Wnek, Debra J. Trantolo, Thomas M. Cooper, and Joseph D. Gresser*

Structure and Properties of Multiphase Polymeric Materials, *edited by Takeo Araki, Qui Tran-Cong, and Mitsuhiro Shibayama*

Plastics Technology Handbook: Third Edition, Revised and Expanded, *Manas Chanda and Salil K. Roy*

Photonic Polymer Systems: Fundamentals, Methods, and Applications, *edited by Donald L. Wise, Gary E. Wnek, Debra J. Trantolo, Thomas M. Cooper, and Joseph D. Gresser*

Handbook of Radical Vinyl Polymerization, *edited by Munmaya K. Mishra and Yusef Yagci*

OLIGOMER TECHNOLOGY AND APPLICATIONS

CONSTANTIN V. UGLEA

Institut de Recherches Biologiques
Iasi, Romania

MARCEL DEKKER, INC. NEW YORK • BASEL • HONG KONG

ISBN: 0-8247-9978-X

The publisher offers discounts on this book when ordered in bulk quantities. For more information, write to Special Sales/Professional Marketing at the address below.

This book is printed on acid-free paper.

MARCEL DEKKER, INC.
270 Madison Avenue, New York, New York 10016
http://www.dekker.com

Current printing (last digit):
10 9 8 7 6 5 4 3 2 1

PRINTED IN THE UNITED STATES OF AMERICA

Preface

The industrial-scale production of polymers, together with the harnessing of atomic energy and the use of planetary ocean resources, represent ongoing challenges of technical progress. However, the further development of polymer science calls for numerous and significant changes in the approaches and methods of polymer research. This volume focuses on processing devices and techniques for obtaining industrial-scale production of some new oligomer types.

One of the primary thrusts in oligomeric research has been in developing oligomers as intermediates in preparative polymer chemistry. Research in the chemistry of chain-ended functional oligomers is important to preparative polymer chemistry because of the use of oligomers as macromonomers and telechelics, and for their other reactive properties [1,2].

Although macromonomers and telechelics are oligomers, the three terms can be distinguished from one another by the functionality of their chain ends and by the nature of the products resulting from reactions of their chain ends. An oligomer does not necessarily contain any reactive groups. However, the reactive chain ends, if present, are statistically distributed, such as in oligomers produced by condensation polymerizations. A telechelic is an oligomer containing at least one reactive group, while a macromonomer is generally considered to be an oligomer containing homopolymerizable end groups. In contrast, the end groups of a telechelic may undergo chain extension through either condensation polymerization, addition reactions, or ring-opening reactions.

IUPAC defines an oligomer as a substance composed of molecules containing one or more species of atoms or groups of atoms (constitutional units) repetitively linked to each other. This definition does not specify an absolute degree of polymerization or molecular weight to distinguish an oligomer from a polymer, but it further states that the physical properties of oligomers vary with the addition or removal of one or a few of the constitutional molecular units. This structure–property definition is perhaps the most meaningful part of the definition of oligomer. Analogously, oligomerization is defined as the conversion of a monomer or a mixture of monomers into an oligomer. Although monodisperse oligomers provide more valuable information than polydisperse oligomers, the latter are still important.

The term *oligomer* originated from the Greek words *oligos*, meaning few, and *meros*, meaning parts. The first use of *oligo-* with reference to low-molecular-weight polymers dates back to 1930, when Burckard [3] proposed that the term *oligosaccharide* be used to designate carbohydrates having molecular weights between those of monosaccharides and polysaccharides. *Oligomer* was then adopted by Kern [4] and Zahn [5]. Their extensive work in this field led to their current widely accepted meaning.

The use of oligomers as models for the understanding of polymerization mechanisms and structure–property relationships and as intermediates in preparative polymer chemistry represents the significant contribution of oligomers to the progress of polymer science [6].

The science of oligmer applications has opened new perspectives in the technology of polymer processing based on the direct transition of liquid oligomers into solid articles. This is performed by using reaction injection molding (RIM), a process that utilizes rapid chemical reactions, allowing the production of solid parts from liquid oligomers, with reaction times on the order of seconds to minutes. Formulations are adjusted to give a range of products from elastomers to more rigid structural parts.

Over the last two decades, oligomers have thus become crucial to the development of the industrial synthesis of adhesives, lacquers, dyes, and coatings as well as to the formation of elastic or rigid components and structural parts.

No doubt, the numerous and various applications of oligomers will continue to grow. That is why the development of oligomer science as a separate discipline represents an attractive and fascinating objective.

Constantin V. Uglea

REFERENCES

1. Harris, F. W., and Spinelli, H. J. (eds.), *Reactive Oligomers ACS Symp. Ser.*, 1985, p. 282.
2. Rempp, P., and Franta, E., *Adv. Polym. Sci.*, 58:1 (1984).
3. Burckard, H., Bohn, E., and Winkler, S., *Berichte*, 63B:989 (1930).
4. Kern, W., Oligomere und pleinomere von synthetischen faserbildenden, *Chem. Ztg.*, 76:667 (1952).
5. Zahn, H., and Gletsman, G. B., Oligomere und pleinomere von synthetischen faserbildenden Polymeren, *Angew. Chem.*, 75:772 (1963).
6. Percec, V., Pugh, C., Nuyken, O., and Pask, S. D., "Macromers, Oligomers and Telechelic Polymers," in *Comprehensive Polymer Science*, Eastmond, G. C., Lednith, A., Russo, S., and Sigwalt (eds.), vol. 6, Pergamon Press, Oxford, 1988, p. 281.

Contents

1

Oligoethylenes and Oligopropylenes

I. INTRODUCTION

The boom in petrochemistry following World War II and especially the building up of a large pyrolysis plant led to the increased yield of ethylene and propylene, as well as of polyethylene, polypropylene, and corresponding oligomers.

Olefins, particularly ethylene and propylene, are the basic building block of the petrochemical industry. They are easily available, cheap, reactive, and readily processed into a range of useful products. The last 20 years have witnessed increasing importance of higher linear C_6–C_{20} α-olefins, which are today a source of biodegradable detergents, new kinds of polymers, lubricants, and many other industrially useful chemicals.

II. SYNTHESIS

The synthesis of oligoethylenes (OEs) and oligoprophylenes (OPs) is based on two categories of methods: *direct oligomerization* of monomers and *separation of oligomers* formed during the process of polyolefin production. We discuss direct oligomerization first and for most of the chapter. Separation of oligomers is covered in the final section of this chapter. In some cases, degradation of polyolefins may be a valuable source of the OEs and OPs.

A. General Aspects

The olefin oligomerization reactions involve the use of transition metal complexes as catalysts. The catalytic cycle of this reaction consists of two steps. The first, the chain growth (propagation) step is

$$\text{Cat—R} + H_2C{=}CH_2 \xrightarrow{r_p} \text{Cat—}CH_2\text{—}CH_2\text{—R} \quad (1)$$

where Cat stands for catalyst; R, alkyl or hydrogen; and r_p, propagation rate.

The *second* step is the hydrogen elimination from β-carbon to the catalytic center:

$$\text{Cat—CH}_2\text{—CH}_2\text{—R} \xrightarrow{r_t} \text{Cat—H} + \text{CH}_2\text{=CH—R} \quad (2)$$

where r_t equals chain-transfer rate.

Depending on the catalyst, β-elimination of the hydride occurs when a metal–hydrogen bond is restored, or β-elimination of a proton occurs when an acidic center is formed again. The relative reaction rates r_p and r_t determine the molecular weight of the product obtained. If $r_p > r_t$, many propagation steps begin before the β-hydrogen transfer occurs, and a higher molecular weight is formed; when $r_t > r_p$, dimers are obtained; and finally, when $r_t = r_p$, oligomers are produced. Many factors determine the ratio of the propagation to the chain transfer rate, and thereby, the molecular weight of oligomerization product. These factors are the kind of metal and its oxidation state, electronic properties and steric volume of the ligands attached to metal; reaction temperature and pressure; monomer concentration; and nature of solvent and molecular weight moderator.

B. Direct Oligomerization

A variety of catalysts, both homogeneous and heterogeneous, have been utilized to convert olefins into olefinic products of higher molecular weight compounds, that is, to dimer and trimer as well as higher oligomers.

Depending on the mechanism of catalyst action, the catalysts for olefins oligomerization may be divided into the following categories: transition metal complexes in homogeneous and heterogeneous systems, organoaluminum compounds, and inorganic salts and acids.

III. OLIGOMERIZATION OF ETHYLENE

The dimerization of alkenes is an important method for the production of higher olefins as well as oligomers that find extensive application as industrial intermediates. The stimulus in this direction was provided by the pioneering studies of Ziegler in the early 1950s. He explored the use of organoaluminum compounds in the selective dimerization of alkenes [1,2].

Three types of mechanisms are reported for the dimerization of olefins: *degenerated polymerization, concerted coupling, and reductive dimerization* [1].

The three important steps involved in the *first mechanism* are (1) initiation reaction (formation) of an activated complex, (2) insertion of a monomer into the activated complex, and (3) transfer reaction (deactivation of the chain).

Coordination of the olefin at the metal hydride center and subsequent insertion of the carbon–carbon double bond of the coordinated olefin into the metal–hydride bond can be related to the initial step of a classical polymerization. The metal–carbon bond formed in this way inserts a second monomer molecule previously coordinated into the same metal center (the propagation step). The dimer is formed by a β-hydride subtraction, a common cleavage reaction of transition metal–carbon bonds. The β-hydrogen of this alkyl group attached to the metal is transferred to the latter with formation of the metal hydride, and the organic residue leaves the metal center as a vinylic olefin [3].

The ease of β-hydrogen abstraction depends of the metal, its valency state, and the ligand environment. The metals of the extreme and of the transition series are prone to β-hydrogen abstraction easily from an attached alkyl group. The complexes based on these metals are good catalysts for the dimerization of olefins.

Selectivity of dimerization is related to the ratio of the rate of β-elimination to the rate of insertion or, in other words, to the ratio of the rate of chain transfer to the rate of propagation. The temperature apparently does not influence this ratio. The steric course of insertion is sometimes ambiguous. The factors that are vital in deciding the products are not understood. Metal–hydrogen and metal–carbon bonds react with olefins either like hydride ions and carbanions or like protons and carbonium ions:

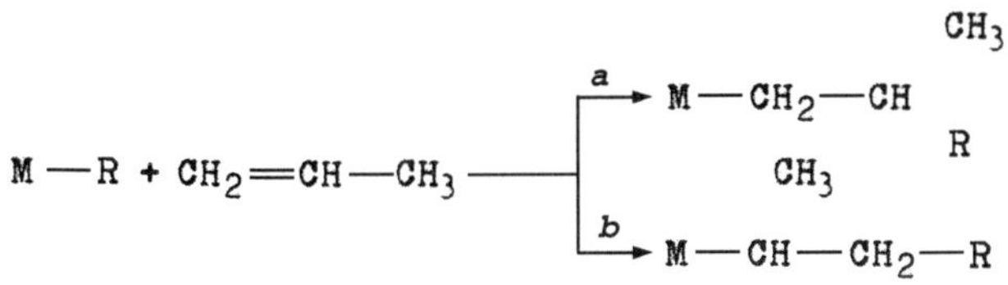

(3)

where M = metal and R = H or alkyl. The factors influencing the ratio *a*/*b* are yet to be ascertained. The ratio may depend on the nature of olefins, the complex involved, and the temperature.

The influence of the ligand on the strength of metal–carbon bond and the course of the reaction is well understood [3]. It is known that the growth step starting from a configuration in which a monomer is coordinated to a transition metal proceeds via a polar, four-center transition state. There is kinetic evidence that both chain growth and β-hydrogen transfer reaction start from the same configuration [3].

A polar six-centered transition state including a monomer, α-carbon, β-carbon, and one β-H of the growing chain attached to the same metal center leading to β-hydrogen abstraction is suggested. This is supported by the ligand effect. Electron-withdrawing ligands withdraw electron density from the metal, thus increasing the positive charge. This polarizes the adjacent bonds including the β-hydrogen, enabling the inclusion of the β-hydrogen into the polar, six-centered transition state:

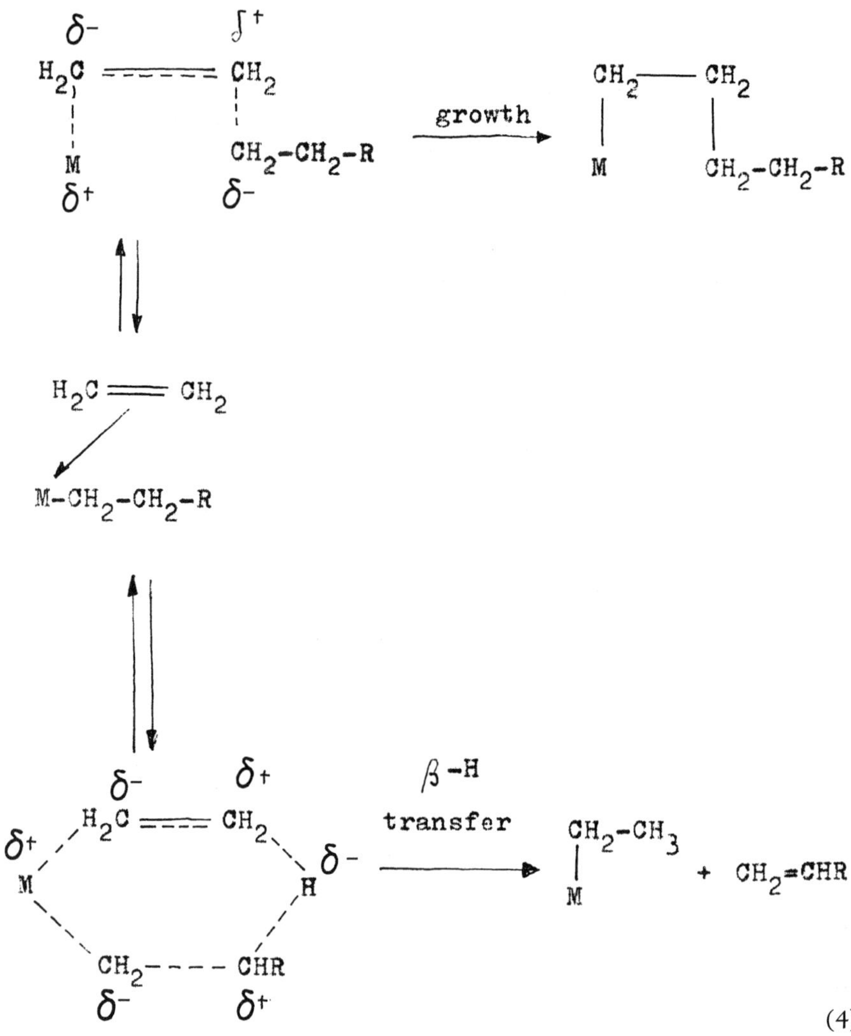

(4)

Donor ligands, on the other hand, reduce the positive charge on the metal. Polarization of the adjacent bonds is less intensive and hence reaches only the α-carbon, thus favoring the four-centered transition state. Hence the chain growth is favored. Therefore, the ligands can control the nature of reactions. The catalytic cycle for the dimerization of ethylene based on this mechanism is presented in the following [4]:

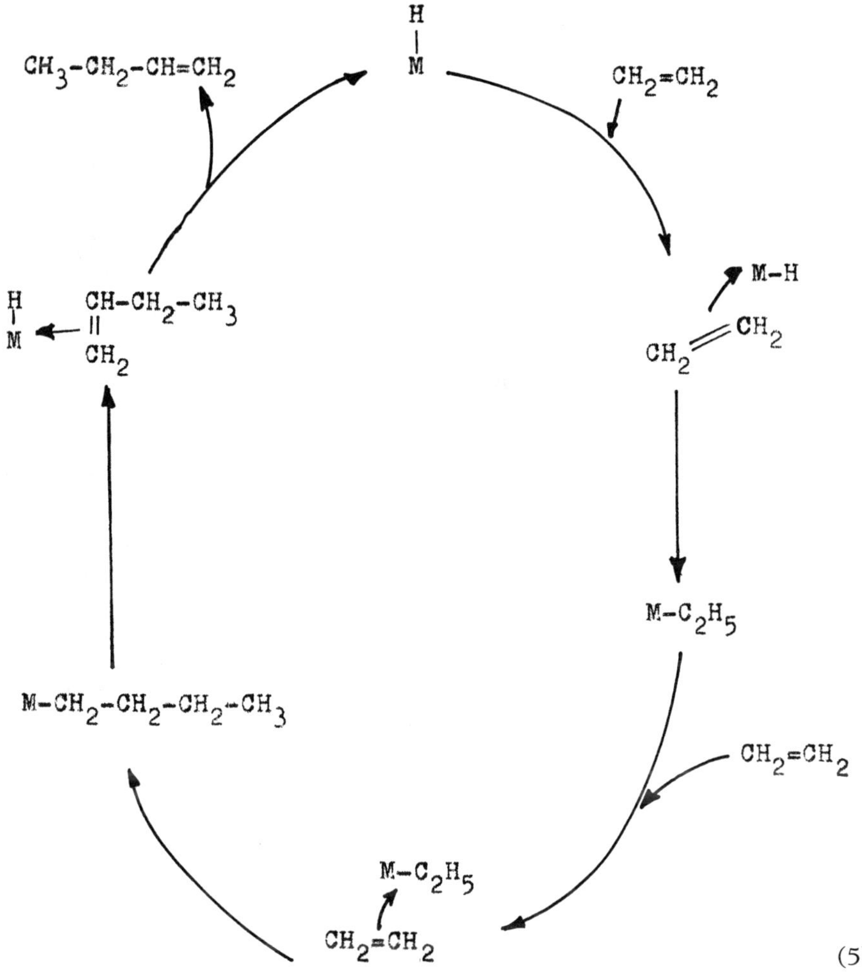

(5)

The *second mechanism* proceeds by the stepwise addition of monomers to the metal followed by the formation of carbon–carbon bonds in a multicentered bond process:

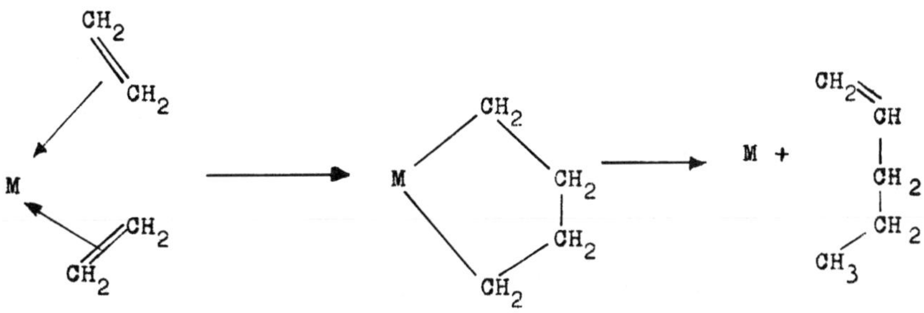

(6)

In the case of monoolefins, it has been formally represented as an activation of the hydrogen–carbon bond followed by coupling:

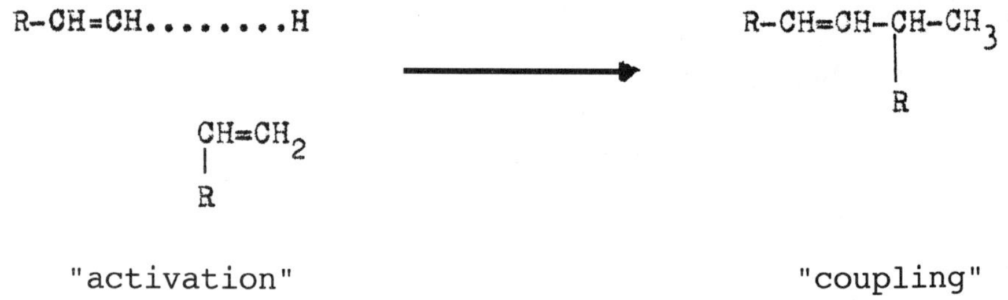

(7)

In a subsequent proposal, the hydrogen abstraction is represented as an oxidative addition on the metal followed by a reductive elimination and then coupling of the attached elements on a coordinated monomer [5]. The catalysts that follow this route have generally no isomerizing activity and exhibit high selectivity to the dimer product.

In the *reductive dimerization*, hydrogen pressure or presence of hydrogen donors or reducing agents can facilitate the dimerization. The products formed are dihydro dimers.

The following hypotheses have been suggested for the mechanisms of these reactions: (1) hydrogen may act as a chain-transfer agent by hydrogenolysis of the metal–carbon bond when β-elimination is not favored:

$$M{-}CH_2{-}CH(R){-}R' + H_2 \rightarrow M{-}H + CH_3{-}CH(R){-}R' \qquad (8)$$

$$2\,M\begin{matrix} / R \\ \backslash R' \end{matrix} \rightarrow M{-}M + R{-}R'$$

and (2) hydrogen may generate or regenerate a hydride after coupling of two alkyl groups:

$$M{-}M + H_2 \rightarrow 2\,M{-}H \qquad (9)$$

The insertion steps of olefins into metal–hydrogen or metal–carbon bonds are of same type as depicted in the preceding mechanisms.

Computer-derived pathways for the formation of butenes from the dimerization of ethylene using various organometallic systems are shown in the reaction series (10) and (11). In these reactions compounds 2 and 3 show the formation of free butenes, whereas compound 4 indicates the computer did not find the formation of free olefins [6a].

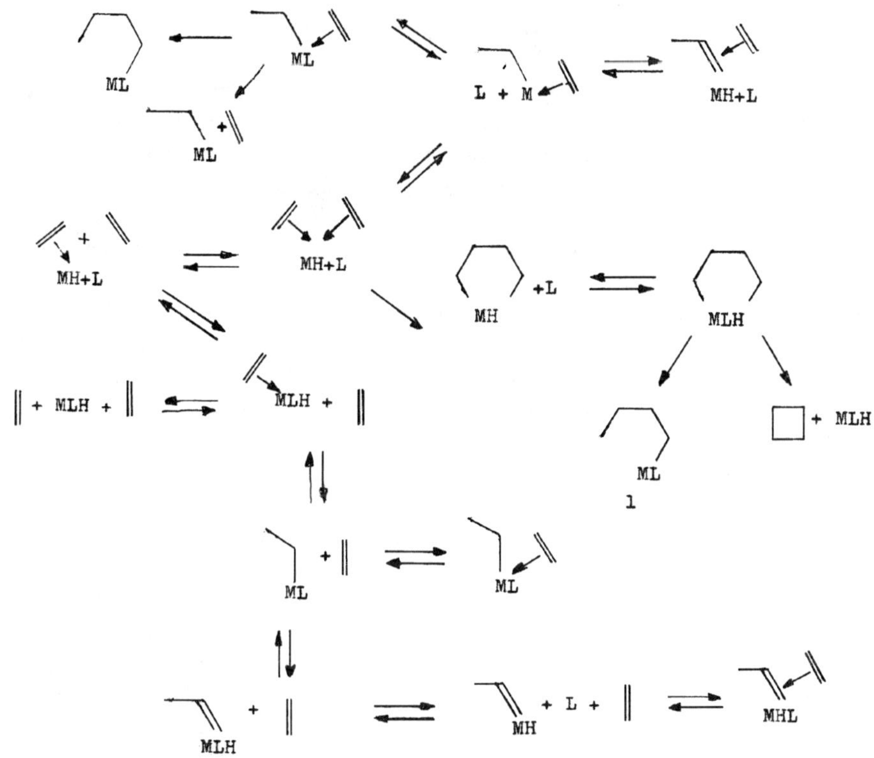

(10)

Computer-derived pathway starting with MLH and two ethylene molecules; M = metal, L = ligand.

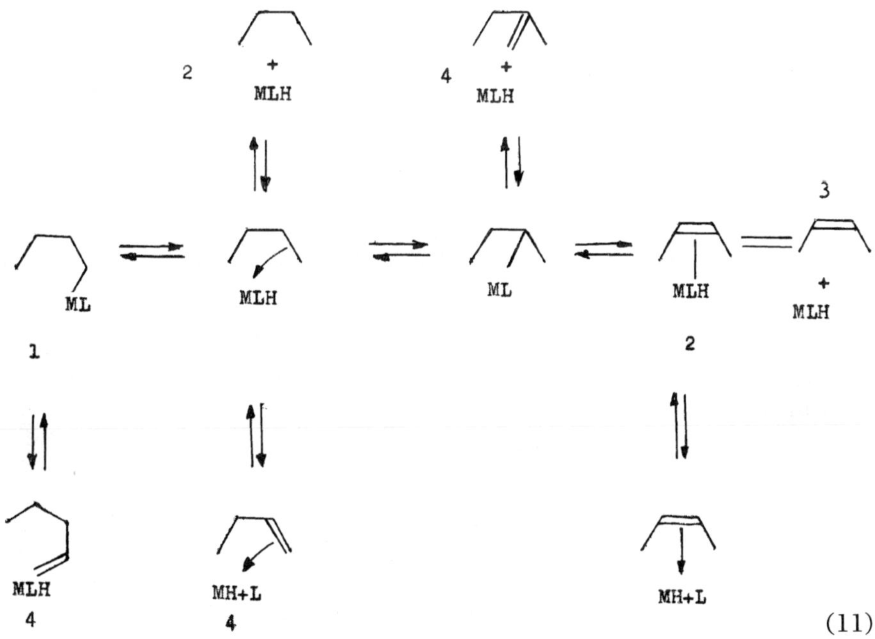

(11)

Computer-derived pathway starting from metal alkyl complex 1; M = metal, L = ligand.

A large number of oligomerization catalysts for alkenes differing in the composition of products and in reaction conditions have been described [1]. Many papers [6–14] on the mechanism of this reaction under the influence of the homogeneous and gel-immobilized catalysts have appeared.

Among the group 1A elements, the alkali-metal compounds like Na_2O, K_2CO_3, and Na_2CO_3 are very versatile dimerization catalysts. Many industrial processes are based on these catalysts.

There are two reports [15,16] on the oligomerization (up to 100%) of ethylene in the presence of catalysts based on elements of group 1B. These catalysts consist of Et_2AlCl_2 or $EtAlCl_2$ and $CuAlCl_4$, Cu_2Cl_2, or $AgAlCl_4$. These catalysts have higher activities at moderate temperatures and pressures compared with those of Friedel–Crafts catalysts.

The only known catalyst for the oligomerization of ethylene from group IIB elements is $ZnAl_2Cl_8$ [15,16]. The cocatalysts employed are Et_2AlCl and $EtAlCl_2$.

Aluminum alkyls and related aluminum compounds are well-known catalysts for the oligomerization of ethylene. The catalytic potential is a consequence of the readiness with which a trivalent aluminum atom forms electron-deficient or so-called "half," bonds. The kinetics of the homogeneous oligomerization of ethylene into 1-butene catalyzed by triethyl aluminum in the gas phase has been studied in Teflon-coated reactors at temperatures ranging between 160 and 230°C [17]. The catalyst is recovered quantitatively from the product mixture. The homogeneous reaction mechanism is summarized in the following reactions:

Propagation

$$\begin{array}{l} \quad /\,(CH_2CH_2)_nH \\ Al - (CH_2CH_2)_mH \\ \quad \backslash\,(CH_2CH_2)_pH \end{array} + CH_2{=}CH_2 \longrightarrow \underset{a}{\begin{array}{l} \quad /\,(CH_2CH_2)_nH \\ Al - CH_2 - CH_2(CH_2CH_2)_mH \\ \quad \backslash\,(CH_2CH_2)_pH \end{array}} \qquad (12)$$

Termination

$$a \longrightarrow \begin{array}{l} \quad /\,(CH_2CH_2)_nH \\ Al - H \\ \quad \backslash\,(CH_2CH_2)_pH \end{array} + CH_2{=}CH(CH_2CH_2)_mH$$

Thermal decomposition of the aluminumalkyl bond [18] yields the Al–H bond and α-olefin. At the end of the process, the hydroaluminum compound reacts very quickly with ethylene, as follows:

$$\begin{array}{l} \quad /\,(CH_2CH_2)_nH \\ Al - H \\ \quad \backslash\,(CH_2CH_2)_pH \end{array} + CH_2{=}CH_2 \longrightarrow \begin{array}{l} \quad /\,(CH_2CH_2)_nH \\ Al - CH_2 - CH_3 \\ \quad \backslash\,(CH_2CH_2)_pH \end{array} \qquad (13)$$

The Al—CH_2CH_3 bond can initiate oligomer chain growth by inserting the next ethylene molecule and thus beginning a cycle of ethylene oligomer production. The chain growth occurs through a four-center intermediate [19]:

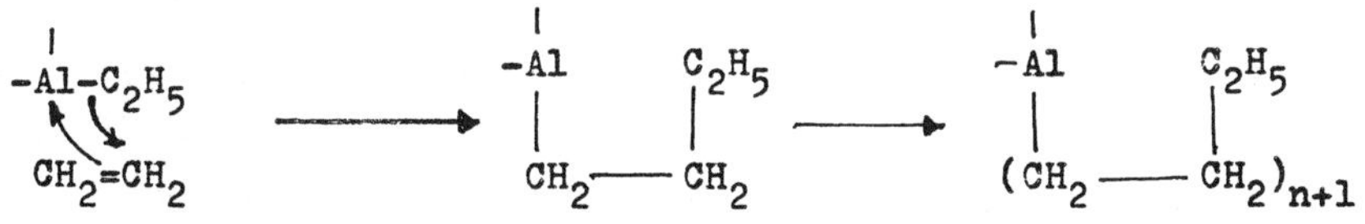

(14)

The detailed mechanism involves the establishment of a preequilibrium between compound *a* (reaction 12) and ethylene to form a complex prior to the formation of the four-center cyclic transition state.

The chain termination reaction involves a six-center transition state:

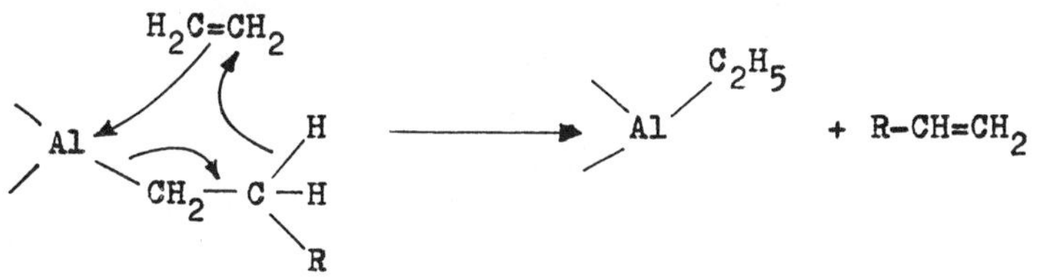

(15)

Thermal decomposition (reaction 12) of the Al–R bond is considerably slower than the ethylene insertion into the Al–H bond (reaction 13). Thus, at lower temperatures, the oligomer growth reaction gives mainly high molecular weight products. At higher temperatures (190°C), thermal decomposition of the Al–R bond begins to dominate and the C_{12} olefin becomes the main product [20–22].

An increase in the reaction pressure increases the ethylene conversion and the amount of linear long-chain oligomers. To avoid polyethylene formation, sulfur or nitrogen compounds are added to the reaction mixture.

Triethylaluminum was heterogenized by bonding to polymers possessing hydroxy groups, such as polyvinyl alcohol [21] or phenol resins [21,22]. The ethylene oligomers obtained in their presence are C_8 to C_{18} olefins with 75% α isomer content.

It is claimed that very high yields of pure 1-butene are obtained at low conversions of ethylene [23–26]. Thus it is possible to dimerize ethylene to 1-butene to a 10% extent and use the resulting mixture for the production of ethylene copolymers [25]. At higher conversions of ethylene (>20%), appreciable amounts of C_6 and C_8 alkenes are produced along with 1-butene [27].

$PbAl_2Cl_8$ with Et_2AlCl or $EtAlCl_2$ is reported as a catalyst for the dimerization of ethylene [15,16].

Titanium compounds, especially halides, in conjunction with aluminum compounds known to be good catalysts for polymerization of ethylene. These give dimers and oligomers by minor changes in the constituents of the catalytic mixture or experimental conditions [28].

The oligomerization reaction of ethylene in the presence of organotitanium compounds takes places via the following steps: active center formation, olefin coordination on the titanium atom and chain propagation, and termination of the chain growth.

A. Formation of a Titanium Catalytic Center

An unoccupied coordination site and titanium alkyl bond in the titanium complex are fundamental requirements for its catalytic activity in olefin oligomerization. The titanium–alkyl bond is formed in the reaction of a titanium compound with alkyl or alkylchloroaluminum compounds when, for example, halide atoms or alkoxy groups of titanium compounds are replaced by alkyl groups of aluminum derivatives. There are many proposals concerning the structure of the Ziegler–Natta catalyst active centers. These are presented in structural formulas (16)–(21) [29].

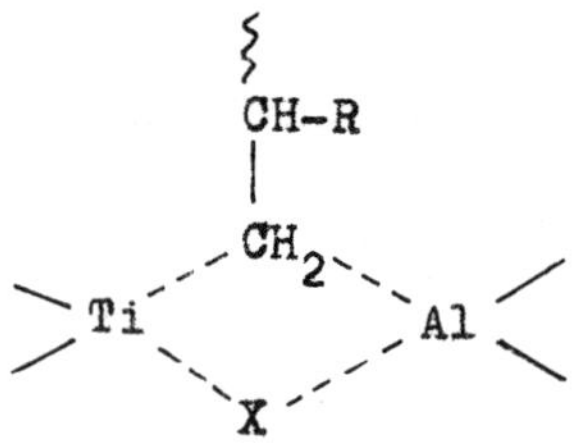

16

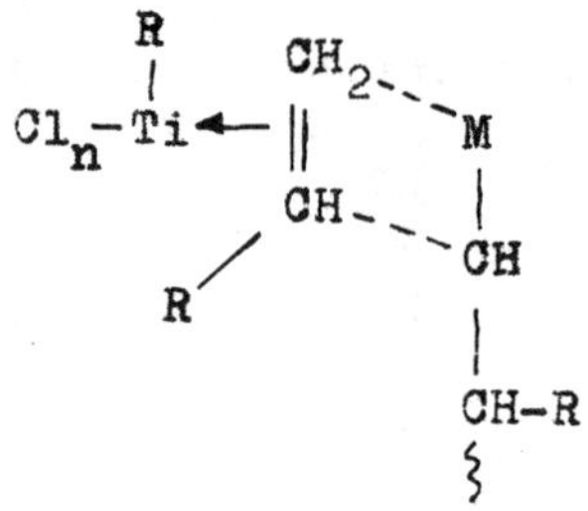

17

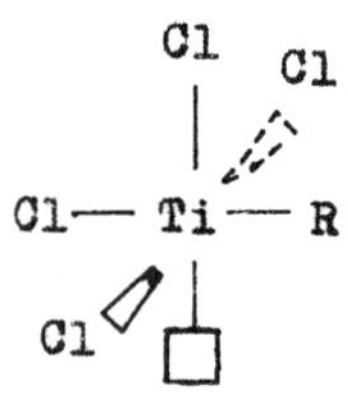

18

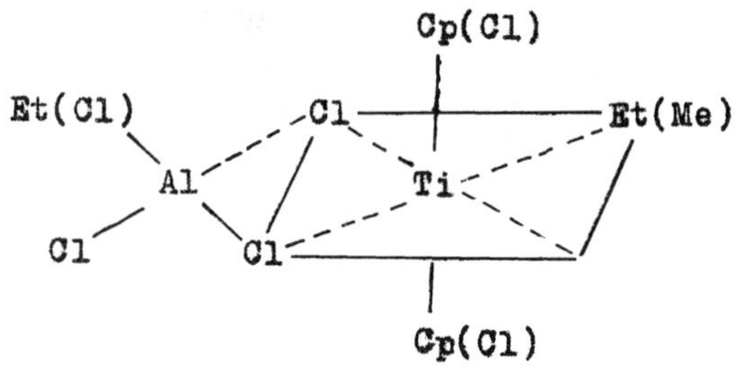

19

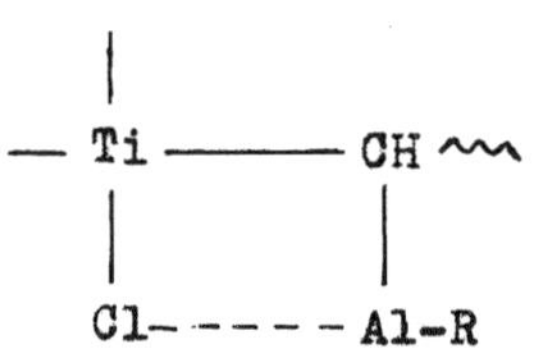

20

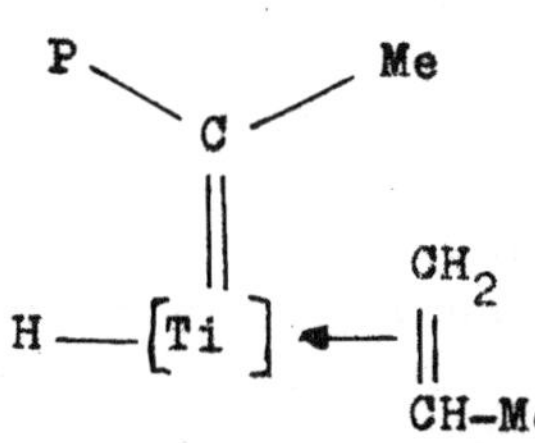

21

(16–21)

They consist of bimetallic systems in which titanium is bonded to aluminum through halide or halide–alkyl bridges, or of monometallic systems where only a titanium ion is the catalytic center [30–36].

B. Oligomer Chain Growth

Oligomer chain growth occurs as the result of olefin coordination on a titanium atom and consecutive olefin insertion into a Ti–C bond. Olefin insertion into a titanium alkyl bond is the fundamental step in the olefin oligomerization reaction in the Ziegler–Natta catalyst. Extensive studies of the olefin insertion mechanism support the bimetallic or monometallic models of catalytic centers.

The olefin oligomerization mechanism on a bimetallic center proposed by Rodriguez, van Looy, and Gabant [37–41] for ethylene oligomerization on the $(MeO)_4Ti/Et_3Al$ system is as follows:

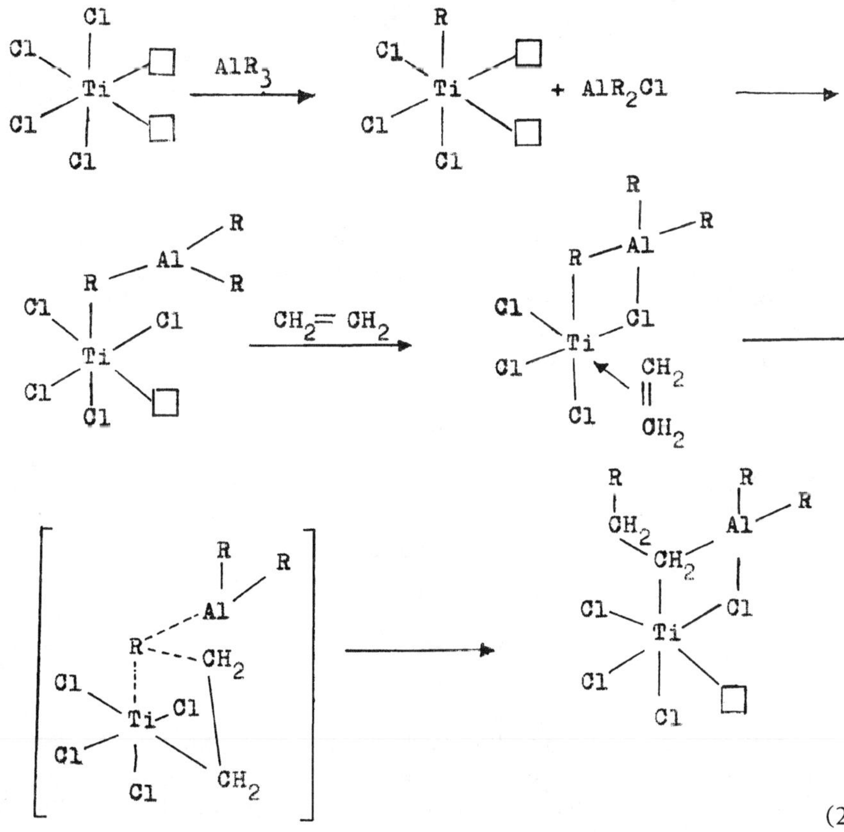

(22)

In this mechanism the first step of the reaction is the alkylation of the titanium center.

The organoaluminum compound plays a dual role. It exchanges the ligands with the titanium ion, thereby forming a labile titanium–alkyl bond and forms, via the Al–O–Ti bridges, the adequately geometrical structure and reduces the titanium (IV) ion to the titanium (III) one. Ethylene complexation causes the change in the complex structure to an octahedron in which ethylene is coordinated and then inserted into the Ti–C bond.

The metathesis type mechanism of olefin oligomerization proposed by Green [36] is given in the following reactions:

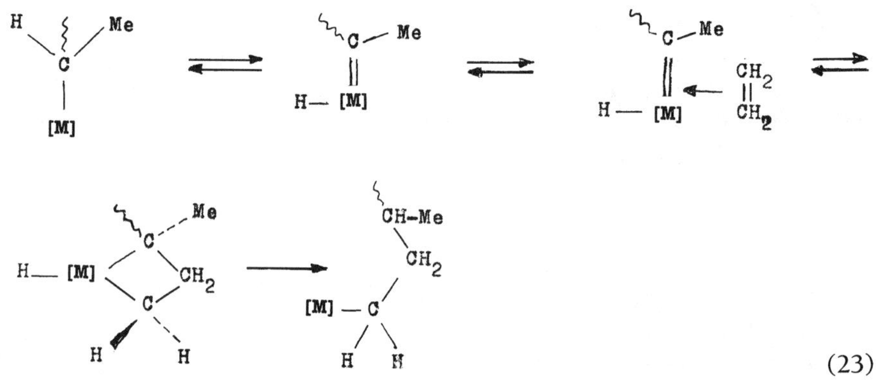

(23)

A metal carbene, Ti═CHR, is the catalyst centered here. This is formed as a result of α-carbon shift from the growing oligomer chain to the titanium ion.

Brookhart and Green [42,43] have proposed a mechanism that is a minor variation of the earlier example. In this mechanism the cleavage of the C–H bond does not occur exactly as in the metathesis type mechanism. The "agostic" hydrogen structure is postulated as

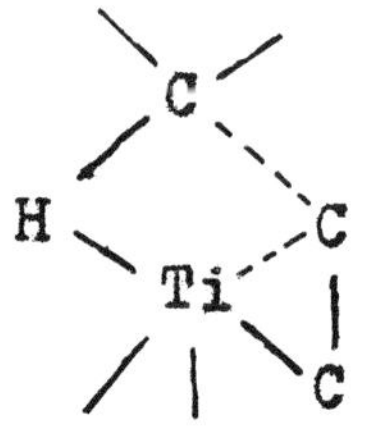

(24)

Both the mono- and bimetallic mechanism are supported by theoretical calculations [44]. These calculations show that the bimetallic mechanism is energetically favorably, but the Jolly and Marynick [45] calculations for $CpTiCH_3$ (Cp = cyclopentadienyl) claim that olefin insertion into the titanium–alkyl bond should occur in the absence of organoaluminum compounds.

For the $TiCl_4$/alkylchloroaluminum system, Langer [46] has proposed an ionic structure of three equivalent forms:

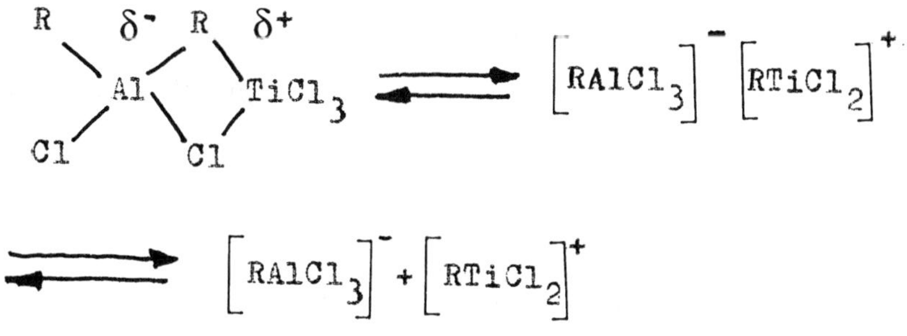

(25)

C. Chain Transfer Reaction

As mentioned above, the relative rates of chain growth and chain transfer determine the molecular weights of the oligomers. This process on the titanium complex catalysts occurs as a result of the β-hydrogen elimination from the oligomer chain to the coordinated olefin [47], after which the α-olefin oligomer is liberated and the catalytic center, namely, the titanium alkyl bond, is restored.

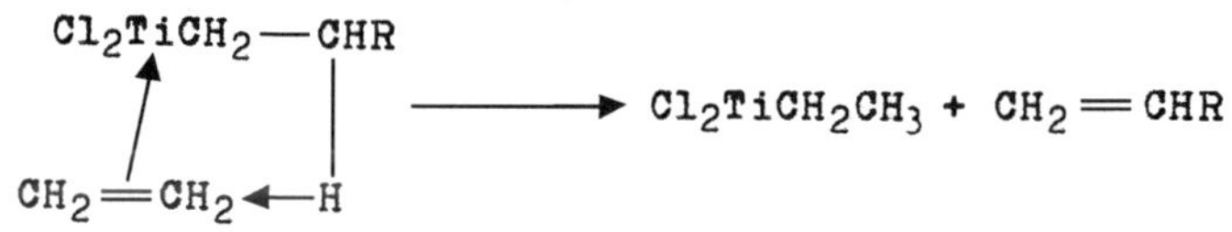

(26)

Most data concerns ethylene oligomerization on the $TiCl_2/Et_nAl_{3-n}$ systems. The influence of reaction conditions on selectivity, Schultz–Flory molecular weight oligomer distribution, and yields has been investigated. These factors are catalyst composition—Ti/Al ratio, catalyst concentration, reaction time, and kind ot the titanium and aluminum ligand.

D. Al/Ti Ratio

When the ratio Al/Ti $\leq$ 1, the ethylene oligomerization does not occur. At Al/Ti > 2, the catalyst efficiency improves [48].

E. Catalyst Concentration [49,50]

An increase in catalyst concentration causes an increase in the oligomer chain-transfer rate—more short-chain olefins are obtained. Increasing the catalyst concentration gives more polymer and decreases the reaction selectivity. This polymer formation is a result of co-oligomerization of ethylene with oligomers.

F. Pressure [49–51]

With increasing reaction pressure, the molecular weight of oligomers as well as conversion to linear α-olefins increases. An increase in the ethylene pressure also inhibits the ethylene co-oligomerization with higher α-olefins.

G. Temperature [49–51]

Increasing the ethylene oligomerization temperature increases the molecular weight of the oligomer and the amount of nonlinear α-olefins in the products.

The rise of the molecular weight of the oligomers obtained is in this case a result of higher activation energy of the chain growth compared with that of the chain transfer to the monomer (β-hydrogen transfer). Under these conditions, ethylene co-oligomerizes with the preformed α-olefins, resulting in a significant quantity of branched hydrocarbons. Also, the reduction of Ti(IV) ions to Ti(III) with organoaluminum compounds is accelerated. This in turn gives more catalytically active polymerization centers (Ti(III) ions) and, therefore, more polyethylene in the product. At higher temperatures, the ethylene alkylation of solvents is also accelerated.

H. Solvent [52–55]

Increasing solvent polarity activates titanium catalyst in the ethylene oligomerization and decreases the molecular weight of the oligomers obtained. More branched olefins exist in the product, because in more polar solvents the cationic character of the catalyst increases.

I. Reaction Time [48]

Prolonged reaction time leads to an increase in oligomer yield and an increase in the amount of branched olefins.

J. Ligands of Titanium and Aluminum

An increase of acidity of organoaluminum compounds causes a decrease of the oligomer molecular weight. When titanium halide (acceptor) ligands in the system $TiX_4/EtAlCl_2$ are changed for alkoxy (donor), more polyethylene is obtained in the product. Halide ligands, having acceptor properties, polarizes Ti–X bonds by giving a positive charge to the titanium ion. This helps six-center intermediate formation and termination of the oligomer chain, which is a result of strong polarization of neighboring bonds (including the polarization of C_β—H). Donor ligands, on the other hand, which decrease positive charge on a titanium ion, favor the four-center intermediate formation and the olefin insertion into the Ti–C bond according to the equation [56]

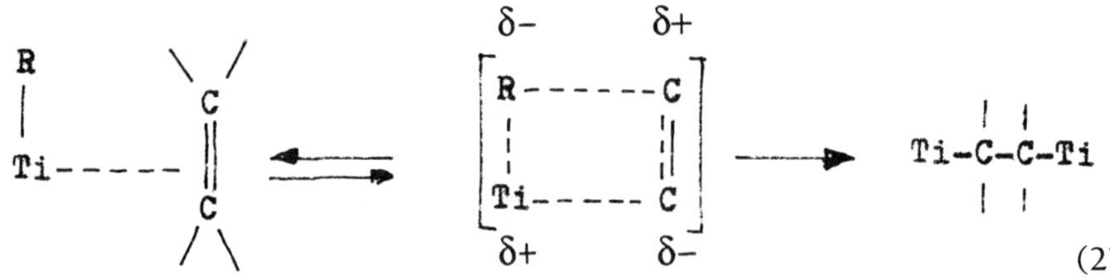

(27)

Donor ligands added to the system $TiCl_4/Et_nAlCl_{3-n}$ (n = 1, 2) increase the catalyst selectivity to linear α-olefins. These catalysts, modified by an addition of ketones, amines, nitriles, phosphines, and sulfur compounds, yield 70–80% C_8–C_{20} olefins by ethylene oligomerization.

Zhukov and his team [57] have reported optimal conditions for the dimerization of ethylene using titanium alcoxides: ethylene dissolved under 2.75 atm pressure in heptane containing 5 g/L of $Ti(O-n\text{-}Bu)_4$ and 20 g/L of $AlEt_3$ and kept for 6–8 h at 60°C. The resulting solution yielded 99.5% pure 1-butene. A copolymer of 1-butene with ethylene was formed during this process. Another report [58] claims that the dimerization of ethylene is favored with a ratio of AlR_3/$Ti(OR)_4$ < 10. The dimerization selectivity is high (90%); a major part of the dimers formed is 1-butene with a small amount of 2-butene. This means that the catalyst system is not a very isomerizing one, and it behaves like triethylaluminum alone, except that the experimental conditions are much less severe. The physicochemical aspects of the catalyst systems have been investigated in detail [59–64].

Belov and his group [65] have observed the formation of polymers at high pressure of ethylene with titanium alkoxide-trialkyl-aluminum systems. Operation at low temperatures (ranging from 10 to 40°C) increases the activity. However, this involves difficulties in heat removal. Use of low-boiling solvents like ethyl chloride and diethyl ether are preferable, because they facilitate heat removal, isolation of 1-butene, and rectification of the solvent. Modifiers like organic esters or H_3PO_4, diphenylamine, and phenothiazine have been used. The inclusion of these compounds in amounts of 0.1–1.0 mol/mol of alkylaluminum leads to reduction in polymer formation associated with loss of activity of the catalyst. Modification of the catalyst $Ti(OR)_4$–AlR_3 by addition of oxygen allows a 10–20% increase in the yield of 1-butene with simultaneous enhancement of the process selectivity. The quantity of oxygen in the reaction vessel is varied within 0.2:1 to 2:1, depending the quality of alkylaluminum present therein.

The dimerization carried out with a mixture of ethylene–hydrogen with a concentration of 5–95% of volume results in enhanced 1-butene formation [66].

It has been reported that with use of $AlEt_3$–$Ti(O-n\text{-}Bu)_4$, the conversion to the dimer is maximal for a Al/Ti ratio of 4:1. The activity can be increased by the addition of triphenyl phosphine or phenylacetilene [66].

Angelescu and coworkers propose the following mechanism for the dimerization of ethylene on the bimetallic titanium–aluminum complex [67,68]:

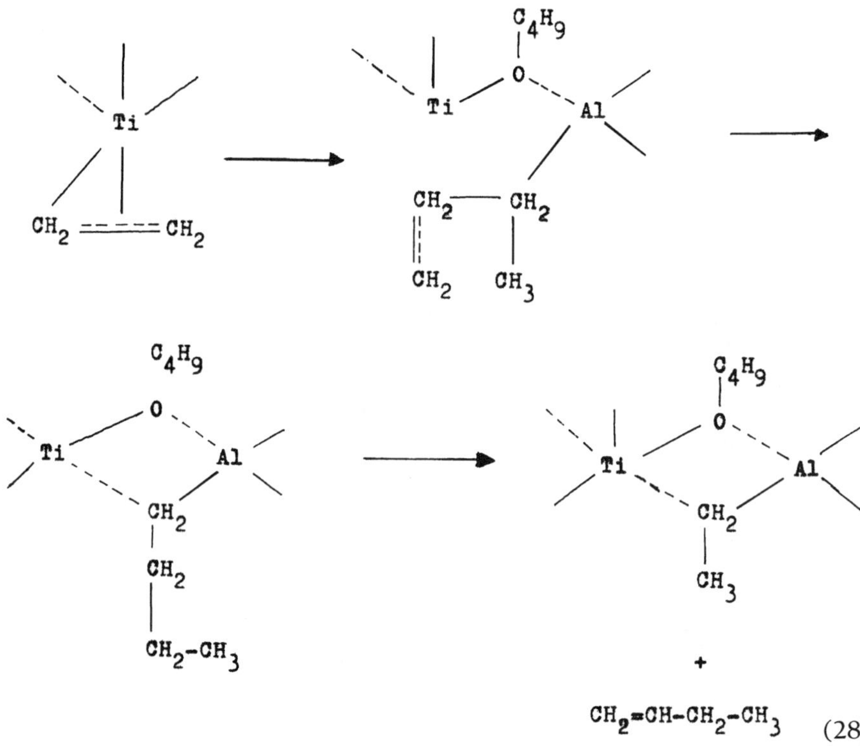

(28)

Typical reactions conditions for the dimerization of ethylene using titanium aryl oxide–Et_3Al systems are summarized in Table 1 [69].

Table 1 Typical Reaction Conditions for the Dimerization of Ethylene Using Titanium Aryl Oxide and Triethylaluminum

	Catalyst		
Parameter	$Et_3Al–Ti(OPh)_4$	$Et_3Al–Ti(OPhMe)_4$	$Et_3Al–Ti(OPhAm)_4$
Al/Ti	3.0	3.0	5.0
Solvent	*n*-Heptane	*n*-Heptane	*n*-Heptane
Pressure (Kg/cm^2)	15	15	10
Ti Salt (g)	0.2280	0.2770	0.429
Temperature (°C)	40	40	75
Time (min)	40	40	60
Product (g)	21.2	22.6	35.7
Butenes (wt%)[a]	91.5	91.3	80.2
Hexenes (wt%)	5.6	5.7	17.1
Polymer (wt%)	2.9	3.0	2.7
Catalyst activity (g of butenes/g of Ti)	85.1	74.4	67.1

[a]Selectivity to 1-butene is approximately 99%.
Source: Ref. 69.

Cossee [70] originally proposed a mechanism for the dimerization of ethylene by the $Ti(OMe)_4$–$AlEt_3$ system. The generation of an active catalyst is given in the following reactions:

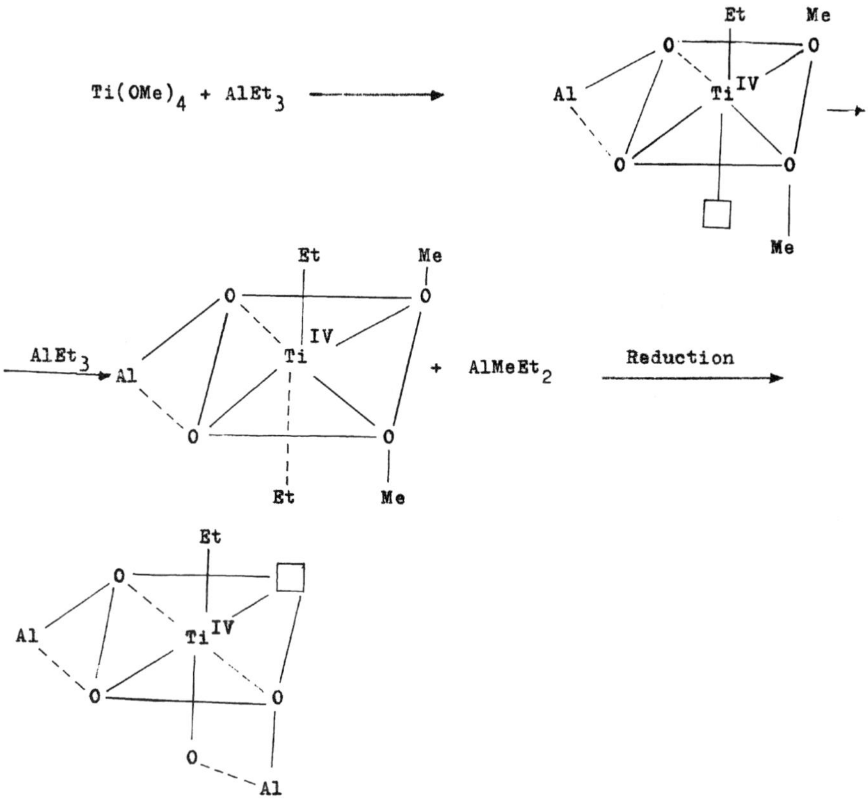

A (29)

Even though the original catalyst A is depicted as an octahedral complex, the self-consistent all-valence-electron molecular orbital calculation shows that the most stable complex is a trigonal–bipyramidal one with titanium ethyl at an intermediate position between two octahedral sites [71–73].

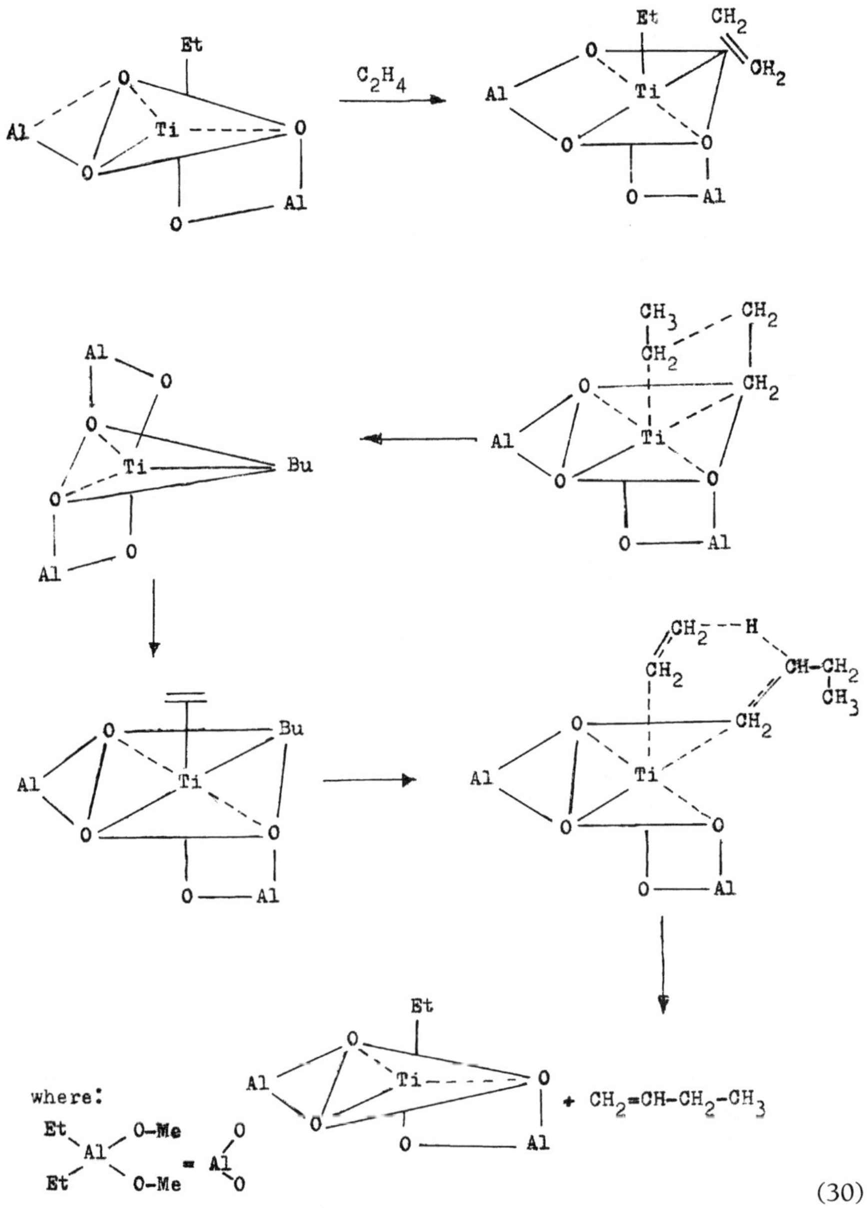

(30)

The same theoretical study shows that both chain propagation and β-hydrogen transfer responsible for the alkene liberation are favored decisively by the titanium *d* orbitals.

Belov [74,75] and his group have proposed formation of intermediate complexes containing $\geqslant Ti-CH_2-CH_2-Ti \leqslant$ groups for the dimerization of ethylene by $Ti(O{-}n{-}Bu)_4AlEt_3$:

$$Ti(OR)_4 + Al(C_2H_5)_3 \rightarrow C_2H_5Ti(OR)_3 + Al(C_2H_5)_2OR \quad (31)$$

$$C_2H_5Ti(OR)_3 + Al(C_2H_5)_3 \rightarrow (C_2H_5)_2Ti(OR)_2 + Al(C_2H_5)_2\,OR \quad (32)$$

$$(C_2H_5)_2Ti(OR)_2 \rightarrow C_2H_4Ti(OR)_3 + C_2H_6 \tag{33}$$

$$Ti(OR)_4 + C_2H_4Ti(OR)_2 \rightarrow \geqslant Ti—CH_2—CH_2—Ti \leqslant \tag{34}$$

$$C_2H_5Ti(OR)_3 + C_2H_4Ti(OR)_2 \rightarrow 2\ Ti(III)[Ti(OR)_3 + C_4H_9Ti(OR)_2] \tag{35}$$

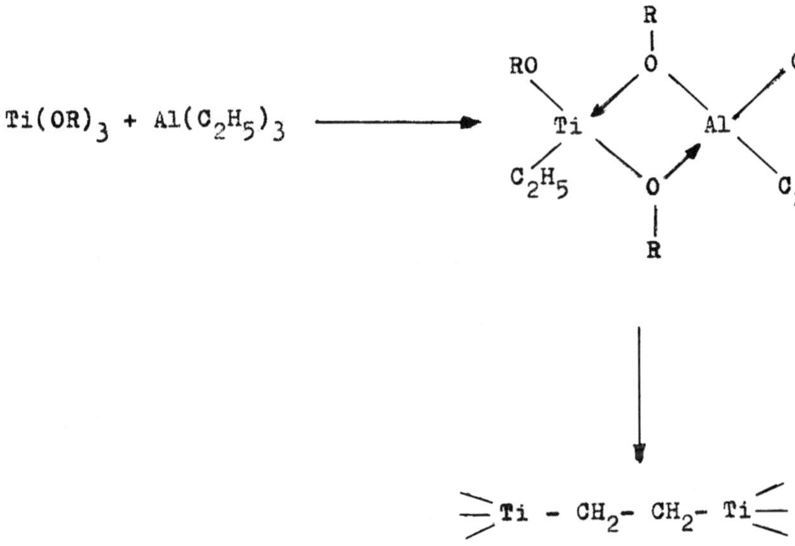

(36)

These behave as binuclear active centers:

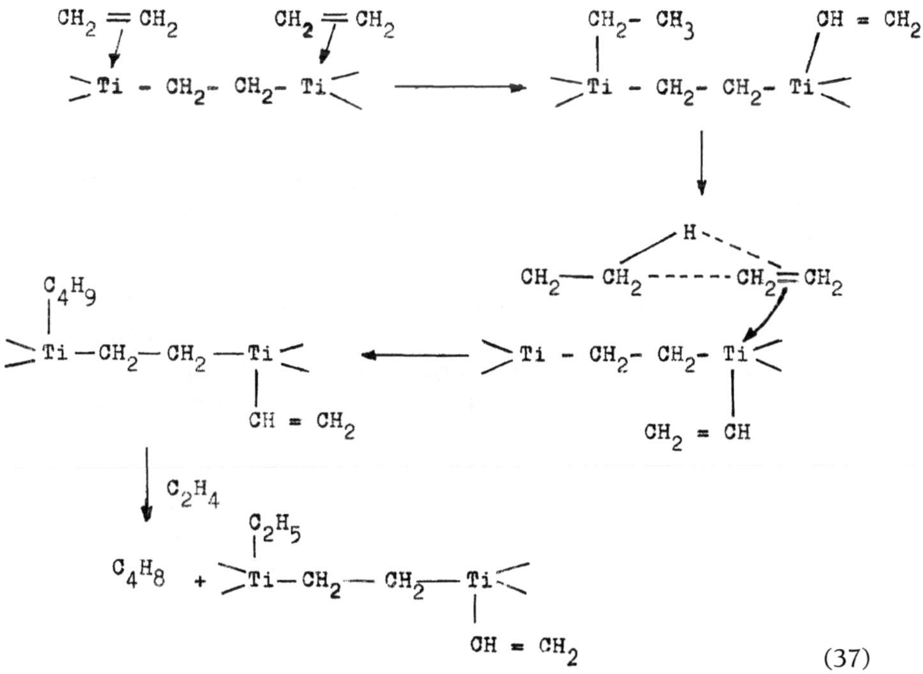

(37)

The same authors have reported that bititanium ethylene-bridged complexes in the presence of $AlMe_3$ are produced as a result of the recombination of carbenium intermediates:

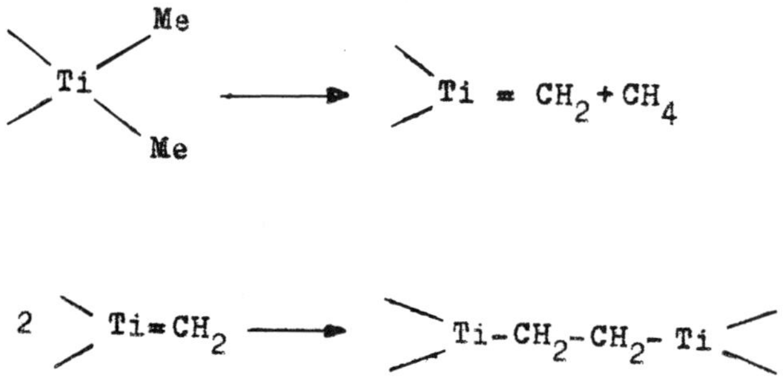

(38)

Conversion of carbenium species into metallacyclobutane and similar compounds (structure *A*, reactions 39) are more facile [77,78] than the recombination of carbene species as shown in reactions (38):

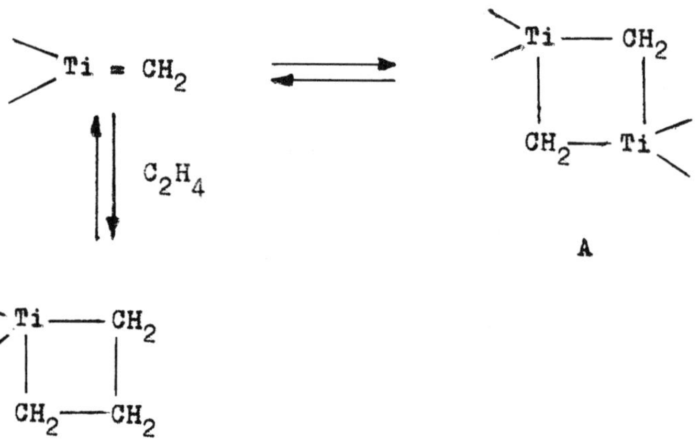

(39)

Thus, the proposal is an unusual one and requires further investigations.

Detailed reports [79–89] on the kinetics and mechanism of ethylene dimerization by the $Ti(OR)_4$–AlR_3 system in various solvents are available. Different values of the activation energy are reported [81,83,87].

The two-component organometallic catalyst $MeTiCl_3$–$MeAlCl_2$ [90,91], which is distinct from other Ti–Al systems, shows the following order of activity for the dimerization of ethylene in organic solvents: chlorinated hydrocarbons > aromatic hydrocarbons > aliphatic hydrocarbons. It exists as the following structural formulas:

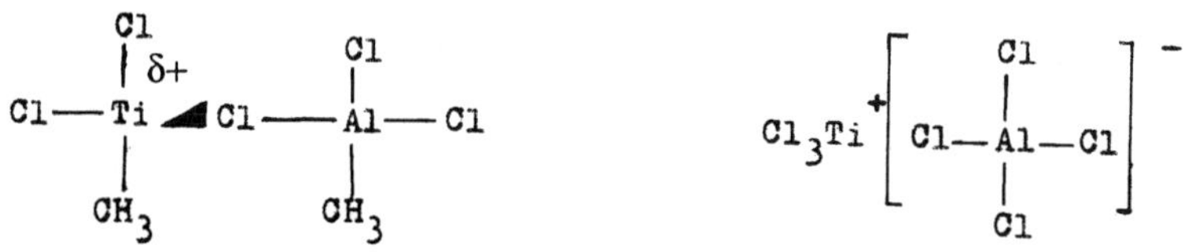

40 41 (40,41)

The selectivity of the catalyst is poor in these solvents. The catonic part of the complex coordinates with the olefin, and the mechanism of the process is given in the following reactions:

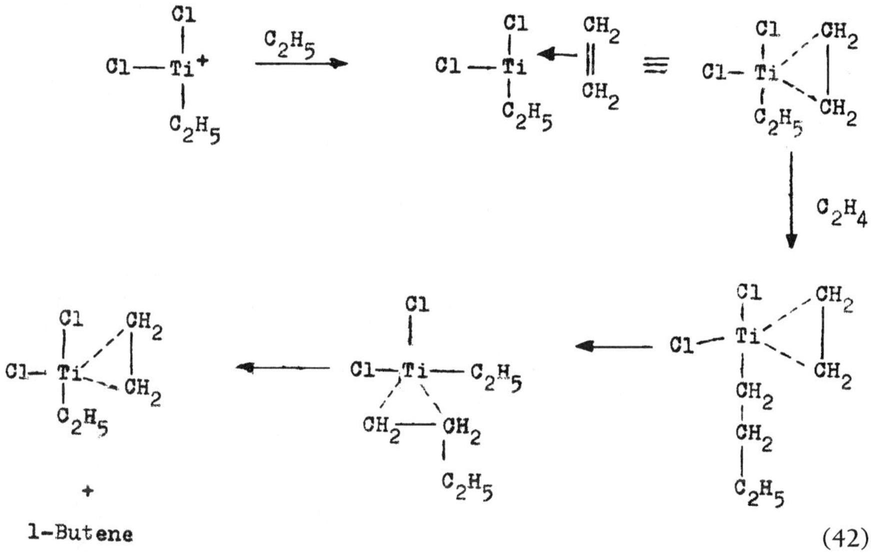

(42)

Apart from the alkylaluminum compounds employed in previous investigations, *N,N*-dialkylaminoalcanes can also be used as a cocatalyst [92]. With this system 95% selectivity to 1-butene is achieved.

Cyclopentadienyltitanium trichloride associated with amalgams or alkali metals dimerizes ethylene to 1-butene [93].

Wreford and his coworkers [94] have reported that $(n\text{-}C_4H_6)_2Ti$–dmpe (dmpe = 1,2-bis(dimethylphosphino)ethane) catalyzes the dimerization of ethylene. A mechanism involving formation of a metallacyclopentane complex is proposed:

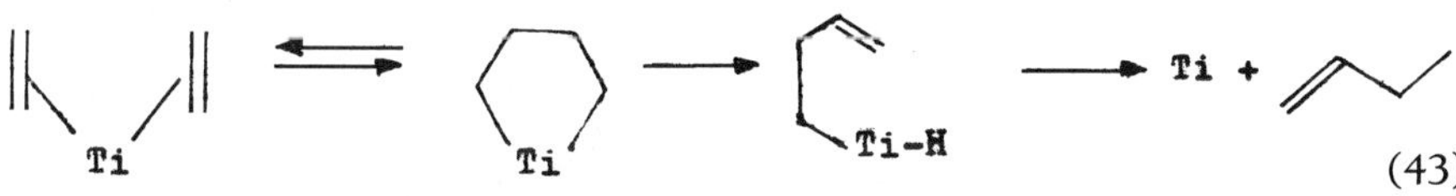

(43)

This metallacyclopentane complex decomposes to give 1-butene. This is not necessarily by the β-elimination sequence. Intermolecular hydrogen transfer and α-elimination have been suggested as alternative decomposition paths leading to 1-butene. The kinetics of the reaction shows first-order dependence on the catalyst and the olefin. Formation of a monoolefin complex is the probable rate determining step.

In Cosee's mechanism [70], the stabilization of a titanium *d* orbital by interaction with an empty antibonding orbital of the alkene was suggested to be important for catalytic activity. There is no theoretical evidence for the $d \rightarrow \pi$ back-donation [95]. Belov's binuclear active centers are ambiguous in light of the formation of metallacyclobutane and similar compounds [82–89]. The oxidation number of titanium is important in these reactions; Ti(III) is known to be involved in polymerization. The coordination of olefin is favored when the metal is in a higher oxidation state [95–97]. Wreford's proposal [94] of a concerted coupling of two molecules of ethylene to a titanium atom affording a titanium (IV)-cyclopentane species, which then decomposes to 1-butene by β-hydrogen transfer, explains the high selectivity for the formation of dimers. The presence of free H^+ or H^- species in the catalyst system is responsible for the isomerization of dimers [98]. The absence of any such species ensures high selectivity to 1-butene and the absence of isomerization to 2-butene.

K. Heterogenized Titanium Catalyst

Reaction of titanium compounds with surface hydroxyls of inorganic gels, their strong adsorption on the inorganic salts or oxides, and their reaction with active functional groups of polymers lead to heterogenized titanium catalysts, which after modification with alkyl metal derivatives are active in the olefin oligomerization reactions [99–102].

Zirconium compounds are highly active in ethylene oligomerization [103,104]. Kinetic investigations of the ethylene oligomerization in the presence of the catalyst $Zr[OOCCH(CH_2)_2]_4/Et_3Al_2Cl_3$ have shown that (1) the oligomerization does not occur when the ratio Al/Zr < 8; (2) the average oligomerization degree decreases when the ratio Al/Zr increases; and (e) the oligomerization degree is not influenced by the kind of solvent used.

Condition (1)—that the ratio Al/Zr must be at least 8—is explained by the necessary presence of the four dimeric sesquichlorosesquiethylaluminum compounds during the oligomerization center formation in the following systems:

$$ZrY_4 + 4\ Et_3Al_2Cl_3 \rightarrow \underset{\text{I}}{\overset{\text{I}}{}}\overbrace{EtZrCl_3 \cdot EtAlCl_2 \cdot EtAlClY}^{\text{I}} + 3[\overbrace{Et_2AlCl \cdot EtAlClY}^{\text{II}}] \tag{44}$$

where Y = OOCR, $CH_2C_6H_5$, or $C_5H_7O_2$.

If an excess of organoaluminum compounds is used, then the following reaction occurs:

$$I + Et_3Al_2Cl_3 \rightarrow \overbrace{EtZrCl_3 \cdot EtAlCl_2 \cdot EtAlCl_2}^{\text{III}} + II \tag{45}$$

Complexes I and III are ethylene oligomerization–active centers. Compound II is inert in ethylene oligomerization.

The oligomerization reaction succeeds when ethylene molecules insert into Zr–C bonds in the active complexes I and III:

$$I + nCH_2{=}CH_2 \rightarrow \overbrace{C_2H_5(C_2H_4)_nZrCl_3 \cdot EtAlCl_2 \cdot EtAlClY}^{\text{IV}} \tag{46}$$

$$III + nCH_2{=}CH_2 \rightarrow \overbrace{C_2H_5(C_2H_4)_nZrCl_3 \cdot EtAlCl_2 \cdot EtAlCl_2}^{\text{V}} \tag{47}$$

The oligomer chain termination reaction occurs as a result of the oligomer chain reaction with monomer:

$$IV + CH_2{=}CH_2 \rightarrow C_2H_5(C_2H_4)_{n-1}CH{=}CH_2 + I \qquad (48)$$

$$V + CH_2{=}CH_2 \rightarrow C_2H_5(C_2H_4)_{n-1}CH{=}CH_2 + III \qquad (49)$$

The rate of co-oligomerization, solvent alkylation, and oligomer chain transfer on chloroethylaluminum derivative were found to be very small. Simultaneously, the ethylene oligomerization reaction rate in presence of the above catalysts decreases with the reaction time. The EPR (electron paramagnetic resonance) investigations of the system $Zr(i\text{-}PrCOO)_4/Et_3Al_2Cl_3$ have shown that the reduction $Zr(IV) \rightarrow Zr(III)$ does not occur, but the following reaction takes place [105]:

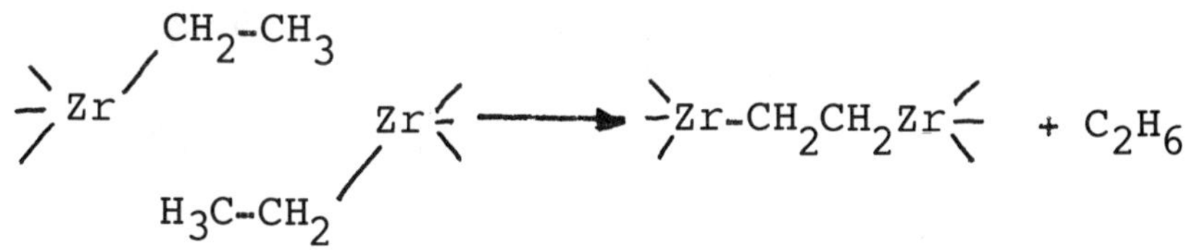

(50)

This disproportionation reaction is responsible for this catalytic system's deactivation.

Additional information concerning this catalyst's deactivation was obtained from electrical conductivity investigations in toluene. During oligomerization, ethylene and other α-olefins were found to form cationic type centers R^+A^-, which show little activity under reaction conditions as they dissociate to ions and disappear in subsequent reactions in the order heptane < ligroin < cyclohexane < olefins < toluene; the α-olefins increase oligomerization rate modestly but does not change the oligomer molecular weight.

The system $Zr(i\text{-}PrCOO)_4/Et_nAlCl_{3-n}(n = 1.5\text{–}2)$ served as a model for the study of the influence of reaction conditions on ethylene oligomerization [106–109]. The maximum oligomerization rate was observed for the system with n – 1.7. A change of n from 1.5 to 1.7 resulted in an increase of branched olefin (with internal C=C bonds) content in the products. The main reason for this phenomenon is the higher concentration of α-olefins formed in comparison with that of ethylene in the reaction medium, caused by the very high activity of catalysts. This enhances the catalyst reaction with higher α-olefins. Increasing n from 9.5 to 17.3 decreased the yield and the molecular weight of the oligomers.

An increase in the alkyl (R) chain length in the $Zr(RCOO)_4$ salts decreases the catalytic activity [109]. Raising the reaction temperature from 100 to 150°C results in increases in the reaction rate, the reaction yield, and the average molecular weight of the α-olefins. A higher reaction temperature (200°C) did not increase the reaction yield, but it did increase the length of α-olefin chains [109].

When reaction pressure is raised during the ethylene oligomerization, a higher reaction rate and selectivity with unchanged oligomer molecular weight is achieved.

Catalyst activity and the selectivity of the $Zr(RCOO)_4/Et_nAlCl_{3-n}$ system is altered when the solvent is changed and increases in the order toluene > n-heptane

> cyclohexane > ligroin. This is related to the lower ethylene solubility in these solvents, resulting in competitive, higher α-olefin co-oligomerization. The same relationships are observed for other zirconium catalysts. For the $ZrCl_4/Et_2AlCl$ system, increasing Al/Zr ratio from 0.25 to 4.0 gave better yields, lower average molecular weight, and higher α-olefin selectivity [110].

Catalysts giving C_6-C_{20} olefins in very high yield and purity (95%), containing Zr halides, organoaluminum compounds and sulfides, disulfides, tiophenes, thiourea, phosphines, or primary amines are described in Idemitsu Petrochemical Co. Ltd. patents [111–113].

The $ZrCl_4/EtAlCl_2$ system is not active in the ethylene oligomerization, and the catalysts $ZrCl_4/BuLi$ and $ZrCl_4/Bu_2Mg$ exhibit very weak activity in this reaction [114]. When Et_3Al is used as modifier for $ZrCl_4$, a very active but less selective oligomerization catalyst is obtained.

Similar catalytic behavior is observed in the case of $Zr(acac)_4EtAlCl_2$ (acac = acetylacetonate) and $(C_3H_7)_4Zr/Et_3Al_2Cl_3$ in the solution by the hydride complex *b*; compound *d* is generated via compound *c*:

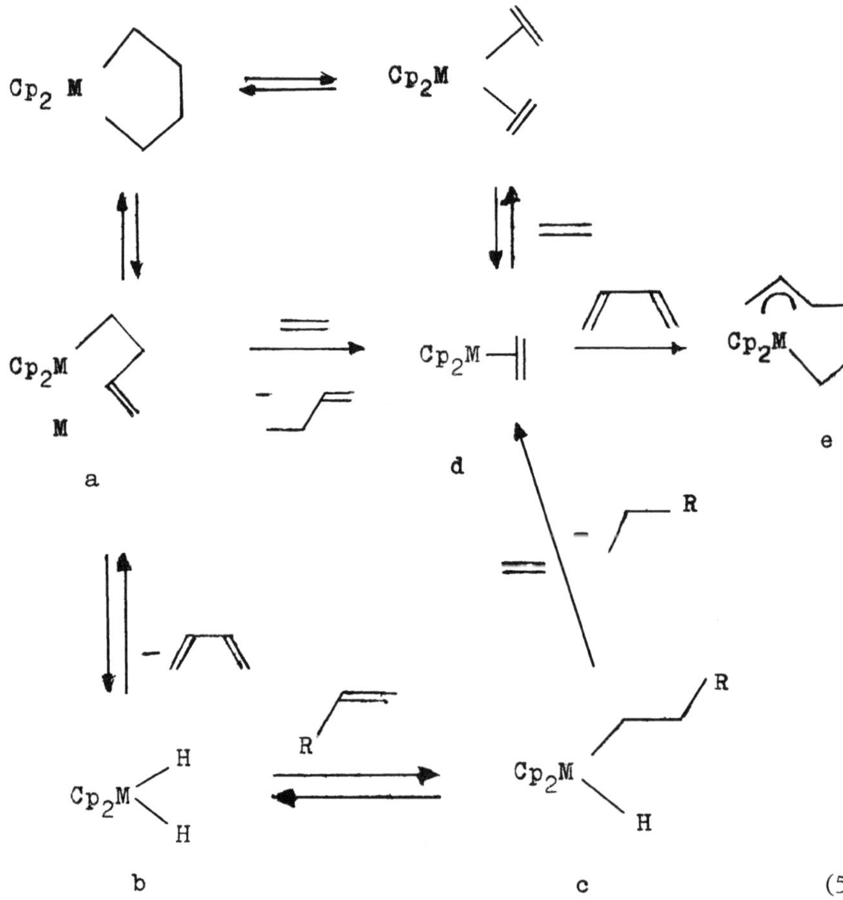

(51)

Dimerization of ethylene by a bis(cyclopentadienyl) metallacyclopentane complex (M = Zr, Hf; Cp = η^5-C_5H_5).

The olefinmetallocene complex *d* is trapped by the butadiene formed in the reaction cycle with formation of the catalytically inactive compound *e*.

The complexes of group 4B elements under optimum conditions can serve as catalysts for the dimerization and oligomerization of ethylene [115–122]. Note that Ti(IV), Zr(IV), and Hf(IV) do not have any 3*d*, 4*d*, and 5*d* electrons, respectively, to coordinate strongly with the olefin. One possibility is that the olefin behaves simply as a weak base with a lone pair electrons. Consequently, the alkene is only weakly bonded, probably by overlap of s orbitals of the metal with the π orbitals of the alkene [123]. This may be an important requirement for catalysis as distinct from compound formation. The weak coordination allows the olefins to rotate to form a nonpolar transition state, which is conducive to catalytic reactions.

In conclusion, zirconium catalysts are generally more active than titanium ones. The catalyst selectivity for ethylene oligomerization in the presence of zirconium compounds is in the range 15–30 Kg/g of zirconium/h. Selectivity to linear α-olefins approaches 99% [124].

A constant ethylene supply is necessary to avoid the formation of branched olefins with internal C=C bonds. Such olefins are the products of the higher olefins' co-oligomerization and isomerization when the ethylene concentration is too low in the reaction medium.

The use of vanadium trichloride for the dimerization of ethylene is patented [125]. VCl_3 and isobutylaluminum sequichloride at 60°C and 3 atm pressure gives 1-butene (7%), *trans*-2-butene (3%), octane (0.6%), and higher-boiling distillates (1.4%).

The preparation of $PhNb(COD)_2$ (COD = cyclooctadiene) [126] and $(dmpe)_2NbH_5$ [121], and their uses as catalysts for conversion of ethylene to 1-butene have been reported in patients.

Schrock and coworkers have investigated the dimerization of ethylene catalyzed by tantalocyclopentane complexes [127–129]. The advantage of these complexes is that the isomerization of primary product is negligible. $CpTaCl_2(C_4H_8)$ (Cp = cyclopentyl) at the 40 psi of ethylene pressure gives 1-butene selectivity (3% 2-butene) [130]. On prolonged reaction, ethylene and 1-butene codimers are formed by the decomposition of mixed metallacycle. The authors have observed [131] that $Ta(CH_3CMe_3)$ $(CHCMe_3)$ and two moles of trimethylphosphine produce a homogeneous catalyst that "rapidly" dimerizes ethylene selectively to 1-butene. The rate constant of this reaction is reported as approximately $k = 10^{-4}$ s^{-1} at 36°C. The active component is $(C_4H_9)Ta(C_2H_4)_2(PMe_3)_2$ (reaction 52, compound *a*), which has trigonal bipyramidal geometry:

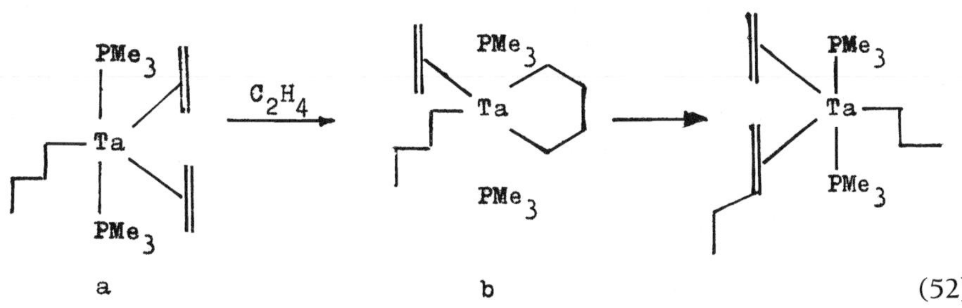

(52)

The trimethylphosphine methylphosphine ligands are axial, and ethylene ligands are equatorial and perpendicular to the trigonal plane. The mechanism for the dimerization reaction is shown in reaction (52). A metallacyclopentane complex forms when ethylene attacks compound *a* and one of the metallacycles. TaC(α) bonds are cleaved by a β-hydrogen atom from the butyl ligand to give compound *c*. The alternative possibility, namely, transfer from metallacycle to the butyl and the ethylene ligand, should be slow. The rate determining step in this mechanism is *a* → *b* transformation. This kind of metallacyclopentane mechanism is a plausible mechanism for insertion of ethylene into a metal–ethyl bond whenever such insertion is not expected to be fast and/or when the metals in lower oxidation states are formally oxidized by forming a metallacyclopentane complex. The insertion mechanism can be viewed as proceeding through two steps, which include the formation of a metallacyclopentane complex as shown in the following reaction [132]:

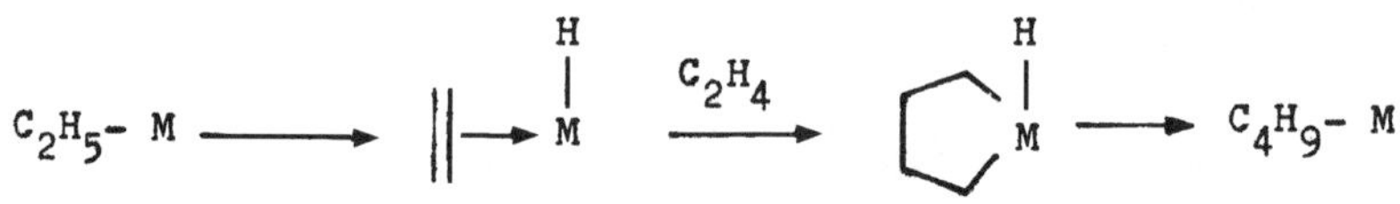

(53)

Interestingly, the analogous niobium system fails to dimerize ethylene to 1-butene because of its inability to form a metallacyclopentane complex.

L. Chromium, Molybdenum, Iron, Tungsten, Cobalt, and Manganese Complexes

The chromium complexes are $CrCl_3L_3$ and $CrCl_2L_2(NO)_2$, wherein the ligand L are pyridine and tri-*n*-butylphosphine in conjunction with ethylaluminum dichloride, effect simple dimerization of ethylene at 50°C [133–135]. A conversion of 4700 g butenes per g of chromium complex is achieved with the catalyst $CrCl_3$(4-Etpy)$_s$. The butene fraction consists of 1-butene (50%), *trans*-2-butene (32%), *cis*-2-butene (18%), and isobutene (0.1%). Cr^+ or Cr^{2+} species may be involved in the reaction. Here chromium atoms are probably associated with the organoaluminum halides to form bridged chromium–halogen–aluminum species.

A chloro-bridged molybdenum complex, [(π-C_6H_6)–Mo(π-allyl)Cl_2]$_2$, with ethylaluminum dichloride catalyzes the dimerization of ethylene in benzene medium at 20°C [136].

Wideman [137] has reported a catalyst based on tungsten. The active catalyst is prepared by heating tungsten hexachloride and 2,6-dimethylaniline in a minimum amount of chlorobenzene and subsequently treating with diethylaluminum chloride. A conversion of 176,000 mol of ethylene/mol of tungsten is achieved in 1 h at 40°C and 27 atm of ethylene pressure. The yield of 1-butene is 92% in the process. An increase in pressure from 27 to 34 atm resulted in the conversion of 184,000 mol of ethylene/mol of tungsten, thereby achieving a better yield of 1-butene (98%) at a faster rate.

Manganese chloride, manganese malonate, and manganese acetylacetonate [138] with organoaluminum halides are reported to dimerize ethylene in chlorobenzene as solvent. The optimum aluminum to manganese ratio is 3:1. At 45 atm of ethylene pressure and 80–85°C, butenes consisting of 36.6% 1-butene, 42.2% *trans*-2-butene, and 21.2% *cis*-2-butene are formed.

Ferrocene and ferrous chloride along with organoaluminum halides are known to convert ethylene to a mixture of butenes at 1–20 atm and −10 to +50°C [139,140].

In the case of cobalt, modified Ziegler systems have limited activity toward ethylene dimerization and oligomerization [141]. Cobalt(II) or cobalt(III) and a reducing organometallic compound system have been specially proposed for this reaction. Tris(acetylacetonato)Co(III) and triethylaluminum convert ethylene at 30°C into *n*-butenes with a selectivity of 99.5% [142]. The product consists of a mixture of 95% 2-butenes and 5% 1-butenes. The molar ratio of AlR_3/Co must be between 2 and 5; beyond that, the activity decreases. On the other hand, the addition of triphenylphosphine decreases the rate of the reaction.

(Dinitrogen)hydrodotris(triphenylphosphine)Co(I) has also been used to dimerize ethylene [143]. Here the dimerization of ethylene takes place at room temperature without the use of a Lewis acid. Ethylene conversion decreases with time, presumably due to partial decomposition of the catalyst. However, the decomposition is slow at 0°C. A mechanism has been proposed in which the olefin is first inserted in the cobalt–hydride bond and then a second molecule of the olefin is inserted in the cobalt–alkyl bond. Displacement of the dimer by the olefin regenerates the cyclic process.

The dimerization of ethylene is selectively catalyzed by halotris(triphenylphosphine)Co(I) in halobenzene containing boron trifluoride etherate [144,145]. The catalytic activity is significantly affected by the solvent, bromobenzene being the most effective. The rate of ethylene absorbtion decreases in the order bromobenzene > iodobenzene > *o*-dichlorobenzene > chlorobenzene > *o*-chlorotolene. The optimum ratio of boron to cobalt for the borontrifluoride etherate/bromotris(triphenylphosphine)Co(I) system is 1. This suggests a strong 1:1 interaction between the two components:

$$CoBr(PPh_3)_3 + BF_3 \cdot OEt_2 \rightarrow [Co(PPh_3)_3]^+[BF_3 \cdot Br]^- \qquad (54)$$

Lewis acids other than aluminum chloride and boron trifluoride etherate slow no oligomerization activity with cobalt complexes. Addition of triphenylphosphine and water stops the oligomerization almost completely.

Speier and Kabanov [146,147] reported the kinetics and mechanism of the dimerization of ethylene using bis[(ethylene)tris(triphenylphosphine)cobalt] as a catalyst. The predominant product is *trans*-2-butene (63%). The mechanism is given in the following reactions:

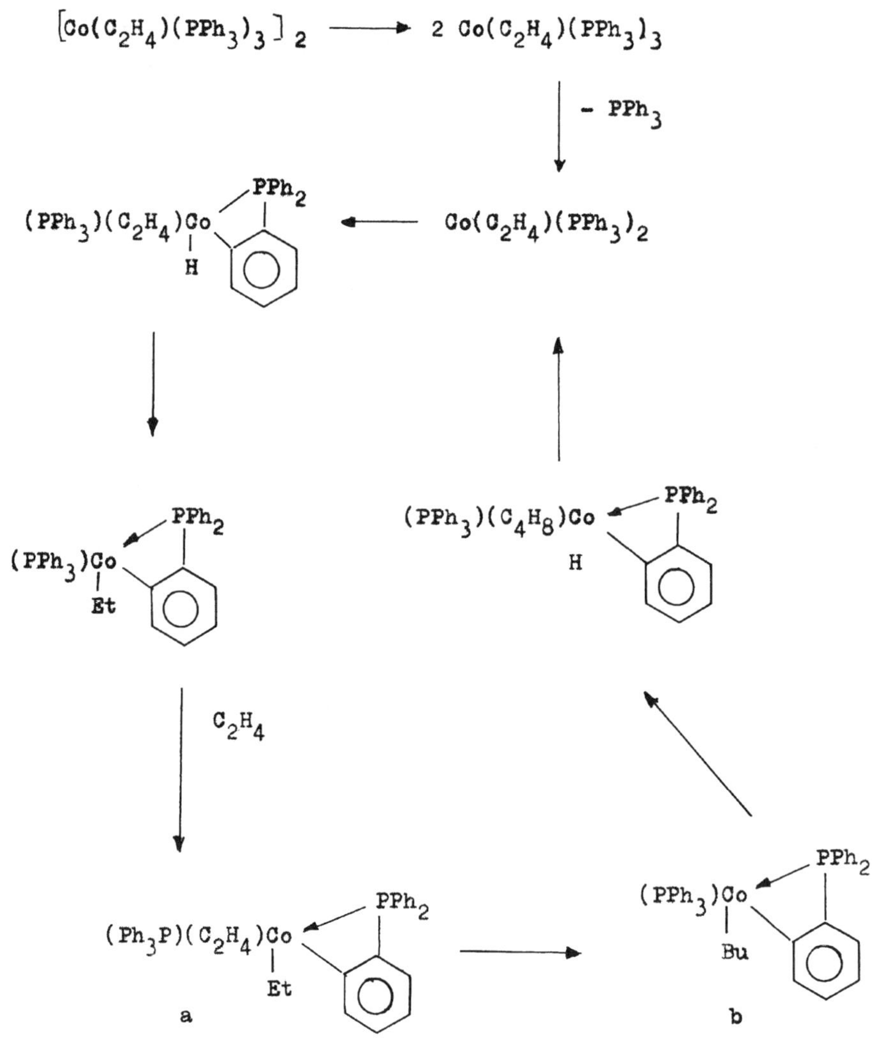

(54A)

Dimerization of ethylene by $[Co(C_2H_4)(PPh_3)_3]_2$.

Conversion of compound *a* to *b* is the rate determining step in this mechanism.

M. Nickel-Based Catalyst Systems

Nickel-based catalyst systems constitute one of the important catalysts for olefin oligomerization, because nickel is the only metal that can control the mode of linking of olefins. The specific activity is high, and it is one of the less expensive transition elements. Because of these advantages, much research has gone into the study of nickel-catalyzed oligomerization [148–181].

The oligomers obtained with these systems follow a Schulz–Flory type distribution. The nickel catalyst complexes can be divided into two groups: nickel(II) compounds modified with alkyl or hydride main-group metal derivatives (Ziegler–Natta type catalyst) and chelated nickel compounds with a Ni–C bond.

The discovery of "nickel effect" represents the starting point for the development of the Ziegler–Natta catalysts [182]. A basis for the elucidation of the effect was provided by studies on the reduction of nickel compounds by organoaluminum compounds, the existence of nickel bridges, and the interactions between nickel (0) and Lewis acids as well as organic compounds of main-group metals. Formation of multicenter bonding systems involving trialkylaluminum compounds and nickel atoms has been demonstrated from these studies, for example

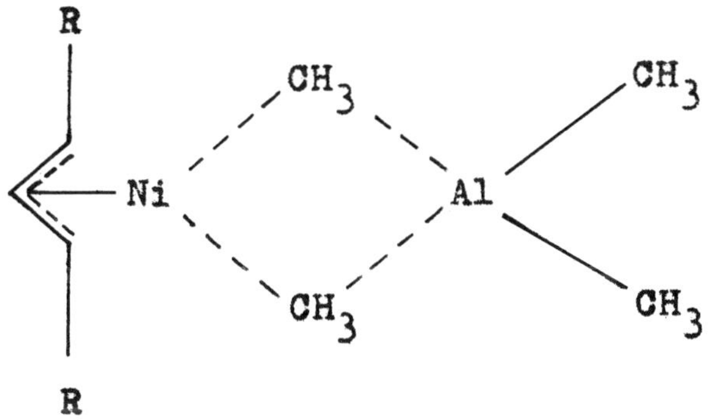

(55)

These systems further react with coordinated ethylene molecules in a concerted manner to give butenes or higher olefins.

The first reports of the dimerization of alkenes by nickel complexes were published almost simultaneously in Germany [183,184], France [185], and Russia [186,187]. Ewers has used a highly active homogeneous catalyst prepared by heating $Ni(acac)_2$, $Ni(\pi\text{-allyl})_2$, or $Ni(\pi\text{-allyl})Cl$ with dialkylaluminum chloride in a toulene medium. A paper by Arakawa [188] describes the oligomerization of ethylene under the influence of (π-allyl)nickel chloride in conjunction with Lewis acids and tertiary phosphines. The allyl group does not participate in the reactions [189]. It is neither displaced as in cyclooligomerization nor inserted as in polymerization; rather, it is found on the nickel at end of the reaction. Its role is probably that of a stabilizer of certain electronic states.

Chauvin et al. [185] have used the $Ni(acac)_2/EtAlCl_2$ system effectively for the oligomerization of ethylene.

An important feature of nickel-catalyzed oligomerization is that a Lewis acid is not necessarily a reducing agent when the nickel is bonded to less than two "hard" anions (π-allyl, nickel halide, etc.) and is a reducer when the bivalent nickel is bonded to two "hard" anions. $Et_2Al(OEt)$ is a weak Lewis acid, and hence the system $Ni(acac)_2$-$Et_2Al(OEt)$ is less active. But in the system $NiCl_2$–PR_3–$AlCl_3$ the phosphine can play the role of reducing agent.

The ethylene oligomerization catalyst cycle in the presence of nickel catalyst, HNiYL (where, Y = acyl, carboxyl, halogen; L = phosphine), is shown in the following reaction series [71]:

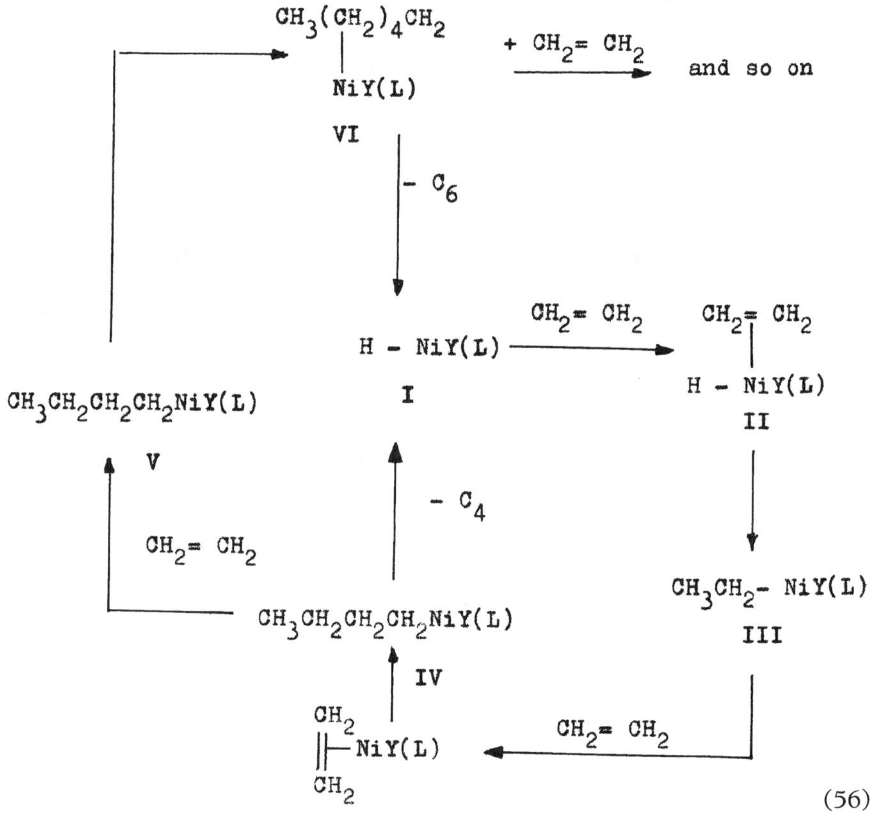

(56)

Catalytic cycle of ethylene oligomerization in the presence of a nickel catalyst.

Ethylene coordination to complex I with the Ni–H bond, which is the catalyst's active center, gives complex II. The coordinated ethylene insertion into the Ni–H bond with formation of a Ni–C_2H_5 group (complex III) begins the catalytic cycle via complexes IV, V, and VI. The chain transfer occurs through the β-hydrogen elimination from the oligomer chain in complex IV by the nickel atom, whereby the Ni–H bond is restored and oligomer liberated:

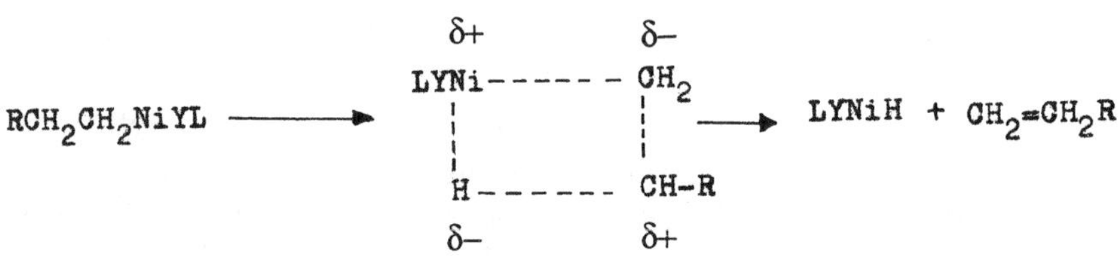

(57)

Formation of the Ni–H or Ni–alkyl bonds in nickel compounds is responsible for their catalytic activity in olefin oligomerization. This process for the system (π-C_3H_5)NiCl/Et_nAlCl_{3-n} (n = 1, 2) is proposed to be as follows:

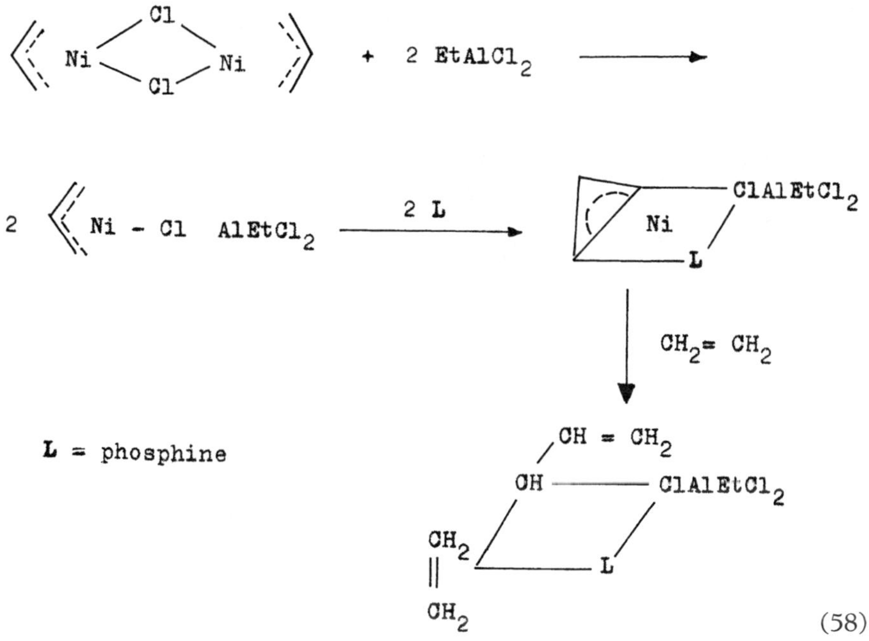

(58)

In the system Ni(acac)$_2$/Et_2AlOEt (1:1), nickel alkylation occurs according to the following reaction:

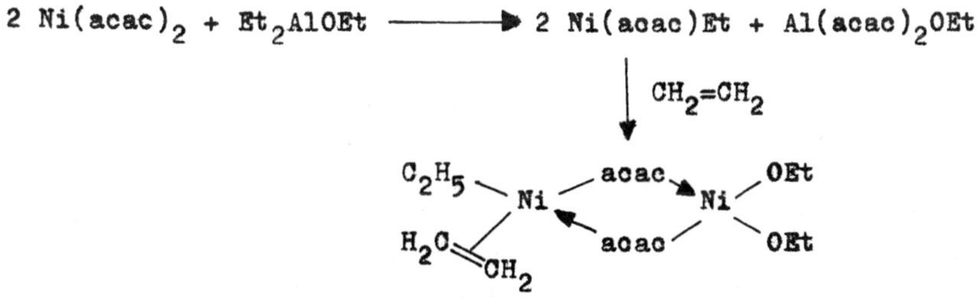

(59)

The catalytic species formed "in situ" by the reaction of Ni(II) with $R_{6-x}Al_2X_x$ (X = Cl) are generally more active. Many attempts have been made to isolate intermediates [191,192]. Experimental observations are in agreement with the ionic structure $(L_n\text{–Ni–H})^+A^-$, where A^- is a noncomplexing anion derived from a Lewis acid [185]. The L may be one of the following: (1) the monomer or the dimer product, (2) a solvent molecule, (3) a component added to the catalytic system, like a phosphine, or (4) any compound having a heteroatom capable of coordinating with the nickel.

It is possible to oligomerize ethylene by nickel in the absence of a Lewis acid. Butenes are formed by heating nickelocene to 200°C in the presence of ethylene [193]. In this process, homolytic decomposition of nickelocene produces excited nickel atoms, which catalyze dimerization. This is similar to heterogeneous catalysis. To explain the nickelocene-catalyzed dimerization of ethylene, Tsutsui [194] has proposed a three-step mechanism represented in the following reactions:

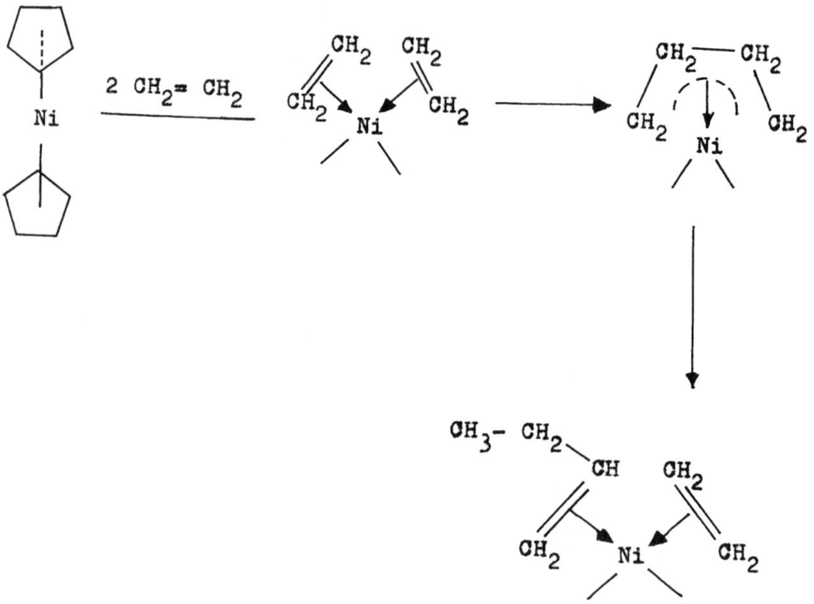

(60)

Nickelocene-catalyzed dimerization of ethylene.

$(\eta^5\text{-}C_5H_5)Ni(\eta^3\text{-}C_5H_7)$ [158] is a highly active unicomponent catalyst for the conversion of ethylene to higher olefins at 145–150°C. At high conversion of ethylene (70–90%), the dimeric product (80–86% yield) contains a high percentage (82–90%) of 1-butene. The cyclopentadiene group remains bonded to the nickel during catalysis while the cyclopentyl group is labile. A possible mode of activation is the reversible elimination of cyclopentadiene from $(\eta^5\text{-}C_5H_5)$-$Ni(\eta^3\text{-}C_5H_7)$ to generate (π-cyclopentadienyl)nickel hydride as a catalytically active intermediate:

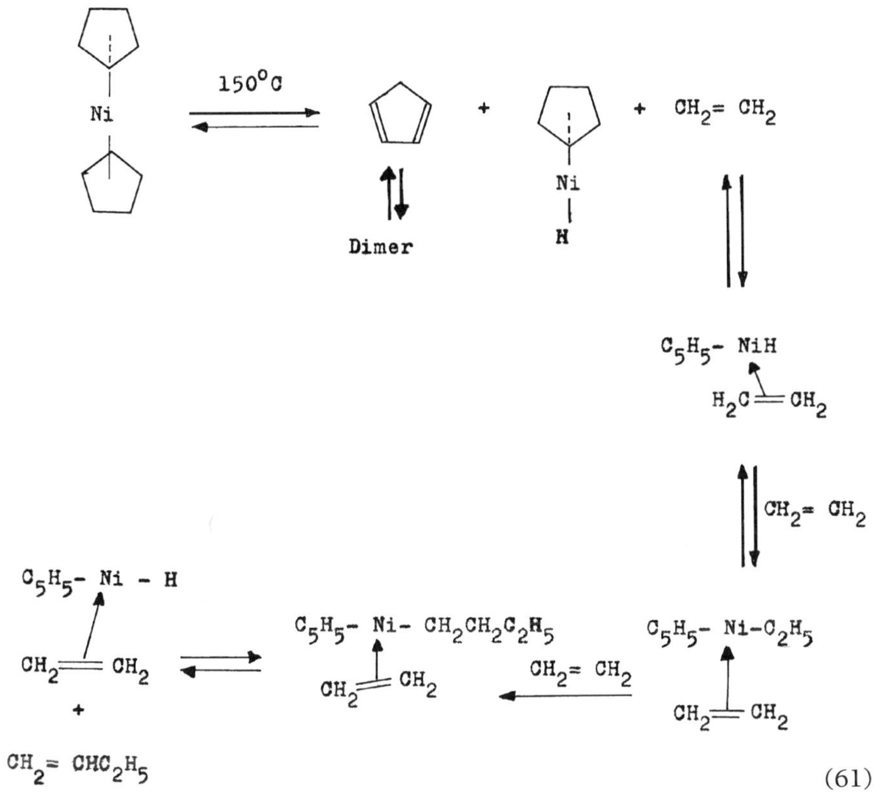

(61)

(η^5-Cyclopentenyl)(η^3-cyclopentadienyl)nickel-catalyzed dimerization of ethylene.

Kalke [193] has reported dimerization of ethylene catalyzed by bis(triphenylphosphine)(pentachlorophenyl)chloronickel (II) activated with silver salts like $AgClO_4$ or $AgBF_4$. The activity is enhanced by adding a catalytic amount of PPh_3 ($PPh_3/AgClO_4 \leq 1$). An excess of PPh_3 ($PPH_3/AgClO_4 \leq 2$) stops ethylene oligomerization. Separate runs show that $AgNO_3$, $NaClO_4$, and $NaBF_4$ do not activate the catalyst. Enhancement of the activity of the complex by a catalytic amount of PPH_3 may be attributable to an increase in the solubility of $AgClO_4$. In contrast to the PPh_3 complex, the phenyldimethylphosphine complex is much less active for ethylene oligomerization.

$(C_6H_5)Ni(PPh_3)_2Br$ gives maximal activity for ethylene oligomerization when the molar ratio of added $AgClO_4$ to the complex reaches 2 [195].

Phosphorus-31 nuclear magnetic resonance (NMR) study of the reaction between $AgClO_4$ and the complex reveals that the variation in oligomerization activity becomes parallel with the concentration of $(C_6H_5)Ni(PPh_3)ClO_4$ formed in the reaction. The coordinately unsaturated state of the complex will be one of the reasons for its catalytic activity. The role of $AgClO_4$ is to remove of halogen and PPh_3 from the original complex, although excessive removal of ligands results in the formation of an inactive complex.

Nickel phosphine metallacyclopentanes, especially tris(triphenylphosphine)-tetramethylnickel(II), catalyze the production of cyclobutene and 1-butene from ethylene [196]. The course of the reaction is depicted in the following reactions:

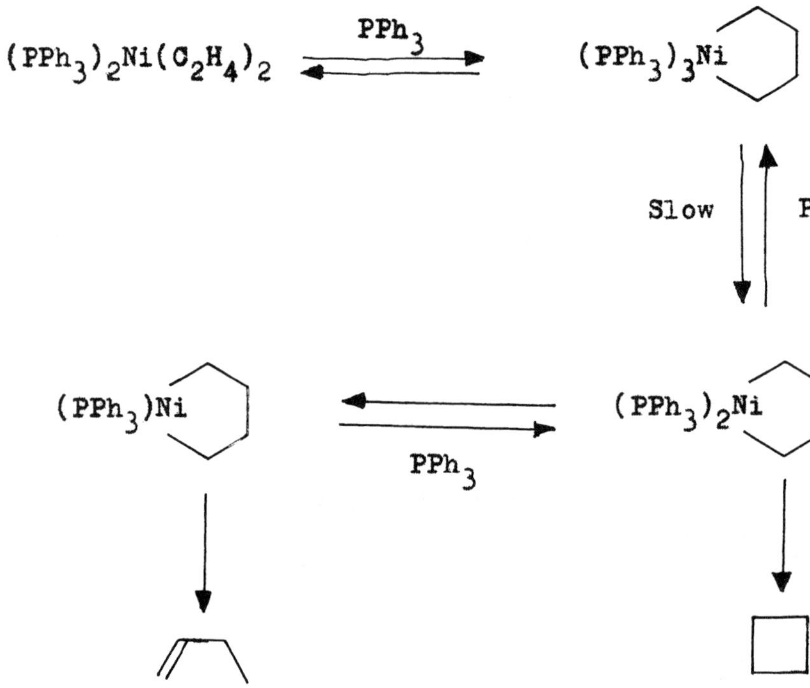

(62)

Dimerization of ethylene catalyzed by nickelacyclopentane phosphine complexes.

Muzio and Löffler have proposed the following mechanism of the Ni–H bond formation for the catalyst [197]:

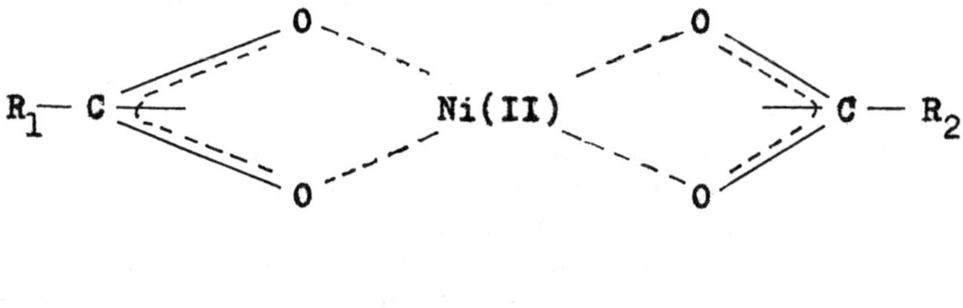

(63)

$EtAlCl_2$ and Et_2AlCl as a mixture is the necessary cocatalyst in this system. The mechanism of its action is presented in the following reactions:

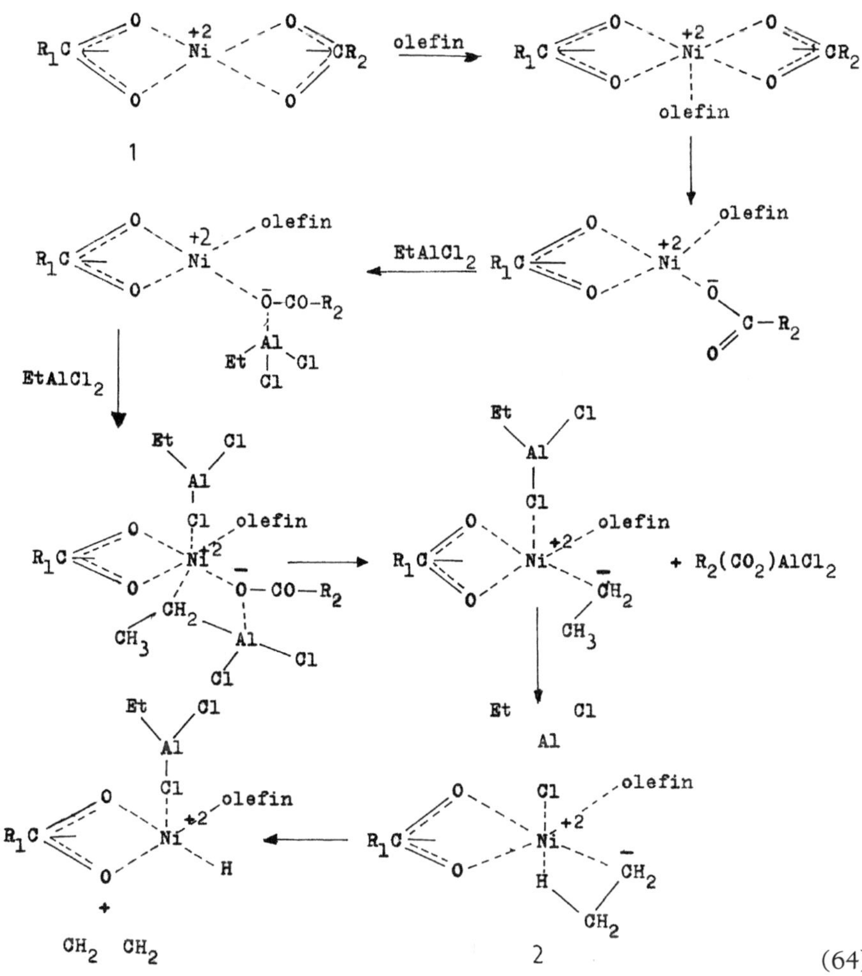

(64)

Mechanism for hydride formation in a nickel-catalyzed olefin oligomerization.

The reaction was carried out by adding at the beginning the nickel salt *1* to the olefin. Next the organometallic compound was added. If the nickel salt were mixed with the organometallic compounds with no olefin present, the nickel salt would be reduced and lose its catalytic activity. The first step of the reaction consists of the olefin coordination to the nickel atom. This helps to avoid reduction of the nickel in the complex *1* with an added organoaluminum compound. At first $EtAlCl_2$ complexes with one of the oxygen atoms in the complex *1*. Additional aluminum compound is coordinated to the nickel through the chlorine bridge; thus, it can neutralize negative charge on the oxygen atom, which in turn helps the salt $R_2COOAlCl_2$ to split off. Thus the nickel in the complex *2* (reaction 64) is coordinately unsaturated. Internal rearrangement leads to formation of Ni–H bond.

Many chelated nickel complexes are highly active and selective in the ethylene oligomerization without added organoaluminum compounds [198–201]:

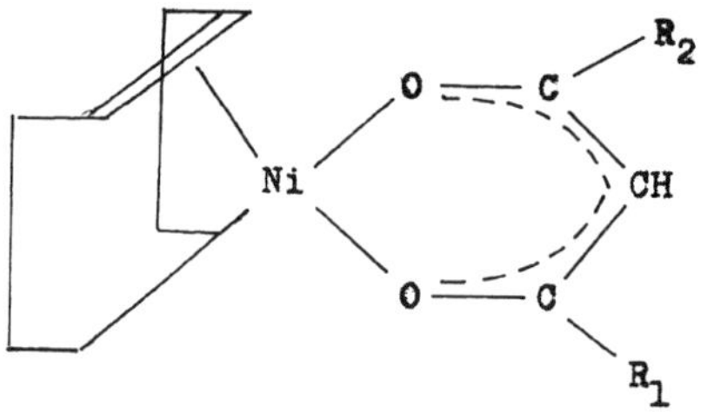

1

$R_1 = R_2 = CH_3, CF_3$

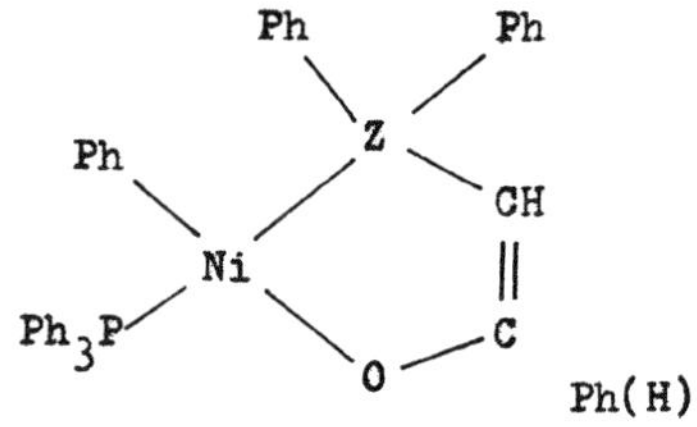

2

Z = P, As

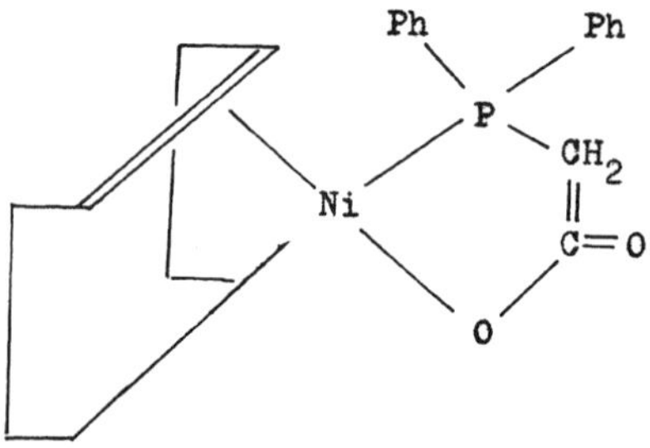

3

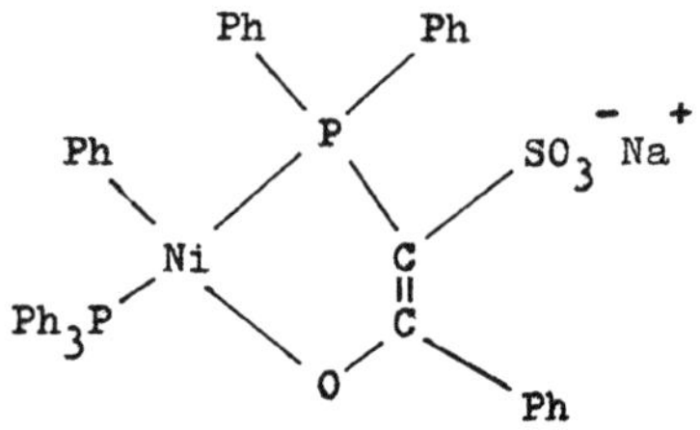

4

(64A)

Complexes of type *1* are very active catalysts for the oligomerization of ethylene. Linearity of 80% can be obtained in the C_8 oligomers (complex *1*). Complex *2* has been found to be the most interesting ethylene oligomerization catalyst. It converts ethylene at 50°C and 10–100 bar to α-olefins possessing 99% linearity and >95% α-olefin content [202]. The geometric distribution of olefins can be modified by pressure and addition of Ph_3P. Increasing the Ph_3P/Ni ratio in complex *2* by trialkylphosphine inactivates the system. When $(\alpha\text{-naphtol})_3P$ substitutes for Ph_3P, 70% of linear polyethylene is obtained in the product of ethylene oligomerization [203]. Polyethylene is obtained in the product of ethylene oligomerization [203].

Complexes similar to compound *2* (structural formulas 64A) may be formed on reacting bis-1,5-cyclooctadiene nickel (0) with Ph_2PCH_2COOH [204], thiolactic acid [205], *o*-mercaptobenzoic acid [205], phosphorus ylides [205], or Ph_2-$PCH_2C(CF_3)_2OH$ [206]. All these systems have been patented by Shell and are effective in the Shell higher olefin process (SHOP) [207].

Complexes *2* and *4* do not have Ni–H bonds. These bonds are formed during the oligomerization reaction, probably through ethylene insertion into Ni–Ph bond and the immediate split off of a styrene molecule:

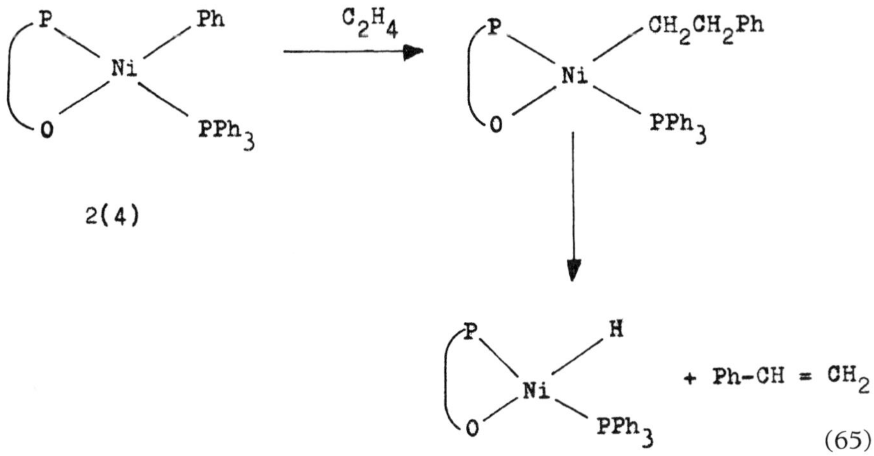

(65)

In the chelated complex *1* (structural formulas 64A) the Ni–H bond is formed as a result of the following reactions:

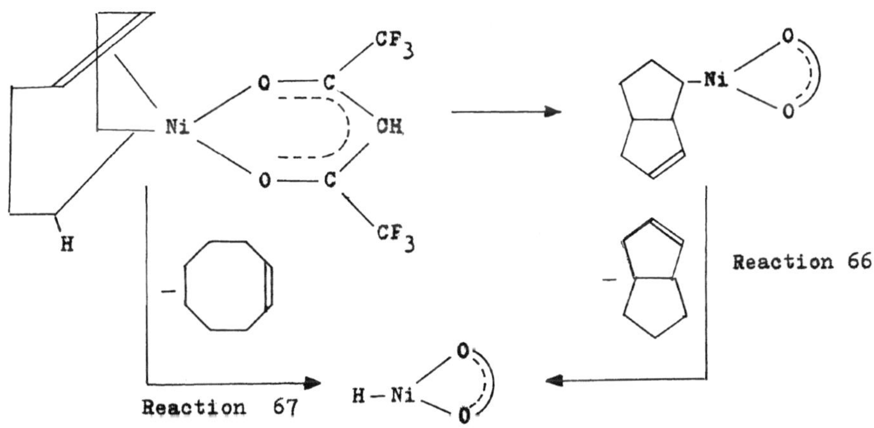

(66,67)

As stated earlier, reaction (66) is favored over reaction (67). During ethylene oligomerization in presence of complex *1* (structural formulas 64A) mainly bicyclo(3.0.0)octene-2 (beside a small amount of 1,5-cyclooctadiene) was found [206].

The Ni–H bond in the reaction of ethylene with complex 3 (structural formulas 64A) is formed as follows [208]:

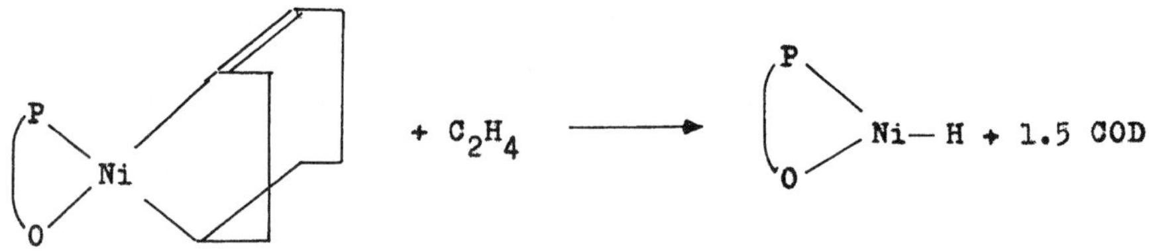

3 (see structural formulas 64A)

COD = cyclooctadiene (68)

Active nickel complexes in ethylene oligomerization generally have a square-planar structure. Their activity and selectivity depend on the nature of the ligands surrounding the nickel ion. The more basic the phosphine (L) ligands and greater their bulk in the complex (π-C_3H_5)NiBrL/$Et_3Al_2Cl_3$, the higher the oligomer molecular weight [172]. This phenomena is explained by the β-hydrogen elimination from the five or more coordinate intermediate:

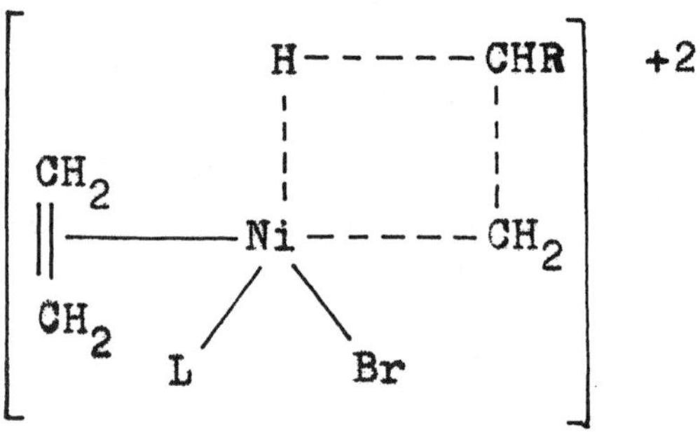

(69)

whereas the olefin insertion into the nickel–alkyl bond occurs through the less-hindered intermediate:

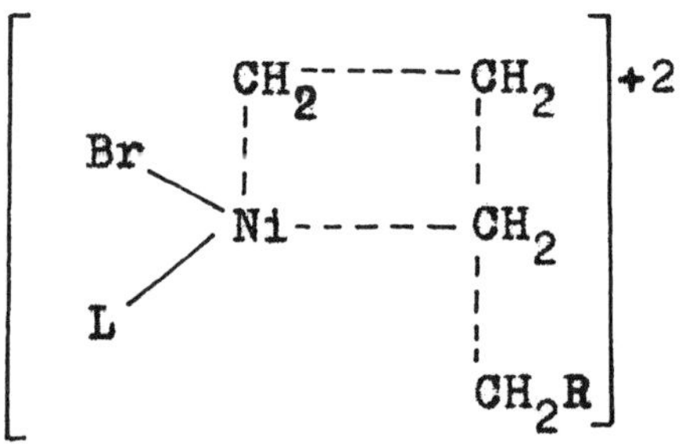

(70)

An increase in steric volume of the phosphine L helps in the formation of the structure (70) and the growth of the oligomer chain. A similar effect, an increase in the oligomer molecular weight, is observed when there is a decrease in the electron density in the nickel ion in complex *1* in (64A). An exchange in this complex of donor alkyl ligands for CF_3, which causes the withdrawal of an electron from the nickel ion, diminishes the insertion reaction barrier and, therefore, the growth of the oligomer chain and its linearity [199].

Using strong polar solvents such as glycols [209–211], pentanol [212], metanol [213], and methanol/H_2O (0.5–20%) [214] for the catalytic system similar to compound *a* (structural formula 65) allows the catalyst solution to separate in a better grade from the oligomer phase. The complexes *a* and *b* (structural formula 68) were heterogenized on alumina–silica and polystyrene supports.

Nickel salts with fluoroorganothiol or sulfide followed by addition of organic aluminum halide or alcoxide or borohydride may be also used in the oligomerization processes of olefins [215]. Catalytic systems based on nickel salts can be activated by active ligands, for example, 1,3-diketone [181,216].

Nickel bis-trifluoroacetylacetonate in combination with aluminum alkyls [217], 1,1,1,5,5,5-hexafluoropentane-2,4-dione, nickel salts, and $HAlR_2$ [218], Ni incorporated in synthetic mica-montmorillonite [219], and Ni-(2,4-pentanedithionate)–$[P(C_4H_9)_3]Cl/(C_2H_5)_2AlCl$ in chlorobenzene were used in the oligomerization of ethylene or propylene [220]. The selectivity of these catalytic systems depends on the Al/Ni ratio (optimal value 0.72–2.0). The presence of traces of water in the system did not affect the oligomerization process.

Nickel(II) complexes with chelating phosphines are effective catalysts for the oligomerization of ethylene and double bond shift isomerization of olefins, with high specificity for dimeric compounds rich in linear isomers [221,222].

Earlier processes [223,224] achieve ethylene oligomerization by means of a catalytic system made up of a nickel salt, a ligand based on phosphinonic acid, and a reducer. Organic (nickel formiate or acetate) or inorganic (nickel chloride or bromide) salts, the ligand with the $(C_6H_5)_2P{-}CH_2CH_2{-}COOMe$ (M = Na, K) structure, and sodium or potassium borohydride as reducer have been preferred.

Ethylene can also be oligomerized in the presence of a catalyst formed as a result of a reaction between metal aluminum, partially chlorinated paraffinic hydrocarbons, and nickel acetylacetonate [225].

Sulfonated nickel ylide complex is a potent single-component catalyst for ethylene oligomerization in aromatic and polar solvents [226]. The modification of this complex with various organoaluminum compounds ($AlEt_2Cl/AlEtCl_2$ or AlEt–O–Et) results in the formation of a new catalytic system with 10–20 times higher activity in ethylene oligomerization in aromatic solvents compared with that of the original ylide [201,227–229].

Ethylene could also be oligomerized to normal α-olefins in aqueous CH_3OH containing 0.5–20% water in the presence of nickel ylide [214]. The water improved the separation of the catalyst phase from the olefin-rich product phase, increased the purity of the α-olefin product, shifted the product distribution toward higher α-olefins and not affect the catalytic activity.

The effects of various organoaluminum compounds on ethylene oligomerization with Ni-ylide were evaluated in toluene solution at 30–120°C and ethylene at pressure 6.6–14.6 atm. An excess of $AlEt_3$ (20:1) reduces Ni-ylide to Ni(0) under these conditions, resulting in a complete loss of oligomerization activity. Tetraethylaluminoxane (produced in the reaction between $AlEt_3$ and water in toluene solution) also deactivates Ni-ylide in ethylene oligomerization.

Aluminum alcoxides (produced in reaction between $AlEt_3$ or $Al(i\text{-}Bu)_3$ and appropriate amounts of alcohols in toluene solution at 20°C) produce, in combination with Ni-ylide, highly active catalyst for ethylene oligomerization to mixtures containing predominantly 1-alkenes.

The major product of ethylene oligomerization in the presence of Ni-ylide–aluminum alcoxides are linear 1-alkenes with even carbon atom numbers, from C_4 to C_{40}. In addition to 1-alkenes, linear alkenes with internal double bonds (*trans* and *cis*) are always present in the reaction product.

Oligomerization of ethylene is also achieved in the presence of heterogeneous catalysts obtained by supported nickel complexes on various organic or inorganic materials.

The oligomerization of ethylene with a heterogeneous catalyst of a polystyrylnickel phosphorus ligand complex is carried out, and the result is compared with that of the oligomerization of ethylene with a homogeneous catalyst $Ni(acac)_2/Et_3Al_2Cl_3$/phosphorus ligand. When PPh_3 is used as a phosphorus ligand, the homogeneous catalyst gives a considerable amount of trimer together with the dimer, but the heterogeneous polystyrylnickel catalyst yields only butenes with mainly 1-butene. The results are discussed in terms of the steric and electronic effect of a phosphorus ligand and the steric effect of the catalyst support (i.e., polystyryl chain) [230].

On activation by $BF_3 \cdot OEt_2$, $Ni(PPh)_4$ anchored on brominated polystyrene exhibits high dimerization catalytic activity for ethylene [230]. The catalyst is reused without loss of activity.

The reaction of halogenated polystyrene with tetrakis(triphenylphosphine)nickel gives a supported complex. This system, when associated with boron trifluoride etherate, is a catalyst for the dimerization of ethylene. Indeed, at 0°C and atmospheric pressure of ethylene, a suspension in toluene of the polymer–nickel complex, to which has been added $BF_3 \cdot Et_2O$ (BF_3/Ni = 20/1 mole), absorbs ethylene selectively, giving butene [231].

Six recycles have been performed without readdition of boron trifluoride. The activity remains constant, apart from a small activation effect observed at first recycle, which is not clearly understood. The toluene solution, when separated from the catalyst, was proved to have catalytic activity. These observations mean that the dimerization is affected by a nickel species that remains attached to the polystyrene and is not soluble in toluene. Furthermore, the readdition of BF_3, not being needed at each recycle, indicated that the active species is a $Ni–BF_3$ anchored species complex.

More detailed studies on the influence of the repartition of the nickel on the polystyrene have been performed [232]. It has been found that the initial rate of dimerization of ethylene is almost independent of the nickel concentration in the polystyryl complex, as long as the same total amount of nickel is utilized in the reaction. This may be understood if the nickel sites are all equivalent, and equally accessible, either to BF_3 or to ethylene; there is no additional interaction between BF_3 and the halogenated polystyrene.

The influence of the nature of the solvent cannot be readily explained. Hexane is the most effective, twice as much as benzene or chlorobenzene, whereas diethylether has a pronounced inhibiting effect. This goes against what would be expected from the swelling ability of these solvents on polystyrene. It must be assumed that there is some sort of competition between the olefin and the solvent for coordination to the nickel. In the case of ethylene, which is known to have a much higher coordinating power than propylene, there is no more competition, and solvent effects are then leveled.

The direct addition of BF_3 and water onto the solid nickel–polymer complex gives a catalyst that exhibits an induction period before the dimerization of ethylene is effective. However, the preformation of the catalyst in a solvent like hexane gives a catalyst that shows no induction period. This can be understood in terms of a physical effect on the support. Indeed, in the latter case the solvent remaining in the polymer leaves it swelled, thus allowing the immediate access of ethylene to the catalytic sites. In the former case, in contrast, the formation of some dimer is required to play the role of a solvent and to promote the progressive swelling of the support.

Nickel sulfate associated with porous, inorganic oxide such as alumina is used for catalytic purposes in olefin oligomerization [174]. The ethylene oligomerization activity of Ni–alumina catalyst is approximately proportional to the Ni concentration in the catalytic system. Increasing the Ni concentration resulted in a marked shift to a higher product (comparable at some level of ethylene conversion). The effects on product distribution and catalyst deactivation are due to the Ni being associated with sites of lower basic strength [233,234].

Various compounds of the complex organo-Ni types show activity in olefin oligomerization processes. Ni-chelates [235] and bis(1,5-cyclooctadiene)Ni(0) deposited on omega zeolites or Y-type zeolite support [236] have high reactivity in the oligomerization of ethylene to form higher olefins.

The systems supported on alumina–silica gels were less active than their homogeneous analogues. However complex 2 (structural formulas 64A), bonded to the polystyrene chain through phosphine or PO (bifunctional phosphine) ligands, forms active and highly selective (99%) catalysts, giving high molecular weight linear α-olefins [237].

N. Kinetic Aspects of Oligomerization and Co-oligomerization of Ethylene in the Presence of Ni-Containing Catalysts

The kinetic mechanism of ethylene oligomerization in the presence of Ni-containing catalysts includes the following steps [200]:

Initiation:

$$Ni{-}H + CH_2{=}CH_2 \xrightarrow{k_i} Ni{-}CH_2{-}CH_3 \tag{71}$$

Chain propagation:

$$Ni{-}(CH_2{-}CH_2)_n{-}C_2H_5 + CH_2{=}CH_2 \xrightarrow{k_p} Ni{-}(CH_2{-}CH_2)_{n+1}{-}C_2H_5 \tag{72}$$

Spontaneous chain termination through β-elimination reactions:

$$Ni{-}(CH_2{-}CH_2)_n{-}C_2H_5 \xrightarrow{k_t} Ni{-}H + CH_2{=}CH{-}(CH_2{-}CH_2)_{n-1}C_2H_5 \tag{73}$$

The molecular weight distribution (MWD) of oligomers corresponding to this chain reaction scheme is described by the equation

$$P_{(n)} = A\gamma^n \tag{74}$$

where $P_{(n)}$ is the number distribution function of oligomers of oligomerization degree n (the fraction of oligomer molecules containing n monomer units), A is the combination of the rate constants of reactions (71–73), and γ is the probability of the chain propagation step [198,238].

$$\gamma = \frac{k_p C_m}{k_p C_m + k_t} \tag{75}$$

where C_m is the monomer concentration in solution.

The number distribution function $P_{(n)}$ is proportional to $Q_{(n)}/n$, where $Q_{(n)}$ is the weight portion of the oligomer molecules with the oligomerization degree n estimated from gas chromatograms of oligomerization products. Combining this expression and Eqs. (74) and (75) gives

$$\log \frac{Q_{(n)}}{n} = \log A + n \log \gamma \tag{76}$$

that is, linearization of experimental gas chromatography data in olefin distribution products in the coordinates log $[Q_{(n)}/n]$ vs. n should produce straight lines with slopes log γ.

When co-oligomerization of ethylene and higher α-olefin takes place, all these steps (reactions 71–73) result in the formation of heterogeneous reaction products. The structure of these products depends on the position of the higher olefin unit in an oligomer chain and on the mode of olefin insertion into Ni–H of Ni–C bonds, primary or secondary.

If a higher α-olefin participates in reaction (71), two types of reaction products can be formed depending on the olefin orientation with respect to the Ni–H bond.

1. Primary Insertion

$$\text{Ni—H} + \text{CH}_2\text{=CH—R} \xrightarrow{k_i'} \text{Ni—CH}_2\text{—CH}_2\text{—R} \tag{77}$$

Subsequent ethylene insertion into the Ni–C bond (reaction 72) followed by chain termination reaction (73) produces linear α-olefins,

$$\text{CH}_2\text{=CH—(CH}_2\text{—CH}_2)_{n-1}\text{—CH}_2\text{—CH}_2\text{—R}$$

If the α-olefin CH_2=CH—R involved in the primary insertion into the Ni–H bond (reaction 77) has an even number of carbon atoms, the products formed in the co-oligomerization reaction are indistinguishable from linear α-olefins formed in ethylene homooligomerization (reactions 71–73), which always proceed parallel with co-oligomerization. However, existence of the primary α-olefin insertion into the Ni–H bond is obvious from the product analysis in the case of co-oligomerization of α-olefins with odd numbers of carbon atoms (propylene, 1-pentene, 1-heptene, etc.).

2. Secondary Insertion of α-Olefins

$$Ni{-}H + R{-}CH{=}CH_2 \xrightarrow{k_i''} Ni{-}CHR{-}CH_3 \qquad (78)$$

In the initiation reaction produces, after subsequent ethylene insertion reactions (reaction 72) and a chain termination reaction (reaction 73), α-olefins with methyl-branched chains:

$$CH_2{=}CH{-}(CH_2{-}CH_2)_{n-1}{-}CHR{-}CH_3$$

Gas chromatography data provide quantitative information on probabilities of primary and secondary insertions in the initiation reactions for propylene and ethylene. The estimates are based on the assumption that, to first approximation, the probabilities of β-elimination reactions (reaction 73) do not depend on the structure of the first monomer unit in an oligomer chain if the unit is separated from the reaction point in the β-elimination reaction by several ethylene units, for example, that k_t' and k_t'', values in reactions

$$Ni{-}(CH_2{-}CH_2)_n{-}CH_2{-}CH_2R \xrightarrow{k_t'}$$
$$Ni{-}H + CH_2{=}CH{-}(CH_2{-}CH_2)_{n-1}{-}CH_2CH_2R \qquad (79)$$

$$Ni{-}(CH_2{-}CH_2)_n{-}CHR{-}CH_3 \xrightarrow{k_t''}$$
$$Ni{-}H + CH_2{=}CH{-}(CH_2{-}CH_2)_{n-1}CHR{-}CH_3 \qquad (80)$$

are equal if $n > 2$–3. In this case the yield ratio for two olefins, the odd linear α-olefin and the methyl-substituted olefin $CH_2{=}CH{-}(CH_2{-}CH_2)_{n-1}{-}CHR{-}CH_3$, correspond to the $k_i':k_i''$ ratios for (77) and (78).

One important particular case of reactions (77) and (78) is insertion of an olefin into the Ni–H bond, with the immediately following β-elimination resulting in regeneration of the olefin. However, the secondary type insertion of the olefin molecule will favor the formation of an olefin with the double bond in the second position in the chain due to higher reactivity of the CH_2 group in β-elimination compared with that of the CH_3 group:

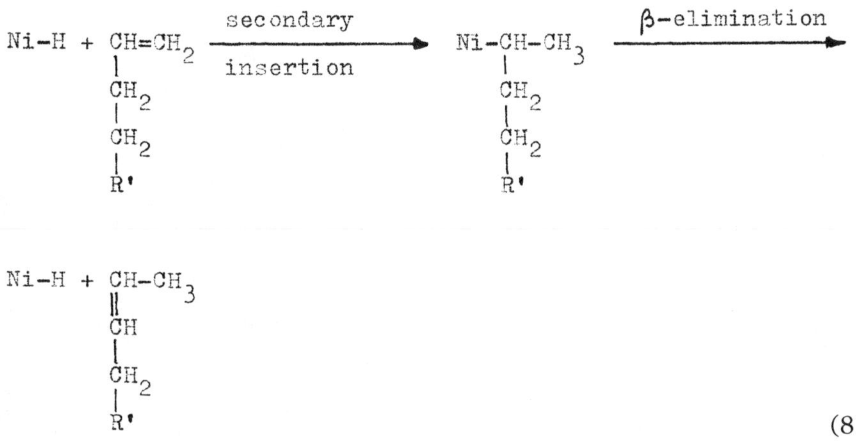

(81)

This reaction occurred in all cases of ethylene–higher α-olefin oligomerization: the added α-olefin was partially isomerized to *cis*- and *trans*-2-olefins. Conversions in these isomerization reactions varied from 2.5 to 6%. *Cis*- and *trans*-2-olefins with linear chains are also formed in ethylene homooligomerization reaction [227]. Their presence in ethylene oligomers can be similarly explained as a secondary reaction (reaction 81) with the participation of the ethylene oligomers. Olefins with other internal double bonds cannot be formed in reaction (14) from α-olefins. Their possible formation from other olefins with internal double bonds (e.g., formation of 3-olefins from 2-olefins, etc.) in reactions analogous to reaction (81) is severely restricted.

Olefins can participate in chain propagation reactions (reaction 3) according to the same two mechanisms as in initiation reactions:

$$Ni{-}(CH_2{-}CH_2)_n{-}CH_2{-}CH_3 + CH_2{=}CHR \xrightarrow[\text{insertion}]{\text{primary}}$$

$$Ni{-}CH_2{-}CHR{-}(CH_2{-}CH_2)_2{-}CH_2{-}CH_3 \qquad (82)$$

$$Ni{-}(CH_2{-}CH_2)_n{-}CH_2{-}CH_3 + CH_2{=}CHR \xrightarrow[\text{insertion}]{\text{secondary}}$$

$$Ni{-}CHR{-}CH_2{-}(CH_2{-}CH_2)_n{-}CH_2{-}CH_3 \qquad (83)$$

If the *chain termination* occurs at the point where the last unit in the growing oligomer chain is the ethylene unit, α-olefins with vinyl double bonds and branched alkyl groups are formed. The type of branching in such olefins depends on the type of the higher olefin involved and on the position of the olefin unit in the oligomer chain. If the chain termination occurs immediately after primary olefin addition to the Ni–C bond, it results in the formation of olefins with vinylidene double bonds. When ethylene homooligomerization with the catalytic system containing Ni-ylide and $AlEt_2OEt$ takes place [227], ethylene oligomers participate in auto co-oligomerization reactions with ethylene.

If the chain termination occurs immediately after secondary olefin insertion into the Ni–C bond, it results in the formation of linear olefins with internal double bonds:

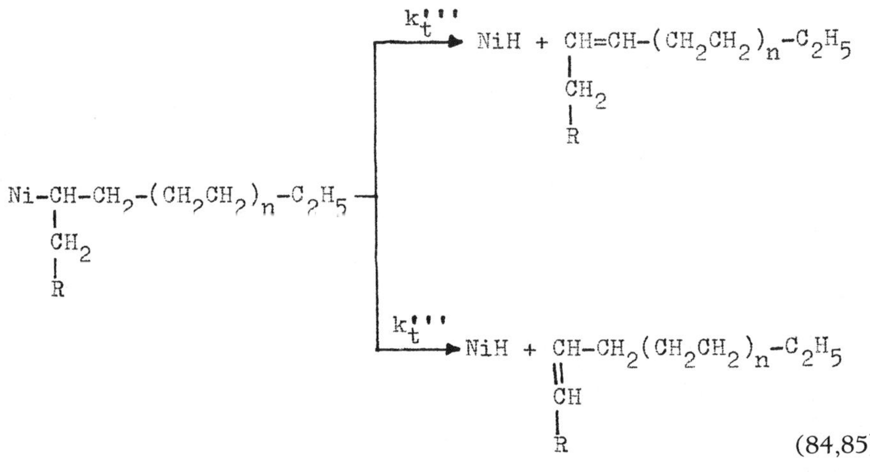

(84,85)

These olefins are also present in significant concentrations in the co-oligomerization product, indicating the importance of the secondary olefin insertion into the Ni–C bond.

Higher olefins have much lower reactivity in co-oligomerization reactions compared with that of ethylene. Chain propagation reactions (insertion into the Ni–C bonds) with participation of α-olefins exhibit poor regioselectivity, primary insertion being 60% more probable than the secondary insertion. Ethylene is significantly more reactive in chain propagation reactions: 50–70 times compared with olefin primary insertion and 100–120 times compared with olefin secondary insertion. Reactivities of α-olefins in chain propagation reactions decrease slightly for higher alkyl groups. Reactivities of Ni–C bonds in chain propagation and chain termination reactions strongly depend on the structure of the alkyl group attached to the Ni atom. The Ni–CHR–CH_2–R bond has very low reactivity in ethylene insertion reactions and usually decomposes in the β-hydrogen elimination process. Kinetic analysis of olefin co-oligomerization reactions provides numerous analogies with olefin copolymerization reactions in the presence of Ziegler–Natta catalysts.

O. Ruthenium- and Rhodium-Based Catalysts

$RuCl_3$ in methanol medium is reported to be a dimerization catalyst for ethylene, but with poor selectivity and activity [239]. When $RuCl_3$ is heated with 500–800 atm of ethylene at 130°C for 10 h, 2-butene is formed along with hexenes, octenes, and higher olefins [240].

Cramer [241] has made a detailed study of the dimerization of ethylene under the influence of *rhodium chloride* in alcoholic hydrochloric acid solutions and proposed the mechanism as shown in the following reactions:

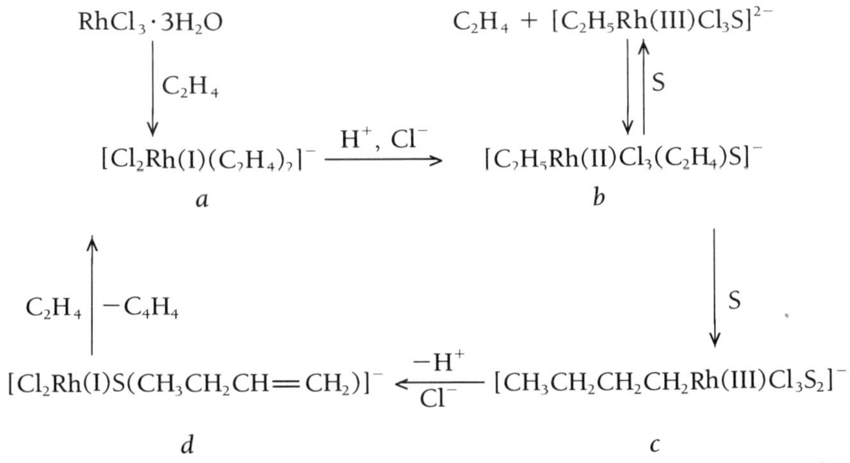

(86)

where S = Cl^-, H_2O or solvent.

Activation of the catalyst occurs by the reaction between $RhCl_3$ and ethylene with formation of a complex anion (compound *a*, reaction 86) of univalent rhodium with two ethylene ligands. There occur a fast protonation of a given ethyl complex *b* and the insertion of a coordinated ethylene at the rhodium carbon σ-bond (the rate determining step gives the butyl complex *c*). This is rapidly converted into a complex compound *d*, from which a molecule of butene is replaced by ethylene, regenerating the starting complex *a*. The solvent molecules are introduced into the reaction pathway in order to satisfy the coordination number of Rh(I) and Rh(III).

However, Shrock and coworkers [126–128] have proposed that the crucial intermediate in Cramer's ethylene dimerization system could be a metallacyclopentane complex and the butyl complex is formed as a result of its protonation:

$$[Cl_3(C_2H_4)\underline{RhCH_2CH_2CH_2CH_2}]^{2-} \xrightarrow[\text{solvent (S)}]{H^+} [Cl_3(C_2H_4)Rh(C_4H_9)S]^- \qquad (87)$$

The rate of dimerization of ethylene is described by Cramer [241]:

$$\frac{d[C_4H_8]}{dt} = k[C_2H_4][H^+][Cl^-][Rh] \qquad (88)$$

It is also possible to dimerize ethylene by a nonionic reaction using $(\pi\text{-}C_2H_4)Rh(C_2H_4)_2$ as a catalyst, but the reaction is not fast.

Rhodium chloride supported on silica gel is found to be more active than the homogeneous catalyst for ethylene dimerization [242,243]. As for the homogeneous catalyst, hydrogen chloride remarkably enhanced the catalytic activity. 1-Butene formed in the initial stage is isomerized to 2-butene. The dimerization activity per unit weight of catalysts increases in the order silica gel supported > silica alumina supported > alumina supported.

The active rhodium species of the supported catalysts is the surface compound *a* formed from rhodium chloride and the surface silanol according to the following reaction:

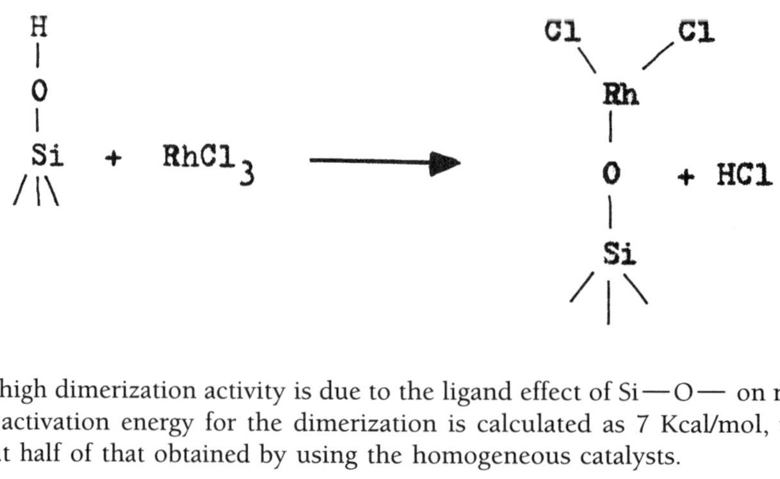

(89)

The high dimerization activity is due to the ligand effect of Si—O— on rhodium. The activation energy for the dimerization is calculated as 7 Kcal/mol, which is about half of that obtained by using the homogeneous catalysts.

$[Rh(SnCl_3)_2Cl_4]^{3-}$ immobilized on AV-17-8 anion-exchange resin is a highly active, stable, and selective catalyst for the dimerization of ethylene to *cis*- and *trans*-2-butene in acidic media [244]. A mathematical model has been derived to describe the process in terms of the ethylene pressure, temperature, and reaction time. The optimum yield of butenes is 88%. A *cis/trans* ratio of 1:2.7 is obtained after 6 h at 75°C and 42 atm of ethylene.

P. Palladium and Other Platinum Group Complexes

Use of palladium complexes in ethylene dimerization is less prevalent, presumably because of palladium's activity and higher cost.

Palladium chloride dimerizes ethylene at 20–70°C and 1–40 atm to yield butenes with 90% selectivity [245]. It has been suggested that this compound initiates the dimerization in the presence of hydrogen chloride by a mechanism analogous to that of Friedel–Crafts catalysis:

$$
\begin{aligned}
&CH_2{=}CH_2 \xrightarrow[HCl]{PdCl_2} CH_3CH_2^+PdCl_3^+ \xrightarrow{CH_2{=}CH_2} \\
&(CH_3CH_2CH_2)^+PdCl_3^- \rightarrow (CH_3CH_2^+CHCH_3)PdCl_3^- \rightarrow \\
&PdCl_2 + HCl + CH_3CH{=}CHCH_3 \qquad (90)
\end{aligned}
$$

The dimerization of ethylene into *n*-butenes by means of tetrachlorobis(ethylene)palladium in nonhydroxylic media (benzene or dioxane) has been attempted [246]. Other palladium salts (fluoride, bromide, iodide, nitrate) tested in the dimerization of olefins do not form complexes of the type $(C_2H_4)_2Pd_2X_4$. Palladium cyanide dimerizes ethylene twice as slowly as $PdCl_2$, probably on account of deactivation of the catalyst by a polyethylene deposit formed along with the dimer.

The solvent used in these dimerization reactions dictates the activity of the catalyst. Besides benzene and dioxane, acetic acid [246,247], halogenated hydrocarbons [245,247], nitroderivatives [247], sulfones, tetrahydrofuran (THF), ethyl acetate, phenol, dimethylformamide, hydroquinone, catechol, benzyl alcohol, salicylic acid, anisole, and acetone have been used [247]. In solvents like benzene and chlorohydrocarbons and in highly polar solvents like dimethylsulfoxide and dimethylformamide, dimerization does not occur. The reaction proceeds in solvents containing oxygen atoms. The dissociative solvents like phenol and acetic acid show high solvent effect. In carboxylic acids the rate of dimerization decreases with an increase in pK_a.

Use of deuterated acetic acid as a solvent shows that the dissociated anion is taking part in the reaction. The rate of reaction follows the order weakly dissociative solvent > dissociative solvent > nondissociative solvent.

An important contribution to the mechanism of the dimerization by palladium is due to Ketley [248]:

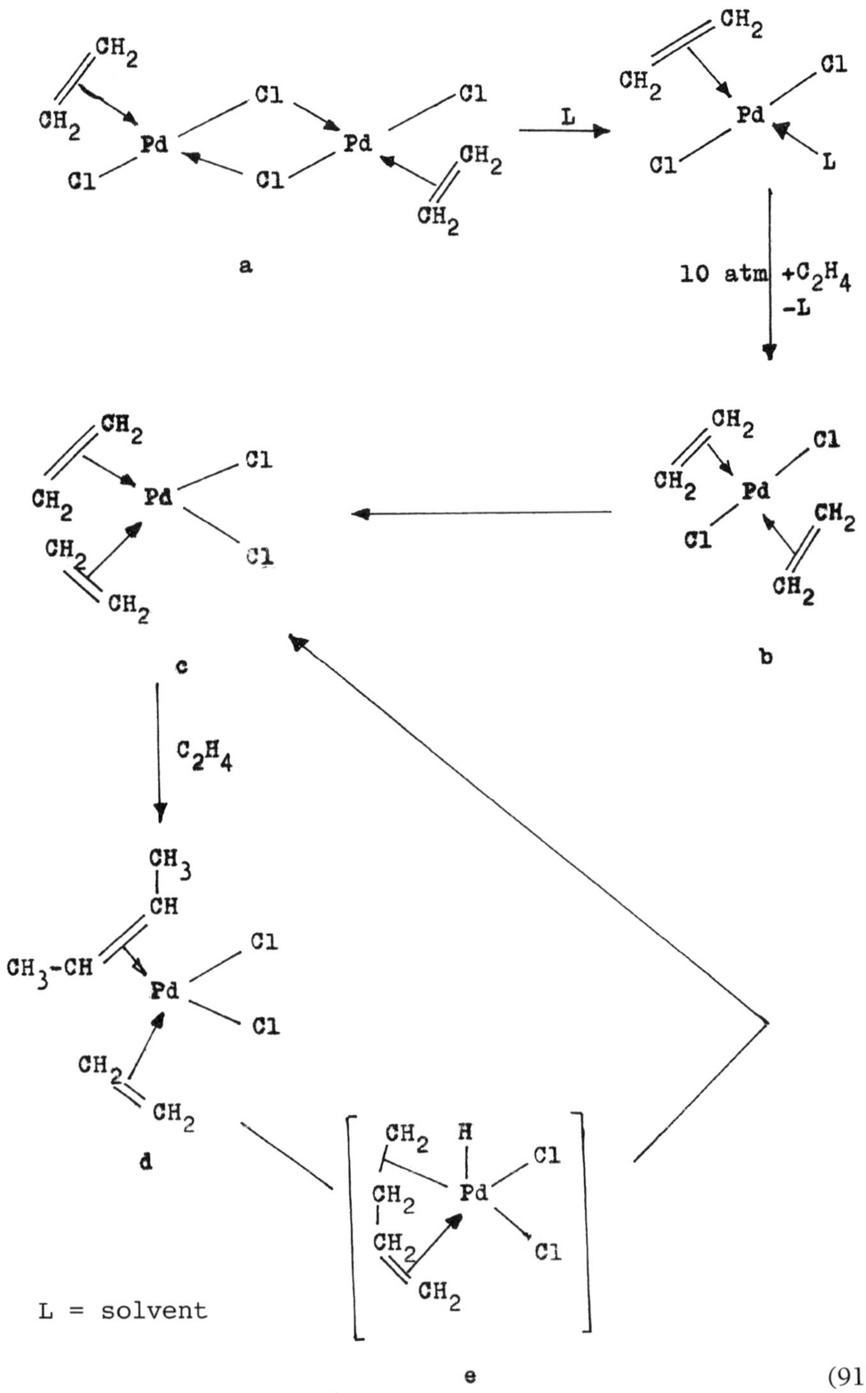

(91)

According to Ketley, in the first step solvents like ethanol acting as weak ligands or additives will cause the opening of the chloride bridges of the Kharasch complex *a* (reaction 91), resulting in the formation of complex *b* under a positive pressure of ethylene. Now complex *b* can undergo geometric isomerization bringing the two ethylene molecules into position next to each other. The detailed mechanism of the conversion of *c* to *d* is far from clear, no evidence having been obtained on the intermediate metal hydride. Ketley has postulated the transient formation of a Pd–H species. This can arise from a vinylic hydrogen abstraction by the metal (*c* → *e* → *d*).

The dimerization of ethylene by $Pd(BzCN)_2H_2$ has been reported by Barlow et al. [249]. Dimerization and the accompanying isotopic exchange of ethylene have been studied with $Pd(BzCN)_2Cl$ in benzene [250]. Both ethylenes take place after an induction period, the extent of which is determined by the presence of a hydrogen releasing olefin such as 3-methyl-1-butene. The kinetics of dimerization is first order in ethylene, and the rate does not immediately respond to changes in ethylene pressure. The reaction product is exclusively butenes with a composition of 1-butene (4%), *cis*-2-butene (36%), and *trans*-2-butenes (60%), which are close to the values corresponding to the equilibrium of isomerization.

(η-Arene)$PdAl_mCl_{n-2}$ (arene = benzene, toluene, *p*-xylene, $m = 1$, $n = 7$; arene = benzene, $m = 2$, $n = 7$) catalyzes the dimerization of ethylene (85–90%) at room temperature [251]. The suggested mechanism for these reactions involves the substitution for one arene unit by ethylene followed by coupling of two ethylene molecules:

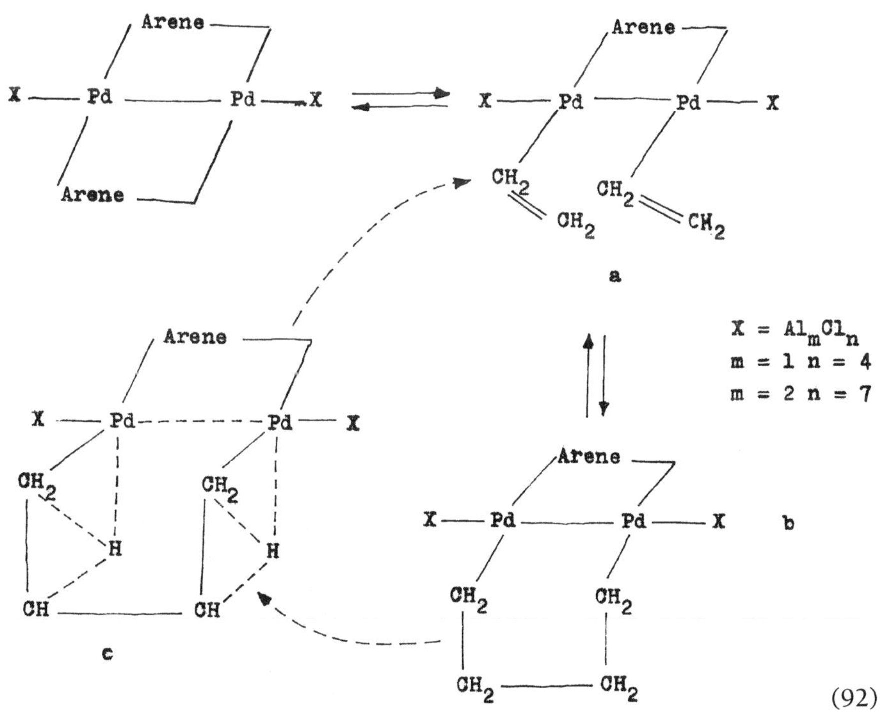

(92)

The evolution of *b* can be supported to proceed through a hydride transfer process (*c*) producing butene and, in the presence of ethylene, again compound *a*.

Palladium chloride supported on silica gel and γ-alumina are active for the dimerization of ethylene [252]. $PdCl_2(Me_2SO)_2$ and $K^+PdCl_3(MeSO)$ on KSH-2 silica gel catalyze ethylene dimerization. The yield of dimer and selectivity are high with use of 0.1% dimethylsulfoxide complex precipitated from acetone at 90°C and with 0.3% ionic complex precipitated from chlorobenzene at 100°C [252–255].

$IrCl_3$ in alcohol medium is capable of dimerizing ethylene to 95–99% 2-butene and 1–5% 1-butene [256].

$Pt(PPh_3)_4$ supported on heterogenized polystyrene [256] or *p*-chlostyrene–divinylbenzene copolymers along with $BF_3 \cdot OEt_2$ in hexane medium dimerizes ethylene to a mixture of butenes.

Q. Other Catalytic Systems for Oligomerizing Ethylene

Catalytic systems employed for ethylene oligomerization can also contain various organic or inorganic compounds. Thus, processes have been noted in which catalytic systems are made up of alkyl aluminum dichloride and organic dithiometal-activators combined with Mg salts at a molar ratio 1.0:0.1–10:0.5–10 [257]; zirconium carboxylate–alkyl aluminum chloride (low temperature radical oligomerization of ethylene [238]; and a ternary catalyst system with a small amount of BuONa as the third component [113,124,258–262]. This ternary system was more active than the binary system. A catalytic activity of 12,935 g Zr/h after 2 h reaction at 90°C and a selectivity of 80% for C_4–C_{10} olefins with 84% linearity were attained using a ternary catalyst system. The infrared spectra of $ZrCl_4$ and BuONa in cyclohexane indicated that $(BuO)_nZrCl_{4-n}$ (n = 1–4) existed in the ternary catalyst system.

Vanadium and molybdenum compounds in combination with $EtAlCl_3$ and Al_2O_3, respectively, are also used in the oligomerization of ethylene [263,264].

Various zeolites, for example, HZSM-5, omega zeolites, ZSM-25, HZSM-11, and HZSM-48, have been investigated as ethylene oligomerization catalysts [265–277]. These catalysts yielded gasoline and diesel fuels. The process occurred not by the classical carbonium ion mechanism, but by the formation of alkoxy structures as intermediate products. Not all of the acid OH groups (~20%) were active in this reaction, due to the structural heterogeneity caused by the presence of nonequivalent Si atoms.

Zeolites possess some of the useful heterogeneous catalyst properties like the possibility of carrying out reactions in a flow system, easy catalyst regeneration, and product separation, but they have shown poor selectivity to α-linear olefins, and a large amount of branched and cyclized olefins, paraffins, and aromatics in the products, especially when ethylene oligomerization is carried out at higher temperatures (>200°C) [275,277]. Hydrocarbon cracking is also observed at these temperatures. Only in presence of high-silica zeolites possessing strong Bronsted acid sites are linear oligomers obtained. In this case, the mechanism of the ethylene oligomerization assumes a cationic character [278].

Well-defined oligoethylenes were also synthesized using organolithium compounds and a bifunctional Lewis bases with the following structures:

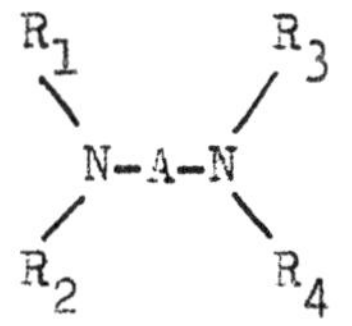

and

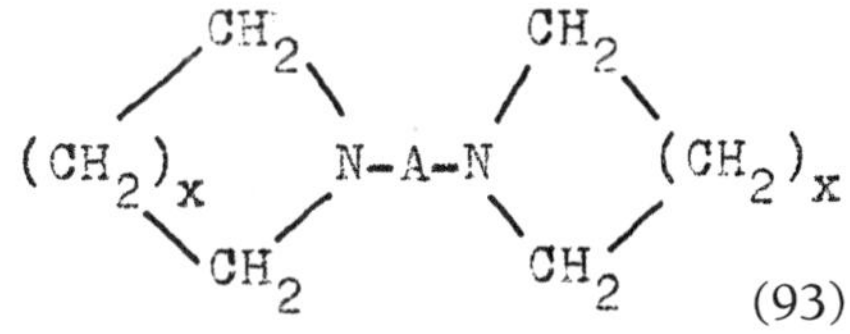

(93)

where R_1-R_4 are same or different alkyl radicals of 1 to 4 carbon atoms, A is a nonreactive group, namely, $(CH_2)_n$ where $n = 1-4$, and x is an integer from 0 to 3 inclusive [279].

Oligomers of tetrafluoroethylene (particularly pentamer) are prepared by oligomerization of tetrafluoroethylene in an organic solvent in the presence of a fluoride compound and a crown ether [280].

Organic peroxides are used for the radicalic oligomerization of ethylene to prepare waxy, low molecular weight polyethylene [281,282]. Examples of suitable peroxides are organic perioxidicarbonates such as diisopropyl peroxydicarbonate, dicyclohexyl peroxidicarbonate, and peresters such as tertiary butyl peroxypivalate. The latter peroxide gives particularly good results. The peroxides may be used in the process either alone or in mixture with each other.

R. Industrial Ethylene Oligomerization Processes

Mel'nikov [171] shows a pilot plant for ethylene oligomerization through a continuous process in the presence of a catalytic system made up of $Al(C_2H_5)_{1.5}Cl_{1.5} \cdot TiCl_2$. The obtained product is made up of higher alkenes of the class C_6 to C_{30}. Figure 1 illustrates the flow chart of the pilot plant. $TiCl_4$ and $Al(C_2H_5)_{1.5}$ in benzene or toluene with concentrations of 0.6 to 5% and 0.2 to 4%, respectively, are supplied from vessels 1 and 2 via pumps 3 in reactor 4 made of stainless steel and provided with cooling jacket 5, through which flows a mixture of glycerine and water, cooled in cooler 6. Ethylene is supplied in the reactor from a pressure vessel. Reaction temperature is $-5°C$ up to $+5°C$, and working pressure 20 to 25 kg/cm^2. Reaction mixture is passed through reactor 7 and vessel 8, where deactivation of the catalyst takes place with an alcoholic NaOH solution of the 3 to 10% concentration at 15 to 18°C and 3 to 25 kg/cm^2 pressure. NaOH:$(Al^{+3} + Ti^{+4})$ molar ratio = 6:1. After reaction mass deactivation, it is passed into vessel 10 for decantation, after which it is filtered in press filter 12. The filtrate is kept in vessel 13, where an organic phase is separated that contains oligomers, and from which an aqueous phase is eliminated. The process and the plant permit one to obtain a one-tone olefin C_2-C_{32} from 1.01 to ethene, 3.5 kg $Al(C_2H_5)_{1.5}$, 2 kg $TiCl_4$, and 6.3 kg NaOH.

Large industrial-scale production of α-olefin oligomers is carried out by Gulf Oil Chemical Co. [283–287], Ethyl Corp. [288], and by Shell Oil Co. [288–291]. The Gulf and Ethyl processes use Et_3Al as catalyst although they differ in some details. In the Ethyl process, in contrast to the Gulf process, the products are recycled to increase the amount of the most desirable linear α-olefins C_6-C_{14}. This leads, unfortunately, to higher branched olefin content in the yield as a result of higher α-olefins reactions with alkylaluminum compounds.

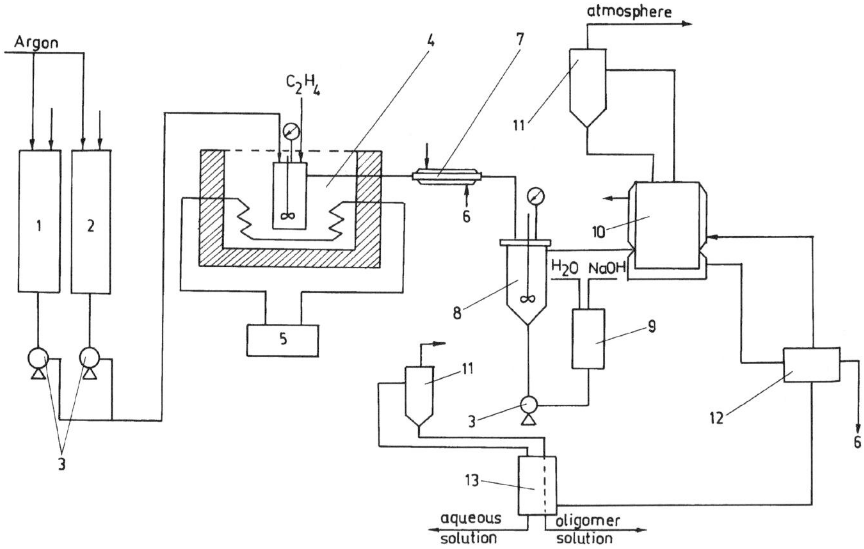

Figure 1 Basic layout of a plant for the production of oligoethylenes. 1: solution vessel for $TiCl_4$; 2: solution vessel for $Al(C_2H_5)_{1.5}Cl_{1.5}$; 3: pumps; 4: oligomerization reactor; 5: vessel for freezing agent; 6: freezing agent; 7: heat exchanger; 8: deactivation vessel; 9: tank for alkaline solution; 10, 13: separators; 11: condenser; 12: filter. (Adapted from Ref. 171.)

The Ethyl process [292] is carried out at 100°C under 210 psig pressure, and the Gulf process at 190°C under 4000 psig uses 14% Et_3Al in heptane solution as catalyst. The ethylene conversion reaches 85%.

The most modern ethylene oligomerization process—SHOP (Shell higher olefin process)—consists of three steps: ethylene oligomerization, oligomers isomerization and cometathesis (Figure 2) [291].

Ethylene oligomerization is a catalytic process, with a homogeneous catalyst consisting of nickel chlorite and the potassium salt of *o*-(diphenylphosphino)benzoic acid in 1,4-butanediol. The oligomerization reactors are operated at 80–120°C and 1000–2000 psig. The rate of reaction is controlled by the rate of catalyst addition. High partial pressure of ethylene is required for a good rate of reaction and high linearity of the α-olefin product. The catalyst for oligomerization is largely immiscible with the α-olefin product. After the reaction, a product is separated from catalyst solution and excess ethylene gas. Catalyst and ethylene are recycled. The oligomerization reaction provides a range of α-olefins with even numbers of carbon atoms from C_4 to C_{40}. These are fed into a distillation column and split into three fractions: a C_4–C_8 fraction, the desired C_{10}–C_{14} fraction, and a heavy C_{16}–C_{40} fraction. The light and heavy fractions are fed to an isomerization reactor where α-olefins are isomerized to internal olefins:

$$CH_3CH_2CH{=}CH_2 \xrightarrow[80-140^\circ C,\ 40-250\ \text{psig}]{MgO} CH_3CH{=}CHCH_3 \qquad (94)$$

The internal olefins pass to the metathesis reactor, where the short- and long-chain internal olefins disproportionate [291]:

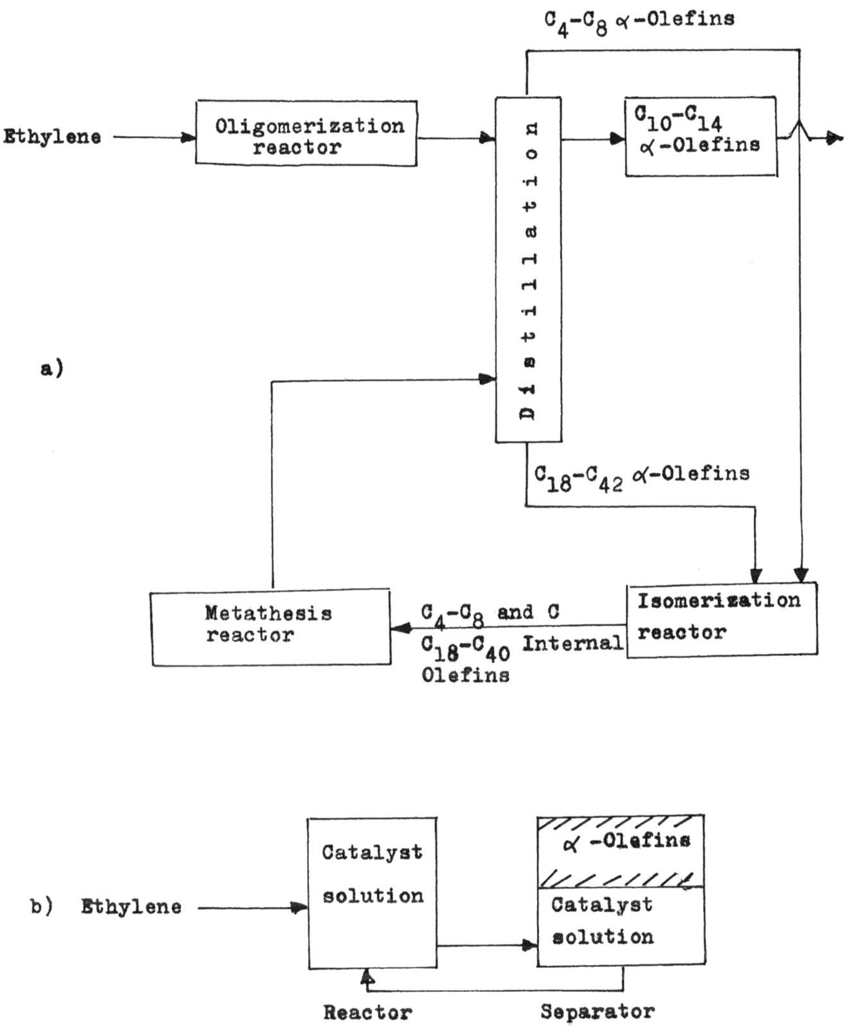

Figure 2 Shell higher olefin process (SHOP). Part (a) shows the Shell higher olefin process. Part (b) shows catalyst recycling, making the most of the two-phase solvent system. (Adapted from Ref. 291.)

$$CH_3CH{=}CHCH_3 + CH_3(CH_2)_8CH{=}CH(CH_2)_8CH_3 \xrightarrow[80-140^\circ C,\ 50-250\ psig]{MnO_2/Al_2O_3} 2\ CH_3CH{=}CH(CH_2)_8CH_3 \qquad (95)$$

When higher internal olefins are reacted with ethylene, α-olefins are obtained:

$$RCH{=}CHR_1 + CH_2{=}CH_2 \rightarrow RCH{=}CH_2 + H_2C{=}CHR \qquad (96)$$

The C_{10}–C_{14} fraction is fed into a hydroformylation reactor, where it is converted to the corresponding straight-chain aldehydes and then to alcohols. When the cobalt carbonyl complex catalyst is used, causing the internal C=C bond to migrate to the α-position in the olefins, the C_{10}–C_{14} internal olefins are passed to the hydroformylation reactor.

α-Olefin oligomerization products are characterized by a geometric molar growth factor, defined as

$$K = \frac{\text{moles of } C_{n+2} \text{ olefin}}{\text{moles of } C_n \text{ olefin}} \tag{97}$$

The weight distribution of various α-olefin oligomerization product fractions is a function of K [289]. Control of the K factor is the key to the process, since K not only sets the product distribution in α-olefin oligomerization, it also determines the average carbon number of the entire SHOP product. The K factor can be readily varied by adjusting the cocatalyst composition. The SHOP process gives highly pure linear α-olefins in the range C_6–C_{18}.

A new ethylene oligomerization process, using as catalyst the nickel salt of an organic acids–Et_3Al system [293], has been patented in the United States. Also, in the Soviet Union, the systems $Zr(OPr)_4$–$Et_3Al_2Cl_3$ and $TiCl_4$–$EtAlCl_2$ have been patented as catalysts for industrial ethylene oligomerization processes [294]. In this case the following conditions must be maintained: 20–30 atm, 100–110°C, and toluene as solvent. After 1 h of reaction, 515 kg α-olefins (18% C_4–C_6, 30% C_8–C_{10}, and 38% C_{12}–C_{18}) per kilogram of Zr are obtained. For the titanium catalyst, where the reaction was carried out at 20–50°C under 20–25 atm, the product consists of 18% C_4–C_8, 42% C_8–C_{10}, and 35% C_{12}–C_{18} α-olefins. The C_4–C_6 fraction recycled with unreacted ethylene resulting in an increase of the desirable C_{12}–C_{18} fraction to 93–96%, with the content of α-olefins reaching 95% in the final products. Also, the Zr catalysts described in the Idemitsu patents provide a base for commercial use [111–113].

Table 2 summarizes the available published and patent literature.

IV. OLIGOMERIZATION OF PROPYLENE

The propylene oligomerization literature is considerably poorer than that for the ethylene oligomerization. The catalysts for oligomerization of propylene are also active in ethylene oligomerization. Also in this case, the titanium, zirconium, and nickel complexes are the most popular catalysts. Propylene oligomers are more structurally diverse than ethylene oligomers.

The versatility of the oligomerization of propylene is very much evident from the extensive applications the individual products and their mixtures find. The oligomerization of ethylene essentially involves the addition of a C–H bond of one olefin molecule across the double bond of a second one. On the other hand, in the oligomerization of propylene four products are possible if only the vinylic C–H bonds are considered, two products involving a C–H bond of methylene carbon and another two involving a C–H bond of the methine carbon. The major products of the initial step of the oligomerization reaction are *n*-hexenes, 4-methylpentene, 2-methylpentene, and 2,3-dimethylbutene. The selectivity of the products depends on the metal, the ligands, and the mode of activation of the olefin.

Pilot-scale as well as industrial-scale (IFP dimersol process) oligomerization of propylene using organometallic catalysts have been reported [540,541]. The product composition, mechanism, and reaction conditions are discussed together with the various homogeneous and supported metal complex catalysts [542–548].

The propagation step of propylene oligomerization in the presence of transition metal complexes occurs through four intermediates, wherein propylene inserts into the metal–hydrogen and metal–alkyl bonds. Also, in the chain-termination reaction, β-hydrogen can be eliminated from CH_3, CH_2, or CH groups of the oligomer chain. All these possibilities give rise to various olefin isomers.

$CuAlCl_4$ and Cu_2Cl_2 in the presence of Et_2AlCl [15] in isooctane medium catalyzes the dimerization of propylene at atmospheric pressure to give *trans*-4-methyl-2-pentene (30.4%) and hexene (27.4%), with 2-methyl-2-pentene (4.5%), 2-methylpentane (1.3%), 4-methyl-1-pentene (1%), and 2-methyl-1-pentene (0.6%) as minor products. $AgAlCl_4$ and $ZnAl_2Cl_8$ with $EtAlCl_2$ or Et_2AlCl is also used for conversion of propylene to *trans*-4-methyl-2-pentene and 2-methyl-2-pentene.

A patent report [547] is available on the oligomerization of propylene in the presence of $BBu_3MR_{n-m}H_m$ (M = Al, Ga, In, Be, Mg, Zn; R = monovalent hydrocarbon radical; n = valence of metal; m = 0–3) at 195–200°C. Organoboranes promote the isomerization of the product to an α-olefin, which is stabilized by shifting an alkyl group to selectively form 2-methyl-1-pentene.

The dimerization of propylene by alkylaluminum proceeds *via* carbanion intermediate [548]. In the reaction between propylene and Al–i-Pr_3, initially an unstable alkylaluminum compound is formed:

$$CH_3{-}CH_2{-}CH_2{-}Al\langle \; + \; C_3H_6 \longrightarrow CH_3{-}CH_2{-}CH_2{-}CH(CH_3){-}CH_2{-}Al\langle \qquad (98)$$

This then reacts with propylene according to

$$CH_3{-}CH_2{-}CH_2{-}CH(CH_3){-}CH_2{-}Al\langle \; + \; C_3H_6 \longrightarrow$$

$$CH_3{-}CH_2{-}CH_2{-}C(CH_3){=}CH_2 \; + \; CH_3{-}CH_2{-}CH_2{-}Al\langle \qquad (99)$$

If the propylene dimer 2-methyl-1-pentene is taken out of the reaction mixture, the mechanism of the process becomes catalytic and a small amount of the alkylaluminum compound will be sufficient to dimerize the propylene [549]. In the dimerization of propylene at high temperature and pressure using $AlPr_3$ (Pr = propyl) catalyst, the integrated contact number K_0 is reported to be the most convenient and reliable control parameter for automatic process control [550]. It determines both the selectivity of the process and the degrees of the conversion.

Table 2 Selected Literature on Oligomerization of Ethylene

Catalytic system	Products	Ref.
$(C_2H_5)_3Al$–Ni salts	1-butene	295
$(C_2H_5)_3Al$–$(C_2H_5)AlX$; X = halogen or alkoxy group	up to 20 carbon atoms, 1-butene containing α-olefins	296
$(C_2H_5)_3Al$	1-butene (48.9%), 1-hexene (31.1%), 1-octene (16.4%), C_4–C_{14} α-olefin (3–5%)	297
$R_xAlR'_{3-x}$–$R_2SnR'_2$; $x = 2$; R = alkyl; R′ = hologen	1-butene (31%), *trans*-2-butene (36%), *cis*-2-butene (20%)	298
R_3Al; R = C_1–C_6 alkyl or aryl	1-butene, hexenes	299, 300
$(C_2H_5)_3Al$	oligomers	301
$(C_2H_5)_3Al$–*i*-Bu_3Al	α-olefins C_4–C_{40}	302
$(C_2H_5)_3Al$–Bu_3Al	olefins C_{10}–C_{40}	303
R_3Al–(*i*-$Bu_2Al)_2O$; R = alkyl	α-olefins C_{10}-C_{40}	304
$(C_2H_5)_3Al$	olefins C_6–C_{12}	305–308
$(C_2H_5)_3Al$–Ph_3N	olefins C_4–C_{40}	309
$(C_2H_5)_3Al$–Ni, Co, Pt	olefins C_2–C_{24}	310–312
$(C_2H_5)_3Al$-polyethers	oligomers	313
Al-CCl_4	oligomers, M_n = 100	314, 315
$Ti(OR)_4$–$(C_2H_5)_3Al$	1-butene (66–99%), 2-butene	316–319
$Ti(OR)_4$–R_3Al–additives	1-butene, 2-ethyl-1-butene, 3-methyl-1-butene	320
$Ti(OBu)_4$–R_3Al, R = Me, Et, Me_2CHCH_2	1-butene, hexene, polymers	321
Ti(O–*i*-Pr$)_4$–$Al(C_2H_5)_3$	1-butene, 2-butene, hexenes, polymers	322
$Ti(OBu)_4$–$(C_2H_5)_3Al$	1-butene	323–330
Ti(O–*n*-Bu$)_4$–$[(CH_3)_2CH]_3Al$	1-butene, 2-butene, polymer	331
$CpTiCl_2$–amalgams of alkali metals; Cp = cyclopentadienyl	1-butene (88%), 1-hexene (12%)	332–334
Ti–Cp	*n*-butane (7%), 1-butene (15.4%), *cis*- and *trans*-2-butene (42%), butadiene (31.6%)	335
$Ti(OR)_4$–R'_3Al; R = Bu; R′ = hexyl, Et, Bu	1-butene	336–338, 118
Ti(O–*n*-Bu$)_4$–R_3Al	1-butene	339, 340
Titanium alcoholate–R_3Al	1-butene	341
$Ti(OBu)_4$–$(C_2H_5)_3Al$–$CH_3OCH_2CH_2OCH_3$	1-butene, mixture of hexenes	342
$Ti(OBu)_4$–R_3Al–$C_3H_7OH/C_4H_9OH/C_6H_5OH$	1-butene	343
Ti(O–*n*-Bu$)_4$–$(C_2H_5)_3Al$–$P(OBu)_3$	1-butene	344
$Ti(OR)_4$–R_3Al–oxygen/amines	1-butene, higher olefins, polymer	345–347
$Ti(OR)_4$–R_3Al–$CpTiCl_2$/*o*-phenylendiamine	1-butene	348–350
$Ti(OR)_4$–$H_2AlNMe_2/H_2AlNBu_2/HAlClNBu_2$	1-butene	351
$Ti(OR)_4$–$(C_2H_5)_3Al/AlH$(*i*-$C_4H_9)_2$	1-butene (94.7%), higher olefins, polyethylene (5.4%)	352
$Ti(OC_2H_5)_2(acac)_2$–$(C_2H_5)_3Al$	1-butene	353
$Ti[N(CH_3)_2]_4$–$(C_2H_5)_3Al$	1-butene	353
(π-$C_5Me_5)Ti(OR_3)$–$(C_2H_5)_3$–Al	1-butene (90.9%), hexene (7.9%), butane (61%)	354
$(OR)_3Ti(acac)$–R_3Al–R_2HAl	1-butene	355

Table 2 Continued

Catalytic system	Products	Ref.
$Ti(OR)_4$–organoaluminum–$PR''(OR')_2$; R′ = alkyl; R″ = thionyl	butenes	356
$TiCl_4$–alkylaluminum	olefins C_4–C_{24}	357
$TiCl_4$–$(C_2H_5)AlCl_2$–$AlCl_3$	olefins, M_n = 120	358
$TiCl_4$–$(C_2H_5)_3Al_2Cl_3$	olefins, C_6–C_{28}	171
$TiCl_4$–$(C_2H_5)AlCl_2$	oligomers, M_n = 70–300	359
TiX_4–R_nAlX_{3-n}; R = alkyl, X = halogen	linear α-olefins C_{12}–C_{18}	360, 361
TiX_3Y–$(C_2H_5)_2AlCl$–MCl_3; Y = Cl, OR, OOCR; M = Fe, Sn, Ti; X = 0	linear α-olefins C_{12}–C_{18}	362
TiX_3Y–R_nAlX_{3-n}–M, M = Fe, B, Sn; Y = Cl, OR; X = Cl, n = 1, 2	olefins C_{12}–C_{20}	363
$TiCl_4$–$C_2H_5AlCl_2$–L; L = acetylacetonate	olefins C > 30 (25–55%) and olefins C > 30 (25–50%)	364
$TiCl_4$–$(C_2H_5)_3Al$	α-olefins C_{18}–C_{22}	365
$TiCl_4$–$C_2H_5AlCl_2$–styrene	—	366
$TiCl_4$–$C_2H_5AlCl_2(C_2H_5)_2AlCl$–L; L = anthracene	oligomers	367
$TiCl_4$–chloroorganoaluminums–L; L = amines, nitriles, esters	olefins C_4–C_6	368
$TiCl_4$–chloroorganoaluminums–mercaptanes	olefins C_8–C_{20}	369
$TiCl_4$–$D(C_2H_5)_3Al_2Cl_3$–$C_2H_5AlCl_2$–L; D = acetone; L = ethylmercaptane	olefins C_6–C_{20}	370
$TiCl_4$–R_nAlX_{3-n}–PX_3; X = Ph, alkyl, n = 1, 2	olefins C_8–C_{20}	371, 372
$TiCl_4$–$C_2H_5AlCl_2$–PCl_3	olefins C_4–C_{20}	373
$TiCl_5^-R_3C^+$–$(C_2H_5)_3Al_2Cl_3$; R_3C^+ = carbonium salt	oligomers	374
$(C_2H_5O)_3TiCl$–$C_2H_5AlCl_2$	olefins C_8–C_{20}	375
$(BuO)_4Ti$–$(C_2H_5)_3Al$	olefins C_4	340
TiX_4–$C_2H_5AlCl_2$–MY_n–MgZ_m; X = Cl, Br, acac; n = 2, 3; M = Ni, Co; Y = A_2NC-(=S)S—; A = C_2H_5; Z = Cl; m = 1, 2	olefins C_4–C_{10}	376
$TiCl_4$(surface ligands)–$(C_2H_5)_3Al_2Cl_3$; surface ligands = 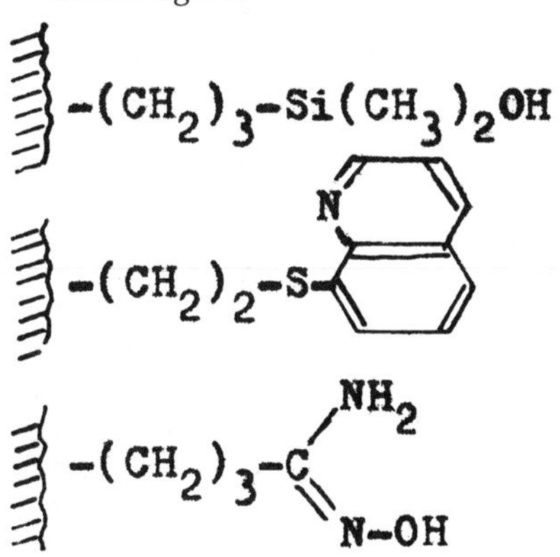	olefins C_2–C_{18}	377
$Co(acac)_3$–$(C_2H_5)_2Be$	2-butene (96%)	378

Table 2 Continued

Catalytic system	Products	Ref.
$Co(acac)_2$–BuLi	butenes	379, 380
$Co(acac)_2$–$(C_2H_5)_2Al$–OC_2H_5–BuLi	butenes and hexenes	381
$Co(acac)_2$–$(C_2H_5)_3Al$	butenes	382
$Co(acac)_2$ supported on $Al_2(PO_4)_3$–$(CH_3)_3Al$	2-olefins	383
$CoBr_2(PPh)_2$–$(C_2H_5)AlCl_2$–bochmite–bayerite–γ-Al_2O_3 impregnated with bis(*n*-butylsalicylideniminato)–cobalt(II)–$(C_2H_5)_3Al_2Cl_3$	higher olefins	384, 385
(π-allyl)nickel halide–$TiCl_4$–$VOCl_3$	2-butene	386, 387
(π-allyl)Ni halide–R_3Al	1-butene (21%), *trans*-2-butene (52%), *cis*-2-butene (27%), hexenes	388, 389
(π-allyl)Ni complex–$AlCl_3$–alkyl benzene complex	butenes, hexenes, polymer	390
(π-allyl)Ni halide–$AlCl_3$–$AlBr_3AlRCl_2$–PR_3	butenes	391–395
(π-C_2H_5)(π-C_5H_7)Ni	butenes (78%), hexenes (15%), octenes (3%)	396
(π-C_2H_5)(π-C_2H_7)Ni supported on SiO_2–Al_2O_3	butenes, hexenes, octenes	397
(π-C_2H_5)(π-C_5H_7)Ni supported on SiO_2–Al_2O_3	hydrocarbons C_4–C_{22}	167, 398
$Ni(acac)_2$–$R_{6-x}Al_2Cl_x$–PPh_3 ($x = 2$–4)	butenes	399–403
$Ni(acac)_2$–$(C_2H_5)_2Al$–BCl_3	butenes	404
$Ni(acac)_2(C_2H_5)_3Al$–$(C_2H_5O)_3Al$–PhC≡CH	1-butene (85%)	405, 406
$Ni(acac)_2$ supported on brominated polystyrene–$BF_3 \cdot OEt_2$–$(i\text{-Bu})_3Al$	butenes	407–409
$Ni(acac)_2$ supported on inorganic oxide–$(CH_3)_3Al$	1-butene, 2-butene, 1-hexene	410
$Ni(acac)_2$ attached to graft copolymer of ethylene–propylene–dicyclopentadiene with 4-vinylpyridine–$(i\text{-}C_4H_9)_3Al$	1-butene (37%), *cis*-2-butene (26%), *trans*-2-butene (37%)	411
$Ni(acac)_2$ supported on (ethylene–propylene–vinyl–norbornene)copolymer–$ClAl(i\text{-}C_3H_7)_2$–PPh_3	1-butene (80%), *cis*-2-butene (9%), *trans*-butene (11%)	412
Nickel fluoroacetylacetonate–$(C_2H_5)_2Al(C_2H_5)$	butene	413
Nickel salts–halogen–aluminum–$B(OR)_3$	2-butene	414
$NiCl_2$–$AlCl_3$–PR_3–amine	butenes (78%), C_6 olefins (20%)	415
$NiCl_2$ or $Ni(OAc)_2$–$NaBH_4$–phosphinobenzoic acid	C_4 and higher olefins	410
$NiCl_2$–$6H_2O$–$Ph_2PCH_2CO_2K$–$NaBH_4$	butenes (45%), C_6–C_{10} olefins (48%), C_{12}–C_{20} olefins (6.3%)	209
NiX_2–Ph_2PCH_2COOH–$NaBH_4$–PPh_3 (X = Cl)	C_4–C_8 (41%), C_{10}–C_{18} (40.5%), C_{20} (18.5%)	210, 211, 416
Polymeric gel immobilized $NiCl_2$–$Ni(NO_3)_2$–$(C_2H_5)AlCl_2$	butenes	417
trans-$(PPh_3)_2Ni$(σ-aryl)–Br	butenes	418

Table 2 Continued

Catalytic system	Products	Ref.
$(C_6H_5)Ni(PPh_3)_2Br-AgClO_4$	butenes	419
(*o*-Tolyl)$Ni(PPh_3)Br-$(1-naphthyl)$Ni(PPh_3)_2Br-BF_3$-$BF_3 \cdot OEt_2$	butenes	420
$NiX_2(PR_3)_2-Al_2R_{6-x}Cl_x$–haloalkane, $x = 2-4$ (X = Cl)	butenes	421–423
$NiCl_2(PBu_3)_2-(C_2H_5)_2-AlCl(C_3-C_{12})$alkyl halide	butenes	424
$LiNiPh_3P{=}CH_2$	α-olefins	425
$NiX_2-PR_3-R_n'AlX_{3-n}$ supported on solid carrier (X = halogen; R = alkyl, cycloalkyl; R′ = R; $n = 1-3$)	α-olefins	426
$NiCl_2(PBu_3)_2$ supported on $SiO_2-Al_2O_3-(C_2H_5)AlCl$	1-butene (12%), 2-butene (88%)	427
NiX_2-PR_3–pyridine–Bipyridyl–$RAlX_2$ (X = halogen, R = alkyl)	butenes	428
$Ni(PPh_3)_4-AlCl_3-AlBr_3$	1-butene (4.4%), *trans*-2-butene (68.5%)	429
$Ni(PPh_3)_4$ supported on polymer–$BF_3 \cdot OEt_2$	butenes	430
$Ni(PCl_3)_4-AlBr_3-LiBu$	butenes	431, 432
$Ni(PPh_3)_2(NO)_2-(C_2H_5)AlCl_2$	butenes, hexenes	433
$Ni(CO)_4-PPH_3-(C_2H_5)_3Al_2Cl_3$ supported on Al_2O_3	C_4 hydrocarbons	434
$Ni(CO)_4(C_2H_5)_2AlCl-(C_2H_5)AlCl_2-PPh_3$	butene, higher olefins	435
$Ni(CO)_2(PPh_3)_2$ supported on $Al_2O_3-SiO_2-(C_2H_5)_nAlCl_{3-n}$ (n = 1, 2)	butenes	436, 437
$Ni(PR_3)_2$(ethylene)$-AlCl_3$	butenes	438
$NiL_2-(C_2H_5)AlCl_2$; L = picoline	butene (59%), hexene (35.7%), octene (6.9%)	439
$NiL_2-(C_2H_5)_2AlCl$; (L = *o*-phenylene-diamine)	butenes	440
$Ni(\pi\text{-}L)_2X_2-Al_2R_{6-x}Cl_x-PR_3$ ($x = 2-4$, L = tetramethylcyclobutadiene)	C_6 olefins (17%), 1-butene (78%)	441–445
$Ni(COD)_2-CF_3COCH_2COCF_3$ (COD = cyclooctadiene)	butenes	446
$Ni(COD)_2-CF_3COOH$	butenes, higher α-olefins	447, 448
$Ni(COD)_2-Ph_2PCH_2COOH$	C_4 to C_{22} olefins	449
$Ni(COD)_2-PPh_2CH_2COOH$	C_4 to C_{24} olefins	199, 450
$Ni(COD)_2-PPh_3-Ph_3P{=}CHCOPh$	C_4 to C_{30} olefins	202
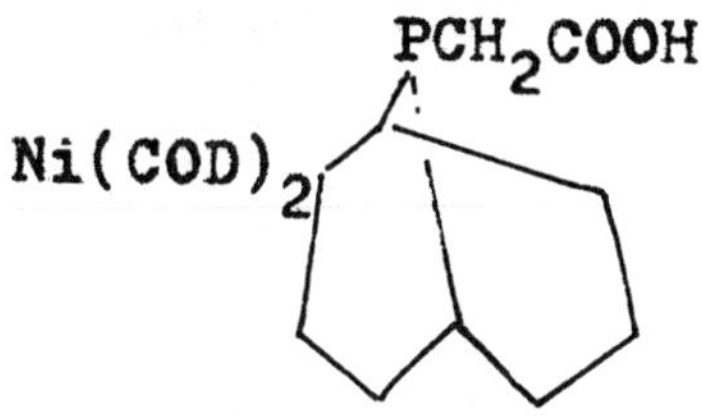	α-olefins	451
$NiX_2L_2-Al_2R_{6-x}Cl_x$ (L = hexamethyl-phosphoramide, $x = 2-4$)	butenes	452
$NiBr_2-NiCl_2$–phosphoric tris-(diamide)–AlR_3-AlRX_2	butenes, hexenes	453

Table 2 Continued

Catalytic system	Products	Ref.
Nioleate$-(i\text{-}C_4H_9)_2AlCl-(t\text{-}Bu)C_6H_4Me-$oxygen	butenes, hexenes	454, 455
Nioleate$-(C_2H_5)_3Al_2Cl_3-PPh_3$	butenes (80%)	456
Ni palmitate$-(C_2H_5)_3Al-PhC{\equiv}CH$	1-butene, *cis*-2-butene, *trans*-2-butene	457–459
Ni diisopropylsalicylate$-Al_2R_{6-x}Cl_x$ (x = 2–4)	1-butene, *cis*-2-butene, *trans*-2-butene	460
$Ni(RCOO)_2-R_mAlX_{3-m}-PX_3$ (R = alkyl, aryl; m = 1, 2; X = halogen)	butenes	461, 462
Bochmite$-\gamma Al_2O_3$ impregnated with 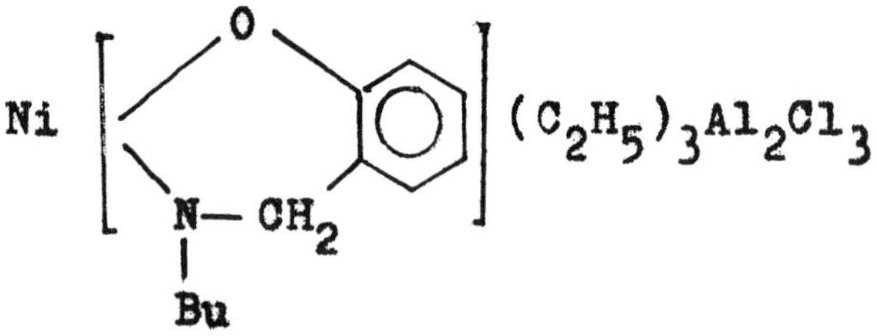	butenes	463, 464
$Ni(Me_2SO)_6\cdot NiCl_4-(C_2H_5)_3Al_2Cl_3-(C_2H_5)AlCl_2$	butenes (61%), hexenes (39%)	465
$CpNi(C_2H_4)(C_2H_5)$	butenes, hexenes	466
$Zr(OR)_4-Al(C_2H_5)_3$ (R = Bu or C_2H_5)	1-butene (90%), hexene (10%)	467
$Zr(O-n\text{-}Bu)_2(OC_6H_4Cl)_2PPh_3-(C_2H_5)_2AlCl$	1-butene (58.4%), 1-hexene (29%), 1-octene (12.6%)	468
$(\pi\text{-allyl})ZrBr_3-C_6H_5CH_2ZrBr_3-(C_2H_5)_2AlCl-(C_2H_5)AlCl_2$	C_4 and higher olefins	469
$Ru(CO)_2Cl_2$ bonded to SiO_2 or silica gel or Al_2O_3 through silane derivatives containing amino groups	butenes	470
$RhCl_3\cdot 3H_2O-HCl$	1-butene, 2-butene	471–473
$RhCl_3\cdot 3H_2O$	1-butene, 2-butene	474
$RhCl_3-LiAlH_4-PPh_3$	1-butene, 2-butene	475
$Rh_2Cl_2(SnCl_3)_4-HCl$	butenes	476
$[Rh(SnCl_3)_2]^{-3}$ anchored on AV 17-8 anion-exchange resin–HCl	*cis*-2-butene, *trans*-2-butene	477
$PdCl_2-HCl$	2-butene (95–99%)	478, 479
$PdCl_2-$sulfones	butenes	480
$PdCl_2-(CH_3)CH(NO_2)CH_3$	butenes	481
$PdCl_4$(ethylene	butenes	482
$[(\pi\text{-methylallyl})PdCl]_2-AgF_4-PBu_3$	butenes, hexenes	483
$CpTaCl_2(C_4H_8)$ (Cp = cyclopentadienyl anion)	1-butene	127, 130
WCl_4^- 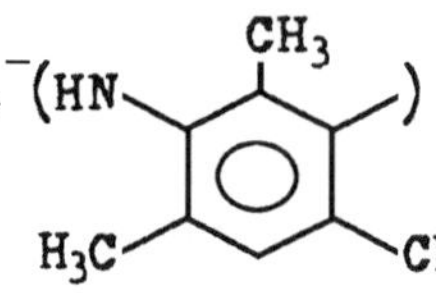 $-(C_2H_5)_2AlCl$	butenes	484

Table 2 Continued

Catalytic system	Products	Ref.
$ZrCl_aBr_b(donors)R_{3-n}AlX_n$; R = C_1-C_{20} alkyl; X = Cl,Br; *a*,*b* = 0–4; *a* + *b* = 4	α-olefins $C_{13}-C_{20}$	485
$ZrCl_4-(C_2H_5)_3AlCl-C_2H_5AlCl_2-n\text{-}BuOH$	olefins, M_n = 120	486
$Zr(OR)_4-(C_2H_5)_2AlCl-C_2H_5AlCl_2$; R = C_3H_7, C_4H_9	olefins C_{20}	487
$ZrCl_4-R_2AlR'$; R = alkyl, R′ = RO	α-olefins	488
$Zr(acac)_4$–alkylbenzene–C_2H_5AlCl	olefins	489
ZrR_4; R = OC_4H_9	olefins	490
$Zr(OCOR)_4$; R = alkyl	olefins, C_4-C_{30}	491
$Zr(O-n\text{-}C_4H_9)_2-(OC_2H_4Cl)_2-(C_2H_5)_2$-AlCl–$PPh_3$	olefins, C_4-C_8	492
$MX_n-R_2AlCl-R_3CX$; X = Cl,Br; R = alkyl; M = Ti, Zr, V, Cr	olefins $C_{12}-C_{22}$	493
$NiCl_2\cdot 6H_2O-NaBH_4-Ph_2PCH_2COOH$	olefins C_4-C_{20}	494, 495
$NiCl_2-NaBH_4-2Ph_2PC_6H_4COOH$	olefins C_{40}	496, 497
$Ni[OP{=}O)(OR)_2]_2-(C_2H_5)_3Al_2Cl_3$; R = C_1-C_6 alkyl	—	498, 499
$NiCl_2\cdot 6H_2O-NaBH_4$-*o*-diphenylphosphinophenol	olefins C_8-C_{18}	500, 501
$NiCl_2\cdot 6H_2O$–sodium bis[diphenylphosphino]-4-methyl-benzenesulfonate–$NaBH_4$	olefins C_4-C_{18}	502
$L_nNi-[O-(R)HOC_6H_4PR_2]$; L = COD; R = alkyl; n = 2–4	oligomers	503, 504
$Ni(acac)_2$–haloorganoaluminums–halohydrocarbons	oligomers, *M* = 30–300	505, 506
$Ni(CO)_4-(C_2H_5)_3Al_2Cl_3-PPh_3$	olefins C > 8	507
$Ni(acac)PR_3Cl$; R = C_2H_5	olefins	508
$Ni(acac)_2$ supported on $(SiO_2/Al_2O_3)-(C_2H_5)_3Al_2Cl_3$	olefins C > 6	509, 510
$Ni(acac)_2-(SiO_2/Al_2O_3)-(C_2H_5)_3Al_2$-$Cl_3PPh_3$	olefins C_8-C_{14}	511
π-C_3H_8NiCl anchored on γ-$Al_2O_3-(C_2H_5)_3Al_2Cl_3$	olefins C_8-C_{14}	512–514
$Ni(COD)_2-P(CH_2COOH)_3$	α-olefins	515
$Ni(COD)_2-CH_3P(C_6H_4COOH)_2$	olefins C_4-C_{10}	516
$Ni(COD)_2$–9-(carboxymethyl)–9-phosphabicyclo[3.3.1]nonane	linear α-olefins	517
$Ni(COD)_2-RPCH_2c({=}O)NA_2$; R = alkyl, A = H, aromatic group	α-olefins	518, 519
$Ni(COD)_2-Ph_3P{=}CHC({=}O)X$; X = OH	olefins C_4-C_{20}	520
$Ni(COD)_2-Ph_2PCH_2COOH$	α-olefins C_4-C_{27}	521
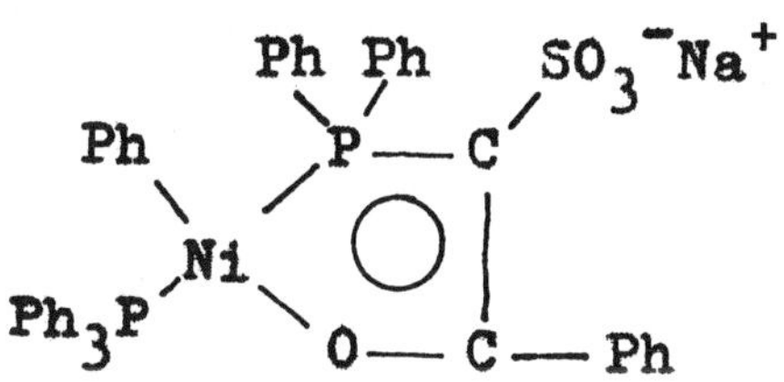	α-olefins	522–525

Table 2 Continued

Catalytic system	Products	Ref.
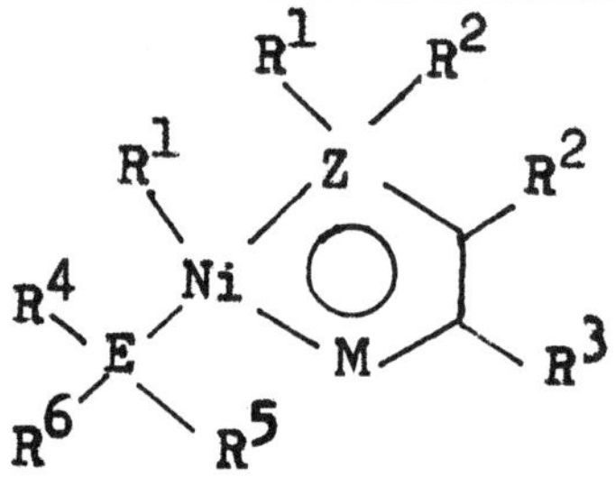 R^1–R^6 = H; M = S, O; E = P, As; Z = P, As	linear α-olefins	525
$(COD)_3Ni_2(CO)_2$ supported on Al_2O_3/SiO_2	olefins C_4–C_{14}	524–527
$[Pd(CH_2CN)_4](BF_4)$–CH_2CN	olefins C_4–C_{10}	528
BF_3–Al_2O_3	olefins C_4–C_{10}	529
Hydrogen mordenite	olefins	530
Ni–Al_2O_3/SiO_2	olefins	531
MoO_3–Al_2O_3	olefins	532
Ni–NaY zeolite	olefins	533
NiO–Al_2O_3–SiO_2	oligomers	534, 535
K-graphite	olefins C_8–C_{14}	536
H-ZSM-5	olefins, paraffins	537, 538
OGO-1 zeolite	olefins	539

Calculations and experimental data show that optimum process conditions are achieved at $K_0 \sim 2$-4.

Schmidt and coworkers [551] have reported the effect of organoaluminum compounds on the catalytic properties of complex catalysts used in the oligomerization of propylene. The selectivity for the process depends strongly on temperature, pressure, and contact time. In a continuous process of oligomerization of propylene at 200 atm with $AlEt_3$ activation it is observed that the yield of oligomers increases with temperature [552]. Studies on the oligomerization kinetics [553,554] have shown that the reaction order with respect to propylene is close to unity. Activation energies of 11.7 and 14 kcal mol^{-1} have been reported [553,554].

The Ziegler oligomerization of propylene is used as the first stage in the production of isoprene from propylene [555]. The Goodyear Tire and Rubber Co., which produces 2-methyl-1-pentene on an industrial scale, has carried out extensive research on the use of a propylene dimer as a starting material in the industrial synthesis of isoprene [556].

There is a patent [15] on $PbAl_2Cl_8$–$EtAlCl_2$ or Et_2AlCl catalyzed oligomerization of propylene to *trans*-4-methyl-2-pentene and *n*-hexenes. Higher activities at moderate temperatures and pressures are observed when compared with those of the Friedel–Crafts catalysts.

Propylene shows outstanding inclination toward the cationic oligomerization. Therefore, the acidic system $TiCl_4/EtAlCl_2$ oligomerizes propylene to irregular strongly branched olefins [56]. A change of the chlorine atoms in this system, whether the one next to titanium or aluminum for the donor ligands, for example, leads to yield of linear olefins [56]. Both actions increase the electron density in the titanium ion and the linearity of obtained oligomers.

Investigations of the $TiCl_4/EtAlCl_2$ system during the co-oligomerization of ethylene with propylene indicated that, depending on the reaction temperature and solvent, anionic coordination and cationic center coexist.

$AlCl_3$ formed as a product of the $TiCl_4$ alkylation or a reaction of $EtAlCl_2$ with halogenohydrocarbon solvent can form the cationic center for propylene oligomerization [557,558]:

$$TiCl_4 + EtAlCl_2 \rightarrow EtTiCl_3 + AlCl_3 \leftrightarrows (EtTiCl_2)^+(AlCl_4)^- \quad (100)$$

$$EtAlCl_2 + RCl \rightarrow AlCl_3 + RH + C_2H_4 \quad (101)$$

It is assumed that the $TiCl_4/EtAlCl_2$ system oligomerizes olefins to a lesser or greater degree according to a cationic mechanism, in which the oligomer chain has a positive charge and forms an ionic pair with a negative counterion, for example, $AlCl_4^-$ [559]. A decrease in Ti(IV) ion electron affinity decreases its acidity and increases a coordination mechanism contribution to the whole oligomerization reaction, which in turn causes an increase of the product linearity.

During the investigation of the propylene oligomerization in the presence of the $TiCl_4$–*i*-Bu_2AlCl system in toluene, it was proved that the solvent participated in the oligomer chain-termination reaction as the chain-transfer agent. This is the reason that a part of the product obtained has oligomers with toluene moiety and end groups [560].

Propylene dimerization has been achieved by using an AlR_3–$Ti(OR)_4$ system with an $AlR_3/Ti(OR)_4$ ratio < 10. $Ti(acac)(OR)_3$, where R = butyl or isopropyl, with $AlEt_3$ at 60°C under 8.5 atm of propylene gives 4-methyl-1-pentene (43%), 4-methyl-2-pentene (1.6%), *n*-hexene (9%), and 2-hexene (46.4%) [561].

The dimerization of propylene occurs as a side reaction during propylene polymerization using $TiCl_3$ catalyst [562]. Several possible mechanisms of dimer formation have been discussed. In the presence of a commercial catalyst, dimer formation is 50–100 times higher than that in the presence of microspherical $TiCl_3$ catalyst obtained in the laboratory by the reduction of $TiCl_4$.

In the case of propylene oligomerization in the presence of zirconium compounds, the molecular weight of propylene oligomers depends on the kind of the zirconium catalyst used. In presence of the $Zr(acac)_4$–$Et_3Al_2Cl_3$–R_3P (R = alkyl or aryl) system, low molecular weight C_4–C_9 oligomers are obtained [563]. When the system bis(pentamethylcyclopentadienyl)dichlorozirconium activated with methyl aluminoxanes (see structure 102) is used, longer-chain oligomers, having more than C_{16} in the chain are found in products [564]:

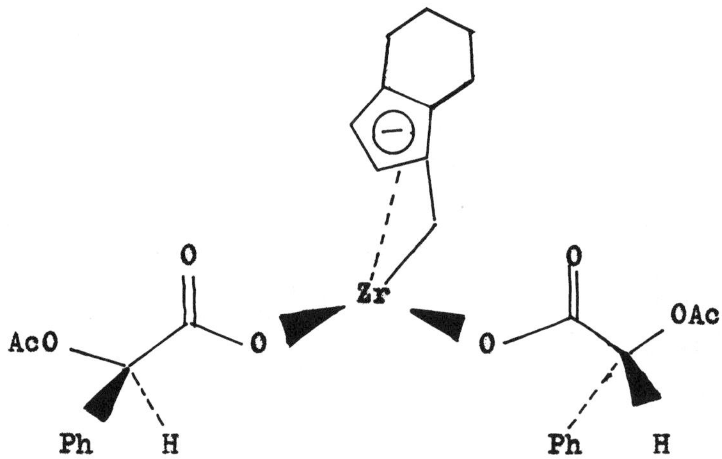

Methylaluminoxane (102)

(*S*)-[1,1′-ethylene-bis(4,5,6,7-tetrahydro-1-indenyl)] zirconium-bis(*o*-actyl(*R*)-mandelate) activated with methylaluminoxane (structural formula 102) is a stereospecific catalyst of the propylene and 1-butene oligomerization [565].

The $Zr-CH_3$ bond is the active center of zirconium catalysts. It is formed in the methylation reaction of the zirconium ions in the complexes with, for example, methylaluminoxanes. The next steps of the catalyst action are olefin complexation and olefin insertion into the $Zr-CH_3$ bond:

$$\text{Cat-H} + nCH_2{=}CH_2{-}R \longrightarrow \text{Cat-}(CH_2CHR)_{n-1}CH_2\underset{\displaystyle CH_3}{\underset{|}{C}H}{-}R \longrightarrow \qquad (103)$$

The chain termination reaction occurs as a result of the β-hydrogen elimination reaction and the Zr–H bond restoration:

$$\text{Cat-H} + CH_2{=}CR{-}(CH_2{-}CHR)_{n-2}CH_3\underset{\displaystyle CH_3}{\underset{|}{C}H}{-}R \qquad (104)$$

(π-C_4H_6)$_2$Zr(dmpe) dimerizes propylene to 2,3-dimethyl-1-butene by a process similar to the titanium analogue. A process for the conversion of propylene to C_6 and higher α-olefins by (α-allyl)$ZrBr_3$ and (benzyl)$ZrBr_3$ has been reported [566]. Zirconium(IV)acetylacetonate with $Et_3Al_2Cl_3$ and PPh_3 or PBu_3 at 65°C catalyzes the oligomerization of propylene. Molar ratios of monophosphine to zirconium of about 1–10:1 and Lewis acid to zirconium of 1–40:1 are preferred. Use of multifunctional phosphines such as 1,2-bis(diphenylphosphino)ethane in place of the unidentate phosphine in the catalyst composition shows no catalytic activity for the oligomerization of propylene. Apart from zirconium(IV)acetylacetonate, zirconium salts of cyclic and aromatic carboxylic acids like zirconium cyclohexene carboxylate, zirconium phenylacetate, zirconium benzoate, and zirconium phthalate can be used.

There is a patent on the $RR^1_nR^2NbAm$ (R = cyclopentadienyl; R^1 = benzyl; R^2 = alkyl; n = 1, 2) catalyzed dimerization of propylene in decane medium [567].

β,β'-Disubstituted and α,β-disubstituted tantalacyclopentane complexes are intermediates in the selective catalytic oligomerization of propylene to a mixture of tail to tail (tt) and head to tail (ht) dimers [568]:

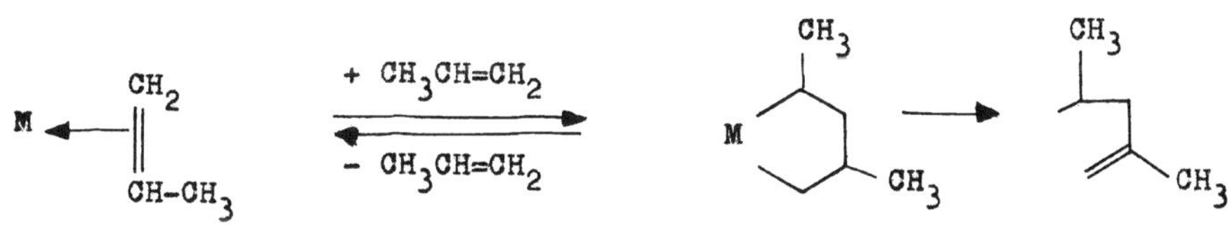

(105)

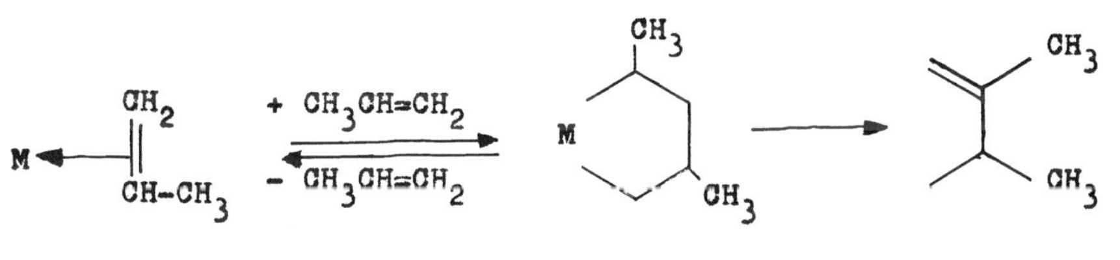

$M = [\eta^5\text{-}C_5H_5]TaCl_2$ (106)

The investigators of these reactions have observed that (η^5-C_6H_5)Cl_2-$TaCH_2CH_2MeCHMeCH_2$ is the crucial intermediate in the catalytic dimerization of propylene to largely 2,3-dimethyl-1-butene. Unfortunately, this catalyst becomes inactive after ~20 turnovers, possibly because Ta(η^5-C_6H_5)Cl_2(C_3H_6), which must be formed as an intermediate, is unstable at 25°C [568].

Deuterium labeling studies show that each tantalcyclopentane ring contacts to the tt or ht dimer [129]. An alkenyl hydride complex is an intermediate in each pathway:

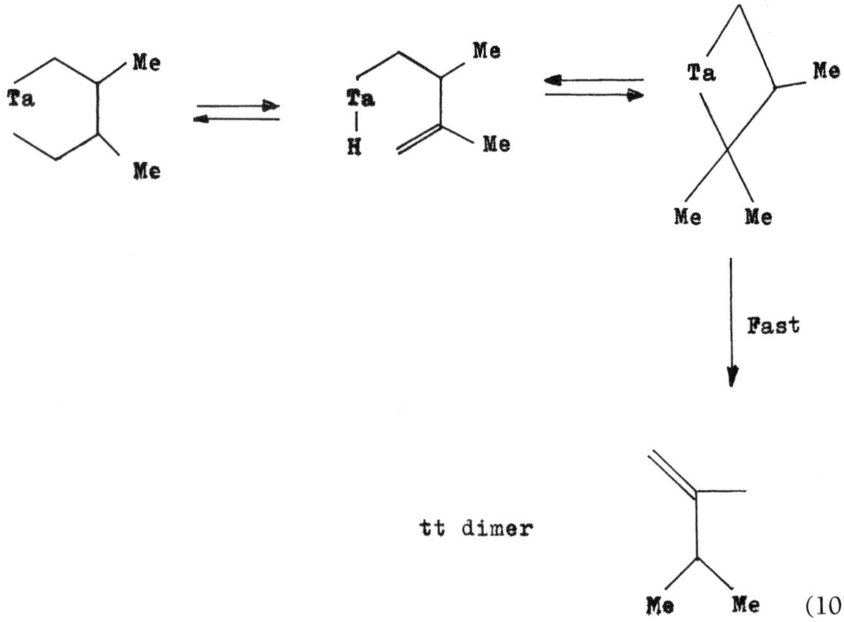

(107)

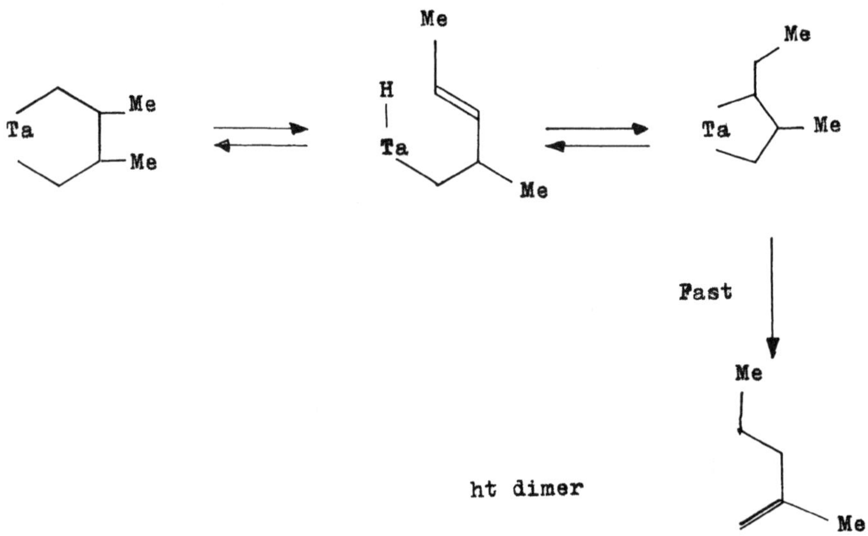

(108)

It forms reversibly from the tantalacyclopentane complex. The rate of the ring contraction step may be kinetically important, and decomposition of the tantalacyclopentane complex is fast relative to the rate at which it forms its alkenyl hydride precursor. In the second pathway, the β-hydrogen abstraction can lead to compound *a* (reaction 109) and subsequently to an α,α,α′-trisubstituted metallacycle:

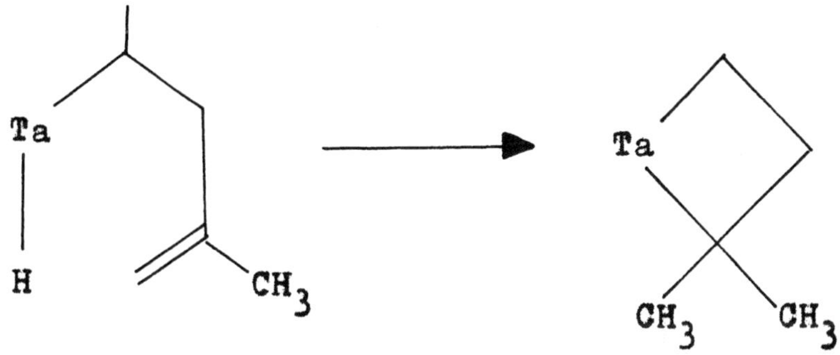
(109)

This is much more sterically crowded than the α,β-disubstituted metallacycle given in the original pathway. This choice is applicable only to α,β′-disubstituted metallacyclopentane rings, since the β,β-disubstituted metallacyclopentane must collapse to the much more crowded α,α,β-trimethyltantalacyclopentane complex.

$CrCl_3(py)_3$, $CrCl_3(4\text{-}Etpy)_3$, and $[CrCl_3(PBu_3)_2]_2$ complexes have been used as catalysts for dimerization of propylene [567,568]. In addition to dimers, trimers and tetramers are formed. The C_6 fraction is composed mainly of 2-methylpentene (69%) and *n*-hexane (31%). $EtAlCl_2$ is used as the cocatalyst in these systems.

Treatment of WCl_6 with aniline (ligand) and $Et_3Al_2Cl_3$ gives $HWCl_3 \cdot H_2NPh$, which on reaction with propylene is converted into a series of reaction intermediates [569]. The order in which the reagents are combined is important to the reaction. As the Al/W ratio decreases, the proportion of dimerization also decreases. WCl_6 or WBr_5 heated at 132°C with aniline in conjunction with $EtAlCl_2$ is also used as a catalyst for the oligomerization of propylene at 60°C. Additives like phenols [570], carboxylic acid esters [571], and diketones [572] with the WCl_6–$PhNH_2$–$Et_3Al_2Cl_3$ system influence the conversion and selectivity of oligomerization of propylene considerably. 2,3-Dimethyl-1-butene and 2-methyl-1-pentene are the major products in all these reactions. A WCl_6–HOC_6H_4–$COOCH_3$ [570] system gives 94% conversion and 99% selectivity to hexenes. On the other hand, the WCl_6–C_6H_5–CH_2OH system produces 77% conversion and 90% selectivity to dimers [571].

There is a patent report [572] on a manganese-based metal complex as a catalyst for the oligomerization of propylene. When $AlCl_3$ is added to $MnCl_2$ it forms a 1:1 complex initially. Here $AlCl_3$ acts as a Lewis acid. If $AlPr_3$ and propylene are added to this, 2- and 4-methyl-2-pentene and 2-hexenes are formed. The activity of the catalyst increases with an increase in complex concentration.

The $FeCl_3$–$AlCl_3$ (1:1) system with $AlPr_3$ cocatalyst is a good catalyst for propylene oligomerization [573]. It has higher activity than the $MnCl_3$–$AlCl_3$–$AlPr_3$ system.

The $Fe(acac)_2$–$(i\text{-}Bu)_2AlCl$ system is also known to catalyze the dimerization of propylene [574]. The dimerization of propylene to 2-methyl-2-pentene in heptane, toluene, or chloroform solvents in the presence of the tricomponent system $Fe(acac)_n$–Et_2AlCl–1,5-cyclopentadiene (where $n = 2, 3$) has also been patented [575].

The two-component cobalt catalysts such as $CoCl_2$–Et_2AlCl with methylene chloride have high activity for propylene dimerization. The following types of catalyst have also used [576]: CoX_2–AlR_3; CoX_2–AlX_3 and related systems; CoX_2–AlX_3–AlR_3 (R = alkyl, X = halogen). Many of these systems have very high activities. The transitional metal salts give 95% selectivity to dimerization in some cases. The dimer products are similar in all cases and consist of 30% hexenes and 70% methylpentenes. Selectivity to *n*-hexenes is increased to 50% at 150°C. It is presumed that cationic catalysis is involved.

The $CoCl_2$–$AlCl_3$ (1:1) complex system with $AlPr_3$ is effective for homogeneous oligomerization of propylene. It has higher activity compared with the $MnCl_2$–$AlCl_3AlPr_3$ and $FeCl_3$–$AlCl_3$–$AlBr_3$ systems [574].

$HCo(N_2)(PPh_3)_3$ has been used to oligomerize propylene [577]. 2-Methyl-1-pentene is the main product. The addition of 3 moles of tri-*n*-butyl-phosphine/mol of cobalt complex considerably reduced the oligomerization rate. A mechanism as shown in the following reactions is proposed [577]:

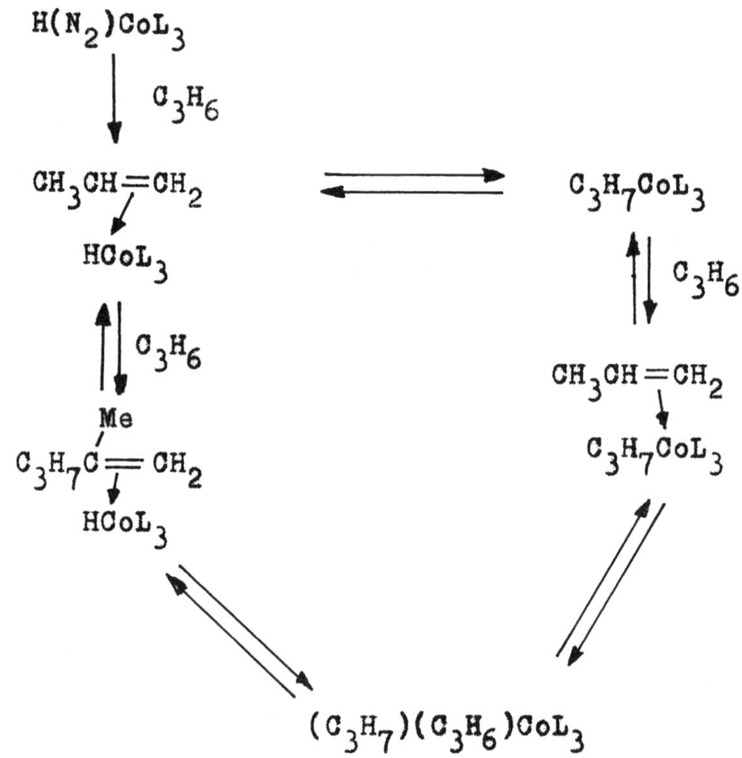

L = tertiary phosphine (110)

The kinetic of homogeneous propylene dimerization with $HCo(N_2)(PPh_3)_3$ [578] or $Co(N_2)(PPh_3)_3$ [579] as catalysts under mild conditions can be described by

$$\text{Rate} = \frac{K_1K_2K_3[\text{Co}][C_3H_6]}{[\text{PPh}_3] + K_1 + K_1K_2[\text{C}_3\text{H}_6]} \tag{111}$$

where K_1, K_2 and k_3 are constants.

The structure and mechanism of the nickel complex action in the propylene oligomerization are the same as in the ethylene oligomerization. The Ni–H and nickel–alkyl bonds are the active centers of the catalyst. The catalytic propylene oligomerization begins with olefin insertion into the Ni–H bond. Hydrogen from the Ni–H bond can be bonded to the carbon C_1 or C_2, forming, respectively, branched or linear alkyl groups bonded to the nickel ion. These alkyl groups can migrate to the propylene molecule in two modes: to the propylene carbons C_1 or C_2. These possibilities are responsible for the branched products. It is possible, however, to control these reactions with the help of the phosphine ligands. Detailed investigations [71] of this kind of catalyst have unequivocally established that the phosphine's influence on the oligomer structure has steric, not electronic, character. More bulky phosphines in these catalysts are responsible for formation of more linear oligomers.

Generally, nickel complexes do not oligomerize propylene to higher molecular weight olefins. Dimers and trimers are mainly the products of this reaction. The predominant literature propylene dimerization is based on the nature of the activity of complex organometallic catalysts based on nickel compounds and the effects of ligands attached to the nickel on the selectivity of the process [544,580–582].

The IFP dimersol process (Dimersol G) for the single-stage dimerization of propylene yields hexenes present in high-octane gasoline [583,584]. The process takes place under the rigorously controlled conditions of reduced pressure and ambient temperature in the presence of catalytic amounts of a nickel and aluminum alkyl complex, which is immediately destroyed in an NH_3–H_2O system upon completion of the dimerization.

Catalysts based on (π-allyl)nickel halide synthesized at the Max Planck Institute (Germany) [184] are used for the oligomerization of propylene on a laboratory scale as well as on a pilot plant scale. (π-Allyl)nickel halides with various cocatalysts like $AlCl_3$, $TiCl_4$, $MoCl_5$, $VOCl_4$, and WCl_3 [585] are active for the formation of *n*-hexenes, 2-methylpentene, and 2,3-dimethylbutenes from propylene. When these catalysts are modified by adding tertiary phosphines, the dimerization of propylene can be diverted, depending on the phosphine used, to give either 2-methylpentenes or 2,3-dimethylbutenes. These products are precursors of isoprene and 2,3-dimethylbutadiene, respectively. The mixtures of them can be used as motor fuel constituents. The Lurgi-Ruhrgas Co. (Germany) has set up an industrial plant for the dimerization of propylene with a capacity of about 100,000 tons/year [586].

It has been suggested that the activity of complex Ziegler catalysts toward the dimerization of alkenes arises from the formation of complex hydrides, for example, HNiCl, coordinated to the alkylaluminum halide.

The influence of phosphines in propylene dimerization has been studied extensively [159,587,588]. In the series PPh_3 to $P(i\text{-}Pr)_3$ the yield of *n*-hexenes decreases gradually from 21.6 to 1.8% according to the order $PPh_3 > Ph_2P(CH_2)_3\text{–}PPh_2 > Ph_2PCH_2Ph > Ph_2P(i\text{-}Pr) > Ph_2PCH_2\text{–}PPh_2 > P(CH_3)_3 > PEt_3 > P(n\text{-}Bu)_3 > P(CH_2Ph)_3 > P(NEt)_3 > P(NEt_2)_3 > P(i\text{-}Pr)_3$. However, the yield of 2,3-dimethylbutene increases up to 67.9% in the preceding series. Still higher yields of 2,3-dimethylbutene are achieved by using phosphines with *tert*-butyl groups. However, with the introduction of two *tert*-butyl groups in combination with one isopropyl group in the phosphine, the yield of 2,3-dimethylbutenes drops to 29.1%. There are two effects operating, namely, the inductive effect of the group R attached to the phosphorus and the steric hindrance of the *tert*-butyl groups. Better insight into the nature of the effect of phosphine on the catalytic reaction can be obtained when the mechanism and composition of dimers are taken into account.

The possible routes leading to the formation of different dimers are shown in the following reaction:

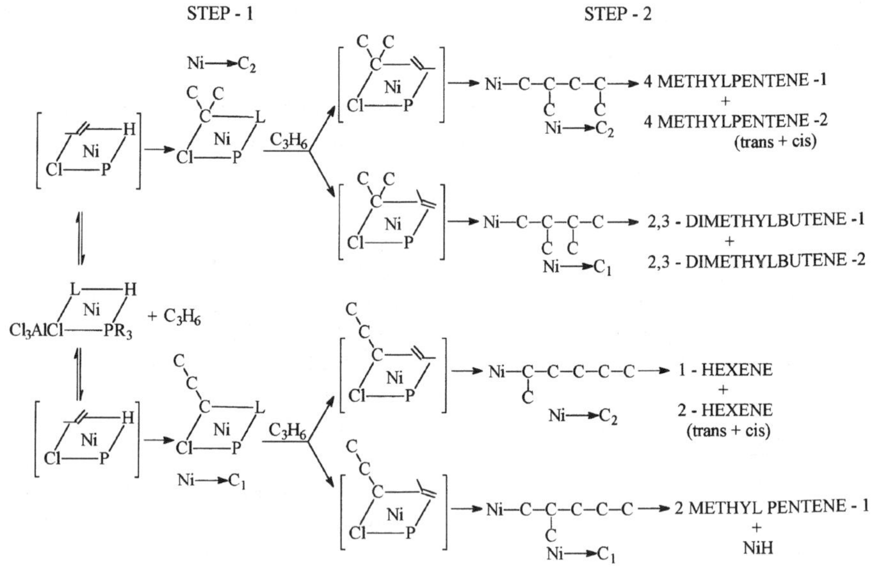

(112)

Possible routes leading to different propylene dimers in the presence of nickel complex.

A square-planar nickel hydride complex is suggested as the catalytic species [589]. In the first step, the nickel hydride catalyst adds across the double bond of propylene to give two intermediates, namely, a propyl nickel and isopropyl nickel complex. Both of these intermediates can react further with propylene by insertion of the double bond into the nickel–carbon bond, resulting in formation of four more intermediates. β-Elimination of nickel hydride from these intermediates produces the possible products of propylene dimerization, namely, 4-methyl-1-pentene, *cis*- and *trans*-4-methyl-2-pentene, 2,3-dimethyl-1-butene, *n*-hexene, 2-hexene, and 2-methyl-1-pentene. Terminal unbranched olefins are rapidly isomerized under the influence of catalyst by a process of repeated nickel hydride addition and elimination to the internal olefins. Therefore, under ordinary reaction conditions the yield of 4-methyl-1-pentene is low.

The influence of phosphines on the direction of addition of nickel hydride or nickel alkyl to propylene is debatable. If it is assumed that the isopropyl nickel complex and the propyl nickel complex have the same reactivity toward propylene, the direction of addition in the first step in the reactions (112) is given by the ratio

$$\frac{\text{Ni} \rightarrow \text{C}_2}{\text{Ni} \rightarrow \text{C}_1} = \frac{\text{\% of 4-methyl-1-pentene + 4-methyl-2-pentene + 2,3-dimethyl-2-butene}}{\text{\% of hexenes + 2-methyl-1-pentene}} \tag{112A}$$

Similarly, the average direction of addition in the second step is given by the ratio

$$\frac{\text{Ni} \rightarrow \text{C}_2}{\text{Ni} \rightarrow \text{C}_1} = \frac{\text{\% of 4-methyl-1-pentene + 4-methyl-2-pentene + hexenes}}{\text{\% of 2,3-dimethyl-1-butene + 2,3-dimethyl-2-butene + 2-methyl-1-pentene}} \tag{113}$$

To obtain these ratios for different phosphines, one must avoid the isomerization of 2-methyl-1-pentene and 4-methyl-2-pentene. With strongly basic phosphines, the rate of isomerization as well as of dimerization decreases in the order PMe_3 > PPH_3 > PEt_3 > (*i*-Pr)P(*t*-Bu)$_2$. The isomerization can be suppressed at low conversion.

The quantity of higher oligomers formed during the dimerization of propylene is also influenced by the nature of the phosphines. Their yield increases with the basicity of phosphines. Also, the propylene dimers obtained with basic phosphines are isomerized only to a slight extent.

The dimerization of propylene on a (π-C_3H_5)Ni–Br(PCy_3) (Cy = crotyl) [590] in the presence of $EtAlCl_2$ proceeds with a turnover number of 60 at −75°C and 230 at −55°C. Extrapolated to 25°C, this gives a turnover number of 60,000, which is comparable to the activity of catalase. At higher conversions the catalyst has diminished activity. The product formed included 10–15% higher olefins, 18% 4-methyl-1-pentene, 1–3% *cis*-4-methyl-2-pentene, 76% 2,3-dimethyl-1-butene, and 4% 2 methyl-1-pentene as well as traces of other isomers.

With the [(π-Cy)NiCl]$_2$–$TiCl_4$ catalyst system the activation energy for the dimerization of propylene is observed to be 15.2 kcal/mol [591]. The kinetics and product distribution for this and several other (π-alkyl)nickel catalysts have been determined with PPh_3 as proton acceptors: the yield of 2-methyl-2-pentene varies from 13 to 56.5%.

There are reports [592–594] on the dimerization of propylene catalyzed by heterogenized (π-allyl)nickel halides. Polymer-anchored π-allylic nickel complexes similar to nonsupported complexes are found to be effective catalysts for propylene dimerization after activation with a Lewis acid such as $EtAlCl_2$ (molar ratio of Al/Ni = 15:5). Using a crosslinked resin as a support, the dimerization can be performed continuously, since the catalytic centers remain active for a long time without any further addition of aluminum cocatalyst. The release of metals during this reaction is low. The reactions are carried out either in bulk propylene or in chlorobenzene solution. The conversion reaches 95% at room temperature. The product has the composition of 2% dimethylbutenes, 67% methylpentenes, and 31% hexenes. Hexene content obtained with polymer-anchored nickel catalyst is higher than that observed with the (π-allyl) nickel triphenylphosphine complex with $EtAlCl_2$. The drawback of the polymer-anchored catalyst is that it loses 40% of its initial metal content after the reaction.

$Ni(acac)_2$–$Et_2Al(OEt)$ at 40°C dimerizes propylene selectively to linear olefins, preferred solvents being ethers like diglyme. The optimum Al/Ni ratio is found to vary with different aluminum alkyls like $AlEt_3$, $Et_2Al(OEt)$, and $EtAl(OEt)_2$ [595]. With $AlMe_3$ and $Me_2Al(OEt)$ the optimum activity is observed at Al/Ni = 2:1. Maximum productivity is achieved between 20 and 40°C. Investigation of the reaction with various α-diketones reveals that the nickel complex of dibenzoylmethane shows an improvement over nickel acetylacetonate on a molar basis. For this complex the optimum Al/Ni ratio is 1:1 for the $AlMe_3$ cocatalyst. Linear dimerization activity is also observed with alkyl compounds of lithium, boron, and magnesium. Of these, the highest activity is observed by using BEt_3.

$Ni(acac)_2(i\text{-}Bu)_2AlCl$ is reported to catalyze the dimerization of propylene to yield a mixture of dimers. The dimer yield per unit weight of $Ni(acac)_2$ is improved by a factor of 5–8 when the homogeneous catalyst is replaced by $Ni(acac)_2$ on a solid carrier like Al_2O_3, SiO_2, or K_2CO_3. Catalytic activities of $Ni(acac)_2$ with Et_2AlCl, $Et_3Al_2Cl_3$, $EtAlBr_2$, and $EtAlCl_2$ are also observed [574].

Thermodynamic constants have been calculated for 12 dimerization reactions of propylene at 5, 25, 127, and 227°C, and the equilibrium composition of the dimerization products at these temperatures has been determined. The influence of various phosphine additives in the $Ni(acac)_2$–$(i\text{-}Bu)_2AlCl$ catalyst system on the composition of dimerization products has been investigated and a dimerization scheme presented [596].

The same reaction is studied in the presence of gel-like catalytic systems containing $Ni(acac)_2$ and $RAlX_2$ (X = Cl). The activity of gel-like catalytic systems in heptane, 2,2,4-trimethylpentane, and propylene dimers remains constant for hundreds of hours [597].

The activity of nickel salts supported on Al_2O_3 or SiO_2 in combination with the $(i\text{-}Bu)_2AlCl$ cocatalyst toward propylene dimerization decreases in the order $Ni(acac)_2 > NiCl_2 > Ni(NO_3)_2 > NiSO_2$ [598].

In the dimerization of propylene with a 1:45:16 Ni:Al:P mole ratio for the $Ni(acac)_2$–$Et_2Al_2Cl_3$-PPh_3 system, raising the reaction temperature from −50 to 40°C decreases the yield of 2-methylpentenes [599]. The yield of 2,3-dimethylbutenes increases, and the yield of hexenes remain constant. An increase in the P/Ni ratio from 0 to 8 at Al/Ni ratio of 45 is associated with a decrease of hexenes and an increase of 2,3-dimethylbutenes.

The effect of reaction parameters like catalyst concentration, temperature, and pressure on the selectivity of the dimerization of propylene in the presence of the catalytic system $Ni(acac)_2$–$(i\text{-}Bu)_2Al$–Cl–L (L = PPh_3, PCy_3, or $(Me_3C)_2PBr$) has been studied [600]. The isomerizing activity of phosphorus ligands is PPh_3 > catalyst without L > PCy_3 > $(Me_3C)_2PBr$. An increase in the pressure decreases the isomerization and increases the dimer yield. Optimum dimer yield is obtained at temperatures from 1–10°C to 20°C. At low catalyst concentration in the presence of $(Me_3C)_2PBr$, 23% of thermodynamically unstable 4-methyl-1-pentene is obtained. Compounds like $P(OR)_3$ are recommended as modifiers for the foregoing system in place of PPh_3 [601].

The nature of the phosphorus ligand in the $Ni(acac)_2-Et_2Al_2Cl_3-PR_3$ catalyst [589] on the dimerization of propylene in toluene has been studied. The electronic effect of substituents in triarylphosphine, $P(PhX)_3$ (X = *p*-Cl, *p*-H, *p*-Me, *p*-OMe, *o*-Me, etc.), on dimer distribution was measured. The catalytic activity decreases in the series *p*-Cl > *p*-H > *o*-Me > *p*-Me > *p*-OCH_3 > *p*-Et > *p*-Bu. Both the basicity and molecular bulkiness of the phosphines favor formation of 2,3-dimethylbutenes rather than hexenes or 2-methylpentenes.

Propylene dimerization in toluene over 1:8:3 $Ni(acac)_2-PPh_3-Et_3Al_2Cl_3$ catalyst is second order in propylene and second order in catalyst [602]. The thermodynamic parameters have been calculated for the catalyzed and uncatalyzed gas phase reactions. The foregoing catalytic system shows higher activity at 10°C than at 26°C in the isomerization of 2-methyl-1-pentene. This suggests that the catalytic species is thermally unstable even at 26°C. Very low temperatures, −50 to −20°C, have been employed to prevent deactivation of the catalyst. The kinetics of dimerization carried out in a static system containing propylene of comparatively high concentration (propylene/Ni molar ratio = 240:2400) have been studied. The rate law is given by

$$\frac{d[C_6H_{12}]}{dt} = k[\mathrm{Ni}][C_3H_6]$$

$$k = 10^{10.1}e^{-13{,}000/RT}\ \mathrm{mol^{-1}\ L\ s^{-1}} \qquad (114)$$

The structure and yield of the propylene dimerization products have been studied as functions of the catalyst composition and solvent [603]. The highest yield is obtained in toluene with a relative molar composition of $Ni(acac)_2-AlEt_3-PPh_3-BF_3\cdot OEt_2$ at 2:1:4:3.5. This corresponds to a B/Ni molar ratio of 15. At a B/Ni molar ratio <5 the system is catalytically inactive. When $BF_3\cdot OEt_2$ is preconditioned in anhydrous toluene for a few days, the optimal ratio decreases. At the optimum catalyst composition the yield is the highest in toluene, then in benzene, and much lower in chlorobenzene, yet the relative distribution of products in toluene and chlorobenzene is closer than that in benzene. The Bronsted acids activate the catalyst containing nickel(0).

Complexes of $Ni(acac)_2(C_4H_8)$ supported on polystyrene containing PEt_2, P(*i*-Bu)$_2$, $P(OEt)_2$, $P(NEt_2)_2$, and $PCl_2\cdot AlCl_3$ along with Et_2AlCl, $EtAlCl_2$, or (*i*-Bu)$_2$AlCl activators catalyze the oligomerization of propylene, giving 1300–1700 g of oligomers/h/g of nickel [604]. The catalysts are stable for >1000 h in alkane solvent. The precise oligomer composition depends on the catalyst components, solvent, and temperature.

Nickel oleate with $Al_2R_{6-x}Cl_x$ (x = 2–4) bring about the dimerization of propylene [390]. $Ni(O_2C-C_{15}H_{31})_2-Bu_2AlCl-L$ (L = dibenzo-18-crown-6, polyethyleneglycol 2000, $X(CH_2CH_2O)_2CH_2CH_2X$ where X = piperidine, PhS, or PhO) containing a 50:1 ratio of Al/Ni catalyzes the reaction at 0–10°C and 1 atm. Using crown ethers and polyethylene glycol ligands with the complex, ≤90% selectivity for methylpentenes can be achieved [605].

$Ni(O_2CR)_2$–(*i*-Bu)$_2$AlCl–piperidine or morpholine systems (R is undefined) are also known to dimerize propylene. Conversion of propylene is the highest (80–85%) with piperidine as additive [606].

Study of the dimerization of propylene in the presence of phosphine complexes of nickel(0) is available [607]. The $Ni(PPh_3)_4$–$AlBr_3$ or $AlCl_3$, $Ni(PPh_3)_4$–$BF_3 \cdot OEt_2$ [608], and $Ni(PPh_3)_4$–Et_2AlCl or $Et_3Al_2Cl_3$ [609] systems are a few of them. Addition of HF to a Ni–$(PPh_3)_4$–BF_3 catalyst system for propylene dimerization up to a Ni/HF ratio of 1:1 significantly increases the rate. Analogous results have been obtained upon the addition of HF and H_2SO_4 to the $Ni(PPh_3)_4$–$BF_3 \cdot OEt_2$ catalytic system [610]. The composition of the propylene dimers is hardly affected by the introduction of a Bronsted acid into the catalytic system. The order of addition of these acids has a significant effect on the yield of hexenes. If HF is added to the $Ni(PPh_3)_4$ solution before the Lewis acid, the yield of hexenes is doubled. There is also a report on the dimerization of propylene catalyzed by $Ni(PPh_3)_4$ in conjunction with CF_3COOH [611].

A polystyrylnickel complex prepared by oxidative addition [490] of brominated polystyrene to $Ni(PPh_3)_4$ and activated with $BF_3 \cdot OEt_2$ and a catalytic amount of water acts as an efficient catalyst for the dimerization of propylene at room temperature and atmospheric pressure. Solvents like *n*-hexane, toluene, benzene, methylene chloride, and chlorobenzene increase the rate of reaction. One role of the solvent is to swell the matrix polymer to allow access of the substrate olefin to the interior of the polymer gel. Some dipole–dipole interaction between the nickel site, the olefin, and the solvent molecule may be prevailing, so that competitive coordination of the olefin and the solvent to the nickel site may be possible. The effect of temperature shows that the rate of the dimerization reaction decreases with an increase in temperature, while selective formation of methylpentanes increases up to 90% at 40°C. 2-Methyl-2-pentane is the major C_6 olefinic product.

$Ni(PPh)_2(C_2H_4)$ in conjunction with CF_3COOH of H_2SO_4 [611] catalyzes the dimerization of propylene. Ni–$(i\text{-}PPr_3)(C_2H_4)_2$ and Ni–$(i\text{-}PPr_3)_2(C_2H_4)_2$ with $BF_3 \cdot OEt_2$ also convert propylene into dimers. Treating the same complexes with $AlBr_3$ or HCl gives complexes that dimerize propylene. However, the bis(ethylene) complex with $TiCl_4$ or WCl_6 gives rise to *trans* C_6 products from propylene. Ni–$(i\text{-}PPr_3)_2(C_3H_6)_2$ reacts with BF_3 at low temperature like −78°C. During this reaction, coordinated propylene is dimerized selectively to 2,3-dimethyl-1-butene [612]:

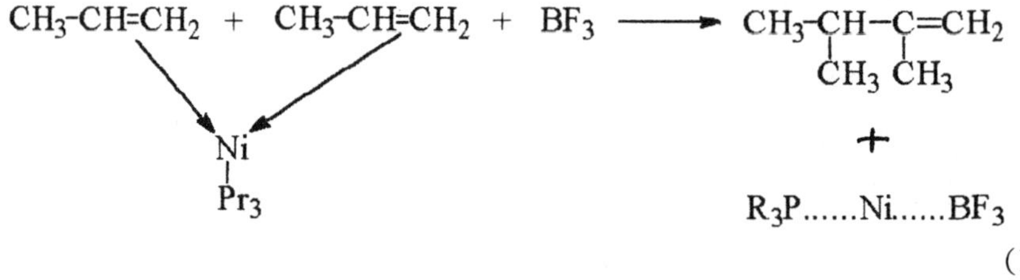

(115)

This observation leads to the conclusion that propylene ligands in the starting complex are coordinated rigidly in a "methyl to methyl" orientation and are not capable of free rotation relative to the coordination axis at −78°C. The final adducts are active catalysts of olefin dimerization.

Addition of $AlCl_3$ in chlorobenzene solution to $Ni(CO)_2(PPh_3)_2$ results in removal of two triphenylphosphine ligands and produces an active nickel catalyst for propylene dimerization. With $BF_3 \cdot OEt_2$ also, the same carbonyl complex can convert propylene into dimers. The activity of various cocatalysts decreases in the order $NbF_3 > SbF_3 > TaF_5 > BF_3 \cdot OEt_2 > AlCl_3 > InBr_3$. The surprising fact is that even in the presence of a hundredfold excess of Lewis acid no free acid remains in the medium, possibly because the Friedel–Crafts behavior of these compounds may be suppressed by trace amounts of nickel.

$Ni[P(OPh)_3]_4$ with $AlCl_3$ or $AlBr_3$ [613], $Ni[P(OEt)_3]_4$ [614,615], and $Ni(CO)_2[P(OPh)_3]_2$ [612] are active for the conversion of propylene to C_6 and substituted C_5 olefins. $Ni(C_2H_4)[P(OC_6H_4R)_3]_2$ (R = *o*-OMe) are the most active [611].

Studies of the dimerization of propylene in the absence of catalytic systems based on phosphine complexes of nickel(I) have been reported [607]. Formation of active complexes containing alkylaluminum compounds like $AlEt_3$, Et_2AlCl, and $Et_3Al_2Cl_3$ with $NiCl(PPh_3)_2$ have been observed for propylene dimerization [609]. $Ni(PPh_3)_2Cl$ or $NiCl(PPh_3)_3$ and the $BF_3 \cdot OEt_2$ system are also efficient for the conversion of propylene to dimers. Addition of Bronsted acids increases the catalytic activity.

Nickel(II) phosphine complexes like $NiCl_2(PPh_3)_2$ and $NiCl_2(PBu_3)_2$ with $R_{6-x}AlCl_x$ (x = 2–4) form active complexes for the dimerization of propylene. The degree of isomerization of propylene dimers [616] to 2,3-dimethyl-1-butene and 2-butene during propylene dimerization in the presence of $NiCl_2(PCy_3)_2$–$(i\text{-}Bu)_2AlCl$ increases in the order of solvents $C_6H_5CH_3 < m\text{-}Br_2C_6H_4Pr < o\text{-}Cl_2C_6H_4$. Halogenated hydrocarbons participate in the regeneration of the catalytic complexes and also allow secondary cationic oligomerization of propylene.

The 1:2 HNiCl and $P(i\text{-}Pr)_3$ complex with $Et_3Al_2Cl_3$ is reported to catalyze the dimerization of propylene. $Et_3Al_2Cl_3$, apart from participating in the formation of active catalytic species, maintains an adequate concentration of HNiCl. Apart from the catalytic conversion of propylene into linear dimers, tris(triphenylphosphine)tetramethylenenickel(II) can produce cyclodimers by oxidation of the complex prepared form the olefin.

The dimerization of propylene by means of the $NiCl_2L$–$EtAlCl_2$–$0.5PBu_3$ (L = tetramethylcyclobutadiene) complex [192] is faster in chlorobenzene than in benzene. The rate of the reaction also depends on the nature of Lewis acid. Moreover, the ratio of the rate of isomerization to dimerization increases with the dielectric constants of solvents. So when unisomerized dimers are wanted, the dimerization is better conducted without solvent:

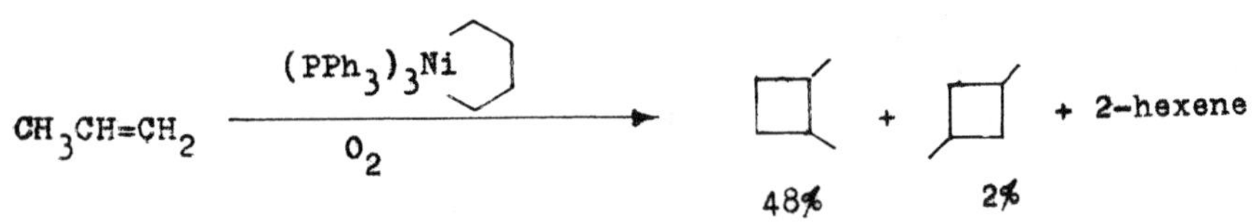

(116)

A selective dimerization of propylene to 2,3-dimethylbutene catalyzed by $R_4P[(i\text{-}Pr_3P)NiCl_3]$ with $Et_3Al_2Cl_3$ in a toluene medium has been reported [617]. The increasing temperature (−20 to +20°C) leads to the formation of C_9 olefins at the expense of 4-methyl-1-pentene. This suggests a secondary codimerization of the product olefin with propylene. Most of the olefins are the thermodynamically less favored α-olefins, indicating the absence of double-bond isomerization under these conditions. The Al/Ni ratio, although having a predominant effect on reaction rate and yield at low values, has no influence on the catalyst selectivity.

Different cocatalysts such as $R_2Al_2Cl_4$ or $R_4Al_2Cl_2$ produce a comparable catalyst with a somewhat decreased activity. The high rates of reaction in nonpolar solvents make the presence of ionic intermediates unlikely. Use of other solvents such as benzene or chlorobenzene do not produce any major change in the product composition. The following active species are formed by the reaction of the monomer with a coordinated nickel hydride:

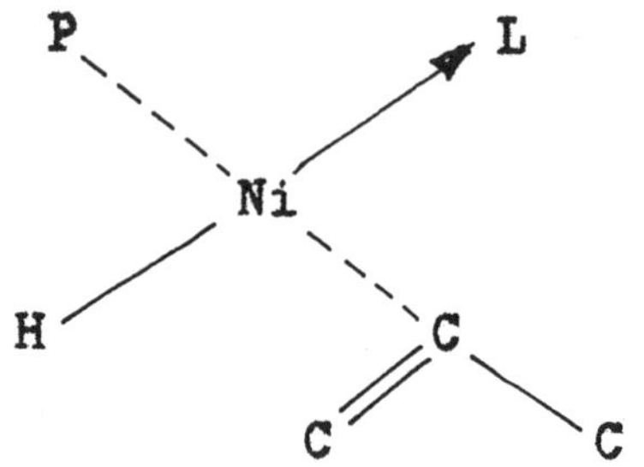

(117)

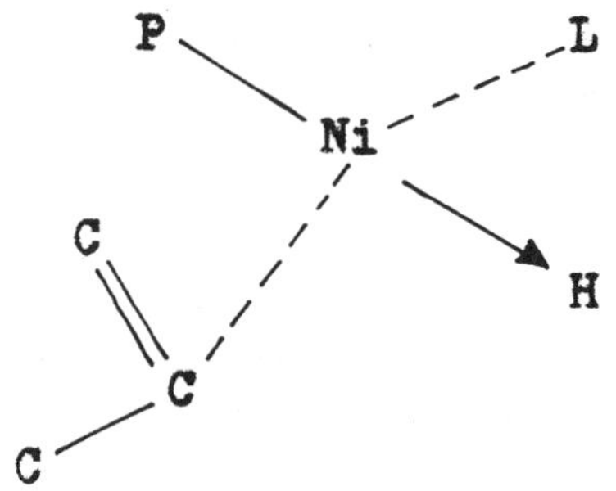

(118)

where P = triisopropylphosphine and L = Lewis acid complexed halide ligand.

The nickel–butadiene complex deposited on PCl_3-treated Al_2O_3 and activated with $AlEt_3$ shows good catalytic activity for the dimerization of propylene, giving predominantly 2-methyl-2-pentene in halogenated hydrocarbons and 4-methyl-2-pentene in pentane [618].

Diethyl(bipyridyl)nickel(II) activated as a catalyst for propylene dimerization by the addition of a Lewis acid such as $EtAlCl_2$ or Et_2AlCl with an Al/Ni ratio of 2–4 shows high catalytic activity in chlorinated aromatic solvents only. However, acry(bipyridyl)nickel halide complexes activated by the addition of $EtAlCl_2$ do not require such chlorinated aromatic solvents in order to exhibit high catalytic activity [619]. *o*-Chlorodiphenyl(bipyridyl)nickel chloride shows the highest catalytic activity among the various aryl(bipyridyl)nickel halide complexes tried. The coordination of 2,2′-bipyridyl in the active species is not clear.

The requirements for selectivity toward a particular product isomer in the dimerization of propylene have been investigated by using a square-planar nickel(II) complex with a chelating Schiff base and similar ligands and an alkylaluminum in the presence of a phosphinous atom containing additives [620]. Steric and electronic effects are separated. The observed electronic preference for the anti-Markovnicoff mode of reaction is discussed on the basis of Chatt model of olefin coordination, taking into account the assymmetrical nature of the propylene π and π^* molecular orbitals.

By mixing $Ni(CO)_4$, various phosphines, and $AlCl_3$ in a suitable solvent, catalysts that are active toward the dimerization of propylene are formed [621]. The maximum activity is obtained with a phosphine/aluminum ratio of 2. In a trigonally hybridized nickel(phosphine)$_2$–$AlCl_3$ complex that is formed, the coordinated $AlCl_3$ serves to lower the electron density of the nickel, thereby facilitating the π-bonding of two propylene molecules to the central nickel atom. The complexed olefin moieties are allowed to rotate freely around the metal–olefin bond, enabling them to attain a sterically suitable orientation that will allow the dimerization to proceed.

The two orientations favored are the "transoid" and "cisoid" geometries:

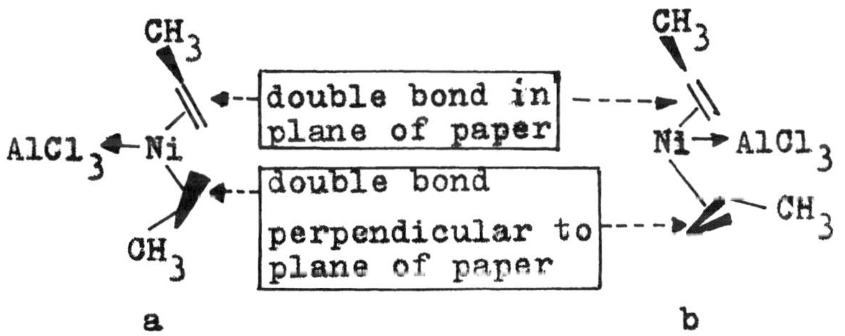

(119)

The "transoid" (a) and "cisoid" (b) orientations of the nickel–bis(propylene)–aluminum chloride complex.

The dimerization of the two coordinated propylene units can take place through a concerted electronic rearrangement with a concomitant hydride shift. The electronic rearrangement can take place *via* metal *d* orbitals of suitable geometry.

The mechanisms for the two orientations are shown in the following reactions:

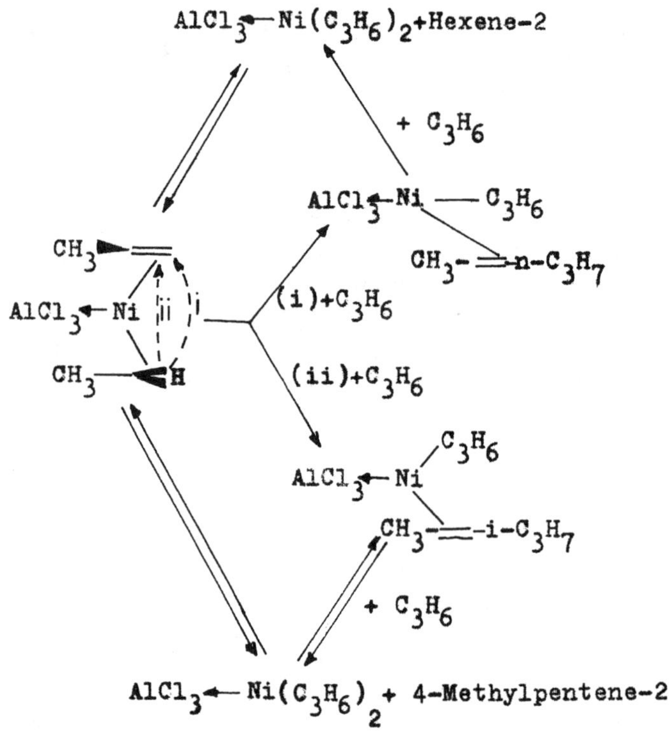

(120)

Dimerization mechanism for the "transoid" orientation.

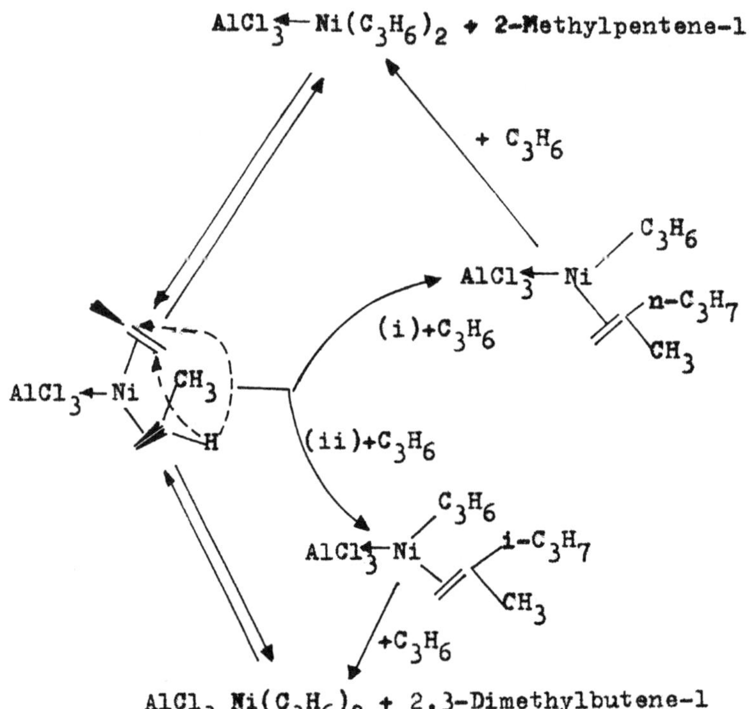

(121)

Dimerization mechanism for the "cisoid" orientation.

The $NiCl_2$–$AlCl_3$ (1:1) complex in the presence of $AlPr_3$ is an effective homogeneous dimerization catalyst for propylene [573]. Treating $NiCl_2$ on a solid carrier with $AlPr_3$, $Et_3Al_2Cl_3$, or $(i\text{-}Bu)_2AlCl$ gives a heterogeneous dimerization catalyst. Catalysts prepared from nickel poly(4-vinylpyridine) complexes [166] in an ethylene–propylene–norbornene polymer and alkylchloroaluminum also have been used for the dimerization of propylene. These catalysts are stable over a wide temperature range. Their activity increases with the increasing ability of the solvent to swell the polymer.

Hydrated iron group metal salts supported on porous carriers after drying acid calcination modified with organoaluminum compounds are used as catalysts in Huels–UCP–"Hexall" and "Octol" processes for propylene oligomerization and propylene–butene co-oligomerization [622–624].

$RhCl_3 \cdot 3H_2O$ is reported as a propylene dimerization catalyst in the presence of additives like alcohol and nitrobenzene [240,476]. $RhCl_2(SnCl_3)_4$ is effective at 50–70°C and 1.5–40 atm of propylene. However, addition of 2–3 mol/L of HCl decreases the pressure required and increases the rate and selectivity of the reaction.

$PdCl_2$ is used as a catalyst for the dimerization of propylene in the presence of various solvents like chloroform, dichloromethane, and anisole [476]. In all cases, high proportions (65–90%) of straight-chain hexenes are formed. However, in the presence of a cocatalyst like $EtAlCl_2$ and additives (PPh_3, $AsPh_3$, or $SbPh_3$), 2-methylpentene is the predominant product [619]. Pd(olefin) Cl_2 catalyzes the reaction to give 100% *n*-hexene selectivity. When $Pd(BzCN)Cl_2$ is used, straight-chain products are formed. This is associated with the isomerization of olefin [476]. A mechanism involving a hydropalladium(II) compound as the catalytically active species is suggested.

A homogeneous catalyst solution made up of $Pd(acac)_2$, PR_3, and $EtAlCl_2$ in 1,2-dichloroethane dimerizes propylene selectively up to 95% linear hexenes, although at a relatively low reaction rate. Replacement of the phosphine by a phosphite increases the rate but considerably lowers the selectivity [625]. The ratio of P/Pd is critical (P/Pd = 2). At P/Pd $\geq$ 5 no catalytic activity is observed, presumably due to the blocking of free sites. Relatively high ratios of Al/Pd are required, indicating that the catalytic species is formed in an equilibrium reaction. Activity is found for $5 \leq$ Al/Pd ≤ 25 with an optimum around 20. The following structural formula is proposed for the active species, where P is the phosphine:

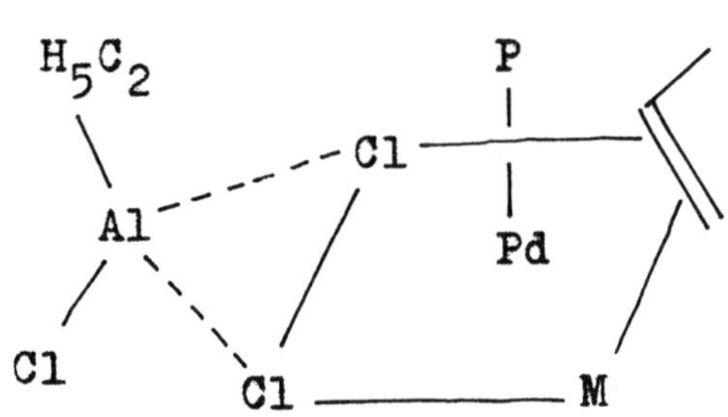

(122)

There are two patents on the use of an acetyl-acetonate complex of the first member of lanthanides for propylene dimerization [626,627]. According to them, a mixture of $Ce(acac)_3$, PPh_3, and chlorobenzene is stirred at 24–29.5°C. To this is added $Et_3Al_2Cl_3$, propylene is fed at higher (600 psi) and lower (<150 psi) pressures, and the mixture stirred to give the dimers. The yield of dimers is less in the absence of PPh_3.

Thorium nitrate, upon treatment with PPh_3 in chlorobenzene solvent and subsequent mixing with $Et_3Al_2Cl_3$, generates the active catalyst, which can dimerize propylene at 150–500 psi and 115–120°C to give the dimerized products [628,629].

In 1933 Universal Oil Products Co. developed the supported catalytic system H_3PO_4/SiO_2 for propylene oligomerization [630]. Also, asbestos can be used as carrier in this reaction. Use of a neutral carrier such as active carbon leads to the formation of alkyl phosphates, which volatilize from the catalyst bed. Thus H_3PO_4 is removed from the catalytic system. This does not occur when acidic carriers are used. Silicon [631–636], kaolin [637], silica, silicon, carbide, glass, quartz [638], and alumina–silica [639] were investigated as supports of H_3PO_4.

Also, some processes for propylene oligomerization using only H_3PO_4 as a catalyst have been described [639]. The reaction mechanism for this kind of oligomerization is presented in the following reactions [640]:

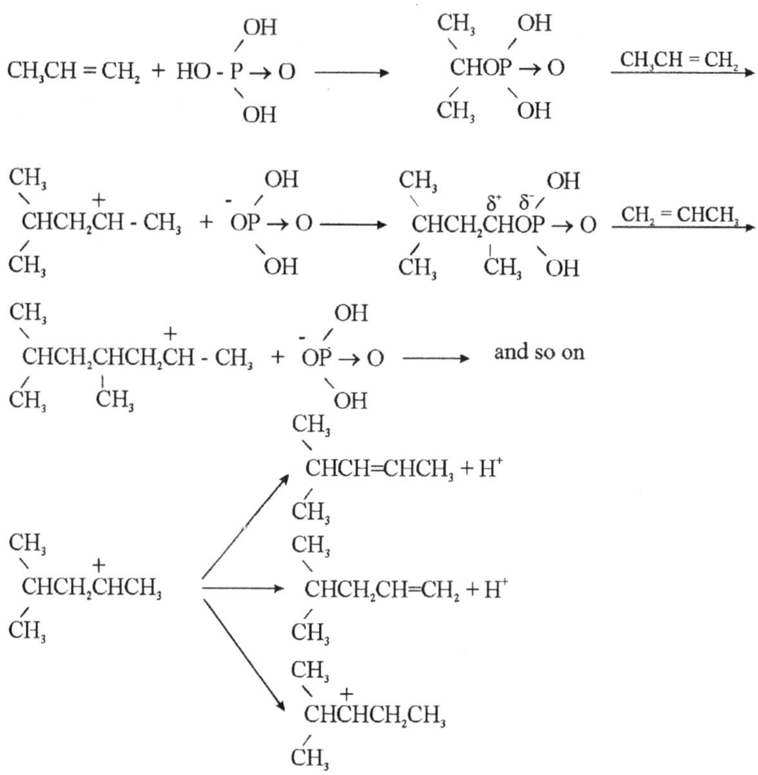

A (123,124)

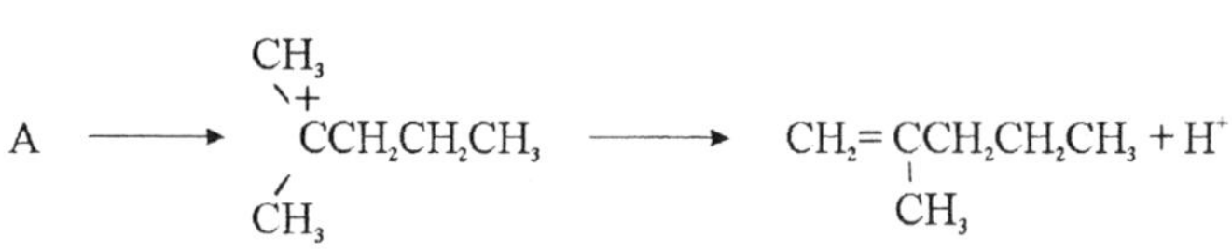

(125)

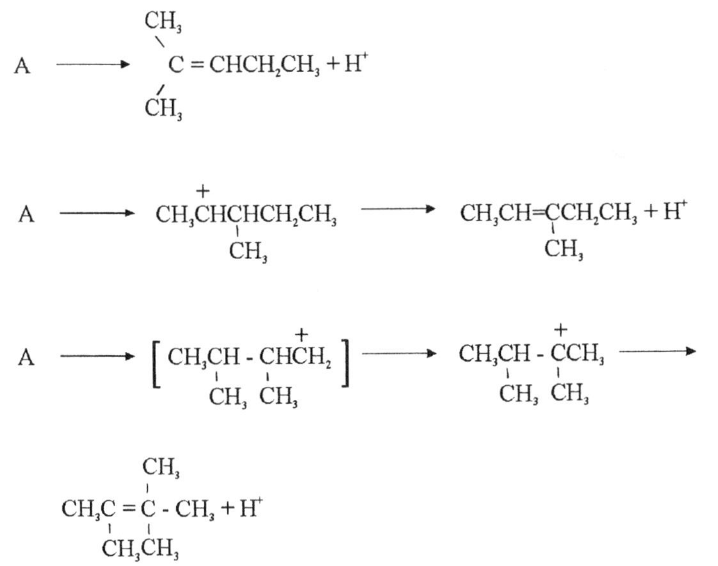

(126–128)

Dimers, trimers, and tetramers are formed when the reaction is carried out at lower temperatures (~250°C). At higher temperatures, longer-chain oligomers are formed. Propylene hydropolymerization reactions, yielding paraffins, begin to occur when the reaction temperature is higher than 280°C. The addition of steam to the reaction mixture at this temperature reduces the paraffin content in the products. The molecular weight of oligomers grows when the reaction pressure is raised.

Repas [631] has proved that addition of water to the system H_3PO_4/SiO_2 increases the propylene oligomerization product yield. He has also proposed a mechanism for this reaction in which surface propylene phosphate esters are the catalytic centers for propylene oligomerization. This catalytic system is often modified with cupric [641–643], nickel [642,644], and calcium salts [641], manganese derivatives [645], and amines [646,647].

Many other acidic catalysts possessing a Lewis acid as a component are used in propylene oligomerization. The most popular are BF_3/H_3PO_4 [648–651], BF_3/MeOH [652], BF_3/BuOH [653], $AlCl_3$ [654], and $ZnCl_2$ [655,656], and they are listed later in Table 3. The cationic mechanism of propylene oligomerization in the presence of Lewis acids is represented in the following reaction series:

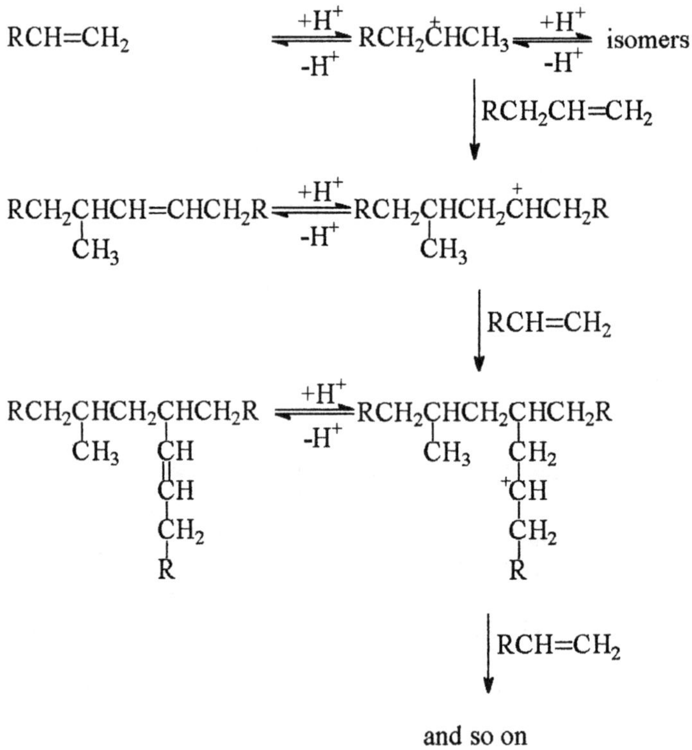

(129)

Cationic oligomerization of α-olefins in the presence of the system BF_3/BuOH [649].

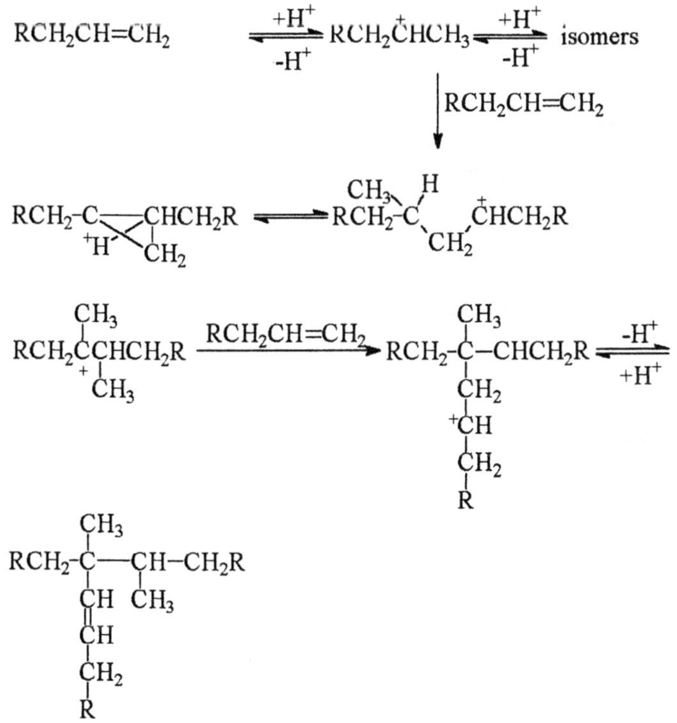

(130)

Mechanism of cationic α-olefin oligomerization in the presence of a Lewis type catalyst [647].

In these cases, an increase in reaction temperature causes a decrease in oligomer molecular weight [651].

Propylene oligomerization in the presence of alumina–silica gels and zeolites, for example, HNaY [657], are initiated on the surface of acidic Bronsted type sites after propylene adsorption. An increase in the amount of the acidic centers on the carrier surface and their acidic strength results in increasing of the propylene oligomerization rate and greater branching of the product:

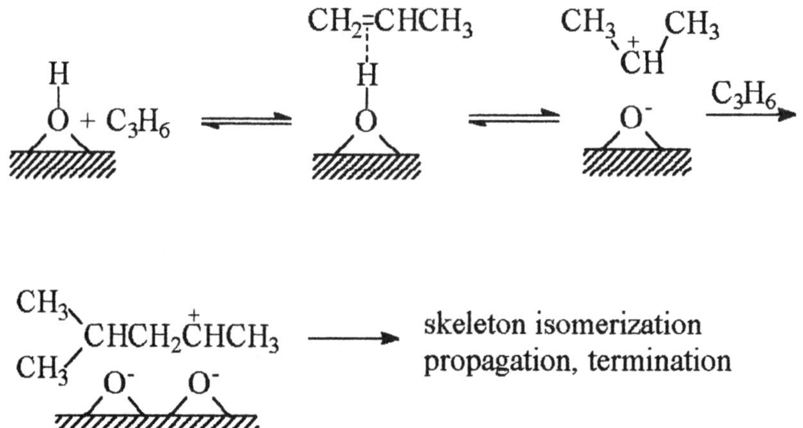

(131)

The oligomer chain branching depends on the carrier porosity. The following order of decreasing branching with pore size was established for zeolites: omega > HY > mordenite > ZSM-5 > offretite > boralize [658–663]. Omega with the largest (10 Å) pore opening gives the most branched oligomers.

In the presence of these catalysts, low molecular weight oligomers are mainly formed. Some paraffins and aromatics are formed as products of the side reactions, namely, isomerizations, cracking, aromatization, and hydrogen transfer.

The propylene oligomerization on HY zeolites at 480–580°C occurs through surface alkoxy derivatives formed in the reaction between protonated alkenes and surface oxygens of zeolites. Carbocations are not intermediates in this case. The mechanism of this reaction is presented in the following [664]:

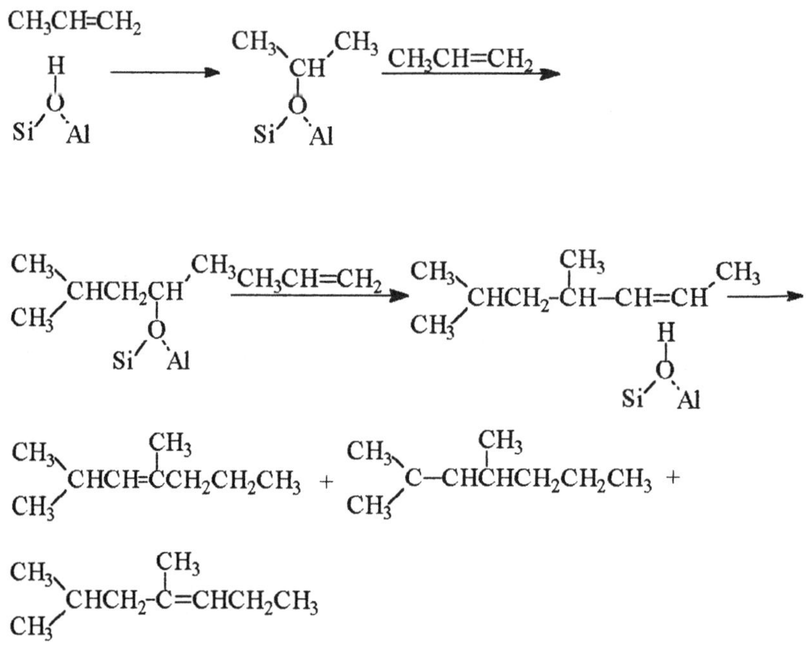

(132)

A point of some interest in propylene oligomerization is the selective poisoning of sites on the outside of zeolite HZSM-23 particles with trialkylpyridine or organophosphite compounds having an effective cross section larger than that of the pores. In this case, propylene is oligomerized in zeolite pores, in this way improving the linearity of the product [270].

Table 3 summarizes the other published and patent data on the oligomerization of propylene.

Olefin reactivity in the oligomerization reaction decreases in the following order: ethylene > propylene > 1-butene 1-hexene > 1-octene > 1-decene. This order results from the fact that higher α-olefins can form a greater amount of isomers than ethylene and propylene in the oligomerization reaction. Catalysts similar to those used for ethylene and propylene oligomerization can be used for higher α-olefin oligomerization.

V. OLIGOMERIZATION OF HIGHER α-OLEFINS

The catalytic systems used for higher α-olefin oligomerization are presented in Table 4. Among the most popular catalysts used for oligomerization of higher α-olefins are titanium, zirconium, and nickel complexes, homogeneous and heterogeneous Lewis and Bronsted acids, and inorganic oxides.

The system $TiCl_4$–R_nAlCl_{3-n} (n = 1–3, R = alkyl) oligomerizes 1-hexene [940]. Modification with halogenohydrocarbons, phosphoro- and sulfuroorganic compounds, alkali metal hydrides, and nickel salts gave relative catalysts in the 1-decene [783], $CH_2{=}CH{-}(CH_2)_n{-}CH_3$ (n = 1–5) [941], C_6–C_{14} α-olefins [942], and C_3–C_6 α-olefin oligomerizations [943].

The mixed catalytic system $TiCl_4$–$Ti(OBu)_4$–$Zr(OBu)_4$–$Et_3Al_2Cl_3$ (1:2:1:10) oligomerizes with high-yield 1-decene to a product having an average molecular weight of M_n = 356 [784,838].

Octene trimers are the main products of the 1-octene oligomerization in the presence of the system $TiCl_4$–$Et_3Al_2Cl_3$, and $ZrCl_4$ modified with $AlCl_3$ oligomerizes 1-decene to trimers and tetramers with a yield of 93% [944]. All of these titanium and zirconium catalysts have one deficiency: they promote alkylation of olefins with aromatic solvents [944].

The sulfonated ylide nickel complex 4 (see structures 1–4, reactions 64A) activated with Et_2AlOEt co-oligomerizes ethylene with higher olefins, for example, 1-hexene, 1-heptene, 1-octene, and 1-decene. Olefin insertion into the Ni–H bond is nonselective. The chain propagation step, α-olefin insertion into the nickel–alkyl bond, is also nonselective. The Ni–C bond reactivity in the olefin insertion reaction depends on the structure of an alkyl group bonded to the nickel ion.

The system $Ni(acac)_2$–$Et_3Al_2Cl_3$ is active in 1-hexene oligomerization. Its modification with donor ligands, for example, PPh_3, increases the yield and molecular weight of the products obtained, mainly trimers. Similar systems with trifluoroacetylacetonate ligands and with PPh_3 or PBu_3 phosphines [842], and the system $RCOONiOOCR'$–$EtAlCl_2$ where R = CF_3, R′ = 2-ethylhexanoate [845,847], gave dimers and trimers in 1-butene oligomerization.

Table 3 Selected Literature on Oligomerization of Propylene

Catalytic system	Products	Ref.
$(C_2H_5)_3Al$	2-methyl-1-pentene	665, 666
$(C_2H_5)AlCl_2$	propane (50%), hexene (36%), nonene (3–5%), dodecene (0.5%)	532, 667–669
$(C_2H_5)Al–H_3Al–(CH_3)_2$-$(CH_3)_2AlH$	2-methyl-2-pentene	532
$(C_2H_5)_2AlH–(CH_3)_3Al–Ph_3Al$	2-methyl-2-pentene	532
$(C_2H_5)_3Al_2Cl_3$–P(O-dodecyl)$_3$	hexenes, 4-methylpentenes	670
$R'_nAlX_{3-n}BiR_3$ (R = R′ = alkyl; X = halogen; n = 1, 1.5, or 2)	hexenes, 4-methylpentenes	671
$Al(OR)_3$ (R = propyl or hexyl)	hexenes, nonenes (4–5%)	672
$(C_2H_5)_3Al$ or $[(CH_3)_2CHCH_2]_3Al$	2-methyl-1-pentene (85–90%), 2-methyl-2-pentene, higher olefins	673
$Ti(OR)_4–(C_2H_5)_3Al$	4-methyl-1-pentene, 4-methyl-2-pentene	341
$CoCl_2–(C_2H_5)_{6-x}Al_2Cl_x$	hexenes, 2,3-dimethylbutenes	674–680
$CoCl_2$–charcoal–$(C_2H_5)_2AlCl$	hexenes	681
$CoNH_2(PPh_3)_3Co(N_2)(PPh_3)_3$	2-methyl-1-pentene (65.3%), 4-methyl-2-pentene (17%), 2-methyl-2-pentene (2.7%), *trans*-3-methyl-2-pentene (2.2%)	682
$[Co(PBu)_3]_2Cl_2–R_{6-x}Al_2Cl_x$	dimethylbutenes	683
$L_2CoX_2–(C_2H_5)AlCl_2$ (L = quinoline, isoquinoline, PPh_3, pyridine; X = Cl, Br)	2-methylpentenes (70.8%), hexenes (25.3%), 2,3-dimethylpentenes (5.7%)	684
$L_2CoX_2–(C_2H_5)AlCl_2$ (L = pyridine, 4-ethylpyridine, quinoline, PPh_3; X = Cl, Br)	2-methylpentenes (67.3%), hexenes (30.7%)	685
$Co(acac)_2–HAl(C_2H_5)_2$–L-COD; 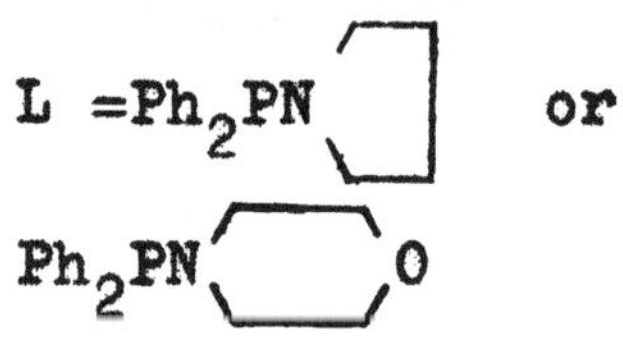	2-methyl-1-pentene (85%)	686
Nickel salt–$R_{6-x}Al_2Cl_x–PR_3$ (x = 2–4)	2-methyl-1-pentene, hexene, *cis*-2-hexene, *trans*-2-hexene, 4-methyl-2-pentene	687–689
(π-allyl)NiBr–$AlBr_3–PPh_3–PCy_3$	n-hexenes (20%), 2-methylpentenes (75%)	690
Ni[Hf(acac)]$_2$–$(i\text{-}C_4H_9)_3Al$	hexenes (77.8%), 4-methylpentenes (4.3%), 2-methylpentenes (17.2%), 2,3-dimethylbutenes (0.8%)	691
$Ni(acac)_2–(C_2H_5)_3Al–(C_2H_5)_2(OC_2H_5)Al–(C_2H_5)(OC_2H_5)_2Al$	1-hexene (4.2%), *cis*-2-hexene (32.7%), *trans*-2-hexene (28.9%), *cis*-3-hexene (6.0%), *trans*-3-hexene (5.8%), 4-methyl-1-pentene (1.7%), *cis*-4-methyl-2-pentene (2.9%), *trans*-4-methyl-2-pentene (10.7%), 1-pentene (5.6%)	692, 693

Table 3 Continued

Catalytic system	Products	Ref.
$Ni(acac)_2-(C_4H_9)_3Al$	2-methylpentenes	694
$Ni(acac)_2-(C_2H_5)_2AlCl-H_2O$	dimers, higher oligomers	695, 696
$Ni(acac)_2-(C_2H_5)_3Al-PhC{\equiv}CH$	2-methyl-1-pentene (53%), 2-methyl-2-pentene (41%)	697
$Ni(acac)_2-R_{6-x}Al_2Cl_x-PPh_3$	*n*-hexenes (31%), 2-methylpentenes, 2,3-dimethylpentenes	698–701
$Ni(acac)_2-(C_6H_5)_2Al_2Cl_3-PPh_3$	C_6 alkenes, C_9 olefins, higher oligomers	702
$Ni(acac)_2-R_3Al-BuPCl_4$	methylpentenes, 2,3-dimethylbutenes, hexenes, higher oligomers	703
$Ni(acac)_2-(C_2H_5)_3Al$	2-hexenes (56%), 3-hexenes (13%), 4-methyl-1-pentene (2.0%), 4-methyl-2-pentene (4.0%), 2-methyl-2-pentene (12.0%), 2-methyl-1-pentene (6.5%)	704
$Ni(acac)_2-(C_2H_5)AlCl_2-PCy_3$	C_6 olefins, C_9 olefins	705, 706
$Ni(acac)_2-R_{6-x}Al_2X_x-BrP(CH_3)_2$; $x < 6$	4-methyl-1-pentene	707
$Ni(acac)_2-(C_2H_5)_3Al_2Cl_3-PPh_3$	C_6 olefins	708
$Ni(acac)_2-(C_2H_5)_3Al_2Cl_3-PCy_3$	methylpentenes (36.9%), 2,3-dimethyl-1-pentene (60%), 2,3-dimethyl-2-butene (1.2%), 4-hexene (1.0%)	709
$Ni(acac)_2-(C_2H_5)_2AlCl-P_2O_5$	methylpentenes, dimethylbutenes, hexenes	710
$Ni(acac)_2 \cdot xH_2O-(C_2H_5)_2Al(OC_2H_5)$	hexenes (80%)	711, 712
$Ni(acac)_2-(C_2H_5)_2Al(OC_2H_5)-PPh_3$	2-hexene, 1-hexene, 2,3-dimethylbutenes	429
$Ni(acac)_2-(C_2H_5)_2AlCl/(C_2H_5)AlCl_2-PBu_3$	methylpentenes (69%), *n*-hexene (11%), 2,3-dimethylbutene (20%)	713
$Ni(PhCOCHCOPh)_2-(C_2H_5)_2Al(OC_2H_5)$	linear hexenes (77%)	714
$Ni(acac)_2$ and an inorganic Ni salt imported on Al_2O_3/SiO_5^{4-}	C_6 olefins	715
$Ni(acac)_2$ on alkaline earth metal carbonate or oxide–$R_{6-x}Al_2Cl_x$	hexenes	716
$Ni(acac)_2$ supported on Al_2O_3 or aluminosilicate–$(C_2H_5)_3Al_2Cl_3-(i\text{-}Pr)_3P$	C_6 olefins (67–85%), C_9 olefins (13–27%), higher oligomers (2–6.4%)	717
$Ni(acac)_2$ supported on $SiO_2/Al_2O_3-PR_3-R_{6-x}Al_2X_x$; $x < 6$	2-methylpentene (35.4%), hexenes (12.5%), 2,3-dimethylbutene (0.6%), higher oligomers (44.5%)	718, 719
$Ni(OOCR)_2-(C_2H_5)_2AlCl$ (RCOOH = octenoic acid, oleic acid)	hexenes, nonenes, dodecenes	720
Ni oleate–$(i\text{-}Bu)_2AlCl$	2-methyl-2-pentene (72.7%), *cis*- and *trans*-4-methyl-2-pentene (11.9%), 1-hexene (7.8%), *cis*-dimethyl-2-butene (7.4%), 4-methyl-1-pentene (0.2%)	721
$Ni(OOCR)_2-(C_2H_5)AlCl_2-CF_3COOH$ (RCOOH = C_9-C_{13} carboxylic acids)	hexenes, nonenes	722

Table 3 Continued

Catalytic system	Products	Ref.
$Ni(OOCR)_2-(C_2H_5)_2AlF$ (RCOOH = oleic acid, isopropyl or salicylic acid)	*n*-hexene (40–51%), 2-methylpentene (45–57%), 2,3-dimethylbutenes (2–4%)	723
$Ni(OOCR)_2-(C_2H_5)_3Al_2Cl_3$ (RCOOH = *p*-toluidine-*m*-sulfonic acid, *o*-aminobenzenesulfonic acid)	4-methyl-1-pentene (18%), 4-methyl-2-pentene (53.1%), 2-methyl-1-pentene (3.8%), 2-methyl-2-pentene (18.5%), 1-hexene (17.7%), 3-hexene (4.1%), 2,3-dimethylbutene (1.2%)	724
$Ni(COC)_2-(C_2H_5)_{6-x}AlX_x-PPh_3/PCl_3/P(i\text{-}Pr)_3$; $x < 6$ (RCOOH = naphthenoic acid)	*n*-hexene, 2-methylpentene, 2,3-dimethylbutene, 4-methylpentene	725–728
$Ni(OOCR)_2-AlCl_3-(C_2H_5)_3Al$ (RCOOH = octenoic acid)	2-methyl-2-pentene	729
$Ni(OOCR)_2-PCl_3-(C_2H_5)_3Al$ (RCOOH = octenoic acid)	2-methyl-1-pentene (6%), 2-methyl-2-pentene (55%), 4-methyl-1-pentene (0.4%), 4-methyl-2-pentene (16%), 2,3-dimethyl-1-butene (3%), 2,3-dimethyl-2-butene (14%), *trans*-2-hexenes (5.5%)	730
$Ni(octenoate)-(C_2H_5)_3Al-BF_3 \cdot OEt_2$	dimers	731
$Ni(OOCR)_2-(C_2H_5)_3Al-SnCl_4$ (RCOOH = octenoic acid)	dimers	732
$Ni(OOCR)_2-(C_2H_5)_3Al/[(C_2H_5)_2Al]_2O-PCy_3-$isoprene (RCOOH = naphthenoic acid)	mixture of dimers	733, 734
$NiCl_2(PPh_3)_2-R_{6-x}Al_2X_x$ (x = 2–4); $x < 6$	*n*-hexene, 2-methylpentene, 4-methyl-2-pentene, C_9 olefins	735
$NiCl_2 \cdot 2[PPh(cyclohexyl)_2]-(C_2H_5AlCl \cdot H_2O)$	4-methyl-1-pentene (10%), *cis*-4-methyl-2-pentene (13.2%), *trans*-4-methyl-2-pentene (15.2%), 2,3-dimethyl-1-pentene (43.1%), 2-methyl-1-pentene (13.3%), *n*-hexenes (3.3%)	736
$NiX_2(PR_3)_2$ $R_{6-x}Al_2X_x$ (X = Cl, Br; R = Bu, Ph; x = 2–4)	C_6 olefins	737
$NiCl_2(PPh_3)_2/NiCl_2(PPh_2A)/NiCl_2(PCy_3)_2$	C_6 olefins	738
$NiCl_2(PBu_3)_2-(C_2H_5)_2AlCl/(C_2H_5)AlCl_2-(CH_3)_3CCl$	2,3-dimethylbutenes, 2-methylpentenes, *n*-hexene	739, 740
$NiCl_2(PPh_3)_2-(C_2H_5)AlCl_2-(PPh_3)_2Mo(NO)_2Cl_2$	C_6 olefins (30.3%) and higher olefins	741
$NiCl_2(Ph_2CH{=}CHPPh_2)-(C_2H_5)AlCl_2$	C_6 olefins	742
$Ni(PPh_3)_2SO_4-CH_3AlCl_2/(C_2H_5)AlCl_2$	2-methyl-2-pentene	743
$NiCl_2(i\text{-}PPr_3)_2-(C_2H_5)_3Al_2Cl_3-C_7H_{16}N$	2,3-dimethylbutenes	744
$NiCl_2(PPh_3)_2-(C_2H_5)AlCl_2$	2,3-dimethylbutenes (13.2%), 2-methylpentene (72%)	428
$NiCl_2(i\text{-}PPr_3)_2-(C_2H_5)_3Al/(C_2H_5)_3Al_2Cl_3$	C_6 olefins (90%), C_9 olefins (10%)	745

Table 3 Continued

Catalytic system	Products	Ref.
$NiCl_2(PPh_3)-AlCl_3-(C_2H_5)_3N-PPh_3-Bu_3P$	4-methyl-1-pentene, 4-methyl-2-pentene, 2-methyl-2-pentene (major product), 2,3-dimethylbutene	746
$NiCl(PPh_3)_2/NiCl(PPh_3)_3-BF_3\cdot OEt_2$–Bronsted acids	hexenes, methylpentenes	747
$NiClCPh_3-AlBr_3-PPh_3$	dimer and trimer	748
$Ni(PEt_3)_2(CF_2{=}CF)Cl-(C_2H_5)AlCl_2$	C_6 olefins	749
$Ni(COD)_2-(C_2H_5)_3P$/2-Cl-1,3-butadiene/*p*-ClC_6H_4COCl	C_6 olefins	750, 751
$Ni(CH_2{=}CHCN)_2-R_{6-x}Al_2Cl_x-PR_3$	4-methyl-1-pentene, 2-methyl-2-pentene, 2-methyl-1-pentene, 2-hexene, 3-hexene, 2,3-dimethyl-1-butene, 2,3-dimethyl-2-butene	752, 753
Ni salt–$RAlX_2$; (X = Cl)	2,3-dimethyl-2-butene	754, 755
$Ni(acac)_2-(C_4H_9)_3Al-BCl_3$	2-methyl-1-pentene	756
Ni naphthenate–$(C_2H_5)_3Al-PPh_3$–2,4,6-$Cl_3C_6H_2OH$	2-methyl-1-pentene, 2-methyl-2-pentene, 4-methyl-1-pentene, 4-methyl-2-pentene, 2,3-dimethyl-1-butene, 2,3-dimethyl-2-butene, 1-hexene, *cis*-2-hexene, *trans*-2-butene	757
$Ni(CO)_2(PR_3)_2-NbF_5/MoF_6/SbF_5/TaF_5/AsF_5/WF_6/PF_5/VF_5$ (R = Ph, cyclohexyl)	4-methyl-1-pentene (8.55%), 4-methyl-2-pentene (51%), 2,3-dimethyl-2-butene (2%), 2-methyl-1-pentene (3%), 2-methyl-2-pentene (10%), *n* = hexenes (25%)	758
$Ni(CO)_2(PPh_3)_2$ supported on acidic, calcined, Al_2O_5/SiO_2	2,3-dimethylbutenes (15%), 2-methylpentenes (53%), hexenes (16%), C_9 olefins (15%)	759
$Ni(PPh_3)_4$ attached to polystyrene–$BF_3\cdot OEt_2-H_2O$	methylpentenes (70.3%), 2,3-dimethylbutene (15.8%), *n*-hexene (13.9%)	760, 761
$NiBr(NO)(PPh_3)-(C_2H_5)AlCl_2$	*n*-hexene, 2-methylpentenes, 2,3-dimethylbutene	762
$(R_4P)_2NiCl_4-(C_2H_5)_3Al_2Cl_3$	2-methyl-1-pentene (4%), 2-methyl-2-pentene (48%), 4-methyl-1-pentene (1%), 4-methyl-2-pentene (34%), 2,3-dimethylbutene (5%), *n*-hexenes (8%)	763
$(R_4P)(NiBr_3PR_3)-R_{6-x}Al_2Cl_x$	*n*-hexene, 2-methylpentene	764
$NiX_2-AlCl_3-PR_3$	2,3-dimethylbutene, *n*-hexene, 2-methylpentenes, 2,3-dimethylbutenes	765
$NiCl_2-Et_3Al_2Cl_3-(i\text{-}C_3H_7)_3P$	2,3-dimethylbutenes	766
$NiCl_2/NiBr_2-(C_2H_5)_2AlCl-PPh_3/PBu_3$	2,3-dimethylbutenes	767
$NiCl_2-(C_2H_5)_3Al$-HCl/$(i\text{-}C_3H_7)_3P$/butadiene	C_6 olefins and C_9 olefins	768
$NiCl_2/NiCl_2$ supported on a solid carrier–$(C_3H_7)_3Al/(C_2H_5)_3Al_2Cl_3/(i\text{-}C_4H_9)_2AlCl_3$	2-hexenes, 2-methyl-2-pentene, 4-methyl-2-pentene	769
$NiBr_2L_2-(C_2H_5)_3Al_2Cl_3$ (L = hexamethylphosphoramide)	*n*-hexene (68%), 4-methyl-1-pentenes (28%)	770

Table 3 Continued

Catalytic system	Products	Ref.
$NiCl_2L_2$ (L = *N,N'*-dicyclohexyl-2-(cyclohexylamino)-malonamide)	4-methyl-1-pentene, *n*-hexene, 2,3-dimethylbutene, 2-ethylbutenes, methylpentenes	771
$NiCl_2(BuNH_2)_4-(C_2H_5)AlCl_2$	2-methyl-2-pentene	772
$NiL_2-(C_2H_5)AlCl_2$ (L = *N,N*-dimethyl-2-mercaptoethyl-amine)	2-methylpentene (60.3%), hexene (33.1%), 2,3-dimethylbutane (5.2%), 3-methylpentene (1.4%)	773
$Ni(sacsac)PPh_3Cl-(C_2H_5)_2AlCl$ sacsac = 2,4-pentanedithionatol	dimethylbutene (10%), methylpentene (74%), hexene (16%)	774
Ni[poly(4-vinylpyridine)]–$(C_2H_5)AlCl_2/(C_4H_9)_2AlCl$	2-methylpentane, *n*-hexene, 2,3-dimethylbutene, 2-ethylbutane	775
$[Ni(Me_2SO)_6][NiCl_4]-(C_2H_5)_3Al_2Cl_3$	2-methyl-2-pentene (70%)	776
$[Ni(R_2SO)_6][MX_4]-(C_2H_5)_2Al_2Cl_3$ (R = CH_3, C_2H_5, C_6H_5; M = Fe, Ni, Co; X = Cl, Br, or I)	hexenes, methylpentenes	777
$NiCl_2(Ph_2SO)-(C_2H_5)_3Al_2Cl_3$	methylpentenes (75%)	778
$NiL_2-(C_2H_5)_2AlCl/(C_2H_5)_2$-$AlCl_2$–KOH	hexene, nonene	779
$PdCl_2-(C_2H_5)AlCl_2-PPh_3/As$-$Ph_3/SbPh_3$	2-methylpentenes	780
PdL_2-$(C_6H_5)AlCl_2$ supported $Pd(CN)_2$	hexenes (92%)	781
	dimers (66%), trimer (11%), oligomers (8%)	782
$TiCl-_4-R_2AlCl$	olefins, $\overline{M}_w = 92$	783
$TiCl_4-(C_2H_5)_3Al-RCl$	oligomers	783
$TiCl_4-(ONi)_5(OH)_n-C_2H_5AlCl_2$; $n = 1-3$	oligomers	784
$TiCl_4-Ti(OBu)_4$-$Zr(OBu)_4-(C_2H_5)_3Al_2Cl_3$	olefins, $M = 114$	785
$(C_2H_5O)_2TiCl_2-C_2H_5AlCl_2$	olefins, $M = 340$	786
$(\pi\text{-}C_3H_7)NiBr-(C_6H_{11})_3P-C_2H_5AlCl_2$	olefins C > 12	589
$(\pi\text{-}C_3H_5)_2Ni-SiO_2$	olefins C_6-C_{12}	594, 787–789
$Ni(RCOO)_2-(C_2H_5)_2AlCl$; RCOO = octanoate, oleate	branched olefins	790, 791
$Ni(acac)_2$ supported on $SiO_2-(C_2H_5)_3Al_2Cl_3-Ph_{3-n}PCl_n$	olefins C > 9	718, 792–795
$[CH[C(CF_3)O]_2]_2Ni-(C_2H_5)_2Al(OEt)$	olefins C_6-C_{12}	796
$Ni(acac)(Bu_2P)Cl-(C_2H_5)_2AlCl$	oligomers	797
$NiX_2-(C_2H_5)_2AlCl$	oligomers	179
$NiCl_2 \cdot H_2O$–alumina spheres–$(C_2H_5)_2AlCl$	olefins C > 6	622

Table 3 Continued

Catalytic system	Products	Ref.
Ni–microfibrous carbon fibers–haloalkylaluminum	oligomers	798
$Lu[\eta^5\text{-}C_5(CH_3)_5]_2CH_3$–$(C_2H_5)_2O$	high oligomers	799
$[[(CH_3)_2Si[\eta^5\text{-}C_5(CH_3)_4]_2LuH]_2$	high oligomers	800
R_nAlX_{3-n}–chlorohydrocarbon; $n = 1$–3; R = hydrocarbyl	oligomers $\bar{M}_w = 592$	801
R_3Al; R = alkyl C_2–C_{15}; x = Cl	α-olefins C_7–C_{14}	802
BF_3–HF	oligomers C_{12}–C_{15}	803
BF_3–polyolefins	oligomers	804
BF_3–TiO_2	oligomers C_6–C_{24}	805
$AlCl_3$–$CH_3C_6H_5$–TiO_2	oligomers $\bar{M}_w = 1074$	806
$AlCl_3$–graphite	oligomers	807
$AlBr_3$	oligomers	808–810
$TiCl_3$	oligomers	811
TiF_4–Al_2O_3	oligomers	812
H_3PO_4	oligomers	813
H_3PO_4–SiO_2	oligomers C_9–C_{12}	814
$H_4P_2O_7$–SiO_2 NaA zeolite	oligomers C_{12}	815
HF	oligomers C_7–C_{14}	816
Mineral acids	oligomers	817
CuP_2O	oligomers	818
Cu/Zr–aluminum phosphate	oligomers C_5–C_{12}	819
Al_2O_3–SiO_2–MoO_3	oligomers	820
10% NiO–Al_2O_3–SiO_2	oligomers C_6–C_{10}	821
TiO_2–ZrO_2–Ni	oligomers C_{10}	822
Sulfided Ni–mica–montmorilonite (N-SMM)	oligomers C_{12}–C_{21}	822
Ni-HZSM-5	oligomers C_6–C_{15}	266, 642
Ti/Al–silicalite (SiO_2/Al_2O_3 = 161; SiO_2/TiO_2 = 46)	oligomers C > 7	271
SiO_2–Fe_2O_3–TiO_2	oligomers C > 6	823
ferrerites $M_{x/n}(TO_2)_x(SiO_2)_{36-x}$; M = cation, T = B, Al, x = 0.05–6.6; n = valence of M	oligomers C_6–C_{17}	824
Zeolites: H, NaHy, NaMgY	oligomers, aromatics	825, 826
Zeolite NaX–30% Ni(II)	oligomers C_6–C_{12}	821
Zeolite 25M	oligomers C_3–C_{11}	827
HZSM-5	olefins, paraffins	828–830
Zeolite Y	oligomers	831
HZSM-12	oligomers C_9–C_{15}	832
Zeolites (ZSM-5, bovalites, offretite-HY, mordenite or omega)–20% sepionite	oligomers	269
SiO_2		833

Table 4 Oligomerization of Higher α-Olefins

Catalyst	α-Olefin	Products	Ref.
$TiCl_4-(C_2H_5)_{3-n}AlCl_n-3RCl$; n = 0,1,1.5	olefins C_3-C_{14}	oligomers $C_{35}-C_{350}$	783
$TiCl_4-(C_2H_5)_3Al_2Cl_3$	1-hexene	vinyliden olefins	834
$TiCl_4-AlCl_3-LiH$	1-butene, 1-hexene	oligomers	835
$TiCl_4$–Ni salt–organoaluminums (2–20:1–5.5:5–86)	α-olefins C_3-C_6	oligomers	836
$TiCl_4$/P,N,S-hydrocarbons/$R_3Al_2Cl_3$ (1:0.1–4:1–5)	1-decene	oligomers	837
$TiCl_4(OBu)_4Zr-(C_2H_5)_3Al_2Cl_3$ (1:2:1:10)	$CH_2{=}CH(CH_2)_nCH_3$	oligomers	838
$TiCl_4$–propylene oxide–$(C_2H_5)_3Al_2Cl_3$ (Al/Ti = 1.25)	1-octene	olefins C_{24}	839
$Cp_2ZrCl_2-(CH_3)_3Al-H_2O$	1-hexene	oligomers	840
$ZrCl_4(HfCl_3)-AlCl_3$	1-decene	$C_{30}-C_{50}$	841
$Ni(CF_3COCHCOCF_3)_2-(C_2H_5)_3Al_2Cl_2$-$PPh_3$	1-butene	C_8-C_{16}	842, 843
(COO)Ni(1,2-diketones)$RCO_2-NiO_2-CR_1$ (R_1 = hydrocarbyl C_5, R_2 = C_{1-3} hydrocarbyl)	1-butene	oligomers	179, 844, 845
Ni(octanoate)$_2$–$C_2H_5AlCl_2$	butene	oligomers	791
Ni(acac)$_2$ supported on SiO_2/Al_2O_3/ $(C_2H_5)_3Al_2Cl_3$/monoolefins, diolefins	α-olefins C_2-C_{10}	olefins C_8-C_{10}	846
Ni(carboxylate 11% Ni)/$C_2H_5AlCl_2$–CF_3COOH	butenes	olefins C_8-C_{16}	847, 848
Ni(acac)$_2$ supported on SiO_2/Al_2O_3/ $(C_2H_5)_3Al_2Cl_3/PPh_3$	butene	olefins C_8-C_{12}	794
$(COD)_3Ni_3(CO)_2$ supported on Al_2O_3/ SiO_2	1-hexene	olefins C_{12}	849
5% NiO-$Al_2O_3-SiO_2-C_2H_5AlCl_2$	butenes	olefins C_8-C_{18}	850
$Ni/Al_2O_3/AlCl_3/(C_2H_5)_2AlCl$	butenes	olefins C_8-C_{12}	851, 852
$W[OCH(CH_2Cl)_2]_2Cl_4-(C_2H_5)_2AlCl-(C_4H_9)_2O$	2-methyl-2-pentene	oligomers	853
$WCl_4-LiAlH_4$	1-hexene	oligomers	854
WCl_6–branched-chain aliphatic alcohols	1-butene	olefins	855
Cp_2M supported on $Al_2O_3/SiO_2/(C_2H_5)_3$ Al_2Cl_3, M = Cr, Fe, Co, Ni	butene	olefins C_8-C_{18}	856
$C_2H_5AlCl_2-RCl$	1-hexene	high branched-chain oligomers	857
$R_nAlX_{3-n}-RCl$; X = halogen, R = H, alkyl, n = 0–3	olefins C_3-C_{11}	oligomers	801, 858
$AlCl_3$–alkylaluminum halide	α-olefins C_6-C_{16}	oligomers	859
$R_3Al-RCl$; R = hydrocarbyl	α-olefins	oligomers	860
AlX_3 or $RAlX_2$–haloalkanoic acid; X = Cl, F; R = alkyl	1-hexene	olefins $C_{12}-C_{24}$	861, 862
$(C_6H_5)_2AlOC_2H_5$–zeolite	isobutene	olefins	863
$AlCl_3$–i-C_3H_7OH (19:0.7)	1-dodecene	oligomers	864
$AlCl_3$	olefins C_4-C_{10}	oligomers	865–867

Table 4 Continued

Catalyst	α-Olefin	Products	Ref.
$(C_2H_5)_2Al_2Cl_3$ or $C_2H_5AlCl_2$	olefins C_4	oligomers	868
$AlCl_3$–Al–$AcO(CH_2)_nOAl$, $n = 2$–4	olefins C_6–C_{10}	oligomers	869
$AlCl_3$–alkyl aromatic hydrocarbons containing O and N ligands	1-decene	oligomers	870
$AlCl_3$–Al_2O_3	olefins C_4–C_8	oligomers	106, 871, 872
AlH_3–Davison SiO_2	isobutene	oligomers	872
BF_3–alcohols C_6–C_8	1-decene	olefins	873, 874
BF_3–BuOH	linear olefins C_3–C_4	dimers, trimers	875–878
K_2O–CaO–Fe_2O_3–Na_2O_2–Al_2O_3–SiO_2	olefins C_9–C_{14}	oils	879
$MSO_4 \cdot n$HCl; M = Zn,Sn; $n = 0.2$–1.7	isobutene	olefins C_8–C_{24}	880
$CuSO_4$–2HCl	isobutene	oligomers	881
Cr–Al_2O_3–SiO_2	1-hexene	oligomers	882
$TaCl_6$–SiO_2	1-butene	olefins C_6–C_20	883
Ta_2O_5–Al_2O_3	1-butene	olefins C_8–C_{12}	884
Zn–HZSM-5 zeolite	olefins C_6–C_9	olefins	885
Ni–HZSM-5	butene	olefins	886
SiO_2–H_3PO_4	*n* butene	olefins	887
Ce–Y zeolite	butenes	oligomers	888
Wyoming bentonite	olefins C_2–C_{10}	alkenes	889
Zeolite HKL	isobutene	olefins	890
HZSM-5/Cr(V)	isobutene	oligomers	891
HZSM-5	1-hexene	oligomers	892
HZSM-12	α-olefins C_5–C_{12}	oligomers	893
ZSM-5	olefins C_2–C_6	liquid oligomers	665
L and Y zeolites: mordenite, crionite, and pentasil	isobutene	oligomers	894
HM mordenites	1-butene	oligomers	895
BF_3–C_2H_5OH	1-decene	olefins C_{30}–C_{40}	896
BF_3–SiO_2–H_2O	1-decene	olefins	897, 898
BF_3–O_2–SiO_2	1-decene	olefins	899, 900
BF_3–Al_2O_3	1-decene	olefins C_8	901, 902
BF_3–CH_3OH	2-decene	oligomers	903–905
BF_3–H_2O–BF_3 (gas)	1-octene	oligomers	906
BF_3–H_2O or alcohols	1-alkenes	oligomers	907
BF_3–mannitol	1-decene	oligomers	908
BF_3–H_2O–alkanoic acids	1-hexene, 1-decene, 1-tetradecene	high branched-chain dimers to pentamers	909, 910
BF_3 (gas)–$BF_3 \cdot 2H_2O$	1-hexene	olefins	911
BF_3–NiO	α-olefins C_4–C_{14}	oligomers	912
BF_3–Naftion 501	α-olefins C_{13}–C_{15}	dimers, trimers	913
Ni-mordenite	butenes	liquid oligomers	914
Ga–ZSM-5	olefins C_2–C_5	oligomers	915
SiO_2–H_3PO_4	isobutene	oligomers	916
Zn–Al_2O_3–SiO_2	olefins C_2–C_6	oligomers	917
Al_2O_3–TiO_2–F	butenes	oligomers	918
WO_3–Al_2O_3–F	butenes	oligomers	919

Table 4 Continued

Catalyst	α-Olefin	Products	Ref.
ZrO_2–MoO_3	butenes	oligomers	920
WO_3–γ-Al_2O_3	butenes	oligomers	921
Zeolon 100H (H-type mordenite)	isobutene	oligomers	922
NaHY zeolite	butene	highly branched oligomers	923, 924
NaY zeolite	butene	oligomers	925
NaCaNiY zeolite	isobutene	oligomers	926
Na_2O–SiO_2–Al_2O_3–H_2O	isobutene	branched olefin oligomers	927, 928
SiO_2–Al_2O_3	isobutene	branched oligomers	928–930
Ca,Zr,Al orthophosphate	butenes	oligomers	819
$AlPO_4$	2-butene	oligomers	931
BPO_4	1-decene	oligomers	932
Naftion 425 (perfluorosulfonic acid resin)	1-decene	oligomers	933
Amberlite 15	1-decene	oligomers	934
Naftion 501	α-olefins C_{14}–C_{40}	oligomers	935
Ion exchanger–SO_3H	isobutene	oligomers	936
Lewis and proton acids	2,4,4-trimethyl-2-pentene		937, 938
H_3PO_4–C_{act}	isobutene		938
$(COD)_2Ni$–PR_3	methylene-cyclopropene	dimers	938

Heterogenized catalysts have also been tested in this reaction. Calcined nickel salts supported on aluminum and activated with (*sec*-BuO)$_3$Al or an $AlCl_3$ + Et_2AlCl mixture oligomerized butenes to C_8–C_{12} olefins [622,623,844].

Lewis and Bronsted acids [945] and modified or unmodified inorganic oxides [946] with various ions are the most popular catalysts for higher α-olefin oligomerization. The reactivity in the oligomerization reaction decreases in proportion to the increase of the olefin length (1-butene > 1-hexene > 1-octene > 1-decene) and the C=C bond position (1-butene > 2-butene).

In conclusion, olefin oligomerization catalysts can be divided into three groups: homogeneous transition metal complexes, trialkylaluminum compounds, and heterogeneous and homogeneous Lewis and Bronsted acids.

The Ti(IV) and Zr(IV) and, particularly, the Ni(II) and Ni(I) complexes exhibit the highest activity and selectivity in olefin oligomerization. Their activity and selectivity depend on electronic and steric factors of the central metal ion in these complexes.

An increase in positive charge on the central metal ion caused by acceptor ligands generally increases the catalyst's activity but decreases molecular weights of the oligomers obtained.

Donor ligands make olefin insertion into the metal–carbon bond much easier. This results in an increase of the oligomer length and linearity.

Large, bulky ligands favor the formation of intermediates from which the β-hydrogen elimination becomes more difficult, and therefore the higher oligomers are obtained. The presence of such ligands in these complexes contributes to a relatively high linearity of the products.

Nickel complexes, particularly ylides, are most selective and active in the ethylene oligomerization to higher linear α-olefins. They have been applied industrially.

There are many papers concerning titanium and zirconium complexes, which are also very active and selective as catalysts for olefin oligomerization. Their main deficiency is polyethylene and branched oligomer formation as side products.

The systems consisting of other metal complexes, for example, lanthanides, are less active or are being investigated at present.

Cationic catalysts, homogeneous and heterogeneous, for example, $AlCl_3$, metal oxides, zeolites, inorganic acids), are generally used for oligomerization of olefins possessing three or more carbons in the chain. These catalysts are less selective than transition metal complex catalysts. In their presence, mainly highly branched oligomers and products of the alkylation, cracking, aromatization, and isomerization reactions are obtained, particularly at higher temperatures.

Halogenoalkylaluminum compounds and the organoaluminum–halogenohydrocarbon systems are the catalysts of higher olefin oligomerization via a cationic mechanism.

VI. SEPARATION OF OLIGOOLEFINS FROM INDUSTRIAL POLYOLEFINS

Very little work has been published dealing with the light products in high-pressure free radical polymerization of ethylene. Kobayashi [947] has examined the liquid fraction produced during high-pressure polymerization of ethylene in a stirred autoclave at pressures from 1200 to 1600 atm. He reported the liquid fraction to be a complex mixture and identified several individual compounds.

A partial analysis of the low molecular weight "oil" fraction produced in the free radical initiated polymerization of ethylene is given by Von Dohlen and Wilson [948]. In this study three samples of liquid fractions were examined. These samples were all extracted with water to remove the acetone or isopropanol chain transfer agent and then dried over calcium sulfate. The third sample is obtained using propylene chain transfer agent. The outstanding feature of the gas chromatograms of these samples is the peak-for-peak correspondence. The same components were also found in a run made without chain transfer agent. Consequently, the same mechanism must be functioning in all cases to produce the small molecules observed, and this mechanism does not involve the chain transfer agent except in a minor way.

Chromatographic analysis of the liquid residue extracted from a resin and of the oils condensed from the vapor phase again reveal a peak-for-peak correspondence showing the continuity of these materials. To determine how the low molecular weight fractions fit into the overall polymer MWD, a commercial low-density resin was analyzed by GPC (gel permeation chromatography). The gel chromatogram shown represents about 30 wt% of the total sample. A closer examination of the low molecular weight and of the MWD clearly reveals that the quantities of oils with a degree of polymerization less than 10 are too great for them to be considered part of the same population as the solid polymer. Thus, the oil and the resin are considered as constituting two distinguishable most probable distributions.

On the other hand, Takahashi et al. [949–951], making use of gas chromatography and mass spectroscopy, claim that more than 40 oligoethylenic homologues are present in commercial high-pressure polyethylene. These compounds consist of a variety of alkanes, alkenes, cycloalkanes, aromatics, alkanones, alkanols, and esters. The oligoethylenic compositions vary greatly among the resins, and the oligoethylenes fall into three types. The *first type* appear in only particular resins, the *second* in almost all the resins, and the *third* is a hybrid of the other two—it appeared in almost all the resins, but its quantity was unusually high in particular resins.

The differences in the structures and quantities of oligoethylenes represent the differences in the reactions producing them. The first type of oligomers is considered to be due to the particular chain transfer agent used in the polymerization process, the second to monomer impurities, and the third to both these origins.

The main consequence of these studies is that the formation of the oil in industrial high-pressure polyethylene is determined by the structure of the chain transfer agent used in the polymerization process and by the nature of the impurities of the monomer.

A partial analysis of the low molecular weight "oil" fraction produced in the low-pressure polyethylene obtained with Ziegler catalysts is given by Turcu et al. [952]. This study reveals that the impurities of the monomer improve the formation of the oligoethylenes and that the increase of the partial pressure of the monomer results in the decrease of the oligomer amount in the industrial polyethylene.

Most oligomer polyolefins have the appearance of wax (hence the denomination of polyolefinic waxes). Technological parameters like monomer partial pressure, the presence of some comonomers and catalyst concentration influences the shape of molecular weight distribution of the "oils" produced during industrial polymerization of ethylene. Thus, in a continuous industrial reactor, ethylene polymerization in the presence of a Ziegler catalyst, making use of common industrial experimental conditions, led to the formation of considerable oligomer amounts (3 to 5% vs. polymer amount) whose structure depends on the working conditions and initial composition of the system, taking into account polymer destination (injection, blowing or film processing).

Occurrence of oligomers in polyolefins is largely determined by the nature and intensity of the growing chain termination reactions. Transfer with H_2 (formation of CH_3 end groups), transfer to $Al(C_2H_5)_3$ (less probable), disproportionation reaction (formation of vinyl or vinylidene end groups), and transfer to monomer can quantitatively influence oligomer occurrence [953].

In copolymerization of ethylene with propene or 1-butene, the tendency of α-olefins to act as chain transfer agents can be seen. In this case, hydrogen from β-carbon (tertiary) is more accessible and leads to occurrence of supplementary amounts of low molecular weight species.

Oligoolefins as such, many of them commercialized (see Table 5), have found significant applications [954]. In plastics processing, oligoethylene can be used as lubricant, demolder, pigment concentrate (master batch), or dyestuff. Light industry makes use of oligoolefins in obtaining cleaning substances, polishing waxes, and shoe polishes, and in the formulation of some emulsions. Metallurgy uses these products as greases, anticorrosive agents, and lubricants in the process of cold-drawing. The timber and paper industries use oligoethylenes for obtaining hydrophobic emulsions, coatings, wax molders and soakers, adhesives, mat lacquers, and printing inks; these are quite resistant to friction. In the long run, the textile industry applies oligoethylenes as finishes and soakers [954].

Oligoalkenes obtained by a continuous radicalic process may be added to lubricating oils. They show activity as low-temperature sludge dispersants and as detergents. Since oligomers produced by such a process have relatively uniform molecular weight and composition, the effect of a given amount of the polymer as a sludge dispersant or detergent can be heightened by suitable control of molecular weight [955–957].

Oligomers of certain 1-olefins are highly useful as base fluids for preparing lubricants, and the like, generally described as functional fluids, by the use of appropriate additives. α-Olefin oligomers can, in some instances, also be used as functional fluids without the use of property-modifying additives. Those oligomers that are useful, either directly or with additives, as functional fluids generally have at 100°C viscosity within the range of about 1–15 centistokes. The dimer is generally removed from the reaction product to avoid volatilization. Therefore, the primary oligomer product is a mixture of trimer, tetramer, and pentamer. But since

Table 5 Main Types of Commercialized Oligoolefins

Producer	Trademark	Types	Structure
Allied Chem. Corp.	AC-PE	6A, 316A, 392A, 615A, 617A, 629A	Oligoethylenes
BASF	A-Wachs	AH6, AL6	Oligoethylenes
Crowley Chem. Co.	Polyfin		Oligoethylenes
	Polystar, Polyper		Atactic oligopropylene
Eastman	Epolene	C10P, E14P, N14P	Oligoethylenes
	Eastobond	M500S	Atactic oligopropylene
Hoechst	Hostalub lubricant	XL223, PA130, PA190, PA520	Oligoethylenes
NL Industries Inc.	Plastflow	POP	Oligoethylenes
Petrolite Corp.	Polywax		Oligoethylenes
VEBA Chemie	VEBA Wachse	A616, A415, A310, A217, AX1539, AX2729, AV1550, AV1551	Oxidized high-density oligoethylenes

Source: Ref. 954.

the different viscosity specifications for the broad range of products for which a trimer, tetramer, and pentamer mixture can be used varies widely, the proportions of trimer, tetramer, and pentamer must vary substantially from (functional) product to product in order to meet these viscosity specifications. Because the amount of pentamer resulting from the oligomerization is generally less than about 15 wt% and is more difficult to quantify the trimer, the behavior ratio has been used as a convenient indicator of the direction of changing product composition. Because the oligomer composition that is generally obtained by direct oligomerization processes is not the composition that is required to meet the viscosity specifications for a desired product, a separation procedure is required to obtain a fraction having the desired trimer to tetramer ratio.

The low molecular weight polymers obtained by oligomerization of α-olefins have a high viscosity index and a low pour point and are superior in acid stability, and so they can be used as a base for high-grade lubricants, cosmetics, or fiber-treating agents. Liquid oligomers having low viscosity are employed as gas turbine oil and as hydraulic fluid for aircraft. Oligomers having medium viscosity are used as hydraulic fluid for general machine and ice-machine oil, and oligomers having high viscosity may be employed as grease and gear oil [958].

Foaming or frothing is frequently encountered at the surface of liquids due to the entrapment of gases, such as air. This foaming occurs especially when the liquid is heated under reduced pressure or agitation. The textile, paint, and paper industries are places where typical foaming problems are encountered.

Defoamers, which, when added to a foaming liquid, break down or inhibit the formation of bubbles in the foam, are of course old in the art. Some of these defoaming compositions are a mixture of compounds such as organosiloxane polymers, including silicone oil and/or normally hydrophilic materials that have their surfaces rendered hydrophobic, such as silica.

One of the problems with these old compositions is that their effective shelf lives are relatively short, because a "settling out" of the various components, such as the silica, occurs. Before it can be used, the defoamer usually has to be agitated, as in a mechanical shaker, so that the components can again be in an intimate admixture.

Another problem that affects the effective shelf life of these silica-based defoamers is their sensitivity to water. The normally hydrophilic, or water loving, silica is treated to be predominantly hydrophobic or water repelling before it is used in a defoaming composition; nevertheless, some of the silica is still hydrophilic in nature. As a result, when water enters the silica-based defoaming composition, the hydrophilic silica is attracted to the hydroxyl groups of the water molecules, causing a settling out of the silica from the mixture. The composition has to be agitated to break up the complex formed by the silica and the water.

Pure organic defoamers have been produced in the past, but they have been found to be ineffective in that they require a large amount of the defoaming agent to accomplish the defoaming process.

The silica-based defoamers are expensive because such a large quantity of silica, or other expensively treated particles, is used in the composition.

Oligoethylenes with molecular weights of 300 to 3000 may be used as defoaming components in the defoaming composition of a surfactant, a normally hydrophilic agent (silica or alumina), and a mineral oil (paraffin oil). A defoamer composed of 100% oligoethylenes (M_n = 300–3000) has proved to be a very effective defoaming agent. However, the effectiveness of the oligoethylenes as defoamers is increased further if various components are mixed with it. These components can be predominantly hydrophobic materials (fatty acids, polyolefin resins, silicone oil) having relatively high melting points and hydrophilic compounds (silica, alumina, talc) the surfaces of which have been treated to be hydrophobic [959].

During high-pressure polymerization of ethylene, oligoethylenes are also obtained in a proportion of 1%. These latter components have the appearance of viscous liquids, pastes, waxes, or solids. In the case of propylene, atactic oligomers formed during industrial polymerization achieved with Ziegler–Natta catalysts represent 10% of the yield. Within worldwide polyolefin yield (approx. 30 million tons per year in 1994), oligomers represent on average ca. 1 million tons per year. Such data justify the desire to reevaluate these secondary products.

VII. MODIFIED OLIGOOLEFINS

Oligoolefins produced by means of industrial processes are separated and purified by extraction with solvents in the presence of decolorants or by melting and filtering [960].

Notwithstanding all this, in some types of applications oligoolefin properties do not lead to optimal results, and there are cases when their application is practically impossible due, in the first place, to the nonpolar character of oligoethylenes. Although this drawback can be partially eliminated, (e.g., by oxidation), grafting of oligoethylenes presents a more attractive way to vary oligoolefin properties and, thus, application range. The many monomers that can be grafted allows one to obtain myriad grafted oligoolefins.

In principle, the methods employed for oligoolefin grafting do not differ from those known for grafted copolymer synthesis [961]. As with these latter methods, synthesis usually employs radicalic initiation.

Among radicalic methods, grafting in the melt is most frequent. Oligomer and monomer are stirred at a temperature that exceeds their melting temperature in the presence of a radicalic initiator (as a rule, peroxides or hydroperoxides) able to decompose into free radicals at the reaction temperature. Active grafting positions are formed *via* the direct attack of free radicals generated through the initiator, thermal decomposition, and transfer reactions between oligoolefins and growing macroradicals present in the system from monomer homopolymerization. Due to oligomers' low molecular weights and high grafting temperatures (over 100°C), reaction mixture viscosity is low enough not to cause special problems for the stirring system.

The most commonly used grafting monomer is maleic anhydride (MA), which is grafted on low-density polyethylene (LDPE) oligomers or on high-density polyethylene (HDPE) oligomers, with 4 to 30% grafted MA, the unreacted monomer being eliminated from the reaction mixture by melt purging with a N_2 stream [962–964]. Grafting of oligomers produced at the yield of industrial LDPE (M_n = 7000, d = 0.907) is achieved by thermal initiation with 5% MA (240 to 275°C) in the absence of O_2, for 45 min [962]. The use of an initiator (azoisobutironitril

[AIBN] or benzoyl peroxide [POB]) in oligomer grafting, resulting from secondary products in industrial yield of HDPE, allows one to lower the reaction temperature to 150°C [963]. Oligomers obtained in LDPE synthesis or by common polyolefin degradation have a wide MWD and a high content of double bonds. Oligomers obtained in the yield of HDPE, or of ethylene–propylene (EP) elastomers *via* a Ziegler–Natta process, have a linear chain, a narrow MWD, and a reduced double-bond content. Thus oligomers (M_n = 1500) can be grafted with 20% MA at 180°C, for 4 h, using dicumyl peroxide as an initiator.

Bronstert and Mueller [965–968] suggest another way of maintaining initial oligomer viscosity: performing oligoolefin grafting with a mixture of MA and styrene (St), in the absence of the initiator and by heating. In the oligomer melt, initial MA is added, and later an equimolecular amount of St. Under the grafting conditions used (10 to 50% monomer in the reaction mixture, 120–180°C, 1–4 h), due to a stressed tendency of alternating copolymerization of the two monomers, a total conversion is obtained, exceeding 95%, but with a particularly high grafting output (>85%) and a reduced content in statistic copolymer. The equimolecular St–MA ratio provides an alternating structure of grafted chain, with, however, the probability of also grafting a mixture in which this ratio is 1:0.7–1.40. As a support, oligoethylene and oligopropylene are used with M_n = 2000 to 15,000; these are obtained through direct synthesis or degradation of the corresponding polymers. Occasionally, MA can be substituted for by acrylic acid (AA) and/or methacrylic acid (MAA). Thus, by grafting at 140°C of 110 parts MA and 117 parts St on 1000 parts oligoethylene (M_n = 5400), a grafted oligomer is obtained with melt viscosity (at 140°C) of 1200 cP. By performing grafting under the same conditions, but in the presence of an initiator, ditertiarybutylperoxide (DTBP), the viscosity of the reaction product is 25,300 cP [965].

Besides MA, other monomers have been used for grafting. Crotonic acid (*trans*) grafting on oligomers obtained through thermal degradation of polyethylene is performed under similar conditions as for MA grafting.

Oligoethylenes (d = 0.918, η = 250 cP at 125°C) are grafted with 10% crotonic acid by heating at 180°C in the presence of 2.7% DTBP for 1.5 h. After the reaction is over, melt viscosity grows significantly [969]. Under similar conditions, poly([*a*]-propylene-g-crotonic acid) is obtained [970]. Both MA and crotonic acid are monomers that homopolymerize with difficulty, so that in grafted oligomers the homopolymer content is low.

Another monomer employed is acrylic acid. By stirring a mixture of 100 parts atactic propylene, 10 parts AA, and 1 part butylperoxide (BP) at 100°C for 40 min, a copolymer grafted with 8.1% AA is obtained, that is, 90% grafting output [971]. Acrylic acid is grafted on oligoolefins (obtained by direct synthesis or thermal degradation, having 0.05 to 0.3% double bonds per 1000 carbon atoms and 2000 cP viscosity at 190°C). The reaction is initiated by BP or perbenzoate of tertiarybutyl, at 70 to 250°C. Under these conditions, 4 to 13% AA is grafted. Similarly, poly(1-butene-g-methylmethacrylate) can also be obtained [972].

For grafting into melt, other monomers have also been employed. Allylglycidyl ether can be grafted on oligoethylene (M_n = 1700), so that a copolymer is obtained with 10.6% grafted monomer [973]. Butyl acrylate is grafted (100°C, for 2 h, *tert*-butylperoxypivalate as initiator) on an admixture of poly(1-butene) with

M_n = 430 and polybutadiene with M_n = 1700, so that a liquid grafted copolymer is obtained (η - 4800 cP at 25°C) [974].

In grafting of vinyl acetate (VA) on oligoethylene at 180°C for 4 h in the presence of DTBP, the saponification index of the reaction product (determining product properties and applications) can be set between 2 and 75 OH/g of grafted oligoethylene by modifying the content of grafted VA [975].

A grafting reaction in the melt can be achieved in the processing equipment, too. Thus, by supplying the Brabender plastograph extruder drum with an admixture of polypropylene and BP solution in AA, copolymers with 6 to 12% grafted AA are obtained, the reaction time being 2 min [972]. Various vinylic monomers can be grafted in a mixer (nitrogen atmosphere, 140 to 190°C, and 10 to 60 min reaction time) on chlorinated oligoolefins with 5 to 25% Cl in the presence of some transitional metal complexes (Co, Fe, Mn acetylacetonate), inducing homolytical dissociation of C–Cl bonds, thus forming graft initiating macroradicals. In monofunctional monomers (MMA, lauryl methacrylate, 2-methyl-hexyl methacrylate, or styrene), only one grafting reaction takes place, whereas with bifunctional monomers (allyl methacrylate, vinyl methacrylate, or divinyl benzene), grafting is accompanied by crosslinking [976].

A special technique of grafting in the melt is claimed by Räetzsch et al. [977], making use of a polymerization plant with tubular reactor, at high pressure of ethylene. A copolymer ethylene + 8% ethyl acrylate (EA), emerging from this tubular reactor at 235°C, is passed through a cooler into a stirring vessel, which also has the role of average pressure separator. Into this vessel (200 atm, 170°C), the same comonomer is inserted. Grafting is initiated by free radicals left from the statistic copolymerization phase. The poly(ethylene-co-EA)-g-EA oligomer thus obtained contains 9.5% polymerized EA, that is, 1.5% remains grafted.

Oligoethylene grafting can also be performed in a stirring vessel, located after the pressure separator, at 1 to 50 atm and 250°C. Initially, an oligomer is obtained by ethylene copolymerization with 0.1 to 5% triallylcyanurate, occasionally with VA, too. This copolymer is later grafted with VA, EA, and MA, the reaction being initiated with *tert*-butyl perbenzoate, which is injected (dissolved in oil) in the stirring vessel. In this way oligomers with 4 to 30% grafted monomer are obtained. The triallylcyanurate content of copolymer determines the number of active grafting positions, this occurring preferably with side double bonds inserted through statistic copolymerization. This procedure allows the grafting of some monomers of low reactivity [978].

Oligoolefins may also be grafted in solution, making use of solvents that, at the reaction temperature (80 to 90°C), provide for the homogeneity of the system. Solvents used are benzene [979,980], toluene [981], or *n*-heptanol [979]. The grafting reaction is radicalically initiated and lasts for 5 to 10 h. The reaction product is separated by cooling or precipitation with nonsolvent. This is how oligopropylenes were grafted (M_n = 8000) with MA [979] or MMA [982], and oligoethylene with EA, St, or methacrylic acid [980]. At the same time, grafting of some acrylamide derivatives (*N*-(1,1-dimethyl-3-hydroxybutyl)acrylamide, methacrylamide) on oligoethylenes and oligopropylenes has been reported [980].

Similar conditions are also used for grating in *o*-dichlorobenzene solution of EP or EPD (copolymers ethylene–propylene or ethylene–propylene–diene) with

2-vinylpyridine, *N,N*-dimethylaminoethylmethacrylate, or 2-methyl-5-vinylpyridine [982].

Suspension grafting has been used to obtain resistant materials. Admixtures of ethylene–vinyl acetate copolymer and various oligoolefins are grafted with vinyl chloride (VC), styrene (St), or acrylonitrile (AN) to obtain highly impact-resistant products [983–985]. Ethylene–VA copolymer, oligoolefins, and monomer are stirred in water at 60 to 80°C in the presence of tensioactive agent (water/monomer/polymer ratio = 15–30/7–13/1) for 0.5–1.5 h, enough to effect suspension and monomer diffusion in polymer particles. Grafting is radicalically initiated (T = 60 to 80°C) and last for 5 to 8 h.

Natta [316,986] obtained impact-resistant materials by grafting St, methyl methacrylate, or VC on oligoolefins *via* a hydroperoxidation method. Thus, by barbotage of an air stream in a 15.5% solution of oligoolefins in isopropylbenezene at 70°C for 4 to 8 h, the oligomer chain oxidizes, so that a product with 2 to 4 hydroperoxidic groups is obtained and little degradation by splitting up is seen. Thermal decomposition of the hydroperoxidic groups that are formed generates radicals that initiate grafting.

Oligoolefin grafting can be done through the effect of ionizing rays in an oxidizing medium. This is how peroxidic or hydroperoxidic groups are formed on the chain, the decomposition of which allows initiating macroradicals to be formed. Thus, oligoolefins with M_n = 2000 are irradiated in the air with γ radiation (source Co^{60}, dose flow 0.4 Mrad/day, total dose 1.5 Mrad). The product obtained is introduced in an autoclave, together with VA, for 6 h at 100°C (N_2 atmosphere). Monomer conversion is 100% [987,988].

Ionic grafting of oligoolefins is less frequently employed. Thus, a solution in *n*-hexane of oligopropylene (2 to 10% concentration) is saturated by barbotage at 40°C with anhydrous formaldehyde. Formaldehyde grafting is initiated with tributylamine. The process lasts 3 h at 40°C [989].

Copolymer of olefins with carbon oxide and an unsaturated grignard compound followed by hydrolysis allows one to obtain a co-oligomer with side vinyl groups wherein the chain is made up of structures of the type $—CH_2—CH(R)—$ and $\begin{array}{c}—C(CH)—\\ |\\ CH{=}CH_2\end{array}$. This product becomes quite accessible to the grafting reaction [990].

Other grafting processes are also known, among which the mechanochemical one has the advantage of being able to graft saturated compounds on oligoolefins [991–998].

Grafted oligoolefin properties and applications depend on the molecular weight and especially on the type, frequency, and length of grafted chains. In describing grafting methods, it became evident that most of the monomers are polar, so that nonmodified oligoolefins cannot be used. Grafted polar groups allow for easy emulsification of grafted oligoolefins, so that only stable emulsions are obtained with a high solid body content, which later can be diluted. Emulsions obtained from oligoethylenes grafted with MA or MM (methylmethacrylate) (1 to 15% grafted monomer) are applied as dispersants for dyestuffs, as antidirt agents for textile fibers, and for curing glass fibers used for reinforcement of plastics

[964]. As textile curing agents, oligo(ethylene-g-crotonic acid) emulsions are used (with an acidity index of 20 to 40 mg KOH/g) [969].

Emulsions intended for shiny coatings are obtained from oligo[(ethylene-co-ethyl acrylate)-g-ethyl acrylate], containing 8% copolymerized and 1.5% grafted EA, and from ethylene–triallylanurate–VA ternary copolymer grafted with 4 to 30% EA, VA, methacrylic acid, or MA [977,978]. Emulsions for coatings and adhesives are obtained by neutralization with NaOH aqueous solution of oligo([*a*]-propylene-g-methyl acrylate) or of solutions in heptane of oligoethylene grafted with ethyl acrylate, St, or methacrylic acid [979]. Polishing pastes and waxes are obtained from emulsions of various grafted oligoolefins: oligo(ethylene–g-VA) with saponification index of 39 to 75 KOH/g [975], oligo(ethylene-g-crotonic acid) with acidity index of 20 to 40 mg KOH/g [969], oligo(ethylene-g-acrylic acid), oligo([*a*]propylene–g-acrylic acid), and oligo(1-butene-g-methacrylic acid) having 4 to 12% grafted monomer and an acidity index of 20 to 800 mg KOH/g [922]. Oligo(ethylene-g-MA) with acidity index 50 mg KOH/g is used for improvement of the dispersing capacity of polyolefins during their chlorination in aqueous dispersion.

Also, because of polar groups, oligoolefins have adhesive properties. Oligoolefins grafted with larger amounts of acrylic or methacrylic acid having an acidity index under 20 mg KOH/g [972], and oligomers from LDPE manufacturing grafted with MA show good adhesiveness [962,963]. Adhesives may also be obtained by grafting butyl acrylate on oligodiene mixtures [974]. The product of Eastman Kodak Corporation under the trade name Epolene C-16, obtained by grafting 4% MA on Epolene C-10 (oligo(LDPE) with M_n = 7000), is used in adhesive compositions of the "hot melt" type, for the improvement of polyamide properties. Incompatibility between polyamide and oligoolefins vanishes by the latter grafting. The same company is producing "hot melt" adhesives for various purposes by grafting oligopropylenes with crotonic acid [980]. Oligoethylenes capped with γ-aminopropyltriethoxysilanes are used in the process of polyethylene reinforcement (M_n = 200,000) with glass fibers. Silanic groups make the link between glass fibers and polyethylene chains. In this way a material with resistance at increased traction (1.5 times greater) compared with that of the similar reinforced product untreated with silanized oligoethylenes is obtained [995].

The use of oligoolefins to produce certain coatings and films is based mainly on the adhesive properties of the former. Oligo(ethylene-g-styrene-alt-MA) and oligo[(*a*)-propene-g-(St-alt-MA)] with 20 to 25% grafted monomers (MA/St ratio = 1/1) is used in the form of powder, paste, or solution as "flating agents" in a proportion of 0.1 to 10% in the production of some filmogene materials with excellent properties (mechanical strength, flexibility, reduced reflexion) [966]. Oligo([*a*]propylene-g-methyl methacrylate) is used for waterproofing of textiles, timber, and cements [981].

Grafted oligoolefins are also used as lubricants in processing various plastics. Oligomers with M_n = 2000–20,000 (obtained through thermal degradation of LDPE and especially of HDPE) grafted with 1 to 10% MA or crotonic acid are used in a proportion of 0.1 to 2% for external lubrication of poly(vinyl chloride), with better results than those obtained with ungrafted oligoolefins (type Hoechst Wax OP) [998].

In injection processing of reinforced polyamides with glass fibers, common lubricants (e.g., metal salts, fatty acids esters) although improving flow and, occasionally, the quality of object surfaces, decreases their resistance and rigidity. This latter drawback is eliminated by using compositions of 0.1 to 2% oligo[ethylene-g-(St-alt-MA)] as a lubricant [967].

Polar groups in grafted oligoolefins improve dyeing power of inks for carbon paper [975] and of polypropylene [971].

Grafted oligoolefins can be used in a proportion of 0.1 to 5% as additives in lubricants for internal combustion engines, as antifreezes, and as agents for the

Table 6 Physicochemical Characteristics of Well-Defined Oligoolefins

I. Oligo(ethylenes): H—$(CH_2)_n$—H

n	M	m.p. (°C)	b.p. (°C/mm)	$d_{4/°C}$	Ref.
2	30	−183.3	−88.5	0.5462/−89	1000
4	58	−138.4	−0.5	0.6011/0	1000
6	86	−94.0	86.7	0.6594/20	1001
8	114	−56.8	125.7	0.7026/20	1001
10	142	−29.7	174.1	0.7301/20	1001
11	156	−25.6	195.9	0.7402/20	1001
12	170	−9.7	216.3	0.7487/20	1002
13	184	−5.2	235.5	0.7563/20	1002
17	240	22.1	302.6	0.7767/22	1001
23	324	47.7	354.0/15	0.7631/70	1003
25	352	53.7	262.0/15	0.7693/70	1003
28	394	62.4	263.0/15	0.7755/70	1003
35	492	74.7	331.0/15	0.7814/74	1003
36	506	76.1	298.4/3	0.7783/90	1003
40	562	81.5	$150.0/10^{-4}$	—	1004
50	702	92.3	420.0/15	—	1005
70	982	105.5	$300.0/10^{-4}$	—	1006

II. Oligo(isobutylenes)

Structure	M	m.p. (°C)	b.p. (°C/mm)	d_4^{20}	Ref.
Dimers					
2,4,4-trimethyl-1-pentene	112	−93.6	101.4	0.7150	1007
2,4,4-trimethyl-2-pentene	112	−106	104.9	0.7211	1008
3,4,4-trimethyl-2-pentene	112	—	112.3	0.7392	1009
2,3,4-trimethyl-2-pentene	112	−113	116.3	0.7434	1010
2,3,4-trimethyl-1-pentene	112	—	108.0	0.7290	1011
2,3,3-trimethyl-1-pentene	112	−69	108.4	0.7352	1012
Trimers					
1,1-dineopentylethylene	168	—	85.0/40	0.7599	1012

Table 6 Continued

III. Oligo(dienes)

Structure	M	m.p. (°C)	b.p. (°C/mm)	n_D^{20}	Ref.
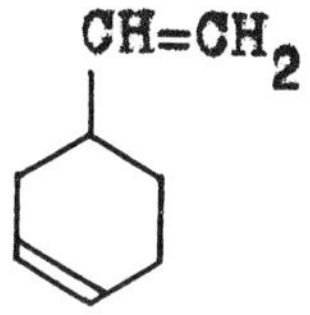	108	–	129	—	1013
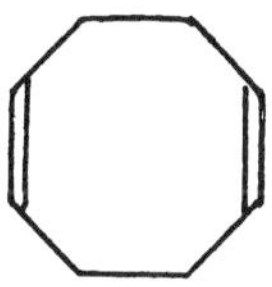	108	−70.1	150.8/755	1.4926	1014
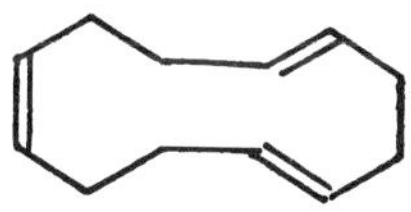	162	—	—	—	1014
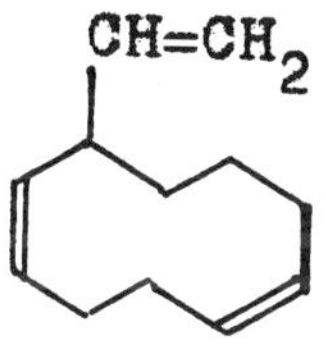	162	—	100/14	—	1015

improvement of viscosity. Their solubility in lubricants is obtained by grafting with monomers containing nitrogen (2-vinylpyridine, 2-methylvinylpyridine, etc.) [980,982].

A ololitivation to bitum of 5 to 7% poly([*a*]propylene-g-acrylic acid), with their COOH groups from grafted chains esterified with glycerine, improves its qualities (flexibility, ductibility, plasticity, resistance to aging), allowing its use as thick foils for hydroinsulation in building engineering [999]. Some properties of epoxy or phenolic resins (water absorption, temperature of heat deformation, chemical stability, and mechanical strength) are improved by mixing with 2% oligo(ethylene-g-allylglycidylether) [973].

Table 6 summarizes the physicochemical characteristics of well-defined oligoolefins.

REFERENCES

1. Lefevre, G., and Chauvin, Y., in *Aspects of Homogeneous Catalysis*, Ugo, R., Ed., Carlo Manfredi, Milan, 1970, p. 107.
2. Wilke, G., in *Fundamental Research in Homogeneous Catalysis*, Tsutsui, M., Ed., Plenum Press, New York, 1979, Vol. 3, p. 1.
3. Henrici-Olive, G., and Olive, S., Olefin oligomerization, *Adv. Polym. Sci.*, *15*, 1, 1974.
4. Parshall, G. W., *Homogeneous Catalysis: The Application and Chemistry of Catalysis by Soluble Transition Metal Complexes*, Wiley, New York, 1980, p. 61.
5. Bestian, H., and Clauss, K., Olefin Reaktion mit der titan kohlenstoff Binding, *Angew. Chem.*, *75*, 1068, 1963.
6. Carbonaro, A., Greco, A., and Dall'Asta, G., Dimerization of ethylene. Kinetic aspects and mechanism, *J. Organometal Chem.*, *20*, 177, 1969; 6a. Barone, R., Chanon, M., and Green, M. L. H., *J. Organometal Chem.*, *185*, 85, 1980.
7. Langer, A. W., Olefin dimerization by transition metal complexes, *J. Macromol. Sci.-Chem.*, *A4*, 775, 1970.
8. Henrici-Olive, G., and Olive, S., α-Olefin aus Äthylene *Angew. Chem.*, *82*, 255, 1970.
9. Negurenko, V. M., and Mamedova, B. A., Deposited Doc. VINITI 185 pp., 1975; *Chem. Abstr.*, *86*, 18839, 1977.
10. Hetflejs, J., and Langova, J., Catalytic dimerization of ethylene, *Chem. Listy*, *67*, 590, 1973.
11. Feldblyum, V. Sh., *Dimerization and Disproportionation of Olefins*, Mir, Moscow, 1979, p. 208 (Russian).
12. Sidorchuk, I. I., Negurencu, V. M., Mamedova, B. A., Filipova, L. V., and Kheldova, W. K., Deposited Doc. VINITI 2195, 1975; *Chem. Abstr.*, *87*, 104020, 1977.
13. Belov, G. P., and Dyacheviskii, F. S., *Dokl. Soobshch.*, 1974, p. 304; *Ref. Zhur.*, *5B*, 1235, 1976 (Russian).
14. Jolly, J. K., in *Comprehensive Organometallic Chemistry*, Wilkinson, G., Stone, F. G. A., and Abel, E. W., Eds., Pergamon Press, New York, 1982, Vol. 8, p. 621.
15. Johnson, B. H., U.S. Patent 3,475,347, 1969.
16. Johnson, B. H., U.S. Patent 3,642,932, 1972.
17. Egger, K. W., Gas phase ethylene dimerization in the presence of organoaluminum compounds, *Trans. Faraday Soc.*, *67*, 2638, 1971.
18. Fernald, M. B., Hay, R. G., and Kresge, A. N., U.S. Patent 3,510,539, 1970.
19. Gruening, H., and Luff, G., Mechanism of ethylene dimerization in the presence of transition metal complexes, *Angew. Makromol. Chem.*, *142*, 161, 1986.
20. Fernald, H. B., U.S. Patent 3,444,263, 1969.
21. Fernald, H. B., U.S. Patent 3,478,124, 1970.
22. Comp. Fr. Raffinage, NL Pat. Appl., 6,400,403, 1964; *Chem. Abstr.*, *62*, 2706c, 1965.
23. Chernikova, I. M., Thesis, Dimerization of olefins, All-union Scientific Research Institute of Olefins, Baku, 1967.
24. Kalashnikova, Z. S., Menaylo, A. T., and Sladkov, A. M., Study of oligomerization process of alkenes, *Tr. Nauchno Isled. Inst. Spirtov Org. Prod.*, 1960, 262 (Russian).
25. Chernikov, I. M., Pisman, I. I., and Dalin, M. A., Dimerization of ethylene. Mechanism and composition of reaction product, *Azerb. Khim. Zh.*, 1965, 35 (Russian).
26. Alan, A. Y., and James, H. K., G.B. Patent 896,822, 1962.
27. Chernikov, I. M., Pisman, I. I., Dalin, M. A., Taktarov, P. K., and Agadzhanov, Yu. A., *Khim. Prom.*, 1967, 328 (Russian).
28. Martin, H., Ethylene oligomerization by titanium compounds, *Angew Chem.*, *68*, 306, 1956.

29. Pino, P., and Lorenzi, G. P., in *Preparation and Properties of Stereoregular Polymers*, Reidel, Amsterdam, Holland, 1980, p. 21.
30. Patat, F., and Siun, H., *Angew. Chem.*, *70*, 496, 1956.
31. Pino, P., Mechanism of ethylene oligomerization by Ziegler Natta catalysts, *Adv. Polym. Sci.*, *4*, 394, 1965.
32. Cosee, P. J., and Arlman, E. J., Active centers in the titanium-aluminum complexes, *J. Catal.*, *3*, 99, 1964.
33. Henrici-Olive, G., and Olive, S., Die aktive Species in homogenen Ziegler-Natta Katalysatoren für die äthylene Polymerisation, *Angew. Chem.*, *79*, 764, 1964.
34. Armstrong, D. R., Perkins, P. G., and Stewart, J. S., Oligomerization of alkenes in the presence of Ziegler-Natta catalysts, *J. Chem. Soc., Dalton Trans.*, 1972, p. 1972.
35. Pino, P., Consiglio, G., and Ringger, J., Ethylene oligomerization by transition metal complexes, *Liebigs Ann. Chem.*, 1977, 509.
36. Green, M., Active species in ethylene oligomerization in the presence of transition metal complexes, *Pure Appl. Chem.*, *50*, 27, 1978.
37. Rodriguez, L. A., Van Looy, H. M., and Gabant, J. A., Studies on Ziegler-Natta catalysts. Part I. Reaction between triethyl-aluminum and α-titanium trichloride, *J. Polym. Sci.*, *A1*, 4, 1905, 1966.
38. Rodriguez, L. A., Van Looy, H. M., and Gabant, J. A., Studies on Ziegler-Natta catalysts. Part II. Reaction between α- or β-$TiCl_3$ and $AlMe_3$, $AlMe_2Cl$ or $AlEt_3$ at various temperatures. *J. Polymer Sci.*, *A1*, *4*, 1917, 1966.
39. Rodriguez, L. A., Van Looy, H. M., and Gabant, J. A., Studies on Ziegler-Natta catalysts. Part III. Composition of the nonvolatile product of the reaction between $TiCl_3$ and Et_3Al or Et_2AlCl, *J. Polymer Sci.*, *A1*, *4*, 1927, 1966.
40. Rodriguez, L. A., Van Looy, H. M., and Gabant, J. A., Studies on Ziegler-Natta catalysts. Part IV. Chemical nature of active sites, *J. Polymer Sci.*, *A1*, *4*, 1951, 1966.
41. Rodriguez, L. A., Van Looy, H. M., and Gabant, J. A., Studies on Ziegler-Natta catalysts. V. Stereospecificity of the active centers, *J. Polymer Sci.*, *A1*, *4*, 1971, 1966.
42. Brookhart, M., and Green, M. L. H., Carbon-hydrogen-transition metal bonds, *J. Organometal Chem.*, *250*, 395, 1983.
43. Brookhart, M., Green, M. L. H., and Pardy, R. B. A., The role of transition metal complexes in olefin oligomerization, *J. Chem. Soc., Chem. Commun.*, 1983, 691.
44. Minsker, K. S., Kasparas, M. M., and Zaicov, C. E., Kinetics and mechanism of ethylene oligomerization in the presence of transition metal complexes, *J. Macromol. Sci., Rev. Macromol. Chem. Phys.*, *C27*, 1, 1987.
45. Jolly, C. A., and Marynick, D. S., Olefin insertion into the titanium-alkyl bond, *J. Amer. Chem. Soc.*, *111*, 7968, 1989.
46. Langer, A. W., Active species in ethylene oligomerization in the presence of titanium compounds, *J. Macromol. Sci. Chem.*, 4, 775, 1970.
47. Zubanov, B. A., and Zavorohin, M. D., *Coordination and Catalysis of Olefin Polymerization*, Nauka, Moscow, 1987, pp. 120–134 (Russian).
48. Brookhart, M., Green, M. L. H., and Pardy, R. B. A., Experimental conditions of ethylene oligomerization reaction. The influence of Al/Ti ratio, *J. Chem. Soc. Chem. Commun.*, 1982, 1410.
49. Russiyan, L. N., Matkovskyi, P. E., Dyachovskyi, F. S., Brikenshtein, H. M. A., Starcheva, G. P., and Gerasina, M. P., The study of the structure of the metal-polymer complex: Polyacrylic acid-Cu(II)-poly-4-vinyl pyridine, *Vysokomol. Soedin.*, *A*, *21*, 1891, 1979 (Russian).
50. Melnikov, V. N., Sycheva, D. A., Matkovskyi, P. E., Golubiev, W. K., and Chekryi, P. S., The influence of catalyst concentration in ethylene oligomerization, *Khim. Prom (Moscow)*, 7, 389, 1985 (Russian).

51. Russiyan, L. N., Matkovskyi, P., Dyachkovskyi, F. S., Brikenshtein, H. M. A., and Gerasina, M. P., Kinetics of oligomerization of ethylene to higher α-olefins by the system $TiCl_4$—$C_2H_5AlCl_2$ in aromatic solvents, *Vysokomol. Soedin., A*, *19*, 619, 1977 (Russian).
52. Dyachkovskyi, F. S., Matkovskyi, P. E., Russiyan, L. N., and Semenov, A. A., Chain termination mechanism in oligomerization of ethylene with the system $TiCl_4$—$C_2H_5AlCl_2$, *Vysokomol. Soedin., A*, *20*, 746, 1978 (Russian).
53. Matkovskyi, P. E., Belova, V. N., Brikenshtein, H. M. A., Dyachkovskyi, F. S., Denisova, Z. A., and Kissin, Yu. K., Mechanism of ethylene oligomerization. The role of solvent, *Vysokomol. Soedin., A*, *17*, 252, 1975 (Russian).
54. Langer, A. W., U.S. Patent 4,396,788, 1983.
55. Langer, A. W., U.S. Patent 4,434,312, 1984.
56. Henrici-Olive, G., *Coordination and Catalysts*, Verlag, Chemie, Weinheim, 1977, p. 146.
57. Zhukov, V. I., Shestat, N. P., Dyadynova, M. N., Shilov, L. A., Sherbyakov, I. D., Denilov, R. Kh., and Tsvetkov, N. S., Oligomerization of ethylene in the presence of titanium alcoxides. Determination of the optimal experimental conditions, *Izv. Sev.-Kavk. Nauchn. Tentra Vyssh. Shk. Nauki*, *3*, 93, 1975 (Russian); *Chem. Abstr.*, *83*, 193844, 1975.
58. Farina, M., and Ragazzini, M., The influence of Al/Ti ratio upon the ethylene dimerization reaction, *Chim. Ind. (Milan)*, *40*, 816, 1958.
59. Brown, C. E. H., and Symcox, R., Oligomerization of ethylene in the presence of organotitanium compounds, *J. Polymer Sci.*, *34*, 139, 1959.
60. Natta, G., Porri, L., Carbonaro, A., and Stoppa, G., Polymerization of conjugated diolefins by homogeneous aluminum alkyl-titanium alkoxide catalyst systems. I. *cis*-1,4-Isotactic poly(1,3-pentadiene), *Makromol. Chem.*, *77*, 114, 1964.
61. Dawes, D. H., and Winkler, C. A., Polymerization of butadiene in the presence of triethylaluminum and n-butyl titanate, *J. Polymer Sci., A*, *2*, 3029, 1964.
62. Angelescu, E. M., and Moldovanu, S., Etude spectrophotometrique des catalysateurs organo-metalliques solubles dans les hydrocarbons, *Rev. Roum. Chim.*, *11*, 541, 1966.
63. Takeda, M., Physico-chemical aspects of the ethylene oligomerization reaction, *J. Polymer Sci., C*, *2*, 741, 1969.
64. Hirai, H., Hiraki, K., Noguki, I., and Nakashima, S., Oligomerization of ethylene. The influence of the catalyst structure, *J. Polymer Sci., A*, *8*, 147, 1970.
65. Belov, G. P., Dzabiev, T. S., and Matkovskii, P. E., U.S. Patent 3,879,485, 1975.
66. Zhukov, V. I., U.S. Patent 4,101,600, 1978.
67. Angelescu, E., and Pîrulescu, I., Ethylene oligomerization by heterogeneous coordinated catalysis, *Rev. Chim. (Bucharest)*, *20*, 521, 1969 (Romanian).
68. Natta, G., and Carbonaro, A., Polymerization of conjugated diolefins by homogeneous aluminum alkyl, *Makromol. Chem.*, *77*, 129, 1964.
69. Yamada, S., and Ono, I., Ethylene oligomerization in the presence of titanium-aluminum organic compounds, *Bull. Jpn. Pat. Inst.*, *12*, 160, 1970.
70. Cosee, P., Dimerization of ethylene in the presence of $Ti(OMe)_4$-$AlEt_3$ system. *J. Catal.*, *3*, 80, 1964.
71. Masters, C., *Homogeneous Transition Metal Catalysis: A Gentle Art*, Chapman and Hall, New York, 1981, p. 135.
72. Navaro, O., Chow, S., and Magnoust, P., Study of the ethylene dimerization in the presence of titanium catalyst, *J. Catal.*, *41*, 91, 1976.
73. Navaro, O., Chow, S., and Magnoust, P., Study of the ethylene dimerization in the presence of titanium catalyst, II. *J. Catal.*, *41*, 131, 1976.

74. Belov, G. P., Dzabiev, T. F., and Dyachovskii, F. S., Intermediate complexes in ethylene oligomerization in *Symposium on the Mechanism of Hydrocarbon Reactions*, Marta, F., and Kallo, D., Eds., Elsevier, Amsterdam, 1973, p. 507.
75. Dzabiev, T. S., and Karfova, N. D., Oligomerization of ethylene. Mechanism and kinetical aspects, *Kinet. Kataliz*, *15*, 67, 1974 (Russian).
76. Belob, G. P., and Dzabiev, T. S., Intermediate complexes in the dimerization reaction of ethylene. *J. Mol. Catal.*, *14*, 105, 1982.
77. Puddephatt, R. J., Intermediate species in the reaction of ethylene dimerization, *Comments Inorg. Chem.*, *2*, 68, 1982.
78. Teble, F. N., Mechanism and intermediates in ethylene oligomerization, *J. Amer. Chem. Soc.*, *101*, 5074, 1979.
79. Matkovskii, P. E., Russiyan, L. N., Dyachkovskii, Kh., Brikenshtein, A., and Gerasina, M. P., Mechanism of ethylene oligomerization, *Vysokomol. Soedin.*, *A*, *19*, 263, 1978 (Russian).
80. Belov, G. P., *Kompleksn. Metalloorg. Katal. Polim. Olefinov.*, 1977, p. 123; *Ref. Zhur.*, *15B*, 1159, 1978 (Russian).
81. Belov, G. P., and Susnov, V. I., Ethylene oligomerization in the presence of catalytic system Ti(n-$C_4H_9O)_4$-AlR_3 in n-heptane medium, *Neftekhimia*, *18*, 891, 1978 (Russian).
82. Belov, G. P., About interaction between phenol and endodicyclopentadiene in KU-2 cationite, *Neftekhimia*, *14*, 1978, 1974 (Russian).
83. Dzhabiev, T. S., Dyachkovskii, F. S., and Shiskina, I. I., Kinetics of ethylene oligomerization, *Izv. Akad. Nauk. SSSR, Ser. Khim.*, 1973, p. 1238 (Russian).
84. Dzhabiev, T. S., Dyachkovskii, F. S., and Ishtryakov, I. A., Kinetics of ethylene dimerization in the presence of tetrabutoxytitanium-trimethylaluminum system, *Neftekhimia*, *14*, 700, 1974 (Russian).
85. Belov, G. P., *Tezis Dokl. Vses. Chugaevskoe Soveshch. Khim., Kompleksn. Soedin., 12th*, 2, 201, 1978; *Chem. Abstr.*, *85*, 159010, 1976 (Russian).
86. Dzhabiev, T. S., Mechanism of butene formation in the process of ethylene dimerization, *Neftekhimia*, *16*, 706, 1976 (Russian).
87. Belov, G. P., Smirnov, V. I., Soloveva, T. I., and Dyachovskii, F. S., Upon the intermediate species in the dimerization reaction of ethylene, *Zhur. Fiz. Khim.*, *51*, 2132, 1977 (Russian).
88. Belov, G. P., Dzhabiev, Z. M., and Dyachkovskii, F. S., *Komplecsn. Metaloorg. Polim. Olefinov*, 1978, p. 119; *Ref. Zhur.*, *20B*, 1174, 1978 (Russian).
89. Belov, G. P., Smirnov, V. I., Soloveva, T. I., and Dyachkovski, F. S., Deposited Doc. VINITI, 1978, p. 1478; *Chem. Abstr.*, *90*, 86645, 1979 (Russian).
90. Bestian, H., and Clause, K., Dimerization of ethylene in the presence of two-component organometallic catalyst $MeTiCl_3$-$MeAlCl_2$, I. *Angew. Chem., Int. Ed. Engl.*, 2, 32, 1963.
91. Bestian, H., and Clause, K., Dimerization of ethylene in the presence of two-component organometallic catalyst $MeTiCl_3$-$MeAlCl_2$, II. *Angew. Chem., Int. Ed. Engl.*, 2, 704, 1963.
92. Cesca, S., Marconi, W., and Santostasi, M. L., Alpha-olefins reactions with systems based on H_2AlNR_2 and $Ni(acac)_3$ or $Ti(OR)_4$, *J. Polymer Sci.*, *B*, *7*, 542, 1969.
93. Shikato, K., Miura, Y., Nakaro, S., and Azuma, K., *Kogyo Kagaku Zashi*, *68*, 2266, 1965; *Chem. Abstr.*, *64*, 15721, 1966.
94. Datta, S., Fisher, M. B., and Wreford, S. S., Dimerization of ethylene in the presence of $(n\text{-}C_4H_6)_2$Ti[1,2-bis(dimethyl-phosphino)ethane)], *J. Organomet. Chem.*, *188*, 353, 1980.
95. Wailes, P. C., Coutts, R. S. P., and Weigold, H., in *Organometallic Chemistry of Titanium, Zirconium, and Hafnium*, Academic Press, New York, 1974, p. 188.

96. Carrick, W. L., Chauser, A. G., and Smith, J. J., Transition metal catalysts. IV. Role of volume in low pressure catalysts. *J. Amer. Chem. Soc.*, *82*, 5319, 1960.
97. Clark, R. J. H., and McAlles, A. J., The influence of the oxidation state of the metal upon the activity of catalyst in the ethylene oligomerization reaction, *J. Chem. Soc.*, 1970, 2026.
98. Commerenc, D., Chauvin, Y., Gaillard, J., Leonard, J., and Andrews, J., Ionic intermediates in the reaction of ethylene dimerization, *Hydrocarbon Process.*, *63*, 120, 1984.
99. Aliev, V. S., and Aliev, A. B., SU Patent 1,154,258, 1985; *Chem. Abstr.*, *103*, 215956, 1985.
100. Kalouj, I. N., Popov, W. G., and Kabanov, W., Elemental organic compounds. Part XXI., *Vysokomol. Soedin.*, *B*, *23*, 368, 1981 (Russian).
101. Fuhrman, H., GER(EAST) Patent 124,623, 1977.
102. Fuhrman, H., GER(EAST) Patent 124,047, 1977; *Chem. Abstr.*, *88*, 51832f, 1978.
103. Longi, P., Greco, F., and Rossi, U., Linear alpha-olefins from ethylene by means of alkylation halides and zirconium compounds as catalysts, *Chim. Ind.* (*Milan*, *55*, 252, 1973.
104. Belova, N. N., Markovskyi, P. E., Gerasina, M. P., Brikenshtein, H. M. A., and Dyachkovskyi, F. S., *Komoleksnyie Metaloorgan Cheskie Katalisatory Polimerizacji Olefinov*, Nauka, Moscow, 1980, p. 30.
105. Mel'nikov, V. N., Matkovskyi, P. E., Russiyan, L. N., and Bucheva, Z. G., Mechanism of interaction between zirconium isobutyrate and sesquiethylaluminum chloride in catalysts for ethylene oligomerization, *Kinet. Kat.*, *29*, 124, 1988.
106. Burinina, N. A., Piechatnikov, E. L., Brikenshtein, H. M. A., Gerasina, M. A., and Mel'nikov, W. N., Kinetic model and mathematical representation of ethylene oligomerization process to linear alpha-olefins upon the influence of Zr(OCO-iso-$C_3H_7)_4$-$(C_2H_5)_3Al_2Cl_3$ system, *Zh. Prikl. Khim.*, *59*, 364, 1986.
107. Pino, P., Ciardelli, F., and Lorenzi, G. P., Induzione asimetrica nella polimerizatione di albune alpha-olefine, *Chim. Ind.* (*Milan*), *46*, 313, 1964.
108. Mel'nikov, V. N., Matkovskyi, P. E., Sycheva, O. A., Chernyh, S. P., Chekruyi, P. S., Belova, W. N., Russiyan, L. N., and Gerasina, M. P., Oligomerization of ethylene to higher alpha-olefins in the presence of zirconium catalysts, *Khim. Prom.* (*Moscow*), 5, 261, 1986.
109. Mel'nikov, V. N., Sycheva, O. A., Matkovskyi, P. E., Mireva, T. A., Chernyh, S. P., Chekruyi, P. S., Belova, W. N., and Russiyan, L. N., Effect of catalyst composition on rate and selectivity of ethylene low-temperature oligomerization, *Khim. Prom.* (*Moscow*), *6*, 823, 1986.
110. Langer, A. W., U.S. Patent 4,442,309, 1984.
111. Shiraki, Y., and Kono, S., JP Patent 6,259,225, 1987; *Chem. Abstr.*, *107*, 58489h, 1987.
112. Shiraki, Y., and Kono, S., JP Patent 62,000,430, 1987; *Chem. Abstr.*, *107*, 7812c, 1987.
113. Shiraki, Y., Eur. Pat. Appl. 241,596, 1987; *Chem. Abstr.*, *108*, 56784z, 1988.
114. Langer, A. W., U.S. Patent 4,361,714, 1982.
115. Kuliev, R. Sh., Olefin oligomerization catalysed by Zr(acac)-$EtAlCl_2$ and $(C_3H_7)_4$Zr-$Et_3Al_2Cl_3$, *Nefterab. Neftekhim.*, *4*, 15, 1985 (Russian).
116. Kuliev, R. Sh., Ethylene oligomerization in the presence of organozirconium compounds, *Nefterab. Neftekhim.*, *6*, 14, 1986 (Russian).
117. Attridge, C. J., Jackson, R., Maddock, S. J., and Thomson, D. T., Oligomerization of ethylene in the presence of Zr-Al compounds, *J. Chem. Soc.*, *Chem. Commun.*, 1973, 132.
118. Ziegler, K., GB Patent 787,438, 1957.

119. Dzhemilev, U. M., SU Patent 857,097, 1981; *Chem. Abstr.*, *96*, 6143, 1982.
120. Langer, A. W., U.S. Patent 3,441,630, 1970.
121. Tebbe, F. N., U.S. Patent 3,133,876, 1976.
122. Dorf, U., Engel, K., and Erker, G., Ethylene oligomerization by zirconium and hafnium organic compounds, *Angew. Chem. Int. Ed. Engl.*, *21*, 914, 1982.
123. Haleard, D. G. M., Oligomerization of olefins. Insertion of hydrocarbon into metal-carbon bond, *J. Polymer Sci.*, *A*, *13*, 2191, 1978.
124. Tanaka, S., Nakamura, Y., Hatta, J., Shimizu, T., and Uchida, N., JP Patent 7,001,441, 1970; *Chem. Abstr.*, *72*, 110750, 1970.
125. Langer, A. W., U.S. Patent 4,377,720, 1983.
126. Schrock, R. P., U.S. Patent 3,932,477, 1976.
127. Schrock, R. P., Sancho, J., and McLain, S., Dimerization of ethylene in the presence of organotantalum compounds, *Pure Appl. Chem.*, *52*, 729, 1980.
128. Schrock, R. P., McLain, S. J., and Sancho, J., Tantacyclopentane complexes: A new catalyst of the ethylene dimerization, I., *J. Amer. Chem. Soc.*, *101*, 5451, 1979.
129. McLain, S. J., Sancho, J., and Schrock, R. P., Tantacyclopentane complexes: A new catalyst of the ethylene dimerization, II, *J. Amer. Chem. Soc.*, *102*, 5610, 1980.
130. McLain, S. J., and Schrock, R. P., Dimerization of ethylene in the presence of $CpTaCl_2(C_4H_8)$, *J. Amer. Chem. Soc.*, *100*, 1315, 1978.
131. Feldman, J. P., Rupprecht, G. A., and Schroch, R. P., Dimerization of ethylene in the presence of $Ta(CH_3CMe_3)$-$(CHCMe_3)$-$P(CH_3)_3$ system, *J. Amer. Chem. Soc.*, *101*, 5099, 1979.
132. Feldman, J. D., Schrock, R. P., and Rupprecht, G. A., Ethylene dimerization. The insertion mechanism, *J. Amer. Chem. Soc.*, *103*, 5732, 1981.
133. Zuech, E. A., Polymerizations with homogeneous chromium catalyst, *J. Polym. Sci.*, *Polym. Chem. Ed.*, *10*, 3665, 1972.
134. Zuech, E. A., U.S. Patent 3,627,700, 1971.
135. Zuech, E. A., U.S. Patent 3,726,939, 1973.
136. Green, M. L. H., and Knight, J., Dimerization of ethylene in the presence of organomolibdenum complexes, *J. Chem. Soc.*, *Dalton Trans.*, 1974, 311.
137. Wideman, L. G., U.S. Patent 3,813,453, 1974.
138. Suaeki, K., and Arakawa, T., JP Patent 7,319,601, 1973; *Chem. Abstr.*, *79*, 91559, 1973.
139. Feldblyum, V. Sh., Krotova, L. S., Leshcheva, A. I., and Kononova, L. D., FR Patent 1,588,167, 1970.
140. Saeki, K., JP Patent 7,322,682, 1973; *Chem. Abstr.*, *79*, 145944, 1973.
141. Feldblyum, V. Sh., and Obeschalova, N. V., Dimerization mechanisms of olefins under the influence of R_2AlCl + $(R'COO)_2Ni$ catalysts, *Dokl. Akad. Nauk. SSSR*, *172*, 368, 1967 (Russian).
142. Natta, G., Dimerization of alkenes in the presence of organocobalt compounds, *Chim. Ind. (Milan)*, 1965, 233.
143. Sun Pu, I., Yamamoto, A., and Ikeda, S., Dimerization of ethylene in the presence of dinitrogenhydridotris (triphenylphophine)cobalt(I), *J. Amer. Chem. Soc.*, *90*, 7170, 1968.
144. Kavakami, K., Mizoroki, T., and Ozaki, A., Dimerization of ethylene in the presence of halotris(triphenylphosphine)cobalt(I), *Chem. Lett.*, 1975, 903.
145. Kavakami, K., Mizoroki, T., and Ozaki, A., Dimerization of ethylene in the presence of cobalt(I) complexes: The mechanism of the reaction, *Bull. Chem. Soc. Jpn.*, *51*, 21, 1978.
146. Speier, G., Dimerization of ethylene. Kinetical aspects of the reaction, *Hung. J. Ind. Chem.*, *3*, 449, 1975 (Hungarian).

147. Kabanov, V. A., and Smetanyuk, V. I., The mechanism of the ethylene dimerization in the presence of bis[(ethylene)tris(tripenylphosphine)cobalt] as catalyst, *Chem. Phys. Suppl.*, *5*, 121, 1981.
148. Maruya, K., Ando, N., and Ozaki, A., Dimerization of ethylene catalyzed by σ-arylnickel(II) compounds, *Bull. Chem. Soc. Jpn.*, *43*, 3630, 1970.
149. Dixon, C. G. P., Duck, E. W., and Jenkins, D. K., Dimerization of ethylene in the presence of organonickel compounds, *Organomet. Chem. Synth.*, *1*, 77, 1971.
150. Maruyama, K., Kuroki, T., Myzoroki, T., and Ozaki, A., Dimerization of C_2D_4 catalyzed by bis(triphenylphosphine)-σ-1-naphthyl nickel(II) bromide, *Bull. Chem. Soc. Jpn.*, *44*, 2002, 1971.
151. Abasova, S. G., Krentsel, B. A., Leshcheva, A. I., Mushina, E. A., and Feldblyum, V. Sh., Nickel catalysed oligomerization of ethylene, *Izv. Akad. Nauk SSSR, Ser. Khim.*, 1972, 644.
152. Schmidt, F. K., Tkach, V. S., and Kalabina, A. V., Ethylene oligomerization and ethylene-propylene codimerization in the presence of complex metallo-organic catalysts on the base of nickel compounds, *Neftekhimiya*, *12*, 819, 1972.
153. Maruya, K., Miziroki, T., and Ozaki, A., Dimerization of ethylene catalysed by σ-aryl nickel compound in the presence of trifluoroboron etherate, *Bull. Chem. Soc. Jpn.*, *45*, 2255, 1972.
154. Kawata, N., Mizoroki, T., Ozaki, A., and Ohkawara, M., Dimerization of ethylene in the presence of organonickel compounds, *Chem. Lett.*, 1973, 1165.
155. Petrushanskaya, N. V., Kurapova, A., and Feldblyum, V. Sh., Oligomerization of ethylene in the presence of organonickel compounds, *Zhur. Org. Khim.*, *9*, 2620, 1973.
156. Petrushanskaya, N. V., Kurapova, A., and Feldblyum, V. Sh., Dimerization of olefins under the influence of bis-(ethylene)-triisopropyl-phosphinenickel(0) and ethylene-bis-(triisopropylphosphine)-nickel(0) in conjunction with Lewis acids, *Dokl. Akad. Nauk SSSR*, *211*, 606, 1973.
157. Petrushanskaya, N. V., Kurapova, A. I., Rodionov, N. M., and Feldblyum, V. Sh., Olefin dimerization under the influence of nickel hydride complexes, *Zhur. Org. Khim.*, *10*, 1402, 1974 (Russian).
158. McClure, J. D., and Barnett, K. W., Organonickel compounds as catalysts of the ethylene oligomerization, *J. Organomet. Chem.*, *80*, 385, 1974.
159. Schmidt, F. K., Tkach, V. S., Levkovskii, Yu. S., and Sergeeva, T. N., in *Katalich. reaktii Vzaidk. faze*, *Nauka*, 1974, Ch. 3, p. 675; *Chem. Abstr.*, *83*, 113.
160. Petrushanskaya, N. V., Kurapova, A. I., Rodionova, N. M., and Feldblyum, V. Sh., Kinetical aspects of ethylene dimerization in the presence of organonickel compounds, *Kinet. Katal.*, *17*, 262, 1976.
161. Angelescu, E., Bisceanu, T., and Nicolescu, I. V., Oligomerization of ethylene, *Rev. Chim. (Bucharest)*, *27*, 745, 1976 (Romanian).
162. Petrushanskaya, N. V., Kurapova, A. I., and Feldblyum, V. Sh., Catalysts of ethylene dimerization based on tris(ethylene)nickel, *Neftekhimiya*, *18*, 58, 1978.
163. Uchino, N., Tanaka, K., Sakai, M., and Sakakibara, Y., *Nippon Kagaku Kaishi*, 400, 1978; *Chem. Abstr.*, *89*, 107889, 1978.
164. Borisova, N. A., Mushina, E. A., Karpacheva, G. P., and Kreutsel, B. A., *Nauchn. Osn. Pererab. Neftigaza Neftekhim. Tezisu Dokl. Vses. Konf.*, 156, 1977; *Chem. Abstr.*, *92*, 40971, 1980.
165. Uchino, N., Tanaka, K., Sakai, M., and Sakakibara, V., *Kyoto Kogei Seni Daigaku Senigakuba. Gakujutsu Hokoka*, *9*, 90, 1979; *Chem. Abstr.*, *91*, 124082, 1979.
166. Kabanov, V. A., Martynova, M. A., Pluzhnov, S. K., Smetanyuk, V. I., and Chediya, R. V., Kinetical aspects of ethylene dimerization in the presence of organonickel compounds, *Kinet. Katal.*, *20*, 1012, 1979.

167. Angelescu, E. A., Anca, N., and Ioan, V., Ethylene dimerization by heterogeneous coordinated catalysis, *Rev. Chim. (Bucharest)*, *30*, 523, 1979.
168. Echmaev, S. B., Ivieva, I. N., Bravaya, N. M., Pomogailo, A. D., and Borodko, Yu. G., Mechanism and kinetical aspects of ethylene in the presence of nickel organo compounds, *Kinet. Katal.*, *21*, 1530, 1980.
169. Andrews, J. W., Bonnifay, P., Cha, B. J., Barbier, J. C., Douillet, D., and Raimbault, J., Dimerization of alkenes, *Hydrocarbon Process.*, *55*, 105, 1976.
170. Bossaert, B., Malatesta, A., and Mourand, J., Eur. Pat. Appl. 82,729, 1983.
171. Mel'nikov, V. N., Ethylene oligomerization, *Khim. Prom.*, 1979, 592 (Russian).
172. Bogdanovici, B., Controlled selectivity in olefin oligomerization catalysed by nickel compounds, *Adv. Organometallic Chem.*, *17*, 105, 1979.
173. Takaoka, C., and Nishikubo, T., JP Patent 72,115,892, 1976; *Ref. Zhur.*, *16C*, 3492, 1978.
174. Miller, J. S., U.S. Patent 4,465,788, 1975.
175. Taylor, J. L., U.S. Patent 4,117,235, 1971.
176. Brossas, J. A., and Friedman, G., FR Patent 2,536,752, 1982.
177. Gartner, K., Gurtler, E., and Horn, H., GER(EAST) Patent 218,109, 1984.
178. Riew, K. C., U.S. Patent 4,481,148, 1978.
179. Gilles, J. C., U.S. Patent 4,238, 1975.
180. Threnkel, R. S., U.S. Patent 4,677,241, 1980.
181. Behr, A., Falbe, V., Frenndenberg, U., and Klim, W., 1,3-Diketones as active ligands in the nickel-catalysed linear oligomerization of olefins, *Isr. J. Chem.*, *27*, 277, 1986.
182. Fischer, K., Jonas, K., Misbach, P., Stubba, R., and Wilke, G., Oligomerization of olefins by Ziegler-Natta catalysts, *Angew. Chem. Int. Ed. Engl.*, *12*, 943, 1973.
183. Ewers, J., Dimerization of alkenes by nickel complexes, *Angew. Chem.*, *78*, 593, 1966.
184. Wilke, G., and Bogdanovici, B., Alpha-olefins oligomerization, *Angew. Chem.*, *78*, 157, 1966.
185. Chauvin Y., Phung, N. H., Genchard, N., and Lefevre, G., Application de l'equilibre mobile d'isomerisation a la polymèrisation des melanges d'olefines. I. Deplacement protrotopique par les catalyseurs organometaliques mixtes, *Bull. Soc. Chim. Fr.*, 1966, 3223.
186. Feldblyum, V. Sh., Obeshchalova, N. V., and Leshcheva, A. I., Dimerization of olkenes, *Dokl. Akad. Nauk SSSR*, *172*, 111, 1967 (Russian).
187. Feldblyum, V. Sh., Obeshchalova, N. V., Leshcheva, A. I., and Baranova, T. I., Dimerization of propylene in the presence of R_2AlCl + $(RCOO)_2Ni$ catalytic system, *Neftekhimyia*, *7*, 379, 1967 (Russian).
188. Arakawa, T., and Saeki, K., *Kogyo Kagaku Zasshi*, *71*, 1028, 1968; *Chem. Abstr.*, *69*, 110199, 1968.
189. Wilke, G., *Proc. R. A. Welch Foundation, Conf. on Chemical Research. IX. Organometallic Compounds*, 1966.
190. Bogdanovici, B., Henc, B., and Karmann, H. H., Mechanism of the ethylene oligomerization under the influence of $Ni(acac)_2$-Et_2AlOEt, *Ind. Eng. Chem.*, 62, 35, 1970.
191. Obeshchalova, N. V., Feldblyum, V. Sh., and Bashchenko, M. M., Intermediates in the Ni-catalysed ethylene dimerization, *Zhur. Org. Khim.*, *4*, 104, 1968 (Russian).
192. Feldblyum, V. Sh., and Osokin, Yu., V., Investigation of the structure of dimers of isohexenes by IR and NMR methods, *Neftekhimyia*, 7, 878, 1967 (Russian).
193. Kolke, T., Kavakami, K., Maruya, K., Mizoraki, T., and Ozaki, A., Dimerization of ethylene in the presence of nickelocene, *Chem. Lett.*, 1977, 551.
194. Tsutsui, M., and Koyano, T., Elemental organic compounds. Part XX. Ethylene dimerization to butene-1, *J. Polymer Sci.*, *A-1*, *5*, 681, 1967.

195. Yshimura, Y., Maruya, K., Nakamura, N., Mizoroki, T., and Czaki, A., Ethylene dimerization catalysed by $(C_6H_5)Ni(PPh_3)_2Br$. Mechanism and kinetical aspects, *Chem. Lett.*, 1981, 657.
196. Grubbs, R. H., and Miyashita, A., Dimerization of ethylene to cyclobutene and 1-butene under the influence of tris(triphenylphosphine)tetramethylenenickel(II), *J. Amer. Chem. Soc.*, *100*, 7416, 1978.
197. Muzio, F. J., and Löffler, D. G., A mechanism for hydride formation in nickel catalysed olefin oligomerization, *Acta Chim. Hung.*, *124*, 403, 1987.
198. Keim, W., Hoffmann, E., Lodewick, R., Panckert, M., Schmidt, G., Fleischauer, J., and Meier, U., Ethylene oligomerization by nickel complexes, *J. Mol. Catal.*, *6*, 79, 1979.
199. Benckert, M., Keim, W., Storp, S., and Weber, R. S., Organonickel complexes as catalyst in the ethylene oligomerization, *J. Mol. Catal.*, *20*, 115, 1983.
200. Kissin, Y. V., Beach, D. L., Co-oligomerization of ethylene and higher linear alpha-olefins. I. Co-oligomerization with the sulfonated nickel ylide-based catalytic system. *J. Polym. Sci.*, *Polym. Chem. Ed.*, *27*, 147, 1989.
201. Kissin, Y. V., Co-oligomerization of ethylene and higher alpha olefins. II. Structure of oligomers, *J. Polym. Sci.*, *Polym. Chem. Ed.*, *27*, 605, 1989.
202. Keim, W., Ethylene oligomerization to higher alpha-olefins, *Angew. Chem.*, *90*, 493, 1978.
203. Keim, W., *Compend. Dtech. Ges. Mineralwiss. Chem.*, *78*, 453, 1978; *Chem. Abstr.*, *93*, 7580t, 1980.
204. Page, N. M., Young, L. B., U.S. Patent 4,855,527, 1989.
205. Binger, P., Ethylene oligomerization under the influence of 1,5-cyclooctadiene nickel(0) and thiolactic acid, *Liebigs Ann.*, 1977, 1065.
206. Muller, U., and Belz, P., Ethylene oligomerization catalysed by nickel(0) complexes and $Ph_2PCH_2C(CF_3)_2OH$, *Angew. Chem.*, *101*, 1066, 1989.
207. Keim, W., Ethylene oligomerization in the presence of nickel complexes, *J. Chem. Ed.*, 1986, No. 63, p. 203.
208. Keim, W., The formation of Ni-H bond in the ethylene oligomerization reaction, *Chem. Eng. Tech.*, *256*, 850, 1984.
209. Mason, R. F., U.S. Patent 3,676, 523, 1972.
210. Mason, R. F., U.S. Patent 3,686,351, 1972.
211. Mason, R. F., U.S. Patent 3,737,475, 1975.
212. Knudsen, R., and Reiter, S., U.S. Patent 4,482,640, 1985.
213. Beach, D. L., and Harrison, J. J., U.S. Patent 4,382,153, 1983.
214. Beach, D. L., and Harrison, J. J., U.S. Patent 4,711,969, 1987.
215. Threlkel, R. S., U.S. Patent 4,677,241, 1980.
216. Gao Zhanxian, Effect of betaketonate ligands and cocatalysts on linear oligomerization of ethylene catalysed by nickel complexes, *Shiyon Huagong*, *15*, 464, 1986; *Chem. Abstr.*, *106*, 120254g, 1987.
217. Zhang Kai and Lu Junmin, Linear oligomerization of propylene catalysed by nickel trifluoroacetylacetonate and aluminum alkyls, *Gongxueyuan*, *25*, 117, 1986 (Chinese); *Chem. Abstr.*, *107*, 7689t, 1987.
218. Liu Zhanxian, Linear oligomerization of propylene with chelated nickel complex catalyst, *Gaodeng Xuexiao Huaxue Xuebao*, *6*, 1127, 1985 (Chinese); *Chem. Abstr.*, *104*, 20778u, 1986.
219. O'Connor, C. T., Jacobs, L. L., and Kajima, M., Propene oligomerization over synthetic mica-monmorilonite (SMM) and SMM incorporated nickel, zinc and cobalt. *Appl. Catal.*, *40*, 277, 1988.
220. Brown, S. J., Chaterbrock, L. M., and Master, A. F., Kinetic and mechanism studies of nickel-catalysed olefin oligomerization, *Anal. Catal.*, *48*, 1, 1989.

221. Cavell, K. J., and Masters, A. F., Olefin oligomerization and isomerization catalysts. The preparation, properties and catalytic action of dithionickel(II) complexes with chelating phosphines, *Aust. J. Chem.*, *39*, 1129, 1986.
222. Lutz, E. F., U.S. Patent 4,528,416, 1985.
223. Lutz, E. F., U.S. Patent 3,825,615, 1974.
224. Singleton, D. M., U.S. Patent 4,472,522, 1984.
225. Nadyr Mir-Ibragimoglu Seldov, RO Patent 70,798, 1986.
226. Beach, D. L., and Harrison, J. J., U.S. Patent 4,293,727, 1981.
227. Beach, D. L., and Kissin, V. V., U.S. Patent 4,686,315, 1987.
228. Beach, D. L., and Harrison, V. V., U.S. Patent 4,310,716, 1982.
229. Kissin, V. V., Co-oligomerization of ethylene and higher linear alpha-olefins. III., *J. Polymer Sci., Polym. Chem. Ed.*, *27*, 623, 1989.
230. Mizoreki, T., in *Catalytic Proc. Int. Symp.*, Delmon, D., and Jannes, G., Eds., Elsevier, Amsterdam, 1975, p. 319.
231. Mizoreki, T., *Catalysis: Heterogeneous and Homogeneous*, Elsevier, Amsterdam, 1975.
232. Kawata, N., and Mizoreki, T., Ethylene oligomerization in the presence of Ni-polystyryl complex, *J. Mol. Catalysis*, *1*, 275, 1975.
233. Espinoso, R. L., Nicolaides, C. P., Korf, C. J. S., and Snel, R., Catalytic oligomerization of ethylene over nickel exchanged amorphous silica-alumina; effect of the nickel concentration, *Appl. Catal.*, *31*, 259, 1987.
234. Behr, A., and Klim, W., The nickel complex catalysed synthesis of alpha-olefins, *Arabian J. Sci. Eng.*, *10*, 377, 1985.
235. Keimi, W., Oligomerization and polymerization of monoolefins via homogeneous and heterogeneous nickel catalysts, *Proc. Int. Symp. Relat. Homogeneous and Heterogeneous Catalysts*, Ermakov, Yu. I., Ed., VNU Sci., Utrecht, 1986, p. 499.
236. Barnett, K. W., U.S. Patent 4,628,139, 1986.
237. Penckert, M., and Keim, W., Supported catalysts for the olefin oligomerization, *J. Mol. Catal.*, *22*, 289, 1984.
238. Kissin, V. V., in *Proc. Symp. on Transition Metal Catalysed polymerization; the Ziegler-Natta and Metathesis Polymerization*, Cambridge University Press, New York, 1988.
239. Alderson, T., Ruthenium chloride as catalyst in ethylene dimerization reaction, *J. Amer. Chem. Soc.*, *87*, 5638, 1965.
240. Thomas, A., U.S. Patent 3,013,066, 1961.
241. Cramer, R., Ethylene dimerization by rhodium chloride: Mechanism and experimental conditions, *J. Amer. Chem. Soc.*, *87*, 4717, 1963.
242. Takahashi, N., Supported rhodium catalysts for ethylene dimerization, *J. Amer. Chem. Soc.*, *97*, 9849, 1975.
243. Takahashi, N., Mechanism of ethylene dimerization by supported rhodium chloride catalyst, *J. Mol. Catal.*, *3*, 277, 1978.
244. Antonov, P. G., *Izv. Vyssh. Uchebn. Zaved. Khim. Technol.*, 22, 952, 1979; *Chem. Abstr.*, *92*, 41220, 1980.
245. Kusunoki, Y., The dimerization of ethylene using palladous chloride as the catalyst, *Bull. Chem. Soc. Jpn.*, *39*, 2021, 1966.
246. Van Gemert, J. T., and Wilkinson, P. K., Dimerization of ethylene in the presence of tetrachlorobis(ethylene)-dipalladiumn, *J. Phys. Chem.*, *68*, 645, 1964.
247. Kawamoto, K., Imanaka, T., and Taraganishi, S., *Nippon Kagaku Zashi*, *91*, 39, 1970; *Chem. Abstr.*, 72, 99767, 1970.
248. Ketley, A. D., Dimerization of ethylene in the presence of palladium complexes. Mechanism of reaction, *Inorg. Chem.*, *6*, 657, 1967.
249. Barlow, N. G., Bryant, M. J., Haszeldine, R. N., and Mackie, A. G., Dimerization of ethylene by $Pd(BzCN)_2Cl_2$, *J. Organomet. Chem.*, *21*, 215, 1970.

250. Kitamura, T., Maruya, K., Moroka, Y., and Ozaki, A., Dimerization and isotopic mixing of ethylene by a palladium complex, *Bull. Chem. Soc. Jpn.*, *45*, 1457, 1972.
251. Pertici, P., and Vitulli, G., Dimerization of ethylene in the presence of palladium-aluminum organocompounds, *Tetrahedron Lett.*, 1979, 1897.
252. Usov, Yu. N., Ethylene dimerization in the presence of palladium catalysts, *Neftekhimiya*, *18*, 369, 1978 (Russian).
253. Chekurevskaya, E. D., Usov, Yu. N., and Kuvshinova, N. I., *Vses. Chunguev Sovetsch. Po Khimii Komplecs. Soedin.*, 1978, 436; *Chem. Abstr.*, *90*, 120917, 1979.
254. Usov, Yu. N., and Kuvshinova, N. I., Dimerization of ethylene using as catalyst $PdCl_2(Me_2SO)_2$. Aspects of the kinetics and mechanism, *Kinet. Catal.*, *19*, 1606, 1978 (Russian).
255. Usov, Yu. N., Deposited Doc. SPCTL 898 KhP-D80, 1980; *Chem. Abstr.*, *97*, 91661, 1982.
256. Ikeda, S., GER Patent 2,412,105, 1975.
257. Mamedaliev, G. A., SSR Patent 1,234,397, 1986.
258. He Ren, Wang Yu, and Jiang Jingyuang, Linear oligomerization of ethylene. I. Synthesis of lower alpha-olefins catalysed by zirconium catalyst, *Cuihua Xuebao*, *9*, 58, 1988; *Chem. Abstr.*, *109*, 55301y, 1988.
259. He Ren, Wang Yu, and Jiang Jingyuang, Linear oligomerization of ethylene. IV. Catalytic oligomerization of ethylene by IVB compounds, *Fenzi Cuihua*, *2*, 283, 1988; *Chem. Abstr.*, *109*, 55302, 1988.
260. Mamedaliev, C. A., SSSR Patent 1,351,912, 1982.
261. Kukes, S. G., and Novak, G. P., U.S. Patent 4,613,719, 1986.
262. Mel'nikov, V. M., and Gerasina, M. D., Oligomerization of ethylene to higher alpha-olefins on zirconium containing catalysts, *Khim. Prom.*, 1986, 261 (Russian).
263. Mitsui Petrochemical Ind., JP Patent 57,123,205, 1982; *Chem. Abstr.*, *98*, 4894r, 1983.
264. Anshita, A. G., Oligomerization of olefins on an Al-Me catalyst, *Kinet. Katal.*, *29*, 1926, 1986 (Russian).
265. Beronane, E. G., and Nagy, J. B., Channel network effects in ethylene oligomerization, *J. Mol. Catal.*, *53*, 293, 1989.
266. Miller, S. J., U.S. Patent 4,608,450, 1986.
267. Mravec, D., Reactions of propylene on zeolite H-ZSM-5, *Ropa Uhlie*, *28*, 283, 1986.
268. Tabak, S. A., Conversion of ethylene and butylene over ZSM-5 catalyst, *AIChEJ*, *32*, 1526, 1986.
269. Ocelli, M. L., Propylene oligomerization with pillared clays, *J. Mol. Catal.*, *33*, 371, 1985.
270. Blain, D. A., Eur. Pat. Appl. 311,350, 1989.
271. Giusti, A., Eur. Pat. Appl. 293,950, 1988.
272. Borovkov, V. Yu., and Kazanska, V. B., Mechanism of ethylene oligomerization on ZSM-type zeolites, *Dokl. Akad. Nauk SSSR*, *286*, 914, 1986 (Russian).
273. Mravec, D., Balko, J., and Ilavski, J., Oligomerization of ethylene on synthetic zeolites ZSM-5, *Chem. Prum.*, *36*, 82, 1986 (Slovenian).
274. Chang, C. D., U.S. Patent 4,554,396, 1985.
275. Van der Berg, J. P., Olefin oligomerization under the influence of zeolites, *J. Catal.*, *80*, 130, 1983.
276. Van der Berg, J. P., Zeolites as catalysts in olefin oligomerization, *J. Catal.*, *80*, 139, 1983.
277. Novakova, J., Secondary reactions in olefin oligomerization catalysed by heterogeneous catalysts, *Coll. Czech. Chem. Commun.*, *44*, 3341, 1979.
278. Kustov, L. M., Cationic oligomerization of olefins in the presence of zeolites, *Stud. Surf. Sci. Catal.*, *28*, 241, 1984.

279. Langer, A. V., U.S. Patent 3,790,541, 1974.
280. Ozawa, M., U.S. Patent 4,042,638, 1977.
281. Wistosky, N. J., U.S. Patent 4,087,255, 1978.
282. Stark, K., U.S. Patent 3,835,107, 1974.
283. Gulf Research Co., GB Patent 1,180,610, 1970.
284. Gulf Research Co., GB Patent 1,180,609, 1970.
285. Gulf Research Co., U.S. Patent 3,655,809, 1972.
286. Gulf Research Co., U.S. Patent 3,721,719, 1973.
287. Gulf Research Co., U.S. Patent 3,751,515, 1973.
288. Turner, A. H., Industrial scale production of olefin oligomers, *J. Amer. Oil Chem. Soc.*, *60*, 623, 1983.
289. Lutz, E. F., Technological aspects of olefin oligomerization, *J. Chem. Ed.*, *10*, 377, 1985.
290. NienwenHuis, R. A., Olefin processing, *Petrol Technol.*, *46*, 268, 1980.
291. Reuben, B., and Wittcoff, H., Industrial process of olefin polymerization, *J. Chem. Ed.*, *65*, 605, 1988.
292. Lanier, C., U.S. Patent 3,789,081, 1974
293. Motz, M., U.S. Patent 4,435,606, 1984.
294. Mel'nikov, V. M., SU Patent 1,211,249, 1986; *Chem. Abstr.*, *107*, 97342b, 1987.
295. Ziegler, K., GER(EAST) Patent 94,642, 1957; *Chem. Abstr.*, *54*, 967, 1960.
296. Ziegler, K., BELG Patent 610,302, 1962.
297. Shewbart, W. E., U.S. Patent 4,314,090, 1980.
298. Joseph, R. K., and Leon, M. A., U.S. Patent 3,168,590, 1965.
299. Shevlayakov, I. D., FR Patent 2,341,540, 1977.
300. Ziegler, K., GER(EAST) Patent 964,410, 1957.
301. Aliev, V. S., Mirzoeva, Z. S., and Soldatova, V. A., *Sb. Tr. Inst. Neftekhim. Protesesov im. Yu. G. Mamedaliov, Acad. Nauk. Az. SSSR*, *13*, 69, 1986; *Chem. Abstr.*, *106*, 17827t, 1987.
302. Fowler, A. E., U.S. Patent 4,380,684, 1983.
303. Ziegler, K., Gelbert, H. G., Holzkamp, E., Wilke, C., and Kroll, W. H., Ethylene oligomerization with metal complexes, *Liebieg, Ann.*, *629*, 172, 1960.
304. Hunt, M. W., U.S. Patent 3,217,058, 1964.
305. Aliev, V. S., Mirzoeva, Z. Sh., and Kagramanov, E. E., *Sb. Tr. Inst. Neftekhim. Protsesov im. Yu. G. Mamedalieva, Akad. Nauk. Az. SSSR*, *13*, 167, 1982; *Chem. Abstr.*, *99*, 23011m, 1983.
306. Gulf Research Co., U.S. Patent 3,702,345, 1973.
307. Ethyl Corp., U.S. Patent 3,663,647, 1972.
308. Serratore, J., U.S. Patent 3,499,057, 1970.
309. Gulf Research Co., U.S. Patent 3,502,741, 1970.
310. Ziegler, K., GB Patent 777,152, 1952.
311. Ziegler, K., Wilke, G., and Holzkamp, E., U.S. Patent 2,781,411, 1957.
312. Pol, P. Z., and Rayer, D. J., U.S. Patent 4,455,289, 1985.
313. Raffinage, C. F., NL Pat. Appl. 6,405,301, 1965; *Chem. Abstr.*, *62*, 7888e, 1965.
314. Seidov, M. N., JP Patent 7,857,289, 1978; *Chem. Abstr.*, *89*, 75663, 1978.
315. Sedov, M. N., GER Patent 2,650,585, 1977.
316. Natta, G., Ethylene oligomerization by titanium complexes, *J. Polym. Sci.*, *34*, 151, 1959.
317. Gerasini, M. P., and Brikenshtein, Kh. A., *Osnovn. Org. Sin. Neftekhim.*, 1978, p. 15; *Chem. Abstr.*, *91*, 174751, 1979.
318. Zhukov, V. I., Shestak, N. P., Shumovskii, V. G., and Livin, A. P., SU Patent 600,132, 1978; *Chem. Abstr.*, *88*, 170761, 1978.

319. Jiri, S., and Kveturi, J., CZECH Patent 189,984, 1983; *Chem. Abstr.*, *99*, 70179, 1983.
320. Belov, G. P., *Nauch. Osn. Pererab. Nefti Gaza Neftekhim.*, *Tezisu Dokl. Vses. Konf.*, 1977, p. 203; *Chem. Abstr.*, *92*, 59261, 1980.
321. Izawa, S., JP Patent 7,201,444, 1972; *Chem. Abstr.*, *76*, 85363, 1972.
322. Beach, D. L., and Kissin, V. V., Ethylene oligomerization. IV. Olefin reactivities with titanium based systems, *J. Polymer Sci.*, *Polym. Chem. Ed.*, 22, 3027, 1984.
323. Matkovskii, P. E., and Russian, L. N., Ethylene dimerization on the system Ti(n-$C_4H_9O)_4$-$Al(C_2H_5)_3$ in the media of diethylether in the presence of hydrogen, *Neftekhimia*, 22, 317, 1982 (Russian).
324. Matkovskii, P. E., SU Patent 496,258, 1975; *Chem. Abstr.*, *84*, 73618, 1976.
325. Vybihal, J., CZECH Patent 179,726, 1979; *Chem. Abstr.*, *92*, 129603, 1980.
326. Vybihal, J., Preparation of 1-butene by ethylene dimerization, *Khem. Prum.*, *31*, 640, 1981 (Czech).
327. Vybihal, J., CZECH Patent 179,182, 1979; *Chem. Abstr.*, *92*, 77131, 1980.
328. Zhurkov, V. I., SU Patent 681,032, 1979; *Chem. Abstr.*, *92*, 75808, 1980.
329. Belov, G. P., SU Patent 485,101, 1975; *Chem. Abstr.*, *84*, 44934, 1976.
330. Zhukov, V. I., Optimal conditions of ethylene dimerization on catalyst $Ti(OR)_4$-AlR_3, *Neftekhimia*, 22, 598, 1982 (Russian).
331. Knec, T. E. C., CAN Patent 653,326, 1962.
332. Azuma, K., and Yomo, K., JP Patent 6,904,961, 1969; *Chem. Abstr.*, *71*, 101270, 1969.
333. Azuma, K., and Yomo, K., JP Patent 6,904,962, 1969; *Chem. Abstr.*, *71*, 101271, 1969.
334. Azuma, K., JP Patent 6,904,963, 1969; *Chem. Abstr.*, *71*, 101272, 1969.
335. Pez, G. P., Ethylene dimerization catalysed by organotitanium compounds, *J. Chem. Soc.*, *Chem. Commun.*, 1977, 56C.
336. Montecatini, S. A., ITAL Patent 586,452, 1957; *Chem. Abstr.*, *56*, 7608, 1962.
337. Izawa, M., GER Patent 1,803,434, 1969.
338. Ziegler, K., and Heinz, M., U.S. Patent 2,943,125, 1960.
339. Mitsubishi Comp., JP Patent 82,169,430, 1982; *Chem. Abstr.*, *98*, 88801, 1983.
340. Mitsubishi Comp., JP Patent 77,113,903, 1977; *Chem. Abstr.*, *88*, 120589, 1978.
341. Belov, G. P., SU Patent 485,102, 1975; *Chem. Abstr.*, *84*, 44933, 1976.
342. Vybihal, J., CZECH Patent 179,248, 1979, *Chem. Abstr.*, *92*, 110598, 1980.
343. Zhukov, V. I., and Ivanenko, P. F., SU Patent 495,451, 1975; *Chem. Abstr.*, *83*, 79936, 1975.
344. Yamada, S., and Tago, K., GER Patent 2,026,246, 1971.
345. Sukov, V. I., GER Patent 2,707,830, 1977.
346. Belov, G. P., U.S. Patent 3,911,042, 1975.
347. Belov, G. P., GER Patent 2,462,771, 1980.
348. Belov, G. P., U.S. Patent 3,969,429, 1975.
349. Cambell, I. A., GB Patent 1,447,812, 1976.
350. Belov, G. P., FR Patent 2,274,583, 1976.
351. Marconi, W., and Cesca, F., GER Patent 2,022,658, 1970.
352. ICI Ltd., GB Patent 1,447,811, 1976.
353. Otsu, K., and Iawama, Y., JP Patent 7,238,402, 1972; *Chem. Abstr.*, *78*, 15478, 1973.
354. ICI Ltd., BELG Patent 634,232, 1963.
355. Izawa, S., Yamada, S., and Kunimoto, N., JP Patent 7,143,366, 1971; *Chem. Abstr.*, *76*, 45693, 1972.
356. Nevzorov, V. E., SU Patent 739,044, 1976; *Chem. Abstr.*, *93*, 185724, 1980.
357. Xu, Z., *Shion Huagong*, *14*, 711, 1985; *Chem. Abstr.*, *104*, 151122, 1986.
358. Langer, A. W., U.S. Patent 4,486,615, 1984.

359. Langer, A. W., *Prepr. ACS Div. Petrol. Chem.*, *17*, B119, 1972.
360. Langer, A. W., U.S. Patent 3,662,021, 1972.
361. Langer, A. W., U.S. Patent 3,647,912, 1972.
362. Langer, A. W., U.S. Patent 3,655,812, 1972.
363. Cull, N. L., and Bearden, R., U.S. Patent 3,637,897, 1972.
364. Buben, D., Bearden, R., and Wristers, H. J., U.S. Patent 3,862,252, 1975.
365. Faris, D. V., GER Patent 1,948,992, 1971.
366. Matkovskii, P. E., SU Patent 536,155, 1976; *Chem. Abstr.*, *86*, 139378, 1977.
367. Cull, N. L., and Bearden, R., GER Patent 1,960,793, 1970.
368. Arakawa, T., Salki, K., Sato, K., and Kitasawa, Y., U.S. Patent 3,725,497, 1973.
369. Arakawa, T., Salki, K., Sato, Y., and Kitasawa, Y., U.S. Patent 3,652,705, 1972.
370. Iwakumi, T. A., and Sato, Y., U.S. Patent 3,660,519, 1972.
371. Cull, N. L., Bearden, R., and Mertzweiller, J. K., GER Patent 1,960,778, 1970.
372. Arakawa, T., Sato, Y., and Kitasawa, Y., GER Patent 1,949,878, 1970.
373. Butter, S. H., U.S. Patent 3,981,941, 1976.
374. Matkovskii, P. E., Pomagailo, A. D., Russiyan, L. N., Lisitskaya, A. P., Dyachkovskyi, P. S., and Brikenshtein, K. A., SU Patent 491,404, 1975; *Chem. Abstr.*, *84*, 89577y, 1976.
375. Henrici-Olive, G., and Olive, S., GER Patent 1,924,427, 1970.
376. Mamedaliev, G. A., Aliev, V. S., Aliev, C. M., Azizov, A. G., Aliev, A. B., and Askierova, E. O., SU Patent 1,234,392, 1986; *Chem. Abstr.*, *105*, 173231r, 1986.
377. Fuhrman, H., Semikolenov, V. A., and Yarmakov, Y., Unhomogeneous organotitanium compound as catalyst of ethylene oligomerization, *Kinet. Catal. Lett.*, *11*, 301, 1979 (Russian).
378. Hata, G., JP Patent 6,902,203, 1963; *Chem. Abstr.*, *70*, 114570, 1969.
379. Butte, W. A., U.S. Patent 3,542,896, 1970.
380. Butte, W. A., U.S. Patent 3,542,899, 1970.
381. Rousser, R. E., U.S. Patent 3,364,278, 1968.
382. Nicolescu, I. V., Angelescu, A., Angelescu, E., and Nenciulescu, E., Ethylene oligomerization by heterogeneous coordinated catalysis. II., *Rev. Chim.*, *(Bucharest)*, *30*, 411, 1979 (Romanian).
383. Bwrnham, D. R., U.S. Patent 4,032,590, 1977.
384. Tekeshi, H., and Akihisa, M., JP Patent 7,103,162, 1971; *Chem. Abstr.*, *74*, 124768, 1971.
385. Ozaki, Y., JP Patent 8,181,524, 1981; *Chem. Abstr.*, *95*, 168512, 1981.
386. Altridge, W. G., GB Patent 1,058,680, 1963.
387. Petrushanskaya, N. V., and Feldblyum, V. Sh., SU Patent 290,764 1971; *Chem. Abstr.*, *74*, 140893, 1971.
388. Wilke, G., BELG Patent 651,596, 1965.
389. Van Hoos, D. G., NL Pat. Appl. 6,612,735, 1967; *Chem. Abstr.*, *67*, 53608, 1967.
390. Feldblyum, V. Sh., SU Patent 686,753, 1979; *Chem. Abstr.*, *92*, 41304, 1980.
391. Bogdanovici, B., and Wilke, G., *Proc. of VIIth World Petroleum Congress*, Mexico, 1968, Vol. 5, p. 351.
392. Bogdanovici, B., and Wilke, G., Olefin oligomerization in the presence of organonickel compounds, *Breunstr. Chem.*, *49*, 323, 1968.
393. Ahymes, G., Grabler, A., Simon, A., Kada, I., and Andor, I., *Magy. Kem. Lapja*, *20*, 570, 1965; *Chem. Abstr.*, *64*, 6471, 1966.
394. Borisova, N. A., Karpacheva, G. P., Mushina, B. A., and Krentsel, B. A., ESR study of catalytic systems of olefins dimerization on nickel-π-allyl complexes, *Izv. Akad. Nauk. SSSR, Ser. Khim.*, 1978, 2131.
395. Furman, D. B., Kharitonova, T. M., Thukovski, S. S., Nechaeva, T. I., Fedosov, B. M., Tabor, A. M., Lipovich, V. G., Bragin, O. V., Kalechits, I. V., and Vasserberg, V.

E., Regularities in kinetic conversions of ethylene into butenes on the catalytic system of π-allylnickelhalide-Al_2O_3-$Et_2Al_2Cl_3$, *Izv. Nauk. SSSR, Ser. Khim.*, 1981, 1962.
396. McClure, J. D., and Barnett, K. W., U.S. Patent 3,424,816, 1969.
397. Barnett, K. W., and Raley, J. H., U.S. Patent 3,532,765, 1970.
398. Barnett, K. W., and Glockner, P. W., GER Patent 2,000,102, 1969.
399. Chauvin, Y., FR Patent 1,497,673, 1965.
400. Bergen, N., Beindheim, U., Onsager, O. T., and Wang, H., FR Patent 1,519,181, 1968.
401. Angelescu, A., and Angelescu, E., Ethylene oligomerization by heterogeneous coordinated catalysts. IV., *Rev. Chim.*, *(Bucharest)*, *32*, 653, 1981 (Romanian).
402. Nicolescu, I. V., Angelescu, E., and Angelscu, A., Ethylene oligomerization by heterogeneous coordinated catalysis. III., *Rev. Chim.*, *(Bucharest)*, *33*, 137, 1982 (Romanian).
403. Ewers, J., Olefin oligomerization by supported nickel compounds, *Erdol, Kohle, Erdgas, Petrochemie*, 1968, 763.
404. Izawa, S., JP Patent 7,222,205, 1972; *Chem. Abstr.*, *77*, 87809, 1972.
405. Krepper, H., and Schloamer, K., GER Patent 1,178,419, 1961.
406. Van Looy, H. G., NL Pat. Appl. 7,016,039, 1971.
407. Angelescu, E., Angelescu, A., Nenciulescu, S., and Nicolaescu, I. V., Ethylene oligomerization by heterogeneous catalysis. II., *Rev. Chim.* *(Bucharest)*, 32, 559, 1981.
408. Bacharev, I. N., BELG Patent 852, 473, 1977.
409. Ozaki, K., JP Patent 7,509,592, 1975; *Chem. Abstr.*, *84*, 106354, 1976.
410. Bernham, D. R., GER Patent 2,538,142, 1976.
411. Antonov, A., BELG Patent 852, 472, 1977.
412. Kabanov, V. A., BELG Patent 852,471, 1977.
413. Cannell, L., Magoon, E. F., and Raley, J. H., U.S. Patent 3,424,815, 1967.
414. Toltikov, G. A., SU Patent 539,860, 1976; *Chem. Abstr.*, *87*, 5358, 1977.
415. Van der Boer, L. H., NL Pat. Appl. 6,809, 1967.
416. Bargen, R., Blum, S., Glöckner, P. T., and Kein, W., FR Patent 316,449, 1972; *Chem. Abstr.*, *71*, 49204, 1973.
417. Ogawa, T., and Sakeki, K., JP Patent 7,222,208, 1972; *Chem. Abstr.*, *77*, 100694, 1974.
418. Kabanov, V. A., Deposited Doc., VINITI, 2292, 1980, *Chem. Abstr.*, *95*, 80000, 1981.
419. Maruya, K., Mishio, K., Kawata, N., and Ozaki, A., Oligomerization of alkenes by organometallic compounds, *Nippon Kagaku Kaishi*, 1385, 1973; *Chem. Abstr.*, *74*, 131292, 1973.
420. Ozaki, S., Mizorogi, T., and Marutani, K., JP Patent 7,405,516, 1974; *Chem. Abstr.*, *82*, 111550, 1975.
421. Chauvin, Y., FR Patent 1,547,921, 1966; *Chem. Abstr.*, *71*, 50646, 1969.
422. Chauvin, Y., BELG Patent 707,477, 1966.
423. Draguez, T. H., GER Patent 2,410,851, 1974; *Chem. Abstr.*, *82*, 3739, 1975.
424. Draguez, T. H., FR Patent 2,220,493, 1974; *Chem. Abstr.*, *82*, 3739, 1975.
425. Keim, W., and Kowaldt, F. H., Oligomerization of alkenes in the presence of organonickel compounds, *Erdol, Kohle, Erdgas, Petrochem.*, *78*, 453, 1978.
426. Feldblyum, V. Sh., SU Patent 382,598, 1973; *Chem. Abstr.*, *79*, 65770, 1973.
427. Smith, C. E., Czenkusch, E. L., and Bailey, G. C., U.S. Pat. Appl. B428103, 1976; *Chem. Abstr.*, *84*, 150171, 1976.
428. Zuech, E. A., U.S. Patent 3,485,381, 1969.
429. Chauvin, Y., FR Patent 1,576,134, 1969.
430. Kawata, N., Mizireki, T., and Ozaki, A., Supported organonickel compounds as catalysts of alkenes oligomerization, *J. Mol. Catal.*, *1*, 275, 1975/1976.

431. Bergen, N., Blindheim, U., Onsager, O. T., and Wang, H., S. African Pat. Appl. 6,705,408, 1968; *Chem. Abstr.*, *70*, 87132, 1969.
432. Jenkins, D. K., and Dixon, C. G. P., U.S. Patent 3,641,176, 1972.
433. Dunn, H. E., U.S. Patent 3,644,218, 1972.
434. Furman, D. B., C_4 Hydrocarbons obtained by dimerization of ethylene in the presence of organonickel compounds, *Izv. Akad. Nauk. SSSR, Ser. Khim.*, 1983, 573 (Russian).
435. Gene, N., and Harold, D. L., U.S. Patent 2,969,408, 1955.
436. Yoo, J. S., U.S. Patent 3,992,470, 1976.
437. Furman, D. B., SU Patent 981,307, 1981; *Chem. Abstr.*, *98*, 143993, 1983.
438. Chauvin, Y., FR Patent 1,537,550, 1971.
439. Dunn, H. E., U.S. Patent 3,558,738, 1971.
440. Zulfugarov, Z. G., SU Patent 950,706, 1982; *Chem. Abstr.*, *98*, 34212, 1983.
441. Onsager, O. T., Wang, H., and Blindheim, U., Synthesis of C_6 olefins by ethylene oligomerization in the presence of $Ni(\pi\text{-}L)_2X_2\text{-}Al_2R_{6-x}Cl_x\text{-}PR_3$ system, *Helv. Chimica Acta*, *52*, 107, 1969.
442. Onsager, O. T., Wang, H., and Blindheim, U., Characterization of the mixture obtained by olefins oligomerization, *Helv. Chimica Acta*, *52*, 196, 1969.
443. Chauvin, Y., FR Patent 1,535,550, 1966.
444. Van Looy, H. V., NL Pat. Appl. 6,601,770, 1966; *Chem. Abstr.*, *67*, 2738, 1967.
445. Bergem, N., Blindheim, U., Onsager, O. T., and Wang, H., S. African Pat. Appl. 6,404,671, 1967; *Chem. Abstr.*, *70*, 68530, 1969.
446. Komoto, R. C., U.S. Patent 4,069,273, 1978.
447. Van Zwet, H., GER Patent 2,062,293, 1971.
448. Van Zwet, H., Bauer, R. S., and Keim, W., U.S. Patent 3,644,564, 1972.
449. Farley, F. F., U.S. Patent 3,647,906, 1970.
450. Keim, W., Organonickel compounds as catalysts of alkene oligomerization, *Ann. N. Y. Acad. Sci.*, 191, 1981.
451. Van Zwet, H., Bauer, R. S., and Keim, W., NL Pat. Appl. 7,016,037, 1971; *Chem. Abstr.*, *75*, 110727, 1971.
452. Van Zwet, H., NL Pat. Appl. 6,706,533, 1966; *Chem. Abstr.*, *70*, 126171, 1966.
453. Chauvin, Y., GB Patent 1,158,211, 1969.
454. Feldblyum, V. Sh., Ethylene oligomerization under the influence of catalytic system diizobutylaluminiumchloride-nickel oleate, *Zhur. Org. Khim.*, *6*, 213, 1970 (Russian).
455. Feldblyum, V. Sh., Leshcheva, A. I., and Petrushanskaya, N. V., The influence of the metal nature upon the efficiency of catalytic system in the monoolefins dimerization, *Zhur. Org. Khim.*, *6*, 2408, 1970 (Russian).
456. Feldblyum, V. Sh., GB Patent 1,164,882, 1969.
457. Pisman, I. I., Technology of 1-butene eynthesis, *Khim. Prom. (Moscow)*, *43*, 328, 1967 (Russian).
458. Chernikova, I. M., Pisman, I. I., and Dalin, M. A., Optimal experimental conditions for the 1-butene synthesis, *Azerb. Khim. Zhur.*, 1965, 35 (Russian).
459. Pisman, I. I., Chernikova, I. M., and Dalin, M. A., Mechanism of 1-butene synthesis, *Azerb. Khim. Zhur.*, 1968, 73 (Russian).
460. Chauvin, Y., FR Patent 1,385,503, 1965.
461. Feldblyum, V. Sh., SU Patent 290,706, 1972; *Chem. Abstr.*, *78*, 147319, 1973.
462. Feldblyum, V. Sh., GER Patent 2,618,738, 1974.
463. Schultz, W., Mix, H., Krunas, E., Wilcke, P. W., Reihsig, J., Fuhrman, H., and Grassert, I., GER Patent 2,252,856, 1971.
464. Yagi, Y., Kobayashi, A., and Hirata, I., JP Patent 7,334,564, 1973; *Chem. Abstr.*, *81*, 14036, 1974.

465. Desgrandchamp, G., Hemmer, H., and Haurie, M., FR Patent 2,070,554, 1971.
466. Lehmuhl, H., Naydowski, C., and Danowski, F., Ethylene oligomerization by nickel complexes, *Chem. Ber.*, *117*, 3231, 1984.
467. Ziegler, K., and Martin, H., GER Patent 1,039,055, 1958.
468. Ozaki, A., JP Patent 8,820,729, 1983; *Chem. Abstr.*, *100*, 174258, 1984.
469. Attridge, C. J., Jones, E., Pioli, A. J. P., and Wilkinson, P. J., GER Patent 2,219,048, 1973.
470. Mitchel, T. O., and Whitehurst, D. D., U.S. Patent 3,980,583, 1974.
471. Cramer, R., Ruthenium trichloride as catalyst in alkene oligomerization, *Acc. Chem. Res.*, *1*, 186, 1968.
472. Chauvin, Y., FR Patent 1,521,991, 1966.
473. Mix, H., GER Patent 1,193,934, 1957.
474. Ketley, A. D., FR Patent 1,499,833, 1967.
475. Ketley, A. D., FR Patent 1,529,992, 1966.
476. Boranova, N. V., SU Patent 650,983, 1979; *Chem. Abstr.*, *90*, 186337, 1979.
477. Ignatov, V. M., Alkenes oligomerization by $[Rh(SnCl_3)_2Cl_4]^-$ anchored on an AV 17-8 anion exchange resin, *Neftekhimia*, *22*, 749, 1982 (Russian).
478. Mitchel, T. O., GB Patent 887,362, 1952.
479. Ketler, F. I., U.S. Patent 3,361,840, 1963.
480. Harvey, S. K., U.S. Patent 3,354,236, 1967.
481. Maatschsppij, W. V., NL Patent 294,637, 1965.
482. Schmidt, F. K., *Mekanizm, Katalit. Reaktsii Materialy 3 Ves. Konf. Novosibirsk*, 1982, Ch. 2, p. 156; *Chem. Abstr.*, *98*, 88747, 1983.
483. Hattori, S., GER Patent 2,234,922, 1974.
484. Wideman, L. G., Organowolfram compounds as catalysts of ethylene oligomerization, *J. Catal.*, *43*, 371, 1976.
485. Young, D. A., Jones, L. O., Campione, T. J., Eur. Pat. Appl. 295,960, 1988; *Chem. Abstr.*, *110*, 173979u, 1989.
486. Langer, A. W., U.S. Patent 4,409,414, 1983.
487. Langer, A. W., U.S. Patent 4,434,313, 1984.
488. Langer, A. W., Doyle, G., Burkhardt, T. J., and Looney, R. W., U.S. Patent 4,409,409, 1983.
489. Kuliev, R. S., and Akhmedov, V., SU Patent 1,073,279, 1984; *Chem. Abstr.*, *100*, 212821, 1984.
490. Longi, P., Greco, P., and Rossi, U., ITAL Patent 871,096, 1973.
491. Khodakovskaya, V. A., PCT Int. Appl. 8,000,224, 1980; *Chem. Abstr.*, *93*, 72615x, 1980
492. Ozaki, A., JP Patent 58,201,729, 1983; *Chem. Abstr.*, *100*, 174278c, 1984.
493. Davis, D. V., U.S. Patent 2,549,723, 1950.
494. Kister, A., and Lutz, E. P., U.S. Patent 4,020,121, 1977.
495. Lutz, E. P., and Gantier, P. A., Eur. Pat. Appl. 177,999, 1986.
496. O'Donell, A. E., U.S. Patent 4,260,844, 1981.
497. Smidt, K. F., Moronova, L. V., and Tkach, V. S., SU Patent 654,594, 1979; *Chem. Abstr.*, *91*, 85671h, 1979.
498. Meyer, J. G., U.S. Patent 4,111,834, 1978.
499. Meyer, J. G., U.S. Patent 4,117,022, 1978.
500. Singleton, D. M., Eur. Pat. Appl. 128,597, 1985; *Chem. Abstr.*, *102*, 46405, 1985.
501. Singleton, D. M., U.S. Patent 4,503,279, 1985.
502. Murray, R. E., U.S. Patent 4,689,437, 1987.
503. Singleton, D. M., U.S. Patent 4,503,280, 1985.
504. Singleton, D. M., Eur. Pat. Appl. 128,596, 1985.
505. Seidov, N. H., Dalin, M. A., and Kyazimov, S. M., GER Patent 2,640,194, 1976.

506. Seidov, N. H., and Dalin, M., SU Patent 545,141, 1979; *Chem. Abstr.*, *91*, 158368q, 1979.
507. Gene, N., and Harold, D. L., U.S. Patent 2,965,408, 1960.
508. Brown, S. J., and Masters, A. F., *J. Organometal. Chem.*, *367*, 371, 1989.
509. Ono, N., Murakami, T., and Miki, M., JP Patent 62,142,125, 1987; *Chem. Abstr.*, *108*, 76065u, 1988.
510. Zhavoronkov, M. N., Darogochiuskii, A. Z., and Madirov, A. A., Butene-1 transformations on calcium and manganese forms of type Y zeolites, *Neftekhimia*, *23*, 495, 1983 (Russian).
511. Atlantic Richfield Co., JP Patent 7,509,592, 1975; *Chem. Abstr.*, *84*, 106354, 1976.
512. Przheval'skaya, L. K., and Vasserberg, V. E., *Kinet. Katal.*, *19*, 1283, 1978 (Russian).
513. Furman, D. B., and Vasserberg, V. E., Oligomerization of ethylene on π-allylnickelchloride prepared via complexing in the absorption layer on alumina surface, *Izv. Akad. Nauk. SSSR, Ser. Khim.*, 1979, 2278 (Russian).
514. Furman, D. D., and Vasserberg, V. E., Ethylene gas-phase oligomerization in the presence of π-C_3H_5NiCl-Al_2O_3-$Et_3Al_2Cl_3$ heterogenized system, *Izv. Akad. Nauk. SSSR, Ser. Khim.*, 1981, 673 (Russian).
515. Haensle, P., GER Patent 3,027,782, 1982.
516. Van Zwet, H., and Keim, W., U.S. Patent 3,635,937, 1972.
517. Glokner, P. W., Keim, W., and Mason, R. F., U.S. Patent 3,647,914, 1972.
518. Bauer, R. S., U.S. Patent 3,647,915, 1972.
519. Glokner, P. W., GER Patent 2,053,758, 1971.
520. Shell Oil Co., U.S. Patent 3,686,159, 1972.
521. Panckert, M., and Keim, W., *Organometallic*, *2*, 594, 1983.
522. Beach, D. L., and Harrison, J. J., U.S. Patent 4,377,528, 1983.
523. Beach, D. L., and Harrison, J. J., U.S. Patent 3,394,322, 1982.
524. Beach, D. L., and Kobylinski, T. P., U.S. Patent 4,288,648, 1981.
525. Beach, D. L., and Kobylinski, T. P., U.S. Patent 4,272,407, 1981.
526. Beach, D. L., and Kobylinski, T. P., U.S. Patent 4,272,406, 1981.
527. Beach, D. L., and Kobylinski, T. P., Ethylene oligomerization under the influence of organometallic compounds, *J. Chem. Soc., Chem. Commun.*, *19*, 933, 1980.
528. Sen, A., and Lai, T., Alkenes reactivities in the oligomerization reaction catalysed by organometallic compounds, *J. Amer. Chem. Soc.*, *103*, 4627, 1981.
529. O'Hara, M. J., and Imai, T., U.S. Patent 4,490,571, 1984.
530. Lange, J. P., Gutaze, A., Allgeier, J., and Karge, H. G., Catalytic systems of alkenes oligomerization, *Appl. Catal.*, *45*, 345, 1958.
531. Heveling, J., GER Patent 3,741,302, 1988.
532. Aushita, H. G., Kinetics of ethylene oligomerization in the presence of metallic complexes, *Kinet. Katal.*, *29*, 626, 1988 (Russian).
533. Heveling, J., Van der Beck, A., and De Pender, H., Alkenes oligomerization, *Appl. Catal.*, *42*, 325, 1988.
534. Elev, I. V., Shelimov, B. N., and Kazanskii, V. B., Industrial aspects of olefins oligomerization, *Kinet. Catal.*, *25*, 1124, 1984 (Russian).
535. Lapidus, A. L., Slenkin, L. N., Rudakova, T. N., Myshankova, M. I., Loktev, T. S., Papko, Ya., and Eidus, T., Effect of the composition of catalyst system NiC-SiO_2-Al_2O_3 on its catalytic properties in ethylene and isobutylene oligomerization reaction, *Izv. Akad. Nauk. SSSR, Ser. Khim.*, 1974, 1956 (Russian).
536. Podall, H. E., Foster, W. E., and Giraitis, A. P., Oligomerization reactions of ethylene and higher olefins, *J. Org. Chem.*, *23*, 81, 1958.
537. Meclin, A. S., Boreskov, V. V., and Giraitis, V. B., On the mechanism of ethylene oligomerization on ZSM-type zeolites, *Dokl. Acad. Nauk. SSSR*, *286*, 914, 1986 (Russian).

538. Kubelkova, L., Novakova, J., Wichterova, B., and Jiru, P., The effect of dehydrxylation of HNaY zeolites on the interaction with ethylene and propylene, *Coll. Czech. Chem. Commun.*, *45*, 2290, 1980.
539. Konovalchikov, L. D., Radchenko, E. D., Nefedov, A. F., and Rostanin, B. K., Some aspects of ethylene oligomerization on nonhomogeneous catalysts, *Nefterab. Neftekhim.*, 2, 3, 1987 (Russian).
540. Galperin, I. M., Feldblyum, V. Sh., and Petrushanskaya, N. V., *Prom. Sin. Kauch. Nauch. Tekh. Sh.*, 1971, p. 1; *Chem. Abstr.*, *78*, 15405, 1973.
541. Andrews, J., and Bonnifay, A. C. S., *ACS Sym. Series No. 75*, 328, 1977.
542. Pogozhilskii, V. V., *Dokl. Soobsch. Mendeleevskii. Sezd. Obsch. Prikl. Khim.*, 2, 303, 1975; *Chem. Abstr.*, *88*, 151947, 1978.
543. Malakhova, N. D., Tkach, V. S., and Smidt, F. K., *Osnovn.*, *Org. Sint. Neftekhim.*, 1977, 14; *Chem. Abstr.*, *89*, 214819, 1978.
544. Tkach, V. S., Smidt, F. K., and Shevchenko, I. D., *Katal.*, *Prevrasch. Uglevodorodov*, 1976, p. 168; *Chem. Abstr.*, *89*, 196882, 1978.
545. Agadzhanov, Yu. A., Markasov, P. I., and Lutsenko, A. A., *Neftekhim. Sint.*, 1976, p. 61; *Chem. Abstr.*, *87*, 84415, 1977.
546. Korneev, N. N., Popov, A. F., and Krentsel, B. A., *Complex Organometallic Catalysts*, Hasted Press, New York, 1971, p. 151.
547. William, K. J., U.S. Patent 3,009,972, 1959.
548. Beach, D. L., U.S. Patent 4,377,529, 1983.
549. Beach, D. L., Eur. Pat. Appl. 52,930, 1982.
550. Beach, D. L., and Harrison, J. J., U.S. Patent 4,301,318, 1982.
551. Shmidt, F. K., Tkach, V. S., and Sergeeva, T. N., Analysis of propylene dimerization process, *Katal. Prevrasch.*, *Uglevodorodov*, 175, 1974; *Chem. Abstr.*, *86*, 16002, 1977.
552. Ziegler, K., GB Patent 775,384, 1954.
553. Baas, C. J., *De synthesis van isopren uit propen*, Delft, 1963 (Netherlands).
554. Feldblyum, V. Sh., and Farverov, M. I., Synthesis of isoprene from propylene. Kinetics of propylene dimerization, *Neftekhimia*, *5*, 493, 1965 (Russian).
555. Kryukov, S. I., and Farverov, M. I., *Zhur. Prikl. Khim.*, *35*, 2319, 1962 (Russian).
556. *Chem. Eng. News, 41*, 1, 1963.
557. Semenov, A. A., Matkovskii, P. E., and Dyachkovskyi, V. N., *Vysokomol. Soedin.*, *A*, *21*, 2749, 1979 (Russian).
558. Matkovskyi, P. E., Cherngya, L. I., and Russiyan, L. N., Cationic oligomerization of propylene. Formation of active centers, *Izv. Akad. Nauk. SSSR*, *Ser. Khim.*, 1984, 643 (Russian).
559. Matkovskyi, P. E., Russiyan, L. N., Dyachkovskyi, F. S., Khvastyk, G. M., Dzhabiova, Z. M., and Startseva, G. P., Oligomerization of olefins in the presence of $TiCl_4$-$EtAlCl_2$ system. The proposal of a cationic mechanism, *Vysokomol. Soedin.*, *A*, *18*, 840, 1976 (Russian).
560. Rischina, L., and Chirkov, N. M., Oligomerization of propylene in the presence of $TiCl_4$-i-Bu_2AlCl system. The influence of the solvent, *Vysokomol. Soedin.*, *A*, *16*, 1459, 1974 (Russian).
561. Izawa, S., and Kunimoto, N., JP Patent 7,249,564, 1972; *Chem. Abstr.*, *78*, 147312, 1973.
562. Aladyshev, A. M., Lisitsyn, D. M., and Dyachkovskyi, F. S., Formation of propylene oligomers during the polymerization process, *Vysokomol. Soedin.*, *A*, *24*, 377, 1982 (Russian).
563. Motier, J. F., and Yoo, J. S., U.S. Patent 3,855,341, 1974.
564. Watanabe, M., Eur. Pat. Appl. 268,214, 1988; *Chem. Abstr.*, *109*, 21161, 1988.
565. Kaminsky, W., Ahlers, A., and Maller-Lindenhof, N., Propylene oligomerization in the presence of organozirconium compounds, *Angew. Chem.*, *101*, 1304, 1989.

566. Motier, J. P., and Yoo, J. S., U.S. Patent 4,026,822, 1977.
567. Schrock, R. R., U.S. Patent 4,231,947, 1980.
568. McLain, S. J., Wood, C. D., and Schrock, R. R., Tantalacyclopentane complex as catalyst in olefins oligomerization, *J. Amer. Chem. Soc.*, *101*, 4558, 1979.
569. Brown, M., U.S. Patent 3,897,512, 1975.
570. Brenner, G. S., GER Patent 2,306,439,1973.
571. Maly, N. A., GER Patent, 2,306,433, 1973.
572. Maly, N. A., GER Patent 2,306,434, 1973.
573. Feldblyum, V. Sh., Petrushanskaya, N. V., Leshcheva, A. I., and Baranova, T. I., On the nature of the mechanism of propylene dimerization by homogeneous and heterogeneous catalysts based on transition metals, *Zhur. Org. Khim.*, *10*, 2265, 1974 (Russian).
574. Tkach, V. S., Schmidt, F. K., Sergeeva, T. N., and Malakhova, N. D., The mechanism of propylene dimerization, *Neftekhimia*, *13*, 673, 1973 (Russian).
575. Drew, E. H., GB Patent 1,167,289, 1969.
576. Jones, J. R., Two components catalyst of the type CoX_2-AlR_3 used in the dimerization of propylene, *J. Chem. Soc., C*, 1971, 1117.
577. Pu Sun, L., Yamamoto, A., and Ikeda, T., Dimerization of propylene in the presence of $HCo(N_2)(PPh_3)_3$ system, *J. Amer. Chem. Soc.*, *90*, 7170, 1968.
578. Speier, G., Dimerization of propylene in the presence of $HCo(N_2)PPh_3$, *J. Organometal Chem.*, *97*, 109, 1979.
579. Speier, G., and Marko, L., Alkene dimerization in the presence of organocobalt complexes, *Proc. Int. Conf. Coord. Chem.*, *13th*, II, 27, 1970.
580. Tkach, V. S., *Katal. Prevrasch. Uglevodorodov*, 1976, 141; *Chem. Abstr.*, *88*, 6255, 1978.
581. Shmidt, F. K., *Katal. Prevrasch. Uglevodorodov.*, 1976, 1959; *Chem. Abstr.*, *86*, 188730, 1977.
582. Areshidze, Kh. I., *Soobsch. Akad. Nauk. Gruz. SSSR*, *101*, 589, 1981; *Chem. Abstr.*, *95*, 61348, 1981.
583. Andrews, J. W., *Hydrocarbon Process.*, *61*, 110, 1982.
584. Chauvin, Y., Technology of propylene oligomerization in the presence of nickel complexes, *Nuova Chim.*, *50*, 79, 1974.
585. Feldblyum, V. Sh., *Proc. Int. Congr. Catal.*, 4th, *1*, 222, 1971.
586. Bogdanovici, B., and Wilke, G., *Proc. of VIIth World Petroleum Congr.*, Mexico, Vol. 4, p. 351.
587. Shmidt, F. K., Tkach, V. S., Kalabian, A. V., and Saraev, V., *Rol Koord. v Katal.*, 1976, p. 159; *Chem. Abstr.*, *86*, 188730, 1977.
588. Shmidt, F. K., Zkach, V. S., and Kalabina, I. V., Influence of organophosphorous ligands upon the catalytic characteristics of $Al(C_4H_9)_2Cl$-$NiCl_2$ systems in the process of propylene dimerization, *Neftekhimiya*, *12*, 76, 1972 (Russian).
589. Sakakibara, Y., Tagano, T., Sakai, M., and Uchino, N., Catalytic species in the propylene dimerization in the presence of Ni-compounds, *Bull. Inst. Chem. Res. Kyoto Univ.*, *50*, 375, 1972.
590. Beach, D. L., and Harrison, J. J., Eur. Pat. Appl., 52,930, 1982; *Chem. Abstr.*, *97*, 163257f, 1982.
591. Abasova, S. G., Mushina, E. I., Muraveva, L. S., Kreutsel, B. A., and Feldblyum, V. S., Propylene dimerization with bis-π-allyl(crotyl)-nickel halogenides, *Neftekhimiya*, *13*, 46, 1973 (Russian).
592. Zhukovskii, S. S., *Soderzash. Nickel complecsu materialy Simposium*, Tashkent, Novosibirsk, 1980, Ch. 1, p. 185; *Chem. Abstr.*, *94*, 46425, 1981.
593. Dawons, F., and Morel, D., Dimerization of propylene in the presence of π-allyl-nickel halides, *J. Mol. Catal.*, *3*, 403, 1978.

594. Skupinski, W., and Malinowski, S., Polymer anchored π-allylic nickel complex as catalyst of propylene dimerization, *J. Mol. Catal.*, *4*, 95, 1978.
595. Jones, J. R., and Symes, J., Propylene dimerization in the presence of $Ni(acac)_2$-$Et_2Al(OEt)$, *J. Chem. Soc.*, *C*, 1971, 1124.
596. Beach, D. L., and Harrison, J. J., Eur. Pat. Appl. 52,931, 1982; *Chem. Abstr.*, *97*, 163256 e, 1982.
597. Chedya, R. V., Pluzhnov, S. K., Smetanyuk, V. I., Kabunov, V. A., and Areshidze, Kh. I., *Soobshch. Akad. Nauk. Gruz. SSSR*, *91*, 619, 1978; *Chem. Abstr.*, *90*, 120919, 1979.
598. Tkach, V. S., Shmidt, F. K., and Sergeeva, T. N., Thermodynamic analysis of propylene dimerization, *Kinet. Katal.*, *17*, 208, 1976 (Russian).
599. Beach, D. L., and Harrison, J. J., U.S. Patent 4,310,716, 1982.
600. Shmidt, F. K., and Tkach, V. S., *Tr. Irkutsk. Politekh. Inst.*, *69*, 188, 1971; *Chem. Abstr.*, *78*, 29135, 1975.
601. Aliev, V. S., Soldatova, V. A., Ismailov, T. A., and Abdullaeva, R. M., *Sb. Tr. Inst. Neftekhim. Protsesov im. Yu. G. Mamedalieva, Akad. Nauk. Az. SSSR*, *7*, 37, 1976; *Chem. Abstr.*, *86*, 89095, 1975.
602. Beach, D. L., and Harrison, J. J., U.S. Patent 4,301,318, 1982.
603. Tkach, V. S., Mironova, L. V., Malakhova, N. O., Chuikova, M. A., and Shmidt, F. K., *Osnovn. Sint. Neftekhim.*, *6*, 6, 1976; *Chem. Abstr.*, *90*, 61799, 1959.
604. Aliev, V. S., Khanmetov, A. A., Mamedov, R. Kh., Kerimov, R. K., and Akhmetov, V. M., Dimerization of propylene under the influence of nickel containing complexes and phosphorylated polystyrene, *Zhur. Org. Khim.*, *18*, 265, 1982; *Chem. Abstr.*, *96*, 198992, 1982.
605. Aliev, V. S., Mustafaeva, I. F., Nazirov, F. A., and Gradzhiev, R. K., *Azerb. Khim. Zhur.*, 1979, 10; *Chem. Abstr.*, *94*, 46686, 1981.
606. Aliev, V. S., Mustafaeva, I. F., Soldatova, A. V., Ismailov, T. A., Abdullaeva, R. M., Nasirov, F. A., and Gradzhiev, R. K., *Azerb. Khim. Zhur.*, 1977, 7; *Chem. Abstr.*, *89*, 107891, 1978.
607. Mironova, L. V., Tkach, V. S., Shmidt, F. C., Kalabina, A. V., and Tselyutina, M. I., *Katal. Prevrasch. Uglevodorodov*, 1976, 180; *Chem. Abstr.*, *88*, 6261, 1978.
608. Shmidt, F. K., Mironova, L. V., Kalabina, G. A., Proidakov, A. G., and Kalabina, A. V., Propylene dimerization in the presence of catalytic system based on nickel complexes with organophosphinic ligands, *Neftekhimia*, *16*, 547, 1976.
609. Shmidt, F. K., Mironova, L. V., Saraev, V. V., Grazhykh, V. A., Dimitrieva, T. V., and Ratovoskii. G. V., Active species in the propylene dimerization in the presence of nickel complexes, *Kinet. Katal.*, *20*, 622, 1979.
610. Shmidt, F. K., Mironova, L. V., Thach, V. S., and Kalabina, A. V., Propene dimerization in the presence of a $Ni(PPh_3)_4$ complex with Lewis and Bronsted acids, *Kinet. Katal.*, *16*, 270, 1975.
611. Mironova, L. V., Shmidt, F. K., Tkach, V. S., and Omistriev, V. I., Propylene dimerization under action of nickel complexes activated with Lewis type and Bronsted type acids, *Neftekhimia*, *18*, 205, 1978.
612. Petrushanskaya, N. V., Kurapova, A. I., Rodionova, N. M., and Feldblyum, V. Sh., Propylene dimerization in the presence of $Ni(i\text{-}PPr_3)(C_3H_6)_2$, *Kinet. Katal.*, *17*, 1345, 1976.
613. De Haan, R., and Dekker, J., Dimerization of propylene in the presence of organonickel compounds, *J. Catal.*, *35*, 202, 1974.
614. Shmidt, F. K., Mironova, L. V., Proidakov, A. G., Kalabin, G., Batovskii, G. V., and Dmitrieva, T. V., Oligomerization of propylene under the action of nickel complexes, *Kinet. Katal.*, *19*, 150, 1978.

615. Shmidt, F. K., Mironova, L. V., Proidakov, A. G., and Kalabina, A. V., Spectral data and molecular weight of reaction mixture resulted in propylene oligomerization, *Koord. Khim.*, *4*, 1608, 1978.
616. Nikonova, T. N., Tkach, S., Shmidt, F. K., and Moiseev, I. I., Propylene dimerization on nickel catalysts in solvents of different nature, *Neftekhimia*, *20*, 534, 1980.
617. Eberhardt, G. G., and Griffin, W. H., Dimerization of propylene in the presence of R_4P (i-Pr_3P)$NiCl_3$, *J. Catal.*, *16*, 245, 1970.
618. Akhmedov, V. M., Mardanov, V. G., Sultanova, F. R., Kurbanova, F. F., Mamedov, R. Kh., and Agave, T. A., Deposited Doc. VINITI, 4406, 1980; *Chem. Abstr.*, *96*, 104836, 1982.
619. Uchino, M., Asagi, K., Yamamoto, A., and Ikeda, S., The influence of solvents upon the catalytic activity of organonickel compounds in the propylene dimerization, *J. Organometal. Chem.*, *84*, 93, 1975.
620. Henrici-Olive, G., and Olive, S., Catalyst selectivity in the dimerization reaction of propylene. I. Square-planar nickel (II) complex with a chelating Schiff base, *Transition Met. Chem.*, *1*, 109, 1976.
621. DeHaan, R., and Dekker, J., The synthesis of catalysts used in the dimerization reaction of propylene, *J. Catal.*, *44*, 15, 1976.
622. Frame, R. R., U.S. Patent 4,740,652, 1988.
623. Frame, R. R., U.S. Patent 4,613,580, 1986.
624. Frame, R. R., U.S. Patent 4,734,479, 1988
625. Henrici-Olive, G., and Olive, S., Oligomerization of propylene. The influence of the catalyst structure upon the rate and selectivity of the reaction, *Angew. Chem.*, *87*, 110, 1975.
626. Yoo, J. S., and Konkos, R., U.S. Patent 3,641,188, 1972.
627. Yoo, J. S., and Konkos, R., U.S. Patent 3,803,053, 1974.
628. Yoo, J. S., U.S. Patent 3,655,811, 1972.
629. Yoo, J. S., U.S. Patent 3,808,150, 1974.
630. Andreas, F., and Grabe., K., *Propylene Chemie*, WNT, Warszawa, 1974, p. 310 (Pol).
631. Repas, M., Supported Bronsted acids as catalysts of propylene oligomerization, *Chem. Prum.*, *15*, 543, 1965.
632. Chkneidze, O. Y., *Tr. Vses. Nauchno-Isled. Inst. Pererab. Nefti*, *9*, 228, 1963; *Chem. Abstr.*, *60*, 1568, 1963.
633. Piechsczek, F., PL Patent 57,962, 1969; *Chem. Abstr.*, *72*, 48064, 1970.
634. Beach, D. L., and Harrison, J. J., U.S. Patent 4,293,502, 1982.
635. Muchinskyi, D. Y., Propylene oligomerization by inorganic catalysts, *Izvest. Akad. Nauk. Turkmen SSSR*, *6*, 58, 1960; *Chem. Abstr.*, *55*, 15902, 1961.
636. Fabisz, E., POL Patent 104,719, 1982.
637. Morrell, J. C., U.S. Patent 2,586,852, 1952.
638. Shell Div. Co., U.S. Patent 2,619,512, 1952.
639. Shell Div. Co., GB Patent 969,403, 1964.
640. Calif. Res. Corp., U.S. Patent 2,852,579, 1958.
641. Mayer, G., GER Patent 885,701, 1948.
642. Allister, J. H., GB Patent 710,537, 1954.
643. Bielawski, V. S., and Mavity, J. M., U.S. Patent 2,618,614, 1952.
644. Corner, E. S., and Lunch, C. S., U.S. Patent 2,778,804, 1954.
645. Mavity, J. M., and Bielawski, M. S., U.S. Patent 2,692,241, 1954.
646. Mavity, J. M., U.S. Patent 2,537,282, 1946.
647. Kemp, J. D., U.S. Patent 2,826,622, 1958.
648. Tpcheyev, A. V., and Pushkin, J., Propylene oligomerization by cationic mechanism, *Dokl. Akad. Nauk. SSSR*, *58*, 1057, 1947 (Russian).
649. Aragonish, P. J., GB Patent 727,944, 1955.

650. Muessig, C. V., and Lippincott, S. B., U.S. Patent 2,816,944, 1957.
651. Andiso, G., and Priola, A., Structure of imers and trimers obtained in the cationic oligomerization of propylene, *Makromol. Chem.*, *189*, 111, 1988.
652. Lachance, P., and Eastham, A. M., The oligomerization of propylene in the presence of BF_3-MeOH catalyst, *J. Polym. Sci.*, *Polymer Symposia*, *72*, 203, 1976.
653. Larkin, J. M., and Watta, L. W., U.S. Patent 4,417, 082, 1983.
654. Terres, E., Ebert, R., Loebmann, K. H., Rauth, G., Seekirchern R., and Wulf, C., Propylene oligomerization in the presence of Bronsted acids and $AlCl_3$, *Erdöl Kohle*, *12*, 542, 1959.
655. Petrow, A., Propylene oligomerization with H_3PO_4-$ZnCl_2$ system, *Fett-Seifen Anstrichmittel*, *57*, 798, 1975.
656. Anszus, L. J., and Petrow, A. D., Oligomerization of propylene by cationic catalysts: Bronsted acids-$ZnCl_2$ system, *Dokl. Akad. Nauk. SSSR*, *70*, 425, 1950 (Russian).
657. Kubelkova, L., Novakova, J., Dolysek, Z., and Ziras, P., The effect of decationation of hydroxylated HNaY zeolites on their interaction with ethylene and propylene. Hydrogen complexes oligomerization and isotope exchange, *Coll. Czech. Chem. Commun.*, *45*, 3101, 1980.
658. Galya, L. G., Occelli, M. L., and Han, D. C., Oligomerization of propylene catalysed by zeolites, *J. Mol. Catal.*, *32*, 391, 1985.
659. Dimitrov, K., and Dimitrov, R., *Cod. Soffii Univ. "Klement Okhridze," Khim. Fac.*, *73*, 125, 1979; *Chem. Abstr.*, *101*, 24019, 1984.
660. Matreev, G. A., *Monomery Polyprod. Neftekhim. Sint.*, 1983, 36; *Chem. Abstr.*, *102*, 46293, 1985.
661. Garwood, E. W., Intrazeolite Chemistry, *ACS Symp. Series 238*, 383, 1983.
662. Quann, R. J., Green, L. A., Tabak, S. A., and Krambeck, F., Propylene oligomerization in the presence of nonclassical catalysts, *Ind. Eng. Chem. Rev.*, *27*, 565, 1988.
663. Wilaner, K. G., Mechanism and kinetics of propylene oligomerization by zeolites, *Appl. Catal.*, *31*, 339, 1987.
664. Haw, J. F., Richardson, B. R., Oshero, I. S., Lazo, N. D., and Speed, J. A., The role of surface alkoxy derivatives in the oligomerization of propylene in the presence of zeolites, *J. Amer. Chem. Soc.*, *111*, 2052, 1989.
665. Smith, P., and Goddard, R. E., GB Patent 876,680, 1961.
666. Goddard, R. E., and Smith, P., GB Patent 853,187, 1960.
667. Chauvin, Y., GER Patent 2,911,691, 1979.
668. Ziegler, K., U.S. Patent 2,695,327, 1954.
669. Ziegler, K., GER Patent 878,560, 1953.
670. Akio, O., and Takamishi, K., JP Patent 7,397,803, 1973; *Chem. Abstr.*, *80*, 285, 1974.
671. Akio, O., and Takamishi, K., JP Patent 7,397,802, 1973; *Chem. Abstr.*, *80*, 70284, 1974.
672. Mercier, S., and Michiele, M., FR Patent 1,419,670, 1965.
673. Kryukov, S. I., and Farborov, M. I., *Izv. Vyssh. Uchebn. Zaved. Khim. i Khim. Technol.*, *7*, 821, 1964; *Chem. Abstr.*, *62*, 8985, 1965.
674. Chauvin, Y., FR Patent 1,533,588, 1966.
675. Chauvin, Y., FR Patent 1,420,952, 1965.
676. Van der Graas, I., NL Pat. Appl. 6,705,681, 1967.
677. Crassens, I., FR Patent 1,535,201, 1966.
678. Abrahams, J., GB Patent 1,101,657, 1965.
679. Van der Graas, I., NL Pat. Appl. 6,618,467, 1966.
680. Hamberg, J. K., and Jones, J. R., NL Pat. Appl. 6,609,512, 1967; *Chem. Abstr.*, *67*, 53609, 1967.

681. Hamberg, J. K., and Jones, J. R., NL Pat. Appl. 6,609,513, 1967; *Chem. Abstr.*, *67*, 53611, 1967.
682. Petit, F., and Arzoyan, C., Transition metal complexes as catalysts in propylene oligomerization, *J. Organometal Chem.*, *202*, 319, 1980.
683. Eberhardt, G. G., U.S. Patent 3,482,001, 1969.
684. Dunn, H. E., U.S. Patent 3,686,343, 1972.
685. Dunn, H. E., U.S. Patent 3,734,975, 1973.
686. Petit, F., Musatti, H., Peiffer, G., and Biono, G., Nickel complexes. Structure and catalytic activity in olefin oligomerization reaction, *J. Organometal Chem.*, *244*, 273, 1983.
687. Aliev, M. I., and Fischer, S. I., *Sb. Tr. Mamedalieva Neftekhim. Protesovim.*, *Yu. G. Mamedalieva. Akad. Nauk Az. SSSR*, *9*, 130, 1977; *Chem. Abstr.*, *89*, 42275, 1978.
688. Izawa, M., Yamada, S., and Kunimoto, N., JP Patent 7,524,281, 1975; *Chem. Abstr.*, *84*, 30384, 1976.
689. Abasova, S. G., SU Patent 289,102, 1970; *Chem. Abstr.*, *74*, 142621, 1971.
690. Van Brook, S. H., NL Pat. Appl. 6,409,179, 1965; *Chem. Abstr.*, *63*, 5770, 1965.
691. Zhou, K., Gao, Z., and Keim, W., *Proc. Int. Congr. Catal.*, Berlin, 1984, Vol. 5, p. 429.
692. Abrahams, J., GB Patent 1,101,498, 1965.
693. Hambling J. K., and Jones, J. R., GB Patent 1,123,474, 1968.
694. Wolfgang, S., GER Patent 1,541,413, 1968.
695. Takagi, E., Matsui, M., Otsu, K., and Iwana, Y., JP Patent 7,138,766, 1971; *Chem. Abstr.*, *76*, 24668, 1972.
696. Chauvin, Y., Lefevre, G., and Uchino, M., FR Patent 1,540,270, 1966.
697. Higo, K., and Kare, S., GER Patent 1,178,419, 1964.
698. Hamber, V. I., NL Pat. Appl. 6,612,339, 1967; *Chem. Abstr.*, *67*, 63689, 1967.
699. Hamber, V. I., NL Pat. Appl. 6,812,618, 1967.
700. Hambling, J. J., GB Patent 1,106,734, 1966.
701. Chauvin, Y., FR Patent 1,560,865, 1967.
702. Chauvin, Y., GB Patent 1,243,809, 1971.
703. Phung, N. H., FR Patent 1,549,177, 1968.
704. Hambling, J. K., and Jones, J. R., NL Pat Appl. 6,608,574, 1967; *Chem. Abstr.*, *67*, 2739, 1967.
705. Castille, V. P., and Parker, P. T., GER Patent 2,021,523, 1971.
706. Castille, V. P., and Parker, P. T., GER Patent 2,021,524, 1971.
707. Shmidt, P. K., Tkach, V. S., and Kalabian, A. V., SU Patent 379,554, 1973.
708. Chauvin, Y., FR Patent 1,095,635, 1971.
709. Neal, A. H., and Parker, P. T., U.S. Patent 3,686,352, 1972.
710. Tskagi, E., Matsui, M., Otsu, K., and Iwama, F., JP Patent 7,223,286, 1972; *Chem. Abstr.*, *77*, 87819, 1972.
711. Jones, J. R., GB Patent 1,143,863, 1971.
712. Jones, J. R., and Priestley, I., GB Patent 1,159,496, 1969.
713. Drew, E. H., U.S. Patent 3,390,201, 1968.
714. Jones, J. R., and Symes, T. J., GER Patent 1,813,115, 1969.
715. Bercik, P. S., Oligomerization of ethylene on inorganic compounds, *Ind. Eng. Chem. Prod. Res. Dev.*, *17*, 214, 1978.
716. Tkach, V. S., Shmidt, P. K., Malakhova, N. D., and Levkoskil, Yu. S., SU Patent 535,266, 1976; *Chem. Abstr.*, *86*, 105904, 1977.
717. Schulz, W., Voelkner, S., and Kurras, E., GER(EAST) Patent 112,973, 1975; *Chem. Abstr.*, *84*, 121086, 1976.
718. Yoo, J. S., and Milam, R. L., U.S. Patent 3,697,617, 1972.
719. Yoo, J. S., and Milam, R. L., U.S. Patent 3,954,668, 1976.

720. Chauvin, Y., and Gaillard, J., BELG Patent 874,819, 1979.
721. Tyurayaev, I. A., and Bodnaryuk, T. S., GB Patent 1,051,564, 1966.
722. Chauvin, Y., and Phung, N. H., Eur. Pat. Appl. 12,685, 1980; *Chem. Abstr.*, *94*, 16361, 1981.
723. Cannell, L. G., and Majoon, E. F., U.S. Patent 3,355,510, 1965.
724. Takahashi, M., Fujii, Y., and Yamaguki, M., JP Patent 7,627,644, 1976; *Chem. Abstr.*, *86*, 89148, 1977.
725. Raylor, K. A., GB Patent 1,131,146, 1966.
726. Smith, E. T., GB Patent 1,140,821, 1967.
727. Mori, M., Magaoka, I., Hirayanagi, S., and Kihinoki, A., JP Patent 7,249,563, 1972; *Chem. Abstr.*, *78*, 110508, 1973.
728. Sato, H., GER Patent 2,835,365, 1979.
729. Mori, H., JP Patent 7,224,523, 1972; *Chem. Abstr.*, *77*, 87812, 1972.
730. Yoshiura, H., and Arakawa, T., JP Patent 7,537,163, 1975; *Chem. Abstr.*, *85*, 5176, 1976.
731. Mori, H., JP Patent 7,222,207, 1972; *Chem. Abstr.*, *77*, 100695, 1972.
732. Mori, H., and Shimizu, S., JP Patent 7,222,206, 1972; *Chem. Abstr.*, *77*, 87808, 1972.
733. Mori, H., and Shimizu, S., JP Patent 82,169,433, 1982; *Chem. Abstr.*, *98*, 106775, 1983.
734. Mori, H., and Shimizu, S., JP Patent 57,167,932, 1982; *Chem. Abstr.*, *98*, 88802, 1983.
735. Takeshi, H., and Akihisa, M., JP Patent 7,103,161, 1971; *Chem. Abstr.*, *74*, 124757, 1971.
736. Chauvin, Y., and Uchino, M., FR Patent 1,549,202, 1968.
737. Eberhardt, G. G., FR Patent 1,528,160, 1966.
738. Hata, T., JP Patent 7,222,807, 1972; *Chem. Abstr.*, *77*, 87818, 1972.
739. Tripers de Hault, D., GER Patent 2,410,851, 1974.
740. Uchino, M., Chauvin, Y., and Lefevre, G., Propylene oligomerization catalysée par nickel complexes, *Compt. Rend. Hebd. Seances Acad. Sci.*, *265*, 103, 1987.
741. Zuech, E. A., U.S. Patent 3,865,892, 1975.
742. Dunn, H. E., U.S. Patent 3,636,128, 1972.
743. Yoshiura, A., and Arakawa, T., JP Patent 7,522,523, 1975; *Chem. Abstr.*, *84*, 43278, 1976.
744. Griffith, W. P. Jr., U.S. Patent 3,467,726, 1969.
745. Griffith, W. P. Jr., and Butte, W. A., GER Patent 1,810,027, 1967.
746. Rohr Scheid, F., GER Patent 1,643,976, 1971.
747. Born, M., Chauvin, Y., Lefevre, G., and Phung, N. H., Utilisation des combination complexes dans la réaction de polymérisation d'olefins, *Compt. Rend. Hebd. Seances Acad. Sci.*, *268*, 1600, 1969.
748. Arakawa, T., JP Patent 7,222,204, 1966; *Chem. Abstr.*, *77*, 87807, 1972.
749. Dunn, H. E., U.S. Patent 3,689,588, 1972.
750. Fahey, D. R., U.S. Patent 4,101,566, 1978.
751. Fahey, D. R., U.S. Patent 4,123,447, 1978.
752. Fahey, D. R., GB Patent 1,138,575, 1966.
753. Itakura, M., and Ito, E., JP Patent 7,417,722, 1974; *Chem. Abstr.*, *83*, 11139, 1975.
754. Feldblyum, V. Sh., SU Patent 405,849, 1973; *Chem. Abstr.*, *80*, 82043, 1974.
755. Feldblyum, V. Sh., Petrushankaya, N. V., and Kutin, A. M., SU Patent 436,808, 1974; *Chem. Abstr.*, *81*, 135425, 1974.
756. Schaffel, G. S., U.S. Patent 3,104,269, 1963; *Chem. Abstr.*, *59*, 14181, 1963.
757. Schaffel, G. S., BELG Patent 868,377, 1978.
758. Chauvin, Y., GER Patent 2,032,140, 1971.
759. Yoo, J. S., U.S. Patent 4,010,216, 1977.

760. Kawata, N., Mizoroki, T., and Ozaki, A., JP Patent 75,135,001, 1975; *Chem. Abstr.*, *82*, 135075, 1976.
761. Kawamoto, K., Imanaka, T., and Teranishi, S., Nickel phosphine attached on macromolecular compounds as catalyst of propylene oligomerization, *Nippon Kagaku Zasshi*, *89*, 639, 1968.
762. Maxfield, P. L., U.S. Patent 3,427,365, 1966.
763. Eberhardt, G. G., and Griffith, W. P. Jr., U.S. Patent 3,472,911, 1969.
764. Eberhardt, G. G., U.S. Patent 3,459,825, 1967.
765. Carter, C. O., U.S. Patent 4,176,086, 1979.
766. Swift, H. E., and Ching-Young, W., U.S. Patent 3,622,649, 1971.
767. Uchino, M., Chauvin, Y., and Lefevre, G., Halogenures de nickel utilisées dans la catalyse de reactions d'oligomérisation, *Compt. Rend. Acad. Sci., Ser. C.*, *265*, 103, 1967.
768. Reinaecker, R., and Wilke, S., GER Patent 2,493,217, 1978.
769. Dunn, H. E., U.S. Patent 3,872,026, 1975.
770. Dunn, H. E., U.S. Patent 3,760,027, 1973.
771. Dunn, H. E., U.S. Patent 3,651,111, 1972.
772. Yoshiura, H., and Arakawa, T., JP Patent 7,537,162, 1975; *Chem. Abstr.*, *85*, 20584, 1976.
773. Dunn, H. E., U.S. Patent 3,592,870, 1971.
774. Masters, A. F., and Gavell, K. J., PCT Int. Appl. 8,302,907, 1984; *Chem. Abstr.*, *100*, 7932, 1984.
775. Dunn, H. E., U.S. Patent 3,737,434, 1973.
776. Desgrandchamps, G., Hemmer, H., and Haurie, M., GER Patent 1,964,701, 1971.
777. Desgrandchamps, G., Hemmer, M., and Haurie, M., FR Patent 1,602,822, 1971.
778. Desgrandchamps, G., Hemmer, H., and Haurie, M., GER Patent 2,001,923, 1970.
779. Mix, H., Kurras, E., and Fuchs, W., GER(EAST) Patent 99,556, 1973.
780. Phung, W. H., and Lefevre, G., Compléxes metalliques utillisées dans la propylene oligomérisation, *Compt. Rend. Hebd. Seances Acad. Sci.*, *265*, 519, 1967.
781. Dunn, H. E., U.S. Patent 3,709,955, 1973.
782. Smutny, E. J., U.S. Patent 4,436,946, 1984.
783. Loveless, F. C., U.S. Patent 4,642,410, 1987.
784. White, M. A., GER Patent 3,500,638, 1985; U.S. Patent 4,579,991, 1986.
785. Kvisle, S., Blindheim, U., and Ellested, O. H., Propylene oligomerization in the presence of titanium complexes, *J. Mol. Catal.*, *26*, 341, 1984.
786. Immergut E. H., Kollman, G., and Malaleta, A., Mechanism of the propylene oligomerization catalysed by titanium complexes, *J. Polym. Sci.*, *51*, S-57, 1961.
787. Skupinski, W., and Malinowski, S., Nickel complexes supported on silica gel. The role of support in the propylene oligomerization, *J. Organometal. Chem.*, *99*, 465, 1975.
788. Malinowski, S., and Scupinski, W., Propylene oligomerization catalysis and kinetics, *Ann. Soc. Chim. Pol.*, *48*, 359 1978.
789. Skupinski, W., and Malinowski, S., The role of solvents in propylene oligomerization, *J. Organometal. Chem.*, *117*, 183, 1976.
790. Pruvot, A., Commerence, D., and Chauvin, Y., Nickel salts as catalysts in propylene oligomerization, *J. Mol. Catal.*, 22, 179, 1983.
791. Yoo, Y. S., and Milam, R. L., U.S. Patent 3,697,617, 1972; *Chem. Abstr.*, *78*, 3652, 1973.
792. Yoo, Y. S., and Milam, R. L., U.S. Patent 3,954,668, 1976.
793. Yoo, Y. S., and Erickson, H., U.S. Patent 3,992,323, 1976; *Chem. Abstr.*, *86*, 734081, 1977.

794. Yoo, Y. S., and Erickson, H., GB Patent 1,411,692, 1973; *Chem. Abstr.*, *84*, 60247, 1976.
795. Cannell, L., U.S. Patent 3,592,869, 1971; *Chem. Abstr.*, *75*, 98144v, 1971.
796. Jones, J. R., and Priestly, I., GER Patent 1,643,716, 1971.
797. Sachnidis, S. J., Clutterbruck, L. M., Masters, A. F., Brown, S. J., and Tregloan, P. A., A new organometallic complex of propylene oligomerization, *Appl. Catal.*, *48*, 1, 1989.
798. Bugai, E. A., SU Patent 711,041, 1980; *Chem. Abstr.*, *92*, 129656r, 1980.
799. Jeske, G., .Oligomerization of olefins. Mechanism of hydrocarbon insertion, *J. Amer. Chem. Soc.*, *107*, 3091, 1985.
800. Jeske, G., Oligomerization of olefins. The role of aromatic solvents. *J. Amer. Chem. Soc.*, *107*, 8103, 1985.
801. Ferraris, G., Priola, A., and Cesca, S., GER Patent 2,831,554, 1979; *Chem. Abstr.* *90*, 139977a, 1979.
802. Mirasuke, E., U.S. Patent 3,586,734, 1971.
803. Mirasuke, E., U.S. Patent 2,528,876, 1950.
804. Robert, M., U.S. Patent 3,932,553, 1976.
805. Robert, M., U.S. Patent 2,766,312, 1956.
806. Zhang, S., *Faming, Zhuali, CM* 87,101,742; *Chem. Abstr.*, *110*, 135923p, 1989.
807. Imai, T., U.S. Patent 4,463,212, 1984.
808. Ethyl Corp., U.S. Patent 3,071,634, 1963.
809. Ethyl Corp., U.S. Patent 2,525,784, 1950.
810. Fontana, C. M., *Cationic Polymerization and Related Complexes*, *Proc. Conf.*, 1952, p. 121; *Chem. Abstr.*, *48*, 5625, 1954.
811. Fontana, C. S., U.S. Patent 3,193,546, 1965.
812. OP Inc., FR Patent 2,402,478, 1979.
813. OP Inc., U.S. Patent 3,041,386, 1962.
814. Mangasaryan, N. A., *Khim. Technol. Topl. Masel.*, 1988, 698; *Chem. Abstr.*, *109*, 93685j, 1988.
815. Abdrakhimov, Yu. R., *Khim. Technol. Topl. Masel.*, 1975, 16; *Chem. Abstr.*, *85*, 108193p, 1976.
816. Pasky, J. Z., U.S. Patent 4,444,985, 1984.
817. Abrahams, J. V., U.S. Patent 2,870,217, 1959.
818. *Petrol Refiner*, *33*, 193, 1954.
819. Iwamura, Y., U.S. Patent 4,613,719, 1986.
820. Clark, A., Oxides of transition metal, *Ind. Eng. Chem.*, *45*, 1476, 1953.
821. Kusnetsov, O. T., Activity and selectivity of ion exchange forms of zeolites NaX and catalysts NiO-Al_2O_3-SiO_2 in propylene conversion, *Neftekhimia*, *20*, 200, 1980 (Russian).
822. Espinoza, R. L., Scurell, M. S., and Van Walsen, H. J., Inorganic molecules as catalysts in propylene oligomerization, *Actas Symp. Iberieum Catal.*, *2*, No. 9, 1625.
823. Giusti, A., Gusi, S., Bellussi, G., and Fattere, V., Eur. Pat. Appl. 290,068, 1988; *Chem. Abstr.*, *110*, 115529m, 1989.
824. Guth, L. J., Faust, A. G., Raatz, F., and Lamblin, J. M., Eur. Pat. Appl. 269,503, 1988; *Chem. Abstr.*, *109*, 150248f, 1988.
825. Dzwigaj, S., Haber, J., and Romotowski, T., Oligomerization of propylene catalysed by zeolites, *Zeolites*, *4*, 147, 1984.
826. Kiviesi I., Technology of propylene oligomerization. Catalysts and experimental conditions, *Kem Lapja*, *32*, 605, 1977; *Chem. Abstr.*, *89*, 41882, 1978.
827. Garwood, W. E., Synthesis of higher olefins by propylene oligomerization, *Prepr. ACS Div. Petrol. Chem.*, *257*, 563, 1982.

828. Mravec, D., Lacny, Z., and Havsky, J., Propylene oligomerization, *Ropa Uhlie*, *28*, 283, 1986.
829. Garwood, E. V., *ACS Symp. Series*, No. 218, 1983.
830. Chester, A. W., U.S. Patent 4,517,399, 1985.
831. La Pierre, B., and Wong, S. S., U.S. Patent 4,430,615, 1983.
832. Tabak, S. A., U.S. Patent 4,254,295, 1981.
833. Yang, Z., *Shigon Huagong*, *15*, 243, 1986; *Chem. Abstr.*, *105*, 153596k, 1986.
834. Sycheva, O. A., Melnikov, V. N., and Vladimirova, L. I., Higher olefin oligomerization in the presence of metal complexes, *Neftepererab. Neftekhim.*, *9*, 20, 1984 (Russian).
835. Isa, H., JP Patent 7,912,952, 1979; *Chem. Abstr.*, *91*, 177912, 1970.
836. Masagutov, R. M., SU Patent 992,504, 1983; *Chem. Abstr.*, *99*, 5179u, 1983.
837. Hashimoto, M., and Arakawa, T., JP Patent 7,400,802, 1974; *Chem. Abstr.*, *81*, 25021, 1974.
838. White, M. A., U.S. Patent 4,579,991, 1986.
839. Brennan, J. A., Oligomerization of hexenes, Structure of resulted oligomers, *Ind. Eng. Chem., Prod. Res. Dev.*, *19*, 2, 1980.
840. Schoenthal, G. V., and Slaugh, E. H., Eur. Pat. Appl. 257,695, 1988; *Chem. Abstr.*, *109*, 54852e, 1988.
841. Bobsein, R. L., U.S. Patent 4,436,948, 1984.
842. Gao Zhang, A., Liu, B., and Zhou, K., *Dalian Goungxueyuam Xuebae*, *26*, 32, 1987; *Chem. Abstr.*, *107*, 237361, 1987.
843. Gao Zhang, A., and Zhou, K., *Vingyouang Huaxue*, *4*, 40, 1987; *Chem. Abstr.*, *106*, 176924a, 1987.
844. Keim, W., Behr, A., and Kraus, G., Mechanism of olefin oligomerization, *J. Organometal. Chem.*, *251*, 377, 1983.
845. Le Ponnoc, J., U.S. Patent 4,398,049, 1983.
846. Cannell, L., U.S. Patent 3,592,869, 1971.
847. Chauvin, Y., FR Patent 2,481,950, 1981.
848. Chauvin, Y., U.S. Patent 4,387,262, 1983.
849. Beach, D. L., and Kobylinski, T. P., U.S. Patent 4,293,726, 1981.
850. Sauki, K., Ono, N., and Murakami, T., JP Patent 61,151,136, 1986; *Chem. Abstr.*, *106*, 4515b, 1987.
851. Frame, R. R., and Imai, T., U.S. Patent 4,737,480, 1988.
852. Frame, R. R., and Imai, T., U.S. Patent 4,737,479, 1988.
853. Avdeikina, E. G., Tlenkopschev, M. A., and Korshak, Yu. V., Oligomerization of olefins. Selectivity of nickel complexes, *Vysokomol. Soedin.*, *A*, *28*, 552, 1986 (Russian).
854. Yuffa, A. V., Vershinius, L. I., Fuman, D. B., and Bragin, O. V., Synthesis of liquid oligomers by olefin oligomerization, *Neftekhimia*, *25*, 495, 1985 (Russian).
855. Austin, R., Beach, L. D., and Pellegrini, J. P., U.S. Patent 4,319,065, 1982.
856. Ono, N., JP Patent 62,142,125, 1987; *Chem. Abstr.*, *108*, 76065u, 1988.
857. Setartseva, G. P., Olefin oligomerization on Pd-zeolite catalysts, *Neftekhimia*, *28*, 53, 1988 (Russian).
858. Loveless, C. T., Eur. Pat. Appl. 139,343, 1985; *Chem. Abstr.*, *103*, 105433e, 1985.
859. Tsvetkov, O. K., S.U. Patent 711, 044, 1980; *Chem. Abstr.*, *92*, 129655u, 1980.
860. Loveless, C. T., U.S. Patent 4,469,010, 1984.
861. Higashimura, T., Synthesis of oligomers by cationic oligomerization, *J. Appl. Polymer Sci.*, *27*, 593, 1982.
862. Higashimura, T., JP Patent 5,877,827, 1983; *Chem. Abstr.*, *99*, 70181, 1983.
863. Helbig, G., GER(EAST) Patent 154,188, 1982.
864. Mandai, H., GER Patent 2,837,235, 1979.

865. Regov, S. A., *Poluch. Primen. Prod. Neftekhim.*, 1982, 123; *Chem. Abstr.*, *98*, 182222c, 1983.
866. Kitamura, T., JP Patent 7,966,993, 1979; *Chem. Abstr.*, *91*, 124211b, 1979.
867. Akatsu, S., JP Patent 90,195,108, 1989; *Chem. Abstr.*, *111*, 195622v, 1989.
868. Minsker, V., Technology of higher olefins oligomerization, *Khim. Technol. Topl. Masel.*, 1977, 27 (Russian).
869. Isa, H., JP Patent 76,122,002, 1975; *Chem. Abstr.*, *86*, 105905, 1977.
870. Tsvetkov, C. N., Transformation of higher olefins by homogeneous catalysis, *Neftekhimia*, *29*, 224, 1989 (Russian).
871. Berlin, A. A., Functionalized oligomers obtained by higher olefins oligomerization, *Neftepererab. Neftekhim.*, 1988, No. 2, p. 25 (Russian).
872. Prokofiev, K. V., Uses of hydrocarbon oligomers, *Neftepererab. Neftekhim.*, 1988, No. 8, p. 22 (Russian).
873. Morganson, N. E., U.S. Patent 4,409,415, 1983.
874. Morganson, N. E., Eur. Pat. Appl. 77,133, 1986; *Chem. Abstr.*, *99*, 76361, 1983.
875. Watta, W. E., U.S. Patent 4,413,156, 1983.
876. Onopchenko, A., Catalytic oligomerization of higher olefins, *Ind. Eng. Chem. Prod. Res. Dev.*, *22*, 182, 1983.
877. Hammond, K. G., Eur. Pat. Appl. 136,377, 1988; *Chem. Abstr.*, *100*, 71847, 1984.
878. Larkin, J. M., U.S. Patent 4,434,309, 1984.
879. Kuliev, R. Sh., SU Patent 759,529, 1980; *Chem. Abstr.*, *93*, 189046d, 1980.
880. Sangalov, Yu. A., Industrial transformation of higher olefins, *Neftekhimia*, *17*, 865, 1977 (Russian).
881. Rafikov, S. R., SU Patent 690,024, 1979; *Chem. Abstr.*, *92*, 758114, 1980.
882. Panchenov, G. M., Formation of active species in higher olefin oligomerization, *Zhur. Fis. Khim.*, *48*, 884, 1974 (Russian).
883. Johnson, T. H., U.S. Patent 4,476,343, 1984.
884. Brennan, J. F., and Lester, G. B., U.S. Patent 3,373,853, 1973.
885. Miller, S. J., U.S. Patent 4,542,251, 1985.
886. Miller, S. J., U.S. Patent 4,608,459, 1986.
887. Ramus, G., *Proc. Int. Congr. Catal.*, *9th*, p. 4, 1974.
888. Weitkamp, J., *Proc. Int. Conf. Zeolites*, *5*, 858, 1980.
889. Gregory, L. H., U.S. Patent 4,531,014, 1985.
890. Minachev, K. M., 1-Decene transformation in higher oligomers, *Neftekhimia*, *28*, 164, 1988 (Russian).
891. Kucherov, A. W., Catalysis of higher olefin oligomerization, *Kinet. Katal.*, *30*, 193, 1989 (Russian).
892. Anderson, J. R., Chang, V. P., and Western, R. J., Surface transformation of zeolites in order to catalyse olefin oligomerization, *J. Catal.*, *118*, 466, 1989.
893. Tabak, S. A., U.S. Patent 4,254,295, 1981.
894. Minachev, K. M., Olefin oligomerization under the influence of transition metal complexes, *Izv. Akad. Nauk. SSSR, Ser. Khim.*, 1987, 1225 (Russian).
895. Kojima, M., Olefin oligomerization, *Ind. Eng. Chem. Res.*, *27*, 248, 1988.
896. Alpino, N. L., Valentine, F. G., and Granes, G. W., U.S. Patent 4,239,930, 1981.
897. Magdavakar, A. M., and Swift, H. E., U.S. Patent 4,308,411, 1981.
898. Cupples, B. L., and Madgavakar, A. M., U.S. Patent 4,213,001, 1980.
899. Morganson, N. E., and Beroik, P. G., U.S. Patent 4,429,177, 1984.
900. Morganson, N. E., U.S. Patent 4,365,105, 1982.
901. Gutry, C., and Engelhard, Ph., FR Patent 2,504,121, 1982.
902. Cupples, B. L., Heilman, W. J., and Kresque, A. N., GER Patent 2,651,637, 1977.
903. Eastham, A. M., Synthesis of oligomers by olefin polymerization, *Can. J. Chem.*, *59*, 2621, 1981.

904. Pilo, P., Ciardelli, P., and Lorenzi, G. P., Induzione asimetrica nella polimerizatione di alcune alpha olefine, *Chim. Ind. (Milan)*, 46, 313, 1964.
905. Brennan, J. A., Practical aspects of olefin oligomerization catalysed by transition metal complexes, *Ind. Eng. Chem. Prod. Res. Dev.*, *19*, 2, 1980.
906. Okamura, Y., Shiozawa, K., and Kagano, T., JP Patent 7,889,089, 1978; *Chem. Abstr.*, *90*, 39440b, 1979.
907. Nipe, R. N., U.S. Patent 4,225,739, 1980.
908. Bith, R. R., U.S. Patent 4,227,027, 1980.
909. Shubkin, R. L., Characterization of reaction product obtained by olefin oligomerization, *Prepr. ACS Div. Petrol. Chem.*, *24*, 809, 1979.
910. Shubkin, R. L., Baylerian, M. S., and Maler, A. R., Separation by GPC of olefin oligomers, *Ind. Eng. Chem. Prod. Res. Dev.*, *19*, 15, 1980.
911. Shubkin, R. L., and Baylerian, M. S., U.S. Patent 4,376,222, 1983.
912. Larkin, J. M., U.S. Patent 4,395,578, 1983.
913. Darden, J. W., Marquis, E. T., and Watts, L. W., U.S. Patent 4,400,565, 1983.
914. Van der Berg, J. P., Kortbeek, A. G. T., and Grandvallet, P., Eur. Pat. Appl. 281,208, 1988; *Chem. Abstr.*, *109*, 231744, 1988.
915. Yoshimura, T., JP Patent 62,721,785, 1987; *Chem. Abstr.*, *106*, 21685, 1987.
916. Miller, S. J., GER Patent 3,229,829, 1983.
917. Miller, S. J., U.S. Patent 4,423,268, 1983.
918. Nippon Oils and Fats Co. Ltd., JP Patent 8,139,430, 1981; *Chem. Abstr.*, *96*, 35912h, 1982.
919. Subbotin, A. N., Khodakov, V., Anders, K., and Nakshunov, V. S., The influence of solvents upon the selectivity of olefin oligomerization reaction catalysed by transition metal complexes, *Izv. Akad. Nauk SSSR, Ser. Khim.*, 1981, 1724 (Russian).
920. Nippon Oils and Fats Co. Ltd., JP Patent 8,139,428, 1981; *Chem. Abstr.*, *96*, 35913j, 1982.
921. Nippon Oils and Fats Co. Ltd., JP Patent 8,139,429, 1981; *Chem. Abstr.*, *96*, 35914k, 1982.
922. Kamiyama, S., Shiozawa, K., and Kaceko, K., CAN Patent 1,212,659, 1986.
923. Dabka, J., Insertion mechanism in olefin oligomerization *Bull. Acad. Pol. Sci. Ser. Sci. Chim.*, *24*, 137, 1976.
924. Dabka, J., Termination reaction in olefin oligomerization catalysed by transition metal complexes, *J. Chem. Soc., Faraday Trans.*, *1*, *77*, 2633, 1981.
925. Kusnetsov, O. I., Panchencov, G. M., and Tolkseheva, F. I., Olefin oligomerization initiated by radicalic catalyst, *Neftekhimia*, *21*, 849, 1981 (Russian).
926. Tea, K., JP Patent 5,922,783, 1984; *Chem. Abstr.*, *102*, 185685t, 1985.
927. Hashimoto, T., and Kokubo, M., Eur. Pat. Appl. 90,569, 1982; *Chem. Abstr.*, *100*, 34978y, 1984.
928. Nippon Oil Co., JP Patent 59,212,438, 1984; *Chem. Abstr.*, *102*, 167321f, 1985.
929. Le Page, J. F., Cosins, J., Jaquin, M. J., and Miguel, J., BELG Patent 874,821, 1978.
930. Jaquin, B., Le Page, J. F., Cosepus, J., and Miguel, J., GB Patent 2,006,263, 1979.
931. O'Hara, M., and Imai, T., U.S. Patent 4,476,342, 1984.
932. Tada, A., Suzuka, H., and Imazu, Y., Catalytic capacity of transition metal complexes in olefin oligomerization, *Chem. Lett.*, 2, 423, 1987.
933. Le Page, J. P., and Miguel, J., GB Patent 2,006,263, 1979.
934. Cares, W. R., U.S. Patent 4,065,512, 1977.
935. Watts, L. W., and Marquis, E. T., U.S. Patent 4,367,352, 1983.
936. Nippon Oil Co. Ltd., JP Patent 58,213,725, 1983; *Chem. Abstr.*, *100*, 13975, 1984.
937. Gibadulina, Kh. M., and Galimov, Z. F., *Izv. Vyssh. Uchebn. Zaved. Neft. Gaz*, *26*, 38, 1983; *Chem. Abstr.*, *99*, 38833f, 1983.

938. Heublin, G., Albrecht, G., and Knempfel, W., Oligomerization of 1-butene, *Acta Polymerica*, *35*, 220, 1984.
939. Binger, P., Bunkman, A., and McMeeking, J., GPC characterization of higher olefin oligomers, *Angew. Chem.*, *85*, 1053, 1973.
940. Sycheva, O. A., Melnikov, V. N., and Vladimirova, L. I., Oligomerization of higher olefins in the presence of $TiCl_4$-R_nAlCl_{3-n} complexes, *Neftepererab. Neftekhim.*, *9*, 20, 1983 (Russian).
941. Hashimoto, M., and Arakawa, T., JP Patent 7,000,802, 1974; *Chem. Abstr.*, *81*, 25021, 1974.
942. Isa, H., and Nagai, S., JP Patent 7,912,952, 1979; *Chem. Abstr.*, *91*, 177912, 1979.
943. Masagutov, R. M., SU Patent 992,504, 1983; *Chem. Abstr.*, *99*, 5179u, 1984.
944. Bobseen, R. L., U.S. Patent 4,436,948, 1984.
945. Nadgavkar, A. M., and Swift, H. E., U.S. Patent 4,394,296, 1983.
946. Kustov, L. M., Inorganic oxides as catalyst of higher olefins oligomerization, *Kinet. Katal.*, *30*, 169, 1989 (Russian).
947. Kobayashi, S., Ethylene oligomerization, *Kogyo Kagaku Zashi*, *72*, 2511, 1969 (Japanese).
948. Von Dohlen, W. C., and Wilson, T. P., Low-molecular-weight species from high-pressure polyethylene, *J. Polymer Sci., Polym. Chem. Ed.*, *17*, 2511, 1979.
949. Takahashi, M., Satch, T., and Toya, T., Oligoethylenes in high pressure polyethylene. I. Identification of homologues, *Polymer Bull.*, 2, 215, 1980.
950. Takahashi, M., Satch, T., and Toya, T., Oligoethylenes in high pressure polyethylenes. II. Reaction mechanism, *Polymer Bull.*, 2, 643, 1980.
951. Takahashi, M., Satch, T., and Toya, T., Oligoethylenes in high pressure polytehylene. III. Alkylated cyclopentanes, *Polymer Bull.*, *4*, 497, 1981.
952. Tureu, A. T., Toader, M., Boborodoa, C., and Iloisou, F., Oligomers in the ethylene polymerization with Ziegler-Natta catalysts, *Materiale Plastics (Bucharest)*, *17*, 153, 1980 (Romanian).
953. Ritchie, P. D., *Vinyl and Related Polymers*, Iliffe Books, London, 1968, p. 75.
954. *Modern Plastics Encyclopedia, Guide to Plastics*, McGraw-Hill Inc., 1979, p. 193.
955. Leister, N. A., and Picollini, R. J., U.S. Patent 3,994,958, 1976.
956. Madgavikar, A. M., and Swift, H. E., Fixed-bed catalytic process to produce synthetic lubricant, *Ind. Eng. Chem. Prod. Res. Dev.*, 22, 675, 1983.
957. Beresnev, V. V., SU Patent 979,374, 1982.
958. Hirishi, T., Ukigai, T., and Sato, M., U.S. Patent 3,997,623, 1976.
959. Smith, R. L., and Johns, J., U.S. Patent 3,959,175, 1976.
960. Rätzsch, M., and Grundmann, H., RO Patent 51,211, 1968.
961. Sperling, L. H., *Recent Advances in Polymer Blends, Grafts and Blocks*, Plenum Press, New York, 1974.
962. Brunson, M. O., and Gillen, W. D., U.S. Patent 3,484,403, 1969.
963. Vulpe, T., Părăuşanu, V., and Mihăiţă, D., RO Patent 54,317, 1972.
964. Tomoshige, T., GER Patent 2,241,057, 1973.
965. Bronstert, K., and Mueller, D. W., GER Patent 2,303,745, 1974.
966. Mueller, D. W., Klug, H., Bronstert, K., and Grah, L., U.S. Patent 3,883,458, 1975.
967. Wurmb, R., Mueller, D. W., Dorst, H. G., Bronstert, K., and Theysohn, R., U.S. Pat. Appl. B 507,456, 1976; *Chem. Abstr.*, *85*, 22354, 1976.
968. Bronstert, K., GER Patent 2,348,840, 1975.
969. Mainard, K. R., U.S. Patent 3,859,385, 1975.
970. Ames, W. A., U.S. Patent 4,159,760, 1979.
971. Tsotomo, K., and Onami, F., JP Patent 77,850,981, 1967; *Chem. Abstr.*, *68*, 88155, 1968.
972. Connel, R. L., and Meyer, M. F., BELG Patent 742,272, 1970.

973. Tachi, A., GER Patent 2,853,862, 1979.
974. Maruyama, M., JP Patent 13,288,631, 1979; *Chem. Abstr.*, *91*, 109152, 1979.
975. Heintzelman, W. J., and Naiman, M. I., U.S. Patent 3,437,623, 1969.
976. Abrahams, W. J., GB Patent 1,198,706, 1970.
977. Räetzsch, M., Killian, R., and Gladigan, G., GER Patent 2,024,947, 1970.
978. Räetzsch, M., Gladigan, G., Geber, M., Blau, P., and Pabst, P., GER(EAST) Patent 117,077, 1975.
979. Matsumoto, Y., JP Patent 24,175, 1965; *Chem. Abstr.*, *68*, 60633, 1968.
980. Coleman, L. E., FR Patent 2,103,060, 1972.
981. Nakata, N., and Hasegawa, T., JP Patent 77,037,131, 1967; *Chem. Abstr.*, *67*, 12648, 1967.
982. Rohm and Haas, GER Patent 2,634,033, 1977.
983. Mori, H., Kakitani, H., and Sugh, H., JP Patent 77,038,192, 1970; *Chem. Abstr.*, *75*, 88728, 1970.
984. Kosaka, V., JP Patent 19,291, 1972; *Chem. Abstr.*, *78*, 137114, 1973.
985. Alberts, H., Synthesis of oligomers, *Adv. Chem. Series*, *142*, 214, 1975.
986. Natta, G., Kinetic evaluation of olefin oligomerization and cooligomerization, *J. Polym. Sci.*, *34*, 685, 1959.
987. Abrahams, W. J., GB Patent 887,731, 1961.
988. Lombrose, C., FR Patent 1,170,326, 1959.
989. Palvarini, H., U.S. Patent 3,352,945, 1967.
990. Pohlman, H., GER Patent 2,233,799, 1974.
991. Romanov, V., Low-molecular species in LDPE, *Plastic Kautsch.*, *12*, 521, 1965.
992. Protasov, V. L., Applications of oligoethylenes, *Nauch. Tr. Mosk. Technol. Inst. Legk. Prom.*, *36*, 63, 1963 (Russian).
993. Pustovit, N. G., SU Patent 564,309, 1977.
994. Gaylord, N. G., GER Patent 2,308,749, 1971.
995. Prestan, J. E., U.S. Patent 3,505,279, 1970.
996. Union Carbide Corp., GB Patent 1,184,971, 1970.
997. Anker, E. H., GER Patent 2,165,330, 1971.
998. Hoechst, A. G., GB Patent 1,204,655, 1970.
999. Kaneko, H., JP Patent 9,471, 1971; *Chem. Abstr.*, *75*, 22003, 1971.
1000. Downer, J. D., *Chemistry of Carbon Compounds*, Vol. 1, Part A, 2nd Ed., Elsevier Publ. Co., 1964, p. 364.
1001. Mueller, A., *Proc. Roy. Soc. (London)*, *Ser. A*, *120*, 437, 1928.
1002. Parks, G. S., *J. Amer. Chem. Soc.*, *52*, 1032, 1930.
1003. Krafft, F., *Chem. Ber.*, *40*, 4479, 1907.
1004. Sondheimer, F., *J. Amer. Chem. Soc.*, *78*, 6265, 1957.
1005. Templin, P. R., *Ind. Eng. Chem.*, *48*, 154, 1956.
1006. Flory, P. J., and Vrij, A., *J. Amer. Chem. Soc.*, *85*, 3548, 1963.
1007. Whitmore, F. C., Langhlin, K. C., Matszeski, J. F., and Sernatis, J. D., *J. Amer. Chem. Soc.*, *63*, 757, 1947.
1008. Whitmore, F. C., *J. Amer. Chem. Soc.*, *53*, 3136, 1934.
1009. Dixon, J. A., *J. Amer. Chem. Soc.*, *70*, 5369, 1948.
1010. Whitmore, F. C., *J. Chem. Soc.*, *63*, 2035, 1941.
1011. Bartlett, P. D., *J Amer. Chem. Soc.*, *63*, 498, 1941.
1012. Ratuev, M. J., *Dokl. Akad. Nauk. SSSR*, 75, 1958.
1013. Vanghan, W. F., *J. Amer. Chem. Soc.*, *54*, 3863, 1952.
1014. Cauntlyer, J. C., and Smith, J. W., *Ind. Eng. Chem.*, *45*, 1732, 1955.
1015. Duncan, N. J., and Janz, J., *J. Chem. Phys.*, *30*, 1644, 1959.

2

Acrylic Oligomers

In the 1930s compounds of oligoester acrylate type were synthesized by esterification or by transesterification of polyols with metacrylic derivatives [1,2]. It was only in 1946 that Berlin worked out the scientific basis of the general method of obtaining oligoesteracrylates [3].

Oligoesteracrylates can be represented by the following general formula:

$$CH_2{=}\underset{\underset{X}{|}}{C}{-}Y{-}CH_2{-}Z{-}CH_2{-}Y{-}\underset{\underset{X}{|}}{C}{=}CH_2 \qquad (1)$$

where X = H, CH_3, C_2H_5, halogen, CN, OCH_3; Y = —OCO—, $-O\underset{\underset{O}{\|}}{C}O-$, $-HN\underset{\underset{O}{\|}}{C}O-$, $-O\underset{\underset{O}{\|}}{P}O-$; Z = $-(CH_2)_n-$, $-(CH_2O)_n-$, $-(CH_2CH_2O)_n-$, and $n = 0\text{–}50$.

The diversification of synthesis methods and of monomers implied in obtaining oligoesteracrylates created compounds with complex structures not representable by the general formula (1). For this reason, membership of some oligomers in a class of oligoacrylates is classified using the following essential structure characteristics:

Presence in the molecule of at least two acrylic, methacrylic, halogenoacrylic groups or their derivatives. These groups can be marginal or lateral.

Presence between the two marginal acrylic groups on an oligomer block of alkylenic, oxyalkylenic, heterochain, elementochain, heterocyclic, elementocylic, or mixed structures.

The link between the internal oligomeric block and the two marginal acrylic groups is by an esteric type bond.

I. SYNTHESIS

Most of the synthesis methods for acrylic oligomers are based on condensation reactions of di- and polyfunctional compounds. Besides these methods, acrylic oligomers can also be obtained by correspondingly modifying other oligomer categories.

So far, more than 150 types of oligomers from the category of oligoester-acrylates have been obtained; on an industrial scale, only 10–15 products have been obtained [4].

A. Oligoesteracrylates (OEAs)

The most widespread method of obtaining oligoesteracrylates (OEAs) is condensation telomerization [1,2]. In principle, condensation telomerization is based on the stoichiometric reaction of functional groups in the presence of telogene, a monofunctional compound. This contains a —OH or —COOH group, which by its insertion within the growing chain eliminates the possibility of continuing the reaction. The general scheme of this process is represented by the following reaction:

$$nR(COY)_2 + (n + 1)R'OH + 2CH_2{=}C(X)COY' \xrightarrow{-H_2O(HCl,\ ROH)}$$

$$CH_2{=}C(X)COOR'O[OCRCOOR'O]_nOCC(X){=}CH_2 \qquad (2)$$

where R and R′ represents alkyl or aryl radicals, X = H, CH_3, halogen, and so forth, Y = OH, OR, Cl, and Y′ = OH, OR, OROH. The polymerization degree n depends on the molar ratio between reactants and telogene.

The process takes place in inert solvents, under stirring, and in the presence of acid catalysts and inhibitors.

As solvents, aromatic hydrocarbons (benzene, toluene), and chlorinated derivatives of aliphatic or aromatic hydrocarbons can be used. Reference 3 includes some technological details and the factors that influence the development of the synthesis process.

In the last 10 years, various modifications of the synthesis process have been worked out with a view to improving OEA yield technology. The use of a powerful acid catalyst (H_2SO_4) has, besides ecological drawbacks, the problems involving the occurrence of side reactions, especially between H_2SO_4 and OH groups in the system with ester formation [4].

Sulfocationites have also been used as heterogeneous catalysts for the industrial synthesis of OEAs. Later on, a new type of catalyst was used, based on polysulfophenilenquinone. This compound simultaneously fulfills the role of catalyst and inhibitor [5–9].

The degree of polymerization of OEA varies, with experimental conditions, from 2 to 33. The frequency of these species is variable.

Russia produces a wide range of OEAs on an industrial scale; see Table 1.

Interphasic oligocondensation is another procedure for OEA synthesis. In this respect, mono- and dicarboxylic acid chloroanhydrides dissolved in hydrophobic organic acids as well as bisphenol or glycol solutions are used as raw materials. The following reactions takes place:

$$2CH_2{=}C(X)COCl + nR(COCl)_2 + (n + 1)R'(OH)_2 \xrightarrow{NaOH}$$

$$CH_2{=}C(X)COOR'OCORCOOR'OOCC(X){=}CH_2 + 2(n + 1)NaCl \qquad (3)$$

Table 1 Oligoesteracrylates Produced in Russia

Name	Trade name	Formula	Functionality	$\bar{M}$
α-Methacryloyl-ω-methacryloyl-oligo(oxyethylene)	TGM-3	$MO(CH_2CH_2O)_3M^a$	4	286
α,ω-Bis(methacryloyloxy)oligo-(ethyleneglycolphthalate)	MGF-1	$C_6H_4\left<\begin{matrix}COOCH_2CH_2OM\\COOCH_2CH_2OM\end{matrix}\right.$	4	390
α,ω-Bis(methacryloyloxy)oligo-(diethyleneglycolphthalate	MDF-1	$C_6H_4\left<\begin{matrix}CO(OCH_2CH_2)_2OM\\CO(OCH_2CH_2)_2OM\end{matrix}\right.$	4	492
α,ω-Bis(methacryloyloxy)oligo-(triethyleneglycolphthalate)	MGF-9	$C_6H_4\left<\begin{matrix}CO(OCH_2CH_2)_3OM\\CO(OCH_2CH_2)_3OM\end{matrix}\right.$	4	566
α-Methacryloyl-ω-methaacryloyl-eneglycoloxyoligo(diethyl-eneglycolphthalate)	MDF-2	$M(Z)_nO(CH_2CH_2O)_2M^b$	4	714
α,ω-Tetramethacryloyloligo-(glycolphthalate)	TMGF-11	$C_6H_4\left<\begin{matrix}COOCH(CH_2OM)_2\\COOCH(CH_2OM)_2\end{matrix}\right.$	8	586
α,ω-Hexamethacryloyloligo-(pentaeritritadipinate)		$MOCH_2-\overset{\displaystyle CH_2OM}{\underset{\displaystyle CH_2OM}{\overset{\vert}{\underset{\vert}{C}}}}-CH_2O(Y)_nM^c$		
$n = 1$	7-1		12	788
$n = 2$	7-20		16	1172

[a] $M = CH_2{=}C(CH_3)CO{-}$.
[b] $Z = {-}O(CH_2CH_2O)_2OC{-}C_6H_4{-}CO{-}$.
[c] $Y = CO(CH_2)_4{-}COO{-}CH_2{-}C(CH_2OM)_2{-}CH_2{-}O{-}$.

By this method, crystalline OEAs have been obtained, containing aromatic sequences in the main chain. These sequences arise from the use of iso- and terephthalic acid chloroanhydride or of diphenylolpropane as raw materials [10].

Another variant of condensation telomerization envisages the use of a transesterification process according to the reaction [11]

$$2CH_2{=}C(X)COOCH_3 + nCH_3OOCRCOOCH_3 + (n + 1)R'(OH)_2 \rightleftarrows$$
$$CH_2{=}C(X)COOR'OOCRCOOR'COOC(X){=}CH_2 + 2(n + 1)CH_3OH \qquad (4)$$

One of the earliest methods of OEA synthesis was oligoesterification in two steps. In the former stage, methacrylic esters of glycol or polyols that contain free OH groups are obtained. In the second stage, esterification of hydroxyls contained in the glycol methacrylate with dicarboxylic acids is realized. The following reactions takes place:

$$CH_2{=}C(X)COOH + R(OH)_2 \rightarrow CH_2{=}C(X)COOROH + H_2O \qquad (5)$$

$$CH_2{=}C(X)COOROH + HOOC{-}R'{-}COOH \rightarrow$$

$$CH_2{=}C(X)COOR{-}OOC{-}R'{-}COOH \qquad (6)$$

Various OEAs have been obtained by this method [12–14].

OEA synthesis can also be achieved by additive telomerization of cyclic ethers. Typical of this method is the reaction between anhydrides and epoxides. The process takes place in the presence of the telogenes in the category of acrylic acid or of its derivatives. An example is shown in the following reaction [15]:

$$n\ R\begin{matrix} \diagup CO \diagdown \\ \diagdown CO \diagup \end{matrix}O + (n+1)\ \underset{\diagdown O \diagup}{CH_2{-}CH}{-}CH_2OR' + [CH_2{=}C(X)CO]_2O \xrightarrow[(C_2H_5)_3N]{170°C}$$

$$CH_2{=}C(X)CO[OCH_2{-}\underset{\displaystyle CH_2OR'}{\underset{|}{CH}}{-}OOCRCO]_2OCH{-}\underset{\displaystyle CH_2OR'}{\underset{|}{}}CH_2OOCC(X){=}CH_2 \qquad (7)$$

where R = C_6H_4, R′ = CH_3, C_2H_5, $C_6H_4CH_3$, and n = 1–3.

OEAs with regular structure, saturated or unsaturated, have been obtained via the reaction between glycols, cyclic anhydrides, and glycidylmethacrylate [16].

Methods of OEA production also include those that use nonpolymerizable compounds as raw materials. Thus, one can mention modification of chlorinated paraffins with acrylic monomers as an example:

$$\sim CH_2{-}CH(Cl){-}(CH_2)_n{-}CH(Cl){-}CH_2\sim + CH_2{=}C(X)COOH \xrightarrow[-ClK]{120-140°C}$$

$$\sim CH_2{-}CH(Cl){-}(CH_2)_n{-}\underset{\displaystyle OCOC(X){=}CH_2}{\underset{|}{CH}}{-}CH_2\sim \qquad (8)$$

Oligoestermaleinacrylates can be obtained through the reaction between OH groups of oligoestermaleates and methacrylic acid [18].

By applying the aforementioned methods, carbo- or heterochain OEAs with a nitrogen, sulfur, halogen, or metal content can be obtained. These OEAs can contain saturated, unsaturated, or aromatic structural motifs arranged in a linear or branched structure.

By the method of step condensation telomerization, OEAs with triazinic cycle content were obtained. The synthesis includes in the first stage the production of diols with triazinic groups in the main chain and OH end groups with the following structure [19]:

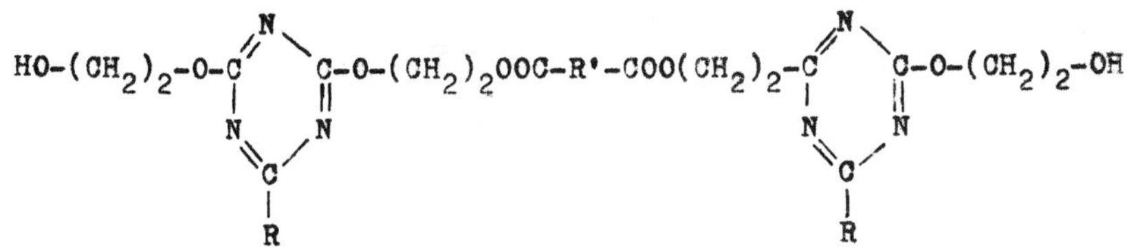

(9)

where R′ = $—(CH_2)_n—$, $n = 4, 8$, and R = C_6H_5.

In the following stage the oligodiol that is produced is esterified with the methacrylic acid.

Numerous OEAs with $n < 4$ are soluble in alcohols, acetone, esters, aromatic hydrocarbons, chlorobenzene, tetrahydrofuran (THF), dichloroethane, and other organic solvents as well as in vinylic or acrylic monomers. OEAs are not water-soluble.

OEAs can participate in processes of homo- or copolymerization with themselves or with other monomers, olgiomers, or copolymers.

B. Oligocarbonateacrylates (OCAs)

Oligocarbonateacrylate (OCA) synthesis uses, in the main, the method of condensation telomerization. In some patents issued from 1945 to 1947 [20–23] experimental results are presented that prove that carbonic acid forms oligomers containing polymerizable groups of the allylic or acrylic type and one or more carbonate groups. The production of dimethacrylic esters of carbonic acid and of glycol is demonstrated. The process develops in several stages, in the presence of acid catalysts.

In Russia OCAs and oligocarbonatemethacrylates (OCMs) were obtained via condensation telomerization and in solution at moderate temperatures or at the interface [24,25].

A typical OCA and OCM formation reaction is the following:

$$2CH_2{=}C(X)COOROH + Cl{—}CO{—}OR'{—}CO{—}OCl \xrightarrow[-2(R)_3N\cdot HCl]{2(R)_3N}$$

$$CH_2{=}C(X)COORO{—}CO{—}OR'O{—}CO{—}ORO{—}CO{—}C(X){=}CH_2 \qquad (10)$$

HCl acceptor organic amines are used, especially the tertiary ones. Their main role is to enhance the reaction speed and prevent destructive processes.

At the beginning of the process, ionic adducts of tertiary amines with chloroformates of glycol or bisphenol are formed.

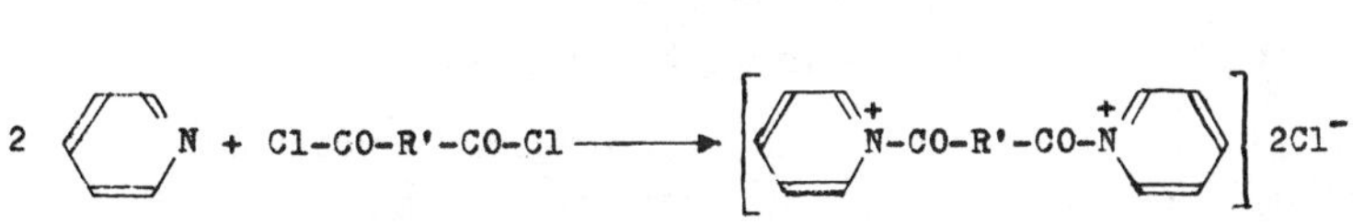

(11)

These adducts represent much more reactive intermediaries than the initial chloroformates. Their reactions with alcohols and glycols form carbonic acid esters and amine chlorohydrates through the following reaction:

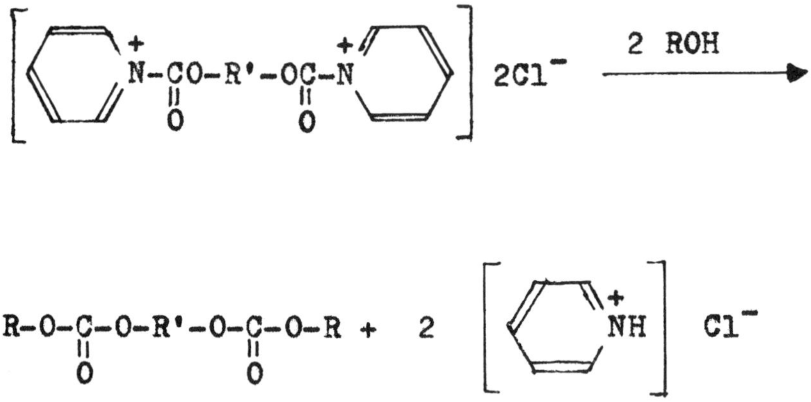

(12)

By synthesis of oligoesters with the general formula given in reaction (10), OCA and OCM are produced with the desired physicochemical properties. With a view to thoroughly achieving the envisaged purposes, the following were particularly taken into account:

Maintaining the ratio of bischloformate or bisphenol and telogene in the vicinity of 0.5.

The use of methylene chlorite as an acceptor of pyridine and telogene, which at room temperature prevents both side reactions and, especially, postpolymerization of oligocarbonates.

Carrying out of the former stage of the process at low temperature (−10 to −5°C). Carrying out this stage at increased temperatures substantially diminishes oligoester formation. In the latter stage, it is desired to work at 40°C (the boiling temperature of the solvents). Under these conditions, a monotone evolution of the reaction and, hence, of the molecular weight of produced oligomers are achieved.

Through the aforementioned methods, more than 20 OCAs and OCMs were obtained with the general formula shown in reaction (10). Most of them contain aliphatic or aromatic polyoxyethylenes or polyoxypropylenes. Likewise, OCAs and OCMs can contain halogens or other organogenic elements [26–28].

In general, OCA and OCM yield technology is based on the following main processes:

Synthesis of oligoesters via ionic adduct yield (first step) and through its condensation with ethyleneglycol acrylic esters (second step)

Filtering of reaction product for chlorohydrate separation

Processing of oligoester solution with salt solution for the removal of tertiary amines that remained unreacted

Azeotrope drying of ester solution in methyl chloride and solvent elimination

OCM synthesis can also be achieved via interfacial condensation telomerization in a single step using glycol chloroformates or bisphenol with monomethacrylic esters of ethyleneglycol in the presence of an inorganic acceptor for HCl (aqueous solution of alkali bases) [29].

The use of aliphatic glycol chloroformates allows the production of OCA and OCM at room temperature and in the absence of organic solvents. The variant yielding OCM based on bisphenol bischloroformates with bischloroformates of aliphatic dihydroxylic compounds in the presence of inorganic acceptors for HCl takes place at temperatures ranging between −10 and 40°C and has been industrialized in Russia.

The first OCM produced industrially in Russia was α,ω-(methacryloiloxyethylenoxycarbonyloxy)ethylenoxyethylene (OCM-2), with the following structure:

$$CH_2{=}\underset{\underset{\displaystyle CH_3}{|}}{C}COO{-}(CH_2)_2{-}O\underset{\underset{\displaystyle O}{\|}}{C}O{-}(CH_2)_2{-}O{-}(CH_2)_2{-}O\underset{\underset{\displaystyle O}{\|}}{C}O{-}(CH_2)_2{-}OOC\underset{\underset{\displaystyle CH_3}{|}}{C}{=}CH_2 \qquad (13)$$

The characteristics of oligoacrylates produced in Russia are listed in Table 2.

To provide stability during OCA and OCM processing, various additives are used. Thus, phenolic derivatives (ionol, diphenylpropane) prevent coloring processes and oxidizing agent effects [30].

OCAs and OCMs show a higher sensitivity than OEAs toward various initiating means. They can easily be polymerized by UV radiation or high-energy radiation (gamma ray, rapid electrons, etc.) [31,32].

Table 2 Physicochemical Characteristics of OCAs and OCMs with General Formula $[CH_2{=}C(X)COO{-}(CH_2)_2O\underset{\underset{\displaystyle O}{\|}}{C}O]_2{-}R^a$

R	Trade name	M_n (determined)	Refractive index	Melting point (°C)
$-(CH_2)_2-$	OCM-1	380	1.4650	—
$-(CH_2)_3-$	OCM-8	380	1.4648	—
$-(CH_2)_4-$	OCM-9	410	1.4649	—
$-(CH_2)_6-$	OCM-10	433	1.4584	38–39
$-(CH_2)_2-O-(CH_2)_2-$	OCM-2	427	1.4670	—
$-(CH_2CH_2O)_2CH_2CH_2-$	OCM-3	—	1.4670	
$-(CH_2CH_2O)_3CH_2CH_2-$	OCM-4	525	1.4682	—
$-CH_2-C(CH_3)_2-CH_2-$	OCM-11	—	1.4642	—
$-CH_2-C(CH_2Cl)_2-CH_2-$	OCM-12	485	1.4842	—
$-C_6H_4-C(CH_3)_2-C_6H_4-$	OCM-5	—	1.5400	—
$-C_6H_4-$ (para)	OCM-6	433	—	67–68
$-C_6H_4-$ (metha)	OCM-7	435	1.5021	—

[a] X = H, CH_3.
Source: Ref. 4.

OCA and OCM crosslinking can also take place in the presence of radicalic initiators. In the main, peroxides and hydroperoxides are used. Cold strengthening is achieved in the presence of some compounds containing transitional metals, for example, cobalt, vanadium salts, or ferrocyan compounds [33].

C. Oligourethaneacrylates (OUAs)

The production of oligoesters with urethanic groups and unsaturated acrylic type and groups is based on the following two reactions:

Interactions of diisocyanates with oligodiols and oxyalkylacrylates [34–36]
Interactions of glycol or bisphenol bischloroformates with diamines and alkyl or alkylarylacrylates that contain chloroformate or amine end groups [37–39].

For the first reaction, oligomers with hydroxylic end groups—for example, oligopropyleneglycol with M_n = 700–2000, oligoethyleneglycoladipateswith M_n = 400–2500, or oligodieneglycols with M_n = 2000–4000—were used [40]. The role of the diisocyanic component is held by aliphatic or aromatic diisocyanates (e.g., 1,6-hexamethylenediisocyanate, 2,4-toluenediisocyanate, and 4,4′-diphenylmethanediisocyanate). Monoacrylic or monomethacrylic esters of ethylene- and propyleneglycol are introduced as unsaturated compounds. For the second reaction, oxyethyleneglycol bisformates, polymethyleneglycols (ethylene-, tetra-, hexamethylene-, neopentyl-) and phenol bischloroformates are introduced. As the aminic component, aliphatic or aromatic polyamines are used; and as the unsaturated component (telogene), aminoalkyl, aminoalkylacryl esters, acrylic or methacrylic acid, or oxyalkylacrylate monoformates are used.

Oligourethane(meth)acrylate (OUM) synthesis through the first reaction can be performed via several procedures. One of these involves the reaction of oligoglycols of different structure and variable molecular weight with diisocyanates (in a NCO:OH ratio = 2:1) in mass or in solution in the presence of butyleneglycol dilaureates [35,40]. Oligoester diisocyanate (macrodiisocyanate) is formed, which then reacts with oxyalkyl(meth)acrylates. If the process is achieved in one step without separation of macrodiisocyanate, unexpected products may form that degrade the final properties of the oligomer.

With a view to eliminating side reactions and obtaining macrodiisocyanate in a pure state, the process can be achieved in two steps; in the first step, macrodiisocyanate separation is achieved, and in the second step, macrodiisocyanate reacts with monomethacrylic ester of ethyleneglycol at room temperature for 6 to 10 h. With this method, in Russia, the OUMs called OUA-2000T were obtained based on ethyleneglycol monomethacrylic ester, 2,4-toluendiisocyanate, and oligooxypropyleneglycol. Using oligodieneglycols, the product called PDI-UAK was obtained.

Another procedure for obtaining OUMs is based on the reaction between oligoethyleneglycols and acrylic monomers containing isocyanic end groups [36]. The process takes place in two stages. In the first one, the following reaction takes place:

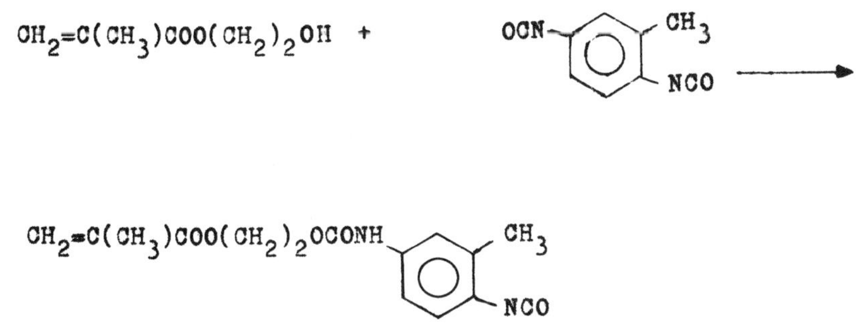

(14)

In the second step, the reaction of the dihydrocyclic compound with chloroform monodiisocyanate in an argon atmosphere occurs, via the following reaction:

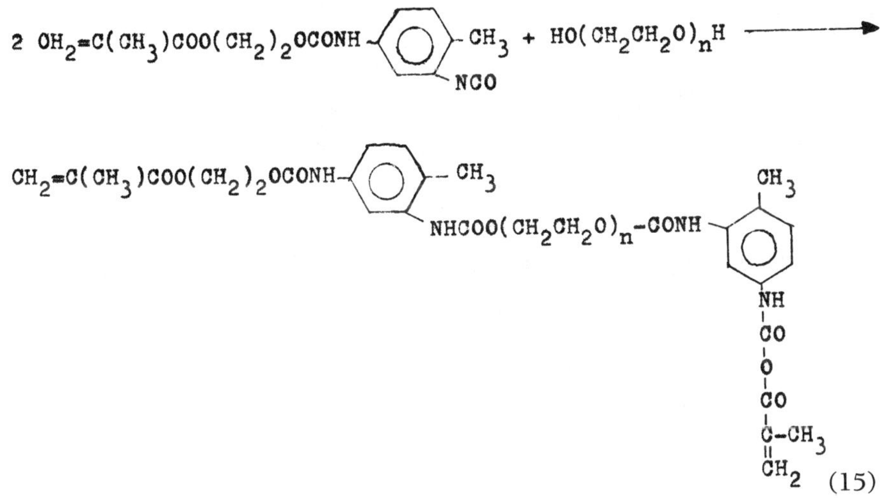

(15)

Table 3 lists the main physicochemical characteristics of OUAs and OUMs produced in Russia.

Some side reactions that can develop during OUA and OUM synthesis give rise to numerous economic and ecological problems. As a consequence, there was a tendency to obtain polyurethanes without using diisocyanates. To reach this goal was possible via the synthesis of polymerizable reactive oligomers containing urethanic groups without diisocyanates as raw materials [41–43]. These oligomers are obtained by condensation telomerization at low temperature of glycol or bisphenol bischloroformates with telogene (2-phenylaminomethacrylate) using the following reaction:

Table 3 Physicochemical Characteristics of OUAs and OUMs Produced in Russia

Trade name	M_n	Refractive index	Melting point (°C)
OUA-550	1,100	1.505	—
OUA-1000	1,550	1.4905	—
OUA-2000	2,550	1.4725	—
OUA-1500	2,100	1.5025	—
OUM-62	660	1.2720	76
OUM-150	730	1.2590	38
OUM-282	890	1.2010	10
OUM-400	970	1.1620	—
OUM-600	1,180	1.1180	6
OUM-1000	1,580	1.1820	19
OUM-1500	2,090	1.1990	41
OUM-2000	2,510	1.2020	43
OUM-3000	3,700	1.2030	56
OUM-6000	6,920	1.2050	63
OUM-15000	14,900	1.2080	66
OUM-2F	570	1.5300	—
OUM-3F	630	1.5270	—
OUM-4F	670	1.5228	—

$$2\ CH_2{=}C(CH_3)COO(CH_2)_2{-}NH{-}C_6H_5 + Cl{-}COO{-}R{-}OOC{-}Cl \xrightarrow[-\ 2HCl]{2.2\ py}$$

$$CH_2{=}C(CH_3)COO(CH_2)_2{-}\underset{\displaystyle C_6H_5}{\underset{|}{N}}{-}OOC{-}R{-}OOC{-}\underset{\displaystyle C_6H_5}{\underset{|}{N}}{-}(CH_2)_2{-}COO(CH_3){-}C{=}CH_2 \qquad (16)$$

where R = $—(CH_2CH_2O)_nCH_2CH_2—$ and n = 1–3.

The reaction takes place in solution (solvent, methylene chloride) in the presence of pyridine or other tertiary amines, which are acceptors for the HCl generated by the reaction.

This procedure has been industrialized in Russia. The products shown in Table 3 have been obtained under the trade name OUM-F.

OUMs can also be obtained by oligocondensation at the interface in water–organic solvent systems in the presence of alkaline bases [42,43].

D. Oligooxyalkyleneacrylates (OAAs)

The original method of oligooxyalkyleneacrylate (OAA) synthesis was based on the principle of cationic oligomerization of cyclic ethers in the presence of acrylic compounds as chain transfer agents. The first compounds obtained were based on

THF [44], dioxolan, and trioxolan [45] in the presence of methacrylic acid via the following reaction:

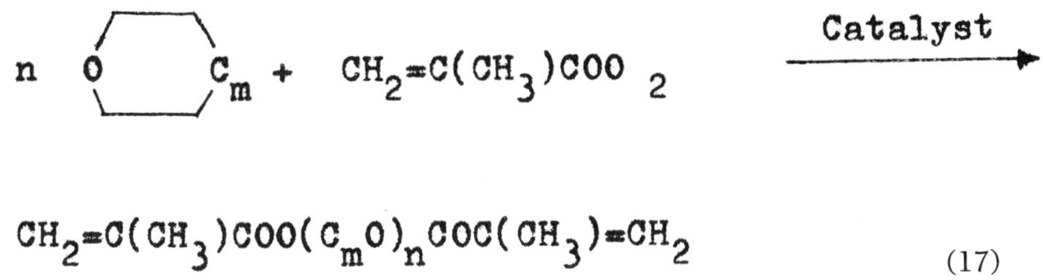

(17)

where

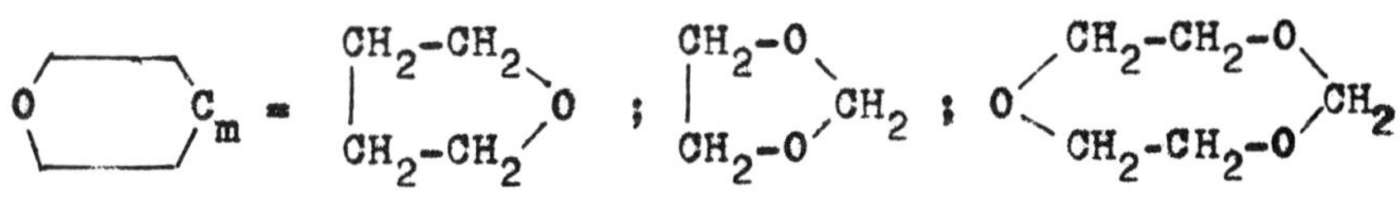

(18)

Oligomerization takes place in an inert atmosphere. In the reaction mixture, which contains cyclic ether and telogene, under stirring at 5°C, in mass or in solution, the catalyst is introduced, which in the case of THF is $SbCl_5$ and in the case of tri- or dioxolane is sulfuric acid. Then the reaction mixture is kept for 24 h at 19 to 21°C. After organic phase separation, this is washed with water until neutral, after which the solvent is eliminated by distillation and drying. The products obtained have M_n of 500 to 1000.

Also interesting is THF copolymerization with propylene oxide in the presence of $SbCl_5$ and methacrylic anhydride [46–48]. This reaction is the basis of the one-step synthesis of OAA without use of glycols. The process mechanism [49] shows that methacrylic anhydride does not participate in the initiation phase. Methacrylic groups appear in the polymer chain as a result of macroradical transfer.

E. Oligoalkylenoxyphosphinatacrylates (OFAs)

This oligomer category gained prominence from its use in fireproof material formuation. Oligoalkylenoxyphosphinatacrylates (OFAs) are produced by oligocondensation of heptylic- and dodecylphosphinic acid chloroanhydrides with oxyalkylmethacrylates in the presence of pyridine as a HCl acceptor. The following reaction takes place:

$$RPOCl_2 + 2CH_2{=}C(X)COOCH_2CH_2OH \xrightarrow[-2HCL]{Py}$$

$$\underset{\displaystyle R}{CH_2{=}C(X)COOCH_2CH_2O\overset{}{\underset{|}{P}}(O)OCH_2CH_2OOCC(X){=}CH_2} \qquad (19)$$

where X = H, CH_3 and R = C_7H_{15}, $C_{12}H_{25}$.

OFAs can also be obtained by condensation telomerization of two bifunctional compounds: heptylphosphinic acid dichloroanhydride, diethyleneglycol, and oxyethylmethacrylate. Oligoesters with a phosphor content can also be produced from diphenylolpropane [50].

F. Oligosiloxaneacrylates (OSAs)

Oligomers with a silicon content are of interest because they can combine inorganic and organic polymer properties advantageously.

Silicoorganic compounds with methacrylic groups linked by aliphatic radicals are produced via the reaction between bis(chloromethyl)tetramethylsiloxane and the potassium salt of methacrylic acid. The reaction takes place at high temperature, and process yield does not exceed 50% [51,52].

Oligomethyl(methacryloylmethylene)dimethylsiloxane synthesis is performed using the following general formula:

$$\begin{array}{l}(CH_3)_2\underset{\displaystyle CH_2OOCC(CH_3){=}CH_2}{\underset{|}{Si}}O\{Si(CH_3)_2O[Si(CH_3)_2O]_n\}_m{-}Si(CH_3)_3\end{array} \tag{20}$$

This was described by Bulatov et al. as early as 1966 [53].

Linear oligomers with silicon and methacrylic end group content can be obtained via the reaction between α,ω-dichlorooligosiloxane and oxyethylenemethacrylate in the presence of tertiary amines as acceptors for HCl. The process takes place in benzene. The main reactions are as follows [54]:

$$2CH_2{=}C(CH_3)COO(CH_2)_2OH + Cl{-}\left(\begin{matrix}R\\|\\SiO\\|\\R\end{matrix}\right)_n{-}\begin{matrix}R\\|\\Si\\|\\R\end{matrix}{-}Cl \xrightarrow[-2(R)_3N\cdot HCL]{2(R)_3N}$$

$$CH_2{=}C(CH_3)COO(CH_2)_2O\left(\begin{matrix}R\\|\\SiO\\|\\R\end{matrix}\right)_n(CH_2)_2OOCC(CH_3){=}CH_2 \tag{21}$$

where R = H, CH_3, n = 3–190, or R = C_6H_5, CH_3, n = 2–4.

II. OLIGOESTERACRYLATE COMPOSITES

Recently, compositions made up of OEAs and various linear polymers have gained importance. In the initial stage, this system is similar to that made up of a plastified polymer [55–62]. Later on the system is transformed because of unsaturated zones found in the oligomer and the polymer. Ideally, the polymerization process is developed in a heterogeneous system made up of the dispersed oligomer and the polymeric dispersion medium.

Analysis of prognoticative literature [63,64] shows that, at least until the year 2000, the consumption of oligomer–polymeric composites will be practically constant. Furthermore, it is foreseen that approx. 40% of polymer industry products will apply mixtures of reactive oligomers and polymers (ROP).

A. Physical Chemistry of ROP Systems

The final properties of materials obtained by ROP solidification can be guided via the nature, molecular weight, and mutual compatibility of the two components. Yet one cannot establish a direct correspondence between a certain final property of the mixture and, for instance, the value of initial oligomer molecular weight.

The behavior of ROP mixtures during a solidification process can be interpreted by means of the temporary plasticizing theory suggested by Berlin [65] and later on detailed by Frenkel and Meshikovskii [66]. In essence, these workers show that by adding a reactive oligomer in the mass of a polymer the initial effect is plasticization. Under these conditions the resulting mixture can be more economically processed, because oligomer addition diminishes polymer viscosity and mechanochemical degradation intensity. In the following stage (curing, strengthening) the oligomer undergoes chemical changes by which it is linked to polymer matrix. Thus, a new product emerges with improved physicochemical properties by simplified processing with low energy consumption. End product properties are determined by the entirety of thermodynamical characteristics and by chemical–colloidal properties of the initial system; this accounts for the lack of direct correspondence between a certain property of the initial oligomer (or of the polymer) and a characteristic of the final material.

Even in 1977 Berlin [67] recognized that interpretation of ROP mixtures was based on the principle of temporary plastification was outdated. It is indeed quite clear that without knowing the thermodynamics of ROP systems, the kinetics of the physicochemical changes that take place in these systems, and the laws governing the manifestation of macroscopic properties of these systems in terms of their microscopic structure it will not be possible to realize a systematic scientific approach to formulating ROP mixtures.

In contrast to oligomer and polymer physicochemistry, as well as that of mixtures of nonreactive oligomers and polymers, the physical chemistry of ROP developed more slowly. Knowing the initial-mixture physicochemical properties and the morphological and chemical changes that take place during processing directly guides the establishment of laws that govern these systems.

Conversion of a liquid ROP mixture into a solid final product passes through three important stages: liquid ROP with high viscosity, intermediate solidifying mixture, and end product (polymer–polymer solid mixture).

As regards reactive oligomer–polymer liquid systems, over the past years much data concerning mutual solubility of the two components has been developed [68–76]. In recent years several insights have been gained regarding the dynamics of the process of achieving thermodynamic equilibrium in ROP systems [77–79]. Thermodynamic equilibrium in ROP systems is supposedly established step by step and in different locations. This hypothesis is supported by the apparent morphological nonhomogeneity between various phases of an ROP system [80–83]. This assumption is based on a series of earlier data concerning the abnormalities of OEA solidification rates [84–86]. Additionally, other studies indirectly confirmed the existence in liquid OEA of some supramolecular formations the space structure of which can be advantageous or unfavorable for the solidification process [87]. The lack of clear proof generated suspicion of the presence of some molecular association in the systems made up of liquid oligomers and of their effects on the kinetics and crosslinking processes of OEA and ROP systems. It is still recognized that the existence of some ordered areas in liquid oligomers represents the decisive factor in the macroscopic properties of solid materials obtained by polymer melt solidification or by liquid oligomer or ROP mixture solidification [88]. The behavior of these systems has been generalized by means of *sibotactic model* (from the Greek *sibos* = ark), which succeeded in scientifically substantiating the dependence on temperature and time of viscosity, thermal conductivity, and diffusion in systems made up of polymer or metal melts [89]. This model has been subsequently applied to the interpretation of liquid oligomeric system properties [90,91]. To understand properly the description of a liquid system by means of this useful model, one must more accurately define the principles on which the sibotactic model is based. The model was initiated by the turn of this century and developed in the 1970s by Frenkel [92].

By sibotaxes one understands the microzones of a liquid that have a certain type of spatial arrangement of particles. The term *cluster* is similar to sibotaxis but does not specify the spatial nature of the arrangement. Sibotaxes are distributed at random within the system and are unstable. The duration of their existence is determined by the composition and temperature of the system, as well as by intermolecular links available in the system.

Sibotaxes look very much like molecular association or like solvates, notions used to describe bicomponent systems. Due to the thermal motion of molecules, sibotaxes are not precisely delimited; they do not evince a physical surface whose crossing is marked by a dramatic change of some properties of the system. The disappearance of sibotaxis is not linked to certain parameters of state of the system. Thus, the sibotactic model reflects realistically the microhomogeneous structure of a monophasic liquid.

The sibotactic model clarified numerous points of confusion in the analysis of phase diagrams of oligomeric and ROP systems [93,94].

Let us analyze via the sibotactic model some data regarding polymerization of some ROP systems (i.e., the OEA–rubber system) [90,91]. Suppose that after polymer–oligomer contact, the mixing of the two components occurs via a diffusion process. The driving force of this process is the tendency to minimize the chemical potential of the system. Further on, the equilibrium state of the system is achieved by means of sibotaxis. This process takes place without changing macroscopic parameters of the state of the system and without modifying the chemical potential gradient. The contradiction between the change of some properties of the system with maintenance of constant thermodynamic functions can be explained by the fact that formation and evanescence of sibotaxes modify in various ways certain components of system entropy without changing the total entropy of the system.

The sibotactic nature of OEA has been confirmed by data obtained for some systems made up of OEA and unreactive oligomers [64].

It is also obvious that the presence of sibotaxes influences the process of ROP system crosslinking [95]. Thus the presence of these formations permitted explanation of the absence of bimolecular recombination of macroradicals during radicalic polymerization of oligocarbonateacrylates. The presence of sibotaxes in liquid OEA increases efficiency of the polymerization process and explains the coexistence in the initial stages of the solidification process of network (rigid zones) and amorphous and friable zones [96]. Experimental data obtained from the behavior of some mixtures made up of OEA and poly(vinyl chloride) [97] and OEA–rubber [98] point to the same conclusions.

Infrared spectra and other optical methods indicated that sibotaxes comprise 6 to 40 OEA molecules [99]. As regards subsequent and final stages of OEA or ROP solidification processes, there is, as far as we know, only one experimental result [100], indicating that the degree of liquid phase ordering is transmitted to the solid material obtained. This observation has special technological implications that require some general comments. Integral passing of OEA blocks (without chemical changes) into the structure of the solid material leads to some novel aspects of polymer processing technology. In other words, it means the introduction into this domain of some elements of chemical cybernetics. Concretely, it is about the introduction into the raw materials (especially OEAs) of some additional informational elements with a view to providing more accessible conditions for the processing technology.

Each technological technique for processing an oligomer or oligomer–polymer mixture can be characterized by the following relation:

$$IT = \text{constant} \tag{22}$$

where I represents the amount of information contained in the raw materials, and T represents the energetic consumption of the processing. Introduction into raw materials of some additional informational quantities (i.e., increased value of I) will cause T to diminish. The information contained in raw materials is represented by their degree of supramolecular organization.

To put it in a nutshell, we can say that the technological cycle of obtaining solid materials from liquid ROP systems comprises a certain sequence of physical and chemical processes the aim of which is to attach to the finite product enough information to provide the complex of desired physicochemical properties. The use of raw materials with high information content diminishes the difficulty and energetic consumption of processing. From this point of view, we are obliged to formulate initial systems by means of some oligomers with a certain degree of structural organization and choose those technological processes that allow us to preserve these organizational forms and produce their complete transposition into the architecture of the solid material ultimately obtained.

B. Uses of Oligoesteracrylate Composites

Various industrial domains use OEAs of various chemical compositions, based on glycol, glycerine, pentaerythrite, and phthalic or adipic acid. For the rubber industry, a special importance is held by polyfunctional OEAs 7-1 and 7-20 (Table 1) [101,102]. Yet the use of polyfunctional OEAs has some drawbacks. These products in their commercial forms include ca. 35% organic solvents. This prevents the use of these technologies in an open system. If the solvent is removed, these OEAs became too viscous and unstable (they self-polymerize) and create difficulties in dosing and transport. As a result, the possibility of using some tetra- and polyfunctional OEA composites was considered [103,104]. Under these conditions, it was found that the properties of resulting composites are not additive and a synergetic effect occurs [105–108]. Likewise, the cured products show lower properties than those yielded with the individual OEAs [109].

Yet the use of OEA composites allow achievement of some cured products with desired properties. Cured products achieved with OEA composites show very good adherence to metal surfaces. This is an advantage yet a technological problem, too. However, the problem has been solved by introducing triethanolamine into the composite [110].

The use of OEA for rubber crosslinking opens new possibilities for the use of all kinds of fillers [111]. Inactive fillers intensify the strengthening process of rubber by means of OEA. Active filler materials participate, however, in the strengthening process, and OEA contributions to improved physicochemical properties of the cured product become less important.

OEA–rubber system solidification is achieved in the presence of physical and chemical radical initiators [112,113]. It is difficult to determine a general criterion for selecting the best initiator system for a given rubber–OEA mixture [114–120]. From the experimental data obtained, it has been discovered that peracids and chloroorganic compounds show the highest efficiency for the crosslinking process.

Of interest is the use of OEA as a binder for reinforced plastics. In this respect, either polyfunctional OEA composites and tetrafunctional OEA, or OEA mixtures with other reactive oligomers might be used [121].

Reinforced materials based on glass fiber, carbon fiber, and boron fiber have been produced in which the binder was made of OEA [122–130].

OEA-based composites are used for the production of electrical insulation, elastic concretes, and abrasive materials and for some medical purposes [131]. By means of OEA, copolymers can be produced that provide hydrophilic membranes permeable to urea and water, some biologically active polymers, contact lenses, and other medical materials [132–135].

Hydrogels are macromolecular materials with a water content of 30 to 90%. They provide fairly good biocompatibility, being adequate materials for catheter sutures, blood detoxificants, artificial sensors, ophthalmologic protection (contact lenses, artificial cornea), conjunctive tissue substitution, burn therapy, dental implants, aesthetic surgery, and immobilization of bioactive agents.

Of remarkable significance in hydrogel systems are anionic vinylic monomers (derivatives of acrylic acid) and β-hydroxyethyl methacrylates with crosslinking properties (ethyleneglycol dimethacrylate derivatives).

The need to avoid trauma produced by surgical sutures focused attention on the use of adhesives for tissues, in the form of cyanoacrylates and acrylic resins [138].

In ophthalmology, the use of acrylic resins and acrylic hydrogels for the manufacture of contact lenses have been mentioned [138].

I. Use in Dentistry

Acrylic resins (ARs) are the most widely used materials in stomatology. They were initially used only as materials to seal cavities in front teeth, where amalgams are not aesthetically desirable. ARs did not succeed with silicate cements.

Classical poly(methylmethacrylate)-type AR properties are far from those necessary for a definitive filling material. They do not adhere to hard dental tissues. As a result their receipts included monomers (methacrylic acid) to improve adhesion. At the same time, comonomers with bulky substituents have been used, as well as special application techniques to reduce volume shrinkage by polymerization. Such a comonomer is cyclohexyl methacrylate, which halves volume shrinkage. Under these conditions, the ARs yielded are more fragile than the classical ones, and their thermal dilatation coefficient is higher. Due to the residual monomer (2 to 5%) ARs have a toxic effect on dental pulp.

Masuhara's investigations [139,140] yielded in 1964 Palakav acrylic resin (Kulzer Co., Hamburg, Germany), then other similar types: Orthomite adhesive (Rocky Mountain Dental Products Co., Richmond, VA), meant for stomatology. The commercial sets contain monomer (methyl methacrylate), catalyst (tri-*n*-butyl boron), acid agent (PO_4H_3, 65%), and coupling agent (3-methacryloyloxypropyl-1-trimethoxysilane). The main innovation in these systems is the compound tri-*n*-alkyl boron, substituting for the organic peroxides.

The new initiators decompose into free radicals in the presence of water, so that resin adhesion improves itself, since total removal of water traces on dental surfaces is not required.

Masuhara [141] also claims the production of composite acrylic resins (CARs). Yet despite all efforts, these types of ARs have not become useful stomatological materials.

Diacrylic resins (DARs) appeared from the need to eliminate some flaws of restorative materials viz. amalgams, silicate cements, and ARs. These materials, besides their good qualities, also have disadvantages: surface nonuniformity and color instability over time. DARs have some advantages: small concentration at hardening, high physicomechanical strength, chemical stability, and adhesion. These materials, too, have dentine toxicity.

The earliest types of DARs appeared in the period between 1964 and 1972: Addent and then Addent-35 in 1964, Addent-12 and Dakor in 1966, Concise and Addent XV in 1967, Adaptic, Blendant, Surgident, and DFR in 1968, Epoxylite and HL-72 in 1970, Enamelite in 1971, and Restodent and Prestige in 1972. These materials have gained a place in stomatological practice and succeeded in eliminating to a great extent silicate cements of ARs. Table 4 lists the most important DARs used in adhesive techniques. The bulk of commercial DARs are composites 15 to 30% by volume organic phase and 70 to 85% inorganic phase. The adhesion between the two phases is achieved by a coupling agent. The organic phase also contains a mixture of monomers and 5% additives—initiators, promoters, polymerization inhibitors, photosynthesizers, absorbents, dyestuff, and so forth.

Commercial products are generally of bicomponent form: liquid–powder or paste–paste. The composition of the two components differs from product to product.

Various diacrylic monomers have been synthesized by means of the reaction between methacryloyl chloride and aliphatic and aromatic diols:

$$2CH_2{=}C(CH_3){-}COCl + HO{-}R{-}OH \rightarrow$$

$$H_2C{=}C(CH_3){-}COO{-}R{-}OOC{-}C(CH_3){=}CH_2 \qquad (23)$$

where R = $-(CH_2)_4-$, $-C_6H_4-$, $-O-CH_2-O-C_6H_4-O-(CH_2)_4-O-$.

Aliphatic diacrylates have been used as the crosslinking agents of monoacrylates.

Table 4 Commercial Types of DARs

Producer	Trade name	Observations[a]
Spofa Dental, Praha, Czech.	Evicrol	C
3M Comp., Detroit, U.S.	Addent 12, 35, XV:	C
	Concise, Bond System, Sylar	C
Lee Pharmaceuticals, Cleveland, U.S.	Expoxylite HL-72, CBA-9080, Enamelite	C
	Restodent, Prestige, Smile, Epoxylite 9075	S
Johnson and Johnson	Adaptic	C
L. D. Caulk Co., Chicago, U.S.	Dakor	C
	Nuva-Fil, Nuva-Seal	S, UV
Amalgamated Dental Co., Ltd., Manchester, England	Alpha Sea	S, UV
Ker Sybran Co., London, England	Blendent, Surgident, DFR	C
Kulzer Co., GmbH, Hamburg, Germany	Estic	C
Kuraray Co., Ltd., Kyoto, Japan	Clearfil Bond System F	C
Espe GmbH, Dortmund, Germany	Espe Fissure Sealent	C, UV
De Trey GmbH, Berna, Switzerland	Cosmic	
Pennwalt SS White Corp., Boston, U.S.	Cervident	C
Vivadent, Lichtenstein, Lichtenstein	Isopast	CO

[a]C = composite resins filled with inorganic compounds; CO = composite resins filled with organic compounds; S = simple resins; UV = UV curable resins.

Aromatic diacrylates, for example, *p*-bis(hydroxyalkyl- or hydroxybenzeneesters dimethacrylates, yield materials with very low water absorption, exceptional transparency, and high hardness and scratching resistance.

Modern resin-based restorative materials used in stomatology originated with the invention of the so-called Bowen's monomer and the introduction of composites [142]. This monomer is known as BisGMA, a label easier to use than chemical name of 2,2-bis[4-(2-hydroxy-3-methacryloyloxypropoxy)-phenyl]propane. Today, both unfilled and composite resins are widely employed in dentistry—as binding agents, pit and fissure sealants, direct filling materials, orthodontic adhesives, and resin cements [143].

Many research efforts have been directed toward improving the properties of DACs, resulting in the development of various monomers and of adhesive compositions that harden when exposed to UV light.

The major components of the adhesive compositions were stated to be three parts by weight of the BisGMA and one part by weight of methylmethacrylate monomer. In the foregoing BisGMA dental restorative compositions, the low molecular weight methylmethacrylate serves essentially as a reactive extender or diluent to reduce the viscosity of the compositions, whereby they can be conveniently used in dental applications. The benzoyl methyl ether is employed in these compositions as a photosensitizer. Another substance typically used in the dental restorative compositions is benzoyl peroxide, or a similar compound, which serve as a free radical initiator.

The monomer BisGMA may be obtained by either the reaction of bisphenol A (BPA) and glycidyl methacrylate (GMA), or the reaction of the diglycidyl ether of bisphenol A (DGEBPA) and methacrylic acid (MA). The following reactions take place:

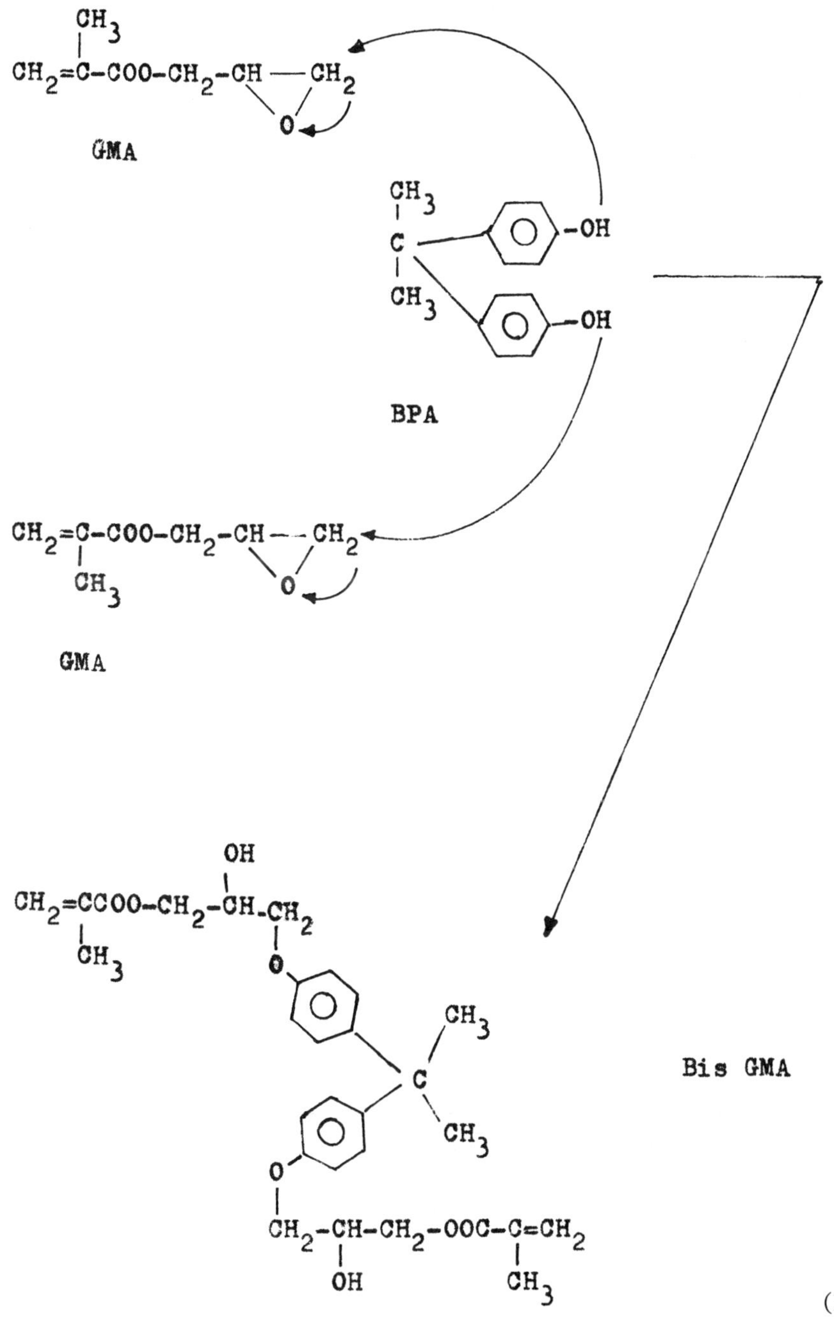

(24)

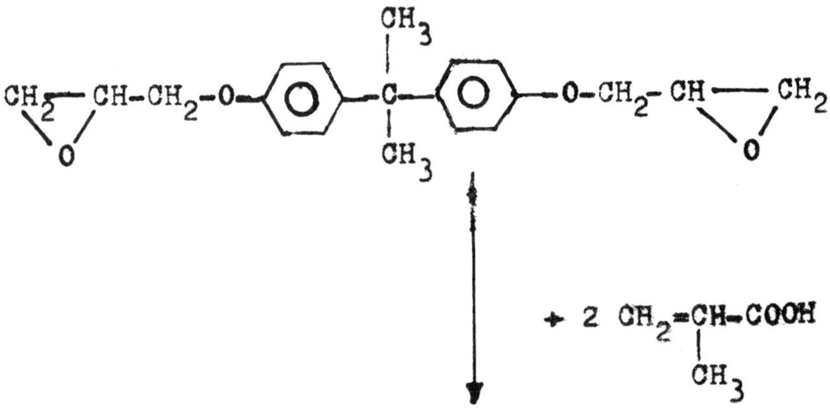

Bis GMA (25)

Many proprietary unfilled and composite dental resins contain mixtures of BisGMA and triethylene glycol dimethacrylate (TEGDMA) as the diluent monomer [144,145]. Thus, in the chromatogram of the Evicrol resin, the dimethacrylate monomers TEGDMA and BisGMA were identified. However, the chromatogram contains two other peaks with longer retention time in high-performance liquid chromatography but with the same UV254/UV280 absorbtion ratios as the BisGMA peak [146]. To identify these unknown peaks the authors took into account that DGEBPA, employed in the synthesis of BisGMA (reaction 25), is the main component of some commercial epoxy resins with low molecular weight ($\bar{M}_n$ = 360–400), usually composed of three individual oligomers ER_x (x = 1–3), the first oligomer ER_1 being DGEBPA. By reacting such a low molecular weight epoxy resin with MA, a mixture of the following oligomers is obtained:

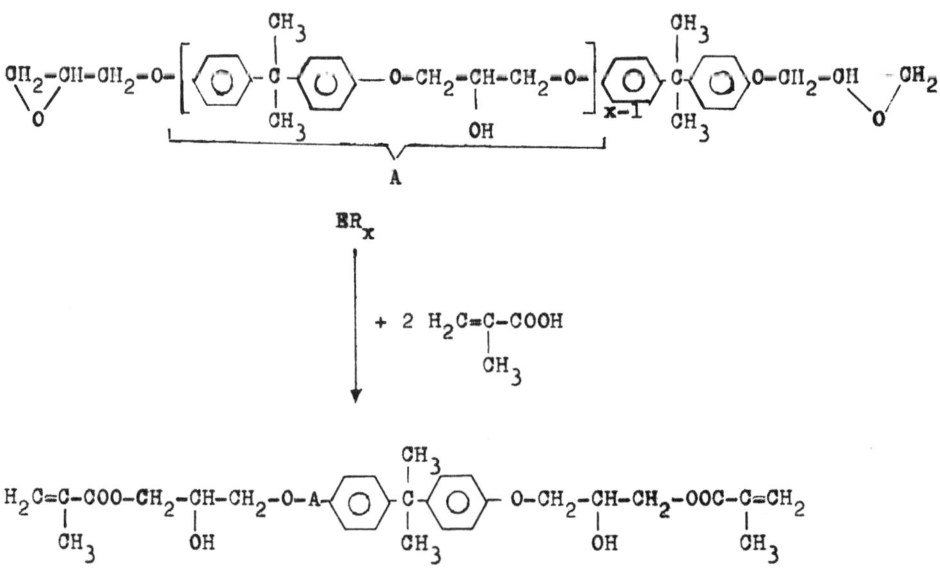

Bis GMA_x (26)

The reaction product contains three BisGMA type oligomers, denoted $BisBMA_x$ ($x = 1-3$), corresponding to the ER_x oligomers. The first oligomer $BisGMA_1$ is in fact the "true" BisGMA monomer. The composition of the reaction product (84.9% $BisGMA_1$ + 13.8% $BisGMA_2$ + 1.3% $BisGMA_3$) matched the composition of the initial epoxy resin.

In the synthesis of the BisGMA monomer by either of the two routes, the ring opening of the epoxy ring proceeds upon attack of the C_1 atom ("normal" ring opening, see reactions (24), (25), so that a "normal" (linear) structure is formed at the end of the molecule. However, attack on the C_2 atom ("abnormal" ring opening) is also possible, so that an "isomeric" (branched) structure is formed at the end of the molecule. Consequently, three isomers may appear: *n,n*- ("normal" end groups at both sides of the molecule), *n,i*- (a "normal" group at one side and a "isomeric" group at the other side of the molecule), and *i,i*- ("isomeric" end groups at both sides of the molecule:

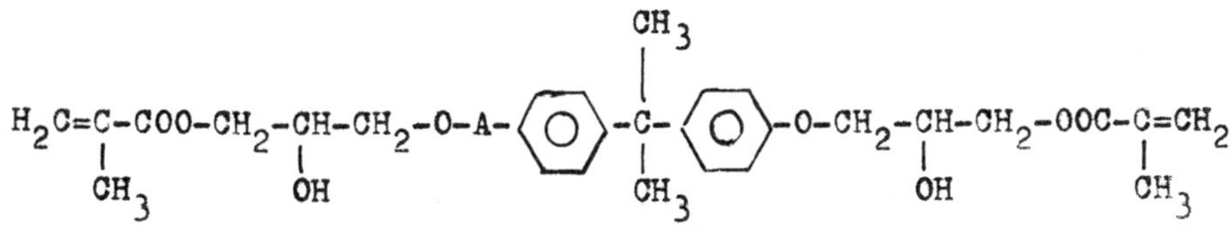

n,n-Bis GMA (27)

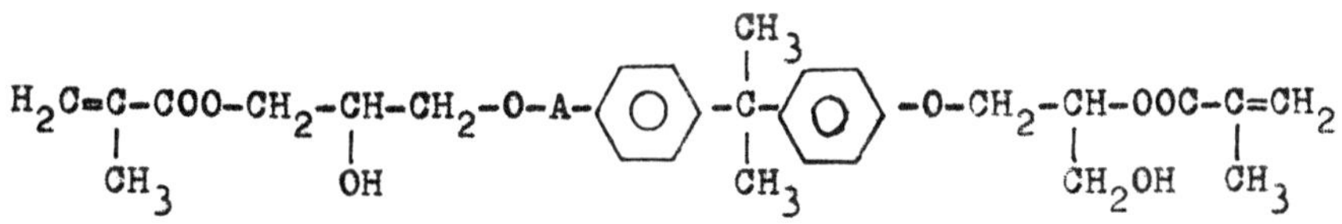

n,i-Bis GMA (28)

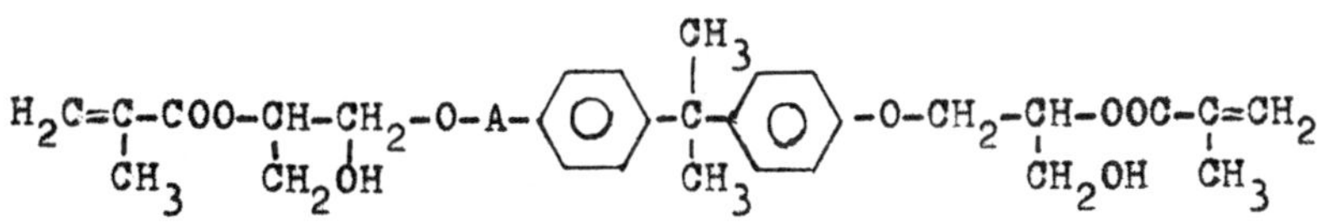

i,i-Bis GMA

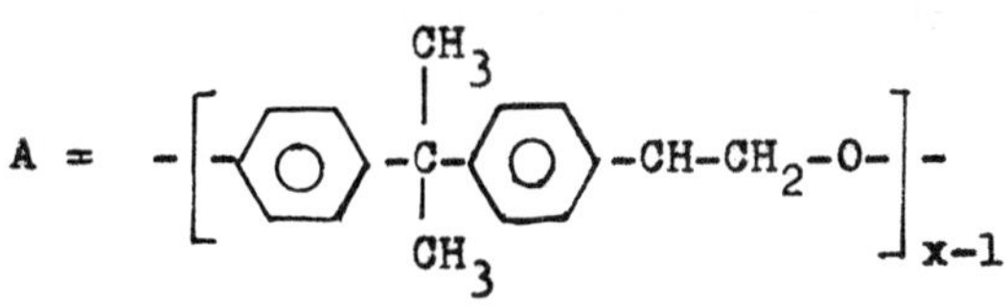

(29)

The main reaction, however, is the "normal" ring opening of the epoxy group (attack on the C_1 atom), so that the reaction product contains mainly *n,n*-isomer.

Due to the diversity of the diacrylate types (aliphatic, alycyclic, and aromatic) and multiple possibilities of material formulation, it is possible, at least theoretically, that a wide range of resins can be obtained by directing the properties of the products toward those required for use in stomatology. The main concepts of this molecular engineering can be summarized as follows:

Passing from linear to crosslinked structure gives chemical resistance, dimensional stability, higher softening point, and the highest content in inorganic filler that can be included.

An increased degree of crosslinking gives increased chemical stability.

Substitution for the methylenic groups with the ether ones increases monomer fluidity.

Elimination of secondary hydroxylic groups increases resistance to moisture.

BisGMA monomer has a series of disadvantages: high viscosity, chemical instability, and hydrophilic character are the main shortcomings. While the foregoing dental restorative compositions are useful, a persistent problem that arises in practice is inhibition caused by the presence of oxygen. Because of this inhibition by oxygen, the desired complete hardening of the resin to the surface to which it is applied is not obtained and, instead, a tacky surface is produced [147].

The attempt to use diacrylic derivatives has achieved a substantial improvement of diacrylic resin properties. Aromatic diacrylates, for example, dimethacrylate-bis(hydroxyalkyl)-bisphenol A; with the structure

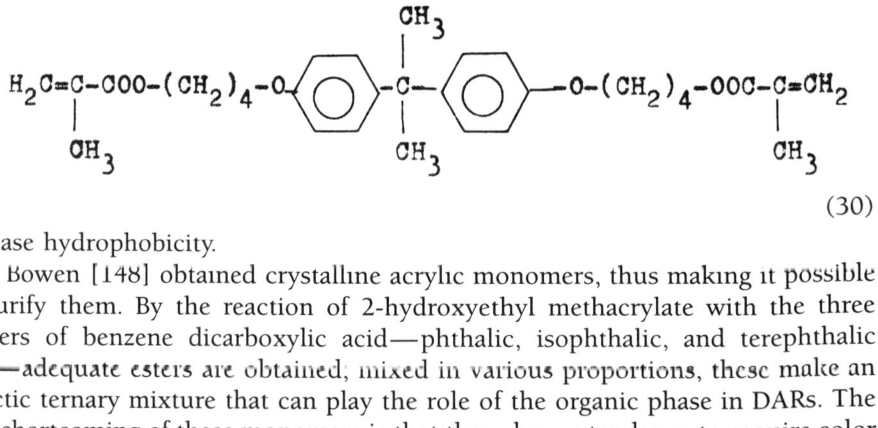

(30)

increase hydrophobicity.

Bowen [148] obtained crystalline acrylic monomers, thus making it possible to purify them. By the reaction of 2-hydroxyethyl methacrylate with the three isomers of benzene dicarboxylic acid—phthalic, isophthalic, and terephthalic acid—adequate esters are obtained; mixed in various proportions, these make an eutectic ternary mixture that can play the role of the organic phase in DARs. The only shortcoming of these monomers is that they show a tendency to acquire color in the presence of the polymerization accelerators (*N*,*N*-dimethyl-*p*-toluidine).

Isomer dimethacrylates (*o*-, *m*-, and *p*-)bis(hydroxyethyl)-dihydroxybenzenether,

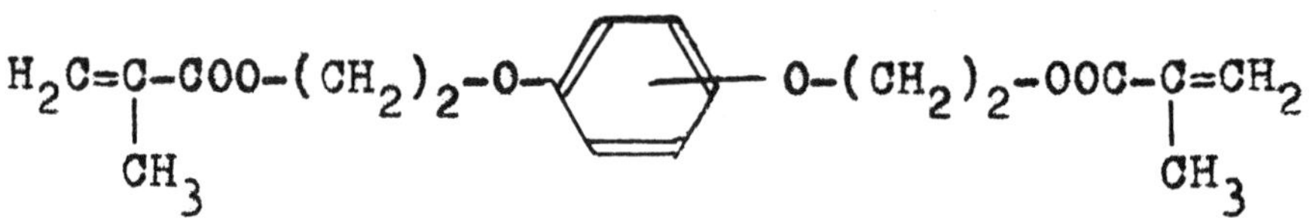

(31)

can make an eutectic ternary mixture that remains as a liquid at 5°C and does not acquire color in the presence of polymerization promoters [149–151].

The synthesis in 1949 of methyl-2-cyanoacrylate and the finding that it can, when applied in a thin layer, even in the presence of water, polymerize in a few seconds, adhering to the support, marked the beginning of the development of a new class of adhesives. Initially, 2-methylcyanoacrylate,

$$CH_2{=}C\begin{array}{l} \diagup CN \\ \diagdown COOCH_3 \end{array} \tag{32}$$

has been tested as an adhesive for soft tissues [152].

The disadvantages of its use (polymer hydrolisis, with formation of toxic products) became obvious, so that higher alkyl 2-cyanoacrylates (ethyl-2-cyanoacrylate, propyl-2-cyanoacrylate, and 1-butyl-2-cyanoacrylate) have been tried, and finally octyl-2-cyanoacrylate or fluorurated cyanoacrylates.

Unlike other adhesive resins, in which polymerization is performed by a radicalic mechanism, cyanoacrylate resins (CNAs) polymerize anionically, according to the reaction

$$CH_2{=}C\begin{array}{l} \diagup CN \\ \diagdown COOR \end{array} \xrightarrow{A:^-} + \; CH_2{-}C:^-\begin{array}{l} \diagup CN \\ \diagdown COOR \end{array} \xrightarrow{A:^-}$$

$$A{-}CH_2{-}C:^-\begin{array}{l} \diagup CN \\ \diagdown COOR \end{array} \xrightarrow{+CNA} A{-}CH_2{-}\underset{\substack{| \\ COOR}}{\overset{\substack{CN \\ |}}{C}}{-}CH_2{-}\underset{\substack{| \\ COOR}}{\overset{\substack{CN \\ |}}{C}}:^-$$

$$\xrightarrow{+(n-2)CNA} A{-}(CH_2{-}\underset{\substack{| \\ COOR}}{\overset{\substack{CN \\ |}}{C}}{-})_n{-}H \tag{33}$$

The process is initiated by an electron donor ($A:^-$), which explains the action of water traces on the CNA layer applied on the surface. This mechanism presents a special advantage for adhesive joints on enamel and dentine surfaces, in vivo, a situation in which water removal, quite necessary for other adhesives, is very difficult. The polymerization reaction is promoted in a basic medium and inhibited in an acid medium. Another characteristic of CNAs, arising from the peculiarities of the synthesis process, is their capacity to bind chemically, by covalent links, to proteins, during polymerization. It was found out that in the presence of blood or of protein solutions (10 mg/ml), CNAs polymerize on the spot [153].

Beech [154] points out that, because of the anionic character of CNA polymerization, the product obtained adheres to dental surfaces. CNA polymerization can be initiated by the NH_2 or OH groups in proteins. Since the initiator remains bound to the monomer, the polymer remains bound to proteinic support. Thus, because of the outstanding importance of organic phase in dentine, this hard tissue represents an exquisite support for CNAs. NH_2 and OH groups in collagen (the main component in dentine organic phase) amino acids represent active positions of CNA polymerization initiation, so that the polymer formed is chemically bound to dentine surfaces. That is why poly(cyanoacrylates) are used as cavity liners.

Although the first utilization of CNAs as adhesive for dental tissues resorted to methyl cyanoacrylates (in the United States) and ethyl cyanoacrylate (in Japan), the results were satisfactory.

The need to resume the application many times prevented the commercialization of methyl and ethyl cyanoacrylates for fissure sealing. A reduced capacity to adhere to enamel, the diminished resistance of the material to hydrolysis, and toxicity of products of hydrolysis accounts for the refusal in 1974 of the application of these materials by the Council on Dental Materials and Devices. To this the scarcity of long lasting chemical and biological investigation was added [155]. A reduced resistance to impact is also notable, and in some cases their adherence to dentine has not been satisfactory [156].

2. Medical Uses

Synthetic polymers have long played a relatively important role in present-day medical practice. Their potential, either as prosthetic materials or as constituents for pharmaceutical uses, has become increasingly apparent. Presently, pharmaceutically active polymers, or macromolecular drugs, constitute a well-established research field in medical chemistry.

Drugs are rarely, if ever, administrated to patients in an unformulated state. The vast majority of available drugs that are potent at the milligram or microgram level could not be presented in a form providing an accurate and reproducible dosage unless mixed with variety of excipients and converted by controlled technological processes. It has been long known that the effect of an active drug substance may depend on the way in which it is formulated. The majority of systematically acting drugs are administrated orally and therefore the drug must traverse certain physiological barriers, including one or more cell membranes. In this way the effects are often greatly weakened; on the other hand, parenteral administration, especially in the case of an important side effect or a strong activity, may lead to deplorable results.

Most pharmacological agents are low molecular weight compounds that readily penetrate into all cell types and are often rapidly excreted from the body. Consequently, large and repeated doses must be given to maintain a therapeutic effect. Due to very limited specificity of the majority of drugs, a range of serious side effects is often observed. A drug can be prepared so that its optimal level remains preserved for a sufficiently long time. Prolongation of action may be achieved by a variety of techniques, which include various encapsulation procedures, dispersion in a hydrophobic vehicle or porous polymer materials, binding of the drug to macromolecules of a polymer or oligomer, and the formation of a drug–carrier complex.

A polymeric drug may be defined as a polymeric substance that can exert a pharmacological activity when administrated to a living organism. This activity may be displayed in different ways. For example, the polymer may be active by itself, without being structurally related to any nonmacromolecular active compound. Alternatively, the polymer may be active because it contains moieties structurally related to nonmacromolecular compounds of known activity.

In the latter case, the active moieties may be present as constituents of the macromolecular backbone, or attached to a macromolecular backbone as side substituents by means of covalent bonds. When these polymers are able to release active molecules after entering the patient's body, by either injection or absorption, the term *polymeric pro-drug* is also used. It is apparent that when the active moieties are among the constituents of the main backbone, the release of active molecules takes place as a consequence of a degradation process involving the whole macromolecule. However, when the active moieties are present as side substituents, the release of the active molecules takes place by cleavage of the bonds holding these moieties, leaving virtually unaffected the main chain of the polymer, which in turn may, or may not, be degraded in time by a separate process. In this case, the macromolecular structure to which the active molecules are attached is usually referred to as the polymer main chain or *carrier*. The term *matrix* is also used to indicate the polymeric constituents of a physical mixture of polymer–drug combinations; these, however, will not be discussed in this book.

A polymeric pro-drug is typically composed of two distinct parts: the polymer backbone or chain, and the active moieties linked to this polymer carrier by means of covalent bonds that are cleavable in the body fluids. When the drug–carrier combination circulates within the body, the carrier can transport the active principles before they exert their activity.

So far, high polymers have been considered in most studies of polymeric pro-drugs. The tremendous amount of research work carried out in this field has been reviewed by several authors [157–165]. Only a few general remarks will be made here. Sustained activity in time of the free drug and selective localization at the preferred site of action (targeting) are the results most often sought. The main drawbacks of polymeric pro-drugs intended to be active systematically are their poor or absent absorbtion through the gastrointestinal walls or through the skin. Thus, they are suitable only for parenteral administration. They are often poorly or not excreted after injection unless their molecular weight is reduced below a certain threshold by degradation processes involving the main backbone. This is possible only for certain types of structures, not including polyvinylic structures. Thus, an accumulation of macromolecular residues after detachment of the drug moieties may be expected. The foregoing picture is not true where oligomeric carriers are concerned; since the molecular weight of oligomers ranges from a few hundreds to a few thousands they are easily eliminated by normal body processes.

One of the main features of high molecular weight polymeric drug carriers is that they contain binding sites to which the drug residues can be attached (along with solubilizing units and targeting groups). These sites must be sufficient in number to allow for a reasonable loading of drug, and they may be randomly distributed along the macromolecular chain. The contribution of terminal groups is considered to be negligible in this respect.

Oligomeric carriers, in contrast, may provide in many cases derivatives with a reasonable number of active drug moieties even if functionalized only at the chain ends. This allows for the utilization of these structures as drug carriers in which the repeating units within the chain, though endowed with the desired physicochemical properties, are not useful as drug binding sites; examples are poly(ethylene glycols) and their homologues. This does not mean, of course, that oligomers with reactive repeating units within the chain cannot be used as well.

Thus, oligomeric carriers offer more opportunities from a structural point of view than high molecular weight polymers. Another element to be considered when dealing with many end-functionalized oligomers is that the reactivity of a chemical function allocated at the end of a flexible chain can be more predictable, since it is practically independent of the segment to which it is attached. These predictions can be made based on known theoretical calculations for polycondensation and stepwise polyaddition processes [166].

In contrast, the nature and length of a macromolecule exerts a great influence on the reactivity of chemical functions directly attached to it. This is exemplified by the slow reaction rate of activated esters on high molecular weight poly(methacrylic acid) with amines [167].

It is well known that molecular size is of paramount importance in the elimination of undegradable polymers from the body. Experimental evidence of this behavior has been reported in the case of poly(vinyl pyrrolidone) and other water-soluble, nonionic polymers [163]. Even though the number of polymers studied in this respect is rather limited, it can be reasonably supposed that at least amphophilic, nonionic oligomers with physicochemical properties similar to those of poly(vinyl pyrrolidone) would show similar behavior. Thus, problems usually associated with the accumulation of residues in the body are not expected in the case of oligomeric pro-drugs, irrespective of any biodegradability question. In principle, a nontoxic, nondegradable oligomer may be safer than a degradable one, since problems arising from the possible toxicity of metabolites would not be encountered.

Shifting from polymers to oligomers does not fully solve the problem of coexistence of the physiological action and product toxicity. Some significant aspects have been depicted in the case of anionic polymers. Note that in this case the molecular weight distribution would be more critical than the average molecular weight. However, a complete molecular weight–pharmacological activity relationship still needs to be determined.

The use of acrylic or methacrylic acid polymers and copolymers as drug carriers has been mentioned in the literature [168]. The toxicology of these compounds is of prime importance in their medical applicability. Initially, acrylic and methacrylic polymers and their biologically active derivatives were found to be too toxic, and clinical application has shown that toxicity is in many instances related to higher molecular weight fractions [169,170]. Experimental evidence of this behavior has been reported [169]. These considerations support the idea that oligoacrylates may be considered as valuable supports for the various drugs (see Table 5).

Telomers with degree of polymerization of 10–50 were obtained by redox oligomerization of acrylic acid:

$$n\mathrm{CH_2{=}CH{-}COOH} + \mathrm{CCl_4} \rightarrow \mathrm{Cl_3C{-}(CH_2{-}\underset{\displaystyle|\atop\displaystyle COOH}{CH})}{-}\mathrm{Cl} \qquad (34)$$

From this oligomeric system, derivatives have been prepared with hydroxy-4-acetanilide (paraacetamol) [170]. This compound exhibits antipyretic, anti-inflammatory, and analgesic activity.

Table 5 Physicochemical Characteristics of Oligomeric Acrylic Derivatives

Structure	M	m.p. (°C)	b.p. (°C/mm)	Ref.
$H[CH_2CH(COOH)]_nCH_3$				
$n = 2$	160	128	—	198
$n = 3$	282	162	—	199
	M	b.p. (°C/mm)	n_D (°C)	Ref.
$H[CH_2CH(CN)]_nH$				
$n = 1$	55	97.1	1.3689/15	200
$n = 2$	108	135/12	1.4312/25	200
$n = 3$	161	195/2.8	1.4609/20	201
$H[CH_2CH(CN)]_nCH_3$				
$n = 1$	69	107	1.3720/20	201
$n = 2$	122	106/2	1.4191/60	201
$n = 3$	175	—	—	201
$n = 4$	22	—	—	201
$H[CH(CN)CH_2]_nCN_2CN$				
$n = 1$	94	287.4	1.4347/20	200
$n - 2$	117	—	1.1611/22	202
$n = 3$	200	95/0.01	—	202
	M	m.p. (°C/mm)	n_D (°C)	Ref.
$H[CH_2—C(CH_3)]_nC(CH_3)_2$ (with COOH on each of the two marked carbons)				
$n = 1$	260	oil	—	203
$n = 2$	346	45	—	204
$n = 3$	432	56	—	203
	M	b.p. (°C/mm)	d^{20}	Ref.
$H_3CO[CH_2—C(CH_3)]_nH$ (with $COOCH_3$ on C)				
$n = 1$	132	147	0.9749	203
$n = 2$	232	241	1.0540	203
$n = 3$	332	116/0.1	1.1045	203
$n = 4$	432	190/0.1	1.1200	203
$n = 5$	632	—	1.565	203
	M	b.p.	n_D	Ref.
$H_3CO[CH_2C(CH_3)]_nH$ (with CN on C)				
$n = 1$	99	—	—	—
$n = 2$	166	92/0.5	1.438	205, 206
$n = 3$	233	165/0.5	1.464	205, 206
$n = 4$	300	255/0.5	1.478	205, 206
$n = 5$	367	300/0.5	1.488	205, 206

3. Other Applications

OEA modification with adequate functional groups allows one to obtain some porous materials or ion-exchange resins [171–175]. Still more important, however, is the use of OEAs for the production of acrylic supports necessary for gel permeation chromatography. Importantly, by means of OEAs hydrophilic and hydrophobic gels can be produced by using methacrylic and dimethacrylic esters of glycol and triethyleneglycol [176,177].

The use of OEAs in the polygraphic industry is driven by the unsophisticated technology, reduced labor, and the possibility of substituting for some nonferrous metals [178,179]. Photopolymerized printing forms based on OEAs allow great progress in polygraphic technology. Commercialized materials like Nyloprint (Hamburg, Germany), Dyna Flex (Baton Rouge, LA), Flexomer, as well as some materials from Russia allow exquisite graphic properties in reproduction and printing. They allow an adequate number of copies to be printed (i.e., 500,000), which stand for a coefficient by 10% as compared with usual forms. Colors are properly rendered, and the forms are quite resistant under the usage conditions employed.

Anaerobic adhesive compositions are blends of dimethacrylates or diacrylates with hydroperoxides and other ingredients. They are stable in the presence of air but polymerize when oxygen is excluded. This characteristic has made them popular for the locking of nuts and bolts in automotive and other product assembly. Originally developed at General Electric Co. [180], they were commercialized by Loctite Corp. [181]. Products with trade names Anaterm, Unigerm, and VAK were commercialized in Russia.

In recent years the problem of *eutrophication*, which can be defined as a slow, natural process of enrichment of water with nutrients such as phosphorus and nitrogen, has received much attention. Uncontrolled or pronounced eutrophication has been found to cause increased algal growth and algal deposits; these are not only unaesthetic, odorous, distasteful, and clog filters or treatment plants, they also create disproportionate demands on the available oxygen in the water. It has been postulated that various human activities have contributed to acceleration of the process through such factors as inordinate enrichment of natural runoff, groundwater, and agricultural drainage, sewage, and waste effluents. It has also been suggested that the phosphorus containing binders present in detergent compositions can be a contributing factor in eutrophication, and therefore substitutes that do not contain phosphorus may decrease the eutrophication problem to some extent.

Sodium polyacrylates and alpha halogen substituted polyacrylates have been suggested as possible detergent boosters [182]. These compounds, however, create as much, if not more, of a problem than they ostensibly solve, since at the molecular weight of the polymer believed contemplated by the U.S. Patent 3,668,230, the molecules would not be biodegradable.

A large proportion of the sanitary treatment performed is done aerobically. If the bacteria in the degradation system cannot consume and degrade a molecule, it may pass through the tank and flow into the surface water and eventually become part of the human water supply. It has been found that bacteria either cannot or have extreme difficulty in degrading long-chain polymers and branched polymers of the type disclosed in Ref. 182. Since so little is known about the effects of so many chemicals, particularly from a carcinogenic and mutagenic standpoint, when ingested by human beings, a compound intended for use in a detergent formulation must be biodegradable or it is dropped from further consideration. As such, the compounds of U.S. Patent 3,668,230, as well as other relatively high molecular weight polyacrylates, would be deemed unacceptable for use in a detergent formulation.

Certain acrylate oligomers have unique and excellent detergent boosting properties. The particular acrylate oligomers found useful are those having an average molecular weight of between about 500 and about 10,000, preferably less than about 3000 [183]. Experimentation with these compounds has shown that the chain terminating moiety plays an important part in both the detergent ability of the oligoacrylates and the relative ease of biodegradability. Although any chain terminating or chain transfer agent that will reasonably function with polyacrylates and does not impair biodegradability of the molecule may be used, such as alkyls, substituted alkyls, hydrogen, and the residue from a free radical initiator, the preferred examples utilize compounds that will terminate at least one end of the chain with a sulfur containing moiety or an hydroxy containing moiety [184]. Preferred cations are alkali metals, ammonium, and substituted ammonium.

With recent concern for volatile solvent pollution in the atmosphere, many studies have been undertaken to determine the safest types of solvents that may be emitted. However, all organic solvents are polluting to some extent and ultimately organic coating systems that utilize a minimum amount of solvent are preferred [185–192]. On the other hand, much importance has been attached in the paint industry to the control of air pollution caused by organic solvents such as hydrocarbons contained in paints, and investigations have been undertaken to develop water-borne paints or powder paints as nonpolluting paints. It is technically difficult in various respects, however, to form a metallic finish coating using water-borne or powder paint. For example, attempts to obtain satisfactory metallic finish coatings from water-borne paints tend to result in coating defects such as the occurrence of pinholes or sagging upon evaporation of water, and this disadvantage is especially prominent when humidity is high in the environment in which coating is performed. To overcome this disadvantage would require extra treatment such as humidity adjustment of the coating environment, a number of coating operations, and repeated baking, which will cause a drastic reduction in productivity. On the other hand, when powder paints are used to form metallic finish coatings, metallic scales become oriented at random, and the appearance of the metallic finish coating obtained differs from that of coating obtained from a solvent-based paint. For this reason, such a coating cannot be repaired with conventional repairing paints (solvent-based air drying paints). Moreover, the powder paint cannot be used for repair because it requires a baking temperature of at least 100°C.

The search for methods of making high solid coatings has been long and difficult. To obtain high solid coatings with adequate spray viscosity, one must start with low molecular weight resins, which on curing produce desirable networks only if the crosslinking reaction employed is fast and proceeds without undesirable side reactions. Several new high solid coating systems have been reported [193,194]. These coatings have excellent physical properties, but better impact strength and stone-chip resistance are desired. Stone-chip resistance is desirable to reduce corrosion and maintain good appearance of auto body panels.

The preparation of various formulations of the high solid coating involved acrylic oligomers or co-oligomers [195,196]. It has been shown that acrylic co-oligomer–melamine formaldehyde resin coatings exhibit acceptable physical properties for high solid coating formulations [197].

REFERENCES

1. Berlin, A. A., Popova, G. L., and Isaeva, E. F., The synthesis of mixed polesters of the acryl series (polyesteracrylates), *Dokl. Akad. Nauk SSSR*, *123*, 282, 1958 (Russian).
2. Sivergin, I. M., Chemistry and technology of macromolecular compounds, *Adv. Sci. Technol.*, *13*, 242, 1980.
3. Berlin, A. A., SU Patent 70,736, 1949; SU Patent 73,031, 1949.
4. Zadontsev, V. G., *Industrial Synthesis of Oligoesteracrylates*, Mir, Moscow, 1978, p. 63 (Russian).
5. Berlin, A. A., Mel'nikov, V. G., and Ivanov, N. S., Catalytic effect and inhibitory activity of some polysulfophenylenequinones on the condensation telomerization reaction, *Plast. Massî*, 1966, No. 1, p. 34 (Russian).
6. Matnisan, A. A., Klashikov, Y. G., and Ilia, A. C., Effects of quinone derivatives on the condensation telomerization reaction, *Vysokomol. Soedin. A 13*, 1009, 1971 (Russian).
7. Berlin, A. A., Mel'nikov, Y. G., and Isaeva, E. F., On the catalytic activity of sulfocation exchangers in etherification, *Kinetica i Kataliz.*, *20*, 1341, 1979 (Russian).
8. Charchischavili, B. I., Platonov, I. Y., and Isaac, I. N., Synthesis of some methacrylate glycol ethers in the presence of sulfocationites, *Plast. Massî*, 1980, No. 9, p. 43 (Russian).
9. Korovin, L. P., Synthesis of some acrylic ethers in the presence of ion exchangers, *Izv. Vuzov Khimia i Khim. Technol.*, *20*, 1847, 1977 (Russian).
10. Bowen, R., Medical application of aromatic oligoetheracrylates, *J. Dental Res.*, *49*, 810, 1970.
11. Bondariuk, F. P., Synthesis of oligoetheracrylates by telomerization condensation, *Zhur. Prikl. Khim.*, *45*, 2500, 1972 (Russian).
12. Pechiney, Saint Gobain, FR Patent 1,509,311, 1968.
13. Hoechst, A. G., GER Patent 2,003,820, 1969
14. Shell Comp., U.S. Patent 3,679,731, 1972.
15. Michailov, M., and Nenkov, A. V., Condensation telomerization of anhydrides and epoxides, *Dokl. Akad. Nauk. Bulgarian Akad.*, *19*, 13, 1966 (Bulgarian).
16. Trebillon, V., FR Patent 1,039,538, 1969.
17. Michailov, S. V., SU Patent 179,919, 1964.
18. Seedov, L. N., et al., Synthesis and polymerization of polymaleinmethacrylates, *Plast. Massî*, 1968, No. 11, p. 25.
19. Kutepov, D. F., and Gloko, S. A., Synthesis of oligoetheracrylates with trazinic content, *Vysokomol. Soedin.*, *B 15*, 203, 1973 (Russian).
20. Smith, R. N., U.S. Patent 2,384,119, 1945.
21. Archibald, L. N., BRIT Patent 606,716, 1948.
22. Fleming, A. H., BRIT Patent 606,717, 1948.
23. Klein, M. V., BRIT Patent 624,019, 1949.
24. Savenkov, Iu. I., SU Patent 215,497, 1966.
25. Afanasiev, L. V., SU Patent 732,291, 1977.

26. Berlin, A. A., Sivergin, Iu. M., and Kefeli, T. Ia., Influence of the structure of oligo(carbonatemethacrylates) on the mechanical properties of their polymers, *Vysokomol. Soedin., A 13*, 2676, 1971 (Russian).
27. Berlin, A. A., Structure-properties relationship of oligocarbonatmethacrylates based polymers, *Plast. Massî*, 1974, No. 5, p. 33 (Russian).
28. Berlin, A. A., Influence of the structure of oligocarbonate methacrylates on the structure and properties of their three-dimensional polymers, *Vysokomol. Soedin., A 20*, 868, 1978 (Russian).
29. Berlin, A. A., Korolev, G. V., and Kefeli, T. Ia., *Acrylic Oligomers*, Khimia, Moscow, 1983 (Russian).
30. Suchov, V. D., Kinetic study of the oxidation of oligocarbonate methacrylate and oligocarbonate acrylates, *Zhur. Prikl. Khim., 50*, 351, 1977 (Russian).
31. Barcalov, I. M., Radiation curing of oligocarbonate methacrylates bis-(methacryloxyethylenecarbonate)diethylenglycol (OCM-2), *Dokl. Akad. Nauk. SSSR*, 222, 1357, 1975 (Russian).
32. Kim, I. P., Radiation curing of bis-(methacryloyloxyethylene carbonate)-propanediol-1,3, *Vysokomol. Soedin., A 20*, 23, 1978 (Russian).
33. Prishchepa, N. D., Khoromskaya, N. D., Shirygaeva, G. V., Kefeli, T. Ia., and Berlin, A. A., Radiation hardening of oligoetheracrylates and oligocarbonate methacrylates compositions, *Vysokomol. Soedin., A 22*, 2311, 1980 (Russian).
34. Berlin, A. A., Influence of modifying additions on the structure and properties of the coatings on the basis of oligocarbonate methacrylate, *Vysokomol. Soedin., A 18*, 2542, 1976 (Russian).
35. Spirin, M. L., Synthesis and characterization of oligoacrylates, *Vysokomol. Soedin., A 10*, 2116, 1968 (Russian).
36. Zemskova, Z. G., Matveeva, N. G., and Berlin, A. A., Oligourethaneacrylates: Synthesis and properties, *Vysokomol. Soedin., A 15*, 724, 1973 (Russian).
37. Berlin, A. A., Synthesis and study of properties of oligourethanemethacrylates and their polymers, *Vysokomol. Soedin., A 22*, 683, 1980 (Russian).
38. Archibald, L. D., BRIT Patent 1,428,454, 1976.
39. Johnson, L. S., U.S. Patent 4,250,322, 1981.
40. Magdinet, V. V., Masliuk, A. F., Rudko, A. R., and Spirin, Iu. L., *Synthesis and Physico-Chemistry of Polymers*, Naukova Dumka, Kiev, 1975, Ch. 15, p. 132 (Russian).
41. Berlin, A. A., A method of synthesis of oligoacrylates, *Vysokomol. Soedin., A 18*, 423, 1976 (Russian).
42. Lester, M. L., BRIT Patent 1,429,405, 1976.
43. Martin, A. L., U.S. Patent 4,259,411, 1981.
44. Berlin, A. A., Matveeva, N. G., and Pancova, E. S., Synthesis of dimethacrylateoligotetramethyleneglycole, *Vysokomol. Soedin., A 9*, 1326, 1967 (Russian).
45. Berlin, A. A., Matveeva, N. G., and Mamedova, E. S., Synthesis of dimethacrylate-oligodiethyleneglycolformals, *Vysokomol. Soedin., A 9*, 1330, 1967 (Russian).
46. Turovskaia, L. N., Matveeva, N. G., and Berlin, A. A., Upon the copolymerization of THF with propylene oxide in the presence of methacrylic anhydride, *Vysokomol. Soedin., A 15*, 1842, 1973 (Russian).
47. Berlin, A. A., Turovskaia, L. N., and Matveeva, N. G., Copolymers of THF with propylene oxide: Kinetics and characterization, *Vysokomol. Soedin., A 18*, 1322, 1977 (Russian).
48. Berlin, A. A., Turovskaia, L. N., and Matveeva, N. G., On the mechanism of cooligomerization of THF with propylene in the presence of methacrylic anhydride, *Vysokomol. Soedin., A 19*, 428, 1977 (Russian).
49. Matveeva, N. G., Turovskaia, L. N., and Berlin, A. A., *Proc. Inter. Conf. Progress in Ionic Polymerization*, Vol. 3, Varna, Bulgaria, 1977 (Bulgarian).

50. Raskina, L. P., Guseeva, R. V., Eltsefon, B. S., and Berlin, A. A., Synthesis of phosphorous containing oligoesteracrylates, *Zhur. Prikl. Khim.*, *41*, 1544, 1968 (Russian).
51. Andrianov, K. A., Leznov, S. S., and Dobagova, A. K., Synthesis and polymerization of methacrylic organosilicon compounds, *Izv. Akad. Nauk. SSSR*, 1957, 459 (Russian).
52. Andrianov, K. A., Leznov, S. S., and Dobagova, A. K., Characterization of the silicon containing oligoacrylates, *Izv. Akad. Nauk. SSSR*, 1959, 1767 (Russian).
53. Bulatov, M. A., Spaski, S. S., Alekseeva, T. V., and Molkanova, T. V., *Elementoorganic Compounds*, Mir, Sverdlovsk, 1966, Ch. 13, p. 3 (Russian).
54. Novitskii, E. G., and Slugina, N. D., Silicon containing oligoesteracrylates, *Plast. Massî*, 1965, No. 4, p. 21 (Russian).
55. Berlin, A. A., Kefeli, T. Ia., and Korolev, G. V., *Polyesteracrylates*, Nauka, Moscow, 1967, p. 372 (Russian).
56. Berlin, A. A., Polymer-oligomer compositions and materials made on their basis, *Plate u. Kautschuk*, *20*, 728, 1973.
57. Berlin, A. A., Zavodcikova, N. N., and Ianovskii, D. M., Oligoesteracrylate modified vinyl chloride copolymers, *Plast. Massî*, 1975, No. 4, p. 47 (Russian).
58. Kuzminskii, A. S., Berlin, A. A., and Arkina, C. N., *Progress in Physical-Chemistry of Polymers*, Khimia, Moscow, 1973, p. 239 (Russian).
59. Dontsov, A. A., Mihlin, V. E., and Dogadkin, B. A., New aspects of oligoesteracrylates composites, *Kauciuc i Rezina*, 1967, No. 1, p. 20 (Russian).
60. Davis, P. D., and Slichter, W. P., NMR relaxation in a radiation cross-linked poly(vinyl chloride) system, *Macromolecules*, 6, 728, 1973.
61. Botschareva, G. G., Poly(vinyl chloride) based composites, *Plast. Massî*, 1975, No. 4, p. 76 (Russian).
62. Lotakov, V. S., Ev'kika, V. S., Markova, L. A., and Nosac, G. D., The influence of oligodienurethaneacrylates upon the behavior of vulcanizates of ethylene-propylene copolymers, *Kauciuc i Rezina*, 1979, No. 11, p. 33 (Russian).
63. Tsuchareva, I. K., and Mezhikovskii, S. M., *Chemical Formulations of Oligomers and Polymers in the Soviet Union*, NIITECHIM, 1979, p. 660 (Russian).
64. Mezhikovskii, S. M., Progress and problems of physical chemistry of polymerizable acrylic oligomers and their transformations into network polymers, *Vysokomol. Soedin.*, *29*, 1571, 1987 (Russian).
65. Berlin, A. A., and Meshikovskii, S. M., Temporary plastification of rubbers by oligoesteracrylates, *Pure Appl. Chem.*, *55*, 2345, 1959.
66. Frenkel, R. S., and Meshikovskii, S. M., Some technological aspects of the modification of rubbers by OEA, *Plaste u. Kautschuk*, *27*, 544, 1980.
67. Berlin, A. A., *Unional Conference on the Chemistry and Physical-Chemistry of Reactive Oligomers*, Khimia, Tschernogolovka, 1977, Vol. 1, p. 8 (Russian).
68. Tager, A. A., The thermodynamical compatibility of *cis*-polyisoprene and nitrile rubbers with oligoesteracrylates, *Vysokomol. Soedin.*, *A 22*, 2234, 1980 (Russian).
69. Frenkel, R. S., Kurilova, T. I., Mezhikovskii, S. M., and Pitschevich, N. A., Structure of rubber-oligoethylacrylates composites, *Kauciuk i Rezina*, 1983, No. 2, p. 7 (Russian).
70. Chalykh, A. E., Avdav, N. N., Berlin, A. A., and Meshikovskii, S. M., Phase equilibrium and natural diffusion in the oligomer-polymer systems, *Dokl. Akad. Nauk. SSSR*, *238*, 893, 1978 (Russian).
71. Chalykh, A. E., Phase equilibrium onto oligoesteracrylates-elastomers, *Vysokomol. Soedin.*, *B 22*, 464, 1980 (Russian).
72. Kotova, A. V., Phase equilibrium in poly(vinyl chloride)-oligoetheracrylates systems, *Vysokomol. Soedin.*, *A 24*, 460, 1982 (Russian).

73. Kotova, A. V., Characterization of poly(vinyl chloride)-oligoetheracrylates systems, *Vysokomol. Soedin.*, *A 25*, 163, 1983 (Russian).
74. Chalykh, A. E., Belokurova, A. P., and Komarova, T. P., Sorption of water and phase structure of systems poly(vinyl chloride)-plasticizers and oligoetheracrylates, *Vysokomol. Soedin. A 25*, 1071, 1983 (Russian).
75. Chalykh, A. E., Belokurova, A. P., and Komarova, T. P., Phase-diagram of systems poly(vinyl-chloride)-oligoetheracrylates, *Vysokomol. Soedin.*, *A 24*, 747, 1982 (Russian).
76. Roginovskaia, G. F., Mechanism of formation of phase structure in epoxide-rubber systems, *Vysokomol. Soedin.*, *A 25*, 1979, 1983 (Russian).
77. Meshikovskii, S. M., Berlin, A. A., Wassiltschenko, E. I., Klynova, V. D., Kalesnikov, V. N., and Kuzminski, A. S., The significance of the solubility of oligoesteracrylates in rubbers during their modification, *Plaste u. Kautschuk*, *24*, 96, 1977.
78. Meshikovsckii, S. M., Morphology and physico-chemical characteristics of the mixtures of rubbers of different nature with oligoester acrylates, *Vysokomol. Soedin.*, *A 18*, 390, 1976 (Russian).
79. Kuleznev, V. N., Klykova, V. D., and Wasiltschenko, E. I., Influence of the solubility of oligoesteracrylates in *cis*-polybutadiene rubber (SKD) on the strength properties of vulcanizates, *Colloid. Zhur.*, *38*, 1976 (Russian).
80. Mezhikovskii, S. M., Study of the rheological properties of polydisperse *cis*-polyisoprene in the presence of oligoesteracrylates, *Vysokomol. Soedin.*, *A 19*, 2719, 1977 (Russian).
81. Mezhikovskii, S. M., Change of structural organization of rubber-oligoesteracrylate solutions during attaining of the therodynamic equilibrium, *Vysokomol. Soedin. B 28*, 53, 1986 (Russian).
82. Maltschevskaia, T. D., Study of the structure and properties of network polymers of the basis of oligocarbonate methacrylates in the process of film formation, *Vysokomol. Soedin. A 18*, 289, 1976 (Russian).
83. Frenkel, S. Ia., *Physics Today and Tomorrow*, Nauka, Leningrad, 1973, p. 176 (Russian).
84. Smets, G., and Van Gorp, R., Acrylic and methacrylic acid copolymerization parameters, *Eur. Polymer J.*, *5*, 15, 1969.
85. Berlin, A. A., Tvorogov, N. N., and Korolev, G. V., Study of solidification rate of oligoacrylates, *Dokl. Akad. Nauk SSSR*, *170*, 1073, 1966 (Russian).
86. Berlin, A. A., Rheological properties of oligoesteracrylates, *Izv. Akad. Nauk SSSR, Ser. Khim.*, 1966, No. 1, p. 193 (Russian).
87. Berlin, A. A., Mechanism of solidification process of oligoesteracrylates, *Vysokomol. Soedin.*, *A 12*, 2313, 1970 (Russian).
88. Vinogradov, G. V., *Polymer Rheology*, Mir, Moscow, 1977, p. 433 (Russian).
89. Wilson, D. R., *Structure of Metallic Melts*, Mir, Moscow, 1972, p. 272 (Russian).
90. Zhil'tsova, L. A., On the relation of kinetics of curing of rubber-oligoester acrylate blends with attaining of thermodynamic equilibrium in a system, *Vysokomol. Soedin.*, *A 27*, 587, 1985 (Russian).
91. Mezhikovskii, S. M., Physical properties of liquid oligomeric systems, *Vysokomol. Soedin.*, *A 28*, 53, 1986 (Russian).
92. Frenkel, Ia. I., *Kinetic Theory of Liquids*, Mir, Moscow, 1975, p. 614 (Russian).
93. Volkov, V. P., Phase-structure in epoxy-rubber systems, *Uspekhi Khim.*, *51*, 1733, 1982 (Russian).
94. Mezhikovskii, S. M., Phase diagram in liquid oligomeric systems, *Cauciuk i Resina*, 1985, No. 11, p. 40 (Russian).
95. Barkalov, I. M., Kurihin, D. F., and Munihs, V. M., Radiation processing of oligocarbonatemethacrylates, *Proceedings of Unional Conf. on the Physical-Chemistry of Oligomers*, Khimia, Tschernogolovka, 1981, Vol. 1, p. 17 (Russian).

96. Kotova, A. V., Tschalih, A. E., Abdrahmova, L. A., and Mezhikovskii, S. M., Irradiation curing of oligoesteracrylates, *Vysokomol. Soedin.*, *A 29*, 1761, 1987 (Russian).
97. Meshikovskii, S. M., Oligoesteracrylates-poly(vinyl chloride) mixtures, *Proceedings of Unional Conf. on the Physical-Chemistry of Reactive Oligomers*, Khimia, Tschernogolovka, 1977, Preprints, Bd. II, p. 362 (Russian).
98. Grishin, V. G., Il'ina, E. A., Shumanov, L. A., Berlin, A. A., Tugov, A. I., and Mezhikovskii, S. M., On the peculiar features of diffusion and solubility in oligomeric plasticizing agents, *Colloid. Zhur.*, *42*, 340, 1981 (Russian).
99. Kuzminskii, A. S., Berlin, A. A., and Arkina, C. N., *Progress in Physical Chemistry of Polymers*, Khimia, Moscow, 1973, p. 239 (Russian).
100. Matveeva, N. G., On the correlation between the structure and morphology of oligomers and network polymers on their basis, *Vysokomol. Soedin.*, *A 20*, 1080, 1978 (Russian).
101. Kuzminskii, A. S., and Arkina, S. N., Mechanical properties of oligoesteracrylates-rubber mixture, *Cauciuc i Rezina*, 1969, No. 6, p. 9 (Russian).
102. Kuzminskii, A. S., Berlin, A. A., and Arkina, S. N., *Progress in Chemistry and Physics of Polymers*, Khimia, Moscow, 1973, p. 238 (Russian).
103. Frenkel, R. S., Physical chemistry of oligoester-rubber mixtures, *Herstellung von Reifen Institute Gummi Technol.*, 1973, No. 2, p. 7 (Russian).
104. Frenkel, R. S., Neue Materialen und Prozess für Gummi Industrie, p. 34, *Verlag der Allunion-Mendelejews-Ges.*, Dnepropetrovsk, 1973.
105. Frenkel, R. S., Rheological properties of oligoesteracrylates *Hochschulnachr. Ser. Chem. Technol.*, *SSSR*, *20*, 626, 1977.
106. Grischin, V. G., Oligoesteracrylate mixtures for rubber processing, *Colloid. Zhur.*, *41*, 192, 1980 (Russian).
107. Ilina, J. A., Oligoacrylates, *SU Conf. on Chemistry and Physical-Chemistry of Oligomers*, Alma-Ata, 1979, p. 137 (Russian).
108. Meshikovskii, S. M., On some specificities of the process of *cis*-polyisoprene modifications by oligoesteracrylates, *Dokl. Akad. Nauk SSSR*, *229*, 410, 1976 (Russian).
109. Frenkel, S. R., Physical-chemistry of oligoesteracrylates, *SU Conf. on Chemistry and Physical-Chemistry of Oligomers*, Alma-Ata, 1979, p. 137 (Russian).
110. Frenkel, R. S., Physical-chemistry of oligoesteracrylates, *Int. Conf. on Rubber*, Prague, 1973, B-76.
111. Pantschenko, W. I., Physical chemistry of oligoesteracrylates, Thesis, Khimia, Volgograd, 1973 (Russian).
112. Berlin, A. A., and Arkina, S. N., SU Patent 411,097, 1973.
113. Arkina, S. N., Herstelung von Reifen, Inst. Gummi Technol., *Int. Conf. Bau.*, Jan 1974, No. 7, p. 3.
114. Frenkel, R. S., *Chemistry of Rubber Modification*, Khimia, Volgograd, 1980.
115. Kurnosova, N. W., SU Pat. Appl. 2,600,363/2305, 1979.
116. Pautschenko, W. I., SU Patent 469,716, 1975.
117. Frenkel, R. S., Rheology of oligoesteracrylates, *Int. Symp. on Rubber and Resins*, Gotwaldow 1979, A-37.
118. Frenkel, R. S., Rubber-oligoesteracrylates systems: Rheology and physico-chemical characterization, *Zhur. Prikl. Khim.*, *46*, 2352, 1976 (Russian).
119. Pautschenko, W. I., Morphology of rubber-oligoesteracrylates mixtures, *Herstelung und Reifen*, 1973, No. 2, p. 5.
120. Pautschenko, V. I., Kinetics of solidification of rubber oligoesteracrylate mixtures, *Herstelung und Reifen*, 1978, No. 4, p. 12.
121. Frenkel, R. S., Mechanism of rubber-oligoesteracrylate solidification, *Cauciuc i Rezina*, 1978, No. 12, p. 12 (Russian).
122. Santoch, S. L., Dearborn, H., and Ares, N. T., U.S. Patent 3,845,010, 1974.

123. Wingler, F., GER Patent 2,703,311, 1978.
124. Slowyc, W., Balle, G., and Munzer, M., GER Patent 2,727,480, 1979.
125. Fidensen, K., GER Patent 2,800,357, 1979.
126. Tatemichi, H., and Ogasawa, T., Problems in oligomer development, *CEER Chem. Econ. Eng. Rev.*, *10*, 37, 1978.
127. Matsui, F., JP Patent 63,122,707, 1988; *Chem. Abstr.*, *109*, 191446b, 1988.
128. Matsui, F., JP Patent 63,122,708, 1988; *Chem. Abstr.*, *109*, 191144c, 1988.
129. Matsui, F., JP Patent 63,122,702, 1988; *Chem. Abstr.*, *109*, 191445a, 1988.
130. Jelew, J., Applicability of oligoesteracrylates, *Plaste u. Kautschuk*, 27, 464, 1980.
131. Brauer, G. M., Oligoesteracrylates, *Polymer Plast. Technol. Eng.*, 9, 87, 1977.
132. Berlin, A. A., and Schutov, F. A., *Reactive Oligomer-Based Materials*, Khimia, Moscow, 1978, p. 296 (Russian).
133. Vacik, I., Permeability of metabolites through hydrophilic membranes, *Coll. Czech. Chem. Commun.*, 42, 2786, 1977.
134. Hrudkova, H., Svec, F., and Kalal, J., Reactive polymers. XIV. Hydrolysis of the epoxide groups of the copolymer glycidyl methacrylate-ethylenedimethacrylate, *Br. Polymer J.*, 9, 238, 1977.
135. Oya, S., and Aras, L., Measurement of transport properties of poly(methylmethacrylate-co-methacrylic acid) ion-containing membranes, *Br. Polymer. J.*, 22, 155, 1990.
136. Refojo, M. F., *Polymers in Medicine and Surgery*, Kronenthal, R. L., and Oser, Z. M., Eds., Plenum Press, New York, 1975, p. 313.
137. Wichterle, O., and Lim, D., U.S. Patent 3,220,960, 1965.
138. Refojo, M. F., *Biomedical Polymers*, Renbaum, A., and Shen, A., Eds., Marcel Dekker, New York, 1971, p. 173.
139. Masuhara, E., Uber die chemie eines neuen haftfähigen kunstoff-füllungs-Materials, *Dtsch. Zahnrztl. Z.*, 24, 620, 1989.
140. Masuhara, E., Studies of dental self curing resins. 4. Bonding of dental self curing resins to dentine surface for the use of alkylboron as initiator, *Reports Res. Inst. Dent. Mater. Tokyo Med. Dent. Univ.*, 2, 511, 1964.
141. Masuhara, E., U.S. Patent 3,829,973, 1974.
142. Bowen, R. L., U.S. Patent 3,066,122, 1962.
143. Bratu, T., Mikulik, L., and Munteanu, D., *Adhesive Techniques in Dentistry*, Facla, Timişoara, 1982.
144. Glenn, J. F., in *Biocompatibility of Dental Materials*, Marcel Dekker, New York, Vol. 3, 1982, p. 97.
145. Ruyter, I. E., in *Posterior Composite Resin Dental Restorative Materials*, Vanherle, G., and Smith, D. C., Eds., Peter Szulc Publishing Co., Amsterdam, Netherlands, 1985, p. 109.
146. Munteanu, D., Isfan, A., and Bratu, D., High-performance liquid chromatographic separation of BisGMA oligomers and isomers in dental restorative materials, *Chromatographia*, 23, 412, 1987.
147. Reaville, E. T., and Streicher, G. M., U.S. Patent 4,097,994, 1978.
148. Bowen, R. L., Crystalline dimethacrylate monomers, *J. Dent. Res.*, 49, 808, 1970.
149. Antonucci, J. M., and Bowen, R. L., Dimethacrylate monomers of aromatic etheresters, *IDREAF*, 54, Abstr. 441, 1975.
150. Antonucci, J. M., and Bowen, R. L., Dimethacrylates derived from hydroxybenzoic acids, *J. Dent. Res.*, 55, 8, 1976.
151. Bowen, R. L., and Antonucci, J. M., Dimethacrylate monomers of aromatic diesthers, *J. Dent. Res.*, 54, 599, 1975.
152. Baume, L. J., Fiore-Donno, G., and Holz, J., La reaction pulpaire à l'egard de l'Addent XI, *Schweiz. Monatschr. Zahnheilkd.*, 81, 1099, 1971.

153. Gollub, S., Schachter, D. C., and Ulin, A. W., Hemostasis blood coagulation and isobutylcyanoacrylate tissue adhesive, *Surg. Gynecol. Obstet.*, *130*, 2, 1970.
154. Beech, D. R., Bonding of alkyl-2-cyanoacrylates to human dentine and enamel, *J. Dent. Res.*, *51*, 1438, 1972.
155. Johnson, D. W., Polymers used in dentistry, Part 1, Cyano-acrylates, *J. Am. Dent. Assoc.*, *89*, 1386, 1974.
156. Newman, G. V., Adhesion and orthodontic plastic attachment, *Am. J. Orthod.*, *56*, 573, 1969.
157. Ringsdorf, H., Structure and properties of pharmacologically active polymers, *J. Polymer Sci.*, *Symposia*, *51*, 135, 1975.
158. Ringsdorf, H., Synthetic polymeric drugs, *Mid. Macromolecular Monogr.*, *5*, 197, 1978.
159. Ferutti, P., Macromolecular drugs acting as precursors in macromolecular active substances. Preliminary considerations, *Pharmacol. Res. Commun.*, *7*, 1, 1975.
160. Ferruti, P., Macromolecular drugs, *Il Farmaco, Ed. Sci.*, *3*, 220, 1977.
161. Batz, H. G., Polymeric drugs, *Adv. Polymer Sci.*, *23*, 25, 1977.
162. Pitha, J., Polymer-cell surface interactions and drug targeting, in *Targeted Drugs*, Goldberg, E., Ed., John Wiley and Sons, New York, 1985, Ch. 6, p. 113.
163. Drobnik, J., and Rypacek, F., Soluble synthetic polymers as potential drug systems, *Adv. Polymer Sci.*, *57*, 1, 1984.
164. Duncan, R., and Kopacek, J., Soluble synthetic polymers as potential drug carriers, *Adv. Polymer Sci.*, *57*, 53, 1984.
165. Ferutti, P., New polymeric and oligomeric matrices as drug carriers, *CRC Critical Reviews in Therapeutical Drug Carrier Systems*, 2, 175, 1986.
166. Flory, P. J., *Principles of Polymer Chemistry*, Cornell Univ. Press, Ithaca, New York, 1956.
167. Ferruti, P., Betelli, A., and Fere, A., High polymers of acrylic and methacrylic esters of N-hydroxysuccinimide as polyacrylamide and polymethacrylamide precursors, *Polymer (London)*, *13*, 462, 1972.
168. Netter, K. J., Ringsdorf, H., and Vilk, H. C., Pharmacologisch aktive polymere, 14. Modellsysteme zur untersuchung der abspaltung niedermolekularer kompnenten von polymeren pharmaka, *Makromol. Chem.*, *177*, 3527, 1976.
169. Bauduin, G., Boudin, D., Martel, J., Pietrasanta, Y., and Bucci, B., Recherche de téloméres à activité pharmacologique potentielle, 2. Téloméres de l'acide acrylique et greffage de composés hydroxylés, *Makromol. Chem.*, *182*,, 773, 1981.
170. Gros, L., Ringsdorf, H., and Scupp, H., Polymeric antitumor agents on a molecular and on a cellular level? *Angew. Chem. Int. Ed. Engl.*, *20*, 305, 1981.
171. Svec, F., Horak, D., and Kalal, J., Reactive polymers. IX. Preparation of macroporous chelate-forming resins from the copolymer glicidyl-methacrylate-ethylmethacrylate containing iminodiacetic groups, *Angew. Makromol. Chem.*, *63*, 37, 1977.
172. Kalalova, E., Sorption of platinum metals on the copolymer glycidyl methacrylate-ethylenedimethacrylate modified with ethylenediamine, *Coll. Czech. Chem. Commun.*, *43*, 604, 1978.
173. Kalal, J., Reactive polymeren. II. Die hestelung von metallchelaten vernetzter polymerrer, *Angew. Makromol. Chem.*, *49*, 93, 1976.
174. Jelinkova, M., Pokorny, S., and Coupek, J., Macroporous hydroxylalkyl methacrylate copolymers containing reactive functional groups, *Angew. Makromol. Chem.*, *52*, 21, 1976.
175. Svec, P., Reactive polymers. VIII. Reaction of the epoxide groups of the copolymer glycidylmethacrylate-ethylene-dimethacrylate with aliphatic amino-compounds, *Angew. Makromol. Chem.*, *63*, 23, 1977.
176. Kubin, M., Hydrophilic gels obtained by the polymerization of oligomethacrylates, *Coll. Czech. Chem. Commun.*, *32*, 3881, 1967.

177. Uglea, C. V., *Characterization of Macromolecular Compounds*, Technical Editure, Bucharest, 1983, Ch. 3 (Romanian).
178. Lasarenko, E. T., Polygraphic materials based on oligoacrylates, *Poligrafia*, 1974, No. 9, p. 26 (Russian).
179. Berlin, A. A., Processing of oligoetheracrylates, *Plast. Massî*, 1977, No. 10, p. 72 (Russian).
180. Krieble, V. H., U.S. Patent 2,895,950, 1959.
181. Murray, B. D., Hauser, M., and Elliot, J. R., Anaerobic adhesives, in *Handbook of Adhesives*, 2nd Ed., Van Nostrand Reinhold, New York, 1977.
182. Dannals, J. A., U.S. Patent 3,668,230, 1972.
183. Lamberti, V., and Chester, R. W., U.S. Patent 3,922,230, 1975.
184. Lamberti, V., and Chester, R. W., U.S. Patent 4,095,035, 1978.
185. Rhum, P., Patrick, W. J., and Aluetto, F., U.S. Patent 4,276,432, 1981.
186. Kison, P., Bryant, G. M., and Carr, F. S., U.S. Patent 4,153,778, 1979.
187. Sakimoto, S., and Yoshida, H., U.S. Patent 4,140,606, 1979.
188. Sperry, P. R., and Wiersema, R. J., U.S. Patent 4,102,843, 1978.
189. Watson, J. D., U.S. Patent 3,880,765, 1975.
190. Emmons, W. D., and Stevens, T. E., U.S. Patent 4,120,839, 1978.
191. Stein, R., BRIT Patent 1,441,814, 1973.
192. Carlson, G. M., and Abbey, K. J., U.S. Patent 4,526,945, 1985.
193. Chatta, M. S., U.S. Patent 4,237,241, 1980.
194. Chatta, M. S., U.S. Patent 4,181,784, 1980.
195. Theodore, A. N., and Chatta, M. S., Modification of acrylic polymers for high solid coatings, *J. Coat. Tech.*, *54*, 77, 1982.
196. Chatta, S. M., and Henk Van Oene, New approach for high solid coatings, *I&EC Prod. Res. & Dev.*, *21*, 437, 1981.
197. Baner, D. R., and Dickie, R. A., Cure response in acrylic copolymer-melamine formaldehyde crosslinked coatings, *J. Coat. Tech.*, *54*, 1982.
198. Anwers, K., and Thorpe, J., *Ann. Chem.*, *285*, 535, 1895.
199. Clark, H. G., *Makromol. Chem.*, *86*, 107, 1965.
200. Zahn, H., and Schaefer, P., *Chem. Ber.*, *92*, 736, 1959.
201. Takata, T., *High Polymers (Japan)*, *18*, 235, 1961.
202. Houtz, C., *Textile Res. J.*, *20*, 786, 1950.
203. Kämmerer, H., and Shkula, J. S., *Makromol. Chem.*, *116*, 62, 1968.
204. Kämmerer, H., and Shkula, J. S., *Makromol. Chem.*, *116*, 72, 1968.
205. Feit, B. A., *J. Polymer Sci.*, *A-1*, *2*, 4743, 1964.
206. Feit, B. A., *J. Polymer Sci.*, *A-1*, *4*, 1499, 1966.

3

Hydrocarbon Resins

I. TERMINOLOGY AND CLASSIFICATION

The term *hydrocarbon resins* refers to hydrocarbon oligomers derived from coke-oven gas, coal-tar fractions, cracked and deeply cracked petroleum stocks, essentially pure hydrocarbon feeds, and turpentines. Typical hydrocarbon resins (HRs) include coumarone–indene resins (CIRs), petroleum resins (PRs), styrene polymers (SPs), cyclopentadiene resins (CPRs), and terpene resins (TRs).

The term *CIR* refers to HRs obtained by oligomerization of the resin formers recovered from coke-oven gas and in the distillation of coal tar and derivatives thereof, such as phenol-modified CIRs.

The term *PR* refers to hydrocarbon resins obtained by the catalytic oligomerization of deeply cracked petroleum stocks. These petroleum stocks generally contain mixtures of resin formers such as styrene, methyl styrene, vinyl toluene, indene, methyl indene, butadiene, isoprene, piperylene, and pentylenes. The so-called "polyalkylaromatic resins" fall into this classification.

The term *SP* refers to low molecular weight homopolymers of styrene as well as copolymers containing styrene and other comonomers such as α-methyl styrene, vinyl toluene, butadiene, and the like when prepared with substantially pure monomer.

The term *vinyl aromatic polymers* refers to low molecular weight homopolymers of vinyl aromatic monomers such as styrene, vinyl toluene, and α-methyl styrene, copolymers of two or more of these monomers with each other, and copolymers containing one or more of these monomers in combination with other monomers such as butadiene and the like. These polymers are distinguished from PRs in that they are prepared from substantially pure monomer.

The term *CPR* refers to cyclopentadiene low molecular weight homopolymers and copolymers derived from coat tar fractions or from cracked petroleum streams. These resins are produced by holding a cyclopentadiene-containing stock at elevated temperature for an extended period of time. The temperature at which it is held determines whether the dimer, trimer, or higher oligomers are obtained.

The term *TR* refers to low molecular weight polymers of terpenes, which are hydrocarbons of the general formula $C_{10}H_{16}$, occurring in most essential oils and oleoresins of plants, and phenol-modified terpene resins. Suitable terpenes include α-pinene, β-pinene, dipentene, limonene, myrcene, bornylene, camphene, and the like. These products occurs as by-products of coking operations of petroleum refining and paper manufacture.

The term *bituminous asphalts* is intended to include both native asphalts and asphaltites such as Gilsonite, Glance pitch, and Grahanite. A full description of bituminous asphalts can be found in Abraham's *Asphalts and Allied Substances*, 6th Edition, Volume 1, Chapter 2, Van Nostrand Co., Inc., particularly Table III on page 60.

The term *coal tar pitches* refers to the residues obtained by the partial evaporation or distillation of coal tar obtained by the removal of gaseous components from bituminous coal. Such pitches include gas-works coal tar pitch, coke-oven coal tar pitch, blast-furnace coal tar pitch, producer gas coal tar pitch, and the like.

The term *rosins* refers to the resinous materials that occur naturally in the oleoresin of pine trees, as well as derivatives thereof including rosin esters, modified rosins such as fractionated, halogenated, dehydrogenated, and polymerized rosins, modified rosin esters, and the like.

The term *rosin-based alkyd resins* refers to alkyd resins in which all or a portion of the monobasic fatty acid is replaced by rosin (a mixture of diterpene resin acids and nonacidic components). Unmodified alkyd resins are polyester products composed of polyhydric alcohol, polybasic acid, and monobasic fatty acid.

The term *chlorinated aliphatic hydrocarbon waxes* refers to those waxes that are commonly called "chlorinated waxes," such as chlorinated paraffin waxes. These waxes typically contain about 30–70 wt% of chlorine.

The term *chlorinated polynuclear aromatic hydrocarbons* refers to chlorinated aromatic hydrocarbons containing two or more aromatic rings, such as chlorinated biphenyls, terphenyls, and the like, and mixtures thereof.

CIRs were obtained for the first time by Kromer and Spilker by 1890 in curing a fraction of coke-oven gas with sulfuric acid [1]. On an industrial scale these products were produced in United States in 1920 [2]. The importance of CIRs increased greatly following 1930.

PRs have been known since 1925. During World War II, Germany made liquid polybutadiene (LPB) under the trade name Plastikator 32, used as a curable and adhesive plastificant for synthetic rubber. A similar product was obtained in United States in 1950, with the aid of a semi-industrial plant at Borger, Texas, under the trade name Butarez [3]. At the same time, petroleum resins based on indene and styrene and their homologues were produced. Their industrial production started in 1950 [2].

TRs have been drawn from some natural products. Compounds thus produced were diversified either by chemical modification or by copolymerization of terpenic monomers with monomers of petrochemical origin [4,5].

The development of petrochemistry after World War II and especially the building up of some large pyrolysis plants as a source of ethene and propene led to the emergence of large amounts of secondary fractions with an increased content of unsaturated compounds. We refer to fractions C_4,C_5 (piperylenes), and C_8–C_{10} containing styrene, vinyl toluene, indene, and their homologues.

The aforementioned situation has determined industrial development and the search for novel consumption fields of PRs. Thus, PRs could come to exceed in importance CIRs and TRs [6]. PR output capacity in 1978 was by 1.5 million tons per year (t/year).

HRs can also be classified from other points of view. In terms of consistency, they can be liquid, semisolid, and solid, while in terms of reactivity they can be thermoplastic and thermoreactive. Nonmodified HR reactivity depends on the degree and nature of unsaturation. Thus, aromatic, mixed (aromatic–aliphatic), and terpenic resins, as well as copolymeric aliphatic resins in the synthesis of which conjugated dienes take a major part, have a thermoplastic character. Resins produced through conjugated-diene polymerization have a thermoreactive character. The same character is held by cyclopentadiene resins and modified HRs [7].

A more scientific classification takes into account resin structure. One can thus distinguish between aliphatic, cycloaliphatic, aromatic, and mixed HRs. Cycloaliphatic HRs include terpenic resins and cyclopentadiene-based resins. A similarity can be noticed, likely to cause confusion, between aliphatic and cycloaliphatic resins. This is because aliphatic resins often contain cyclic structures, arising from the presence of cyclopentadiene and cycloolefins in monomer mixtures subjected to polymerization and from cyclization accompanying some of the polymerization processes. Aromatic HRs include the products produced by polymerization of C_8–C_{10} fractions and CIRs. Mixed resins are made up of complex structures. Their properties will be modulated in terms of component proportion. Thus, these resins can produce various materials that also prove valuable from the technical point of view [8].

HRs can be modified or can be used as such. The unmodified ones are made up of carbon and hydrogen (with the exception of CIRs, which also contain small quantities of oxygen). They are nonpolar, neutral, and without functional groups. The only structural element that is more accessible to the approach of reactants is unsaturated and C–H bonds in the vicinity of double bonds. Modified RHs, except for hydrogenated RHs, are polar and reactive owing to various functional groups.

We shall make a detailed study of HRs according to their structure.

II. ALIPHATIC AND CYCLOALIPHATIC HYDROCARBON RESINS

This category of resins include a broad range of products. Oligoisobutenes and oligobutenes, oligodienes, oligopiperilenes, TRs, cyclopentadiene-based oligomers, oligocyclopentanes, cylcohexenes or dicyclopentadienes, and corresponding oligomers represent the main component of this category of resins.

A. Oligobutenes

It has long been known that normally gaseous olefins can be converted to viscous liquid polymers by means of solid aluminum chloride or by other catalytic systems [9–18]. A well-known and widely used olefin feedstock for such oligomerization is a petroleum refinery butane- and butylene-containing stream, referred to as "B–B stream."

The primary products obtained from the oligomerization of a typical B–B stream are oligomers having molecular weights in the range of from about 900 to about 5000. Oligomers have $\bar{M}_n > 900$, representing from about 10 to 30% of the total oligomers, have been used as alkylating agent.

An improved method [19] for producing oligoisobutylene used fractionated isobutene from a B–B stream. Olefins other than isobutylene of more than two carbon atoms, including butene-2 and both *cis*- and *trans*-butene-1, were considered as "poisons," along with organic sulfur containing compounds, alkyl halides, halogen acids, hydrocarbons of high molecular weight, and in general all sulfur, nitrogen, and oxygen compounds that could form stable complexes with Friedel–Crafts catalysts.

The products produced may be oligomers or high polymers [20,21]. Better known are isobutene copolymers with butenes [22], styrene [23], or cyclopentadienes [24].

Oligomerization of butenes has been realized with various catalytic systems such as $P(C_6H_5)_3$–ethyl aluminum [25], 1,1,1,5,5,5-hexafluoro-2,4-pentanedionatonickel–diethyl-aluminum [26,27], ion-exchanged mordenite [28], Ni-heptanoate–$C_2H_5AlCl_2$-CF_3COOH [29], Ni-*p*-toluene sulfonate–Ni-polystyrene sulfonate [30], $AlCl_3$ [31–34], and iron-group metal compounds in combination with an alkyl aluminum compound and an aluminum alkoxide activator, which improve catalyst activity and stability.

The linear dimerization of butadiene has been effected with a $NiBr_2(PPh_3)_2$ complex associated to sodium borohydride ($NaBH_4$:Ni = 2:1) [34a]. In a mixture of ethanol–tetrahydrofuran (THF) at 100°C, 1,3-butadiene is dimerized to (*E*,*E*)1,3,6-octatriene in 95% or greater yield and with more than 99% selectivity.

The polystyrene-supported analogue, (polystyrene–$PPh_2)_3NiBr_2$, when treated with $NaBH_4$, gave essentially the same results. In pure ethanol, the reaction failed, presumably due to the lack of swelling of the polymer. However, the supported catalyst collapsed after a while, corresponding to about 1500 mol butadiene converted in a batch operation.

Trimers and higher isobutene oligomers were obtained by C_4 fraction oligomerization in the presence of $CuSO_4$ and $AlCl_3$, respectively [16].

It is now well known that $Ni(CO)_2(PPh_3)_2$ is an efficient catalyst for the cyclooligomerization of butadiene to a mixture of 4-vinylcyclohexane (VCH = 30%), cycloocta-1,5-diene (COD = 60%), and cyclododeca-1,5,9-triene (CDT = 10%). This product distribution can be varied by using an excess of CO or phosphites. Particularly, the addition of 1 atm CO in THF or benzene at 115°C, though it slows down the reaction rate, leads to a selective transformation of butadiene to VCH (98%).

The same effect is observed when $Ni(CO)_2(PPh_3)_2$ anchored onto a phosphine-modified styrene–divinylbenzene copolymer is used [34b,34c]. However, the rate using the supported catalyst is three times slower than that using the homogeneous complex, under the same conditions. This is attributed to diffusion limitations. Recycling produced a slow leaching of the nickel, which can be reduced by the use of a more crosslinked copolymer. One may think that this leaching results from the intrinsic mechanism of the oligomerization, which involves more or less the transient decoordination of the phosphine ligands.

A 4 molar excess of *tris-o*-tolylphosphite in the presence of the supported nickel–carbonyl catalyst gave a 92% yield in cyclooctadiene, a result identical to that obtained in homogeneous catalysis. Again, the rate is three times lower than with the supported complex. These catalysts slowly deactivated and collapsed after about 1100–1200 mol of butadiene had reacted [34d]. However, this is not due to the nickel leaching, which either is too slow or may be avoided by the use of a suitable copolymer. The deactivated catalyst no longer exhibited CO stretching absorptions in the infrared, and repressurizing the system with carbon monoxide did not reactivate the catalyst. Nothing more is known about this deactivation process.

Another supported nickel(0) species has been prepared by the reaction of *bis*-(cyclooctadiene)nickel with a phosphinated macroreticular polystyrene. This has a very low activity toward butadiene oligomerization [34e], which is somewhat reminiscent of the deactivated catalyst previously described. However, the addition of $AlE_2(OEt)$ promotes the activity to about 60–100 g of product/g Ni/h. Cyclododecatriene is not found among the products, which consists of 70% COD and 30% VCH, a repartition similar to that obtained with the classical homogeneous systems.

The absence of CDT is indicative of the presence of a phosphine group in the coordination sphere of the nickel throughout the process. The need for activation by aluminum may be attributed to the existence of phosphine-chelated species on the support, which have to be destabilized in order to give the active species. Only an aluminum alkyl compound seems to be able to promote this destabilization. After this initial activation, additional aluminum is no longer required and the activity remains constant, for instance, in a flow system.

The reaction of butadiene with functional molecules to give functionalized oligomers is known to be catalyzed by different palladium complexes including phosphine ligands.

The reaction of trimethylsilanol Me_3SiOH with butadiene, leading to trimethylsilyloctadienes, has been investigated using as catalysts palladium complexes anchored on a variety of modified polymers. Examination of the mechanisms commonly accepted shows that this includes a step where the metal is in the form of a *bis*-(α-allyl) species. In such a species the C_8 chain occupies nearly the whole coordination sphere around the palladium, and it would also have great steric demand. Upon using a supported palladium catalyst, it can be expected, if the mechanism is still operative in heterogeneous phase, either that, due to steric requirements, the fixation of the metal to the polymer–ligand should be loosened and the complex should come out into the solution, or that the catalyst should lead only to C_4 products. The experimental data shows that the former assumption is correct [34f]. The content of the metal on the polymer was always considerably smaller after the reaction than in the freshly prepared catalyst. In no case could the catalyst by recycled. This confirms that a sterically free palladium species is required to effect the reaction. Contrary to the case of the cobalt hydroformylation catalyst, the palladium, once in solution, cannot go back onto the polymer, which then does not act as a reservoir.

Similarly, the reaction of acetic acid with butadiene catalyzed by the same complexes is known to give, in homogeneous phase, a mixture of acetoxybutenes, acetoxyoctadienes, and 1,3,7-octatriene. This reaction has been effected with $PdCl_2$, $Pd(PPH_3)_4$, or $Pd(OAc)_2$ anchored to diphenylphosphinated styrene–divinyl benzene copolymer [34g]. The product distribution is essentially the same as that obtained with heterogeneous catalysts. This distribution varies only slightly as the percentage of palladium or the PPh_2/Pd ratio, is changed.

Once more, palladium leaching is important for anchored $PdCl_2$ or $Pd(PPh_3)_4$ catalysts, but it is less so for $Pd(OAc)_2$. This parallels the observations obtained in the reaction of butadiene with trimethylsilanol, previously described.

The production of oligo- and polyisobutylene was elaborated as early as 1938 [35]. As raw material, C_3–C_4 fraction (90:10 or 50:50 isobutane–isobutene mixture) is used in the presence of $AlCl_3$ solution in alkyl or aryl chloride. The molecular weight of the product and process efficiency depend on raw material composition and system temperature. An increased reaction temperature decreases the average molecular weight of the product.

Oligoisobutenes obtained by cationic process present a wide molecular weight distribution (MWD) ($\bar{M}_w/\bar{M}_n$ = 5–6). To obtain a more homogeneous product, the fractionation of the crude product may be used.

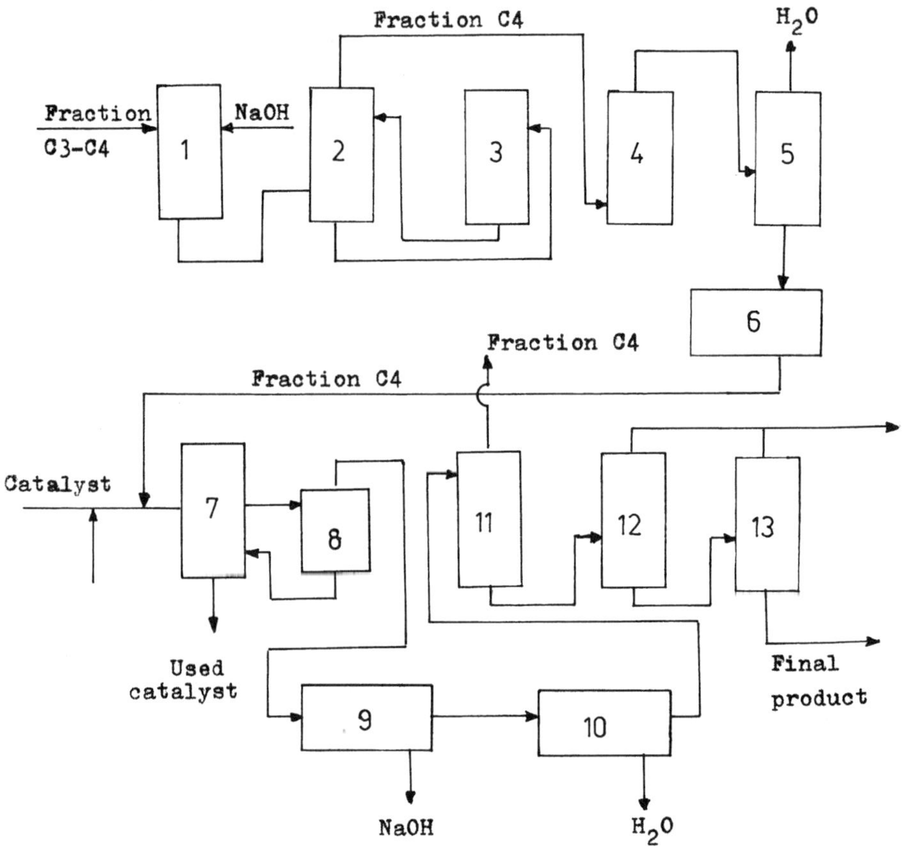

Figure 1 Basic layout of a plant for production of oligoisobutylene. 1: Column for alkaline washing; 2, 3: column for preparation of alkaline solutions; 4: washing columns; 5: column for water distillation; 6: raw material tank; 7: polymerization reactor; 8: filter; 9: alkaline deactivation of polymerization mixture; 10: catalyst separator; 11: vapors column; 12: atmospheric column; 13: vacuum column. (Adapted from Ref. 35a, p. 62.)

Oligoisobutenes obtained by Coden Comp., Baton Rouge, LA, have M_n = 300–2700. These products are obtained at 243 to 273°C by using C_3–C_4 fraction with 15 to 16% isobutene obtained, for example, by cracking and subsequent sulfur and C_3 compound separation. The raw material thus prepared is introduced in vessel 1 (Fig. 1), mixed with the catalyst, and then fed into polymerization reactor 7 made up of three vertical columns (Fig. 2) with a total volume by 5 m^3 (liquid volume 2.5 m^3) and provided at the top with a column condenser. The liquid phase is driven with gaseous ascending stream, leading it into coolers, from which it comes back to the reaction area. The reaction product separates from the catalyst and is introduced into deactivation reactor 9, where it is deactivated with alkaline solution. Purification of the polymerizate from the catalyst traces is carried out by washing with water in vessel 10 and then with steam in column 11. Together with the catalyst, the unreacted monomer is separated and then introduced

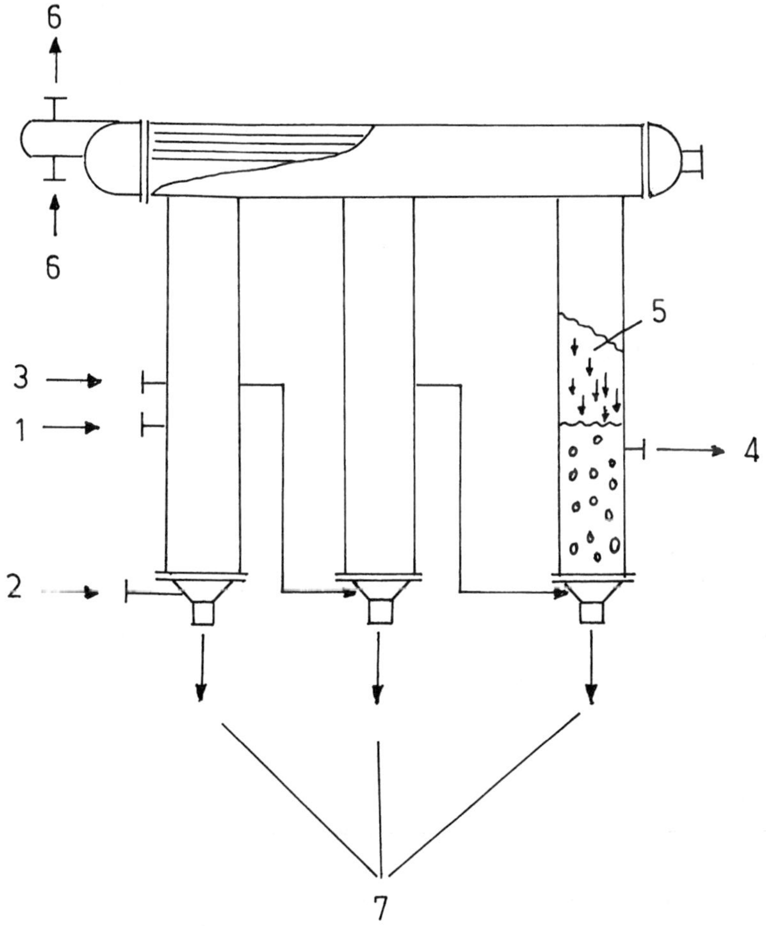

Figure 2 Polymerization reactor. 1: Raw material; 2: catalyst; 3: sludge; 4: ammonia; 5: condensates; 6: polymerization product; 7: exhausted catalyst. (Adapted from Ref. 35a, p. 62.)

into the reactor 7. Further on, the polymerizate is passed through the air column and the vacuum column, where separation from lower oligomers takes place.

Oligoisobutene yield achieved by Amoco Comp., Louisville, KY, uses a similar plant to that of Cosden Comp. What makes it different from the latter is that it uses $AlCl_3$ dissolved in butane (pressurized solution), which later is introduced into the reaction area in dispersed form. As working temperature is high, about 10% pleinomers can be obtained, which are later separated by distillation.

In Russia, oligoisobutenes are produced by means of a specific technology. Thus, a polymerization reactor is heated with a thermal agent, and the average molecular weight is regulated by modifying working conditions. Thus, to obtain octol with M_n = 800–1000, the polymerizate is mixed with alcohol or with alkaline solutions, after which it is degassed. The product with M_n = 600–3500 is obtained by two-step polymerization of C_4 fraction in reactors with intensive stirring. The basic layout of the operating process is given in Fig. 3.

The raw material is introduced into isothermic reactor 3 for the first-step polymerization. At the same time, the catalytic system $AlCl_3$–butylbenzene is also introduced at 297 to 300 K. Under these conditions, isobutene polymerization occurs.

The second step of polymerization takes place in the next polymerization reactor (4), wherein an additional quantity of catalyst is introduced along with the rest of isobutene and *n*-butene resulting from the first-step reactor. At this step, working temperature is 289 to 292 K. After this, product degassing follows (T = 573 K and P = 0.13–0.4 kPa).

As concerns industrial procedures of producing isobutene-based oligomers and polymers, several general observations should be made. Various oligoisobutenes produced at industrial scale, distinguished in the main by the average molecular weight, can be obtained by C_4 fraction polymerization. After preparing the oligoisobutenes by purifying the compounds that poison the catalyst (dienes, sulfur-containing compounds) it still contains butane and butene isomers as well as hydrocarbons with 2–5 carbon atoms. The presence of such compounds in the C_4 fraction depends on the method used to obtain it [36].

By polymerizing isobutene in C_4 fraction, oligomers or pleinomers are mainly obtained along with products resulting from the polymerization of other unsaturated compounds [37].

The aluminoorganic catalytic complex carries out isobutene polymerization in the C_4 fraction. At the same time, the same catalytic system solves the problem of complex reevaluation of this fraction. Catalyst efficiency is so good that the C_4 fraction of isobutene is almost exhausted (remaining monomer 0.5 to 1.5%), and residual gases in fact represent a pure B–B stream.

The practical volume in which polymerization is carried out is also important. In this respect, one should mention isobutene polymerization in a tubular reactor (L = 0.7 m, ϕ = 0.02 to 0.03 m) without stirring under adiabatic conditions—that is, reaction heat is used for preheating reaction mass [38]. The small diameter of the tubular reactor and the small reaction mass circulation rate (0.01 to 0.02 m/s) ensure the laminar character of the flow. Reaction mass constant composition (monomer and catalyst concentration) is ensured by mixer 4 (Fig. 4). Mean residence time of reaction mixture in the reactor is 35 to 70 s. Its adiabatic character does not provide an even temperature throughout the reactor

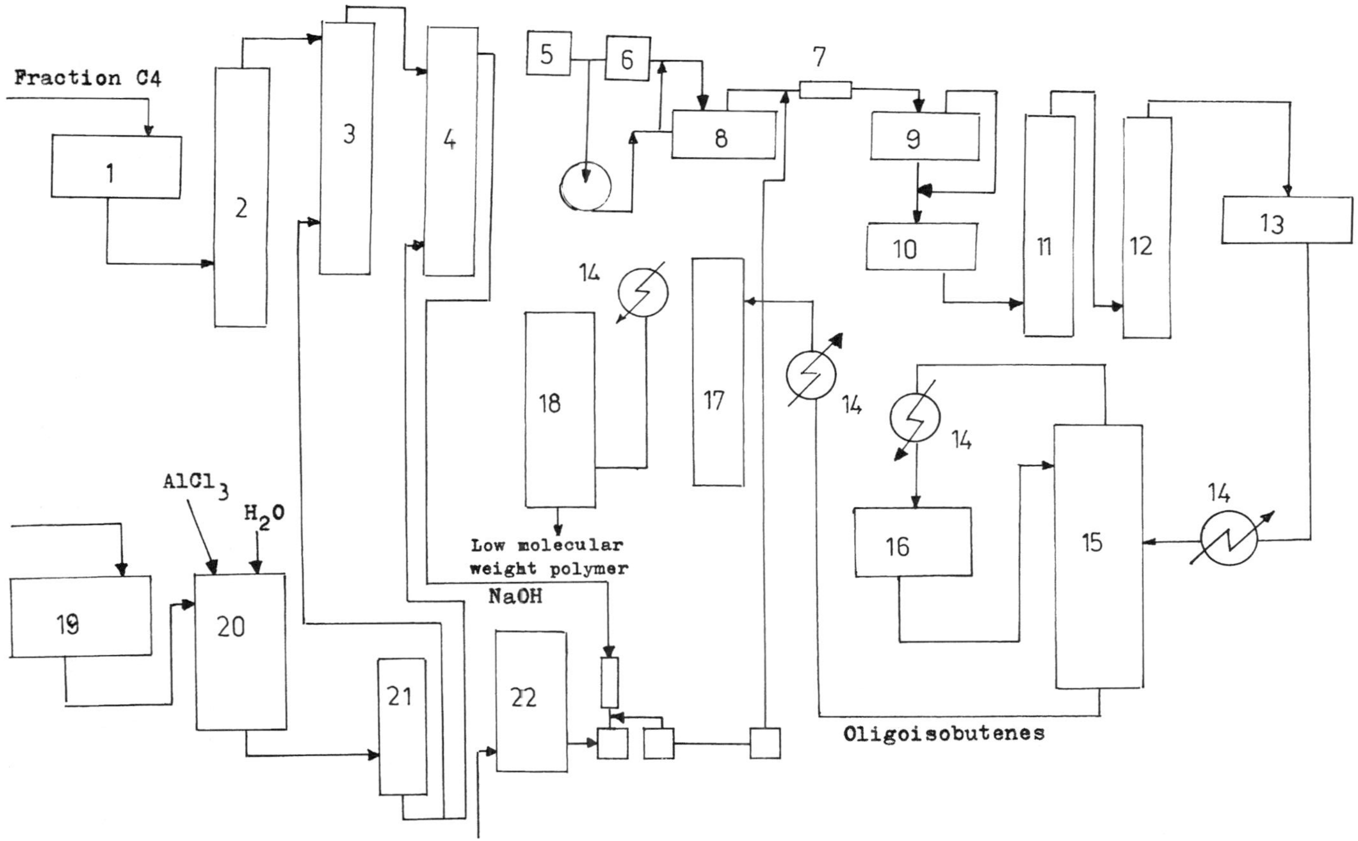

Figure 3 Basic layout of oligomerization fabrication in Russia. 1: Raw material; 2: drying column; 3: first-step reactor; 4: second-step reactor; 5–7: mixers; 8–10: separators; 11: drying column; 12: percolation vessel; 13: filter; 14: heat exchanger; 15: degassing column; 16: collector of working gases; 17: expansion column; 18: collector of oligomers; 19: butylbenzene tank; 20: column for preparation of catalytic system; 21: degasser; 22: tank of NaOH solution. (Adapted from Ref. 35a, p. 64.)

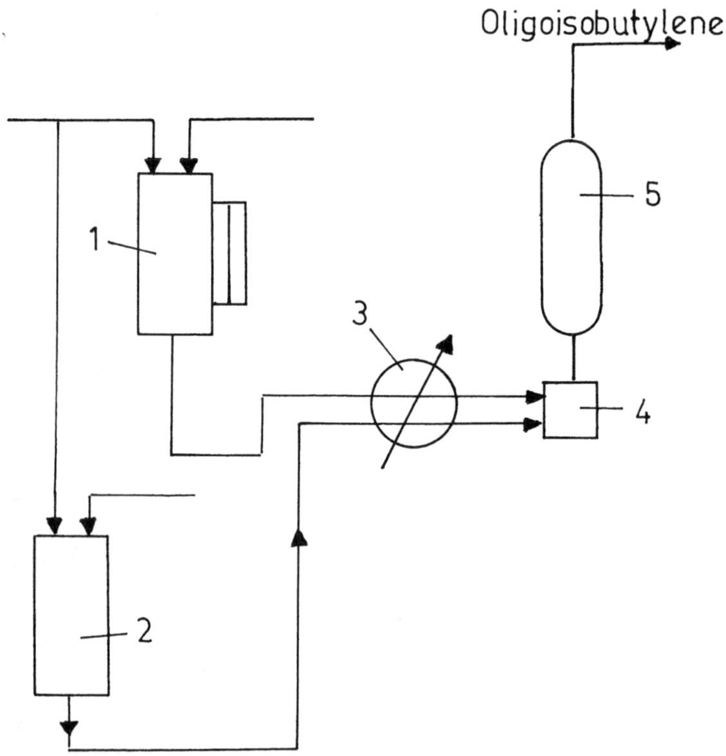

Figure 4 Basic layout of a plant for oligomerization of isobutylene in a tubular reactor. 1: Monomer tank; 2: catalyst tank; 3: heat exchanger; 4: mixer; 5: reactor. (Adapted from Ref. 35a, p. 66.)

length, so that the difference in temperature between reactor input and output is 40 to 100 K.

Optimal reactor sizes are $L = 0.5$ to 800 m and $\phi_{int} = 0.001$ to 0.1 m. A spiral form is preferred. Working temperature and pressure is adjusted so that the process occurs in the liquid phase [39,40]. In general, tubular reactors are more efficient (at equal volume) than the ones commonly used (100 kg/h with $M_n = 500$ to 5000 compared with 1 to 3 kg product/h with the same molecular weight) [40]. On the other hand, one should notice that MWD of reaction product resulting from tubular reactors is wider than that of products resulting from common reactors. This, along with some economic considerations, allowed in some situations (processes based on cationic mechanisms) the use of some small, common reactors to be favored over tubular reactors.

In Russia, the technology of producing oliga- and polyisobutenes in tubular reactors is still used [35]. The plant employs a tubular reactor with volume 5 to 10 m^3, in which the reaction time is 3×10^{-3} h.

The main characteristic of the oligo- and polyisobutene synthesis is that the reaction product is directly obtained and the resulting oligomers have uses such as greases for the axle box of a railroad car and a building adhesive.

Table 1 Oligoisobutenes: Main Procedures and Trade Names

Country	Trade Name
Russia	Oktol
United States	Indopol L, Indopol LV, and Indopol H
England	Hivis
Germany	Oppanol
Japan	Polybutenes HV

Polyisobutylene may be depolymerized on rare-earth-exchanged zeolite at 260 to 280°C. The yield of oligomers and the selectivity of the process are increased by the depolymerization in the presence of water vapors at a 17–30:1 ratio of water to raw material [41–43].

Oligoisobutenes can also be produced by cationic telomerization [44–47].

The presence of marginal double bonds in oligoisobutenes permits an entire sequence of reactions, substantially enlarging the possibilities for using oligoisobutenes. Out of the vast gamut of products obtained based on oligoisobutenes, those obtained via reactions of oligoisobutenes with amines (C-5A, Lubrisol 890 and 894, Ozoa 1200) are notable, along with P_2S_5 (Amoco 12) or those meant for fuels and oils that help prevent internal combustion engine wear [48–50]. Oligoisobutenes with M_n = 200 to 500 are used as antioxidants for obtaining cooling agents. Owing to their dielectrical properties, oligoisobutenes with M_n = 600 to 700 are used to make electroinsulations and condensers. Oktol 600A is used as an oil additive with a view to improve viscosity or heat stability under extreme temperature conditions. Oktol-1000 represents the raw material employed in the production of some adhesives and synthetic oils (Tables 1 and 2).

Chemical modification of oligoisobutenes aims in the main at enhancing the resistance of these products to oxidant agents as well as to mechanical strain (mechanical destruction) and thermal strain (thermal destruction). Thus, oligoisobutenes modified with aromatic compounds as well as products resulting from hydrogenation [51,52] find application in various technical fields and in medical practice. Oligoisobutenes with end groups of the type —C_6H_4—CH are antioxidants or can be used as liquid rubbers [53,54]. The latter products have been made by Enjay Comp., Los Angeles, CA.

B. Oligomers and Co-Oligomers of Acyclic Conjugated Dienes

The first diene oligomer obtained industrially was oligobutadiene, made in Germany in the period between the two world wars under the trade name Plastikator-32 [35]. In 1950–1955 various technologies were worked out and, the ranges of application of liquid butadiene and liquid butadiene–styrene copolymers explored. These products encountered obstacles to commercial recognition. Thus, in this time span it was only such products as "Budium" (Du Pont de Nemours, La Rochelle, France) and "Ricon" (Richardson Co., Richmond, VA) that could be mar-

Table 2 Technical Characteristics of Oligobutenes

Trade name	M_n	Density (kg/m^3)	Kinetic viscosity 373 K (mm^3/s)	Iodine number (g I_2/100 g)
Oktol 200	700–1000	—	190–250	—
Oktol K	700–1500	—	200–600	—
Oktol for adhesive	700–1000	870–890	190–350	—
Oktol 1000	800–1000	870–890	200–300	25–45
Oktol 600A	—	—	—	—
Oktol 600B	—	—	—	—
Oktol A	—	—	75–115	—
Indopol H-50E	720	—	103	—
Indopol H-1000	915	889	210	37.4
HV-15E	700	—	87	—
HV-100	870	886	240	37.7
Hivis-5	780	—	97–120	—
Hivis-10	930	882	240	29.7
Oppanol B3	720	880	230	46.2

keted. It was only by the end of the 1960s and the beginning of the 1970s that a true development of the products in the category of liquid polybutadiene began, especially because of their application as curable coatings and in the manufacture of inks and rubber-based paints. In 1975 the worldwide yield was estimated at 8000 to 15,000 tons.

Data published by the end of the 1960s [36] reported products such as Butarex (liquid polybutadiene made by Phillips Company) [37], KW resin liquid polybutadiene made in Germany in Buna Werke Dresda [38], C-oil, MD-1, MDE-1 (liquid butadiene–styrene copolymer) made by Esso-Research [39], Buton (made by Esso-Research) [40] Baton Rouge, LA, and Dienol-S (butadiene–styrene copolymer) made in Russia [38].

The MWDs of unmodified oligobutadienes range between 1000 and 8000 Da and exhibit different microstructures materialized through *cis*-1,4 or *trans*-1,4 as well as through vinyl structures. Molecular weight distribution is influenced by the method of synthesis. The procedure of live polymerization forms products with narrower MWD, and that of telomerization, products with broader MWD [41]. Oligobutadienes can be easily functionalized so that a wide range of compounds with various applications is obtained [42].

Copolymers of butadiene with styrene and piperilenes represent potential raw materials for obtaining liquid oligomers [43–47].

Copolymerization of vinyl ferrocene with butadiene results in a product widely used as a propellant. The product of the reaction is a liquid copolymer capable of undergoing a curing reaction when used in a propellant composition to form a rubberlike binder material [48,49].

Liquid polymers containing vinylidene groups ($CH_2{=}C\langle$) are known. These polymers are prepared in a number of processes such as cleavage or degradation of high molecular weight dienic elastomers, the free radical polymerization of dienic monomers in the presence of large quantities of chain transfer agents, and the solution polymerization of dienic monomers using lithium catalysts.

These liquid polymers are cured through the vinylidene groups to solid elastomers. This has advantages: compounding ingredients may be simply dissolved or dispersed in the polymer by mixing and the compounded liquid poured or spread into place. The compounded liquid will then quickly cure *in situ* at room temperature or with slight application of heat. Unfortunately, the vinylidene groups of the previously known polymers are not highly reactive at room temperature. Curing to a dry elastomeric state may often take weeks. This has heretofore hindered or prevented the use of these liquid polymers in applications such as commercial caulks and sealants.

The novel liquid polymers are characterized by having reactive terminal vinylidene groups and are prepared from liquid carboxyl-terminated polybutadiene polymers using an amine catalyst [50].

There have long been made many studies on processes for cationic polymerization of unsaturated hydrocarbons, in the presence of a Friedel–Crafts type catalyst, and it is known that rubbery, resinous, liquid, gel-like, or any desired type of polymer may be obtained by the selective use of polymerizing conditions, such as a combination of catalyst and monomeric material.

Typical of these polymers are liquid polybutadienes, petroleum resins, and the like. The supply of hydrocarbon fractions obtained as by-products from ethylene-producing plants has recently increased with the enlargement of the plants in scale or capacity, and in connection with one of the uses of these fractions, studies concerning the cationic polymerization thereof are increasing. Among the fractions making up the by-products: unsaturated hydrocarbon components, conjugated diolefins such as isoprene, 1,3-pentadiene, and cyclopentadiene, and pentenes such as *n*-pentene-1, *n*-pentene-2, 2-methylbutene-1, 3-methylbutene-1, and 2-methylbutene-2. There have already been many reports on processes for the cationic polymerization of the remaining portion obtained by separating the isoprene and cyclopentadiene components from the C_5 fraction. For example, in the presence of boron trifluoride [51], polymerization of the remainder of such a C_5 fraction (the remainder being obtained by the separation or removal of isoprene and cyclopentadiene from C_5 fraction and being composed mainly of 1,3-pentadiene and pentenes) gives a liquid polymer [52]. The thus obtained liquid polymer exhibits a high iodine value since it contains a large proportion of 1,3-pentadiene comonomeric units. Thus, the liquid polymer is useful as a synthetic oil substitute for natural drying oils such as linseed, sunflower, or soybean oil; it is also suitable as a working fluid, such as a lubricating oil, a functional oil, or a heat transfer medium, which is required to be stable. Also, it is not entirely satisfactory to use as a cosmetic base since it is somewhat unsuitable in color and an irritant to human skin.

The oligomer obtained (M_n = 300–1000) by copolymerizing a mixture containing 20–90% by weight of *n*-pentene and 10–80% by weight of at least one other chain pentene entirely eliminates the foregoing drawbacks such as an unsatisfactory color and irritation to human skin if this copolymer is hydrogenated. This new product is excellent in affinity and spreadability (being well attached without stickiness to human skin and well spread thereon). The hydrogenated co-oligomer oil further has nearly the same refractivity and water permeability as squalane, which has been the most recommended material as a base for cosmetics.

Oligomerization of long-chain alkenes or dienes and cycloalkene [53] yields products of higher average molecular weight that are suitable for use as lubricating oil, base oil, gasoline and jet fuel [54].

Applications of anionic processes [55], metathesis reactions [56–58], and cationic or radicalic procedures [60,61] to oligomerization of long-chain alkenes have been presented in the scientific literature.

Various catalytic systems such as boron trifluoride [62,63], aluminum chloride [64–67], tantal halides [68], zirconium halides [69,70], zeolites [71–73], molecular sieves containing metallic oxides [74], ion exchangers [75], fluoroalkane sulfonic acid [76], and montmorilonite [77] have been used to promote oligomerization of long-chain alkenes, cycloalkenes, and dienes.

A method for the evaluation of gel permeation chromatograms obtained by analyzing mixtures of long-chain alkenes or diene oligomers have been proposed [78]. Mathematical relations were derived for proper calibration and calculation of the content of particular oligomers from the shape of the chromatograms.

Co-oligomers of 1-alkenes (C_4–C_{32}) and derivatives of acrylic acids are additives to lubricating oils that are useful as low-temperature and ashless sludge dispersants and detergents [79].

C. Oligocyclodienes

Cyclodienes to be found in C_5 and C_9–C_{10} fractions of pyrolysis gasoline (cyclopentadiene and dicyclopentadiene) are used in HR synthesis [80–82]. The properties of these products and the ease with which they can modify themselves give them considerable importance in the marketplace [83,84].

The presence of polar groups in HRs such as polycyclopentadiene results in improved compatibility of these HRs with high molecular weight substances such as rubbers and synthetic resins, with which the HRs are blended in the preparation of compositions and also to enhance properties of HRs, such as pigment dispersion fluidity, and adhesiveness, that the HRs in their modified form do not readily manifest [85].

III. COPOLYMERIC ALIPHATIC RESINS (CARs)

Copolymeric aliphatic resins (CARs) are obtained through oligomerization of complex mixtures of olefins and dienes with 4 to 6 carbon atoms. This sort of HR prevails in HRs as a whole, in both market importance and range of application. Being developed rather late as compared with other HRs, CARs advanced rapidly due to raw material availability and reduced cost.

CARs can be liquid [86], plastic solids [87], brittle solids [88]. Liquid CARs have smaller molecular weights than brittle CARs [89]. Table 3 lists the main characteristics of liquid CARs from various sources.

Correlation of softening temperature with M_n of brittle CARs is still questionable [89–91].

Table 3 Physical Characteristics of Liquid CARs

Characteristic	Wingtack 10[a]	TPR[b]	Pentacol-1[c]
Aspects		yellow pale liquids	
Color (Gardner scale, 50% solution in toluene)	6	<6	6–7
M_n	—	500–1000	350–500
Viscosity, 25°C. (Pa·s)	14.0–20.0	30.0–100.0	95.0
Volatiles (%)			
3 h at 105°C	—	10	5–10
0.5 h at 115°C	—	—	<10

[a]Product of GoodYear Tire and Rubber Co., U.S.
[b]Product of Toho Chemical Industry Co., Ltd., Kyoto, Japan.
[c]Romanian product.

CARs have been obtained on a laboratory or an industrial scale [92–94]. In this respect, thermal oligomerization is employed [95], the cationic technique [96] or other procedures that combine thermally initiated reactions or Diels–Alder additions with cationic oligomerization [97,98].

Depending on their molecular weight, CARs can have different softening temperatures. As a rule the 100–110°C values are maintained, but products with a high content in polycyclic structures, like those based on Diels–Alder adducts (dicyclopentadiene, cyclopentadiene–isoprene, cyclopentadiene–piperilene, isoprene–butadiene codimers) show higher softening temperatures.

CARs in general show an increased reactivity compared with that oligodiene, distinguishing themselves by their capacity for autooxidative crosslinking (siccativation) and adhesivity [99,100].

IV. TERPENIC RESINS

Terpenic resins are oligomers of some unsaturated dicyclic hydrocarbons with the formula $C_{10}H_{16}$, which are found in coniferous crude oil, the turpentine resulting from the cellulose–sulfate process, and other natural materials. The main compounds of turpentine-sulfate and of some distillation fractions of this are α- and β-pinene (Fig. 5).

Of two components, β-pinene is of significance for the terpenic resin (TR) synthesis (α-pinene is used to obtain camphor [101]).

TRs are thermoplastic products, the production and marketing of which began as early as 1938. They can be considered β-pinene oligomers. Their softening temperatures range between 10 and 135°C. The latter limiting value corresponds to M_n = 1200 to 1500 g/mol [101].

β-Pinene oligomerization is carried out in solution in the presence of Friedel–Crafts catalysts. β-Pinene can be introduced alone in the process or together with other monomers (e.g., isoprene, piperilene, amylene, butadiene).

Figure 5 Structure of pinene isomers. (a) α-Pinene; (b) β-pinene.

TRs have been of particular importance in the manufacture of adhesives from natural and synthetic rubbers. The adhesive compositions obtained with these resins have an excellent balance of adhesion, cohesion, tackiness, and strength. They provide superior adhesives for applying to backings, providing excellent pressure-sensitive adhesive tapes [102]. TRs also have applications in ink or chewing gum formulations, building engineering, and the paper and leather industries [83,103,104].

V. AROMATIC HYDROCARBON RESINS (AHRs)

Aromatic hydrocarbon resins (AHRs) were mentioned as early as the 1930s and were initially obtained from raw materials of carbochemical origin. Later, AHRs were obtained from raw materials from petrochemistry [105–107]. Several thousand tons of AHRs are marketed per year, divided into two subgroups: copolymeric AHRs and homopolymeric AHRs. The first subgroup dominates in importance and is made up of indene–coumarone resins (of carbochemical origin) [108–112] and of indene or styrene–indene resins (of petrochemical origin) [113–116].

The interest in homopolymeric AHRs is growing, since the use of pure monomers allows one to achieve some colorless products with certain higher-performance qualities (increased softening temperature, reactivity, etc.). Styrene homooligomers can be obtained from pure styrene or from styrene concentrates by cationic oligomerization. Styrene co-oligomers can be obtained from styrene and α-methylstyrene, vinyltoluene, cyclopentadiene, or vinylic ethers by cationic oligomerization [113,117–120].

High-temperature oligophenilenes are obtained in the presence of a titanium–aluminum catalyst, conveniently expressed by the generic formula $TiCl_4$/$Al(C_2H_5)_2Cl$, by the following reactions [121–125]:

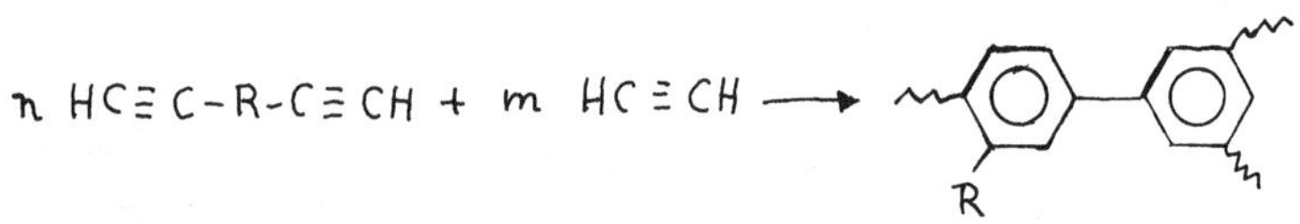

(1)

where R = alkyl,

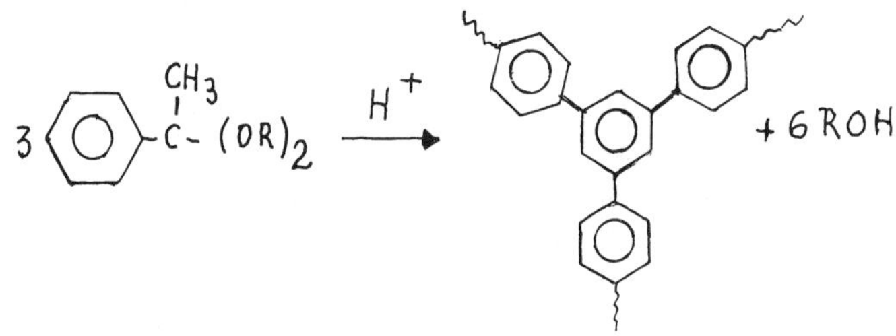

(2)

where R = alkyl, and

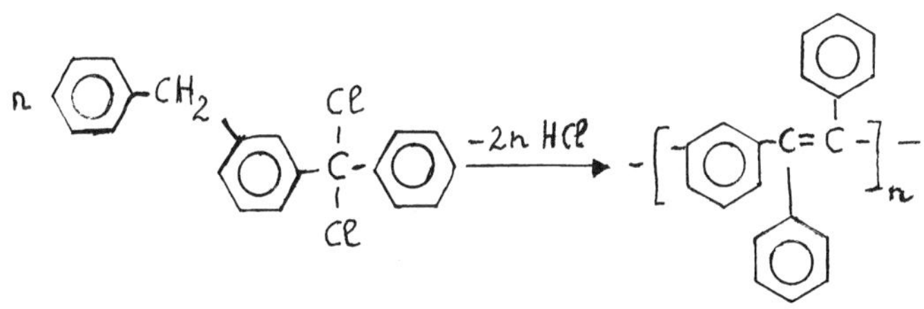

(3)

Oligophenylenes with oxymethylenic groups were prepared in the presence of $P(C_6H_5)_{32}$ or $Ni(CO)_2$ by the following reactions [126]:

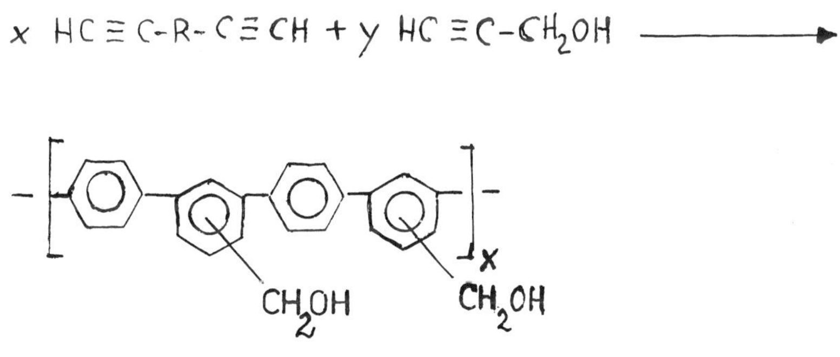

(4)

where R = $—C_6H_4—$.

Metal-containing oligophenylenes soluble in aromatic hydrocarbons were prepared by treating oligomers containing $(X)_m(CH_2)_n$ with Na or K in a hydrocarbon medium, where

X =

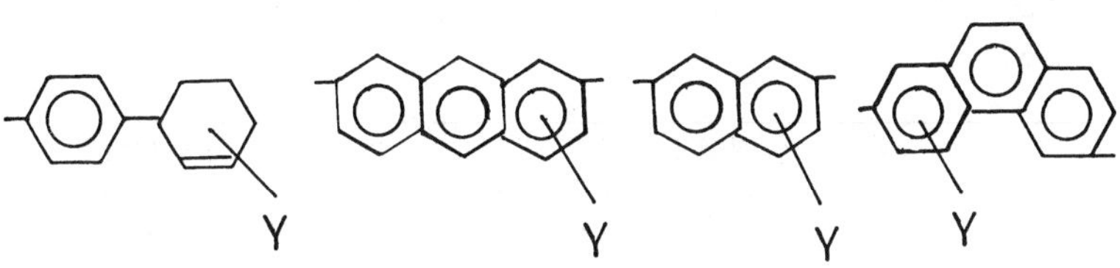

and Y = H, m = 1,2 and n = 1–5 [127].

AHR structure and properties are influenced both by the initial fraction composition and by conditions of synthesis. More frequent is the use of thermal initiation [128–130] as well as cationic initiation. The latter process has used Bronsted or Lewis acids [131–134]. Initiation with Bronsted acids is an outdated process. Catalysts made up of simple or compounded Lewis acids are economically desirable for the industrial manufacture of these products.

Owing to reduced reactivity, obtaining AHRs by radicalic processes requires high temperatures. Thus, the resulting products are dark colored.

Cationic processes catalyzed by the Lewis acids allow more accurate control of molecular weight and softening temperature. The use of Bronsted acids promotes secondary reactions and generates more impure products.

AHRs can be obtained in a wide range of mean molecular weights. By varying the average molecular weight and structure, these resins can be (at room temperature) liquid ($T_{softening} \approx 10°C$) or solid with $T_{softening}$ = 120 to 130°C. They are thermoplastic and have a greater density than other HRs.

VI. MIXED HYDROCARBON RESINS (MHRs)

Mixed hydrocarbon residues (MHRs) are obtained by reevaluation of some unsaturated fractions with a wide compositional spectrum (aliphatic or aromatic) [135–137]. Introduction of aliphatic structures along with the aromatic ones in HRs generally has a favorable effect on the products obtained [138–140].

The aromatic structure/aliphatic structure ratio in MHRs influences some of the physical properties of these resins, as does the nature of the aromatic structure. Thus, aromatic structures originating in styrene and its homologues do not lead to an increased $T_{softening}$ value, whereas indene, coumarone, and their homologues have a significant effect on this parameter. Thus, by co-oligomerization of an aliphatic fraction (33–40% isoprene, 18–23% piperilene, and other C_5 hydrocarbons) with an indene fraction (40% indene, 15% styrene, and 15% coumarone, the rest being made up of other aromatic hydrocarbons) products with $T_{softening}$ = 150°C were obtained (when the aliphatic fraction was dominant [140].

In obtaining MHRs one can also produce products with higher compatibility than various other materials, or with an increased capacity of reaction for chemical modifications.

In general MHRs are obtained by radical oligomerization initiated thermally or with organic peroxides and naphthenates or metal stearates [141], or by cationic oligomerization initiated with Lewis acids [142].

VII. CHEMICALLY MODIFIED HYDROCARBON RESINS

HRs of all types can be chemically modified by various reactions—either polymer-homologous transformations or copolymerization processes (grafted or block). Except for the hydrogenation, these reactions transform HRs from neutral, nonpolar materials to polar, basic or acid products capable of participating in various subsequent reactions. These modifications determine essential changes of HR physical and chemical properties. These resins become water soluble, thermoreactive, and compatible with polar materials.

There are various types of modification reactions. Carboxyl groups introduction, imidization, hydrogenation, oxidation, and epoxidation are (only) some examples. Also known is modification by simple mixing, without chemical reactions. This latter process is characteristic of HR applications, not HR synthesis.

A. Hydrocarbon Resins with Carboxylic Groups

Oligomerization methods for preparing carboxy-functionalized liquid oligomers have been described in the patent and technical literature [143–156]. The main chemical agent employed here is maleic anhydride (MA); thus, the products obtained are called *maleic oils* or *maleic resins* [157]. Also, beta unsaturated mono- or dicarboxylic acids can be used as well as some of their functional derivatives such as acrylates, itaconic anhydride, citraconic anhydride, and fumaric and metharcrylic esters [155,156].

General methods of obtaining functionalized HRs with carboxylic groups include thermal grafting with initiators forming free radicals, or with cationic catalysts of the Friedel–Crafts type. Also included are cationic co-oligomerization of unsaturated hydrocarbons or of some Diels–Alder adducts (obtained from dienic hydrocarbons) with MA or other filodienes with carboxylic functions [155,156].

The reaction with MA is characteristic for the unsaturated compounds. As a result it is applied to liquid polydienes, to cyclopentadiene and dicyclopentadiene oligomers, to copolymeric aliphatic resins, and to cycloaliphatic resins.

1. HR Grafting with MA

Radical grafting of MA on unsaturated polymers has been intensely studied since the 1930s. The first studies envisaged the modification of dienic rubbers. Thus, in 1938 Bacon and Farmer, at the first conference on rubber technology, and in 1939 Dawson and Scott, based on data obtained in the reaction between MA and polyisoprene or polybutadiene in the presence of benzoyl peroxide, suggested two types of structure (taking into account inter- and intramolecular reactions) for modified polyisoprene [158,159]:

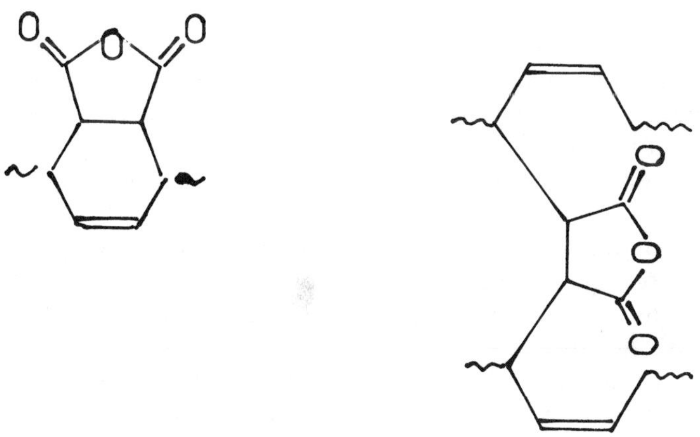

Intramolecular reaction   Intermolecular reaction   (4a)

The substitution takes place in the methylenic groups found in the alpha position to the double bond [160–162].

The reaction does not alter initial unsaturation of the oligomer, both in the case of thermal initiation and of radicalic initiation.

The presence of oxygen enhances crosslinking [162].

The best results are obtained in the absence of oxygen, in the presence of aromatic solvents, and at temperatures in the range 120 to 150°C.

In the absence of radicalic initiators, grafting requires higher temperatures.

In the case of thermal grafting, the presence of isomerization reactions of double bonds was discovered with the resultant formation of vinylidene groups:

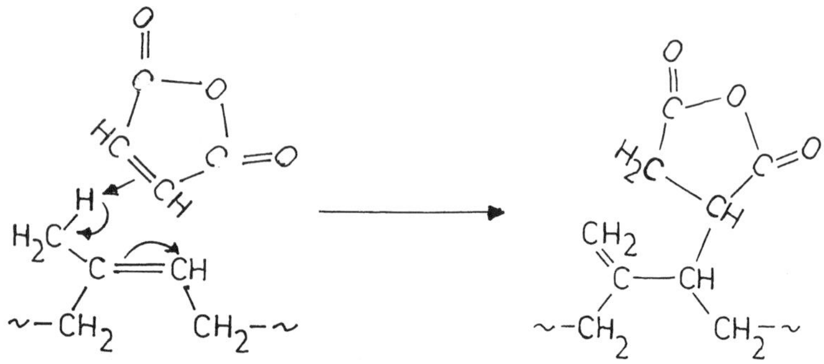

(4b)

Watson [163] has studied the effect of initiators on the reaction of rubbers with MA in order to find out some details of the reaction mechanisms. It was discovered that initiated grafting was inhibited and the thermal grafting was not influenced by the inhibitors. As a result, the two highly radicalic processes are carried out according to different mechanisms.

Supposedly, in the presence of free radicals a hydrogen atom is extracted from a methylenic group of the oligomer located in the alpha position to the double bond, followed by MA binding in the respective position. Secondary reactions can also induce the migration of the double bond from the 2,3-position to the 3,4-position [164].

For thermal grafting of MA, an electronic transfer mechanism was suggested [165].

Experimental conditions for the MA modification of HRs are quite different [143–147]. Organic solvents are preferably employed, and MA is introduced in amounts that vary in terms of maleized degree. The amount of initiator introduced represents 10% of MA quantity.

Liquid polydienes are more reactive to MA than CARs. This behavior is determined by the higher degree of unsaturation and polydiene microstructure. Modification of polydiene with MA occurs at 180 to 190°C in the case of initiated processes, and at 180 to 230°C in the case of thermal processes. The same processes take place in the case of CARs at 180 to 225°C and at 275°C, respectively [144].

HR modification with MA is accompanied by secondary reactions. Branches and crosslinks can appear by intermolecular grafting or by direct radicalic processes between species. Another possible reaction allowed by the presence of some impurities (Si, H_2O, Na^+, K^+, Ca^{2+}, Mg^{2+}) is MA decomposition, which can evolve toward explosion. To avoid secondary reactions, additives are introduced into the system (phenols, di- and trisubstituted phenols, Cu-naphthenate and acetylacetone, halogenated triazine, triazols, phenylenediamine, phenylenecatechol).

There is information regarding cationic initiation of this reaction [166]. The process takes place in the presence of Friedel–Crafts catalysts. Supposedly in this case the mechanism of reaction begins with electrophilic addition of the primary cation R^+ at the oligomer double bond, followed by the attack of carbocation formed on the MA double bond and the transfer toward a species contained in the reaction medium. Also possible are isomerizations of intermediate carbenium ions.

HR modification with MA can be also performed by means of radical-cationic mechanisms [167].

The reactivity of HRs diminishes with lower values of the unsaturation degree.

MA-modified MHRs [168] and aliphatic resins [85] are also known.

The HRs least sensitive to this reaction are indene–coumarone resins and styrene–indenic resins [150].

Grafting of unsaturated hydrocarbon resins with unsaturated mono- or dicarboxylic acids or with functional derivatives of them is less commonly used due to the occurrence of some secondary reactions. Some attempts to obtain HRs modified by this method are known, but the products obtained did not fulfill the performance required [169].

2. Co-Oligomerization of Diels–Alder Adducts of Conjugated Dienes with Filodienes That Contain Carboxylic Groups

Diels–Alder addition of conjugated dienes with filodienes leads to the formation of cyclic, unsaturated codimers:

(5)

Provided that unsaturated acids or their functional derivatives are used as filodienes, unsaturated and substituted cyclic substances are produced by this reaction. A particular case is the synthesis of unsaturated cyclic anhydrides through Diels–Alder reactions between dienes and the anhydrides of unsaturated dicarboxylic acids.

Cyclic conjugated dienes such as cyclopentadiene and the isomeric methylcyclopentadienes undergo uncatalyzed dimerization to yield bicyclic adducts containing unconjugated diolefinic unsaturation [170]. It has been found that a copolymer can be obtained by reacting MA and the dimers of cyclic conjugated dienes in the presence of free radical precursors. The products of the copolymerization process are co-oligomers of MA and the dimers of the cyclic conjugated diene, where the MA:dimer ratio in the copolymer ranges from 1:1 to 2:1. When the MA:dimer ratio is 1:1, the co-oligomer is unsaturated, soluble in polar solvents, and insoluble in hydrocarbons. The 2:1 MA:dimer co-oligomer is saturated, soluble in polar solvents, and insoluble in hydrocarbon solvents.

Although the actual structure of the co-oligomers is not known, it is believed that the unsaturated 1:1 MA/dimer co-oligomer and the saturated 2:1 MA:dimer co-oligomer have the following structural units, respectively:

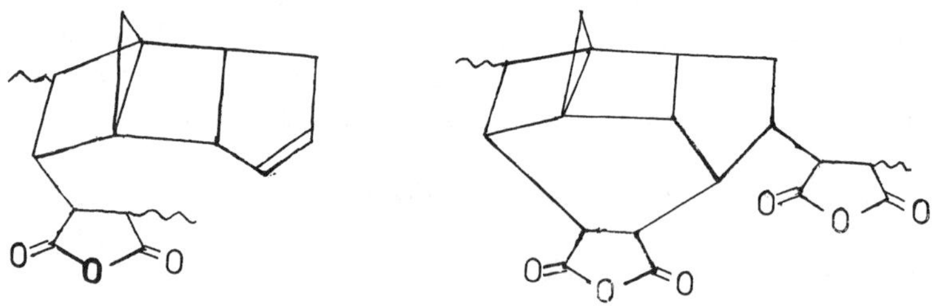

(6a)

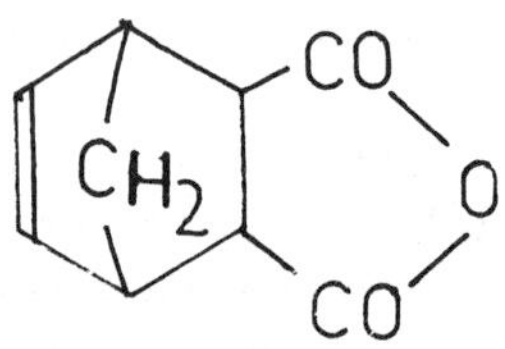

(6b)

The mere heating with MA of unsaturated hydrocarbon mixtures containing conjugated dienes leads to the formation of anhydrides of the following structures:

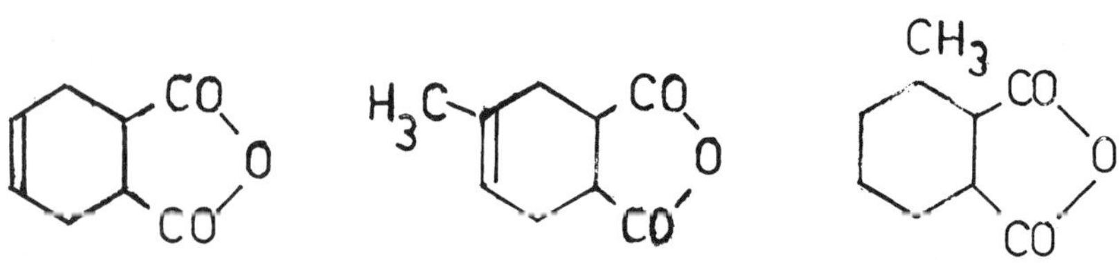

(7)

Reactivity of double bonds in these anhydrides is generally low. Notwithstanding this, modified HRs can be obtained with these anhydrides by using radicalic or cationic processes [171].

The polymerization process of Diels–Alder adducts allows the formation of products with high content in COOH groups with no secondary crosslinking reactions and no chromophore formation. The same approach makes it possible to obtain AHRs with carboxylic functions, since indene–coumarone and styrene–indenic fractions contain dicyclopentadiene, which can be converted into *endo*-methylene-tetrahydrophthalic anhydride by heating with MA.

Functionalized HRs with MA are characterized by the coexistence of some nonpolar and hydrophobic structural motifs with the polar, hydrophilic ones. The same structural motifs show their properties according to their chain frequency and microstructure. Functionalized HRs with MA are acidic, can be saponified, and can form fine, stable aqueous emulsions [172].

Introduction of a large amount of MA into HRs determines water solubility of the HRs. The same structural element determines HR compatibility with polar polymers and intensifies adhesive properties of HRs. In the same circumstances, the softening temperature of the HRs increases.

Before giving a detailed description of the use of functionalized HRs with COOH groups, we mention that these can be used in the paper industry, the automobile industry, and in the production of inks [143–147].

HRs functionalized with COOH groups can be diversified by saponification or esterification. The latter process allows the binding of some polymerizable structures (i.e., alkyl acrylates) [35].

Crosslinked HRs can take part in further polymerization reactions with some natural or synthetic resins (epoxy, phenolformaldehydic, aldehydic, unsaturated polyesters) [173–177].

Low molecular weight polybutadienes (LMWPBs) have been employed in the past for UV curable coatings. However, one must expose these LMWPB formulated coatings to UV radiation for an excessively long period time to achieve an acceptable level of curing.

One means to improve the UV curing time of LMWPBs is to react them with MA before formulation. A particularly useful procedure is the reaction of polybutadiene with MA in the presence of copper inhibitor [178]. The curing spread of this product is still unacceptable for many applications.

The reaction between LMWPBs and the MA results in a shift in the location of the olefinic bond on the LMWPBs. A typical reaction is

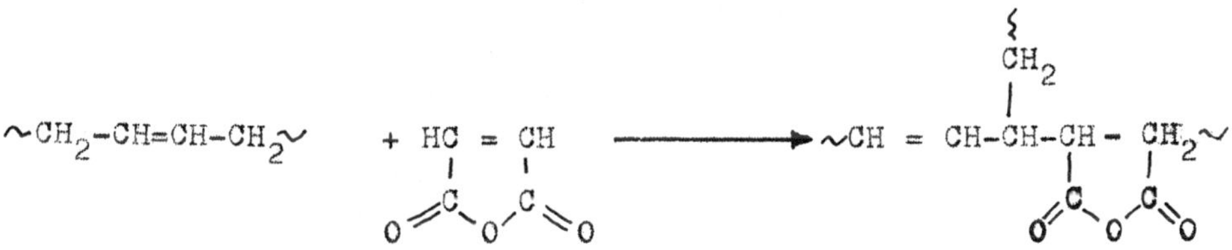

(8)

The modified LMWPB resin as prepared above is then reacted with hydroxyalkyl acrylate [179] or aminoalkylacrylate [180] of the general formulas

$$CH_2{=}C(R_1){-}COO{-}R_2{-}OH \quad \text{and} \quad CH_2{=}CH{-}COO{-}CH_2{-}CH_2{-}N(CH_3)_2 \qquad (9)$$

where R_1 is a hydrogen or methyl group and R_2 is an alkyl group of 1 to 10 carbon atoms. Suitable hydroxyalkylacrylates include 2-hydroxyethyl acrylate, hydroxypropylacrylate, and the like.

The reaction between the hydroxyalkyl acrylate and the modified LMWPB is typified in the following reaction formula:

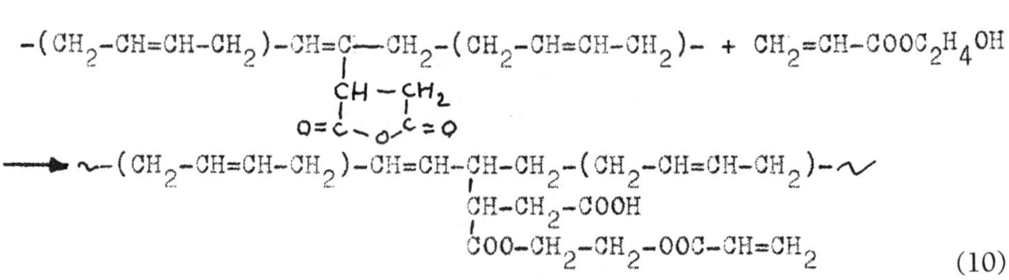

(10)

This new product is cured by any conventional curing technique including UV radiation, thermal baking, metal-catalyzed air drying, and the like. Because of the ease with which it cures thermally the resin could also be applied as a water-borne self-curing coating. Where one wants to employ UV curing system, one must add a photosensitizer.

Water-based paints in electrodeposition coating, brush painting, spray coating, and the like are becoming increasingly popular in the market because such paints are safe and economical and can form smooth, uniform films. It has heretofore been the practice to use natural drying oils such as tung oil and linseed oil and synthetic resins such as alkyd resins, epoxy-ester resins, and the like as starting polymers for vehicles of these water-based paints. After 70 years of technical developments, it has been found that natural drying oils may be substituted for by liquid polybutadienes such as LMWPBs containing 80% or more 1,2-bonds, LMWPBs containing 75–85% 1,4-bonds, and LMWPBs containing *cis*-1,4-bonds, which are reacted with MA and then at least partially neutralized with a basic compound to thereby make the addition product soluble or dispersible in water.

These compositions, however, have both merits and disadvantages for the resultant film properties. For example, when an LMWPB that contains predominantly 1,2-bonds is used as a starting material for such a composition, the resulting mixture forms films of high hardness and resistance to water, but of low resistance to corrosion and of such a high viscosity that there are practical difficulties in workability. When an LMWPB that contains predominantly *cis*-1,4 bonds is used as starting material for such a composition, such a coating composition will offer of high resistance to corrosion but will be of slightly lower hardness and water resistance. Therefore, attempts have been made to overcome these respective disadvantages by using suitable proportions of a mixture of such LMWPBs. A technique has been developed in which LMWPB that contains predominantly 1,2-bonds is used in combination with a second LMWPB that contains predominantly 1,4-bonds. The film obtained shows only intermediate properties of both polybutadienes. It has therefore remained impossible to obtain films that simultaneously possess both the needed resistance to corrosion and high resistance to water.

The addition of an alcohol before, during or after the neutralization results in an aqueous coating composition [181] that can form films that are not only superior in the resistance to corrosion and to water, but also have favorable physical properties such as hardness, adhesion, impact strength, and flexibility. They show excellent draining properties in electrodeposition paint and exhibit a remarkably improved performance in comparison with paints based on other LMWPBs such as *cis*-1,4-polybutadiene as the starting material [182].

Several attempts have been ventured to reduce the viscosity (use of a noncoloring stabilizer such as copper compounds, alkyl amines, ammonium salts, urethane, and urea compounds, etc.). Only a small addition of these compounds causes an undesirable effect on the air drying property and the obtained products are dark colored.

The gelation effect can be completely prevented without causing any ill effect to the reaction if the process is carried out in the presence of *p*-phenylene-diamine derivatives or imidazole [183–185].

A special category is made up of maleated LMWPBs, subsequently converted into succinimidic derivatives by amine curing. The products thus obtained show adhesive properties or can be used as dispersants without ashes for high-speed motor oils.

B. Hydrogenated HRs

The main purpose of hydrogenation is bleaching and increasing the stability of HRs [186]. The process aims the saturation of double bonds that are easily oxidizable and possess chromophoric properties.

Hydrogenation processes for HRs are described in numerous publications and patents [187–192]. As catalysts, Ni Raney, Ni deposited on the inorganic supports, Pt, Pd on alumina powder, and other catalysts based on metallic oxides are used. Pressures of 200 atm, temperatures up to 200°C, and high catalyst concentrations are recommended.

Hydrogenation determines a reduction of HR softening temperatures [193,194]. Interesting effects have been noticed in the case of LMWPB hydrogenation with a high content in 1,2-structures. In this case, hydrogenation determines a modification of LMWPB physical characteristics, and in the case of LMWPB hydrogenation with a high content of 1,4-structures, solid and crystalline products are obtained [35]. Hydrogenation of telechelic polybutadienes determines an increased viscosity [195], and in the case of aromatic HRs an improved stability is seen [188]. In case of indenic, inden-coumaronic, and styren-coumaronic resins, hydrogenation prevents the formations of fulvenes, which produce an intense coloration of these resins.

While being preserved, coumaronic resins modify their color toward darker hues. The cause of these phenomena is the formation of some chromophore with fulvenic type structure as a result of the slow oxidation of indene end groups, with the appearance of carbonylic groups, which condensate with CH_2 groups:

(11)

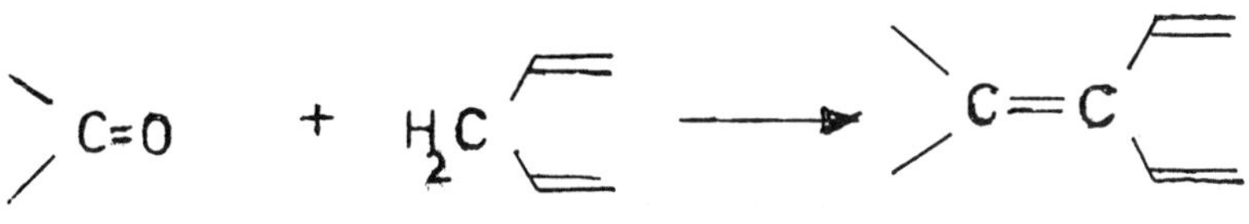

(12)

The strongly colored system of double conjugated bonds from the fulvenic structure can be hydrogenated, but the color reappears over time. That is why it is preferred that the initial unsaturation in indene is diminished.

In indenic terminal units, three areas susceptible to be hydrogenated can be distinguished (Fig. 6). Hydrogenation advances progressively in the three areas in decreasing order of reactivity: I > II > III. The first phase of hydrogenation occurs at low hydrogen pressures and room temperature, in the presence of Ni Raney. The attack of second area requires rather more stringent conditions, while area III is hydrogenated at high hydrogen pressures and a temperature of 200°C. The products obtained by HR hydrogenation find various applications [196].

C. Epoxidized HRs

Due to its unsaturation, LMWPB can be epoxidized. It was shown on the occasion of the first epoxidized LMWPB (Oxiron 2000) that the lowest epoxidic oxygen required for an adequate strengthening is by 3% [197,198]. The process of epoxidation can be applied to a range of oligomers: anionically obtained LMWPB, radicalically obtained oligoisoprene, liquid butadiene–styrene copolymers, cyclopentadienic and dicyclopentadienic resins thermally obtained, liquid or solid unsaturated aliphatic HRs cationically obtained, and mixed HRs cationically obtained. Epoxidized products have a complex structure. They can contain remaining unsaturation, oxiranic groups and other oxygenated functions resulting from secondary reactions of epoxidic cycle opening accompanying the epoxidation process, OH groups, and aryloxy, ketonic, aldehydic, and etheric groups [95].

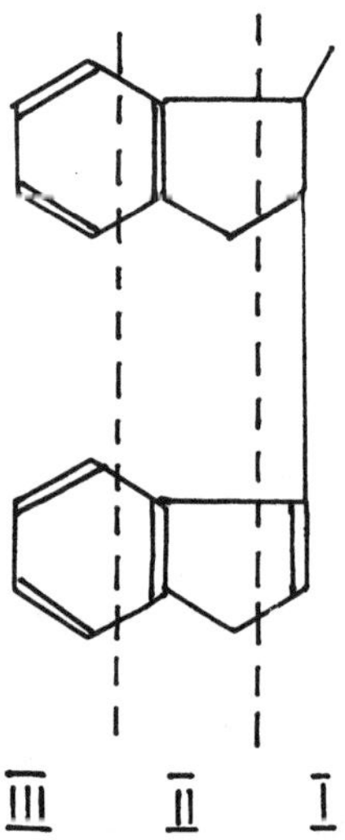

Figure 6 The hydrogenation of indenic terminal units.

Based on economic considerations, in 1962 it was foreseen that the epoxidized LMWPBs would compete with classical epoxy resins and would exceed the yield of these latter ones until 1975. This rapidly resulted in the appearance of a large number of patents and publications with regard to synthesis and uses of expoxidized LMWPBs [199–202].

Epoxidized polybutadiene commercialization was begun in 1960 by Food Machinery and Chemical Corp., which produced three types of modified resins: Oxiron 2000, 2001, and 2002; this was the only manufacturer until 1964, when production ceased.

After 1965 it was discovered that expoxidized polydienes cannot substitute for classical epoxy resins, since they are inferior in some properties, especially a high viscosity brought about by the oxiranic oxygen content [95]. Possessing other valuable properties epoxidized polydienes, represented in the main by epoxidized polybutadiene, have found their own specific uses, in which they cannot be rivaled by epoxy resins. Literature still appears in regard to some liquid polydiene epoxidation, covering applications of manufactured products as well as new commercial products [203–205].

Epoxidation processes are based on the reaction between unsaturated polymers (solubilized in organic solvents) and peracids organically preformed or obtained *in situ* from the respective acids and oxygenated water.

1. Mechanism of Epoxidation by Peracids

Bartlett [206] in 1960 suggested a mechanism of olefin epoxidation with peracids, based on nonionized molecules and containing a three-atom cycle in the transition complex:

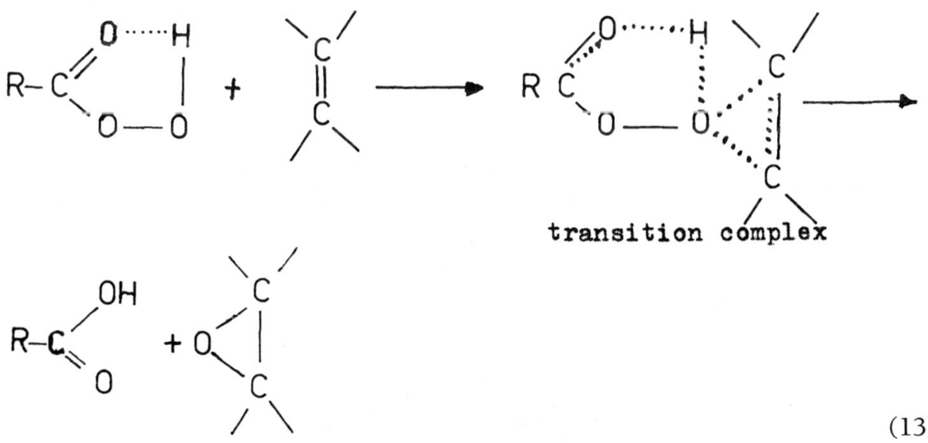

(13)

This mechanism allows a satisfactory interpretation of numerous epoxidation reactions.

Reaction (13) is assumed to be of electrophilic character, with the oxygen atom taking part in the formation of the three-atom cycle in the transition complex. This assumption cannot be proved satisfactorily.

Based on the supposition that peracids can be taken as hydroxy-substituted dioxyranes, Kwart and coworkers [207,208] suggested the following mechanism in which the formation of the transition complex has the character of a 1,3-dipolar addition:

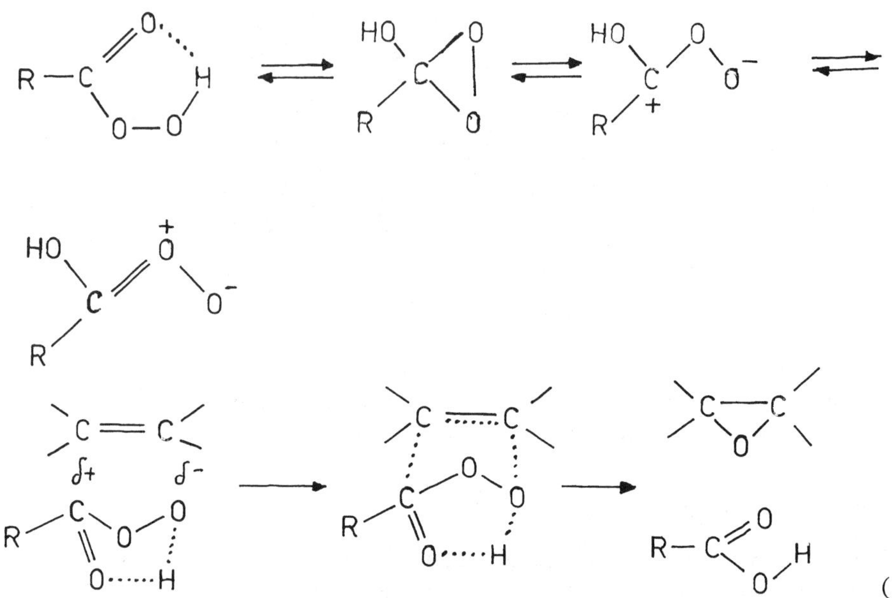

(14)

Mimoun [209] assumes a mechanism based on the formation of a complex π with charge transfer between nucleophilic olefin and electrophilic carbon atom of the peroximethylenic group:

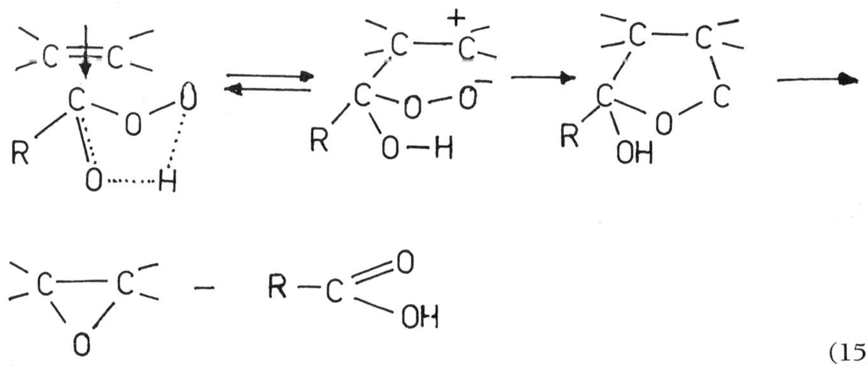

(15)

Because the epoxidation rate is greater in chlorinated solvents than in the aromatic ones, aromatic solvents were considered to compete in their quality of π-donors with double bonds [201]:

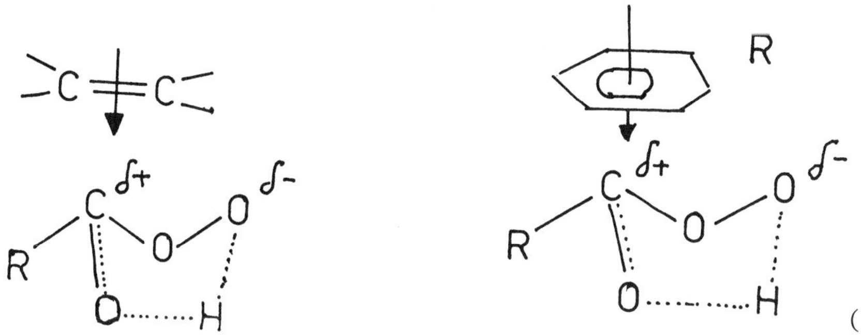

(16)

In the case of epoxidation with peracids formed *in situ*, the process occurs in a markedly heterogeneous environment, with oxygenated water and carboxylic acid to be found in an aqueous phase, and the unsaturated polymer and the solvent forming the organic phase. The peracid is formed in the aqueous phase, and it diffuses in the organic phase, where it attacks the double bond and the resulting carboxylic acid has a tendency to return the aqueous phase. This cycle continues until exhaustion of oxygenated water.

Most aliphatic carboxylic acids show a very low rate of conversion into peracids under the action of oxygenated water at temperatures below 40°C and in the absence of catalysts [211]. Acid catalysts substantially shorten the time to achieve the reaction equilibrium. These catalysts thus become indispensable for the process of epoxidation by the *in situ* method.

Epoxidation with alkylene oxides is a complex process generally accompanied by significant secondary reactions. Less selective are epoxidations with peroxides formed *in situ*. Nevertheless, these processes are preferred in the industry due to their economic advantages and lack of some technical risks [212].

One of the important secondary reactions is the opening of oxiranic cycles under the influences of carboxylic acid [213]:

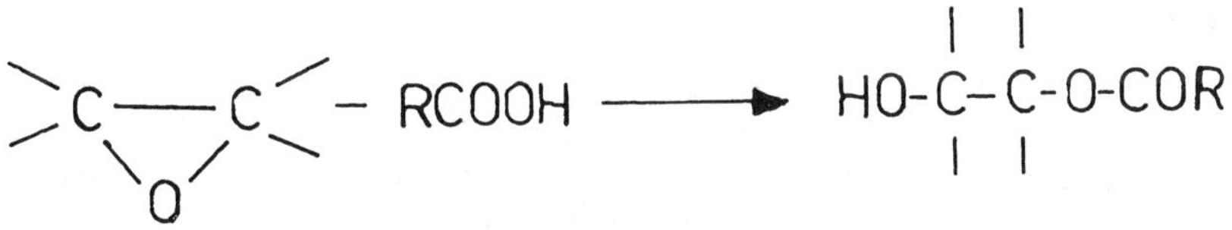

(17)

Hydrolysis of esteric groups is also possible:

$$\text{HO-C-C-O-COR} + H_2O \longrightarrow \text{HO-C-C-OH} + \text{RCOOH}$$

(18)

as well as reactions of oxirans with hydroxylic groups:

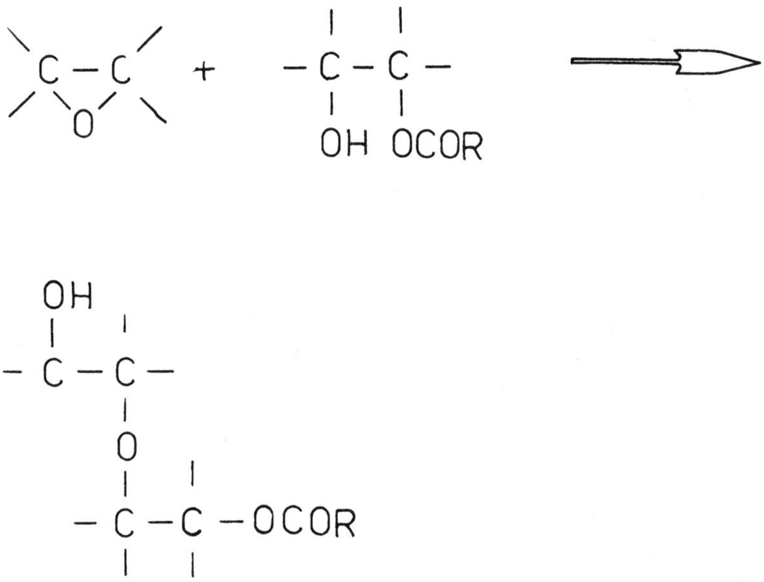

(19)

In media with increased acidity and alkyl substitution of double bonds of the initial unsaturated oligomers, isomerizations of epoxy cycles in ketonic or aldehydic groups may occur. Acid media and high temperatures enhance oligomerizations of oxiranic cycles, with the formation of etheric bonds.

The higher activity of performic acid on the double bond is explained by the highly electrophilic character of the carbon atom bond to the peroxy group, a fact that makes the formation of the σ complex with charge transfer easier. Notwithstanding all this, performic acid gives rise to some difficulties because, in the case of the *in situ* process, it also acts as an agent for the opening of oxiranic cycles. This disadvantage can be eliminated by reducing performic acid concentration, reducing the working temperature, and choosing an optimal duration of the process [214].

The structure of epoxidized oligomers depends on the microstructure of the initial products, on their degree of epoxidation, and on the working conditions employed. In terms of these factors, in the epoxidized polydienes besides oxiranic groups there have been identified unreacted double bonds, too, as well as other structural motifs like acyloxy, hydroxyl, ketonic, and aldehydic groups and etheric links.

Based on these findings, the following conventional structures for epoxidized polybutadiene can be formulated:

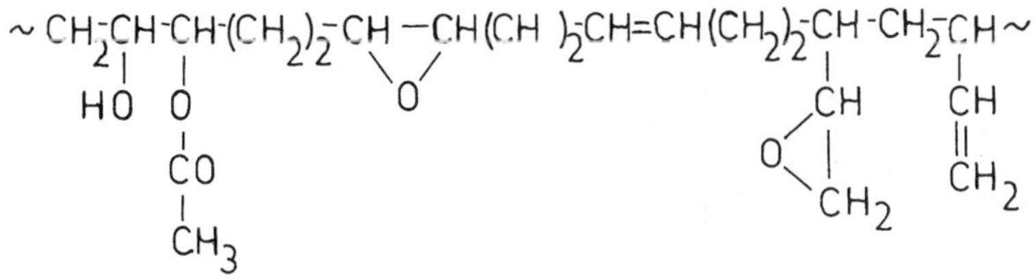

(20)

And for epoxidized polipiperilene:

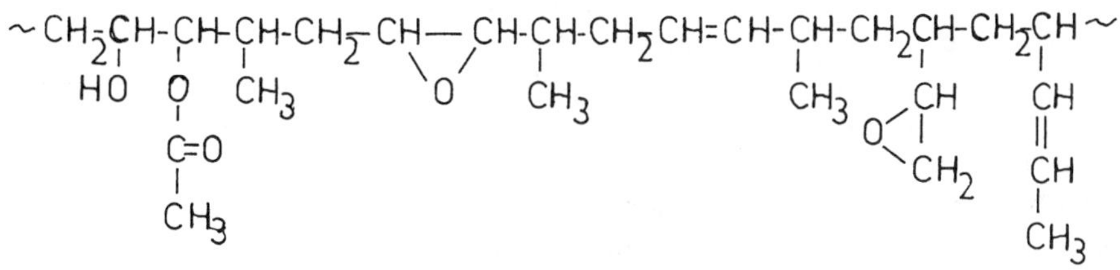

(21)

As regards the origin and the conditions favoring the occurrence of these structural motifs, the following statements have been formulated:

Double bonds come from initial polydiene.

Oxiranic groups represent the main product of the reaction between the double bond and peracids.

The —C(OH)—C(COOH)— structures come from the addition of carboxylic acid to oxiranic groups; their occurrence is favored by the high H^+ concentration, high temperatures, and a long duration of the process.

Ketonic groups, structures that are similar to alylic alcohol $>C{=}C{-}CH_2OH$, and the etheric bonds originate in the polymerization processes of epoxy groups.

2. Properties of Epoxidized Polydienes

The main property of epoxidized polydienes is their reactivity, due in the main to oxiranic groups and remaining unsaturation.

Crosslinking through epoxy groups is done by means of hardeners: polyamines, anhydrides of di- and polybasic acids, phenolic resins, polyphenols, polymercaptans, and carbon sulfur.

The process of crosslinking through epoxy groups can be also done ionically with adequate catalysts (e.g., BF_3), in which case the reaction occurs rapidly and at low temperatures.

The presence of OH groups provides additional possibilities of crosslinking with anhydrides, along with intermolecular reactions that accompany the main hardening reactions:

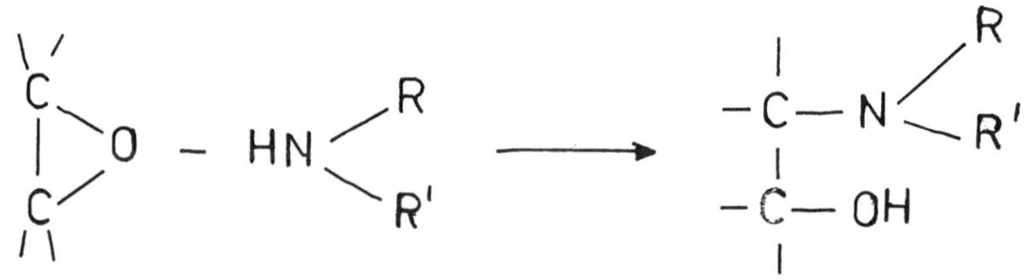

(22)

Unsaturated double bonds in epoxidized polydienes can participate in the hardening process with the aldehydes provided that peroxidic initiators are also introduced into the system. Thermorigid materials appear with satisfactory dielectric constants under high-temperature conditions [215].

The systems made up of epoxidized polydienes and vinylic monomers (e.g., styrene) with mixtures of (especially unsaturated) anhydrides react in ionic catalysis or with radicalic initiators, yielding solid products with various applications. Mixed with other unsaturated polymers, epoxidized polydienes can become solid materials in the presence of radicalic initiators or can participate in sulfur vulcanization of rubber mixtures.

A simple procedure for modifying epoxidized polydienes is their controlled hydrolysis. Thus, hydrophilic products can be produced with applications in the fibers industry [216]:

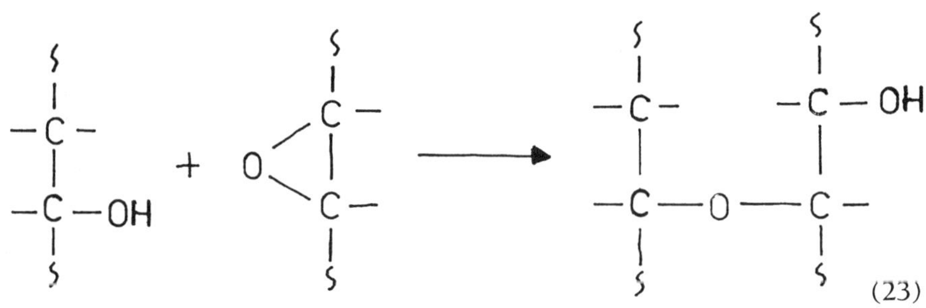

(23)

as well as their modification with the tertiary amines [217–221]:

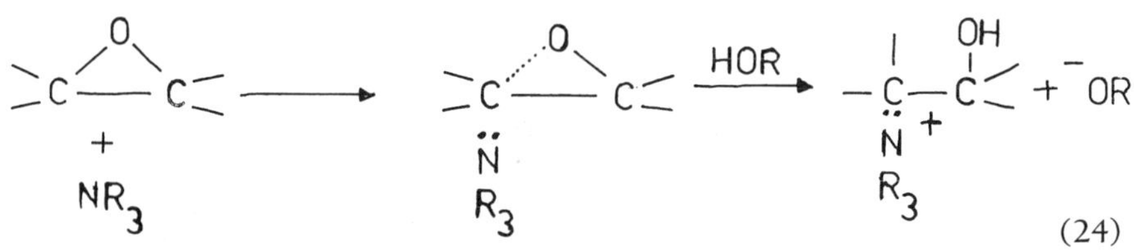

(24)

Numerous methods of modifying epoxidized polydienes aim at introducing into the structure of these resins some unsaturated groups with increased reactivity. One of these is the reaction of oxiranic groups with unsaturated carboxylic acids (acrylic, methacrylic, itaconic) [222] and, more frequently, with acrylic hydroxy esters, like 2-hydroxyethylmethacrylate, 2-hydroxyethylacrylate, and especially reaction products of the semiester type obtained by the reaction between hydroxyalkylmethacrylates and polycarboxylic acid anhydrides [223,224].

In general, epoxidation modifies the physical properties of liquid polydienes. Bleaching, increased specific weight, and increased viscosity are the main effects. If the first is an advantage, the last represents a disadvantage. It was found that this increase in viscosity is less prenounced in polydienes with an increased content in 1,2-structures.

D. Oxidized HRs

The earliest information with regard to oxidized HRs appeared between 1955 and 1966 and relates to oxidized LMWPBs (low molecular weight polybutadienes) launched by Phillips under the trade names Butarez 200 and Butarez 300 [225,226], and to liquid copolymers of butadiene with 20% styrene, obtained by Esso. Research under the trade names Buton 200 and Buton 300 [227–229].

Oxidized LMWPBs can contain 10 to 16% bound oxygen. Oxidation of polybutadienes and of butadiene–styrene copolymers is favored by the presence of hydrogen atoms bound to tertiary carbon atoms and activated by vinyl or phenyl substituents:

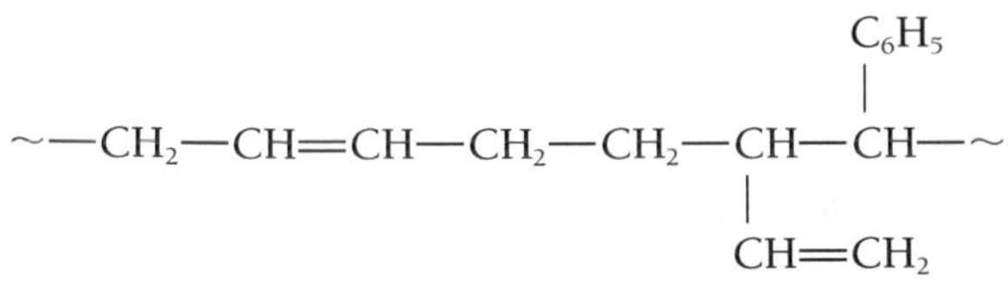

(25)

In the first phase of the oxidation process, hydroperoxides are formed, which subsequently decompose mono- or bimolecularly, generating free radicals:

$$RH + O_2 \rightarrow ROOH \quad (26)$$

$$ROOH \rightarrow RO^{\bullet} \rightarrow {}^{\times}OH \quad (27)$$

$$2ROOH \rightarrow RO^{\bullet} + ROO^{\bullet} + H_2O \quad (28)$$

Further on, the process has a radicalic character:

$$RO^{\bullet} + RH \rightarrow ROH + R^{\bullet} \quad (29)$$

$$R^{\bullet} + O_2 \rightarrow ROO^{\bullet} \quad (30)$$

Interruption occurs by recombination:

$$2\ R^{\bullet} \rightarrow R{-}R \quad (31)$$

$$R^{\bullet} + ROO^{\bullet} \rightarrow ROOR \quad (32)$$

$$2\ ROO^{\bullet} \rightarrow ROOR + O_2 \quad (33)$$

$$RO^{\bullet} + R^{\bullet} \rightarrow ROR \quad (34)$$

In this process, carbonylic functions can also appear:

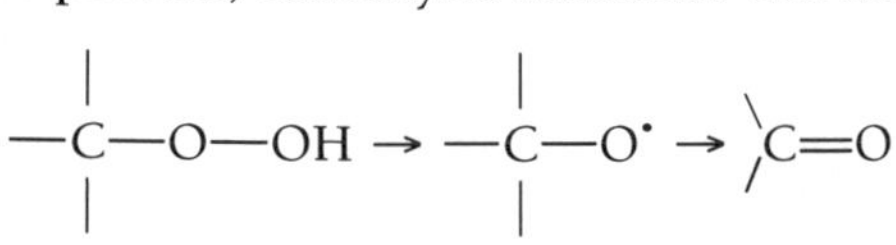

(35)

Oxidized HRs include the products NISSO GQ-1000, oxidized polyisoprene, and oxidized aliphatic and aromatic HRs [230,231].

The oxidation process can be achieved in solution with air in the presence of Co, Mn, Fe ocotates and naphthenates by discontinuous or continuous procedures [232–234].

By oxidation, modification of some characteristics of initial polymers takes place. Table 4 lists several chemical characteristics of liquid copolymer butadiene (80%)–styrene (20%), Buton 100, and its oxidized variants Buton 200 and Buton 300 [227,228].

At the same time, through the same process, HRs gain increased compatibility with other polymers and increase their capacity for oxidative crosslinking, becoming siccative synthetic oils [230].

E. Other Types of Chemically Modified HRs

Hydrocarbon resins can be halogenated by addition or substitution, the reactions taking place in solution by Cl_2 bubbling.

The presence of halogens in the HR structure manifests itself in terms of HR structure and halogen nature. Thus, aliphatic copolymer HR chlorination diminishes their thermal stability, and the elimination of halogens takes place at temperatures higher than 150°C [235]. In the case of cyclopentadiene homopolymers (sensitive to oxidizing agents), the addition of halogens improves their thermal and oxidative stability [236,237].

Chlorinated and sulfochlorinated aromatic HRs are known [231], as well as halogenated LMWPBs and commercial products of this kind. An example is brominated LMWPB marketed under the name Polysar RTV-Liquid Rubber DBBD [35,238].

Halogenated HRs can be employed as raw materials to obtain more sophisticated products. Thus, chlorinated oligocyclopentadiene is applied as a carrier for butadiene or isobutene grafting [239–241].

Maleated HR can later be modified by saponification with alkaline hydroxide [242–245], by copolymerization in emulsion with vinylic monomers, or by treating with ammonia, amines, diamines, and amides [246–251].

In the range of aromatic resins, a special significance is held by the modification with phenolic or phenol–formaldehyde resins. Unsaturated aromatic fractions (styrene–indene, indene–coumarone), mixed with phenol polymerize in the presence of Lewis or Bronsted acids (H_2SO_4, HCl, $AlCl_3$, or activated zeolites). Having removed the nonpolymerizable fraction, two fractions can be separated: a liquid one (isolated by vapors bubbling) and another, solid, one. Both fractions

Table 4 Chemical Characteristics of Buton 100, Buton 200, and Buton 300

Characteristic	Buton 100	Buton 200	Buton 300
Hydroxyl content (mg/KOH/g)	0	100	100
Oxygen content (%)	0	10	16
Unsaturation degree (g I_2/100 g)	330	270	210

can be separated in turn into subfractions that are soluble and insoluble, respectively, in alkaline bases. The values of molecular weights shows that the liquid fraction corresponds to the addition of a phenol molecule to an indene, coumarone, or dicyclopentadiene molecule. The solid fraction corresponds to the addition of a phenol molecule at dimers and trimers of the aforementioned monomers. No water results from these reactions. These data, as well as the observation that alkaline-base-soluble subfractions dissolve cellulose acetate while those that are insoluble in alkaline bases do not dissolve this polymer, led to the conclusion that addition of phenol to the double bond of monomers or oligomers can be realized by the phenolic proton; then, through transposition of the phenolic esters, free phenolic groups are again formed, the substituent being linked at the benzene ring, in *ortho* or *para* positions of the phenol (reaction 36).

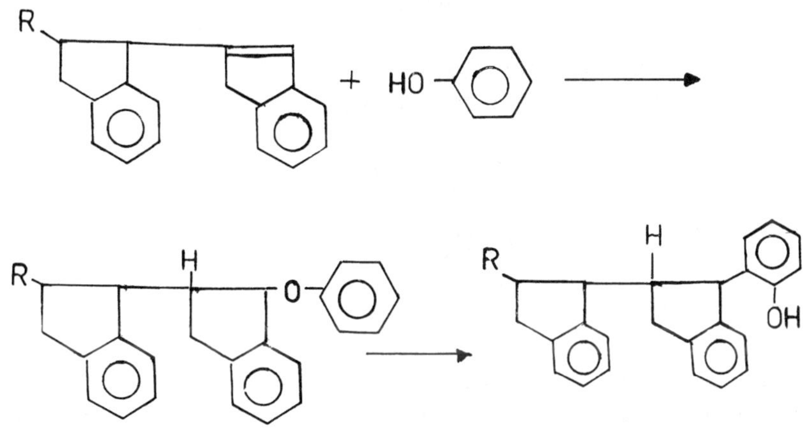

(36)

HR modification with phenol-formaldehyde resins depends on the nature of the latter. Indene-coumaronic resins do not react with resol type resins. Only novolac reacts with aromatic HR in the presence of benzoic acid or HCl, and a product of the following structure is being formed [252]:

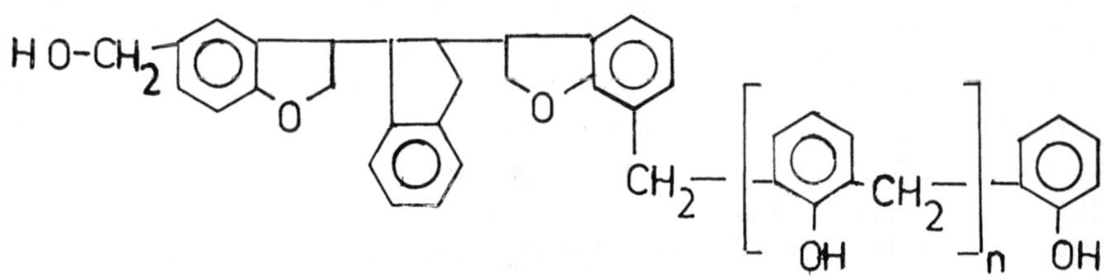

(37)

Other HR modifying methods and agents are known. Thus, aromatic HRs can react with aldehydes due to the reactive CH_2 groups they contain. Colored resins with fulvenic structure are formed [189]. Aldehyde treated inden-coumaronic fractions, in the presence of acid or basic catalysts, can form mixed resins wherein inden-coumaronic structures coexist with sequences specific for hydrocarbon–formaldehydic resins [253].

Volkov [254] studied modification of inden-coumaronic resins with furfurol and acetone, so that in the alkaline catalysis furfurol-acetonic resins bound to indenic resins are obtained.

Unsaturated hydrocarbon resins are able to react with unsaturated polyesters with epoxy–phenolic systems [255].

Liquid polybutadienes, especially those with OH terminal groups, can be modified with isocyanates. Already known is the commercial product NISSO-TP-1000, produced by the reaction with isocyanate of telechelic polybutadiene with OH groups [256]. Copolymeric resins can also be modified with polyisocyanates and urethanic prepolymers [257].

Owing to their double bonds, unsaturated HRs can be grafted with polymers or with various chemical functions. Thus, unsaturated HRs are dissolved in vinyl monomers; then they can be polymerized in solution or emulsion. In the same way, polystystyrene or poly(methyl methacrylate) can be grafted on liquid polybutadiene [35].

HR modification with siccative and semisiccative vegetal oils or fatty acids is frequently mentioned in publications and patents [258–270]. Products thus obtained can be further modified or used as such. Modification of vegetal oils with oligodienes is also possible. Thus, cyclooligopiperilenes can be linked on vegetal oils with a view to improve the physicomechanical properties of the products obtained [271].

F. Telechelic Hydrocarbon Resins

Polymerization methods for preparing telechelic liquid polymers (TLPs) have been described in the patent and technical literature. An excellent review article by D. M. French on functionally terminated butadiene polymers was published in 1969 [272].

Telechelic polymers have several advantages. First, being crosslinked through the terminal groups, their vulcanizates contain no free chain ends. Related to the absence of chain ends is the greater molecular weight between crosslinks that can be achieved with telechelic materials of the same prepolymeric molecular weight. The prepolymer must be made with a low viscosity so that fillers may be mixed into them. To have a sufficiently low viscosity, the prepolymers cannot have a molecular weight much greater than 6000. Granted that highest tensile strength values are obtained with a molecular weight between crosslinks (M_c) of about 10,000, the telechelic prepolymers will be easier to chain-extend with difunctional reactants than those of random type [273–278].

The telechelic polymers also have a greater advantage, which has not been sufficiently appreciated. When cured, the molecules are tied together through their ends; thus, the resulting network has a much narrower distribution of mesh sizes than when randomly located reactive groups are tied together. A number of very short chains between crosslinks are thus avoided, and the elongation at break of the cured stock is increased.

Telechelic prepolymers have been prepared in a number of ways, but since we are concerned with HRs, only free radical and anionic polymerization are pertinent (although some kinds of termination have been introduced by reaction of existing functional groups). Table 5 shows functionally terminated olefin addition

Table 5 Functionally Terminated Olefin-Addition Oligomers

End Group	Oligomer	Producer	Trade name	Ref.
COOH	Oligobutadiene	Thiokol	HC-434	
		Goodrich	Hycar CTB	
		General Tire	Telogen CT	
		Phillips	Butarez CTL	
	Co-oligo(butadiene–acrylonitrile)	Goodrich	Hycar CTBN	
	Co-oligo(butadiene–acrylonitrile-acrylic acid)	Goodrich	Hycar CTBN-X	
	Hydrogenated oligobutadiene	General Tire	Telagen S	
	Oligoisoprene	—	—	278
	Co-oligo(butadiene–isoprene)	—	—	277
	Oligostyrene	—	—	279, 280
	Co-oligo(ethylene–neohexene)	—	—	280
	Oligoisobutene	Enjay	Utrez	
OH	Oligobutadiene	Sinclair	R15, R45 M	
		Goodrich	Hycar HTB-Prim	
		General Tire	Telagen HT-Prim	
			Telagen HT-Sec	
		Phillips	Butarez HT	
	Oligoisoprene	—	—	275, 281
	Hydrogenated oligoisoprene	—	—	275
	Oligochloroprene	—	—	275
	Oligostyrene	—	—	282
	Oligoisoprene	UTC	Utrezdiol	
	Oligoethylene	—	—	283
Mercaptan	Oligobutadiene	—	—	284
	Oligobutadiene	Goodrich	Hycar MTB	
	Co-oligo(butadiene–acrylonitrile	Goodrich	Hycar MTBN	
Amine	Oligobutadiene	—	—	285
	Co-oligo(butadiene–styrene)	—	—	285
	Oligostyrene	—	—	286
Aziridine	Co-oligo(butadiene–styrene)	—	—	287
Epoxide	Oligobutadiene	—	—	288
	Co-oligo(butadiene–styrene)	—	—	288
Bromine	Oligoisobutene	—	—	288

polymers that were available commercially in 1968 or had been described in the literature by this time.

Several types of carboxyl and hydroxyl terminated polybutadienes are made in Japan. The Nippon Soda Co. offers liquid polybutadiene of various molecular weights [273]. These materials have high vinyl content (90%) and are of high viscosity. The carboxy-terminated polybutadienes (CTPBs) of molecular weights 3000 and 4000 are the main products commercially available from Japan at this time.

There are three ways that can be used to regulate the molecular weights of telechelic polymers produced by free radical methods: chain transfer, control of the initiator–monomer ratio, and the temperature of polymerization.

Flory [289] defined chain transfer as the process in which "the active center is transferred from one polymer molecule to another molecule, leaving the former inactive and endowing the latter with the ability to add monomers successively." However, the most efficient chain transfer agents contain active halogen, which adds to the growing polymer chain, and hence are not suitable in the preparation of telechelic polymers.

The instantaneous number average degree of polymerization (DP) is given by the following relation:

$$\overline{\mathrm{DP}}_n = \frac{k_p M}{f(k_d k_t)^{0.5}} I^{0.5} \tag{38}$$

where M = monomer concentration (mol/L), I = initiator concentration (mol/L), f = fraction of initiator radicals that initiate chains, k_d = rate constant for initiator decomposition, k_p = rate constant for chain propagation, and k_t = rate constant for chain termination.

The molecular weight accordingly varies with the monomer-initiator ratio $M/I^{0.5}$. Since both the monomer and initiator are consumed during the reaction and at different rates, this ratio will vary throughout the polymerization reaction. Nevertheless, for a given monomer and initiator the final molecular weight will be a function of the initial charge of monomer and initiator. In most cases the initiator is depleted faster than the monomer. Hence, the higher the amount of initiator added at the start of the reaction, the less $M/I^{0.5}$ will be reduced at a given conversion of monomer and the less the polymer molecular weight will be increased during the reaction. On the other hand, low initial catalyst concentration will lead to higher molecular weight products at the end of the reaction than at the beginning [275].

A more constant molecular weight product would be formed if $M/I^{0.5}$ could be held constant by controlled addition of initiator during the reaction [276,277]. The rate of addition R of initiator solution is given by the following relations:

$$R = \frac{VI(k_d - 2KI^{0.5})}{S} \tag{39}$$

where V = average volume of reactants in reactor in ml, S = concentration of initiator in mol/L, and

$$K = \frac{k_p k_d^{0.5} f^{0.5}}{k_t^{0.5}} \quad \text{in } L^{0.5}/\text{mol}^{0.5} \times \text{sec}$$

Knowing the rate of reaction, one can calculate the value of I needed at any time. At any degree of conversion this value can be inserted into Eq. (39) and the appropriate rate of initiator addition calculated. Typically, the rate of addition is two to four times as great during the first hour of reaction as during the last. To use this method, the kinetic constants must be known. In many cases, the kinetic constants for the polymerizing system are not known.

Free radical initiators that have been employed in the preparation of telechelic HRs are shown in Table 6. The most useful initiators are soluble only in polar solvents, and finding a common solvent for the initiator, butadiene, and polybutadiene has been a problem. Tetrahydrofuran is suitable; t-butanol and acetone containing a small proportion of water [277] have been employed.

The third method of controlling the molecular weight is to vary the temperature of polymerization. From Eq. (38), the DP varies with k_p and $(k_d k_t)^{0.5}$. Each of these rate constants varies with temperature in an Arrhenius manner. When the energy of activation of the processes is included, the combined rate constant expression may be written as

$$a \exp \frac{E_2 + E_3 - 2E_1}{2RT} \tag{40}$$

where a is a constant, E_1 is the activation energy of the propagation constant, E_2 is the activation energy of the termination reaction, and E_3 is the activation energy of initiator decomposition.

Table 6 Initiators employed in Free Radical Production of Telechelic Oligomers

Initiator	Structure	Solvent[a]
Azo-bis-isobutironitrile	$(CH_3—C— N{=})_2$ / \\ H_3C CN	Benzene
4,4′-Azo-bis(4-cyanopentanoic acid)	$(HO—CO—CH_2—C—N{=})_2$ / \\ H_3C CN	THF, 1-butanol acetone, water
Succinic acid peroxide	$(HO—CO—(CII_2)_2COO)_2$	THF
Glutaric acid peroxide	$(HO—CO—(CH_2)_3—COO)_2$	THF, acetone
2,2′-Azo-bis(4-carboxy-2-methylbutyronitrile)	$(HO—CO—CH_2—C—N{=})_2$ / \\ H_3C CN	Dioxane
Diethyl-2,2′-azo-bis-isobutyrate	$(C_2H_5OOC—C—N{=})_2$ / \\ H_3C CN	Toluene, methylcyclohexane
Hydrogen peroxide	H_2O_2	Water
4,4′-Azo-bis-(4-cyano-n-amylalcohol)	$(HO—(CH_2)_3—C—N{=})_2$ / \\ H_3C CN	Ethylene carbonate, DMF

[a]THF = tetrahydrofuran; DMF = dimethylformamide.

The variation with temperature thus depends on the relative magnitude of $E_2 + E_3$ and $2E_1$. The magnitude of E_1 for butadiene is 9.3 kcal/mol [289], whereas E_2 is usually in the range 2 to 6 kcal/mol and E_3 with a few exceptions varies from 23 to 40 kcal/mol [290]. Thus, in the case of butadiene, and indeed in the case of most monomers and initiators, $E_2 + E_3 - 2E_1$ is positive and the molecular weight decreases with increasing temperature of polymerization.

1. HTPB

The preparation of *hydroxyl-terminated polybutadiene* (HTPB) by free radical methods is not well documented. As shown in Table 5, four commercial organizations in the United States supplied these products in 1968. Two of these use ionic polymerization, and of the remaining two, Goodrich and Sinclair, Goodrich is believed to use a free radical method of preparation. The method of production of Sinclair's product is not known.

The Goodrich product might be made using 4,4′-azo-*bis*-(4-cyano-*n*-amyl alcohol) as initiator, since the nitrile band at 2330 cm^{-1} is present in their material. However, a strong ester band is also present, pointing to the possibility that the product is made by addition of an epoxide to a carboxyl-terminated polybutadiene.

Seligman [275] made HTPB by a roundabout method. He prepared an ester-terminated polybutadiene from butadiene and diethyl-2,2′-azo-bis-isobutyrate. This product was reduced to the alcohol with lithium aluminum hydride in refluxing ether for 2.5 h. It was treated with 10% hydrochloric acid and dried.

A process for the preparation of polymers using H_2O_2 as a radicalic initiator has long been well known. However, such a process, when used for the polymerization of conjugated dienes, will give the corresponding polymer in a low yield [291]. Thus, H_2O_2 has heretofore not generally been used as a radical initiator as often as other radical initiators such as benzoyl peroxide and α,α'-azo-bis-isobutyronitrile. Hydroxyl-terminated liquid polymers are in demand for polyurethane elastomers, which have excellent water resistance, and electrical and other properties and may widely be used in various industrial goods such as sealing, potting, coating, and damping materials. Attention has thus been drawn to polymerizing processes wherein H_2O_2 is used as a radical initiator. Therefore, processes for preparing such liquid polymers of a low cost have been sought in the industrial fields concerned. It is well known that the polymers obtained using radical initiators are characterized by molecular weights of wide distribution [292]. The polymers having molecular weights of narrow distribution will be obtained if the monomer and radical initiator are controlled to be always in a constant molar ratio in the polymerizing reaction system [293]. An alternative has been proposed, a polymerizing process using trihydrocarbylorthophosphoric acid ester as a solvent [294]. These proposed processes, however, are disadvantageous in that they use expensive chemicals in large amounts and are therefore uneconomical, and the polymers obtained need to be further subjected to complex purifying treatments.

A process for preparing HTPB having a narrowly distributed and controlled molecular weight in a high yield is achieved by copolymerizing a conjugated diene and a vinyl monomer, using H_2O_2 as the radical initiator, in the presence of at least one additive selected from halogen-containing compounds of nickel, palladium, and platinum and of silver, as well as silver nitrate and silver phosphate [291].

In the preparation of polyurethane elastomers by the reaction of the liquid polymer with a diisocyanate as the chain extender and a diamine or diol as a curing agent, the key factor is the uniformity of the molecular chain length (that is, the chain length between the crosslinking points) of the hydroxyl group–containing liquid polymer. This component plays the role of soft segment, in contrast with a hard segment where the cured material is chemically and physically crosslinked. The formulation of polyurethane with hydroxyl-terminated polybutadiene is therefore considered to be conducive to enhancing the physical properties of the resulting elastomers. In other words, it is desirable that the hydroxyl-terminated liquid polymers have straight-chain molecules, with molecular lengths as uniform as possible, in order to obtain satisfactory physical properties in the resulting polyurethane.

Another method that has been used to produce hydroxyl-terminated HRs is the ozonolysis of butadiene copolymers followed by reduction of the aldehyde produced with lithium-aluminum hydride or sodium borohydride [295–297].

The liquid HR functionalized with OH or epoxy groups may be used for preparing granular composition of carbon black and liquid HRs [298] for electrotechnical uses [299–301].

Poly(alkylene terephthalate)s undergo a gradual degradation due to thermal decomposition during processing. This degradation is due to thermal scission, and it is accompanied by a decrease in melt viscosity and a deterioration of the polyester's physical and mechanical properties.

Surprisingly, however, it has been found that small amounts of HTPB containing a high amount, say 85% mol, of the units of vinyl unsaturation vastly improve the melt stability of polyester resins [302].

2. MTPB and MTBN

Mercaptan-terminated polybutadiene (MTPB) and *mercaptan-terminated butadiene–acrylonitrile copolymer* (MTBN) are commercially available from the Goodrich Chemical Company. They are liquids used as sealants and adhesives. They are compatible with flexibilizing and toughening epoxide resins. The MTPB product has a molecular weight in the neighborhood of 3000 and viscosity of 200 poise. The polymer used in these resins is probably made by a free radical process. Yet it is difficult to see how a mercaptan-terminated product could be produced by a free radical mechanism. Mercaptans are good transfer agents, and if any concentration of mercaptan groups were present in a polymerizing system, the growing chains would remove hydrogen from them and be terminated.

3. Amine-Terminated Polybutadiene (ATPB)

Paracril RF, said to have been produced through a free radical mechanism, was distributed for a few years by the Naugatuck Chemical Co. This material was a liquid of viscosity 50 poise at 25°C and a $\bar{M}_n = 3500$.

The initiation step is very important in the production of liquid HRs by *ionic methods*. To obtain reproducible results, the initiator should be soluble in the butadiene solution. Solubility is easily obtained in the ethers commonly used as lithium solvating agents. To obtain the best mechanical properties and heat aging characteristics, it is desirable that 1,2-addition be kept to a minimum. It has therefore become customary to prepare initiator in an ether, dilute it with a hydrocarbon solvent, and then distill off as much of the ether as possible [303].

Phillip's Petroleum Company has been most successful in preparing initiators soluble in systems relatively free of ethers. They accomplished this by carrying out the polymerization in separate stages. An initiator such as 1-methyl naphthalene was reacted with lithium in diethyl ether for 40 h at −15°C. A slurry was obtained. This mixture was then solubilized by adding an amount of a diene monomer calculated to give a polymer chain 4 to 10 units in length. Dienes employed were isoprene, butadiene, or 2,3-dimethylbutadiene. This first portion of polymer contained a substantial vinyl content. The ether was then distilled from the solubilized initiator and replaced by cyclohexane. Polymerization was continued by addition of the major portion of the monomer to the solubilized initiator. The best description of Phillip's processes are in U.S. Patent 3,274,147, 1967, by Zelinky et al. and in British Patent 921,803, 1963.

Hydroxyl-terminated polybutadiene can be prepared from polymers having two living ends by addition of oxides, aldehydes, or ketones [279,282,285]. HTPB prepared by ionic methods is commercially available from Phillips Petroleum Company and General Tire and Rubber Company. The former supplies a secondary hydroxyl-terminated product and General Tire and Rubber Company produces both primary and secondary hydroxyl-terminated polybutadiene.

Reactions of the secondary material are easier to control. Secondary HTPB can be prepared by capping the living polymer with propylene oxide or aldehydes higher than formaldehyde.

Mercaptan groups have been introduced on living polymer ends by reaction with sulfur [279,286], cyclic disulfides, or cyclic sulfides such as ethylene or propylene sulfide by the following reactions [304]:

$$\sim CH_2^{-}\,Li^{+} + S \longrightarrow \sim CH_2\text{-}S\text{-}Li \tag{41}$$

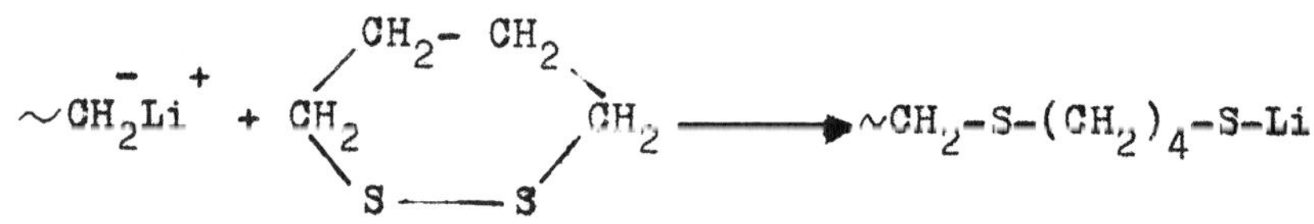

(42)

(43)

The lithium mercaptide groups are then hydrolyzed with acids.

Mercaptan-terminated polybutadiene prepared by anionic means is not commercially available.

4. Amine- and Aziridine-Terminated Polybutadiene

The living polybutadiene end group will remove a hydrogen from a primary or secondary amine and terminate. By including a group within the amine molecule with a greater tendency to add a carbanion than the amine hydrogen, or by the use of tertiary amines containing additive groups, a polymer terminated with amine groups may be obtained [285].

Analogously to their preparation of aziridine- and amine-terminated polybutadiene, Phillips has also prepared epoxide-terminated polybutadiene [288]. Either glycilaldehyde or various diepoxides were added to polybutadiene carbanion:

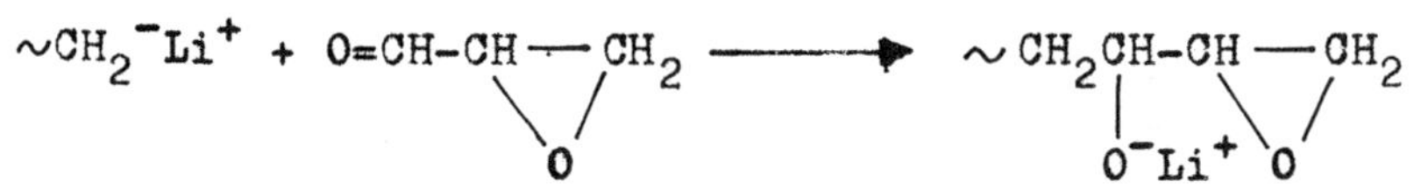

(44)

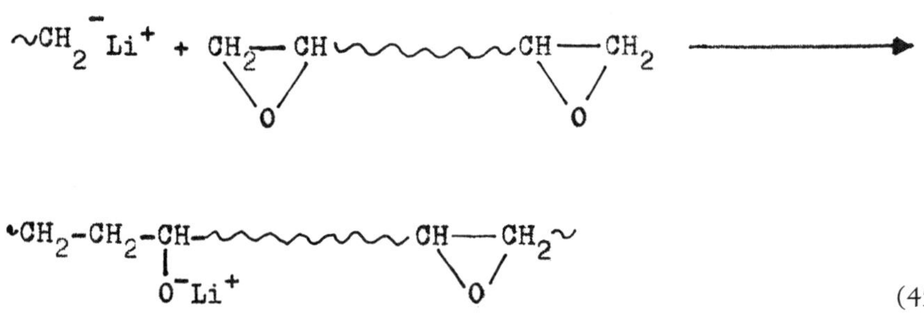

(45)

Characterization of liquid polybutadiene may include measurements of a large number of properties: molecular weight, equivalent weight, viscosity, microstructure, reactive group functionality, molecular weight distribution, plasticizer content, trace metal content, volatiles, and so forth. To these might be added the structure of the end group, whether primary, secondary, or tertiary, and the nature of the structure adjacent to the end group. Only the first six in the foregoing list can be regarded as characteristics of the oligomer itself.

Seven factors influence viscosity of CTPB resin: molecular weight, critical molecular weight for chain entanglement, presence of carboxyl groups, branching and carboxyl functionality, *cis-trans*-vinyl ratio, molecular weight distribution, and viscosity and concentration of solvents.

Dilute solution viscosity of CTPBs and its relation to molecular weight has been determined in different experimental conditions. The results obtained are presented in Table 7.

Intrinsic viscosity measurements [305] in various solvents point out that abnormally high values of Mark–Houwink constants were the rule in nonpolar

Table 7 Values of Constants K and a from the Relation $[\eta] = KM^a$ for Carboxyl-Terminated Polybutadienes (CTPBs)

$K \times 10^{-4}$	a	Solvent	Observations	Ref.
6.75	0.645	Toulene at 25°C	Ionic CTPB	305
7.71	0.634	Toluene at 25°C	Free radical CTPB	305
15.40	0.50	MEK[a] at 30°C	Ionic CTPB	305
20.00	0.50	MEK[a] at 25°C	Sinclair R45M	306
4.48	0.71	Toluene at 25°C	Sinclair R45M	306

[a]MEK = methyl ethyl ketone.

solvents, indicating a high degree of association. Chloroform, dioxane, ethyl acetate, and methyl ethyl ketone were sufficiently polar to break up the associations.

The microstructure, especially the *cis-trans*-vinyl ratio, in CTPB and HTPB depends largely on the method of manufacture. The product obtained in solution by free radical methods ordinarily is of 50 to 60% *trans* configuration and about 20% vinyl structure. The ionic material has a fine structure that depends on the polarity of the solvent during polymerization. In practice, the vinyl structure in CTPB and HTPB prepared by ionic methods is commonly 25 to 30% since some ethers are usually present during at least part of the polymerization reaction. *Trans* configuration will usually be lower than when the material is produced by free radical methods: about 40 to 45%.

French, in his excellent review [272], pointed out that fractions of molecular weight above 20,000 Da in both free radical and ionically produced CTPB and in free radical produced HTPB dramatically influence the physicomechanical properties of the products. The molecular weight distribution of the ionically produced material was narrower than that of the free radical types with the exception that much of Phillip's CTPB had a broader distribution than any of the other samples measured.

There is a discrepancy in the molecular weight ranges for the commercial CTPB resins. Part of the difficulty may be measurement of molecular weight in an unsuitable solvent, methyl ethyl ketone. Another possibility is the presence of high molecular weight material in certain lots of the same manufacturer's product but not in others.

The most useful curatives of CTPB are the aziridines and epoxides, whereas HTPB may be cured with isocyanates. HTPB has also been capped with isocyanates and the resulting prepolymer cured with low molecular weight diols or amines [307].

The Goodrich Company has taken advantage of the fast reaction rate of their CTPB products to devise room temperature curing systems.

Hycar reactive liquid polymers produced by Goodrich Company are homopolymers of butadiene or copolymers of butadiene and acrylonitrile. The isomer content is largely *cis/trans* with vinyl (1,2-addition of butadiene) being 25% percent or less. They have reactive groups in both terminal positions of the polymer chain end and, optionally, may have additional reactive groups pendant on the chain. Hycar contains no solvent or other unreactive components. There are three types of reactive functional groups commercially available. They are carboxyl, acrylate, vinyl, and secondary amine.

Hycar carboxyl-terminated liquid polymers may be considered long-chain dicarboxylic acids having functionalities between 1.8 and 2.4. They are primarily used as modifiers for epoxy resins. Benefits provided include improved impact properties, peel strength enhancement, low temperature shear improvements, and greater crack resistance. Hycar carboxyl-terminated polymers are presently used in many commercial applications; these include epoxy structural adhesives (aircraft, automotive, industrial), epoxy coatings, epoxy potting/encapsulation compounds, solid propellant binder compositions, and plastisol adhesives and sealants. Other applications include elastomeric sealants, epoxy RIM compositions, epoxy powder coatings, advanced epoxy composites, filament winding, and fraction materials.

Hycar amine-terminated liquid polymers have reactive secondary amine groups. They allow the epoxy formulator to achieve rubber modification by replacing a portion of amine curing agent with Hycar and still maintain stoichiometric balance. They are useful in formulating both ambient and elevated temperature curing epoxy systems. Hycar amine-terminated finds use in epoxy adhesives, epoxy maintenance coatings, geophysical cable fillers, foamed products, epoxy encapsulating compounds, automotive adhesives, castable elastomers, epoxy primers, epoxy sealants, and fiberglass-reinforced epoxy compositions.

Hycar vinyl-terminated liquid polymers (Hycar VTBN) have reactive acrylate or vinyl groups and can be reacted into systems involving curing by free radical mechanisms. The reactive vinyl groups are separate from the *cis/trans* vinyl unsaturation contributed by the polymerized butadiene of the polymer backbone.

Diluted with a monomer like styrene, Hycar vinyl-terminated butadiene crylonitrile copolymer, with the structural formula

$$H_2C{=}CH\sim\sim\left[(CH_2{-}CH{=}CH{-}CH_2)_m{-}(CH_2{-}\underset{\displaystyle CN}{\underset{|}{C}H})_n\right]\sim\sim\sim{-}CH{=}CH_2 \qquad (46)$$

and $M_n = 3400$, is a coating fluid of low viscosity that can cure to form rubbery products. In polyester manufacture, Hycar VTBN can modify the molecule by producing flexible domains. In polyester applications, Hycar VTBN can be used to increase toughness and fracture resistance.

Fiber-reinforced plastics fail mechanically in two distinct fashions: gross failure and microcracking. *Gross failure*—separation into two or more parts no longer connected—involves both fiber fracture and pull-out of fibers from the matrix. In this, the principal resistance to crack propagation is provided by inherent fiber strength and by the quality of their matrix bond. High-strength fibers contribute to improved fracture toughness or crack resistance since they require more energy to break. Perhaps unexpectedly, a weak fiber–matrix bond may be preferable to a strong one. If the fiber partially but not completely decouples from the resin, a greater volume of fiber must be stressed to fracture, and thus the energy required to advance the main crack through the composite is increased.

Microcracking, the second failure mode, occurs in the brittle resin phase of the composite. It results from high local stresses in the resin produced by differential thermal contraction and by the electric discontinuities represented by high-modulus fibers and the much lower modulus of the brittle matrix. Tiny cracks tend to propagate along fiber paths for great distances with significant macroscopic consequences: (1) reduction of composite stiffness and strength; (2) reduced resistance to corrosion; (3) poor fatigue resistance; and (4)increased mechanical hysteresis. Flexibilizing the brittle matrix is often done in industry. A flexibilized matrix will reduce or eliminate microcracks, but sacrifice of composite properties—lowered stiffness, strength, and heat resistance—frequently is unacceptable. A preferred remedy is to toughen the matrix resin by the inclusion of small rubber particles as an internal phase. Both modes of failures have been greatly reduced by incorporating reactive butadiene–acrylonitrile liquid rubber (Hycar VTBN) in the polyester resins. This additive forms discrete particles in the cured compound, with size range 5–25 μm. Apparently, these particles toughen the polyester matrix in a manner similar to technique used in high-impact polystyrene. The net result is an improvement in resistance to internal cracking and to impact damage [308].

Elastomer-modified epoxy resins have grown in use in the past 20 years in several application areas connected with structural adhesives, composites, civil engineering/construction, electrical laminates/encapsulants, and corrosion resistance. Some of this increased use has come about through the utility of telechelic butadiene–acrylonitrile liquid polymers.

Both carboxyl and amine liquid polymers have provided chemistries amenable to elastomer modification, with the polybutadiene–acrylonitrile copolymer providing solubility parameters close to if not equaling those of base epoxy resins.

In the mid-1960s CTBNs, with the formula

$$\mathrm{HOOC{-}R{-}[(CH_2{-}CH{=}CH{-}CH_2)_x{-}(CH_2{-}\underset{\displaystyle CN}{\underset{|}{CH}})_y]_m{-}R{-}COOH} \tag{47}$$

were introduced for the purpose of epoxy resin modification. These telechelic polymers are essentially macromolecular diacids. They offer processing ease (and therefore advantage) over the solid carboxylic nitrile elastomers.

Later [309], in 1991 amine reactive versions (ATBN) with the formula

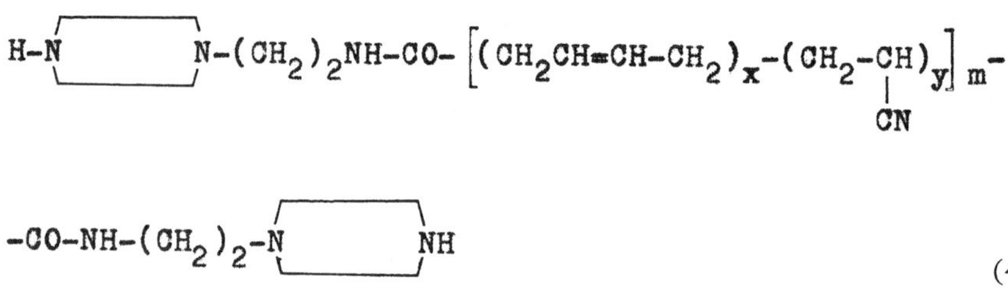

(48)

were issued, thereby offering another way to introduce rubbery segments into a cured epoxy resin network [310–312].

Commercial applications of Hycar–epoxy resin systems were at one time predominantly, indeed almost exclusively, the domain of the structural adhesives industry. Adaptation of some of these systems and the associative epoxy technology surrounding carboxyl-terminated liquid rubber have produced a growing Hycar–epoxy alloy consumption in epoxy applications other than structural adhesives.

A significant consequence of Hycar inclusion in epoxy matrix resin is greatly reduced crack density (surface area cracking per unit volume of sample). These systems have found utility in prepregs for various types of laminates, prepregs having "minimum bonding capability," centrifugal castings, filament windings, and certain types of brand lay-ups [313].

The basic needs for either thermocycling or thermal shock improvements in epoxy resin encapsulants (and, to some extent, in molded insulations and coatings) have occasioned the examination of the utility of elastomer-modified epoxy resin systems. Carboxy-terminated liquid polymers have been shown to provide a route to these improvements while maintaining a good balance between heat distortion temperature and desirable electrical properties.

A German patent [314] disclosed the CTBN-modification of an epoxy-ureaformaldehyde (aminoplast) system used for molding purposes. Rubber modification in this fashion give higher impact strength, tear resistance improvement, and flexural strength enhancement without harming the hardening properties and external appearance of the molded parts.

VIII. HYDROCARBON RESIN PRODUCTION AND CONSUMPTION

HR industrial production began in the United States when in 1920, using as a raw material some coke-chemical fractions, the early products of this category were obtained [315]. In 1927, Neville Cindu Ltd. was one of the largest HR manufacturers [316,317]. Indene-coumaronic resins were marketed in France by HGD (Huiles, Goudrons et Dérivés) in 1927.

The first LMWPB was produced on an industrial scale in 1925 in Germany under the trade name Plastikator 32 [318]. In the United States products of this category were obtained in 1950 [318].

The year 1938 marked the start of terpenic resin production [319], while in 1940, the United States produced the first aromatic HRs using petroleum fractions as raw materials. Europe's aromatic HR production lagged to 1964 [320,321].

Mixed and cycloaliphatic copolymeric resins were achieved in the period between 1960 and 1970 [322]. Thus, in 1965, the main types of HRs were already known and commercialized, with a remarkable production capacity.

For total HRs, one sees that it was the United States alone that in 1950 produced 50,000 t/year, after which the yield volume continually increased to reach 90,000 t/year in 1957, 130,000 in 1963, and approximately 160,000 t/year in 1968. Although petroleum aromatic fractions were used as raw materials to obtain HRs as late as 1940, it was only after 10 years that these fractions covered two-thirds of the whole HR yield volume. Further development of this industry was exclusively based on petroleum-origin raw materials.

Based on published data [323–332], development of HR production in the United States is given in Table 8. From these data, one sees a strong growth until 1963, followed by a standstill at about 150,000 to 160,000 t/year until 1972, followed again by a growth in the year 1978. In the following period, some industrial capacity development for the production of HRs in the United States are

Table 8 HR Production in the United States

Year	1954	1960	1962	1963	1964
Production (t/year)	99,500	120,000	145,000	156,000	156,000
Year	1965	1966	1968	1971	1978
Production (t/year)	147,000	153,000	151,000	161,000	182,000

known. Thus, Exxon Chemical Co. extended to 91,000 t/year the capacity of the plant in Baton Rouge, Louisiana, for Escorez type resins, while Esso Chemicals began the production of Escorez 5000 [333]. A plant for producing polyterpenic resins was at the same time created at Neville Island by Neville Chemicals Ltd. [317].

Goodyear Tire and Rubber Co., extended its plant at Beaumont, Texas, to a 45,000 t/year capacity in 1975 [334], while Hercules Inc., extended its HR yield capacity in the period 1975 to 1977 from 40,000 to 56,000 t/year [335–338].

We do not have recent data available with regard to the entirety of HR production in the United States. Early data estimate a whole capacity of 340,000 t/year [339]. On the other hand, the same source mentions that Pennsylvania Ind. Chem. (PICCO) alone should produce approximately 312,000 t/year [339].

The main manufacturing countries in western Europe are France, Germany, the Netherlands, and England. In France, there are two manufacturing companies: C. d. F. Chimie, specializing in aromatic resins commercialized under the trade name of Norsolene, and Esso Chimie, manufacturing aliphatic, aromatic, and mixed HRs, under the trade name Escorez from raw materials of petrochemical origin. Up to 1972, C. d. F Chimie made use only of raw materials of carbochemical origin; then it began to manufacture aromatic HRs making use of petrochemical raw materials and its own technological process. Initial yield capacity was as much as 10,000 t/year; in the period between 1974 and 1979, this doubled. The process was afterward licensed to Brazil, India, and Spain [340–352].

Esso Chimie based on its own research achieved in 1981–1982 HR yield capacities as high as 7500 t/year sited at Notre Dame de Gravenchon [353]. In 1983 its capacity increased to as much as 14,000 t/year by developing Escorez 5000 with possibilities of further increase to a capacity of 18,500 t/year [354–361]. According to data published in 1985 [357], Esso Chimie initiated a new investment of 50,000,000 francs to increase the capacity of this plant.

In the Netherlands, the HR market is dominated by two companies: Hercules and Neville Cindu. The former built up in 1973 a plant with a capacity of 15,000 t/year based on a Mitsui Petrochem. technology, which uses as raw materials petroleum residues [362–365], and later on C_5 and C_9 fractions [366,367]. In 1984, the same company built a factory for producing hydrogenated HRs [368]. Neville Cindu holds in the Netherlands a capacity of 9,000 t/year profiled on indene-coumaronic resins [369].

In Germany, several companies are involved in HR production. Among these, the most important are Esso and Rütgers Werks [320]. In 1965, a yield capacity for 8000 HR t/year was announced, but for the subsequent period of time we have no information [370]. Hoechst and Reichold Chemie announced acquisition of licenses from Nippon Petrochemical [371–374], and Esso Chemie GmbH has since 1983 increased its yield capacity for Escorez 5000 to 10,000 t/year [333].

Great Britain entered the HR market with ICI in 1973, using a 10,000 t/year plant. As raw materials, petrochemical fractions C_4 and C_5 were employed. The products obtained have been marketed under the trade name Imprez [375–379].

License acquisitions in Australia and Italy have been known [373,378].

For all western Europe in 1974, a 110,000 t/year HR yield is estimated: 45,000 t/year aliphatic and 65,000 t/year aromatic HRs. The raw material sources employed were 20,000 t terpenes, 35,000 t carbochemical fractions, and 55,000 t petroleum fractions. It is thought that at present western Europe produces more than 200,000 t/year HRs.

The oligomer industry, which can produce fine chemicals with high added value, is one of Japan's best industries. It is therefore quite understandable that during the past several years, manufacturers have focused steady attention on oligomers and that various oligomers have been industrialised or are in research and development (see Table 9).

Organic industry, high polymer chemistry, and polymer technology are the background to the oligomer industry, and consequently, this has been developing as an information-intensive industry that gives high priority to technology. It is undeniable, however, that the research and development methods so far employed have leaned rather toward what might be called phenomenalism.

To develop the oligomer industry more effectively and produce new oligomers that can meet societal demands, it will therefore be necessary to achieve full commercialization. When an effort is made to develop a synthesis of a new oligomer, it will in the future also be necessary to develop original molecular designs for oligomers that have so far been regarded as simply intermediate substances. Such designs must be natural extensions of low molecular weight organic compounds and high polymers.

There is also a good possibility that oligomers with original structures, functions, and properties will be created by the positive introduction of such methods as template polymerization and solid-phase polymerization, or by a new concept, "synthetic control."

In addition, there are biologically active substances the development of which the oligomer industry should emphasize in the future.

All these ideas appeared in the minds of Japanese experts as early as the 1960s. Their great merit is that they did not forget them and persevered to achieve them.

To the data listed in Table 9, we can also add that in the early 1960s the Japanese market was dominated by the indene-coumaronic resins (11,000 t/year) produced by four companies: Fuji Iron and Steel Corp., Nittsutsu Chem. Ind. Co., Mikuni Kasei Co., and Mitsubishi Chemical Ind. Co. [381]. Later, the oligomer industry developed in such a way that by the beginning of the 1980s, the total oligomer yield of Japan reached 80,000 t/year [382]. Japanese companies later invested a lot in the production of aromatic HRs by using raw materials of carbochemical and petrochemical origin, reaching in the years of 1980–1990 a total capacity of 160,000 t/year [383,384].

Tables 10 and 11 give data regarding Japan's potential regarding aromatic HRs (obtained from petrochemical sources), and aliphatic, cycloaliphatic, and mixed HRs.

Summarizing the foregoing data shows that the United States was the first to develop HR manufacture at a growing rate maintained as late as 1963, when a standstill appeared, growth resuming in 1972. One of the causes of this standstill is the growth of vinyl resins, which competed with HRs for some applications in building engineering. Later, HRs penetrated into other fields (e.g., adhesives) and reached former capacities that later increased, at a rather lower annual rate, however. This change in the dynamics of yield evolution in the United States can also be due to the emergence of other manufacturers, especially in Japan, as well as in

Table 9 Production of Major Oligomers in Japan

Raw material	Oligomer	Manufacturer	Application
Ethylene	Olefin	Mitsubishi Chem.	Higher alcohol Detergents Epoxy alkanes
α-Olefins	C_7–C_{17} alcohols	Mitsubishi Chem. Mitsubishi Petrochem.	Surfactants Detergents
	α-Olefin oligomers	Lion Fat & Oil Co.	Plasticizers
Ethylene oxide	Polyethylene glycol	Sanyo Chemical Nippon Soda Yokkaichi Chem. Nippon Oil & Fats	Surfactants Cosmetics Drugs Lubricants Paints Adhesives
	Crown ethers	Nippon Soda	Catalysts Reagents
	Liquid polysulfides	Toray Thiocol	Sealants Ahsesives Rocket fuels
Ethylene imine	Polyethylene imine	Nippon Shokubai Kagaku	Paper making Fiber reforming Dyeing industry
Fluoroethylene	Fluorine-containing oligomers	Daikin Kogyo	Lubricating oils Additives for rubber
Propylene oxide	Polypropylene glycol	Asahi Denka Kao-Atlas Sanyo Chem.	Polyurethanes Surfactants Alkyd resins
Glycerine	Polyglycerine	Daicel	Solubilizing agents Emulsifiers
Acrylonitrile	Adiponitrile	Nippon Petrochem.	Nylon intermediates
Butadiene	Cyclododecatriene	Toray Toyo Soda Mitsubishi Petrochem.	Nylon 12 Plasticizers Polyurethanes
	Liquid polybutadiene	Nippon Soda Nippon Zeon Sumito Chem.	Paints Insulating materials
n-Butene, isobutene	Low molecular weight polybutene	Nippon Petrochem.	Lubricants Modifiers Insulating materials
1,3-Pentadiene, isoprene	Liquid polypentadiene	Nippon Zeon	Plasticizers Cosmetics Adhesives

Source: Ref. 380.

Table 10 Evolution of Aromatic Petroleum Resins from C_9 Fraction in Japan in the Period 1976–1983 (in thousands of tons)

Producer	Trade name	1976	1979	1980	1981	1982	1983
Dainippon Inc.	Petrosin	44	40	44	44	44	44
Nippon Petrochem. Co.	Neopolymer	18	18	18	18	18	18
Toho Petroleum	Hiresin	12	12	12	—	—	—
Toyo Soda	Petocal	18	18	18	18	18	18
Fuji Kosan	?	—	—	—	6	6	6

Source: Ref. 385–391.

some European countries, South Korea, and Australia. Thus, South Korea in collaboration with Nippon Petrochemical Co. built a plant for HR production, and Australia by a singular endeavor built a plant for HR yield in order to avoid importing from Pennsylvania Ind. Chem. Corp. [392–403].

Quite significant in the last 15 years is the growing number of HR manufacturing countries, in western Europe, Latin America, Asia, and in eastern Europe. Thus, in 1984, Indian Petrochemical Corp. Ltd. commissioned a plant for aromatic HR production with a capacity of 5000 t/year. The plant is based on the technology worked out by C. d. F. Chimie [404–413]. C. d. F. Chimie's license was granted also in 1978 to the Brazilian company Petrochimica Unidosa with a view to build up a 10,000 t/year plant in São Paolo [409–411]. In South Korea there are two aromatic HR manufacturing companies, each with a capacity of 5000 t/year: Korea Polyester and Kolon Petrochemical Co. Both use the procedure elaborated by Nippon Petrochemical Co. Information is that the Spanish company Union Explosives is negotiating with Nitsui Petrochemical Industries with a view to building a plant of capacity 20,000 t/year petroleum HRs [412,413]. Australia has had since as early as 1977 an Escorez type HR plant.

China developed by itself a plant for inden-coumaronic resins. Its production became well known in 1979, when the export of these products began to some Asian countries [385].

Table 11 Evolution of Aliphatic Copolymeric HRs and Alicyclic HRs in Japan in the Period 1976–1983 (in thousands of tons)

Producer	Trade name	1976	1979	1980	1981	1982	1983
Tonnen Petrochemical	Escorez	10	12	10	12	12	12
Hitachi Chemical	Hirez	5	6	5	8	8	8
Nippon Zeon Co.	Quintone	10	10	8	21	21	21
Sumitomo Chem.	Tackirol	7.5	5	5	5	—	—
Marusen Petrochemical	Marukarez	—	—	—	—	7	7
Toho Petrochemical	Hiresin	—	13.5	—	15	15	15
Arakawa Forest Industries	Arakon	—	7	6	7	7	7

Source: Refs. 385–391.

Table 12 Commercialized Hydrocarbon Resins

Type	Trade name	Producers	Observations
Functionalized liquid polybutadiene	Lithene	Lithium Corp., U.S.	Partial cyclizaiton, M_n = 1000–8000
		Revertex Ltd., England	
		Metallgeselschaft, Germany	
	Ricon	Colorado Chem. Specialities Co., U.S.	Copolymer with styrene, higher content of 1,2-bonds, M_n = 1000–6000
	Intolene 50	International Synthetic Rubber, England	Higher content of 1,2-bonds
	Poliöl	Chemische Werke Hüls, Germany	Higher content of *cis*-1,4- structures, M_n = 1000–4000
	Nisso B	Nippon Soda Co. Ltd., Japan	Structure 1,2 = 90%; M_n = 1000–4000
	Hystl B	Dinachem Corp., U.S.	Structure 1,2 = 90%
Telechelic polybutadiene	Butarez	Phillps Petroleum, U.S.	OH or COOH end groups
	Polybond	Arco Chem., U.S.	OH end groups
	Hystle	Dinachem Corp., U.S.	OH or COOH end groups
	HC 434	Thiokol, U.S.	COOH end groups
	Hycar	BF Goodrich, U.S.	OH, COOH, or halogenated end groups
	Nisso	Nippon Soda Co. Ltd., Japan	OH and COOH end groups, higher content of 1,2-bonds
Modified liquid poly-butadiene	Nisso	Nippon Soda Co., Ltd., Japan	8% epoxy groups, 10% maleic anhydride content
	Butarez	Phillips Petroleum, U.S.	Oxidized hydrocarbons
Low and medium molecular weight polybutadiene	Oppanol	BASF, Germany	Liquid
	Vistar	Advance Solvents & Chem. Corp., U.S.	Liquid
	Indopol	Amoco, U.S.	Liquid
	Cosdene	Cosden Oil & Chem. Corp., U.S.	Liquid
	Oronite	Oronite Chem. Corp., U.S.	Liquid
	Vistanex	Enjay Chem. Corp., U.S.	Liquid and semisolid
	P-10, P-20	Russia	Liquid and semisolid

Aliphatic and copoly-meric resins	Hirez	Hitachi Chemicals, Japan	Good compatibility with other resins, natural oils, and rubbers
	Quintone	Nippon Zeon Co., Japan	Obtained from C_5 fraction
	Marukarez	Maruzen Petrochemical, Japan	
	Imprez	ICI, England	
	Escorez	Esso Chemicals, U.S.	10% content of cyclodienes
	Betaprene	Reichold Chem., U.S.	Obtained by diene or α-olefin oligomerization
	Piccodiene, Piccopole	Pennsylvania Ind. Chem., U.S.	Obtained by diene or α-olefin oligomerization
	Hexalyn, Piccotac	Hercules Inc., U.S.	
	Wing-tack	Goodyear Tire & Rubber Co., U.S.	
Alicyclic and terpenic resins	Quintone (various types)	Nippon Zeon Co., Japan	
	Escorez 8000	Esso Chemicals, U.S.	Cyclopentadienic resins
	Piccodin	Hercules Inc., U.S.	Cyclopentadienic resins
	Neville	Neville Chemicals Co., U.S.	Cyclopentadienic resins
	Piccolyte	Pennsylvania Ind. Chem., U.S.	Terpenic HRs
Modified aliphatic and alicyclic HRs	Escorez 5000	Esso Chemicals, U.S.	Hydrogenated HRs
	Imprez	ICI, England	Maleinized HRs
	Quintone	Nippon Zeon Co., Japan	Maleinized HRs
	Arkon M100	Arakawa, Japan	Hydrogenated alicyclic HRs
Aromatic HRs	Norsolene, various types	C. d. F., France	Indene–coumaronic resins and aromatic petroleum HRs
	Escorez 3000	Esso Chimie, France	Aromatic HRs
	Escorez S 5000	Esso Chimie, France	Copolymer of vinyl toluene and α-methylstyrene
	Petrosin, various types	Dainippon Inc., Japan	Petroleum aromatic HRs
	Nisseki Neopolymer, various types	Nippon Petrochem. Co., Japan	Petroleum aromatic HRs
	Betaprene AR	Teichold Chemicals Inc., U.S.	Aromatic HRs
	Piccotex, Kristalex	Hercules Inc., U.S.	Copolymers of vinyl aromatic monomers

Table 12 Continued

Type	Trade name	Producers	Observations
	Nebony, Nevex, Nevchem, Neville, Necires, various types	Neville Chem. Co., U.S.	Aromatic HRs
	Piccotack, Piccovar, and other types	Pennsylvania Ind. Chem. Co., U.S.	Aromatic HRs
	Coumaron-Inden Harze, various types	Rütgerswerke A. G., Germany	Carbochemical and aromatic HRs
	Piroken	Bulgaria	Aromatic HRs
	Piroplast	Russia	Aromatic HRs
	Arkon	Arakawa, Japan	Aromatic HRs
Aromatic–aliphatic HRs	Petcoal	Toyo Soda, Japan	Mixed HRs
	Tack-ACE	Mitsui, Japan	Mixed HRs
	Hiresin	Toho Petroleum Resin Co., Japan	Copolymeric HRs
	Escorez 2000	Esso Chemicals, U.S.	Copolymeric HRs
	Klyrvel, various types	Velsicol Chem. Corp., U.S.	Copolymeric HRs
	Hercules AR	Hercules Inc., U.S.	Copolymeric HRs
Aromatic and modified mixed HRs	Hiresin 90	Toho Petroleum Resin Co., Japan	Mixed and maleinized HRs
	Piroplast 2	Russia	Maleinized aromatic HRs
	Petrosin 80	Dainippon Inc., Japan	Maleinized aromatic HRs
	Piroken 100 W	Bulgaria	Aromatic HRs
	Arkon P125	Arakawa, Japan	Hydrogenated aromatic HRs
	Neopolymer 140	Nippon Petrochem. Co., Japan	Maleinized aromatic HRs
	Coumarone-Indene Hatze	Rütgerwerke Co., Germany	HRs modified with phenol

Table 13 Hydrocarbon Resin Applications

Field of application	HR role	HR type applied	Observations
Rubber processing	Adhesives, extension agents, active fillers	Aliphatic, copolymeric, liquid polybutadiene, aromatic and mixed	Up to 50 parts in 100 parts rubber are introduced
Plastic processing	Adhesives	Ibidem	Applied in polyolefin processing
Lacquers, paints, printing ink	Substitutes for vegetable oils, protective coatings	Copolymeric HRs, terpenic HRs	Colorless types are preferred [418]
Road markers	Binders and adhesives	Copolymeric HRs	HRs with high softening temperature. Consumption increased by 15–20% in the years 1977–1980 [389]
Adhesives	Main component	Aliphatic, copolymeric, alicyclic, aromatic, and terpenic HRs	Annual rate growth of production is 10–12% [419]
Paper and cardboard industry	Gluing and hydrophobic agents	Aliphatic and maleinized mixed HRs	
Building engineering	Concrete waterproofing agents	Copolymeric and terpenic HRs	
Metallurgical industry	Binders for preparing casting molds	Aliphatic, aromatic, and mixed copolymeric HRs	Types with high softening temperature are preferred
Electronics, electrotechnical industry	Adhesives, sealants	Aromatic HRs	A promising domain of importance [420]
Additives to mineral oils	Viscosity regulators, dispersants	Maleinized or imidized polyisobutene	

Table 14 Physicochemical Characteristics of the Well-Defined Oligomers Based on Aromatic Hydrocarbons

Monomer	Oligomer	m.p. (°C)	b.p. (°C/mm)	n_D^{20}	Ref.
 Benzene	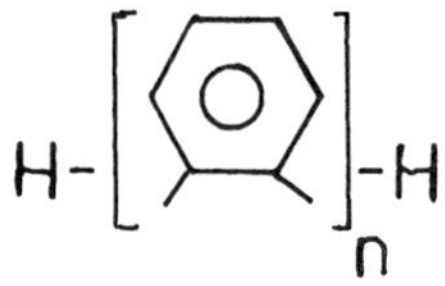				
	Oligo(phenylenes)				
	$n = 3$	59	332	—	422
	$n = 4$	119	420	—	422
	$n = 5$	217	—	—	423
	$n = 6$	520	—	—	424
	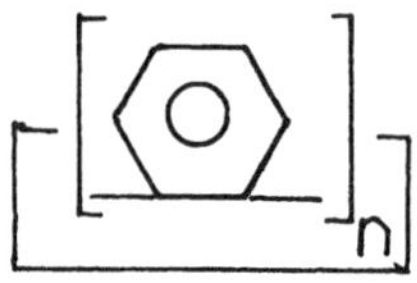				
	Cyclic *o*-oligo(phenylenes)				
	$n = 2$	111	—	—	425
	$n = 3$	196	—	—	426
	$n = 4$	233	—	—	427
	$n = 6$	335	—	—	428
	$n = 8$	425	—	—	428
	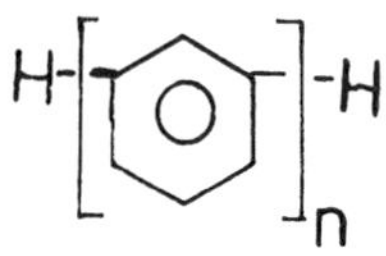				
	Linear *m*-oligo(phenylenes)				
	$n = 3$	89	365	—	429
	$n = 4$	86.5	419	—	430
	$n = 5$	117	—	—	431
	$n = 6$	148	—	—	430
	$n = 8$	159	—	—	430
	$n = 9$	195	—	—	432
	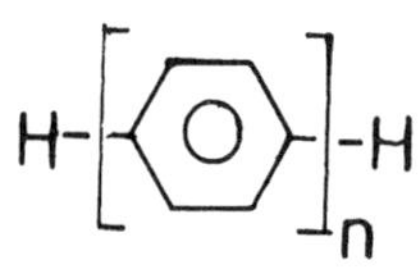				
	Linear *p*-oligo(phenylenes)				
	$n = 2$	71	156	—	433
	$n = 3$	215	376	—	434
	$n = 4$	322	428/10	—	435
	$n = 5$	395	—	—	436
	$n = 6$	465	—	—	437

Table 14 Continued

Monomer	Oligomer	m.p. (°C)	b.p. (°C/mm)	n_D^{20}	Ref.
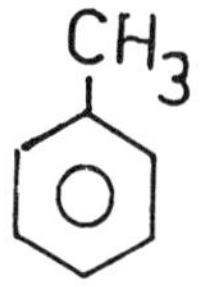 Toluene	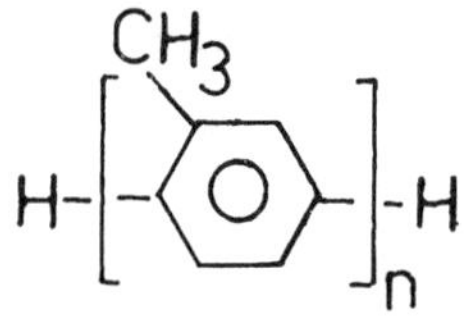				
	Oligo(3-methyl-*p* phenylenes)				
	$n = 2$	—	273	—	438
	$n = 3$	43	—	—	439
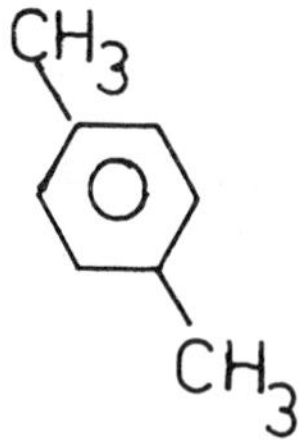 Xylene	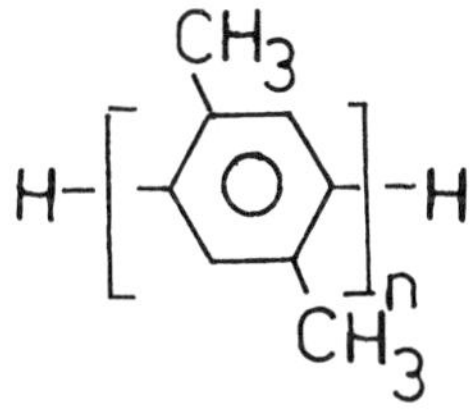				
	Oligo(2,5-dimethyl-*p*-phenylenes)				
	$n = 2$	53	—	—	
	$n = 3$	182	—	—	
	$n = 4$	264	—	—	
	$n = 5$	307	—	—	
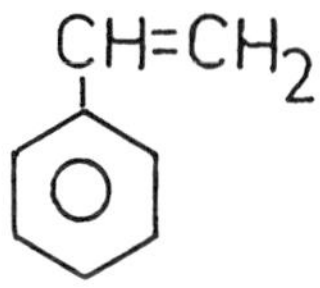 Styrene	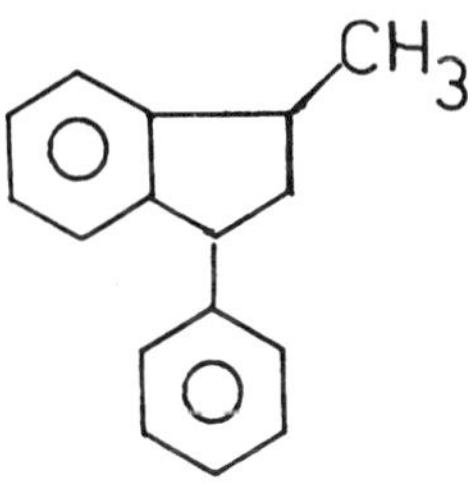 3-Methyl-1-phenylindane (2 stereoisomers)	9.5 25.5	168/16 157/12	1.5810 1.5809	
	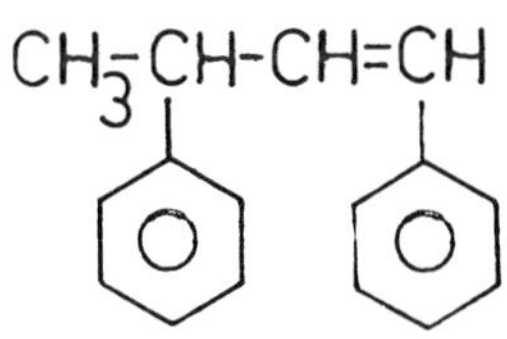 1,3-Diphenyl-1-butene	—	134/1	1.5930	
	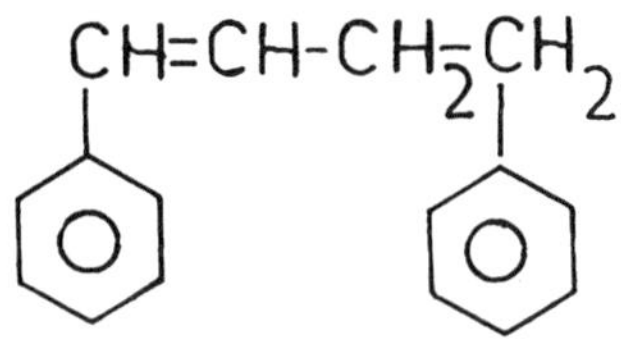 1,4-Diphenyl-1-butene	—	124	—	

Table 14 Continued

Monomer	Oligomer	m.p. (°C)	b.p. (°C/mm)	n_D^{20}	Ref.
	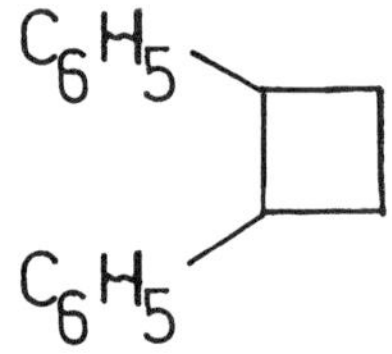 1,2-Diphenylcyclobutane	—	—	1.5913	
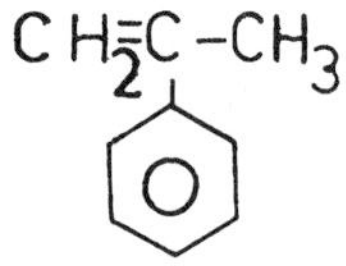 α-Methylstyrene	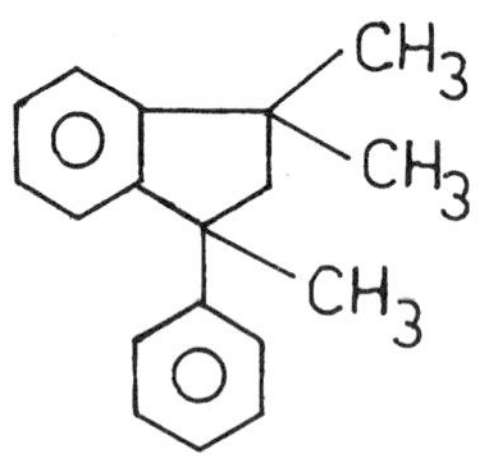 1,3,3-Trimethyl-1-phenylindane		158/10	—	444
	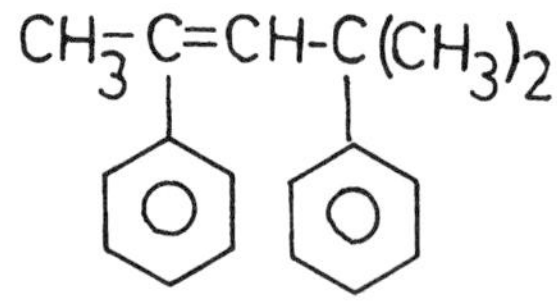 4-Methyl-2,4-diphenyl-2-pentente	52	166/15	1.5728	445
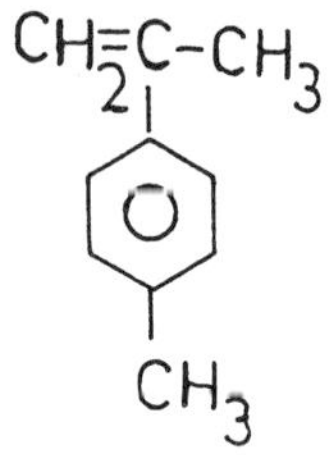 α-*p*-Dimethylstyrene	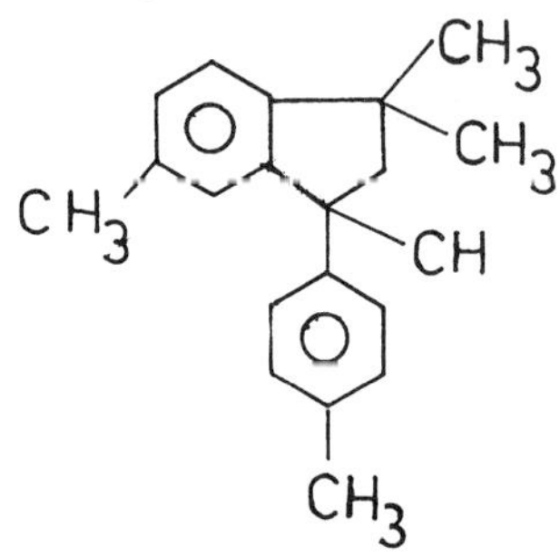 1,3,3,4′6-Pentamethyl-1-phenylindane	40	142/0.8	—	446
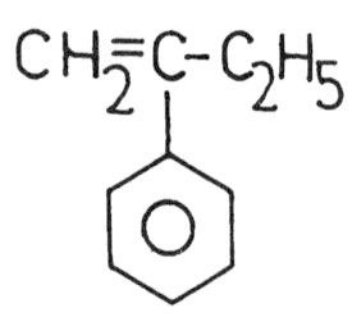 α-Ethylstyrene	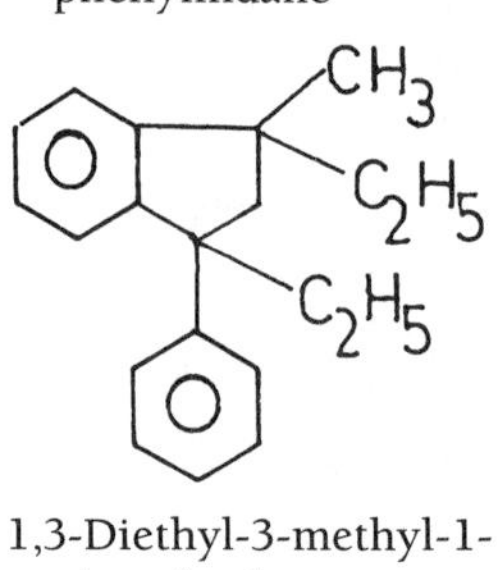 1,3-Diethyl-3-methyl-1-phenylindane	—	104/0.3	1.5642	447

Table 14 Continued

Monomer	Oligomer	m.p. (°C)	b.p. (°C/mm)	n_D^{20}	Ref.
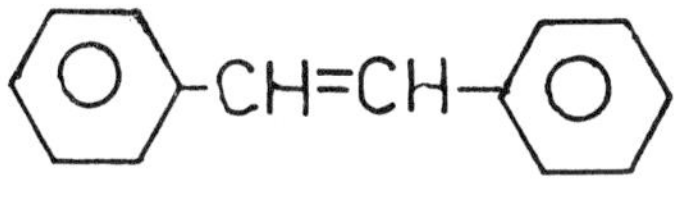 Stilbene	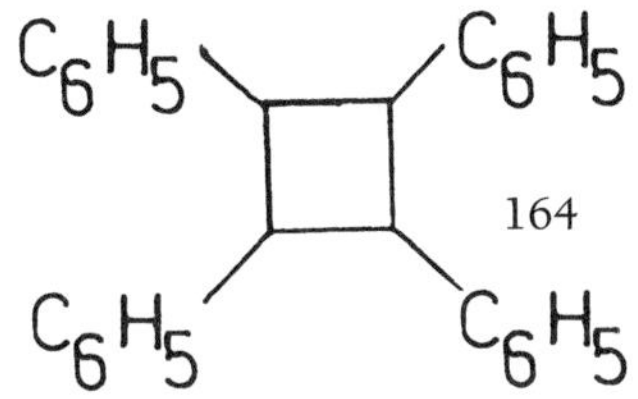 1,2,3,4-Tetraphenylcyclobutane	164	—	—	448
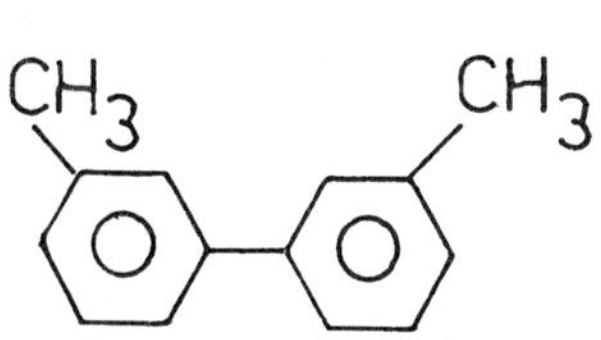 3,3′-Dimethylbiphenylene	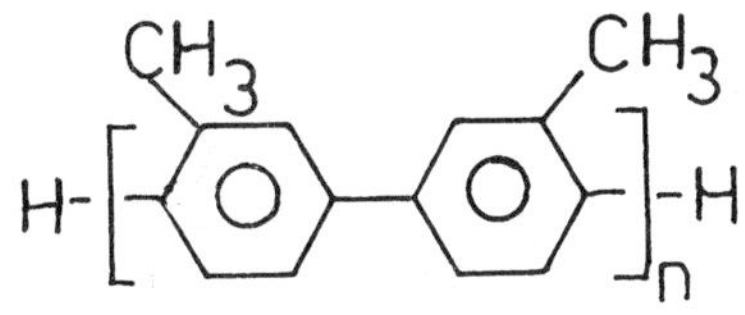 Oligo(3,3′-Dimethylbiphenylenes)				
	$n = 2$	76.5	—	—	449
	$n = 3$	142.0	—	—	450
	$n = 4$	273.0	—	—	451
	$n = 5$	285.0	—	—	451
	$n = 6$	298.0	—	—	451

Bulgaria elaborated a process for obtaining aromatic HRs using as raw materials C_9 fraction and mixed C_5–C_9 fractions. The product is commercialized under the trade name Pirolan and has been initially obtained in a discontinuous plant with a 15,000 t/year capacity.

Russia also manufactures HRs under the name Piroplast [414].

In Romania, technological processes for producing aliphatic and aromatic HRs have been elaborated. Plants with capacities of about 10,000 t/year were built in 1991. Table 12 lists the main products and manufacturers, and some HR characteristics.

The large gamut of HR types and formulations, the large number of manufacturers, as well as the fact that sometimes these deliver not only resins, but also various auxiliary materials, makes it difficult to establish with sufficient accuracy the volume of world production. Public estimates state a level of 500,000 t/year in the years 1975 to 1976 [415], almost 1,000,000 t/year in 1981–1982 [414], and ca. 1,500,000 t/year in the following period [416].

Data for 1967 on the distribution of applications of HRs in the United States showed the following percent distribution:

Rubber processing	21
Protective coatings	14
Floors	31
Adhesives and rolled plates	9
Hard cardboards	4
Impregnations	4
Metallurgical industry	3
Miscellaneous	14

The emergence of crosslinkable vinyl resins was an important moment, as these products successfully competed with HRs in manufacturing floor materials. This led to redistribution of HR consumption and finding new application areas for them. Formulation of new HR-based adhesives and the obtaining of materials for road markers makes up for the diminished consumption in the field of floors. Later, new ranges appeared: for example, thermoplastic polymer (polyolefin) processing, the manufacturing of compounds and protective materials for the electronics industry, and some branches in the machine building industry [417].

Tables 13 and 14 list the main current HR users and the physicochemical characteristics of well-defined aromatic oligomers, respectively.

REFERENCES

1. Mark, H. F., and Gaylord, G. H., Eds., *Encyclopedia of Polymer Science and Technology*, New York, Vol. 4, 1966, and Vol 8, 1968.
2. *Kirk-Othmer Encyclopedia of Chemical Technology*, Interscience, New York, 1980, Vol. 12.
3. Crough, W. W., and Shotton, J. A., Liquid polybutadiene, *Ind. Eng. Chem.*, 47, 2091, 1955.
4. Wheeler, H., U.S. Patent 3,478,225, 1969.
5. Kennedy, J. P., and Makowski, H. S., Reactivities and structural aspects in the cationic polymerization of homo- and diolefinic norbornanes, *J. Polym. Sci.*, C 22, 247, 1968.
6. Eldib, A., Technology of hydrocarbon Resins, *Hydrocarbon Proc.*, 42, 145, 1963
7. Greenlee, O. S., Hydrocarbon resins, *Ind. Eng. Chem.*, 49, 71, 1957.
8. Dumskii, M. V., Berent, A. D., Kozodoi, I. V., and Muhina, T. N., *Petroleum Resins*, Khimia, Moscow, 1983 (Russian).
9. Cai Tianhi, Selective oligomerization of isobutene in C_4 mixture using heteropolyacids as catalysts, *Shigou Huagong*, 15, 207, 1986 (Chinese).
10. Showa Denko, K. K., JP Patent 57,149,232, 1982; *Chem. Abstr.*, 98, 72905v, 1983.
11. Saido, M. J., JP Patent 63,221,847, 1988; *Chem. Abstr.*, 110, 115524f, 1989.
12. Showa Denko, K. K., JP Patent 57,149,233, 1982; *Chem. Abstr.*, 98, 7290w, 1983.
13. M'nachev, Kh. M., Oligomerization of isobutylene on decationized zeolites, *Neftekhimia*, 28, 164, 1988 (Russian).
14. Holpap, P., Continuous liquid-phase polymerization of isobutylene from C_4 hydrocarbon mixture, *Plaste u. Kautsch.*, 33, 6, 1986.
15. Stewart, R., Musser, T. M., and Kothari, G., CH Patent 556,380, 1976.
16. Raficov, S. R., SU Patent 690,024, 1979.
17. Smith, R. L., and Johns, J. W., U.S. Patent 3,959,175, 1976.

18. Beresniev, V. V., Servian, A. K., and Kirpichnikov, P. A., *Oligoisobutene*, Nauka, Moscow, 1974 (Russian).
19. Schammel, W. P., U.S. Patent 4,465,887, 1984.
20. Singleton, W. J., U.S. Patent 3,567,795, 1973.
21. Lapour, P., and Marek, M., Isobutylene oligomerization in the presence of Friedel-Crafts catalysts, *Makromol. Chem.*, *134*, 23, 1970.
22. Watson, J. N., U.S. Patent 3,985,822, 1976.
23. Gorbaty, M. L., BRIT Patent 1,497,643, 1978.
24. Joy, D. R., CER Patent 2,349,530, 1974.
25. Gao Zanxian, Oligomerization of 1-butene based on ethylaluminum sesquichloride/triphenylphosphine catalyst, *Dalran Gongymeyuan Xuebao*, *26*, 32, 1987 (Chinese); *Chem. Abstr.*, *107*, 237361j, 1987.
26. Gao Zhanxian, Oligomerization reaction of 1-butene. I. *Yingyang Huaxue*, *4*, 40, 1987; *Chem. Abstr.*, *106*, 176924a, 1987.
27. Friedlander, R., Make plasticizer olefins via n-butene dimerization, *Hydrocarbon Process.*, *Int. Ed.*, *65*, 31, 1986.
28. Masami, K., Butene oligomerization over ion-exchanged mordenite, *Ind. Eng. Chem. Res.*, *27*, 248, 1988.
29. Gaillard, J., FR Patent 2,504,122, 1982.
30. Gao Zhanxian, Oligomerization of 1-butene. II., *Fenzi Anhau*, *2*, 101, 1988 (Chinese); *Chem. Abstr.*, *110*, 154922d, 1989.
31. Prokof'ev, K. V., Oligomerization of 1-butene in the presence of a complex catalyst, *Neftepererab. Neftekhim.*, 1988, No. 8, p. 22 (Russian).
32. Watson, J. M., and Kysselberge, J. Van, Comments on the structure of low molecular weight butene polymers, *J. Polym. Sci.*, *Polym. Chem. Ed.*, *16*, 1173, 1978.
33. Cotten, J., U.S. Patent 4,113,804, 1977.
34. Frame, R. R., U.S. Patent 4,737,479, 1988; (a) Pittman, C. U., Sequential multistep reactions catalysed by polymer-anchored homogeneous catalysts, *J. Am. Chem. Soc.*, *97*, 341, 1975; (b) Jacobson, E. E., A new polymer anchored catalyst of $Ni(CO)_2(PPh_3)_2$; (c) Delman, B., and Jannes, G., Eds., Elsevier, Amsterdam, 1975, p. 83; (d) Pittman, C. U., Catalytic reactions using polymer-bound vs. homogeneous complexes of nickel, rhodium, and ruthenium, *J. Am. Chem. Soc.*, *97*, 1742, 1975; (e) Allum, K. G., Ni-cyclooctadiene derivatives as oligomerization catalyst of hydrocarbon resins, *J. Organometal. Chem.*, *87*, 189, 1975; (f) Capka, M., Butadiene oligomerization with polymer anchored palladium complex catalyst. Mechanism of reaction, *Coll. Czech. Chem. Commun.*, *38*, 1242, 1973; (g) Pittman, C. U., Wun, S. K., and Jacobson, S. E., Oligomerization reactions catalysed by polymer abchored catalysts, *J. Catalysis.*, *44*, 87, 1976.
35. Luxton, R. A., The preparation, modification and applications of nonfunctional liquid polybutadienes, *Rubb. Chem. Technol.*, *54*, 546, 1981; (a) Minsker, K. S., and Sangalov, Iu. A., Isobutylene and their oligomers, Khimija, Moscow, 1986 (Russian).
36. Hsish, H. I., Structure and molecular weight distribution of oligomers resulted in oligomerization of isobutene, *J. Polym. Sci.*, *A-1*, *7*, 449, 1969.
37. Raficov, S. R., SU Patent 525,707, 1976.
38. Lorkovski, T. H., Isobutene oligomerization; the influence of working conditions upon the structure and composition of oligomers, *Plaste u. Kautschuk*, *13*, 325, 1966.
39. *British Plastics*, 1959, p. 69.
40. Bette, C., Oligomerization of isobutene in tubular reactors, *Chimie et Industrie*, *87*, 246, 1962.
41. Antkowiak, T. A., Synthesis of oligoisobutylene by deploymerization reactions, *J. Polym Sci.*, *A-1*, *10*, 1319, 1972.
42. Schildnecht, E. C., *Vinyl and Related Polymers*, Wiley, New York, 1952, Ch. L, p. 52.

43. Weelock, Ch., U.S. Patent 2,959,531, 1960.
44. Meyer, R., GER Patent 1,111,399, 1960.
45. Gleeson, H. A., U.S. Patent 2,791,631, 1957.
46. Kita, R., Cationic telomerization of isobutene, *J. Coating Tech.*, *48*, 53, 1976.
47. Lizkumerch, A. T., SU Patent 899,574, 1982.
48. Reed, S. F., U.S. Patent 3,886,190, 1975.
49. Baldwin, M. C., and Reed, S. F., U.S. Patent 3,753,812, 1973.
50. Skillicorn, D. E., U.S. Patent 4,129,713, 1978.
51. Sumitomo Chem. Co., Japan, GER Patent 2,228,262, 1972.
52. Motoyuki, Y., and Yaginuma, H., U.S. Patent 3,954,897, 1976.
53. Ionescu, E., Vasilescu, D. S., Oprea, S., Ciobanu, N., Cucinschi, V., and Mărculescu, B., Cyclohexene oligomerization in the presence of Lewis acids, *Materiale Plastice (Bucharest)*, 22, 41, 1985 (Romanian).
54. Miller, S. J., U.S. Patent 4,538,012, 1985.
55. Bozelli, J. W., Dennis, S. K., and Donati, F. A., U.S. Patent 4,482,771, 1988.
56. Reyx, D., Hamza, M., and Compistron, J., Application of the methathesis reaction to the synthesis of prepolymers. 3., *J. Mol. Catal.*, *42*, 1987.
57. Korshak, V. V., and Turov, B. S., Synthesis of unsaturated oligomers by co-metathesis of cycloolefins with linear olefins, *Vysokomol. Soedin.*, *A22*, 723, 1980.
58. Nelson, W., and Heckelsberg, L. F., Synthetic lubricants star-branched oligomers *via* metathesis, *Ind. Eng. Chem. Prod. Res. Dev.*, 22, 178, 1983.
59. Nukina, S., JP Patent 73,14,634, 1973; *Chem. Abstr.*, *79*, 5890f, 1973.
60. Troianker, V. L., Hexene telomerization, *Vysokomol. Soedin. B*, *17*, 181, 1975 (Russian).
61. Höcker, H., and Musch, R., Cyclodecene oligomerization, *Makromol. Chem.*, *176*, 3117, 1975.
62. Larken, J. M., U.S. Patent 4,395,578, 1983.
63. Morganson, N. E., and Vayda, A. V., Eur. Pat. Appl., 77,113, 1983.
64. Startseva, G. P., Oligomerization of hexene catalysed by reaction products of ethylaluminium chloride with chlorine-containing compounds, *Neftekhimiya*, *28*, 53, 1988 (Russian).
65. Onoi, N., JP Patent 62,142,125, 1987.
66. Lychkin, J. P., Kontsova, L. V., and Slashchilina, E. N., Study of the oligomerization of a $C_{20}-C_{40}$ alpha olefin fraction, *Zh. Prikl. Khim. (Leningrad)*, *56*, 844, 1983 (Russian).
67. Miller, S. J., U.S. Patent 4,551,438, 1985.
68. Johnson, T. H., U.S. Patent 4,615,790, 1986.
69. Shinaki, Y., JP Patent 62,001,430, 1987; *Chem. Abstr.*, *107*, 7812c, 1987.
70. Hatekayama, K., JP Patent 60,203,695, 1985; *Chem. Abstr.*, *104*, 132739q, 1986.
71. Cruann, R. J., Green, L. A., Tabak, S. A., and Kramfbeck, F. J., Chemistry of olefin oligomerization over ZSM-S catalyst, *Ind. Eng. Chem. Res.*, *27*, 565, 1988.
72. Miller, S. J., Olefin oligomerization over high silica zeolites, *Stud. Surf. Sci. Catal.*, *38*, 187, 1988.
73. Rabo, J. A., PCT Int. Appl. WO 86,05,483, 1986.
74. Giusti, A., Belussi, S., and Fottora, V., Eur. Pat. Appl. 290,068, 1988.
75. Keyworth, D. A., and McFarlan, C. G., U.S. Patent 4,540,839, 1985.
76. Nippon Oils and Fats Co., JP Patent 58,08,021, 1983; *Chem. Abstr.*, *99*, 23097v, 1983.
77. Jasra, R. V., *Adsorption and Catalytic Conversion of Olefins*, Wiley, New York, 1985.
78. Hudec, P., Analysis of oligomerization products of 1-decene, *Ropa Uhlie*, *31*, 21, 1989 (Czech).
79. Lesister, N. A., and Picollini, R. J., U.S. Patent 3,994,958, 1976.

80. Aso, C., Kumitake, T., and Ito, K., Cationic polymerization of cyclopentadiene. Determination of the structure and its correlation with Friedel-Crafts catalysts, *J. Polym. Sci.*, *B 4*, 701, 1966.
81. Aso, C., Kumitake, T., and Ishimoto, Y., Studies of polymers from cyclic dienes. IV. Determination of the structure of polycyclopentadiene, *J. Polym. Sci.*, *A-1*, *6*, 1163, 1968.
82. Schmidt, A., Tieftemperatur polymerisation des cyclopentadiene mit BF_3-katalysatoren, *Makromol. Chem.*, *130*, 90, 1969.
83. Paulsen, F., and Samuelsen, P., GER Patent 3,148,097, 1983.
84. Lepert, A., Eur Pat. Appl. 85,510, 1983.
85. Kageyama, H., and Maekawa, I., U.S. Patent 4,264,754, 1981.
86. Vasile, C., ROM Patent 77,647, 1981.
87. Ceauşescu, E., Dimonie, I., and Corciovei, A., Study of piperilenes polymerization, *Materiale Plastic (Bucharest)*, *22*, 8, 1985 (Romanian).
88. Moritz, H. C., Pine, A. L., and Ellert, G. H., U.S. Patent 3,442,877, 1969.
89. Vasile, C., Molecular weight distribution in hydrocarbon resins, *Materiale Plastice (Bucharest)*, *20*, 8, 1983 (Romanian).
90. Ceuşescu, E., ROM Patent 86,623, 1984.
91. Zarbov, I., Matveeva, V., and Kabaivanov, V., Physical characteristics-molecular weight relationship in liquid hydrocarbon resins, *Plast. Massi*, 1981, No. 9, p. 54 (Russian).
92. Heurlein, G., and Albrecht, G., Petroleum resins: Synthesis and characterization, *Acta Polymerica*, *33*, 515, 1982.
93. Tekhanovich, M. S., SU Patent 632,721, 1978.
94. Bullard, H., GER Patent 2,014,424, 1970.
95. Feltes, E. M., Ed., *Chemical Reactions of Polymers*, No. XIX, in *High Polymers Series*, Interscience Publ., New York, 1964, pp. 152, 178, 202.
96. Bullard, H., Osborn, R., and St. Cyr, D. R., GER Patent 2,409,430, 1974.
97. Vargin, S., BRIT Patent 1,405,328, 1975.
98. Cupples, B. L., U.S. Patent 4,045,508, 1977.
99. *Japan Chemical Quarterly*, *5*, 11, 1969.
100. Tsukuyia, S., and Sazuki, M., GER Patent 3,132,081, 1982.
101. *Kirk-Othmer Encyclopedia of Chemical Technology*, Interscience, New York, 1954, Vol. 14, p. 700.
102. Gobran, R., U.S. Patent 3,976,606, 1976.
103. Wheeler, H., U.S. Patent 3,478,005, 1969.
104. Jordan, R., Applications of terpene resins, *Coating*, *15*, 335, 1982.
105. Hritakudis, D., Aromatic resins, *Azerb. Khim. Zhur.*, *1*, 108, 1980 (Russian).
106. Akira, W., Application of petroleum resins to hot melt adhesives, *Kami to Parasuchika*, *8*, 31, 1980; *Chem. Abstr.*, *94*, 193256k, 1981.
107. Yashida, T., Applications of hydrocarbon resins resulting from petrochemistry, *Kami to Parasuchika*, *11*, 68, 1983; *Chem. Abstr.*, *98*, 180610k, 1985.
108. Harada, M., and Arakawa, K., *Coal Tar*, *3*, 150, 1951; *Chem. Abstr.*, *46*, 7816, 1951.
109. Anders, H., *Kunstoffe*, *42*, 403, 1952; *Chem. Abstr.*, *47*, 2531, 1952.
110. Bose, S. K., and Basak, N. K., Coumarone-indene resins from crude benzene and coal tar heavy naphta, *J. Sci. Ind. Res.*, *13B*, 422, 1954.
111. Archibald, J. W., BRIT Patent 924,311, 1963.
112. Ceply, J., Czech. Patent 146,963, 1973.
113. Jones, H. W., U.K. Patent 950,602, 1964.
114. Arakawa, M., U.S. Patent 3,504,940, 1971.
115. Shiihara, I., Natural and synthetic oligomers, *Kagaku (Japan)*, *30*, 609, 1975; *Chem. Abstr.*, *84*, 18125p, 1976.

116. Freund, K., GER (EAST) Patent 151,154, 1981.
117. Hokama, T., U.S. Patent 3,803, 079, 1974.
118. Powers, P. O., U.S. Patent 3,000,868, 1959.
119. Cooper, G. D., U.S. Patent 3,749,693, 1973.
120. Raoul, P., FR Patent 2,153,542, 1973.
121. Korshak, V. V., SU Patent 523,118, 1974.
122. Chalk, A. J., and Gilbert, A. R., U.S. Patent 4,108,942, 1978.
123. Roth, C., GER (EAST) Patent 143,914, 1980.
124. Raabe, D., GER (EAST) Patent 209,202, 1984.
125. Chen-Shen Wang, U.S. Patent 3,855,332, 1974.
126. Sergeev, V. A., SU Patent 551,342, 1977.
127. Shatalev, V. P., SU Patent 378,398, 1973.
128. Greenlee, A. J., BRIT Patent 550,798, 1973.
129. Dumskii, I. V., Berent, A. D., Kozodoi, I. V., and Muhina, T. N., *Petroleum Resins*, TsNiiT neftekhim, Moscow, 1983 (Russian).
130. Varsaver, E. M., and Kozodoi, L. V., SU Patent 641,009, 1979.
131. Koliander, L. R., SU Patent 780,472, 1981.
132. Yamada, K., GER Patent 2,417,934, 1974.
133. Jaques, Y., FR Patent 2,233,339, 1975.
134. Kundo, A., and Ushida, S., GER Patent 2,417,912, 1974.
135. Endo, A., and Ushida, S., GER Patent 2,822,786, 1978.
136. Aliev, S. M., SU Patent 988,832, 1982.
137. Karlinski, L. E., and Emelianova, L. P., SU Patent 191,584, 1967.
138. Koliander, L. I., and Sustikov, V. I., SU Patent 648, 567, 1979.
139. Chen Wang, U.S. Patent 3,855,332, 1974.
140. Zarbov, I., Mateeva, R., and Kabaivanov, J., *Mixed hydrocarbon resins*, *God. Viss. Him. Technol. Inst. Sofia*, *24*, 13, 1978 (Bulgarian).
141. Aliev, S. M., SU Patent 861,356, 1981.
142. Benitez, F. M., and English, M. F., U.S. Patent 4,419,503, 1983.
143. Bernard, P., FR Patent 1,368,241, 1964.
144. Perkins, B. R., Jr., and Weis, G. R., U.S. Patent 3,161,620, 1964.
145. Andrewsen, W. H., and Gorda, D. S., U.S. Patent 3,202,679, 1965.
146. Irwin, G. P., and Selwitz, M. C., U.S. Patent 3,412,111, 1968.
147. Zaweski, F. E., and Filby, N. A., U.S. Patent 3,476,774, 1969.
148. Migamoto, S. N., and Watanabe, Y., GER Patent 3,038,573, 1981.
149. Zarbov, I., Pencev, S., and Filipov, L., SU Patent 885,241, 1981.
150. Hultzsch, H., and Rodolphy, A., U.S. Patent 4,401,791, 1983.
151. Vasile, C., ROM Patent 72,122, 1981.
152. Aldred, A. G. C., Eur. Pat. Appl. 81,969, 1983.
153. Snegur, S. A., and Danilenko, T. P., SU Patent 1,049,506, 1983.
154. Strelkov, V. P., SU Patent 604,706, 1978.
155. Heuglein, F. A., GER Patent 1,161,026, 1967.
156. Heuglein, F. A., and Ilgeman, L. R., Synthesis of functionalized hydrocarbon resins, *Makromol. Chem.*, *62*, 120, 1963.
157. Kaiya, A., Kawasaki, Y., and Otsuki, H. H., U.S. Patent 3,952,023, 1976.
158. Bacon, R. G., and Farmer, H. E., *Proc. of 1st Rubber Tech. Conf.*, 1938.
159. Dawson, R. F., and Scott, R. J., Grafting of hydrocarbon resins with maleic anhydride, *Rubb. Chem. Technol.*, *12*, 200, 1939.
160. Alder, K., Pascher, I., and Achmito, A., Oligoisoprene grafted with maleic anhydride. Synthesis and characterization, *Chem. Ber.*, *76B*, 27, 1943.
161. Farmer, H. E., Synthesis and characterization of modified hydrocarbon resins, *Trans. Faraday Soc.*, *38*, 340, 1942.

162. Farmer, H. E., Applications of modified hydrocarbon resins I. Technological aspects, *Rubb. Chem. Technol..*, *15*, 765, 1947.
163. Watson, W. F., Synthesis of modified rubber, *Trans. Inst. Rubber Ind.*, *29*, 32, 1953.
164. Farmer, H. E., Structure of modified rubber, *J. Soc. Chem. Ind.*, *66*, 86, 1947.
165. Mathieu, J., and Valls, J., Thermal grafting of hydrocarbon resins. A proposed mechanism based on electron transfer, *Bull. Soc. Chim. Fr.*, *11/12*, 1509, 1957.
166. Juison, F., FR Patent 1,462,608, 1967.
167. Ceauşescu, E., Corciovei, M., and Dimonie, D., *Chemistry of Modified Hydrocarbon Resins*, Academic House Press, Bucharest, 1988, p. 314 (Romanian).
168. Archibald, B., BRIT Patent 1,090,604, 1967.
169. Tsuchiya, S., and Sasaki, M., GER Patent 3,132,081, 1982.
170. Zarbov, I., Pencev, S., and Filipov, L., SU Patent 885,241, 1981.
171. Ishibe, S., GER Patent 2,409,430, 1974.
172. Grasell, T. G., and Turner, H. A., BRIT Patent 1,031,614, 1966.
173. Mone, J. G., GER Patent 3,801,909, 1978.
174. Moore, T. G., U.S. Patent 4,317,753, 1982.
175. Tsuchiya, S., BRIT Patent 2,061,288, 1981.
176. Strelkov, V. P., SU Patent 604,706, 1978.
177. Tsukiya, S., Sazaki, M., and Hayashi, H., GER Patent 3,081,809, 1981.
178. Heidel, R. B., U.S. Patent 3,546,184, 1970.
179. De La Mare, H. E., U.S. Patent 3,974,129, 1976.
180. Otsuki, Y., Araki, Y., and Kazuho, A., U.S. Patent 4,139,396, 1979.
181. Verdal, J. A., and Patrick, W., U.S. Patent 3,796,762, 1974.
182. Kita, R., U.S. Patent 4,096,106, 1979.
183. Kaiya, A., Kawasaki, Y., and Otsaki, H. H., U.S. Patent 3,952,023, 1976.
184. Yasui, S., and Oshima, T., U.S. Patent 4,080,493, 1978.
185. Kaiya, A., U.S. Patent 4,145,501, 1979.
186. Karsten, M. A., and Delany, L. J., FR Patent 1,448,462, 1966.
187. Ordelt, Z., Jahn, H., and Raubach, H., Hydrogenation technologies of hydrocarbon resins, *Plaste u. Kautschuk*, 15, 350, 1968.
188. Carmody, W. H., U.S. Patent 2,416,904, 1947.
189. Hatzel, H., Wondraczek, R., and Heublein, G., Synthesis of saturated hydrocarbon resins, *Polymer Bull.*, 6, 561, 1982.
190. Girotti, P., Tesei, R., and Telemaco, F., ROM Patent 70,475, 1980.
191. Camberlin, V., Golé, J., and Pascault, J. P., Synthèse et propriétés d'oligomèrec diéniques partiellement et totalement hydrogénés, *Makromol. Chem.*, *180*, 2309, 1979
192. Yasman, Yu. B., SU Patent 1,024,452, 1983.
193. Moritz, H. K., Pine, A. L., and Ellert, G. H., U.S. Patent 3,442,877, 1969.
194. Nakajima N., GER Patent 3,213,948, 1982.
195. French, R. M., Hydroxytelechelic polybutadienes, *Rubb. Chem. Technol.*, *42*, 71, 1969.
196. McGinn, M. W., and Tucker, D. L., Eur. Pat. Appl. 79,225, 1983.
197. Johnston, W. C., and Greenspan, P. F., Epoxidated hydrocarbon resins, *Modern Plastics*, *38*, 135, 1961.
198. Wheelock, Ch., Technology and applications of modified hydrocarbon resins, *Ind. Eng. Chem.*, *50*, 299, 1958.
199. Wheelock, Ch., U.S. Patent 3,022,322, 1962.
200. Kirchof, W., GER Patent 1,173,658, 1962.
201. Seattery, H. G., U.S. Patent 3,410,761, 1968.
202. White, R. W., U.S. Patent 3,382,255, 1968.
203. Tanada, T., Synthesis and application of epoxidated oligobutadienes, *Chem. Economy and Eng. Rev.*, 6, 46, 1974.

204. Shan, Y. Q. J., Modified hydrocarbon resins, *Hecheng Xiangjiao Gougye, 7*, 291, 1984; *Chem. Abstr., 101*, 172762, 1984.
205. Miyazaki, K., and Hara, H., GER Patent 3,238,866, 1983.
206. Bartlett, P. D., Synthesis of epoxidated hydrocarbon resins; mechanism of reaction, *Rec. Chem. Prog., 11*, 47, 1960.
207. Kwart, H., and Hofmann, D. M., Observations regarding the mechanism of olefin epoxidation with per acids, *J. Org. Chem., 31*, 419, 1966.
208. Kwart, H., Starcher, P. S., and Tinsley, S. W., Secondary reactions during olefins epoxidations, *Chem. Commun.*, 1967, p. 335.
209. Mimoun, H., Mechanism of olefins epoxidation, *Angew. Chem., 94*, 751, 1982.
210. Heublein, G., Albrecht, G., and Klopper, S., Mechanism of the epoxidation of olefins, *Acta Polymerica, 36*, 357, 1985.
211. Chadwick, A. F., Technology and uses of modified hydrocarbon resins, *J. Am. Oil Chemist's Soc., 35*, 355, 1958.
212. *Chemical Week*, 1963, April 6, p. 907.
213. Gaylord, N. G., U. S. Patent 4,168,359, 1979.
214. Qi, Y., and Shan, J., *Hechang Xiangjiao Gongre, 7*, 291, 1984; *Chem. Abstr., 101*, 172762, 1984.
215. *Modern Plastics, 38*, 125, 1961.
216. Mayer, R., GER Patent 1,111,399, 1960.
217. Gorke, K., and Bartz, W., GER Patent 2,838,930, 1979.
218. Terry, R. W., and Jacobs, A. F., BRIT Patent 2,008,125, 1979.
219. Mohri, M., and Yoshioka, Sh., FR Patent 2,476,706, 1981.
220. Konietzug, A., and Bartz, W., Eur. Pat. Appl. 43,901, 1982.
221. Zengel, G., Kersten, H., and Malgerlein, H., FR Patent 2,495,965, 1982.
222. Banbo, H., FR Patent 2,496,977, 1982.
223. Yasuno, H., GER Patent 2,910,198, 1979.
224. Yasuno, H., BRIT Patent 2,017,723, 1980.
225. Carmody, W. H., U.S. Patent 2,416,906, 1947.
226. McKay, F. J., U.S. Patent 2,825,662, 1968.
227. Eldib, A., Technology of modified hydrocarbon resins, *Hydrocarb. Process, 42*, 187, 1963.
228. Betts, L. J., Oxydation of hydrocarbon resins, *Chimie et Industrie, 87*, 242, 1962.
229. Kawakami, J. H., and Kwiatkowski, G. T., U.S. Patent 4,275,186, 1981.
230. Jones, H. G., BRIT Patent 802,563, 1958.
231. Kostov, G., and Denev, G., Radicalic mechanism of hydrocarbon resins oxidation, *Kauciuk i Rezina, 11*, 52, 1981 (Russian).
232. Morhard, H. R., and Strzelbicki, G., Modified hydrocarbon resins obtained by radicalic processes, *Kunstoffe, 51*, 751, 1961 (German).
233. Segraves, W. B., and Hillard, G. O., U.S. Patent 2,895,979, 1959.
234. Archibald, H. L., BRIT Patent 802,563, 1958.
235. Antorisin, V. I., Halogenated hydrocarbon resins; structure properties relationship, *Neftepererab. 1 Neftechim., 4*, 82, 1974 (Russian).
236. Kumitake, T., Aso, C., and Ito, K., Halogenated cyclopentadiene resins, *Makromol. Chem., 97*, 40, 1966.
237. Heublein, G., and Frietag, W., Thermal stability of halogenated hydrocarbon resins, *J. Prakt. Chem., 319*, 968, 1977.
238. Ruebensaal, O., *The Rubber Industry—Statistical Report*, 1978.
239. Heublein, G., and Freitag, W., Uses of halogenated hydrocarbon resins, *J. Prakt. Chem., 320*, 725, 1978.
240. Heublein, G., Freitag, W., and Mock, W., Halogenation of hydrocarbon resins. Mechanism of reaction, *Makromol. Chem., 181*, 267, 1980.

241. Heublein, G., and Freitag, W., Uses of chlorinated oligocylcopentadiene, *J. Prakt. Chemi.*, *321*, 544, 1979.
242. Ripley, I. S., GER Patent 2,442,526, 1975.
243. Ripley, I. S., U.S. Patent 3,951,924, 1975.
244. Hayashi, H., GER Patent 2,434,216, 1974.
245. Hayashi, H., U.S. Patent 3,953,407, 1975.
246. Just, C., GER Patent 2,338,643, 1974.
247. Ripley, I. S., and Syson, J. W., GER Patent 2,646,443, 1975.
248. Funaoka, K., and Miwa, T., FR Patent 2,147,279, 1975.
249. Funaoka, K., and Miwa, T., CAN Patent 994,489, 1976.
250. Funaoka, K., and Miwa, T., U.S. Patent 3,804,788, 1974.
251. Ueshima, T., U.S. Patent 3,856,759, 1975.
252. Bajoras, G., *Rev. Liet. TSR*, *10*, 101, 1969; *Chem. Abstr.*, *73*, 5024, 1970 (Lithuanian).
253. Orlov, A. M., SU Patent 258,585, 1969.
254. Volkov, B. V., Modified indene-coumaronic resins with various compounds, *Koks i Khimia.*, *4*, 40, 1966 (Russian).
255. Kempler, T. E., U.S. Patent 4,189,450, 1980.
256. *Chem. Economy and Eng. Rev.*, *6*, 46, 1975.
257. Johnson, A. H., BRIT Patent 1,095,981, 1967.
258. Jofre, L., FR Patent 1,426,206, 1966.
259. Bouduin, A., FR Patent 1,495,182, 1987.
260. Jauvin, L., FR Patent 1,482,925, 1967.
261. Legrande, J., FR Patent 1,495,530, 1967.
262. Marschal, E., FR Patent 1,576,508, 1969.
263. Ives, L., FR Patent 2,154,216, 1973.
264. Carson, J., U.S. Patent 3,523,095, 1970.
265. La Motta, H., U.S. Patent 3,937,674, 1975.
266. Cameron, J. M., U.S. Patent 4,039,414, 1977.
267. Barson, H. V., U.S. Patent 3,450,656, 1969.
268. Orlov, V. I., SU Patent 579,295, 1977.
269. Orlov, V. I., SU Patent 704,976, 1980.
270. Mayer, H., GER Patent 2,246,283, 1973.
271. Richmond, G. H., and Polgaev, A. A., SU Patent 568,667, 1978.
272. French, D. M., Hydroxytelechelic polybutadiene, *Rubb. Chem. Technol.*, *42*, 71, 1969.
273. Furushima, M., Telechelic Polymers, *Japan Chem. Quart.*, *3*, 55, 1967.
274. Istudor, I., RO Patent, 70891, 1980.
275. Seligman, L., U.S. Patent 2,877,212, 1959.
276. Hoffman, R. F., Schreiber, S., and Rosen, G., Telechelic Hydrocarbon resins, *Ind. Eng. Chem..*, *55*, 51, 1964.
277. Thiokol Chem. Corp., BRIT Patent 957,652, 1964.
278. Schoenberg, E., Synthesis of modified hydrocarbon resins, *J. Polymer Sci.*, *54*, 1168, 1956.
279. Uraneck, C. A., Short, J. N., and Zelinski, R. P., U.S. Patent 3,135,716, 1964.
280. Rempp, P., and Loucheiux, M. H., Hydroxytélechelique oligomères, *Bull. Soc. Chim. Fr.*, 1958, 1497.
281. Goldberg, E. J., U.S. Patent 3,055,952, 1962.
282. Fokina, T. A., Modified oligostyrene. Synthesis and characterization, *Vysokomol. Soedin.*, *7*, 946, 1965.
283. Anderson, W. S., Synthesis of modified oligoethylenes, *Polym. Prepr.*, *9*, 773, 1968.
284. Uraneck, C., GER Patent 1,169, 674, 1974.
285. Zelinski, P. P., U.S. Patent 3,109,871, 1963.

286. Loucheiux, M. H., Meyer, G., and Rempp, P., Introduction of amine and groups into oligostyrenes, *Compt. Rend.*, *252*, 2552, 1961.
287. Phillips Petroleum Co., BRIT Patent 944,538, 1963.
288. Phillips Petroleum Co., BRIT Patent 945,851, 1964.
289. Flory, P. J., *Principles of Polymer Chemistry*, Cornell Univ. Press, Ithaca, New York, 1953, p. 111, 132, 145, 158, and 336.
290. Brandrup, J., and Immergut, E. H., Eds., *Polymer Handbook*, Interscience, N.Y., 1965, p. II-1.
291. Wise, R. M., U.S. Patent 3,962,518, 1976.
292. Gaylord, N. G., U.S. Patent 3,959,244, 1976.
293. Thiokol Chem. Corp., BRIT Patent 945,713, 1964.
294. Johnson, R., U.S. Patent 3,733,313, 1973.
295. Fuson, R. C., *Reaction of Oligomeric Compounds*, John Wiley, New York, 1962, p. 226.
296. Rhein, R. A., and Ingham, I. D., U.S. Patent 4,118,427, 1978.
297. Pinazzi, C., FR Patent 2,306,216, 1976.
298. Yamavaki, T., U.S. Patent 3,909,281, 1975.
299. Broecker, B., and Werner, G., CH Patent 581,153, 1973.
300. Schief, H., GER (EAST) Patent 134,444, 1977.
301. Jezierski, A., GER (EAST) Patent 134,230, 1977.
302. Borman, W. F., and Reilly, E. P., U.S. Patent 3,969,306, 1976.
303. Phillips Petroleum Comp., BRIT Patent 972,246, 1964.
304. Richards, D. H., Controlled molecular weight in the free radical oligomerization, *J. Polym. Sci. Polym. Lett. Ed.*, *6*, 417, 1968.
305. French, D. M., Casey, A. W., Collins, C. I., and Kirchner, P., Intrinsic viscosity-molecular weight relationship in carboyl terminated oligobutadiene, *Polymer Prepr.*, 7, 447, 1966.
306. Heuglein, M. B., Whitehurst, D. H., and Sims, D., Intrinsic viscosity measurements of diluted solution of oligomers, *J. Appl. Polym. Sci.*, *12*, 1889, 1968.
307. Verdal, J. A., Curative compounds of carboxyl terminated oligobutadienes, *Rubber Age (NY)*, 1965, pp. 57, 62, 98.
308. McGarry, F. J., Rowe, E. H., and Riew, C. K., Improving the crack resistance of bulk molding compounds and sheet molding compounds, *Polym. Eng. Sci.*, *18*, 78, 1978.
309. Riew, C. K., Modified oligobutadienes with amine end groups, *Rubb. Chem. Technol.*, *54*, 2, 1981.
310. Drake, R. S., and Siebert, A. R., *Adhesive Chemistry*, Leing-Huang Lee, Ed., Plenum Press, 1984, pp. 393–407.
311. Alvino, W. M., Polybutadiene adhesive compositions, *J. Appl. Polym. Sci.*, *24*, 735, 1979.
312. Minoura, Y., Yamashito, S., Okamoto, H., Matsuo, T., Izawa, M., and Kohimoto, S. I., Crosslinking and mechanical properties of liquid rubber. III. Curative effect of aralkyl and alicyclic diols, *J. Appl. Polym. Sci.*, *23*, 1137, 1979.
313. Drake, R., and Siebert, A., Elastomer modified epoxy resins for structural applications, *SAMPE Quarterly*, *6*, No. 4, July, 1975.
314. Hoechst, A. G., GER Patent 2,356,898, 1974.
315. *L'officiel des plastiques et cautschouc*, *25*, 163, 1969.
316. *Chemical Market Abstract*, *67*, 5, 1975.
317. *Chemical & Eng. News*, March 1976, 10.
318. *Chemische Industrie*, 1965, No. 5, p. 269.
319. *Kirk Othmer Encyclopedia of Chemical Technology*, Interscience, 1954, Vol. 14, p. 700.
320. *Eur. Chem. News*, *26*, 659, 1974.
321. *Eur. Chemie*, 1974, No. 21, p. 424.

322. Gessler, N. A., and Sparks, J. V., U.S. Patent 3,070,568, 1962.
323. *Chemische Ind.*, 1965, No. 5, p. 269.
324. *Chemische Ind.*, 1970, No. 6, p. 383.
325. *Chem. Market Abstr.*, *63*, 276, 1971.
326. *Chem. Market Abstr.*, *64*, 106, 1972.
327. *Chem. Market Abstr.*, *64*, 192, 201, 1972.
328. *Chemische Ind.*, 1976, No. 7, p. 404.
329. *Chem. Eng. News*, *44*, 54A, 1966.
330. *Chemische Ind.*, 1981, No. 3, p. 16.
331. *Chemische Ind.*, 1978, No. 3, p. 151.
332. *Plastiques Modernes et Elastomérés*, *17*, 6, 1965.
333. *Eur. Chemie*, *8*, 121, 1985.
334. *Modern Plastics*, *11*, 12, 1973.
335. *Chem. Marketing Reports*, 1974, 21 Oct., p. 7.
336. *Paper J.*, *1*, 19, 1974.
337. *Chem. Market Abstr.*, *66*, 78, 1974.
338. *Chem. Age*, *10*, 9, 1974.
339. Child, E. T., Rambo, M. L., and Nolan, T. T., U.S. Patent 3,125,612, 1964.
340. *L'Officiel des plastiques et du caoutschouc*, *40*, 546, 1972.
341. *Chem. Age*, 1977, Nov. 3, p. 2.
342. *Chem. Market Abstr.*, *69*, 100, 1977.
343. *L'Officiel des plastiques et du caoutschouc*, *200*, 546, 1972.
344. *L'Officiel des plastiques et du caoutschouc*, *251*, 262, 1977.
345. *Chimie Actualité*, 1972, 24, 1977.
346. *Eur. Chem. News*, *32*, 37, 1978.
347. *PROMPT*, *12*, 120, 1978.
348. *Informations Chimique*, *183*, 79, 1978.
349. *Informations Chimiques*, *164*, 98, 1977.
350. *Informations Chimique*, *179*, 83, 1978.
351. *Chim. Actualités*, *1621*, 28, 1978.
352. *L'Officiel des plastiques et du cautschouc*, *287*, 263, 1980.
353. *L'Officiel des plastiques et du cautschouc*, *300*, 540, 1982.
354. *Informations Chimique*, *246*, 76, 1984.
355. *Eur. Chemie*, *18/19*, 322, 1982.
356. *Chimie Actualités*, *1982*, Aug. 7, p. 2.
357. *Information Chimique*, *260*, 79, 1985.
358. *Eur. Chem. News*, Dec. 7, 25, 1982.
359. *PROMPT*, *4*, 67, 1984.
360. *Chimie Actualités*, Jan. 30, 3, 1984.
361. *Eur. Chemie*, *4*, 54, 1984.
362. *Chemical Marketing Reporter*, 1977, July 2, p. 7.
363. *Chemical & Eng. News*, 1972, July 2, p. 7.
364. *Chem. Age*, 1972, Nov. 2, p. 14.
365. *Eur. Chem. News*, 1977, Nov. 2, p. 19.
366. *Rubber World*, 1982, No. 12, p. 8.
367. *PROMPT*, 1983, No. 2, p. 106.
368. *Hydrocarb. Proc.*, *II 64*, 19, 1985.
369. *The Oil & Gas J.*, 1966, Sept. 5, p. 171.
370. Frank, G. H., and Collins, G., *Steinkohlenteer*, Springer-Verlag, Berlin, 1968.
371. *Japan Plastics Age*, 1970, No. 12, p. 16.
372. *Chem. Market Abstr.*, *63*, 145, 1971.
373. *Chem. Week*, 1972, Jan. 19, p. 30.

374. *Chem. Market Abstr.*, 64, 30, 1972.
375. *Chem. Marketing Reporter*, 1972, May 20, p. 16.
376. *Chem. Ind. Notes*, 1980, Aug. 11, p. 35.
377. *Informations Chemie*, 1981, No. 218, p. 95.
378. *Eur. Chemie*, 1978, No. 3, p. 127.
379. *Chem. Market Abstr.*, 64, 533, 1972.
380. Oligomers in Japan, *Technocrat*, 11, 53, 1978.
381. *Chemische Industrie*, 1966, No. 1, p. 26.
382. *BIKI*, 1982, Febr. 10, p. 6.
383. *PROMPT*, 1981, No. 3, p. 82.
384. *Japan Chem. Week*, 1982, Sept. 12, p. 4.
385. *Japan Chem. Week*, 1979, Nov. 1, p. 13.
386. *Japan Chem. Week*, 1980, Dec. 4, p. 4.
387. *Japan Chem. Week*, 1981, May 21, p. 4.
388. *Japan Chem. Week*, 1981, Dec. 24, p. 5.
389. *Japan Chem. Week*, 1983, Nov. 17, p. 6.
390. *Japan Chem. Week*, 1984, Sept. 27, p. 4, 6.
391. *PROMPT*, 1981, No. 2, p. 61, 69.
392. *Chemical Econ. Eng. Rev.*, 1979, Oct./Nov., p. 60.
393. *Chem. Ind. Notes*, 1980, Mars, p. 30.
394. *Chem. Market Abstr.*, 69, 100, 1977.
395. *Japan Chem. Week*, 1976, Nov. 25, p. 4.
396. *Chemische Industrie*, 1965, No. 12, p. 750.
397. *Chemische Industrie*, 1965, No. 11, p. 687, 688.
398. *Japan Chem. Week*, 1975, Jan 25, p. 8.
399. *PROMPT*, 1976, No. 3, p. 88.
400. *Japan Chem. Week*, 1976, Jan. 15, p. 3.
401. *PROMPT*, 1976, No. 3, p. 88.
402. *Japan Chem. Week*, 1978, June 7, p. 2.
403. *PROMPT*, 1978, No. 9, p. 110.
404. *Chemical Marketing Reporter*, 1981, Mars 10, p. 5.
405. *Chimie Actualités*, 1981, Mars 20, p. 7.
406. *Information Chimie*, 1981, No. 212, p. 98.
407. *Information Chimie*, 1985, No. 258, p. 86.
408. *Eur. Chimie*, 1985, No. 6, p. 87.
409. *L'Officiel des plastiques et du caoutschouc*, 1978, No. 263, p. 90.
410. *Eur. Chem. News*, 32, 49, 1978.
411. *Hydrocarbon Process.*, 57, 22, 1978.
412. *Chem. Age*, 1975, Sept. 12, p. 17.
413. *Chemical Market Abstr.*, 67, 128, 1975.
414. Zdenso, D., *Modified Carbochain Resins*, VUSPL, Pardubice, 1982 (Czech).
415. *Ullman Encylcopedie der technischen Chemie*, Verlag Chemie, GmbH, Weinheim, 4 Auflage, Vol. 12, 1976, p. 539.
416. Hronec, M., Kaszanyi, A., Ilavski, J., and Mikulec, J., Hydrocarbon resins, *Ropa Uhlie*, 24, 527, 1982 (Czech).
417. Takeda, A., Uses of hydrocarbon resins, *Chem. Economy and Eng. Rev.*, 8, 26, 1976.
418. *Chemische Technik*, 1984, No. 12, p. 3.
419. *Chemical Week*, 1985, Mars 13, P. SAS 3–38.
420. *Eur. Chem. News, Supplement*, 1983, Oct. 24, p. 20.
421. Allen, C. F. H., and Pingert, F. D., *J. Am. Chem. Soc.*, 64, 1365, 1942.
422. Cade, J. A., and Pilbeam, A., *Tetrahedron*, 20, 519, 1964.
423. Wittig, G., and Lehman, A., *Chem. Ber.*, 90, 875, 1957.

424. Bowden, S. T., *J. Chem. Soc.*, 1931, 1111.
425. Baker, W., Boarland, M. P. V., and McOmie, J. F. W., *J. Chem. Soc.*, 1954, 1476.
426. Copeland, D. G., Dean, K. E., and McNeil, D., *J. Chem. Soc.*, 1960, 1689.
427. Busch, M., and Weber, W., *J. Prakt. Chem.*, *146*, 1, 1936.
428. Gillam, A., and Hey, D. H., *J. Chem. Soc.*, 1939, 1170.
429. Woods, G. F., and Tucker, J. W., *J. Am. Chem. Soc.*, *70*, 2174, 1948.
430. Alexander, R. L., *J. Org. Chem.*, *21*, 1464, 1956.
431. Woods, G. F., and Reed, F. T., *J. Am. Chem. Soc.*, *71*, 1348, 1949.
432. Silverman, L., and Houk, W., *Anal. Chem.*, *27*, 1956, 1955.
433. Clar, E., *Polycyclic Hydrocarbons*, Acadmic Press, New York, 1964.
434. Gerngross, G., and Dunkel, M., *Chem. Ber.*, *57*, 739, 1924.
435. Wirth, H. O., Goenner, K. H., Stueck, R., and Kern, W., *Makromol. Chem.*, *63*, 30, 1963.
436. Campbell, T. W., and McDonald, R. N., *J. Org., Chem.*, *24*, 730, 1959.
437. Kovacic, P., and Lange, R. M., *J. Org. Chem.*, *29*, 2416, 1964.
438. Mayer, F., and Freitag, K., *Chem. Ber.*, *54*, 347, 1924.
439. Wirth, H. O., and Hermann, F. U., *Makromol. Chem.*, *80*, 120, 1964.
440. Marcus, E., Lauer, W. M., and Amold, E. T., *J. Am. Chem. Soc.*, *80*, 3742, 1958.
441. Uglea, C. V., *Makromol. Chem.*, *166*, 275, 1973.
442. Nozaki, T., Tamura, M., Harada, Y., and Saito, T., *Bull. Chem. Soc. Japan*, *33*, 1329, 1960.
443. France, H., *J. Chem. Soc.*, 1939, 1288.
444. Manich, C., *Chem. Ber.*, *40*, 159, 1907.
445. Bennett, M., and Sunshine, N. B., *J. Org. Chem.*, *28*, 2514, 1963.
446. France, H., *J. Chem. Soc.*, 1938, 1364.
447. Woods, G. F., Artsdale, A. L., and Reed, F. T., *J. Am. Chem. Soc.*, *72*, 3221, 1950.
448. Ullmann, F., Meyer, G. M., and Loewenthal, I. M., *Ann. Chem.*, *332*, 38, 1904.
449. Kern, W., *Makromol. Chem.*, *29*, 164, 1959.
450. Kern, W., and Gruber, W., *Makromol. Chem.*, *37*, 198, 1960.
451. Heitz, W., Ullrich, R., and Kern, W., *Makromol. Chem.*, *98*, 29, 1966.

4
Oligoethers

This chapter provides comprehensive information on oligoethers. It deals thoroughly with their synthesis, their chemical and physical properties, and all their applications presently known.

I. OLIGOALKYLENE GLYCOLS (OAGs)

A. Synthesis

Oligoalkylene glycols (OAGs) are produced by addition of epoxides (ethylene oxide, propylene oxide, butylene oxide) to a starting molecule. All the epoxides described can be converted to polymeric substances by opening the ring. Alkali is generally used as catalyst for this reason:

$$n\underset{\diagdown\, O \,\diagup}{CH_2{-}CH}{-}R + R'OH \xrightarrow{NaOH} R'O{-}(CH_2{-}\underset{\displaystyle R}{\underset{|}{CH}}{-}O)_n{-}H \qquad (1)$$

where R = H, CH_3, C_2H_5 and R′ = H, aryl, or alkyl.

Ethylene oxide (EO), propylene oxide (PO), and butylene oxide (BO) can be polymerized to polyethylene glycols (PEGs), polypropylene glycols, and polybutylene glycols. In addition, however, mixtures of the foregoing epoxides can be polymerized with one another, whereupon we obtain a random polymer. In addition to water, alcohols, phenols, amines, and so forth can be used for the starting molecule (i.e., variation of R′ in reaction (1)). When alcohols are employed, polyethylene glycol monoalkyl ether is obtained.

The range of oligoethers is constantly being expanded and comprises tens of types of products. Accordingly, it is usual to employ a batch method and production lines of relatively low capacity. We may, however, single out some grades of oligoethers that make up a considerable proportion of the consumption. These include, first and foremost, oligoethers for the production of elastic urethane foams, including foams for cold forming. Accordingly, it is economically desirable to organize the production of oligoethers using batch and continuous processes. It is, moreover, desirable to use a continuous process to obtain one to three grades of polyols that are in the greatest demand, and use a batch process to give the necessary range of products. This particular idea underlies new developments carried out in Russia in the preparation of oligoethers [1]. These anionic polymerizations of alkylene oxides include, as a rule, the following main stages: the production of the initiator or "starter system," oxyalkylation (the oligomerization of the alkylene oxide), the treatment of the alkaline polymerizate, and removal of the volatiles.

Figure 1 shows the fundamentals of continuous production of Laprol 5003-2B-10. This oligoether, produced in Russia, is a block copolymer of PO with EO of M_n = 5000. It is widely used in the production of elastic or semirigid urethane foams for cold forming.

The polymerization of the PO is effected in column-type continuous bubbling reactors (2 and 5, Fig. 1) in which are reached the necessary molecular weight and desired conversion of the monomer. Because of the particular design of the reactor 5, the issuing alkaline polymerizate contains practically no residual monomer. Oligomer of the desired molecular weight and chemical composition is formed in reactor 6 by the copolymerization of the product prepared in 5 and the EO. The same principle forms the basis of the production of other oligomers that contain fragments of EO in block or statistical distribution. In this setup it is also possible to produce oligomers consisting only of PO, such as Laprol 3003, Laprol 2002, and others.

Another important virtue of this scheme is that part of the alkaline polymerizate of the PO produced in reactor–polymerizer 2 (Fig. 1), of intermediate molecular weight, may be used further in batch operation setups, for the production of other grades of oligoethers. Reactors 2 and 5 are divided into sections by means of perforated partitions, so that each of them functions as an ideal mixing reactor while the number of sections is sufficient to ensure conditions in the liquid phase that yield the same molecular characteristics as those of the corresponding product from a batch operation reactor.

In the bubbling reactors, the necessary contact of the reagents is achieved without the use of mixing devices, with adequate heat transfer. The polymerization

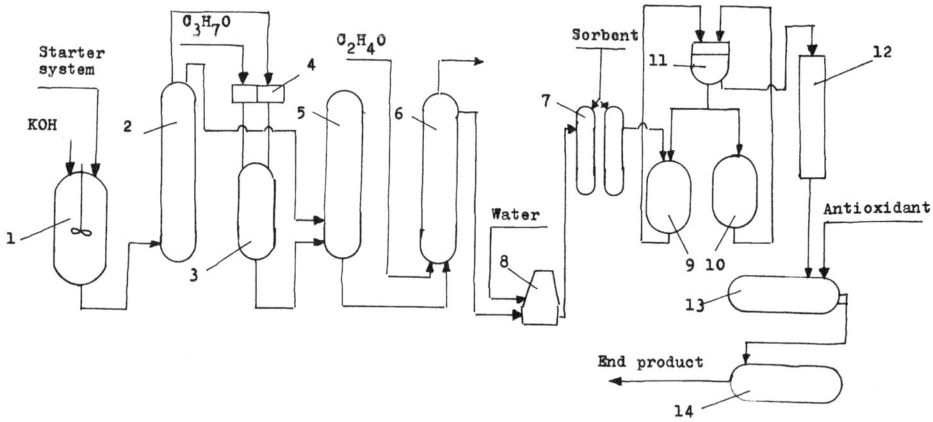

Figure 1 Basic layout of a plant for continuous production of oligo(ethylene glycols), with capacity of 30,000 t/year. 1: Apparatus for preparing the "starter system"; 2: reactor polymerizer; 3: metering vessel for the propylene oxide (PO); 4: condenser; 5: reactor polymerizer for PO; 6: reactor polymerizer; 7,8: mixers; 9,10: collectors; 11: filter; 12: rotary film evaporator; 13: filter; 14: tank for end product. (Adapted from Ref. 1.)

is easily controlled in this case, since it takes place effectively at atmospheric pressure, or slightly exceeding this, and is completely free from risk. The treatment of the alkaline polymerizate is effected, irrespective of its chemical nature, in two mixers (7, 8) operating in series, using easily procured and inexpensive mineral sorbents, which are divided further into automatic chamber-type filter presses with mechanical discharge of the deposit, 11. Then the product is dehydrated on the rotary film evaporator 12, and after filtration is passed to storage or for forming into articles.

The batch process (Fig. 2) is intended for the production of the necessary range of oligoethers. It is versatile process and requires only slight variations in the engineering aspects of the lines for specific products. The production is effected in the reactor 5 with its stirring device, into which particular amounts of the alkylene oxides are fed continuously in the necessary sequence and proportions. The product is processed in the mixer 6 (using mineral sorbents) or else by an acid sorbtion method, in the filter 7 and in the rotary film evaporator 9. The resulting polyol is passed to storage, or for forming into articles. The production schemes complement each other and permit one to manufacture products with good technical and economic characteristics.

The continuous process has significant advantages over the batch process. It allows higher worker productivity, requires less raw material and energy, and the unit cost of polyol produced by the continuous process is almost 20% below that of the batch-produced product. In addition, in the continuous process less wastewater and gas discharges are formed, the level of hazard is lower, and it is possible to produce a large amount of product of uniform quality.

Various types of OAGs are obtained by the foregoing methods. Table 1 lists the most important types of OAGs [2]. In some pharmacopoeias, PEG 1500 is

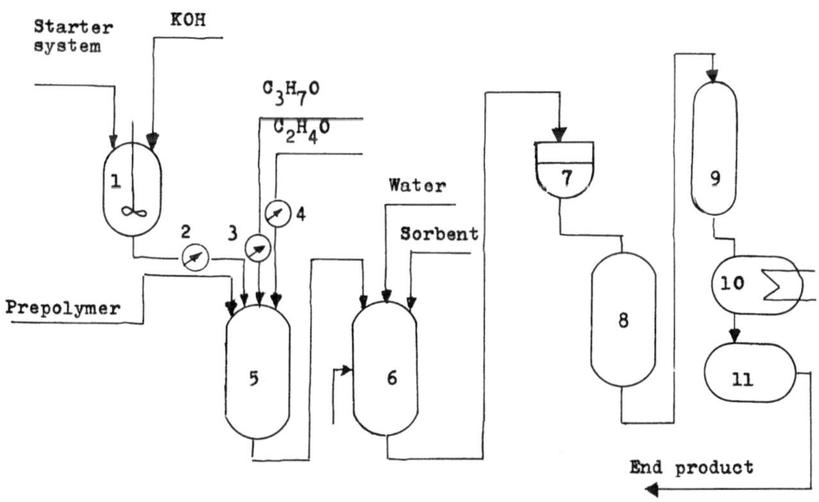

Figure 2 Basic layout of a plant for batch production of oligo(alkylene glycols), with capacity of 10,000 t/year. 1: Apparatus for preparing the "starter system"; 2–4: flow meters; 5: reactor polymerizer; 6: mixer; 7: filter; 8,11: collectors; 9: rotary film evaporator; 10: condenser. (Adapted from Ref. 1.)

Table I Technical Data for PEGs

PEG grade	$\bar{M}_n$	$n_D^{25°C}$	Viscosity,[a] mPa·s (cP)	Description
200	190–210	1.458	46–53[b]	
300	285–315	1.462	66–74	
400	380–420	1.465	85–95	Pale, clear, viscous
600	570–630	1.452[c]	13–15[d]	liquids
1,000	950–1,050	1.453	19–23	Soft wax
1,500	1,400–1,500	1.454	29–34	White, waxy flakes
2,000	1,800–2,200	1.454	40–48	or powder
3,000	2,700–3,300	1.455	60–80	
4,000	3,700–4,500	1.455	90–120	
6,000	5,600–7,000	1.455	170–220	
10,000	8,500–12,500	1.456	450–800	Pale, hard waxy flakes or powder
P_{41}[e]				
41/300	5,000	1.467[f]	1,200	
41/3,000	15,000	1.467	13,000	
41/10,000	20,000	1.467	55,000	
B_{11}				
11/50	1,200	1.457	150	
11/150	2,500	1.459	555	
11/300	3,200	1.459	1,200	
11/700	3,800	1.460	2,800	
B_{01}				
01/20	700	1.446	72	
01/40	1,100	1.448	150	
01/80	1,400	1.450	330	
01/120	2,000	1.450	400	
01/240	2,300	1.451	1,050	

[a]Data on undiluted product.
[b]Experimental data for undiluted products.
[c]Refractive index determined at 70°C.
[d]Experimental data for 50% aqueous solution.
[e]Produced by Hoechst.
[f]The products are identified by a code consisting of a letter and figures. The alcohol initially used is identified by the letter (e.g., B = butanol), while the two figures following indicate the ethylene oxide/butylene oxide ratio. For example, 41, 11, and 01 indicate ethylene oxide/propylene oxide ratios of 4:1, 1:1, and 0:1 respectively. Within the individual series, the products are further identified by their viscosity at 50°C in cSt (mm²/s). Polyglycol $B_{01/80}$ is accordingly a polypropylene glycol monobutyl ether with a viscosity of 80 cSt at 50°C.

also called PEG 1540. In American technical journals, Carbowax is frequently used synonymously with PEG. Carbowax 1540 corresponds to PEG 1500. In the British Pharmacopoeia and Pharm. Nordica, the PEGs are termed "Macrogol." The following designations used in different pharmacopoeias, therefore, all relate to the same PEG grade 1500: PEG 1540 NF XIV (United States) PEG 1500 USP XIX, PEG

1500 DAB 7 (Germany), PEG 1540 Ph. Helv. VI, PEG 1540 Cod. Fr. IX, and PEG 1500 OAB 9. In a number of pharmacopoeias, PEG 3000 is listed as PEG 4000. USP XIX describes for PEG 4000 an average molecular weight from 3000 to 3700. In practice, however, the products used possess a molecular weight much closer to the lower limit of 3000. The following terms, however, all relate to the same PEG 3000: PEG 4000 USP XIX, PEG 300 DAB 7, PEG 4000 BP 73, PEG 3000 Ph. Helv. IV, and PEG 4000 Cod. Fr. IX. PEG 6000 in USP XIX is a product with a molecular weight of 7000–9000. The Hoechst (Berlin, Germany) designation for a product of this type is PEG 6000 USP.

B. Physical Properties

The viscosities in mPa·s (cP) indicated in Table 1 are intended mainly to characterize the individual OAG grades. The dynamic viscosity (mPa·s) can be converted into kinematic viscosity (mm²/s), and vice versa, according to the following formula taking the density into account:

$$\text{mm}^2/\text{s} = \frac{\text{mPa·s}}{\text{density}} \tag{2}$$

None of OAG types up to a molecular weight of 35,000 or their aqueous solutions exhibit thixotropy or other structurally viscous phenomena (dilatancy or rheopexy).

The surface tension of the liquid PEGs 200 to 600 is about 44 mN/m (dyn/cm) at room temperature. There is only a slight difference in the surface tension of liquid and solid PEGs in aqueous solution; a 10% solution of PEG 400 has a value of 64 mN/m (dyn/cm), whereas a 10% solution of PEG 4000 has a value of about 60 mN/m (dyn/cm) at 20°C.

The PEGs possess no characteristic surface-active properties and therefore cannot be included among the surfactants. Nevertheless, they frequently prove to be useful dispersing agents or solubilizers.

The latent heat of fusion of the solid PEGs is 167–188 KJ/Kg (40–45 cal/g), depending to some extent on the degree of crystallinity.

The specific heat of the liquid PEGs at room temperature is about 2.1 KJ/Kg·K (0.5 cal/g deg C). With rising temperature, the specific heat increases steadily and at 120°C reaches about 2.5 KJ/KgK (0.6 cal/g deg C).

One of the major advantages of the PEGs, and frequently a determining factor in their applications, is their excellent solubility in water. The combination of hygroscopicity, viscosity, lubricity, dissolving power, and binding power inherent in the PEGs coupled with their solubility in water makes them ideal in countless different applications.

The liquid PEGs 200 to 600 are miscible with water in any ratio. It must be borne in mind, however, that an addition of water lowers the solidification point compared with that of pure liquid PEG.

When liquid PEGs are mixed with water, a volume contraction takes place. When equal parts by weight PEG 400 and water are mixed together, this contraction amounts to about 2.5%. At the same time, a marked heat effect occurs. The temperature rise taking place when equal parts by weight PEG and water are mixed is about 12°C for PEG 200 and up to 14°C for PEG 600.

Even solid PEG grades have excellent solubility in water. For example, 75 parts by weight PEG 1000 can be dissolved at room temperature in only 25 parts water. Although the solubility in water falls with increasing molecular weight, it does not fall below 50% even for PEG 35,000. The dissolving process can be greatly accelerated by applying heat.

The PEGs exhibit nonionic behavior in aqueous solution. They are not sensitive to electrolytes and are therefore also compatible with hard water.

The liquid PEG grades are markedly hygroscopic, although not to the same extent as diglycol or glycerol, for example. The hygroscopicity decreases rapidly with increasing molecular weight and is clearly related to the available hydroxyl groups.

A rule of thumb is as follows: With a relative humidity of about 50%, PEG 200 has about three-quarters of the hygroscopicity of glycerol. PEG 400 has about half, PEG 600 a third, and PEG 1000 only a quarter.

PEGs with a molecular weight of 2000 and more are virtually nonhygroscopic.

Owing to their solubility in water, PEGs are, however, sensitive to moisture and must therefore be stored in a dry place. PEGs take moisture from the air until an equilibrium is reached. By plotting the water content of the substance in the equilibrium state as a function of the relative humidity, a sorbtion isotherm is obtained.

An adaptable moderate hygroscopicity may be advantageous for a conditioning agent, because products treated with it are less sensitive to climatic changes and have better storage stability.

Occasionally a request is made for a completely anhydrous PEG. This may well be an understandable wish where the product is to be processed with a number of water-sensitive substances. On the other hand, the very hygroscopicity of the liquid PEGs is an obstacle to absolute dehydration, and even the solid grades tenaciously retain a certain water content.

When PEGs are dispatched, care is taken to ensure that the water content is no more than 0.5%. The usual value is about 0.1 to 0.3%. A number of pharmacopoeias, for example, DAB 7, permit a maximum water content of 2%.

If necessary, the water content can be reduced to 0.1% in a drying cabinet at 105°C; with fresh or well-regenerated molecular sieves (pore size: 3–4 Å), it can be reduced to 0.05%.

The PEGs are virtually nonvolatile, a factor of considerable importance in connection with their use as plasticizers and humectants.

If a certain weight loss is established despite the nonvolatility of the PEGs when maintained at a constant temperature of 150°C and above (e.g., when used as heating-bath liquids), this is due not to evaporation, but to loss of volatile products of decomposition.

The thermal decomposition of PEGs leaves neither hard encrustations nor tough sludge deposits. The equipment, heating elements, pipe coils, and so forth, can be easily washed down with water. Moreover, the obnoxiousness of the odors is relatively slight. These advantages have contributed greatly to the use of PEGs in the following sectors:

Immersion baths for shaping thermoplastics
Immersion baths for softening and facilitating the removal of plastic residues from molds, dies, and extruders
Tempering baths for modifying the crystallinity of plastics
Immersion baths for the vulcanization of rubber articles at atmospheric pressure, for filling thermostats, heating baths, etc.
Baths for cleaning spinnerets

In these applications, attempts are frequently made for technical reasons to exploit a material to the limit, and so the behavior of PEGs under thermal stress is described below.

In the PEG 300 to 6000 range there are virtually no differences in thermal stability, and so all grades are equally suitable for heating baths. The only exception is PEG 200; it tends to evolve fumes and to lose weight.

The solid PEGs, too, are good heat transfer media in their molten state. The viscosity is frequently the determining factor in choosing one particular grade; low-viscosity dip baths prevent high carryover losses.

Experience has shown that more fumes and obnoxious odors usually occur when the bath is first heated than in the subsequent stages of treatment. This is due to the low water content, which is expelled initially. Since the lower PEG grades are hygroscopic, moisture may be reabsorbed in the case of fairly long downtimes.

If PEGs are kept constantly at a high temperature, they turn dark fairly quickly, but this does not impair their stability as heat transfer media. At temperatures above 100°C it is, however, essential to add a suitable antioxidant to the PEG.

The type and quantity of antioxidant is governed by the requirements imposed on the PEG. Thus, not only the temperature and dwell time but also the physiological properties of the antioxidant and its solubility or insolubility in water must be taken into consideration. Where exposure to high thermal stress is involved, up to 3% antioxidant should be added.

Among successful antioxidants that dissolve readily in PEGs are the following:

Trimethyl dihydroquinoline polymer (e.g., Vanlube RD, supplied by Lehman u. Voss, Hamburg, Germany)
Diphenylamine derivatives (e.g., Vulkanox DDA, manufactured by Bayer AG, Leverkusen, Germany)
Phenotiazine (supplied by Hoechst Aktiengesellschaft, Verkauf Organische Chemikalen ARV, Frankfurt/Main, Germany)
Phenyl-α-naphthylamine (e.g., Vulkanox PAN, manufactured by Bayer AG)
4,4′-methylene-bis-2,6-di-*tert*-butylphenol (e.g., Antioxidant 702, manufactured by Ethyl Corp., New York)
Butylated hydroxyanisole
Hydroquinone monomethyl ether

The phenolic stabilizers are effective only at lower temperatures—up to about 150°C—but have two advantages: they cause less discoloration, and some of them are water soluble. If the design of the equipment permits, air should always be kept from coming into contact with the bath or the bath should be covered with an inert gas atmosphere. This applies particularly at temperatures between 200 and about 240°C.

Hot PEGs attack iron and steel slightly; as a precaution when the liquid PEGs are used, a certain margin of alkalinity must be created by the addition of about 0.3% hydrated borax or triethanolamine. Because borax dissolves with difficulty in the higher molecular weight PEGs, one also can add 0.2% sodium nitrite to prevent steel corrosion. Other metal combinations should be tested to establish their resistance to corrosion by PEGs.

If PEGs are heated to a very high temperature while exposed to air, a pungent odor is unavoidable. In such cases, adequate ventilation is essential. The pungent odor is due to the evolution of volatile aldehydes. The decomposition products of PEGs may vary, depending on the extent of exposure to the air; in addition to water, carbon dioxide, aldehydes, simple alcohols, and glycol ethers are also formed. There is, however, no known recorded case of injury caused by the obnoxious fumes of the decomposition products. Higher concentrations are, in any case, unbearable because of the pungent smell of the aldehydes.

When air is excluded, the PEG chain breaks down only at temperatures above 250°C. In the molecular weight range of about 300 to 6000, the chain length and the molecular weight are insignificant factors.

The smooth and complete decomposition of the PEGs is the reason behind their suitability as temporary binders for ceramic pressings. When burnt off in a muffle furnace, the PEGs burn with a bluish, fairly short, soot-free flame. In lubricating processes at high temperatures (e.g., in metal foundries), the pyrolytically decomposing PEG may lead to gas lubrication.

Polyethylene glycols with less common or uncommon average molecular weights can be manufactured without difficulty by mixing available PEG grades. In order for the properties of such a mixture not to differ too greatly from those of an unblended product, the molecular weight range must be kept as narrow as possible. In practice, this means that only neighboring grades in the PEG series should be mixed; for example, PEG 500 is made from PEG 400 and 600, PEG 8000 from PEG 6000 and 10,000, and so forth. Where the differences are greater, especially if liquid and solid PEGs are mixed, the result may possibly be soft to ointmentlike products with greatly modified characteristics.

Some frequently required mixtures are made up as follows:

PEG	500	= 40% PEG	400	+ 60% PEG	600
PEG	800	= 38% PEG	600	+ 62% PEG	1,000
PEG	1,200	= 50% PEG	1,000	+ 50% PEG	1,500
PEG	5,000	= 40% PEG	4,000	+ 60% PEG	6,000
PEG	8,000	= 38% PEG	6,000	+ 62% PEG	10,000
PEG	15,000	= 33% PEG	10,000	+ 67% PEG	20,000

Although in formulating such mixtures it is in principle immaterial whether the components used are solid or liquid, homogeneous mixtures are in fact normally obtained straightforwardly only with liquid:liquid combinations. They can be obtained with solid:solid combinations provided both grades are available in an equally fine powder. In all other cases, the products must be mixed in the molten state, particularly the solid:liquid combinations, unless it is possible to work with solutions.

It would appear logical to see PEG ointment grades as the transitional stage between the liquid and the solid grades, that is, somewhere between PEG 600 and PEG 1000. In fact, PEG 800 has the soft consistency of an ointment at room temperature but liquefies even on slight heating. A practical ointment consistency, say, one that spreads in the range −10 to +40°C, is obtained only by mixing liquid and solid PEGs.

A well-trained ointment mixture of this kind is prepared from equal parts by weight PEG 300 and PEG 1500. The preparation of this ointment is extremely simple: the two components are melted together in a suitable vessel by being heated to 60°C with stirring, followed by cooling while stirring continues.

The same ointment base is obtained by allowing the thoroughly mixed molten mass to cool without stirring and then thoroughly stirring or kneading it. The solidified mixture, however, resembles tallow in its dullness and is somewhat harder. This hardness is due a certain crystallinity. After further stirring or kneading, the mixture becomes a smooth ointment deceptively similar to white petroleum jelly.

Hoechst is a supplier of a versatile ready-to-use ointment mixture of PEG 300 and PEG 1500 under the tradename Lanogen 1500 in the form of "fused" material, that is, solidified in the containers. The PEG ointment is DAB 7 (unguentum polyethylenglycol) is identical in composition.

Solid PEGs are used in powder form whenever it is necessary for them to be intimately dry-mixed with components of a different kind. Such is, for example, the case in tablet manufacture, in dry-granulation, and in the preparation of ceramic pressings.

To convey an impression of the particle size distribution of PEG powder, we give here an average screen analysis of PEG powder produced by Hoechst in which the μm measurement indicates the internal mesh width in accordance with DIN 1171 and "mesh" indicates mesh/inch in accordance with the British Standard.

Particle size distribution of PEG 4000 powder:

Less than 90 μm = 170 mesh	About 50%
Less than 150 μm = 100 mesh	About 80%
Less than 200 μm = 60 mesh	About 95%
Less than 300 μm = 52 mesh	About 99%

PEG grades 3000 to 20,000 are available in powder form. PEG 3000 powder is slightly finer than PEG 4000 powder, which in turn is only slightly finer than PEG 6000. Owing to its greater hardness, PEG 20,000 powder is substantially coarser than PEG 4000 powder. The very hard tough-elastic PEG 35,000 is consequently available only in flake form.

For many applications, particularly in pharmaceuticals, cosmetics, and foodstuffs (packaging), the physiological safety of the PEGs is of crucial importance.

When administered orally and cutaneously they are rated as nontoxic. Furthermore, the vapor pressure of PEGs is so low that inhalation is impossible.

Because of their good physiological tolerability, the PEGs were first included in the U.S. pharmacopoeia as long ago as 1950. Since then, they have been listed in the pharmacopoeias of 11 countries.

All PEGs have a very low acute oral toxicity, lower than that of glycerol. The tolerability of PEGs in animals improves as the degree of polymerization rises. In Table 2 we have compiled some of the published values [3,4].

By way of comparison, the mouse LD_{50} for anhydrous glycerol is 26.4 g/kg body. Further details on the physiological tolerability of the PEGs have been published by Rowe [5].

To test chronic oral toxicity, dogs were fed a diet containing 2% of various PEGs (400, 1500, and 4000) for one year. Rats were given 4% PEG 1500 and PEG 4000 for two years [6]. No adverse effects of any kind were recorded; the PEGs proved to be inert.

When administrated singly or repeatedly, the PEGs have no toxic or irritant effect on the skin. For example, Smyth et al. [7] found that a single dose of 20 g/kg of PEG 200, 300, and 400 was well tolerated by rabbits. Because of the low toxicity of PEGs, it was not possible to establish an exact LD_{50} resulting from skin penetration.

As Tusing [8] ascertained, the PEGs in the molecular weight range 300 to 9000 have no adverse effects even after prolonged contact with the skin. The PEGs are tolerated well by the human skin and do not cause any sensitization; they have no macerating action on the areas treated, nor are they absorbed by the skin.

When injected intravenously into rats, undiluted PEG 300 has LD_{50} of 8.0 g/kg [9]. According to investigations by Pfordte [10], the LD_{50} (intravenously in rats) for PEG 600 is 7.8 g/kg. The animals survived a dose of 2.8 g/kg body weight without any pathological changes in the blood serum.

Table 2 Acute Oral Toxicity of PEGs (LD_{50} values[a] in g/kg)

PEG grade	Test animal			
	Rat[b]	Guinea pig[b]	Rabbit[b]	Mouse[c]
200	34	—	20	43
300	39	20	21	—
400	44	16	27	—
600	—	—	—	54
1,000	42	22	—	—
1,500	51	—	—	—
5,000	>50	>51	>50	>50
6,000	>50	>50	—	—
10,000	50	—	—	—

[a] LD_{50} = lethal dose for 50%, i.e., the quantity of the product in g/kg at which 50% of the test animals die.
[b] Data from Ref. 3.
[c] Data from Ref. 4.

Undiluted PEG 300 or 400 injected subcutaneously or intramolecularly into dogs (dose: 2 ml/kg) was for the most part excreted with the urine within 24 h [9,10].

In humans, it was observed that, after an intravenous injection of 1 g PEG 6000 or PEG 1000, 96% and 85%, respectively, had been excreted after 12 h [11]. Similar tests with PEG 400 showed that 77% of the PEG was excreted in the same time [12]. The authors found that no monomethylene glycol is formed as a metabolite.

PEGs are required to be physiologically safe in view of their use in the manufacture of packaging materials for foodstuffs.

In German official publications, PEGs are mentioned—that is, approved—in the assessment of plastics from the health aspect under the Food Law. These specifications are contained in the appropriate "recommendations" of the German Ministry of Health (BGA), which are given in the following.

PEGs may be used as (1) antistatic agents for conveyor belts based on plasticized PVC; (2) mold release agents and lubricants for consumer articles made from unsaturated polyester resins (molecular weight above 1000); (3) humectants for paper, board, and cardboard for foodstuff packaging (max. 7% and max. monoethylene glycol content 0.2%); (4) humectants for artificial sausage skins made from cellulose hydrate; (5) lubricants and mold release agents for modified polystyrene consumer articles; (6) humectants for normal and weatherproof cellophane; (7) processing aids for copolymers of ethylene, propylene, butylene, vinyl esters, and unsaturated aliphatic acids; (8) defoamers for polymer dispersions; (9) lubricants for crosslinked polyethylene; (10) processing aids, lubricants, and mold release agents for consumer articles based on natural and synthetic rubber; (11) lubricants, and antiblocking and antistatic agents for polyethylene; (12) lubricants and mold release agents for consumer articles based on elastomers made from natural and synthetic latex and on dispersions made from solid rubber; and (13) material for externally coating hollow glassware.

It is not possible to give further details here of the use of PEGs under the various national food laws. One noteworthy fact is that in the United States PEGs 200 to 9500 are approved expressly as auxiliaries and additives in the manufacture of consumer articles that come into contact with food in accordance with FDA. In certain instances they are also approved as components of the foodstuff itself. The applications concerned here are those of PEGs as binders and plasticizers for foodstuffs in tablet form, as auxiliaries for tablet coatings, and as carriers for aromatic substances and calorie-free sweeteners.

In view of environmental pollution, the behavior of PEGs in effluent is a matter of crucial importance, especially in their industrial use in the textile sector, in metal processing, and other applications.

In this connection, it must be realized that the rate of biodegradation of PEGs decreases with increasing molecular weight. According to Pitter [13], the lower molecular weight members of the PEG series up to PEG 600 are regarded as biodegradable. It must, however, be borne in mind that the activated sludge requires a certain time to adapt.

The microbiological degradation of other substances is not inhibited by the presence of PEGs; here, too, the PEGs prove to be nontoxic. The toxic inhibition limit for bacteria in the fermentation tube is 5000 mg/L. A summary of the biodegradability of the glycols and polyglycols has been published by Swischer [14].

C. Chemical Properties

Having two terminal primary hydroxyl groups, the PEGs are capable of forming mono-, di-, and polyesters.

The *monoesters* of fatty acids in particular, chiefly those of stearic, oleic, and lauric acid, are very widely used on account of their surface-active properties [15].

By selecting PEG grades with longer or shorter chains, preferably between PEG 200 and PEG 1000, it is possible to control the hydrophilic–lipophilic balance of the desired ester [16]. If the hydrophilic PEG chain of the ester predominates, that ester will tend to be used as emulsifiers in organic (predominant)–water systems, whereas esters with a short PEG chain and a longer fatty acid radical will tend to be used as emulsifiers in water (predominant)–water systems.

Diesters are formed by reacting 1 mol PEG with 2 mol of a suitable monobasic acid, for example, oleic acid. Compared with the monoesters, the diesters are much more lipophilic. Unlike most monoesters of the PEGs, the dioleates of PEGs 200 to 1000 are oil soluble. They are suitable for use as oil additives and cutting oil components, for example.

Being nonionic compounds, the PEG fatty acid esters are largely insensitive to hardness salts, as they are to electrolytes in general. They are broken down only at high temperatures by strong acids and alkalies.

The range of applications of the PEG fatty acid esters covers wide sectors in the textile and leather industries, the chemical industry, and in pharmaceuticals and cosmetics (ointment bases, emulsifiers).

PEG esters are generally manufactured at temperatures of 120 to 150°C, usually with an addition of acid catalysts. Suitable products are sulfonic acids, for example, benzene- or *p*-toluene-sulfonic acids, and also sulfuric or phosphoric acid (0.1 to 0.5% catalyst).

At temperatures of about 200°C, esterification proceeds without a catalyst, but the products obtained are darker in color.

A water jet or steam jet vacuum pump should be connected to a reaction vessel equipped with a stirrer. To continuously remove the water produced in the reaction, nitrogen or carbon dioxide should be used for scavenging. An inert gas blanket also promotes the formation of light-colored products. Under these conditions, simple esterification processes can be completed in a few hours, but the time required lengthens somewhat with increasing molecular weight of the PEG. PEG esters can also be produced by transesterification of PEGs with esters of low-boiling alcohols.

Polyesters of PEGs are obtained by reacting PEGs with dicarboxylic acids. However, the goal is not so much pure PEG polyesters as the possibility of modifying polyesters by inserting a number of PEG chains. Such polyesters are used in the lubricant sector and also in alkyd resins in the paint industry.

PEG polyesters with unsaturated dibasic acids such as maleic acid are used to modify polyester resins for paints so as to improve their elasticity and scratch resistance. Alkyd resins modified with PEGs can be made so hydrophilic that they are dispersible in water [17].

The hydroxyl groups of the PEGs react with aliphatic or aromatic isocyanates to form urethanes (carbamic acid esters). Divalent isocyanates (diisocyanates) and polyhydroxy compounds form high molecular weight polyurethanes by polyaddition:

$$\mathrm{HO{-}R_1{-}OH + O{=}C{=}N{-}R_2{-}N{=}C{=}O \rightarrow}$$

$$\mathrm{{-}O{-}R_1{-}O{-}\underset{\underset{O}{\|}}{C}{-}NH{-}R_2{-}NH{-}\underset{\underset{O}{\|}}{C}{-}O{-}R_1{-}O{-}\underset{\underset{O}{\|}}{C}} \qquad (3)$$

$$\mathrm{{-}NH{-}R_2{-}NH{-}\underset{\underset{O}{\|}}{C}{-}}$$

This is also the basic reaction in the manufacture of polyurethane foams [18]. Linear and branched polyesters or polyethers in the molecular weight range 400 to 6000 are used here as hydroxyl components (polyols). In this field, the derivatives of propylene oxide have achieved greater importance, while the polyethylene glycols are used solely as modifiers, especially in the manufacture of flexible polyurethane foams.

Apart from their reactivity based on the functions of their two terminal primary hydroxyl groups, the PEGs can react in another way by forming addition compounds or complexes on their ether bridges [19].

If liquid PEGs are mixed with water, a marked heat effect occurs. Surprisingly, this is greater than if the same quantity by weight of the much more hygroscopic glycerol, glycol, or diglycol is dissolved in water.

The greater heat effect produced by the PEGs is due to hydration of their ether bridge atoms. It is therefore logical that this heat effect is even greater with PEG 600 than with PEG 200, because PEG 600 contains relatively more ether bridges than PEG 200 in a given quantity by weight.

When the solid PEGs are dissolved in water, the heat of hydration is overcompensated for by the simultaneously occurring negative heat of solution, resulting in a moderate drop in temperature.

The strikingly good solubility in water of PEGs is due to the formation of hydrate complexes. By analogy with the formation of ammonium hydroxide when ammonia is dissolved in water, the process involved in adding water to the ether oxygen of PEGs is referred to as oxonium hydroxide formation. Unlike the aqueous solution of ammonia, however, the PEG solutions react neutrally.

It is well known that the oxygen in an ether bridge possesses two pairs of electrons not involved in the chemical linkage:

$$\mathrm{{-}CH_2{-}CH_2{-}\underset{\cdot\cdot}{\ddot{O}}{-}CH_2CH_2{-}}$$

The high electron density of the ether oxygen characterizes the PEGs as electron donors or proton acceptors (Lewis bases). This explains their tendency to form addition compounds with suitable acids, salts ($HgCl_2$, CdI_2, $MnCl_2$), Lewis acids in general (BF_3), hetero polyacids, urea, and phenol [20].

The obvious explanation for the strikingly good solubility of many phenol compounds in PEGs is complex formation, starting with a hydrogen bond at the basic ether oxygen atom. According to Wurzschmitt [21] the formula of such PEG–phenol complexes should be given as follows:

$$
\begin{array}{l}
HOCH_2\text{—}(CH_2\text{—}O\text{—}CH_2)_x\text{—}CH_2OH \\
\qquad\qquad\quad\;\downarrow \\
\qquad\qquad\quad\; H \\
\qquad\qquad\quad\; | \\
\qquad\qquad\quad\; O\text{—}C_6H_5
\end{array}
\tag{4}
$$

However, the formation of PEG–phenol complexes depends strongly on the concentration and is generally reversible; that is, strong dilution can weaken or sever the complex.

A noteworthy fact is that even the characteristic phenol smell is largely eliminated by the addition of PEGs, but it reappears when the complex is diluted with water.

Pure phenol can be completely masked or inactivated by introducing a certain excess of PEG in the complex concentrate.

According to Schuz [22], the best antidote to percutaneous phenol poisoning or burns is PEG 400. The areas of skin affected by the phenol must be treated as quickly as possible with absorbent soaked in PEG 400. It is therefore advisable to keep PEG 400 (undiluted and without additives) handy for first aid measures wherever there is a risk of accidents involving phenol.

Complex formation is also used for the detection and quantitative determination of PEGs [23]. About 25 different reagents are known that can be used for this purpose owing to formation of precipitates [24].

D. Molecular Weight Distribution

It was pointed out initially that the various PEG grades are not perfectly uniform chemical compounds; rather, they are mixtures of similar polymer species. Information on the molecular weight distribution of PEG grades is given by gel permeation chromatography or by other methods of fractionation [25–27].

The excellent solubility characteristics of the PEGs are of particular importance in relation to their applications. Two advantages are especially significant: first, the ability of PEGs to dissolve many substances, and second, their very good solubility in many solvents.

In the preparation of aqueous solutions, the PEGs sometimes act as specific solubilizers.

Generally, the dissolving power and the solubility of the PEGs decrease slightly as the molecular weight increases. However, both properties are improved by heating.

For the reader's information, there follows a list of some groups of solvents with which the liquid PEGs are very readily miscible and in which the solid PEGs dissolve, although moderate heating is sometimes required to achieve solution: alcohols (methanol, isopropanol, benzyl alcohol); esters (methyl acetate, butyl acetate); glycol ethers (ethylene glycol monomethyl ether, ethylene glycol monobutyl ether, and their acetates); ketones (acetone, cyclohexanone); chlorinated hydrocarbons (ethylene chloride, chloroform); and benzene hydrocarbons (benzene, toluene). The PEGs are insoluble in paraffin hydrocarbons.

Since PEG 400 can be regarded as a typical representative of the liquid PEG grades, while the same is true of PEG 1000 among the solid grades, we have restricted ourselves to these two grades to simplify the following tables. On the basis of this, it is possible to draw conclusions regarding the solubility characteristics of the other PEG grades, and the actual solubility can then be checked fairly easily in each case. The solubility of various substances in liquid PEG 400 at room temperature is given in Table 3 [28]. It can be assumed that substances that dissolve at room temperature in PEG 400 are soluble to roughly the same extent in molten PEG 4000 at 60–70°C.

The solubility of PEG 4000 in various solvents is shown in Table 4. The figures in Table 4 give the approximate percentage of PEG 4000 in the solutions saturated at room temperature.

Generally, the solubility of the solid PEGs increases with rising temperature. This means that a PEG that is virtually insoluble at room temperature can be brought into solution by moderate heating. It is worth noting that solid PEGs are completely insoluble in liquid PEGs at room temperature.

The increase in solubility with rising temperature is sometimes very sharp, as the following example shows [28]. PEG 20,000 is soluble in pure ethanol as follows:

At 20°C	0.1%
At 32°C	1.0%
At 34°C	20.0%

Compatibility is defined here as good miscibility of the PEGs with various substances to form homogeneous mixtures that do not separate even when heated. This is a requirement, for instance, for ointments and other preparations.

Products that are compatible in the sense of being processible together with PEGs include casein, cetyl alcohol, stearic acid, glycerol, polyvinyl pyrrolidone, colophony, vegetable albumin, dextrin, starch, and chlorinated starch. PEGs are gelling agents for nitrocellulose, benzyl cellulose, and various resins. Some ethereal oils are absorbed extremely well by liquid and molten PEGs.

Lack of compatibility is reflected in a tendency to separate, which is particularly marked in the case of ointmentlike mixtures (e.g., mixtures containing beeswax, ceresin wax, ester waxes, or paraffin wax). With mainly solid mixtures, turbidity (gelatin) or brittleness may result or one of the components may exude.

In a number of cases, adequate compatibility exists only with low molecular weight PEGs. For instance, gum arabic yields clear films only with PEG 200; it exhibits marked incompatibility with higher molecular weight PEGs.

Several common types and grades of plastic have good to very good resistance to PEGs, for example, polyethylene, which is frequently used as packaging material and is also suitable for PEG storage containers.

Other plastics that are resistant to PEGs even at elevated temperatures are polypropylene, polystyrene, articles made from natural rubber, polytetrafluoroethylene, polyoxymethylene, and unsaturated polyester resins.

Table 3 The Solubility of Various Substances in PEG 400 at Room Temperature

Substance	Solubility[a]	Substance	Solubility[a]
Acetanilide	16	Eucalyptus oil	10
Acetic anhydride	∞	Formamide	∞
Acetone	∞	Furfural	∞
Acrylic acid	α	Gelatin	i
Acrylonitrile	∞	Glacial acetic acid	∞
Allyl alcohol	∞	Glycerol	∞
Ammonia 25%	∞	Glycerol monostearate	sl.s
Amyl acetate	∞	Glycerol triacetate	∞
Amyl alcohol	∞	Glycol	∞
Aniline	∞	Gum arabic	i
Antipirine	10	Hexachlorophene	sl.s
Azulene	10	Hydrochloric acid 37%	∞
Beeswax	i	Iodine	20
Benzaldehyde	∞	Iron(III) chloride ($6H_2O$)	50
Benzene	∞	Isobutanol	∞
Benzocaine	50	Isobutyl acetate	∞
Benzoic acid	10	Isodecyl alcohol	∞
Benzyl alcohol	5	Isooctyl alcohol	∞
Borax cryst.	0.3	Isopropanol	∞
Brombenzene	∞	Isotridecyl alcohol	∞
Bromofluoresceic acid	10	Lactic acid, 90%	∞
Butanol	∞	Lavender oil	10
Butyl acetate	∞	Lead acetate	1
Butyl amine	∞	Lead stearate	i
Butyl glycolate	∞	Lecithin	i
$CaCl_2 \cdot 2H_2O$	20	Lithium stearate	i
Camphor	10	$MgCl_2 \cdot 4H_2O$	25
CS_2	10	$MnCl_2 \cdot 4H_2O$	40
CCl_4	∞	Menthol	10
Carnauba wax	i	Methanol	∞
Casein	i	$(CH_3COO)_2Hg$	10
Castor oil	i	Methoxybutyl acetate	∞
Ceresin wax	i	Methyl acetate	∞
Cetyl alcohol	sl.s	Methyl ethyl ketone	∞
Cetyl stearyl alcohol	sl.s	Methyl methacrylate	∞
Chloral hydrate	50	Methyl salicilate	∞
Chloramine T	10	Mineral oils	i
Chlorobenzene	∞	Morpholine	∞
Chloroform	∞	Naphthalene	10
Ethylbenzene	∞	beta-Naphthol	40
Ethyl chloride	∞	Nitrobenzene	∞
Ethylene glycol monobutyl ether	∞	Nitromethane	∞
Ethylene glycol monoethyl ether	∞	Octyl alcohol	∞
Ethylene glycol monoethyl ether acetate	∞	Oleic acid	∞
		Paraffin oil	i
Ethylene glycol monomethyl ether	∞	Paraldehyde	50
Ethylene glycol monomethyl ether acetate	∞	PEG laurate	sl.s
		Pentachlorophenol	40
2-Ethylhexanol	∞	Pentachloroethylene	43
Ethyl urethane	50	Phenacetin	10

Table 3 Continued

Substance	Solubility[a]	Substance	Solubility[a]
Phenol	50	Pyrocatechol	50
Phenotiazine	15	Resorcinol	50
Phenyl acetate	∞	Saccharine	10
Phenyl salicylate	50	Salicyladehyde	∞
Piperazine	10	Salicylic acid	30
Potassium iodide	15	Sodium chloride	0.3
Propanol	∞	Sodium cyclamate	3
1,2-Propylene glycol	∞	Sodium nitrite	0.4
Pyridine	∞		

[a]Figures: in % by weight; ∞: miscible in all proportions; sl.s: slightly soluble at room temperature; i: insoluble.
Source: Ref. 28.

Although phenol-, urea-, and melamine-formaldehyde resins may be regarded as resistant, they are less suitable for use at elevated temperatures in continuous contact with PEGs in liquid or ointment form.

PEGs are not likely to attack common synthetic fibers and textiles at normal temperatures.

Plasticized PVC is not resistant, because liquid PEG dissolves out some of the plasticizers, making the material brittle. Cellulose acetate, too, is not resistant. Owing to its camphor content, celluloid is particularly susceptible to attack by PEGs in liquid or ointment form.

Normal paintwork is frequently attacked or partially dissolved by PEGs. Stove-enameled surfaces or stove-enameled coatings inside tubes are, however, immune to such attack.

E. Applications

1. OAGs as Phase Transfer Catalysts

Until 1976 little attention had been paid to oligoethylene glycol dimethyl ethers as potential phase transfer catalysts, despite the knowledge that lower molecular weight analogous glymes were capable of acting as effective alkali metal solvating agents [29–34]. This lack of attention is surprising, since the ability of available ether oxygens to act as phase transfer catalysts generally increases with increasing chain length of the glyme [35,36]. Thus, a considerable acceleration effect could be observed with *oligoethylene glycol dimethyl ethers* (OAGDMs) in several nucleophilic substitutions as well as with the classical permanganate oxidation in benzene solution [35,36].

Taking advantage of OAGDM-induced "purple benzene" [37], the oxidation of stilbene to benzoic acid is enhanced by approximately a factor of 10 [35]. The rate of esterification of benzyl bromide with potassium acetate depends on the OAGDM added and on the solvent present [35]. In ethanol, the reaction proceeded homogeneously even without OAGDM; nevertheless, an additional acceleration effect was observed in the presence of OAGDM. This indicates that the enhance-

Table 4 Solubility of PEG 4000

Solvent	Solubility (% by wt.)	Solvent	Solubility (% by wt.)
Aniline	30	Methanol	20
Benzene	10	Methylene chloride	53
CCl_4	10	Pyridine	40
Chloroform	47	Trichloroethylene	25
1,4-Dioxane	10	Water	55
Ethanol, 60%	50	White spirit	i[a]
Ethylene chloride	46	Xylenol	50
Formamide	30		

[a]i = insoluble.
Source: Ref. 28.

ment cannot be due solely to the increased solubility of potassium acetate and thus requires an increase in anion reactivity resulting from the complexation of OAGDM with the salt [29–33]. For comparison, the rates of esterification caused by OAGDM and by crown ethers under analogous experimental conditions are of roughly the same magnitude.

The effect of OAGDM on the alkylation of sodium diethyl benzylmalonate with butyl bromide was studied in tetrahydrofuran (THF), benzene, and benzene: hexane (1:) [35]. As a result, the rate acceleration with OAGDM was found to be low in THF, fairly high in benzene, and much higher in benzene:hexane, a result that can be explained by different ion-pair aggregation in this solvent system [37]. In THF, where sodium malonate is partly dissociated from the beginning [35], the contribution of OAGDM is only a small one, whereas in benzene and to a greater extent in benzene:hexane, OAGDM addition seems to destroy the ion-pair aggregation to a great extent by complexation. It is also interesting to note that in the case of monoglyme and crown ethers as additives there is a selectivity in the rate acceleration depending on the chain length of the alkyl bromide [35]. Furthermore, it was found that steric bulkiness of the chain causes similar effects—this being observed in the Williamson substitution of alkyl bromides by sodium phenoxide:

$$CH_3(CH_2)_nCH_2Br + NaOC_6H_5 \rightarrow CH_3(CH_2)_nCH_2\text{—}O\text{—}C_6H_5 \quad (5)$$

The investigation of the dependence of the reaction rate on the average molecular weight of the OAGDM employed for the sodium phenoxide reaction (reaction 5), rather surprisingly, yields a sharp increase in the rate with increasing molecular weight [38]. At the present time, no clear explanation can be offered as to why the "polymer effect" [39–41] is influenced to such a large extent by the average molecular weight of OAGDM.

It has been shown that simple PEGs can be also complex with alkali ions, preferably with potassium, and transfer the complexed salt into organic phases, comparably to action of crown ethers [42]. This indicates that PEGs might also be used as powerful phase transfer catalysts for a variety of organic reactions, some of which have already shown good results. The very mild reaction conditions for a high-yield conversion, requiring in most cases room temperature and reaction times of only few minutes, are the main features. This yields well even for phase transfer of potassium fluoride, which is known to be difficult to stabilize in organic media [43]. Thus, in the presence of PEG, aliphatic and aromatic carboxylic acid chlorides

$$RCOCl + KF \xrightarrow[CH_3CN,\ 20°C,\ 3\ h]{PEG} RCOF \tag{6}$$

and aromatic sulfonic acid chlorides

$$RSO_2Cl + KF \xrightarrow[CH_3CN,\ 20°C,\ 3\ h]{PEG} RSO_2F \tag{7}$$

were readily converted by KF into the corresponding fluorides. However, benzil bromide

$$C_6H_5CH_2Br + KF \xrightarrow[6\ h,\ reflux]{PEG} C_6H_5CH_2F \tag{8}$$

and 2,4-dinitrochlorobenzene

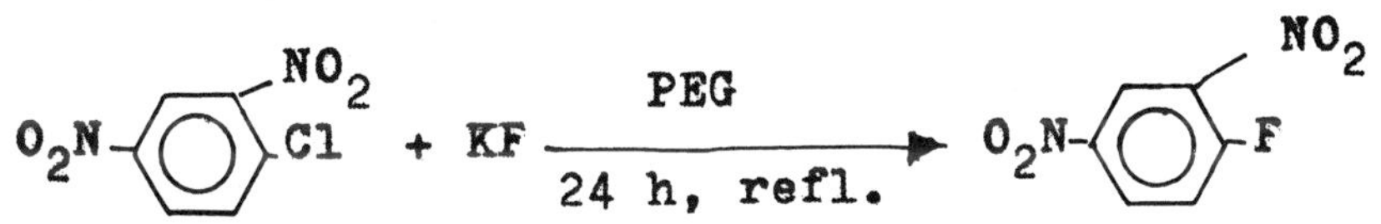

(9)

hardly react under these conditions; cyclohexyl bromide exclusively gave elimination to cyclohexene:

$$C_6H_{11}Br + KF \xrightarrow[6\ h,\ reflux]{PEG} C_6H_{10} \tag{10}$$

Nevertheless, the results reveal that PEGs have a substantial effect in activating the fluoride ion in this reaction system, too.

Esters are essentially inert toward reduction by sodium borohydride [44]. In an extension of this observation, OAG as solvent was found to catalyze the following reaction, which has been successfully carried out with a variety of substances [45]:

$$RCOOR' + NaBH_4 \xrightarrow[65°C,\ 10\ h]{PEG\ 400} R\text{—}CH_2OH \tag{11}$$

An average 10 h reaction time and a reaction temperature of 65°C has been established as optimal. The yields are generally good and lie between 65 and 90%. The results lead one to postulate that specific alkoxyborohydrides are formed under these conditions that may show a higher reactivity than $NaBH_4$ itself.

Among the PEGs' differing molecular weights, the liquid PEG 400 combines the most advantages [46], since it can be directly used as a solvent. PEG 200 is not efficient in ion binding. On the other hand, high molecular weight solid catalysts, such as PEG 4000 or PEG 6000, are almost insoluble in polar organic solvents, and this seems to be of importance for the observed acceleration effect on reactions.

Taken together, lower molecular weight and polymer-supported noncyclic neutral ligands no longer have to be very exclusively classified as weakly activating catalysts. Possibilities for optimizing these ligands as a new tool in organic synthetic phase transfer catalysis have been demonstrated.

2 OAGs as Carriers for Pharmaceuticals and Cosmetics

Synthetic polymers have long played a relatively important role in current medical practice. Their potential, either as prosthetic materials or as constituents for pharmaceutical uses, has become increasingly apparent. Presently, pharmaceutically active polymers, or macromolecular drugs, constitute a well-established research field in medical chemistry.

Drugs are rarely, if ever, administered to patients in an unformulated state. The vast majority of available drugs, which are potent at the milligram or microgram levels, could not be presented in a form providing an accurate and reproducible dosage unless mixed with variety of excipients and converted by controlled technological processes. It has long been known that the effect of an active drug substance may depend on the way in which it has been formulated. The majority of systematically acting drugs are administrated by the oral route, and therefore the drug must traverse certain physiological barriers including one or more cell membranes. In this way, the effects are often greatly weakened; on the other hand, parenteral administration, especially in the case of an individual side effect or a strong effect, may lead to deplorable results.

Most pharmacological agents are low molecular weight compounds that readily penetrate into all cell types and are often rapidly excreted from the body. Consequently, large and repeated doses must be given to maintain a therapeutic effect. Due to very limited specificity of the majority of drugs, a range of serious side effects is often observed. A drug can be prepared so that its optimal level remains preserved for a sufficiently long time. Prolongation of action may be achieved by a variety of techniques, which include various encapsulation procedures, dispersion in hydrophobic vehicles or porous polymer materials, binding of drugs to macromolecules of a polymer or oligomer, and the formation of a drug–carrier complex.

A polymeric drug may be defined as a polymeric substance that can exert a pharmacological activity when administered to a living organism. This activity may be displayed in different ways. For example, the polymer may be active itself, without being structurally related to nonmacromolecular compounds of known activity. In the latter case, the active moieties may be present as constituents of the macromolecular backbone, or attached to a macromolecular backbone by covalent bonds, as side substituents. When these polymers are able to release active molecules after entering the patient's body, either by injection or by absorption, the term *polymeric pro-drug* is also used. It is apparent that when the active moieties are among the constituents of the main backbone, the release of active molecules takes place as a consequence of a degradation process involving the whole macromolecule. However, when the active moieties are present as side substituents, the release of the active molecules takes place by cleavage of the bonds holding these moieties, leaving virtually unaffected the main chain of the polymer, which in turn may or may not be degraded in time by a separate process. In this case, the macromolecular structure to which the active molecules are attached is usually referred to as the polymer main chain or "carrier." The term *matrix* is also used to indicate the polymeric constituents of a physical mixture of polymer–drug combinations; these, however, will not be discussed in this subsection.

A polymeric pro-drug is typically composed of two distinct parts, the polymer backbone or chain, and the active moieties linked to this polymer carrier by means of covalent bonds that are cleavable in the body fluids. When the drug–carrier combination circulates within the body, the carrier can transport the active principle before it exerts its activity.

So far, high polymers have been considered in most studies of polymeric pro-drugs. The tremendous amount of research work carried out in this field has been reviewed by several authors [47–55]. Only a few general remarks will be made here. Sustained activity over time of the free drug, and selective localization at the preferred site of action (targeting) are the results most often sought. The main drawbacks of polymeric pro-drugs intended to be systematically active are their poor or absent absorption through the gastrointestinal walls or through the skin. Thus, they are suitable only for parenteral administration; the fact is that they are often poorly or not excreted after injection unless their molecular weight is reduced below a certain threshold by degradation processes involving the main backbone. This is possible only for certain types of structures, not including polyvinylic structures. Thus, an accumulation of macromolecular residues after detachment of the drug moieties may be expected. The foregoing picture is not valid where oligomeric carriers are concerned; since the molecular weight of oligomers ranges from a few hundreds to a few thousands, they are easily eliminated by normal body processes.

One of the main features of high molecular weight polymeric drug carriers is that they contain binding sites to which the drug residues can be attached (along with solubilizing units and targeting groups). These sites must be sufficient in number to allow for a reasonable loading of drug, and they may be randomly distributed along the macromolecular chain. The contribution of terminal groups is considered to be negligible in this respect.

Oligomeric carriers, in contrast, may in many cases provide derivatives with a reasonable number of active drug moieties even if functionalized only at the chain ends. This allows for the utilization of these structures as drug carriers in which the repeating units within the chain, though endowed with the desired physicochemical properties, are not useful as drug binding sites, such as poly(ethylene glycols) and their homologues. This does not mean, of course, that oligomers with reactive repeating units within the chain cannot be used as well. Thus, oligomeric carriers offer more opportunities from a structural point of view than high molecular weight polymers. Another element to be considered when dealing with many end-functionalized oligomers is that the reactivity of a chemical function located at the end of a flexible chain can be more predictable since it is practically independent of the segment to which it is attached. These predictions can be made based on known theoretical calculations for polycondensation and stepwise polyaddition precesses [56].

In contrast, the nature and length of a macromolecule exerts a great influence on the reactivity of chemical functions directly attached to it. This is exemplified by the slow reaction rate of activated esters on high molecular weight poly(methacrylic acid) with amines [57].

It is well known that the molecular size is of paramount importance in the body elimination of nondegradable polymers. Experimental evidence of this behavior has been reported in the case of poly(vinyl pyrrolidone) and other water-soluble, nonionic polymers [53]. Even though the number of polymers studied in this respect is rather limited, it can be reasonably supposed that at least amphophilic, nonionic oligomers with physicochemical properties similar to those of poly(vinyl pyrrolidone) would show similar behavior. That means that problems usually associated with the accumulation of residues in the body are not expected in the case of oligomeric pro-drugs, irrespective of any biodegradability question. It should be also considered that, in principle, a nontoxic, nondegradable oligomer may be safer than a degradable one, since problems arising from the possible toxicity of metabolites would not be encountered.

Shifting from polymers to oligomers does not solve, on the whole, the problem of coexistence of the physiological action and toxicity of the products. Some significant aspects have been delineated in the case of anionic polymers. It should be pointed out that in this case the molecular weight distribution would be more critical than the average molecular weight. However, a complete molecular weight–pharmacological activity study will needs to be performed.

The suitability of PEGs as carriers for pharmaceuticals was recognized as early as 1935 by Middendorf working in the galenical laboratory at Hoechst [58–60]. In 1939 an auxiliary based on PEG was introduced to the pharmaceutical industry under tradename Postonal.

PEGs can be modified with various monomers to obtain biologically active compounds. Thus, bithionol 2,2′-thio-bis-(4,6-dichlorophenol),

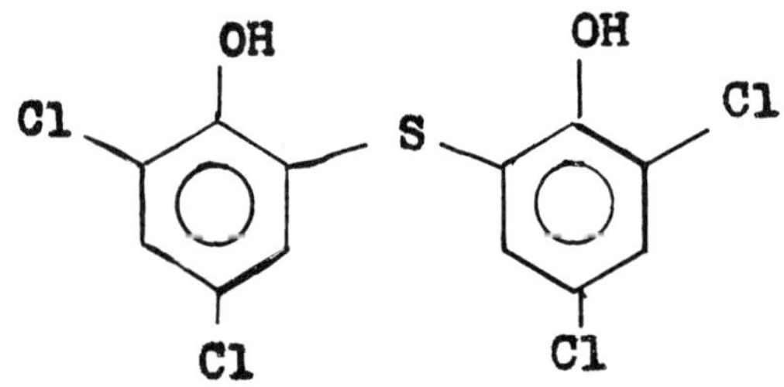

(12)

an effective antibacterial agent, was reacted with phosgene to give bischloroformate, a very desirable intermediate for the preparation of alternating copolycarbonates [61]. Thus, from PEGs of molecular weights 400, 600, 1000, and 4000 and bischloroformate, an alternating copolycarbonate may be obtained. The copolycarbonates prepared from PEG 400 were tough polymers similar in their properties to low-density polyethylene.

Despite the wide range of the PEGs used as segments for the alternating copolymers, only copolycarbonate prepared from PEG 4000 was completely water soluble. The use of the hydrophilic PEG 4000 segment was intended to render the polymer water soluble.

Esterification of PEG with carboxylic acids can be performed by the usual procedures. For instance, esters of PEGs with the anti inflammatory drug Ibuprofen (4-isobutylphenyl-2-propionic acid)

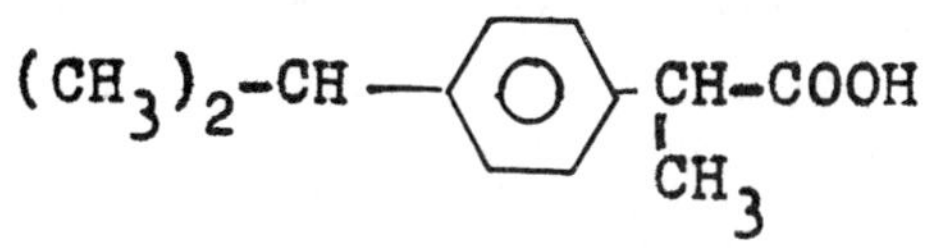

(13)

or with the antimicrobial agent

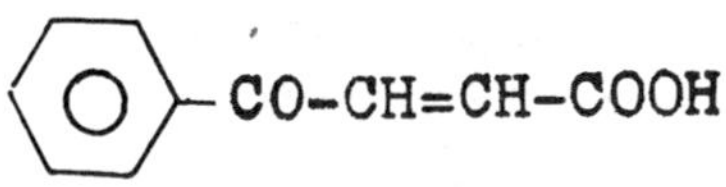

(14)

have been synthesized [62,63].

PEG may be functionalized as PEG–dihydropyridins and PEG–arcylacrylates. These compounds, though pharmacologically active, are not pro-drugs, since they are active by themselves, and do not generate active molecules.

Other end-functionalizations of PEGs have also successfully been produced by the following reactions [64,65]:

$$\mathrm{HO{-}(\underset{\displaystyle R}{\underset{|}{C}H}{-}CH_2O)_n{-}H} \xrightarrow[\text{pyridine, CHCl}_3\text{, 35°C}]{\text{succinic or glutaric anhydride}}$$

$$\mathrm{HO{-}(\underset{\displaystyle R}{\underset{|}{C}H}{-}CH_2O)_n{-}CO{-}(CH_2)_m{-}COOH} \qquad (15)$$

where R = H, n = 10–40, and m = 2,3. This compound may be further functionalized to benzotriazolides or imidazolides.

PEG may also be functionalized with isocyanate groups by the following reaction [66]:

$$\mathrm{HO{-}(CH_2CH_2O)_n{-}H} \xrightarrow[\text{excess}]{\mathrm{ONC{-}R{-}CNO}}$$

$$\mathrm{ONC{-}R{-}NH{-}CO{-}O{-}(CH_2CH_2O)_n{-}CO{-}NH{-}R{-}CNO} \qquad (16)$$

These oligomers are valuable intermediates that can be used for binding hydroxylated, aminated, or carboxylated compounds with urethane, urea, or aminic bonds.

The incompatibility of PEGs with medicaments is often discussed in terms of their ability to form complex compounds. The possibility of forming addition compounds is due to the ether oxygen in PEGs. Since these complex compounds often depend on concentration, are reversible, and in addition, are frequently affected by the presence of water, it is no easy matter to establish the consequences of complex formation. It is necessary to carry out tests in individual cases to establish whether release of these active substances has been inhibited. A delayed release of the active substance from the complex may be desirable from the biopharmaceutical viewpoint. In practice, medicaments known to form complexes with PEGs are therefore used in PEG bases. PEGs are unsuitable as bases for bacitracin and penicillin G and W (complete inactivation), acetylsalicylic acid (release of salicylic acid due to transesterification), and also where discoloration is undesirable [67].

In addition, reports of inactivation of phenobarbital by PEGs have been given, although there is no inactivation in the case of barbituric acid, diethylbarbituric acid, and phentobarbital [68].

The fact that PEGs are readily water soluble, nonvolatile, nongreasy substances acknowledged as being skin-compatible makes them ideal for numerous cosmetic preparations. Because they differ in consistency, ranging from liquid through ointment to solid and hygroscopic, it is possible to select the grade without having to compromise.

The fact that suitable PEG mixtures combine ointment consistency and solubility in water with the ability to dissolve many pharmaceuticals makes them very useful ointment bases.

The dermatological advantages of PEG ointment bases are as follows:

Polyglycols are well tolerated even by sensitive skin and do not cause any sensitization. They have no macerating effect on the areas of skin treated.

Polyglycol ointments can be removed very easily with water, even from very hairy areas and from articles of clothing.

Polyglycol ointments can be distributed easily and evenly on moist areas of skin. When applied they "melt on the skin." A thin film of PEG is almost invisible, and so the products are suitable for discreet treatment.

They do not impede gas or heat exchange from the skin or the exudation of perspiration and other secretions. They soften and detach existing encrustations.

Owing to their osmotic activity, polyglycol ointments have a washing, cleansing, and drying action and are therefore particularly suitable for treating weeping, inflammatory, supurating, ulcerated, and necrotic wounds and dermatoses.

Polyglycol bases have good dissolving power for numerous medicaments commonly used in the treatment of skin ailments and wounds. Molecular dispersion of the active substances in an ointment base that is soluble in water or exudate provides optium effectiveness in many cases. Even undissolved medicaments are usually released very well into the treated area by polyglycol ointments.

Polyglycols are insensitive to electrolytes. They are by nature completely neutral, can be adjusted to any dermatologically acceptable pH, and can be adapted to the pH of the acid mantle of the skin.

The purely technical aspects of PEGs in ointment form were explained earlier. In certain cases, it may be advantageous to make a clear solution of the pharmaceutical in liquid PEG 300 or 400 and then to mix this solution with the higher molecular weight PEG component to produce the finished ointment.

Table 5 Examples of PEG-Compatible Pharmaceuticals

Bismuth gallate, basic	Nitrofurazone
Camphor	Polymixin B
Chloramphenicol	Prophenpyridamine
Diphenylhydramine	Sulfanilamide
Hexachlorophene	Sulfatiazole
Hydrocortisone acetate	Sulfisomidine
Iodochlorohydroxyquinoline	Trypaflavin
Nitrofurantoin	

PEG bases can also be combined with various other bases. They are compatible with cetyl alcohol, cetyl stearyl alcohol, stearic acid, 1,2-propylene glycol, glycerol, glycerol monostearate, and PEG sorbitan monooleate. PEGs are not compatible, however, with paraffin wax, petroleum jelly, oleyl oleates, and hydrogenated peanut oil. Further details on the miscibility and compatibility of PEGs with other substances are given in Table 5.

Typical formulations for finished PEG ointments (e.g., chloramphenicol ointments, antihistamine ointment, chest embrocation) may be found in the literature [69].

Opinions on the suitability of PEGs as suppository masses are by no means uniform. While USP XIX emphasizes the fact that the medicaments are released effectively into the body and also stresses the suitability of PEGs for antiseptic suppositories, there is no lack of critical comments [70].

The PEGs have in practice proved very successful for globules. They are also used as suppository masses when the following properties are desired:

The specific dissolving power of PEGs for numerous active substances
Superior release of the active substance into the body
Resistance to tropical conditions [71]

In choosing the PEG grades the crucial factor is the consistency of the active substances used [72]. In the molten state, all PEG grades are readily miscible with one another and yield homogeneous masses. Table 6 lists a few useful mixtures [73]. Mixture I is particularly suitable for globules. Solid PEG compounds have also been suggested for coating freeze-dried suppositories [73].

Table 6 Useful Mixtures of PEGs

	Mixture			
PEG grade	I	II	III	IV
400	—	—	—	5%
1000	25%	50%	75%	—
1500	75%	—	—	95%
2000	—	50%	—	—
3000	—	—	25%	—

PEGs are used for liquid preparations mainly on account of their dissolving power. Their excellent tolerability was mentioned earlier.

Liquid PEGs have a slightly bitter taste, a factor of some consequence where peroral administration is concerned. The solid grades may be described as having a neutral taste. The taste can be adjusted by suitable additives (sweeteners).

Since PEG 400 does not soften gelatin, it is frequently used for those active substances where the genetable oils commonly employed as carrier substances for filling into gelatin capsules are unsuitable.

Spiegel [74] Anschel [75], and Ritschel [76] give surveys of the use of PEGs as solvents in injections, quoting some practical examples.

For PEG 300 the intravenous LD_{50} in rats is 7.1 ml/kg [77]. PEGs are suitable at the temperatures required for sterilization of injection preparations (e.g., 121°C). Aqueous solutions of barbiturates can be stabilized by the addition of PEGs [78].

If the question of choosing PEG 200 or PEG 400 arises, the greater dissolving power of PEG 200 may be the deciding factor. On the other hand, owing to its molecular weight distribution, PEG 200 contains small quantities of mono- and diethylene glycol, which are virtually absent from PEG 400. USP XIX contains a test specification for PEG 400, by means of which the slightest trace of mono- and diethylene glycol can be checked spectrophotometrically (max. permissible amount is about 0.3%). PEG 400 is accordingly preferable for injection purposes.

If pure PEG 4000 or 6000, which each melt at 65°C, is exposed to high pressure, the resultant tablets are translucent at the edges. The relatively low melting point favors a sintering or compression technique. Owing to these properties, PEG used as a binder in tablet making has at the same time a plasticizing effect, which facilitates the shaping of the tablet mass in the compression process and may counteract capping. An addition of PEG is also recommended in the manufacture of coated tablets [79], as it will give the tablets a certain "elasticity."

The high binding power can lead to an increase in the time taken for the tablet to disintegrate if the amount added is too high. This is because tablets compressed with a high quantity of PEG have a very dense structure with hardly any points of capillary access. Suitable additions to ensure good disintegration times are 2 to 5% of PEG 4000 or 6000. In the case of lozenges, for which PEGs are highly suitable, considerably more PEG can be added, as a slow dissolving rate is desired.

In tablet manufacture, PEG is not only a binder but also acts as a lubricant and furthermore as an antistick agent [80,81].

Often, some magnesium stearate can be replaced by PEG. Frequently, PEGs produce an improved smooth tablet surface with a pleasant gloss and reduced tendency to abrade.

Substances containing additions of PEG are suitable for moist and for dry granulation and also for the direct compression of tablets and sugar-coated tablet cores without prior granulation. PEGs 4000 and 6000 are available in powder form for dry mixing. The particle size distribution of PEG powders has been described in the literature [79].

Finally, another method of granulation with PEG should be mentioned, which is called "melt granulation." If a powder mixture containing, say, 10 to 15% PEG 6000 in powder form is heated to 70 to 75°C, the PEG melts; the mixture then has a paste consistency. If stirred until cold, the powder mixture will form granular particles. Naturally, this labor saving method will be restricted to cases where the active substances withstand heating to about 70°C and where the higher amount of PEG is permissible in view of the disintegration time of the tablet. In this connection, note that PEGs are also suitable as binders for various other pelleted materials, for example, catalysts.

The use of PEGs in tablet coating started with a process of applying a hot solution of PEG 6000 in ethanol [82], thereby saving considerable time as compared with sugar coating. Sugar-free tablets are also suitable for diabetics [83,84]. It was found that PEG 6000 has the additional property of imparting a good gloss to tablets [85].

Various processes involving PEGs as components in sugar-free tablet coatings have been developed [86]. PEGs can also be combined readily with polyvinyl pyrrolidone. In the *Wurster process* (fil coating, fluidized bed process), the addition of PEG to the coating compound is of considerable importance [87–89]. As a specific anchoring agent, it makes gravure print clearly visible when titanium dioxide is used as an opacifier. PEGs also give a smooth appearance to the surface.

In film coating with aqueous acrylic resin dispersions, additions of solid PEG (e.g., PEG 6000) control the permeability of the coating film [90,91].

When sugar-coating suspensions are used, there is an undesirable tendency for the tablets to stick together at the drying stage. This can be prevented by adding fairly small quantities of high molecular weight PEG 6000 to 35,000. If PEG is sprayed as an antistick agent in aqueous solution at an appropriate point in the process, the coating process can be shortened considerably. This tackiness can also be eliminated by adding a few percent PEG directly to the suspension. The tablets then adhere neither to one another nor to the walls of the vessel. A fully automatic high-speed coating method on this principle has been developed [92] and has become widely adopted in the industry [93,94].

The ready water solubility of PEGs makes them ideal for numerous cosmetic preparations. Extensive information on formulations for the production of cosmetics containing PEGs can be found in the literature and in appropriate technical journals. A typical work giving such data is Ref. 95.

The PEGs can be used in the following cosmetic preparations:

Creams, lotions, face lotions
Deodorant, perfume, and insect-repellant sticks
Lipsticks
Toothpastes
Soaps, hand-cleaning pastes, and detergent sticks
Hair care products, face packs, and depilatories
Hair sprays
Bath oils and foam baths
Tabletted shampoos, denture cleaners, and bath cubes

In creams, as in all preparations that tend to dry out, the PEGs have a moisture-stabilizing effect and also a conditioning effect on the skin treated. After application, they leave a pleasant feel on the skin similar to the natural replacement of oils, without producing any sensation of stickiness.

In lotions and face lotions, PEG acts as a cleaning agent. In pre-shave lotions, PEG has the additional functions of a nongreasy lubricant and perfume stabilizer.

PEGs are ideal carriers for sodium stearate and sodium aluminum hydroxylactate. For, unlike ethanol or isopropanol, they are not volatile and thus permit reliable control of the effectiveness of deodorant, perfume, and insect-repellent sticks. The most suitable grades are the liquid PEGs 200 and 600.

In these and other cases, PEGs prove to be outstanding solubilizers for hexachlorophene, dimethyl phthalate, azulene, aluminum hydroxychloride, and so forth.

PEGs can be used in lipsticks as solubilizers for tetrabromo-fluorescein and its derivatives. According to Nowak [95], the solubility in PEG 400 is about 10%. Higher additions of PEG should be avoided because of their good solubility in water, since dyes then tend to bleed.

Since PEGs are nontoxic and nonirritating, they meet the requirement for incorporation in toothpastes, where their main function is to improve the consistency and storage stability. Thus, glycerol and sorbitol can be replaced by PEGs in toothpaste formulations. With increasing molecular weight, the slightly bitter taste of PEGs, which can be easily counteracted by sweeteners, is less pronounced.

PEG has proved highly successful in the production of transparent toothpastes. By using PEG, the refractive index of the mixture, which usually contains a high amount of silicic acid, can be adjusted to achieve good transparency.

PEGs are highly prized as milling aids in toilet soap manufacture. Not only do they facilitate mechanical plastication, they also improve the sharpness of the molded bar contours. They stabilize the perfume and prevent the soap from subsequently drying out and cracking. The intrinsic cleansing action of the soap is enhanced without affecting the lathering characteristics.

PEGs prevent hand-cleansing pastes from drying out and leave a pleasant feel on the skin once they have dried. Very soft smooth shaving creams can also be produced with PEGs.

Soap-free blocks (detergent blocks) can be molded or pressed when PEGs are incorporated. In this application PEGs in the molecular weight range 1500 to 20,000 are suitable as readily water-soluble carriers. The strength and solubility in water can be adjusted by addition of a little cetyl alcohol.

PEGs have proved successful as additives for improving the consistency of nongreasy hair care products, face masks, face packs, and depilatories. It is essential that these products can be washed off after use with clear water, a requirement that is fully met by the PEGs.

The efficacy of aerosol hair lacquers and fixers is based on synthetic resins such as cellulose derivatives, polyvinyl alcohol and acetate, polyvinyl pyrrolidone, and so forth. As a plasticizer and antistatic agent, PEG counteracts the tendency of these substances to dry to a brittle film.

In formulations for bath oils, the PEGs assist the solubilizing action of the active substances for perfume oils. In addition, consistency and skin compatibility are improved.

PEGs are an excellent binder when bath salts, shampoos, denture cleaners, and the like are pressed into tablets. By choosing the appropriate grade, for example, PEG 4000 powder/PEG 20,000 powder, and by incorporating suitable amounts, the dissolving rate can be controlled to a certain extent as required.

A great advantage of PEGs in all cosmetic applications is that they provide no nutrient for microorganisms and are not sensitive to oxidation. Consequently, such formulations have no tendency to go bad. PEGs are virtually free from bacteria when manufactured. Other sensitive components in cosmetic preparations must, if necessary, be protected by antioxidants or preservatives.

When phenol derivatives such as polyhydroxybenzyl derivatives are used as preservatives, they may form complexes with any PEG present, resulting in slight inhibition [95]. A very similar effect is observed in the presence of nonionic emulsifiers, though of course it is much more pronounced. Apart from combination products with a fairly wide range of effectiveness and pH values, quaternary ammonium compounds such as benzalkonium chloride have proved very suitable and effective, provided that there are no anionic emulsifiers present in the formulation [96].

Further details on selecting suitable preservatives together with information on the required concentrations are contained in a review published by Wallhäuser [97].

3. Application of OAGs in the Textile and Leather Industries

The manufacturers of textile and leather auxiliaries use the surface-active nonionic PEG/fatty acid esters in the following sectors:

In the textile industry as softeners, antistatic agents, preparations, skin finishes, conditioning agents, emulsifiers, wetting and cleaning agents, components for lubricants, coning oils, loom oils, and coolants for sewing threads on high-speed machines; finishing of woven and knitted fabrics and nonwovens

In the leather industry as cleaners for degreasing skins and hides; auxiliaries in finishing, impregnating, and dyeing

Their most important characteristic for these applications is certainly their softening and lubricating effect; cleaning and wetting properties are far more pronounced, for example, in the related polyglycol ethers, which are readily obtained by ethoxylating fatty alcohol. Thus, it is the fatty component that largely determines the finishing effect desired on textiles and leathers. The main purpose of esterification with PEG is, therefore, to make the fatty component water soluble or at least dispersible, because virtually all treatments are carried out in aqueous media. The longer the polyglycol chain incorporated, the more hydrophilic the compound is. PEGs with molecular weights above 600, and at most 1000, will therefore be used as esterification components only in rare cases.

Under certain conditions, liquid and solid PEGs can also be used for the treatment of textiles. They possess softening, conditioning, smoothing, and luster-imparting properties and assist the action of antistatic finishes. The effects achieved are, naturally, not washfast, but neither do they impair the absorbency of the goods. For this reason, PEGs are chiefly used as warp lubricants added to sizing liquors.

In textile printing, PEGs are used only in exceptional cases, for example, as solubilizers or dispersing agents. Diethylene glycol is generally the preferred substitute for glycerol.

For use as a softener in finishing liquors, Hoechst Company markets a special formulation of a high molecular weight PEG under the trade name Finishing Wax WL.

4. Application of OAGs in the Rubber Industry

PEGs 4000 to 20,000 are mold release agents that also provide the necessary lubrication during vulcanization processes, thus ensuring that the articles can be easily removed from the molds. They also give such articles an attractive finish, which in the case of black goods is distinguished by a deep shade and velvety sheen.

PEG is suitable as a mold release agent and lubricant for natural rubber and all kinds of synthetic rubber, whether for soft or hard rubber articles. In the tire sector, PEG is often combined with other components to produce special mold release agents. Higher concentrations are used for foam rubber molds, for example, aqueous solution of PEG 4000.

Unlike mineral oils, PEGs are highly compatible with rubber. None of the various rubber compounds is attacked by PEG, nor is the quality of the rubber impaired even if the PEG is directly incorporated. PEGs leave no film on the surface that could cause trouble in subsequent bonding or laquering. They can be readily washed off with water in cases where absolute purity is required, for example, pharmaceutical rubber articles. As a lubricant, PEG is completely nonvolatile even when heated, is physiologically safe, and has excellent skin compatibility when handled.

The liquid PEG grades, particularly PEG 400 and the aqueous solutions of the solid PEG grades, are suitable as lubricants for the air bags and bladders commonly used in the tire industry. PEG as a lubricant is often combined with other components, for example, fine mica powder, that help to release entrapped air. PEG alone or with suitable additives can even be used as a lubricant to facilitate tire fitting.

When textiles are rubberized, PEG can be added as a processing aid directly to the latex. A highly suitable grade is PEG 1500 added in an amount of about 5%, relative to the solid rubber content of the latex. The PEG serves as a lubricant in the rolling process, and as a release agent it prevents the latex from sticking to the rollers.

PEG 6000 is particularly suitable for rubber grades containing pale reinforcing fillers (e.g., highly dispersed silicic acid). They act as dispersants for the filler in the mixing process and even more as accelerators–activators in vulcanization. Appropriate additions in this case are 4–10% PEG, relative to the weight of the reinforcing filler used. The effect of the PEG on the end product is to raise the ultimate tensile strength, the modulus, and the tear resistance.

PEGs 400 to 1500 can be used as heat transfer media in fairly small, electrically heated vulcanizing units and in immersion baths for the vulcanization of rubber articles under atmospheric pressure. The bath liquid can be made thermally stable by adding, say, 1 to 3% phenyl α-naphthylamine.

5. Applications of PEGs in the Manufacture of Polyurethanes

As already described, free terminal hydroxyl groups in the polyglycols can react with diisocyanates to form polyurethanes (PUs). PEGs of various molecular weights can therefore be used as polyol components for foamed and dense polyurethane materials.

Polyethylene glycols are employed in the manufacture of prepolymers for flexible foams and PU elastomers, and as additives to polyols in the manufacture of ultraflexible polyurethane foams.

Prepolymers are produced by the chemical reaction of PEG or PEG–polyether mixtures with an excess of isocyanate. The reaction proceeds in such a way as to leave free isocyanate groups. Not until the second reaction stage does the final reaction (foaming or elastomer formation) take place with considerably less evolution of heat. This two-stage process offers more possibilities for varying the product properties.

Polyol 600 PU is a highly suitable product for this application. It is a polyol pretreated by a special method and has a stepwise reactivity with polyisocyanates; it us used for the manufacture of homogeneous prepolymers with good storage stability. A special data sheet on Polyol 600 PU is available from Hoechst [98].

Polyurethane foam synthesis can be modified by additions of PEG in such a way that an open-cell, very flexible, highly elastic foam is produced, for example, for upholstery.

To produce ultraflexible polyurethane foam, PEG 600 or 1500 or mixtures of these two grades are added to the commercial polyols based on propylene oxide, the amount added being 2 to 10% by weight relative to the usual polyether component.

Extremely flexible and lightweight foams with good resilience can be produced by combining PEG 1500 or 600 with 10 to 15% parts of Frigen 11S (produced by Hoechst), relative to 100 parts by weight polyol including PEG.

Since the PEGs have excellent solubility in water, they can be introduced to the reaction mixture as a concentrated solution in the quantity of water required, preferably before the reaction [98]. The lubricating action of the PEG obviates the need for special lubrication of the metering pump. PEGs are compatible with the conventional activator mixture, but the silicone commonly added must be pumped and metered separately.

The great variety of types of polyurethanes (PURs), ranging from elastomers and rigid plastics to ultrasoft foams, can be attributed mainly to their molecular weight, average functionality, the structure of the end groups, and the structure of the chain of the hydroxyl-containing raw material. Firms outside Russian marketing raw materials for PUR production offer a wide range of oligoethers, 80–90% being composed of hydroxyl-containing oligomers. For instance, Union Carbide offers 40 grades of oligoethers for the preparation of PURs. Accordingly, the widening and improvement of the oligoether range has become one of the main joint undertakings of specialists in Russia and Germany. All polyols intended for PUR production have as a rule a basic company name, including figures or figures and letters to indicate the specific type of polyol [99].

In Germany the family of oligoethers has the basic name Systol, while in Russia the basic name is Laprol, or Lapramol for nitrogen-containing products.

Oligoethers for PUR production are linear or branched products of the polyaddition of alkylene oxides, mainly propylene oxide or ethylene oxide, to polyhydric alcohols, amines, or mono- or polysaccharides. High, and sometimes mutually contradictory, requirements are presented for all polyols: strictly defined and constant content and structure of hydroxyl end groups; complete absence of alkaline, acidic, or any other inorganic impurities; and uniformity in molecular weight, composition, and functionality.

Oligoethers for the production of elastic and semirigid PUR foams have, as a rule, an equivalent mass of 1000–2000, and average functionality of 3 (Table 7). The first and most common polyol of this type, Laprol 3003, is produced by practically all the firms producing oligoethers. It is a product of polyaddition of propylene oxide to glycerol or trimethylolpropane; its molecular weight is 3000. This oligomer is used mainly in the production of PU by hot forming of a number of car components. The quality of Laprol 3003 is not inferior to comparable non-Russian products:

Brand of poly-(oxypropylene glycol)	I	II	III	IV
Hydroxyl group, %	1.76	1.70	1.60	1.62
Acid number, mgKOH/g	0.02	0.05	0.04	0.03
Iodine number, mgI/g	0.76	1.05	1.11	1.00
Moisture, %	0.03	0.07	0.10	0.05
Content of K^+, mg/kg	5	5	5	5

Here, I is Voranol CP 3001 (produced by Dow Chemical), II is Niax Polyol LG-56 (produced by Union Carbide), III is Daltocel F-5601 (produced by ICI), and IV is Laprol 3003 (produced in Russia).

Laprol 3503 belongs to the second generation of polyols for the production of elastic and semirigid PU foams. It has relatively high chemical activity and was specially developed with a view to improving the production of elastic molded

Table 7 Oligoethers for Making Elastic PUR Foams Produced in Russia (Laprol) and Germany (Systol)

Grade	Hydroxyl group content (%)	$\bar{M}_n$	Average functionality	Moisture content (%)	Flash point (K)
Laprol 3003	1.55–1.70	3000	3	0.1	473
Laprol 3017	1.40–1.50	3300	2.8	0.1	373
Laprol 3503	1.55–1.70	3000	3	0.1	473
Systol 2B-5	1.70–1.70	3000	3	0.1	473
Laprol 5003	0.95–1.10	5000	3	0.1	498
Laprol 3203	1.50–1.70	3000	3	0.1	488
Laprol 4003	1.20–1.75	3800	2.8	0.1	483
Laprol 6003	0.75–0.85	6000	3	0.1	473
Laprol 1003S	4.80–5.50	1000	3.6	0.1	488

Source: Ref. 99.

PUR foams. When it is reacted with ethylene oxide, there is block copolymerization at the end groups. Commercial testing of Laprol 3503 has shown that in the production of molded PUR articles there is a reduction in their mass without reduction in strength-related properties, resulting in reductions in the consumption of raw material and the reject rate, and accordingly a reduction in the labor outlay in repair of the articles. Because of the relatively high chemical activity of Laprol 3503, it is possible to dispense with the expensive catalyst DABCO (1,4-diazabicyclo-2,2,2-octane).

The most popular activated polyol in Russia and Germany is Laprol-2B-10. When commercial production of oligoethers was being taken up in East Germany, the product for the comparable purposes did not come up to processing requirement, which made it necessary to replace it with Laprol 5003-2B-10, produced by Syntheswerk (Germany) as Systol T116. Laprol 5003-2B-10 is an oligoether triol of molecular weight 5000, containing upward of 80% of primary hydroxyls, which gives it its high chemical activity. It is used for elastic PUR foams for cold molding, integral semirigid PUR foams, and other PUR materials. With the expansion of Russian production of polyethers, the proportion of this product will fall but, according to preliminary estimations, will make up not less than 20–25% of the total volume of oligoether production.

Further raising of the specifications for the properties of polyols intended for cold or hot forming has made it necessary to develop a product of yet higher molecular weight. The development of such oligomers involves certain difficulties, since with increasing molecular weight of the oligomer there is a rise in the proportion of monofunctional impurities (this is reflected in the iodine number or in the content of unsaturated bonds). Accordingly, in the development of the procedure for making Laprol 6003 special measures became necessary for reducing the proportion of monofunctional impurities (Table 8).

The first generation of oligomers for elastic bulk PU foams were poly(oxypropylene)triols of molecular weight 3000. In the Soviet Union this meant Laprol 3003, in East Germany, the slightly different oligomer Systol T112.

By the joint efforts of Germany and Russia, there has been developed a new oligoether for the production of elastic bulk PUs using modern technology. This

Table 8 Oligoethers with Molecular Weight 600 for Cold Forming Foams

Grade	Producer	Content (%)		Iodine number (g/100 g)
		OH groups	Potassium	
Pluracol 220	BASF-Wyandotte, Memphis	0.8	2	2.56
Thanol SF 6500	Jefferson, Atlanta	0.75	2	2.38
Poly-GX-805PG	Olin, St. Louis	0.81	2	2.60
Daltocel R102	ICI, London	0.75	2	2.40
Desmophen 3073	Bayer, Leverkusen	0.85	5.5	1.72
Lupranol 1371	BASF, Hamburg	0.75	12.5	1.75
Laprol 6003	Russia	0.80	5	1.70

product provides for, along with good processing characteristics, the production of bulk PU foams covering a wide range of apparent densities along with relatively high load-bearing capacity. This is the new polyol Laprol 4003 (Systol T114 in Germany), which is an oligomer with a special, strictly monitored structure of the hydroxyethyl segments of the chain. This makes it easier to meet the processing requirements and produce a foam with prespecified properties. When they use this polyol in elastic bulk PUR foam systems in place of Systol T112, producers can increase the stiffness while keeping the apparent density, or reduce the apparent density without losing stiffness. For instance, according to the requirements of the standard, for an apparent density of 33 ± 2 kg/m^3 the stiffness under 40% compression has to be 325 N, the EB 180%, and for apparent density 30 ∓ 2 the figures are 275 N and EB (elastic behavior) 200%, respectively. According to test data [99], for apparent density 33.9 kg/m^3 the stiffness under 40% compression was 402 N, the EB 299%; and with apparent density 31.9 and 29.00 kg/m^3, 385 N and 245%, and 340 N and 260%, respectively. Systol T114 is one of the oligoethers marketed commercially.

In addition to these most common brands of oligoethers, for making PU foams there are also used such special types of polyols as Laprol 3203-4-80 (a copolymer of tetrahydrofuran and propylene oxide for the production of low-temperature-resistant elastic PU foams that keep their elastic properties down to 223 K), Laprol 3003F (a special grade of high molecular weight oligoether that contains 2–2.5% of phosphorus in order to yield flame-resistant PU foams), Laprol 100/81 (a blend of oligoethers for semirigid bulk PU foams), LANS (a polymer–polyol containing an AN (arylonitrate)/styrene copolymer grafted to an oligoether), which is used to give molded and bulk PU foams with relatively high resistance to loads.

The requirements for oligoethers used to make rigid PU foams are completely different. As a rule, these oligoethers are short-chain (equivalent mass 200), heavily branched polyols (functionality 3–8); much use is made of nitrogen-containing oligoethers. As the raw material for such oligoethers producers use, along with glycerol and trimethylolpropane, different polyhydric alcohols (pentaerytrol, xylol, sorbitol, or mannitol), polyamines (ethylenediamine or diethylenetriamine), saccharides (glucose or saccharose), and so forth.

Russia and Germany produce a quite wide range of oligoethers for making PU foams, with steady updating (Table 9). The products Laprol 263, Laprol 363, Systol T107, and Laprol 503 are all based on glycerol and propylene oxide, differing in molecular weight. They may be used as a basis for compositions with differing pot life and strength in the end product. Laprol 805 is produced by oxypropylation of xylol. This is one of the highest-quality polyols for rigid PUR foams, although it is also the most expensive. Laprol 805 exhibits world-class characteristics. Laprol 564, Systol T106, and Laprol 1258 are all oligoethers based on glucose or saccharose; their high average functionality and moderate viscosity ensure their wide use. Lapramol 294, Lapramol 323, Lapramol 375-2-20, and Systol T103 are nitrogen-containing oligoethers with high chemical activity. They are used above all in a blend with other types of polyols to impart relatively high activity. Lapramol 545 is based on aromatic polyamines.

In Russia only the main grades of oligoethers for the production of rigid PU foams are being marketed commercially. As the demand for these products rises,

Table 9 Oligoethers Used for Making PUR Foams

Grade	Hydroxyl group content (%)	$\bar{M}_n$	Average functionality	Flash point (K)
Laprol 263	18.2–20.4	270	3	513
Laprol 373	13.0–14.5	370	3	493
Laprol 503	10.0–11.0	500	3	475
Systol T107	12.5–13.9	400	3	373
Laprol 564	12.0–13.5	—	4.5	473
Laprol 805	10.5–11.9	800	5	477
Systol T106	12.5–13.5	—	—	373
Laprol 1258	11.0–12.0	1250	8	463
Laprol 503M[a]	7.0–9.0	700	3	—
Lapramol 294	21.0–24.0	290	4	503
Lapramol 323	15.0–17.0	320	3	473
Lapramol 375-2-20	21.0–24.0	370	5	493
Systol T103	13.0–14.5	430	5	373

[a]Chlorine content 12.5–17.5%.
Source: Ref. 99.

the output and available range will increase continuously. In Germany considerable experience in the commercial production of various types of rigid PUR foam has been gained, and at present significant renewal of the commercial types is being attempted, based on jointly developed products. Particular interest is shown in Laprol 564, for which trials of pilot-plant batches have shown economic and technical advantages in its use as compared with Systol T103. According to the preliminary data, the newly developed Lapramol 545 will make it possible to improve the heat resistance and shape retention of potting and integral types of PUR foam, and to improve the processing variables, in particular to shorten the molding cycle in the production of integral PU foam.

At present, there are also oligoethers suitable for making other PUR materials that are used in various industries (Table 10).

6. Applications of PEGs in the Ceramic Industry

PEGs 4000 to 20,000 meet the following requirements in the manufacture of pressings based on steatite and as temporary binders in oxide ceramic bodies:

- Easy, smooth distribution in the ceramic body either by dry blending as PEG powder or as aqueous PEG solution, if necessary followed by spray drying
- Good plasticizing properties
- Good lubricating action in pressing, extrusion, and injection molding without any troublesome sticking to the modulus
- Increased green strength of pressing
- Uniform burning-out without any residue or fumes

Table 10 Oligoethers for the Production of Various PUR Materials

Grade	Hydroxyl group content (%)	$\bar{M}_n$	Uses
Laprol 202	16.0–18.0	200	Modification of isocyanates
Laprol 402-2-100 (Systol T122)	7.5–9.0	400	As chain extenders
Laprol 1103	4.5–5.5	1000	Manufacture of PUR lacquers
Laprol 1102-4-10	2.5–3.3	1100	PUR enamels
Laprol 1502-2-70	2.0–2.3	1500	Isocyanurate foams
Laprol 1052	3.0–3.5	1000	PUR synthetic leather
Laprol 1602	2.0–2.25	1600	PUR synthetic leather
Laprol 2102	1.5–1.7	2000	PUR synthetic leather

Source: Ref. 99.

The electrical properties of ceramic insulators are in no way impaired by the addition of PEG. Usually, as little as 1% PEG 20,000 is sufficient to plasticize the body adequately for extrusion. Spray-dried oxide ceramic material requires a PEG content of 4% on average, while for injection molding the additions may be considerably higher without adversely affecting the electrical properties.

As in the case of ceramic bodies, the plasticizing effect of PEGs is used to good effect in the manufacture of ferrite materials. In this case, PEG 1000 in amounts of up to 5%, as well as the high molecular weight grades, is added [100].

Mixtures of liquid PEGs are suitable as screen printing inks, for example, for printing ceramic titles. By selecting suitable glycols and adding water, the hygroscopicity of the printing ink can be varied within wide limits.

The adhesion of the raw glazes is increased by adding high molecular weight PEGs 6000 to 20,000.

If one part molten PEG 6000 is mixed with five parts glass or ceramic coloring agent, the result is a mixture that, when solidified, can be applied as a paint with water and a brush. Its smooth consistency also makes it suitable for transparent colors. The color layers based on PEG are readily etched and are suitable for scraping techniques.

When heated to about 450°C, the PEGs burn off completely without leaving any residue or carbon deposits.

7. PEGs as Plasticizers for Abrasives and Industrial Cement Products

As in the case of ceramic bodies, PEGs 6000 to 20,000 are suitable for plasticizing and, since they can be readily burned off, as binders for abrasives. They also have a beneficial effect on the structure of the material. Additions of 1 to 2% PEG 6000, relative to the weight of the dextrin used, improve the free-flowing properties of the granules in automatic pressing machines.

In the extrusion of cement and asbestos cement, even fairly small additions of PEG (e.g., 1% and less PEG 6000) have a plasticizing effect and give a fine smooth surface. At the same time, wear and tear on the steel tools is reduced.

Additions of PEG have a good lubricating effect and increase the workability of cement mortar for pressure grouting.

8. Applications of PEGs in the Detergent and Cleanser Industry

Proteolytic enzymes are being used more and more frequently in soaking agents and heavy-duty detergents. By prilling (spray granulating) PEG–enzyme mixtures, it is possible to produce a virtually dust-free enzyme concentrate that is readily metered and presents no hazards when handled. The chief grades used in this application are the higher molecular weight PEGs with melting points of 50–60°C. A highly desirable effect that has come to light in that the PEGs have a distinct stabilizing effect on the enzymes [100,101].

The addition of liquid PEGs or PEG ointments to bioactive enzymatic detergents greatly improves their storage stability [102].

PEGs are a fairly common component of cleaners, even though they have no actual detergent properties, because of the following factors:

The active detergent components in the products concerned are reinforced by the dissolving power of the PEGs; textile fibers are given a pleasant feel.

When added to hand cleansing pastes, the highly skin-compatible PEGs have a very beneficial effect on the skin even if the paste contains aggressive detergents and abrasives; yet they do not impair the effectiveness of the cleaner.

PEGs are not sensitive to electrolytes and are compatible with the acids commonly added to cleaners. As carriers and binders, PEGs offer the advantage that an appropriate selection from the wide range of molecular weights available allows the consistency of the preparations produced to be readily adjusted and the desired conditioning to be achieved. The PEG makes it easy to remove the used cleaner simply by washing, and PEGs of 4000 and above are highly suitable for the manufacture of tabletted or molded rinsing aids with toilet bowl cleaners (toilet blocks). The solubility in water of the finished products can be adjusted to meet requirements by adding small quantities of water-insoluble products such as acid or cetyl alcohol.

Apart from being used in fully synthetic hand-cleansing pastes, PEGs are also added to metal cleaning, grinding, and polishing agents.

9. PEGs in the Lubricant Sector

A large number of requirements are imposed on an ideal lubricant. Both conventional and synthetic products succeed in meeting only some of these requirements in optimal fashion. The requirements imposed on a lubricant are

1. Cost-effectiveness (purchase price, life)
2. Wide range of use (low pour point, high boiling and decomposition point)
3. Adequate stability to temperature, oxidation, and shear
4. Good compatibility with machine construction materials and seals
5. Good viscosity/temperature behavior
6. Low wear
7. Environmental acceptability
8. Low toxicity

We realize from the first glance at this list that there is no lubricant that meets all the given requirements fully. It must also be freely conceded that the conventional mineral-oil-based lubricants represent a balanced compromise.

What is, then, the relationship between conventional and synthetic lubricants? Perhaps this is a suitable point to resurrect the old comparison between the ready-made suit and the made-to-measure suit. Ninety percent of the population are of such a build as to appear well dressed in a ready-made suit. The situation in lubrication technology is similar. Here, it is only in a small minority of cases that a synthetic lubricant with specially "tailored" properties need be used. The aim of this subsection is to give you an idea where the specific strengths of one class of synthetic lubricants, namely PEGs, are to be found [102].

In the production of conventional lubricants, a usable product is obtained essentially by physical refinement from a given blend. Synthetic lubricants, however, are obtained by synthesis from reactive low molecular weight starting products.

As synthetic lubricants, the PEGs belong to the group of polyalkylene glycols that also includes the polypropylene glycols, and the random copolymers of ethylene oxide and propylene oxide and their mono- and diethers [103,104].

A common feature of all polyglycols is their excellent viscosity–temperature characteristics, thus making the products absolutely insensitive to shear stress even under the severest loads. At high temperatures, PEGs decompose completely without leaving any residue such as sludge or coke. With suitable inhibition (aromatic amines, phenols, phenotiazine), lubricants with excellent high-temperature properties can therefore be formulated.

In addition, all polyglycols have good lubricating properties in boundary friction conditions. Information on corrosion and oxidation inhibition and suggestions for additives to combat wear are available in the literature [105].

The use of PEGs in the lubricant sector is basically restricted to the liquid grades PEG 200, PEG 300, and PEG 400. They are used in those cases where solubility in water, absolute physiological safety, and skin compatibility are essential.

Owing to its low solidification point (−50°C), PEG 200 is suitable for lubrication at low temperature (e.g., pneumatic hammers).

The PEGs are particularly suitable for all types of rubber lubrication, since they neither attack nor induce swelling in natural rubber and the numerous types of synthetic rubber.

In the interests of road safety, brake fluid for road vehicles must meet a number of stringent requirements. They should have excellent corrosion protection, lubricating action on metal and elastomers, extremely low congelating point combined with the flattest possible viscosity–temperature curve and high boiling point, which must be depressed as little as possible when moisture is absorbed.

In the United States, the minimum requirements for a brake fluid are laid down in SAE 1703 MVSS 116 (DOT-3, DOT-4). Fluids meeting these requirements are nowadays manufactured by blending the following components: solvents (glycol ethers), lubricants (e.g., polyglycols), and oxidation and corrosion inhibitors together with antiwear additives.

Generally, at least 20% polyglycol-based lubricant is used. The PEGs have excellent viscosity/temperature characteristics and very low congelating points. The products are, furthermore, absolutely insensitive to shear. These properties make the PEGs particularly suitable for the formulation of synthetic hydraulic fluids for extremely low temperatures. Such fluids are used, for example, in construction machinery and in motor vehicle hydraulic systems.

In many sectors of industry (mining, aluminum die casting, metallurgical plant, etc.) fire-resistant hydraulic fluids are used for safety reasons. One way to achieve such fire-resistant hydraulic fluid is to use the water–glycol–polyglycol system. A hydraulic fluid of this type is composed as follows: about 45% water as the nonflammable component, about 25% ethylene or propylene glycol as the antifreeze agent, about 20% highly viscous water-soluble PEG to achieve the desired viscosity and lubricity, about 10% inhibitors to prevent liquid and vapor corrosion, to improve protection against wear, and as defoamer. The additive package and the PEG component determine the quality of such a formulation. Hoechst AG offers Genodyn Hydro 46, a good water–glycol hydraulic fluid with excellent protection against wear [106].

Compared with mineral oil or fully synthetic fire-resistant formulations, water–glycol hydraulic fluids can be regarded as particularly nonpollutive to the environment.

Hydraulic fluids based on water-soluble PEGs represent a much lower environmental risk than mineral oil in the event of spillage. Such hydraulic fluids are used in food factories and in the textile industry [107].

Water-soluble PEGs (e.g., PEG 3000) are suitable for the formulation of mineral-oil-free, water-soluble, fully synthetic cutting fluids. A typical formulation consists of 10–50% PEG P41/3000 (Hoechst AG) and 90–50% Hostacor KS 1. Triethanolamine, metal inhibitors, and if necessary, bactericides are also added to this concentrate to produce the finished lubricant [108]. Depending on the severity of the machining process, 1–5% solutions are used. In the formulation of water-soluble PEGs, very great importance is attached to a minimum of foam, and so defoamers are usually superfluous. Fully synthetic cutting fluids based on polyglycol are noted for longer service life, lower sensitivity to microbiological degradation, and lack of irritating effects on the skin.

Polyglycol P 41/3000 and P 41/10,000 (Hoechst AG) have proved excellent as additives to hardening baths in induction hardening. During the quenching process the formation of a water vapor skin is prevented, and thus weak spots and cracking are minimized [108]. The maximum cooling rate is reduced in the desired manner and operates at higher temperatures. The amount of polyglycol added to the hardening bath is 2–5%.

Polyglycols with suitable additives are noted for excellent aging characteristics, a flat viscosity–temperatures curve combined with absolute insensitivity to shear, and excellent protection against wear. All these properties make polyglycol lubricants particularly suitable for lubricating enclosed heavily loaded industrial gearing, for example, in rolling mills, paper-making machines, mills, kneaders, conveying equipment, and so forth. With due regard for the manufacturer's specifications, these lubricants can also be used for motor vehicle rear-axle drives and tram and motor reduction gears. The PEGs have a lower coefficient of friction than conventional oils, allowing high efficiency in worm gears and thus longer machine service life to be achieved coupled with a lowering of the oil temperature. There are likewise advantages at low temperatures. The lubricant spreads more quickly and energy consumption drops.

In modern plastics calenders, rolling mills, and paper-making machines, lubricants are often exposed to temperatures that are beyond the limits of control via mineral oils. By contrast, polyglycols, when suitably inhibited, can be used at temperatures up to 250°C without any appreciable tendency to carbonization, sludge deposition, or gumming. PEG lubricants incorporating good additives have three to five times the service life of the best mineral oils at extreme temperature conditions [109]. Machine service life is prolonged correspondingly. The excellent viscosity index counteracts difficulties in cold starting. The polyglycols, which decompose without leaving any residue, are also suitable as carriers for graphite or molybdenum disulfide. Such products are used to advantage, for example, as chain link lubricants in furnaces, since the base oil decomposes at high temperatures without forming carbon residues.

The excellent high-temperature properties of correctly inhibited PEG combined with their low congelating point make them particularly suitable as heat transfer media for plastic machinery, chemical plants, wax melting equipment, and so forth. Temperatures up to 250°C may be used without signs of carbonization occurring. PEGs offer the advantages of solubility in water and are therefore widely used in vulcanization at atmospheric pressure, in annealing plastic articles, and in the manufacture of printed circuit boards for the electronics industry [109]. In all these cases, the heating bath liquid must be easily washed off with water.

The optimal decomposition properties of the PEGs make them highly suitable for use in two-stroke engines [110]. While mineral oils tend to thicken and form sludge and deposits, polyglycol-lubricated air-cooled two-stroke engines are much cleaner running. The tendency to sooting up, sparkplug gap bridging, and piston ring sticking is reduced markedly. Even in the combustion chamber there is less deposit. Furthermore, far fewer fumes are emitted.

In the main, castor oil is used for lubricating model engines, but it causes fouling and gumming of the model and engine that is difficult to remove.

The readily methanol- (or methanol/nitromethane-) soluble PEGs do not have this disadvantage while having lubricating properties equal to those of castor oils [110].

Conventional lubricants based on mineral oil tend to increase in viscosity on oxidation or absorption of gases. On decomposition, troublesome sludge formation may occur. The PEGs are free from this drawback and owing to their character have less tendency to absorb methane, ethane, ethylene, propylene, and so forth. They can therefore be used to advantage for lubricating compressors employed in the production of low-density polyethylene. A lubricant for these compressors must meet several requirements:

An adequate lubricant film must be formed on the plunger and piston.
Since the lubricant is dissolved in ethylene and thus enters the end products for food packaging, the toxicological properties are of major importance.
The lower the solubility in ethylene, the lower the lubricant consumption.

Polyglycol B 11/150 (Hoechst AG) is highly suitable as a lubricant for low-density polyethylene compressors, since at 2000 bar it has only 1/10 the solubility of white oil in the compressed ethylene. The maximum permissible polyglycol concentration in the end product was fixed at 1000 ppm in 1973 (3rd Recommendation of the German Federal Ministry of Health, 15 July 1973, 97th Report). In this particular application, PEGs have the following further advantages over white oil:

High load-bearing capacity even without additives
Favorable viscosity–temperature characteristics
Flat pressure–viscosity dependence curve
Low consumption

Since PEGs contain no sulfur, they are also used as lubricants for gas compressors in all sectors of the chemical industry where sulfur compounds can cause catalyst poisoning. Water-soluble grades can be readily removed from a compressed gas by scrubbing with water.

High-grade lubricating greases, which are highly suitable for high- and low-temperature applications, can be manufactured with polyglycols. The thickeners used are modified bentonites or pyrogenic silicic acid. When water-soluble polyglycols are used, it is possible to manufacture greases that are resistant to hydrocarbons. The properties of mineral oil greases thickened with highly dispersed silicic acid are improved by the addition of polyglycol.

Owing to the solubility of PEGs in water, lubricant containing these products can be readily washed out after metalworking, thus enabling surface treatment (painting, electroplating) to be carried out immediately without laborious cleaning.

In the cold working of metals, drawing lubricants containing aqueous solutions of solid PEG grades are advantageous in that they dissipate heat well and can be easily rinsed off. They are suitable for drawing pipes and for drawing sections from mechanically descaled round stock. Aqueous solution of PEG 400 can be used in the hot rolling of aluminum blocks/ PEG 600, for example, is suitable for deep drawing cathode material because the PEGs are free from sulfur and other cathode poisons. The investment casting process commonly used in the precision casting of metals can be varied in an interesting way adding a mixture of water-soluble waxlike PEG 4000 and mica.

Impregnating abrasives with high molecular weight PEG closes the pores and improves the lubricating action for grinding soft metals such as copper and aluminum [111].

PEG 1500 to PEG 4000 are used as binders for grinding and polishing pastes.

The insensitivity of polyglycols to electrolytes makes them suitable as additives in electroplating baths. Even small additions of high molecular weight PEG in tin and copper plating ensure a solid bright metal deposit in the electroplating of aluminum, and stainless steel additions of PEG produce a very attractive surface [112].

10. PEGs in Technology of Wood Treatment

By incorporating PEG in the cell wall of wood, it is possible to prevent wood from shrinking as it dries out and thus produce a dimensionally stable wood [113–115]. Since PEGs do not volatilize, the wood remains permanently and unvaryingly dimensionally stable.

Of the numerous water-soluble compounds that can be used, the PEGs in the molecular weight range 1000 to 2000 have the most advantageous properties. The fairly low hygroscopicity of these products, which are solid at room temperature, means that high dimensional stability can be achieved.

Green wood can be impregnated by immersion in aqueous PEG solution over a period of several weeks. Air-dried wood is best treated in a combined vacuum-pressure process. For example, beechwood can be made almost completely dimensionally stable by immersing it in an approximately 50% solution of PEG 1500 so that it absorbs about 45% PEG. The process has two advantages [116–119]:

The PEG-treated wood can be dried at high temperatures without warping or cracking.

As a result of being treated, the wood becomes much easier to cut and work when it is finally used.

Impregnating fairly large timber such as planks and boards or treating whole tree trunks presents difficulties owing to the long impregnating times required and the associated technical expertise needed. Suitable objects for impregnation are, therefore, those of small dimensions such as parquet flooring blocks, craftsman-made articles, and tool handles. One advantage is that wood preservers, dyes, or curable monomers can be incorporated in the wood together with PEG. In a number of instances it is sufficient to spray just the outer edges of the fresh wood with liquid PEG 200 to 400 to prevent cracking.

Suitable products for bonding PEG-impregnated wood are reinforced poly(vinyl acetate) glues, Kaurit glue, and solvent-free epoxy resins. Since the PEGs have a softening effect on normal surface coatings, reactive polyurethane coatings are preferred for surface treatment [120].

The essence of preserving ancient moist wood is to counteract the danger of disintegration on drying out and to enable the wood to be handled by incorporating PEG [121,122]. The most suitable grades for this are PEG 1500 to 3000, which can also be used in aqueous solution. The wooden finds to be preserved are placed in a 20% aqueous PEG solution, for example, which is kept at about 60°C for several weeks. As the water evaporates, it is continuously replaced by more PEG solution until finally the wood is stabilized by concentrated PEG. Objects treated in this way keep their natural appearance. This method has been successfully used without any great technical expertise for the preservation of ancient woodcarvings on the island of Samos. PEGs have also been widely used in preserving ancient ships [123,124].

11. PEGs as Humectants for Cellulose Film

The liquid PEG grades mainly used for the production of cellulose films have the following valuable properties as humectants or plasticizers: balanced hygroscopicity, complete nonvolatility, absolute transparency, peptizing effect on resin components of the viscose, anchoring effect on nitrocellulose lacquers, and the imparting of dimensional stability.

In accordance with the German cellulose film regulations, cellulose film may contain up to 15% PEG. The PEG can be mixed with other humectants commonly used in cellulose film. PEG 300 is preferred for PUT (plain, untreated, transparent) grades, which are used chiefly for the manufacture of adhesive tape.

For this special application a nonyellowing special grade is available that meets the requirements of the "Sidac test."

12. PEGs as Viscose Modifiers

In addition to the ethoxylated fatty amines, PEGs 400 to 4000 have attained importance as modifiers in production of high-tenacity regenerated cellulose [125,126]. Generally, the two types of modifiers are used together because the PEGs are complementary to the fatty amine ethoxylates, such as Leomine AC80 (Hoechst). As an acceptor, PEG added to the viscose in the spinning process effects limited acidification.

As modifiers, PEGs improve cellulose filaments by giving them an all-skin structure with reduced swelling tendency and higher wet and dry strength. These properties are required for industrial rayon for car tires, for high-tenacity textile rayon, and for high-tenacity rayon staples.

Unless there are special factors determining the choice, it is advisable to consider PEG 1500 first, alternatively PEG 400 or 600. This, however, does not exclude other PEG grades, such as PEG 1000 or 3000, from being considered as suitable modifiers.

The PEG can be added to the viscose during the dissolving stage—preferably in the secondary dissolver—or just prior to spinning. The amount added is about 1 to 3%, relative to the cellulose. The total quantity of modifiers must be within this figure should PEG be used as a modifier together fatty amine ethoxylates. Up to half the total modifier content may consist of PEG. It should be noted that PEG 1500 is advantageous in certain circumstances [126].

13. Applications of PEGs in the Paper Industry

Polyethylene glycols can be used for keeping various grades of paper moist and flat, for example, gravure paper, coated cardboards, and greaseproof paper.

Liquid PEGs 200 and 600 prevent excessive drying out and, in contrast to glycerol, are less hygroscopic and nonvolatile. They do not migrate and have better water retention. Unlike other humectants, PEGs do not give rise to any hardening or yellowing even after several drying operations.

Under the German Food Law, PEGs are approved as humectants for papers, cardboard, and boards for food packaging. As well as liquid grades, mixtures of liquid and solid PEGs (e.g., PEG 400 and PEG 1500 in a 1:1 ratio) are often used and can be applied to a 1 to 5% aqueous solution. The amount applied is governed by the weight per m^2, and the stuff composition. The solution can be applied by brush dampers, size presses, jet dampers, or similar units.

When added to sanitary papers and waddings, mixtures of PEG 400 and PEG 1500 improve softness and tactile properties.

PEGs of various molecular weights (600 to 20,000) influence the consistency and spreadability of coating compounds (modified starch, barite, pigment coatings) and also facilitate the calendering of paper. They improve surface smoothness [127].

Addition of PEG may produce a release effect that, for example, prevents the adhesion of coagulated coatings to the glassing roller or facilitates the removal of transfer backing.

Addition of PEG improves the flatness of gummed papers and photographic papers by counteracting the tendency to curl up. They also make dextrin gums easier to perforate [128]. It must, however, be borne in mind that the adhesion of label adhesives may be reduced by PEGs. Owing to their insolubility in aliphatic hydrocarbons, PEGs increase the resistance of impregnated stuffs to mineral oils and fuels.

14. Applications of PEGs in Production of Dyes and Printing Inks

The properties of PEGs of major importance for the coloring sector are solubility in water, hygroscopicity, dissolving power of dyes, binding power, and antistick characteristics.

These properties, some of which are complex, are utilized in the following applications:

- In steam-set or moisture-set printing inks, PEGs 200 to 400 and Lanogen 1500 (Hoechst) regulate moisture absorption, thus preventing the ink from setting too early. The hygroscopicity of the diethylene glycol normally used for this can be suitably modified by addition of 2–25% PEG, relative to the humectant.
- The highly rubber-compatible PEG 6000 in amounts of 3–5% can be used in flexographic inks. The dispersing PEG improves the contrast, brilliance, and printing properties.
- Setting-off of art prints (i.e., sticking together and transfer of ink) can be prevented by spraying the prints with a fine coating of an aqueous solution of PEG 4000 or 6000. The same effect is obtained by using PEG to treat the interleaves inserted to protect valuable prints.
- In screen printing and textile printing, the PEGs prevent premature drying out.
- As dye carriers and humectants. PEGs can be used, for example, in ball point inks, making inks, and stamping inks [129].
- The good solubility of PEGs in aqueous media can be used to very good effect when water-receptive pigment and shading inks are to be manufactured.
- Higher molecular weight PEGs in powder form can be used as water-soluble wax-link binders for water colors in pellet form and for crayons. Depending of the type of pigment, 3–5% PEG 6000 powder is used along with other additives (such as dextrin or gum arabic). The powder quality is especially suitable for the tabletting of dry blends. Wax crayons for use in water color printing may contain up to 50% PEG 6000 or 10,000. Liquid polyglycols as well are suitable for the manufacture of

paste colors in tubes so as to prevent the paste from drying out and to improve ease of application and flow. The choice of suitable PEG grade is governed by the desired consistency and hygroscopicity. Additions of as little as 5% PEG 600, for example, have good conditioning and plasticizing effects.

15. Applications of PEGs in Biology

As a water-soluble embedding agent for organic preparations, PEG is used in the preparation of histological microtome sections. PEG 4000 is suitable for fixing cytological preparations [130]. Other methods are based on the good compatibility of methacrylates and PEGs [131,132].

In preparing animal specimens and human organ preparations, PEG 1500 can be used instead of paraffin wax. When PEG is used, there are no significant changes from the natural state of the objects [133].

16. Applications of PEGs in Agriculture

Granulated NP and NPK fertilizers have to be protected against caking by a suitable treatment. At the same time, the customer requires the finished granules to be sealed against dusting. Both requirements can be met by coating the fertilizer granules with primary fatty amines in combination with higher molecular weight PEGs.

The following amounts have proved suitable for compound fertilizers: 500–1500 g/t PEG 6000 and 100–250 g/t stearyl amine.

Depending on manufacturing method, NPK fertilizers of identical chemical composition differ in water content, sieve analysis, grain shape, and surface. This accounts for the different amounts required to produce optimal effects. On the basis of results obtained so far, this method is not recommended for conditioning calcium ammonium nitrate.

If there is a cooling drum in the production plant, the fertilizer granules are best coated in this drum. Liquid amine and liquid PEG are added in the chute leading to the cooling drum. The product temperature in this process should be at least 55°C. Amine and PEG can also be used in the form of spray-dried granules or flakes. In this case, however, they must be melted by being added to the hot product before they reach the cooling drum. This can be done on a conveyor belt or in a drum, for example. The product temperature in this case should be higher than the melt temperature of the PEG, that is, at least 70°C [134].

In agriculture, PEGs are used as carriers and solubilizers in pest control. When used with esterified fatty acids, they form dispersing agents and emulsifiers for insecticides and herbicides [134].

17. Applications of PEGs as Absorption Media

Liquid PEGs are used as absorption media in cleaning industrial gases. PEG 400 is used as reaction medium in desulfurizing plants. Toxic solvent vapors, too, can be removed from other waste gas mixtures owing to the selective dissolving properties of the glycols for chlorinated hydrocarbons, aromatics, phenols, and so forth [135]. In this application it is an advantage that the PEGs are virtually nonvolatile and thus do not pollute waste gas in any way.

18. PEGs as Adhesives

Polyethylene glycols, chiefly the liquid grades, are used as humectants and plasticizers for dextrin-based adhesives. The addition of PEG prevents embrittlement and reduces the tendency of the bonded papers to curl up.

In some cases, up to 5% PEG 600 was added to the adhesive compound. The flatness of stamps that were gummed with dextrin-containing adhesives was considerably improved by additions of a little as 0.5% PEG 3000 [136].

In adhesive pens for sticking paper and similar porous materials, the PEGs act as lubricants, plasticizers, and humectants in the pen compound [137]. The most suitable PEGs for this purpose are the higher molecular weight grades from 6000 to 20,000 in amounts of about 10% by weight, relative to the total weight of the pen.

19. PEGs for Water-Free Soldering Flux

An addition of liquid PEG prevents the flux from spattering during soldering. Since the PEGs as solvents neither decompose nor vaporize at raised temperatures, large smooth soldered surface without any encrustations are produced [138–140].

For example, the acid fluxes zinc chloride or zinc ammonium chloride can be used in a 10% aqueous solution in PEG 200. It is advisable to dissolve the metal salts containing water of crystallization in the PEG at about 100°C. Organic soldering resins dissolved in PEG 200 produce a flux that is also suitable for soft soldering in the electronics industry. Care must be taken to remove all traces of flux residues by washing down with water.

20. Applications of PEGs in the Photographic Sector

In the photographic sector, PEGs are suitable for the following products: light-sensitive emulsions, photographic papers, developers, fixing, and stop baths.

The poor compatibility of the PEGs with the gelatin in photographic emulsions may lead to fogging, and so caution is advised in this respect. Nevertheless, PEGs are widely used in the photographic sector. The positive gains resulting from addition of PEG include

Sensitization of emulsions, softening, stabilizing, greater contrast, and as already mentioned, improved flatness of photographic papers.

In photographic developers, PEG additions of 0.1 to 2.5 g/L reduce the induction period and increase the contrast [141].

21. PEG Derivatives as Phase Transfer Catalysts (PTCs)

The use of oligoethers, especially oligo(ethylene glycol)s, as reagents in organic synthesis can be arbitrarily divided into three areas: *phase transfer catalysts, supported catalysts,* and *reaction solvents.*

PTCs are typically used in reactions conducted in a two-phase media where a reagent is used in catalytic amounts to assist in the transfer in one of the reactants from a phase where it is soluble to a phase where it is less soluble.

Note that while PTC chemistry has been the subject of numerous papers and reviews, relatively little has been done on the study of PEG derivatives as catalysts for these processes [142–144].

PEG 400 and PEG 600 were used as PTCs in preparation of various triaryl phosphates and phenytoin by the following reactions, respectively [145,146]:

$$3\ ArONa + POCl_3 \rightarrow (ArO)_3P{=}O + NaCl \tag{17}$$

and

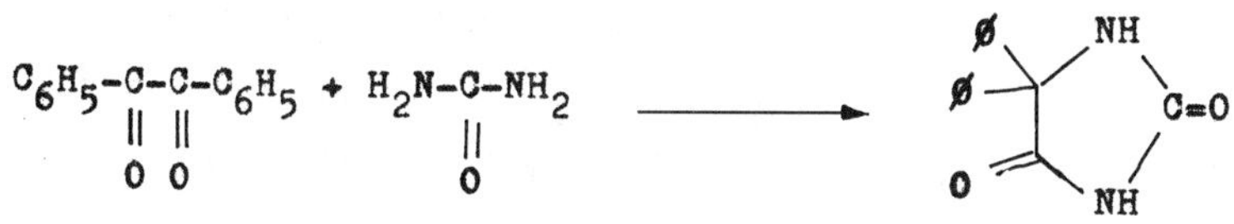

Phenytoin (18)

PEG 1000, diethylether of PEG 1000, dibutylether of PEG 1000, and diethylether of PEG 4000 were used as PTCs in an aqueous–benzene solvent system to *N*-alkylate pyrrole, indole, and several other nitrogen heterocycles by the following reaction [147]:

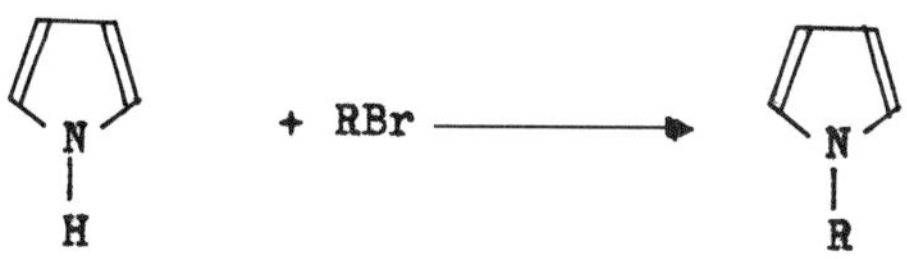

(19)

Similar studies by Regen [148] confirmed that PEG catalysis was superior to that by crown ethers and tetraalkylammonium salts. Alkylation rates decreased with increasing PEG molecular weight and increasing alkyl halide chain length and branching.

Comparative kinetic studies using PEGs as PTCs for the alkylation of imidazole by alkyl halides showed that reaction follows pseudo-first-order kinetics.

Aromatic amines have been alkylated using benzyl bromide and 1-bromododecane in variable yields using a PEG methylether as the PTC [149].

PEG and their α,β-deithylether derivatives were used as PTCs for the alkylation of phenyl-acetonitrile with alkyl halides by the following reaction:

$$C_6H_5{-}CH_2{-}CN + RX \xrightarrow[\text{Benzene, 70°C, 2 h}]{\text{KOH, PEG}} C_6H_5{-}\underset{\displaystyle R}{\underset{|}{CH}}{-}CN \tag{20}$$

The resulting 2-substituted phenyl-acetonitriles were produced in good yields.

N-alkoxyalkyl derivatives of acylanilines are of current interest due to their biological activity. However, up to this time the most facile method for their synthesis utilizes a PEG–PTC catalyzed reaction where yields in excess of 90% were reported [150]. When this reaction was catalyzed by benzyltriethylammonium chloride, a considerable dimer formation was reported [150].

PEGs were reported to exhibit remarkably high activity as catalysts for dehydrohalogenation reactions [151]:

$$C_6H_5{-}CH_2CH_2{-}Br \xrightarrow{\text{KOH}} C_6H_5{-}CH{=}CH_2 + KBr \tag{21}$$

For this reaction, the hydroxy-terminated PEGs exhibited superior performances.

PEGs and their dimethylethers have been utilized as PTCs for the reduction reactions and the synthesis of numerous ethers [152]. In addition, PEGs have been used as PTCs for conventional Williamson ether synthesis conducted in a two-phase liquid reaction media [153].

The dibutylether of PEG 600 was used to catalyze the reaction of *n*-octanol with butyl chloride in 65% aqueous KOH solution to give 95% yield of the corresponding ether [154].

Evans and Berenbaum [155] studied the effect of varying polyether structure on their usefulness as PTCs. In general, of all the polyether derivatives tried, PEG demonstrated the largest rate enhancement for the reaction of sodium phenoxide with *n*-butylbromide in 1,4-dioxane at 45°C.

PEGs with molecular weights ranging from 300 to 1400 were used as PTCs in the preparation of aliphatic polyhydroxyethers,

$$n\mathrm{HO{-}C_6H_4{-}C(CH_3)_2{-}C_6H_4{-}OH} + n\mathrm{CH_2{-}CH{-}CH_2Cl}\ (\text{epoxide, }\overset{\diagdown\ \diagup}{\mathrm{O}}) + n\mathrm{KOH} \rightarrow$$

$$\mathrm{{-}({-}C_6H_4{-}C(CH_3)_2{-}C_6H_4{-}OCH_2{-}CH(OH){-}CH_2O{-})_n{-}} + n\mathrm{ClNa} + \mathrm{H_2O} \qquad (22)$$

and aromatic polyethers [157],

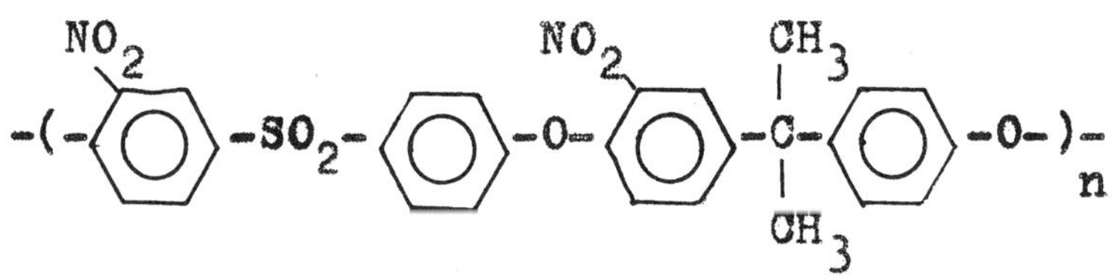

(23)

The last two reactions were conducted at 20–30°C for approximately 24 h. The PEG catalysts evaluated were less active than crown ethers (e.g., dibenzo-18-crown-6); however, they are nontoxic and are substantially lower in cost.

Various aryldiazonium salt reactions, for example, arylbromide, aryliodide, and asymmetrical diaryl formation,

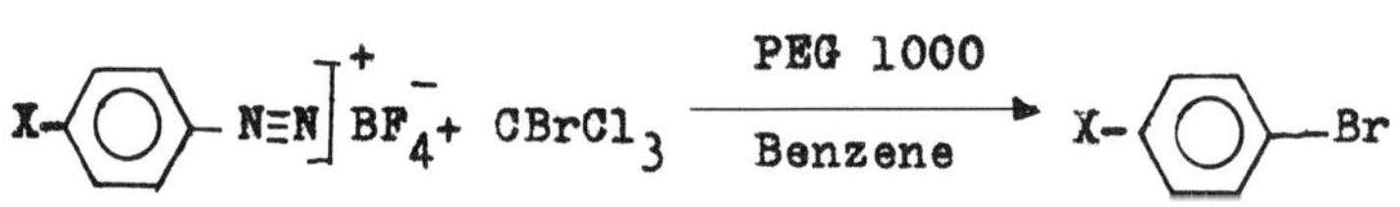

(24)

where X = Br, NO_2, and so forth, and

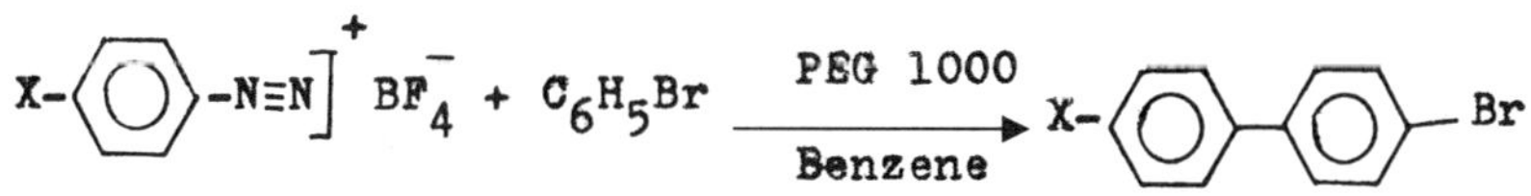

(25)

have been successfully prepared using PEG 1000 as PTC [158].

Also, the dimethylether of PEG 1000 was found to result in 15% lower reaction yields than PEG 1000 itself.

Perhaps one of the most common areas of utilizing PEGs as PTC is for nucleophilic substitution reactions:

$$R'X + MY \xrightarrow[\text{aqueous/organic}]{\text{PEG}} R'Y + MX \qquad (26)$$

These reactions are carried out in aqueous/organic two-phase solvent systems with one of the reagents in an alkali metal salt form. Some examples of the anionic nucleophilic reagents (Y) are hydroxides, halides, cyanides, sulfides, cyanamides, carboxylates, sulfonates, and so forth. In fact, one of the most studied PEG phase transfer catalyzed reaction is the formation of carboxylate esters, such as acetates [159,160].

In summary, the following points can be made in regard to these reactions:

In toluene, PEGs were more effective PTCs than were crown ethers; in *n*-butanol, their effectiveness was about equal; however, in some cases, PEG poisoning was observed in toluene solutions [160].

When compared on an equimolar basis for the catalysis of the reaction of benzyl bromide and potassium acetate, PEG 6000 was more effective than PEG 3000 [161].

The order of reactivity of nucelophiles (Y) in PEG-catalyzed reactions decreases in the following order [158]:

$HS^- > SCN^- > N_3^- > MeOOC^- > CN^- > F^-$

PEGs with molecular weight 400 are usually not efficient for ion binding; therefore, they are typically poor PTCs [162]; in general, the ability of PEG to complex with nucelophilic reagents increases with molecular weight.

Complexation between PEG catalyst and the cation of the nucleophile depends on the terminal group structure of the PEG [163]; for end group structures R, the stability constants decrease in the following order:

$R = —H > —CH_3 > —C_2H_5 > C_6H_5$

PEG 600 has been used in the phase transfer catalysis of the partial oxidation of benzyl alcohol to benzaldehyde [164]. This reaction was conducted in ethylacetate for 30 min at room temperature.

Olefin oxidation, for example, of *trans*-stilbene, was conducted using PEG 600 to yield benzoic acid [165]. In addition to the more conventional oxidations we have described, PEGs have been used for phase transfer catalyzed autooxidation of weak carbon acids [166].

PEGs are also involved as PTCs in other important reactions such as reaction of mercaptans with polychlorobiphenyl [167], reduction of carbonyl functionality to methylene groups [168], isomerization of allylanisole [169], and isoprene synthesis by decomposition of 4,4-dimethyl-1,3-dioxane [170]:

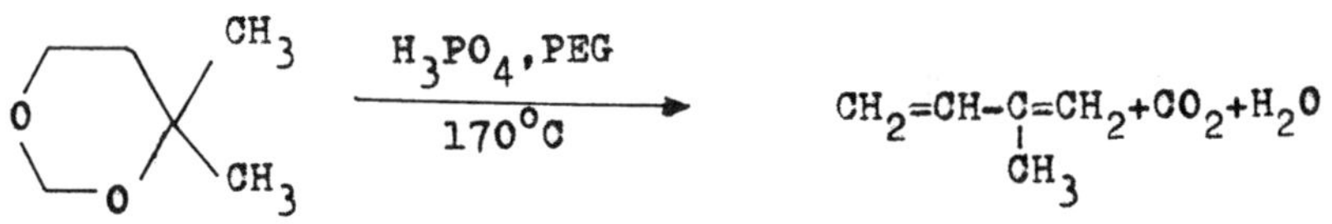

(27)

Various alkyl *p*-tolyl sulfones were prepared under relatively mild conditions using PEG 400 and PEG 1000 and α,ω-diethylether of PEG 1000 as phase transfer catalysts [171].

PEGs were used as solvents for Williamson synthesis [172],

$$H_2N—C_6H_4—O^-Na^+ + CH_3—CHCl—CH_3 \rightarrow$$
$$H_2N—C_6H_4—O—CH—(CH_3)_2 \quad (28)$$

and in a number of redox reactions as the host solvents [173].

In recent years, there have been a number of papers describing the use of solid support bonded PEGs as phase transfer catalysts. One of the major interests in binding PEG derivatives to solid supports is to facilitate the ease of catalyst recovery.

One of the least expensive classes of solid supports is metal oxides, for example, alumina and silica gel. Sawicki [174] reported that PEGs and their monomethylethers may irreversibly bond to the hydroxyl functionality of a silica or alumina surface via a dehydration mechanism

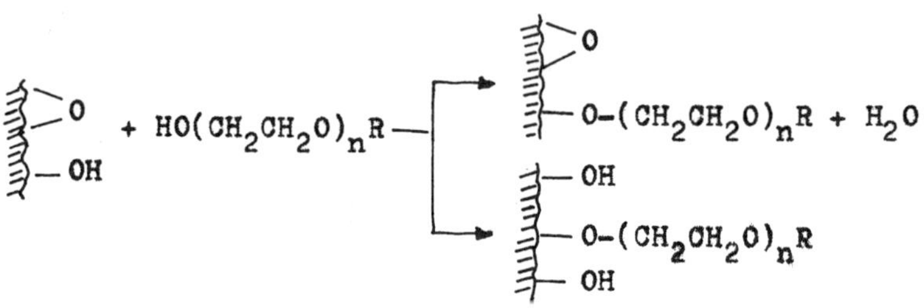

where n = 7–9 and R = H, CH_3 (29)

by simply refluxing in toluene with continuous removal of water by azeotropic distillation. On the other hand, most of the work published on immobilized PEGs as PTCs has utilized various organic polymers, for example, polystyrene and macroporous glycidyl methacrylate copolymers [175–177].

II. CYCLIC ETHERS

Crown ethers were originally defined as macrocyclic polyethers with the following structure [178]:

(30)

where Z = alkylene, arylene, or cycloalkylene.

Typical binding groups are 1,2-ethylene, *o*-phenylene, and 1,2-cyclohexylene, which can also be substituted. In the common crown ethers, *x* runs from 3 to 6. Dioxane is not considered a crown ether.

Hetero analogues, in which oxygen is replaced either partially or completely by other heteroatoms, especially sulfur or nitrogen, are also known [179]. These hetero crown ethers are also called *coronands* or *corands* [180].

Occasionally, open-chain analogues such as compound of the form (*podands*)

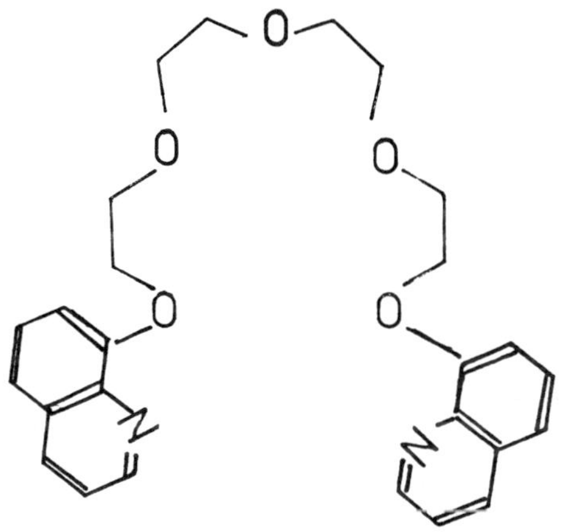

(31)

or oligocyclic compounds (*cryptands*)

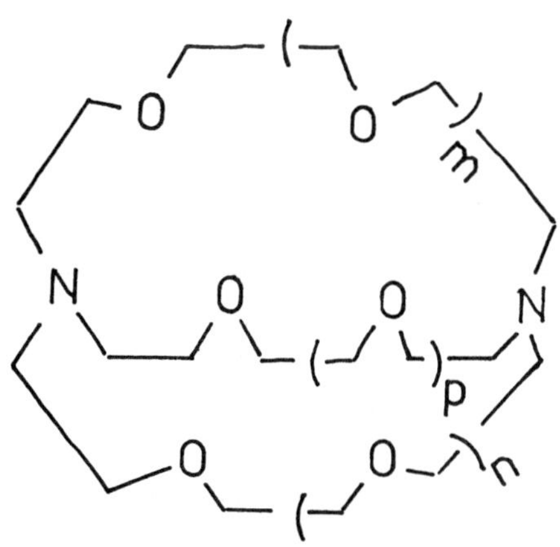

(32)

are also classified as crown ethers. A more detailed description of these compounds is presented in Ref. 181.

Crown ethers of the ethylene oxide oligomer type,

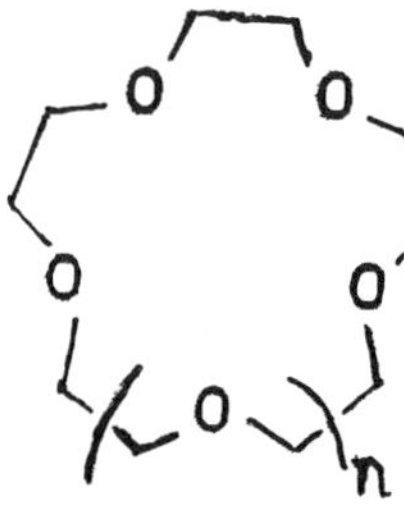

(33)

are colorless, odorless, viscous liquids or solids with a low melting point. They are strongly hygroscopic and readily soluble in most organic solvents and in water [179].

Crown ethers with condensed aromatic rings,

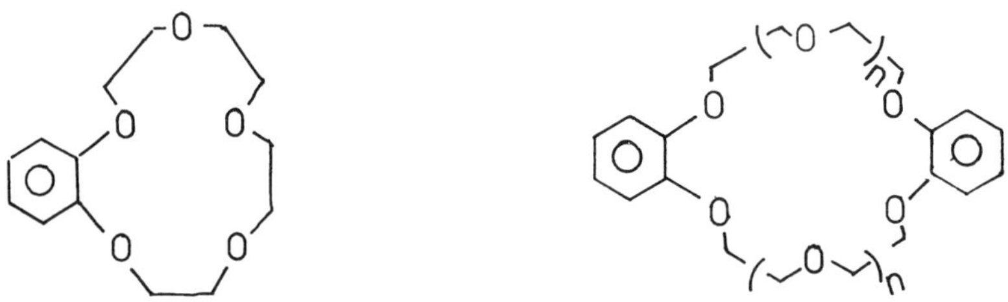

n = 1, 2

(34)

are colorless, rarely hygroscopic, crystaline compounds. At room temperature, they have poor solubility in water, alcohols, and many other common solvents [179]. They are readily soluble only in halogenated hydrocarbons, pyridine, and formic acid.

Crown ethers with alicyclic bridges,

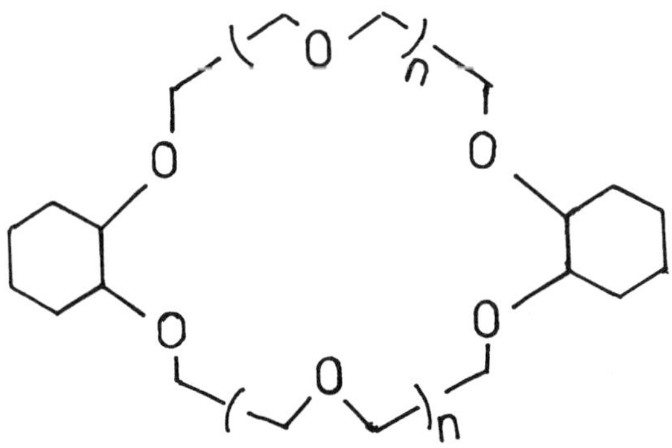

n=1,2 (35)

have much better solubility in hydrocarbons and much poorer solubility in water than the ethylene oxide oligomers [179].

Like simple dialkyl ethers, aliphatic and alicyclic crown ethers are chemically stable [179]. Aromatic crown ethers react like anisole, that is, they can be halogenated or nitrated and they react with formaldehyde [182]. Hydrolysis takes place only in special cases [183]. Crown ethers are also thermally stable; dibenzo-18-crown-6 can be distilled at 380°C without decomposition [184]. With hydrogen ions [185] and in the presence of Lewis acids ($AlCl_3$, $TiCl_4$), oxonium compounds are formed [186].

The heteroderivatives

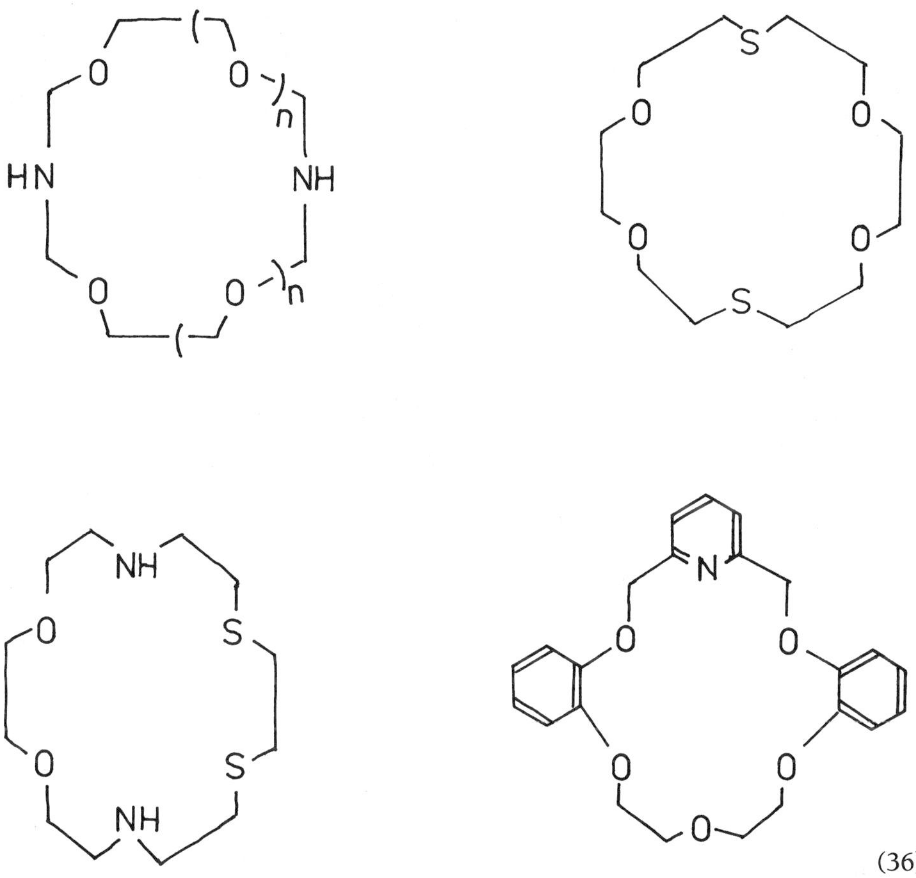

(36)

are usually more reactive than the classical crown ethers. Aza analogues are strong bases and react with acids to form salts.

The most remarkable property of crown ethers is their ability to form stable complexes with alkali and alkaline-earth metal ions and with ammonium ions. Numerous complexes of crown ethers with nonionic organic molecules are also known [187].

Crystalline complexes of crown ethers and metal ions are obtained by mixing two components in a common solvent [182]. The complexes have higher melting points than the free crown ethers. Stoichiometry of the complexes (crown ether: salt = 1:1, 2:1, or 1:2) is largely determined by the fit of crown ether cavity and cation diameter [188]. Structure determination by means of X-ray difraction [189] has confirmed that in the 1:1 complexes the cation is located inside the crown ether ring. For the 2:1 complexes, sandwich structures were found. In the 1:2 complexes, two cations are located inside the crown ether ring. Complexes in which the cation is surrounded by cagelike structure have also been described. In all cases, the cations are coordinated to the ether oxygens, which point into the ring. Hydrogen bridges are effective in complexes with ammonium ions. Frequently, there is also contact with the counterion.

In solution, the complexes exhibit varying degrees of stability [190]. They are most stable under the following conditions:

Optimum ratio of cation diameter to cavity diameter
Optimum number of ether oxygens capable of coordination
Strongly basic ether oxygens
High charge density at the cation
Low polarity of the solvent
Lipophilic, polarizable anion

Nitrogen and, especially, sulfur decrease the stability of alkali and alkaline-earth ion complexes. Conversely, complexation with Ag^+ or other heavy-metal ions and with ammonium ions is enhanced. Pyridine rings increase selectivity for sodium ions [190].

Ion selectivity depends to a large extent on the solvent; in certain cases, it can even be reversed through a change of solvent. In multiphase systems that are suitable for selective salt extractions, ion selectivity is determined by the relative degree of salt transfer. Effective extractions require large anions that are strongly polarizable (e.g., picrate), a highly lipophilic crown ether, and a complex that is highly stable in the organic solvent [191].

Lariat ethers [192,193] have been designed to enhance the cation-binding ability of common crown ethers by introducing a side arm carrying extra donor group(s); they also partly mimic the dynamic complexation process shown by natural macrocyclic ligands. A lariat ether may be represented by the following general formula:

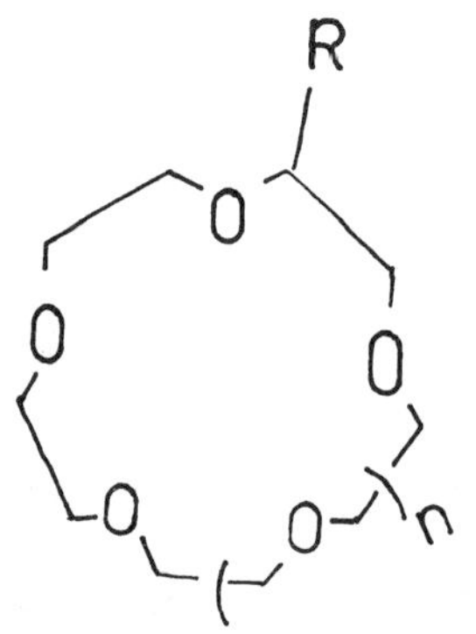

(37)

where R = H, $—CH_2OCH_2CH_2OCH_3$, $—CH_2OH$, $—CH_2(OCH_2CH_2)_n—OCH_3$, and n = 2–5.

Ouchi et al. [194] recently reported a number of new lariat ethers, possessing a variety of single or double side arm(s) and their cation binding abilities. In general, the extractabilities of lariats for mono- and divalent cations increased gradually with extending oxyethylene side arm.

Due to the various applications in metalloenzymes, homogeneous catalysis, electrical conductance, and magnetic exchange processes [195], most of the research focused on multisite ligands, which make possible a close grouping of more than one alkali or alkaline-earth metal ion.

Weber [196] outlined a concept for the synthesis of so-called "multiloop crown-compounds" by which any number and type of crown ethers can be linked together via a spiro skeleton. The ring size varies from 13 to 25 ring atoms, and donor sites vary from exclusively oxygen to combinations of oxygen–sulfur and oxygen–nitrogen. In the following formulas, (38)–(46), some examples of multiloop crown compounds are given [196]:

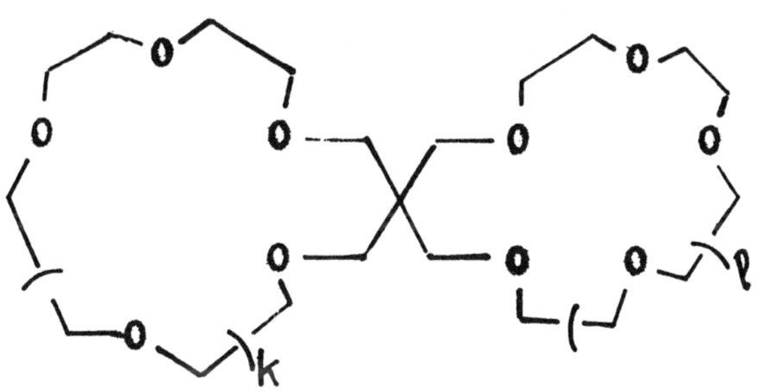

(38)

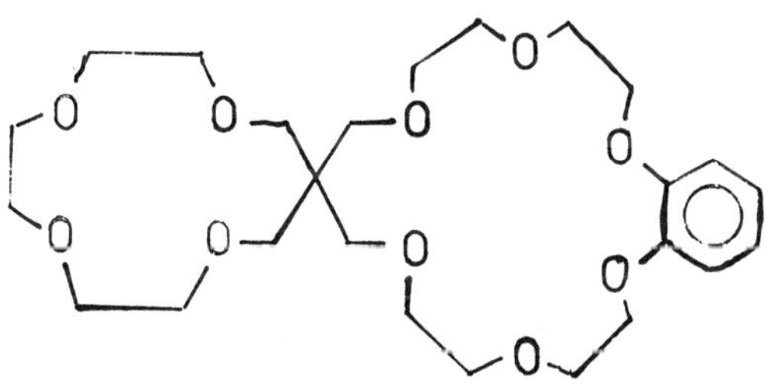

(39)

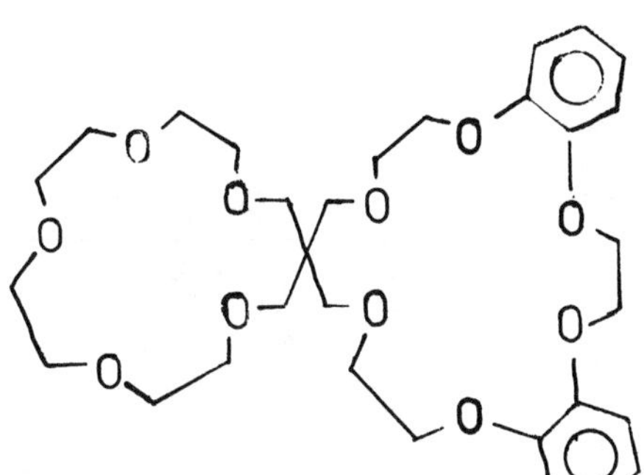

(40)

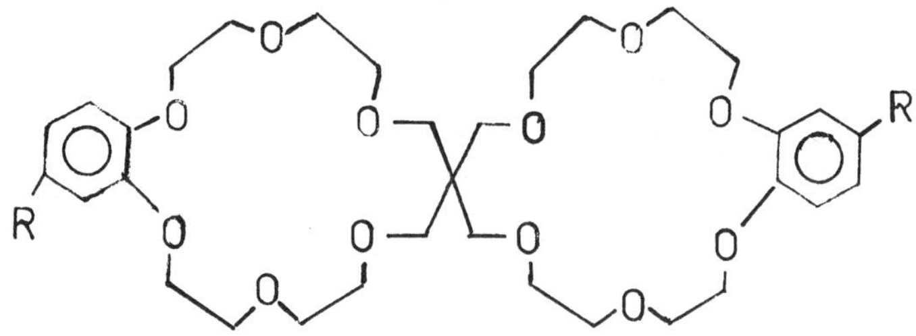

(41)

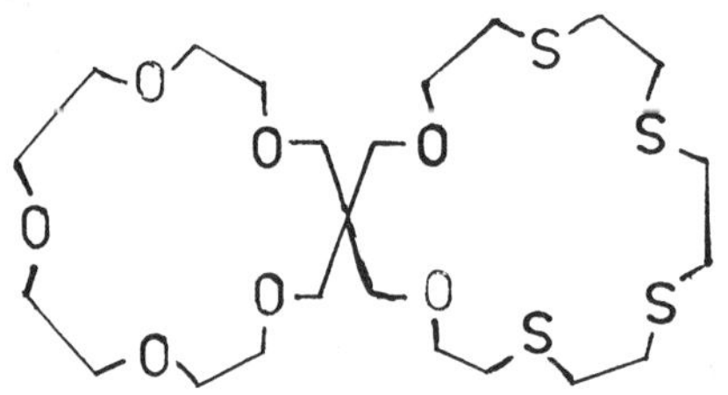

(42)

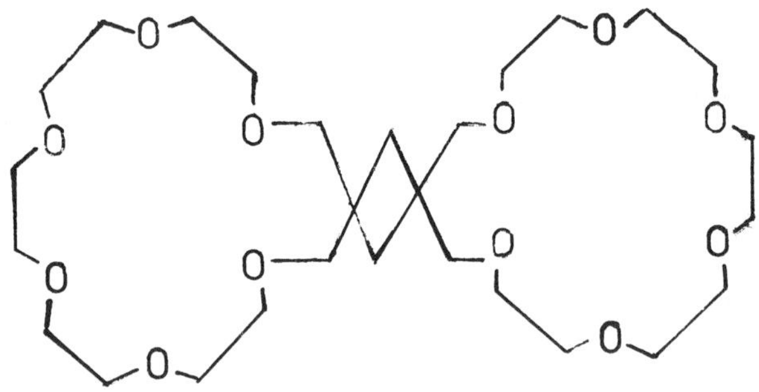

(43)

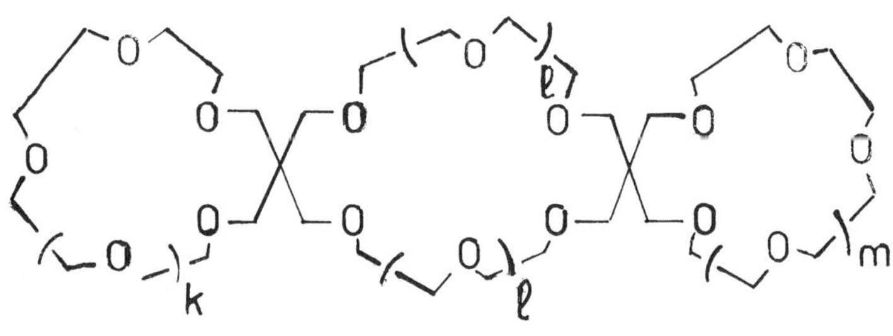

(44)

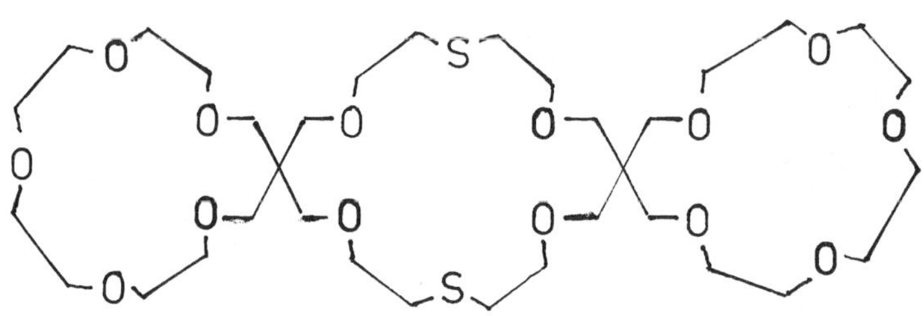

(45)

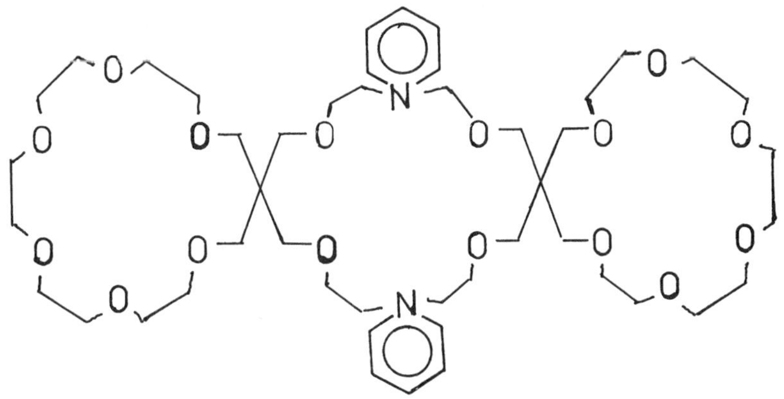

(46)

Variability in the ligand framework is also derived by alteration of the succession of binding compartments and by ring stiffening via benzoannelation as well as by specific expansion of the centers of adjacent crown ring subunits.

The number of the ring subunits ranges from two to four, defining, for example, (38), (41), and (43) as di-, (44) (k = 2, l = 1, m = 2) as tri-, and

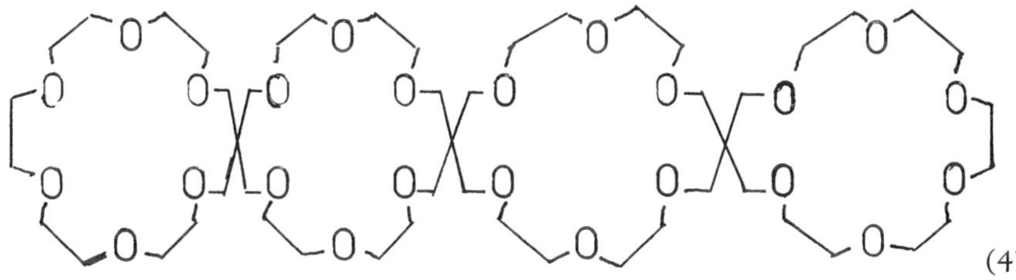

(47)

as *tetra homotropic* molecular receptors.

Oligocoronands, including more than three individual and defined crown subunits, have not been described so far. The ligands (38) (k = 2, l = 0 and k = 4, l = 0), (39), (40), and (42) represent *diheterotropic* and (44) (k = 1, l = 1, m = 3), (45), and (46) *triheterotropic* receptor molecules, respectively. They may successively bind different substrates, yielding cascade complexes.

The synthesis of multiloop ligands is characterized by stepwide cyclization presented in the two following series of reactions:

Synthesis of multiloop compounds (38), (44), and (47):

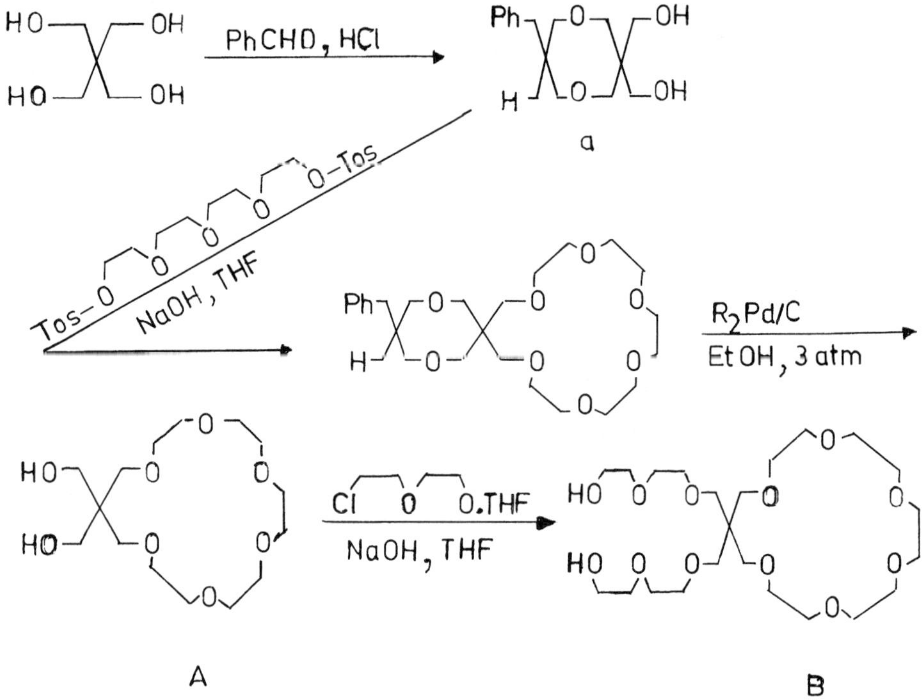

(48)

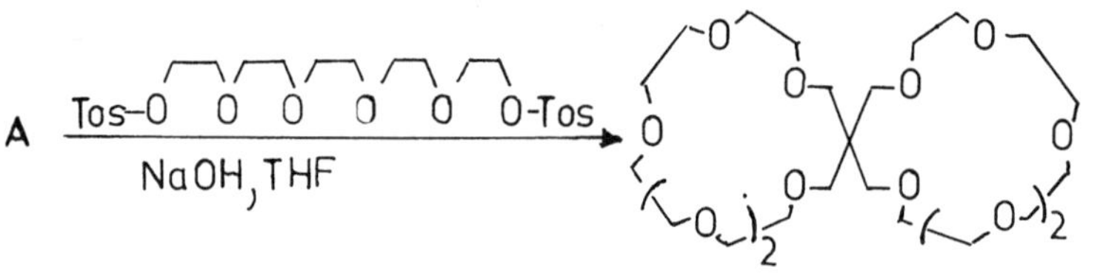

(49)

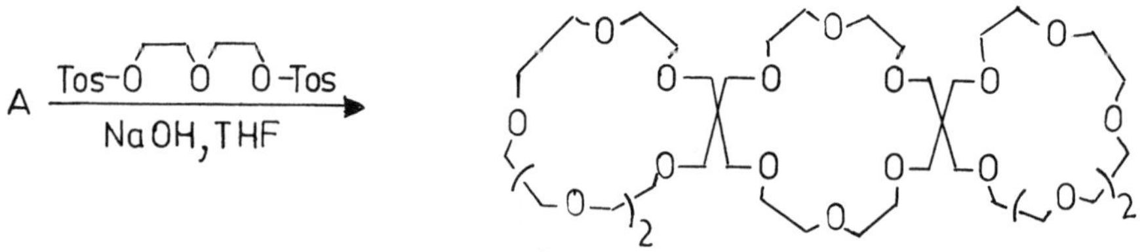

(50)

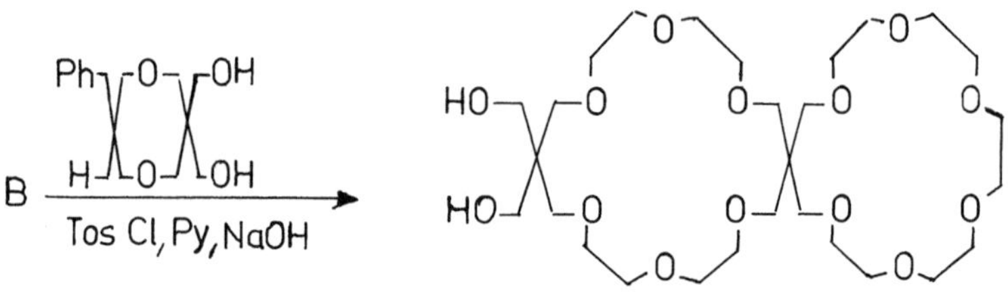

(51)

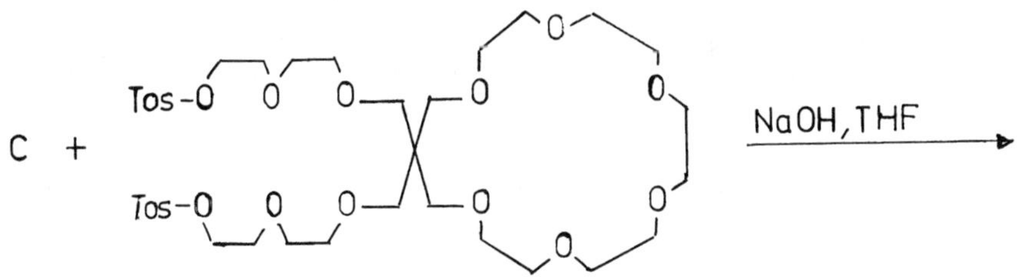

Compound 47

(52)

Synthesis of the multiloop compound (43):

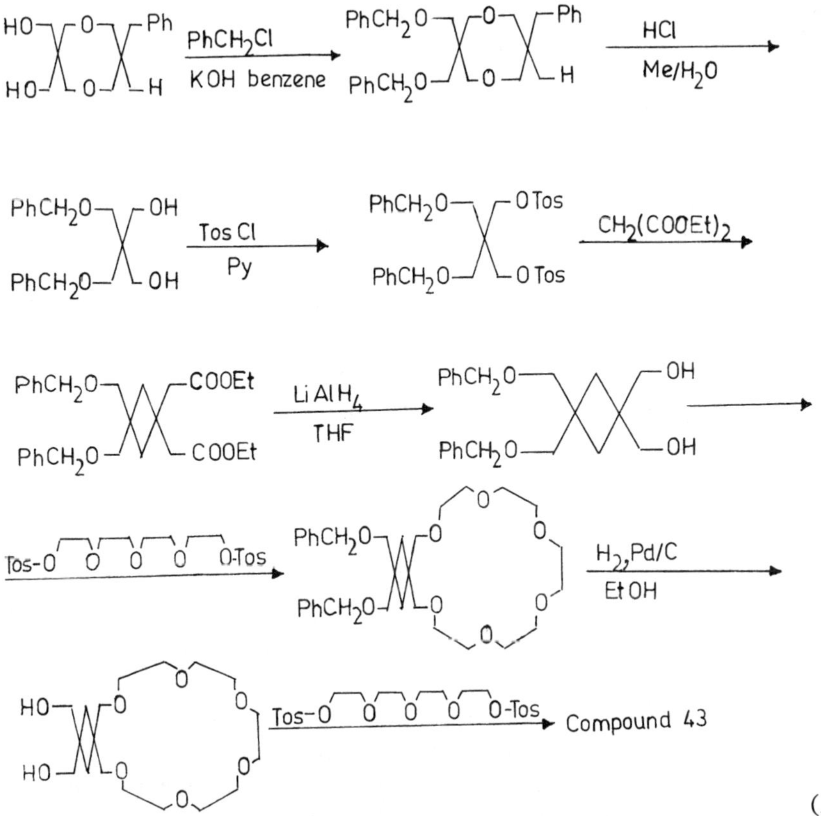

(53)

All these compounds were shown to be generally suitable hosts for the common incorporation of several cations, and they can complex identical or different metal ions (*homo-* or *heteronuclear complexes*, respectively). The symmetrical double loop compounds (38) (k = 0, l = 0; k = 1, l = 1; k = 2, l = 2), (41) (R = H), and (43), which have oxygen donor sites only but which, however, gradually exhibit larger rings and donor numbers, readily form homobinuclear complexes with alkali/alkaline-earth metal ions of suitable size. The asymmetrical diloop compounds

(38) ($k = 2$, $l = 0$; $k = 4$, $l = 0$), (39), (40), and (42) should be able to incorporate several cations of various types.

Most of the crown ethers are commercially available; the preparations are usually of 98% or 99% purity. The most important suppliers are Aldrich (Stedheim), Alfa (London), Fluka (Zürich), Janssen (Berlin), Merck (Berlin), Nippon Soda (Kyoto), Parish (London), PCR (London), Sigma (St. Louis), Suchard (Berna), and Strem (London).

A. Chemical Applications of Cyclic Ethers

Crown ethers are used as complexing agents and phase transfer catalysts in organic synthesis [197–199]. In principle, any chemical reaction where ions or ionic intermediates are involved in any form can be modified or improved through crown compounds. Among others, nucelophilic substitutions (e.g., halogen and pseudohalogen substitutions), O-, N-, S-alkylations, reactions with carbanions, C–C bond formation, additions, eliminations, carbon generation, gas extrusion (N_2, CO_2), oxidations, reductions, rearrangements, isomerizations, and polymerizations are concerned. Further application fields for crown compounds include macrolide synthesis, organometal and protecting group reactions, and problems in phosphorus and silicon chemistry [200–211].

As already explained, crown ethers can complex cations. Thus, cations become compatible with a fatty medium and soluble in organic solvents. To keep the electroneutrality of the solution, the corresponding anion has to be dissolved, too. Not being complexed and only weakly solvated in the organic medium, it is in a very active state and can initiate. In extreme cases, we have even "naked anions" and here chemists jokingly invented the term "ion sex" [212].

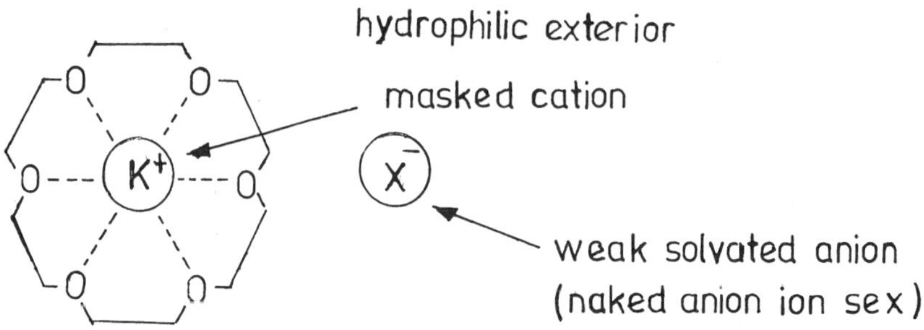

(54)

This applies especially to cryptates, since they lack contact to the complexed cation; whereas in coronates and podates this contact is possible, in principle. Summing up: crown compounds facilitate solution of inorganic salts in organic media and at the same time activate the anion. The preparative usefulness is obvious: the reaction components, which on account of different polarities cannot be united in a common solvent without further means, are now apt to react. Here completely new application fields for saltlike reagents open up. Many reactions that in former times were impossible or only possible with low yields can be carried out in the presence of crown compounds without any difficulty.

A way to overcome the high cost associated with the use of crowns and even more with the use of criptands is to facilitate the separation and recovery of the catalyst, in order to make them reusable. Toward this goal, the development of insoluble catalysts and the new concept of triphase catalysis have been explored [213].

Typical crown and cryptand catalysts were bound as active groups to a polymer matrix of the polystyrene type, crosslinked with 20% or 4% *p*-divinylbenzene.

Polymer-supported podands appear to be more commonly used in solid–liquid systems. A variety of chemical transformations including nucleophilic substitutions at saturated carbon atoms [214,215], esterifications [214], sodium borohydride reduction [214], fluoride-catalyzed Michael addition [214], phenacyl ester synthesis, Darzen's reaction, Wittig alkenylation, and dichlorocarbene reactions can be carried out under these conditions [214].

Extended studies have been carried out in order to achieve further improvement of the catalytic power arising from the ion selectivity of podand resins in solid–liquid triphase systems [216].

Although crown-ether-assisted nucleophilic substitutions are among the most common and also longest-known reactions, further new applications and investigations are still possible. 18C6 solubilized in acetonitrile can be used to accomplish F/Cl-halogen exchange at the silicon atom with a high yield [217].

Several new applications of crown ethers were developed in performing Gabriel-like transformations; for example, high yields of secondary amines are produced, followed by a saponification, when *N*-alkyl or *N*-aryl trifluoroacetamides are alkylated in THF using potassium hydride as the base and 18C6 as the alkylation catalyst [218].

A fairly pronounced 18C6 catalysis is also obvious in the *N*-alkylation of pyrrols, indole, pyrazole, imidazole, benzimidazole, or carbazole via a phase transfer process in diethylether in which potassium *tert*-butoxide is employed as a base [219].

Theoretical as well as practical contributions have been made regarding the substitution chemistry with alkoxy and aryloxy nucleophiles. The former include kinetic investigations dealing with influence of crown ether additives on the reactivity of alkoxides with l-halogene-2,4-dinitrobenzene [220].

On the other hand, the ability of crown ether to complex and thereby to dissolve cations in organic solvents opened the way for the first isolation and characterization of guanosine permethylation products. Methyl iodide–silver oxide/18C6 was used as the methylating agent for this transformation [221]:

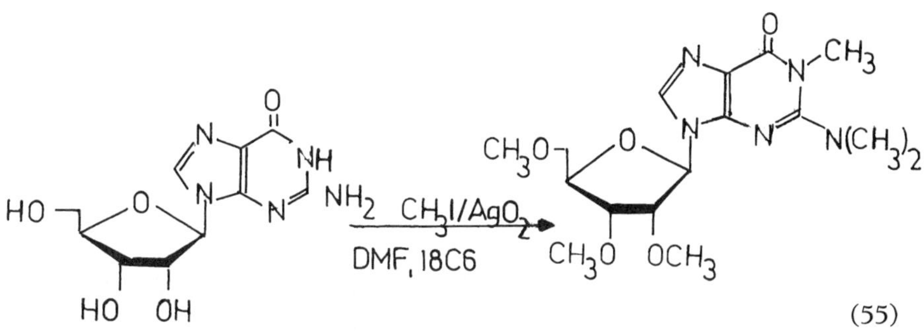

(55)

The preparation of poly (mercaptomethyl)styrene from the corresponding chloromethyl polymer using potassium hydrogen sulfide in DMF and catalytic amounts of 18C6 represents an attractive simplification in the course of the synthesis of transition metal coordination resins [222]:

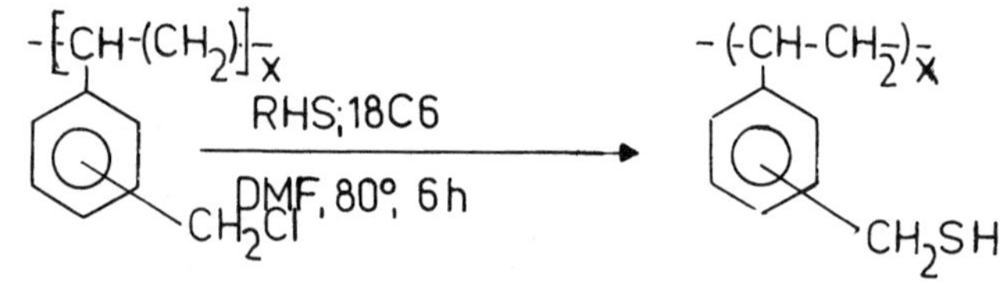

(56)

Following the findings of a Japanese group, polycarbonates can be prepared starting with dihalogenides and potassium or barium carbonate activated by 18C6 [223,224]:

$$x\mathrm{BrH_2C{-}C_6H_4{-}CH_2Br} \xrightarrow[\text{Benzene, 60°C, 24 h}]{\mathrm{K_2CO_3/18C6}}$$

$$\mathrm{{-}(OOCCH_2{-}C_6H_4{-}CH_2)_{\mathit{x}}{-}O{-}} \qquad (57)$$

Reactions involving acid derivative transformations (transesterifications, aminolysis, thiolysis, and halogen and related exchange reactions) have been performed with crown-assisted fluoride ions giving the corresponding acyl [225], silfonyl [226], and phosphoryl fluorides [227] with up to quantitative yields.

The concept of a biologically relevant salt-dependent control of reaction rates could also be realized by an aminolysis, as demonstrated in the following reaction [228]:

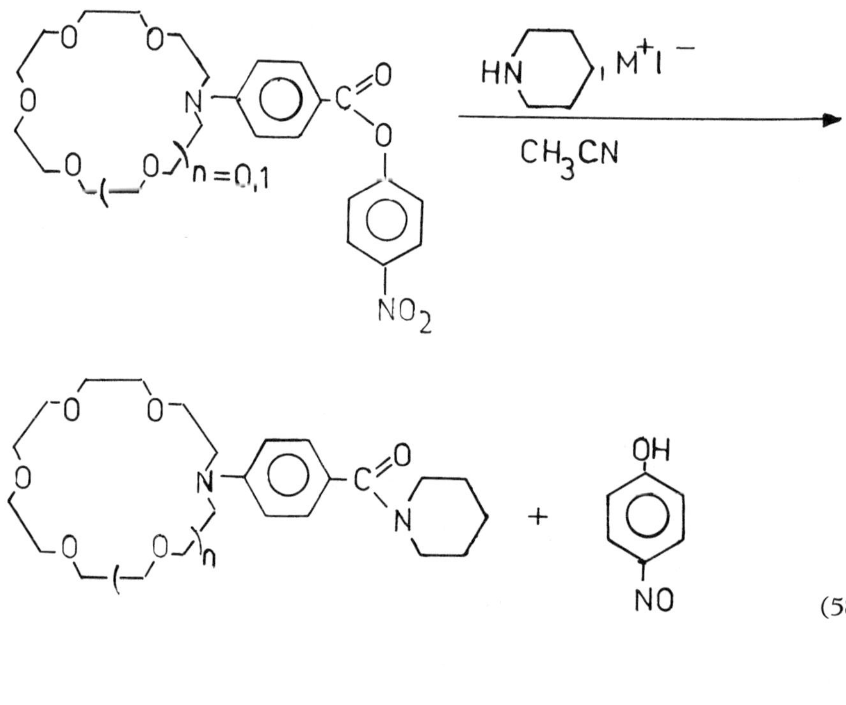

(58)

A significant enhancement is observed in the presence of salts, particularly when the crown ring size matches the ionic radius of the cation and when the cationic charge density is high.

Thiol-bearing chiral crown ethers have been designed as regio- and chirality-recognition sites,

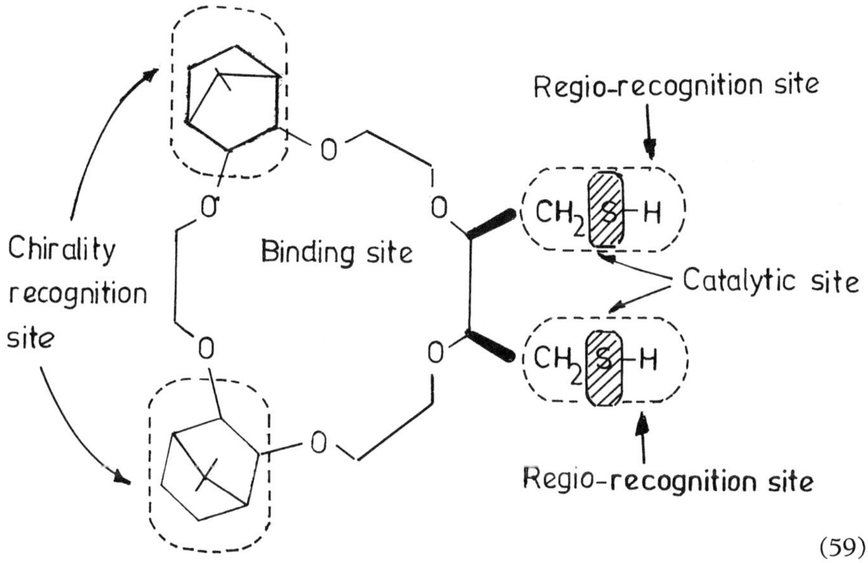

(59)

and are found to exhibit enantiomeric discrimination in the thiolysis of alanine ester salts,

$$\underset{{}^{+}NH_3Br^-}{R-\overset{*}{C}H-COOAr} + Crown-SH \rightarrow \underset{{}^{+}NH_3Br^-}{R-\overset{*}{C}H-CO-S-Crown} \tag{60}$$

by a factor of 1.7 to 1.9 for the rates of *p*-nitrophenol release [229].

In the field of carbanion reactions and C–C bond formation (nucleophilic additions), Cram and Gogan have, in a brilliant work [230], demonstrated that chiral crown compounds

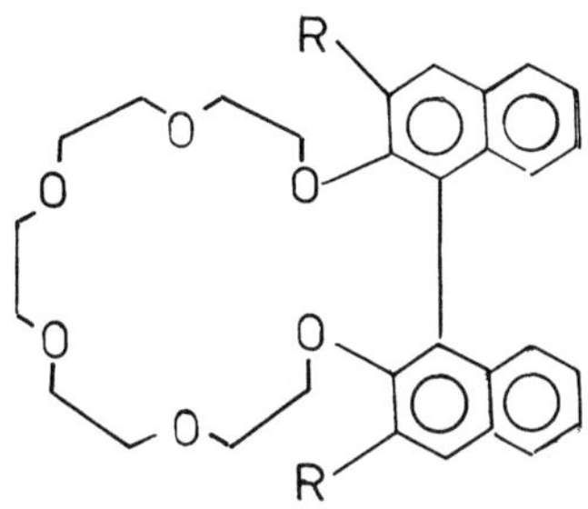

(61)

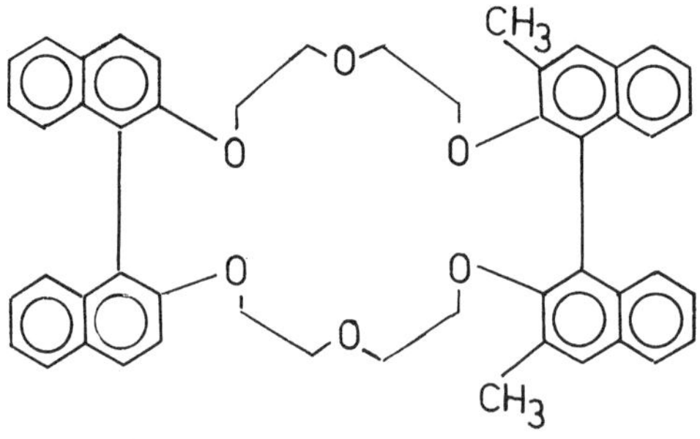

(62)

complexed to potassium bases catalyze, with high turnover numbers, the enantioselective Michael addition of β-ketoester to methyl vinyl ketone,

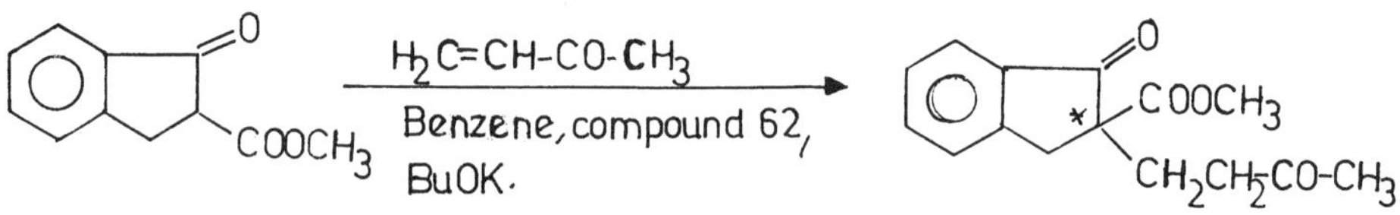

(63)

to yield product of up to 99% optical purity.

The reaction (63) is assumed to proceed by a catalytic chain mechanism:

Initiation:

(64)

Addition:

(65)

Chain transfer:

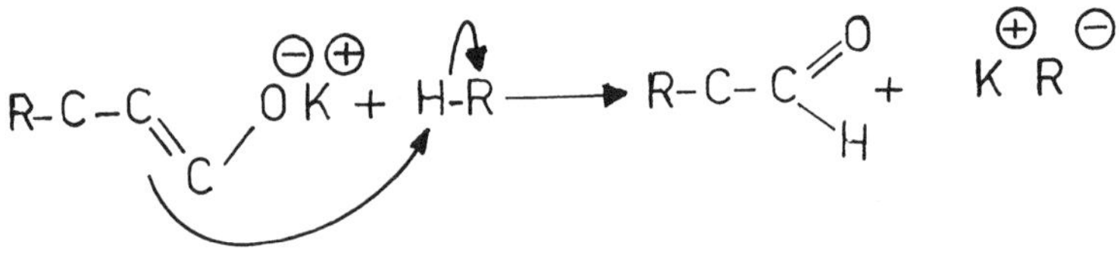

(66)

The configuration-determining step, namely, reaction (65), for the production of the compound resulting from reaction (63) is illustrated by the mechanism given in the following scheme:

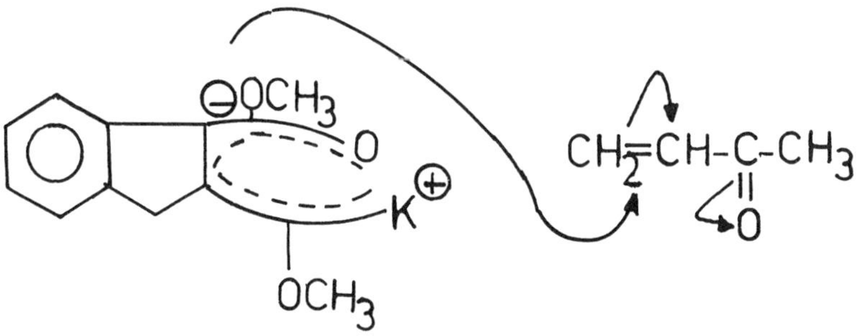

(67)

The direction of the configurational discrimination is explicable on the basis of steric differences in the diastereomeric transition state models for the complex involved. This is sketched by

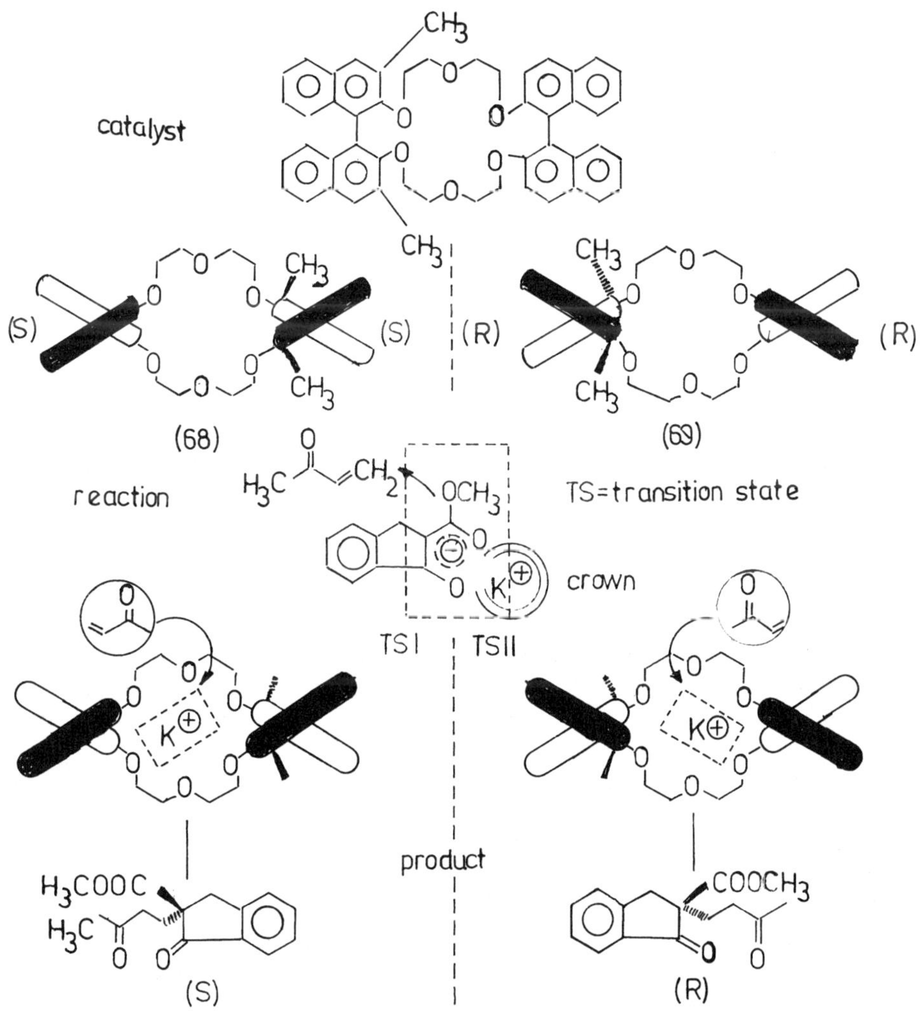

(68, 69)

the (*S,S*)-enantiomer of crown ether (62) predominantly leading to (*R*)-product of reaction (63), and the (*R,R*)-enantiomer of crown ether (62) leading to (*S*)-product of reaction (63).

Octopus molecules

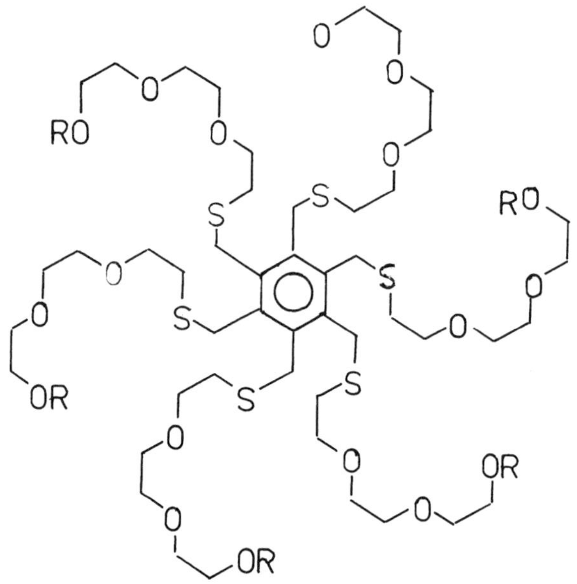

(70)

were employed efficiently as catalysts in the liquid-to-liquid two-phase Darsen condensations of a variety of ketones with chloroacetonitrile and NaOH to give the corresponding oxiranes in high yields [231].

As yet there is no clear understanding of the octopus mode of action, that is, whether a transport of hydroxide into organic phase or a transport of the carbonyl compound into the aqueous phase occurs. The latter, however, seems to be more likely.

It was found that the system *tert*-BuOH/0.1 mol% 18C6/petroleum ether is also very useful in alkyne synthesis, starting from 1,2-dihalides (from alkenes) or 1,1-dihalides (from aldehydes or ketones). Advantages are high yields and fairly mild reaction conditions [232,233].

Organic sulfur compounds, for example, disulfides, thiosulfinates, thiosulfonates, thiols, sodium thilates, and sodium sulfinates, were found to be readily oxidized to both sulfinic and sulfonic acid using the poatassium superoxide/18C6 system [234].

Crown-type compounds have led to considerable progress in the field of metal solutions, since they strongly enhance the solubility of alkali metals in a variety of solvents [235]. As is to be expected, the more strongly complexing cryptands are more efficient than the coronands; however, they differ in the type of counterion formed (metal anions are favored in the case of crown ethers, whereas cryptands mainly produce solvated electrons [236]. These aspects have been reviewed and discussed [237].

Although the reactivity increase caused by crown ethers and cryptands in anionic polymerizations has already found a wide range of application, more details have been reported and a number of questions concerning the type and the behavior of the different species present, both in the initiation and in the propagation steps, have been clarified by, for example, kinetic studies [235]. Special polymerization reactions that were effected in the presence of crown compounds are those starting with butadiene, propene, styrene, 2-vinyl pyridine, ethylene oxide, propylene sulfide, isobutylene sulfide, methyl methacrylate, β-propyllactone, or ε-caprolactone as monomers and alkali metals as initiators [238–246].

The use of crown ethers in photochemistry is still rare. Some work focuses on the photochemical behavior of crown ethers themselves and their cation complexes [247]. The most promising application of crown compounds in this area is the photochemical reduction of complexed Ag^+ to complexed Ag^0 [248].

In summary, crown compounds have been demonstrated to be effective synthetic tools for many types of chemical reactions. New transformations have been made possible, and product yields have been generally improved. A new role for the use of neutral ligands in the regio- and stereoselectivity control of chemical reactions has been outlined. This can be expected to be one of the most important subjects for future research.

A steadily growing application of crown compounds is observed in analytical chemistry and related fields. Fundamentals of their use in this field have already been discussed in a review of Weber [249]. Since then, a great many improvements and further applications have been demonstrated, including ion-selective solvent extraction [250,251], ion chromatography [252], ion-selective electrodes [253–256], chromoionophores and fluoroionophores [257–259], biology [260–262], and medicine [263]. Of the numerous paper dealing with these subjects, only a few important topics have been singled out here in order to make new concepts and trends evidents.

Solvent extraction is one of the most important techniques in analytical chemistry. In this regard, common crown compounds in different solvent systems have been successfully applied to specific ion separation problems.

Bis- and poly(crown ethers) as illustrated in

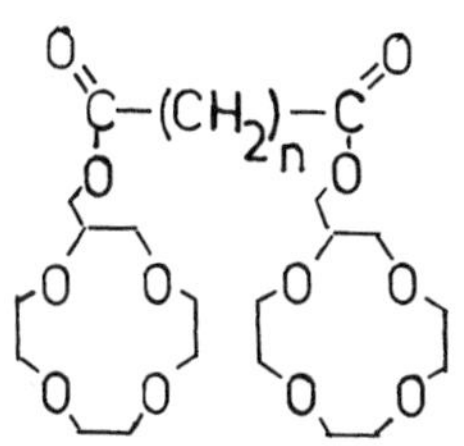

(n=1,3)

(71)

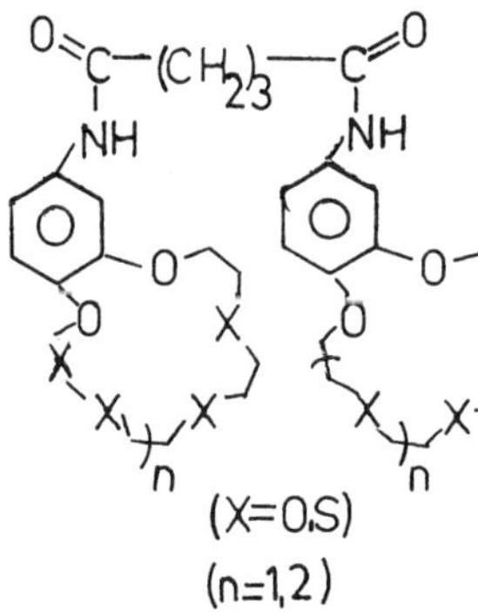

(72)

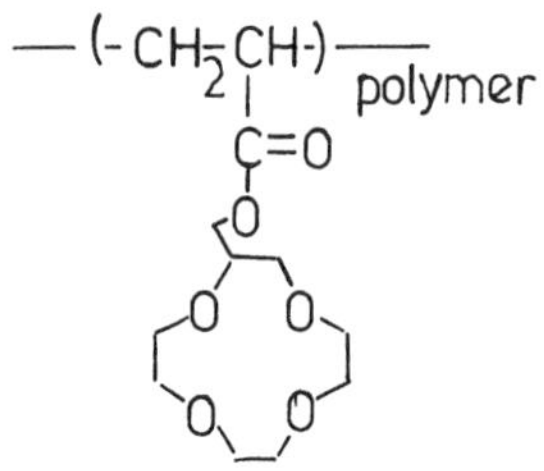

(73)

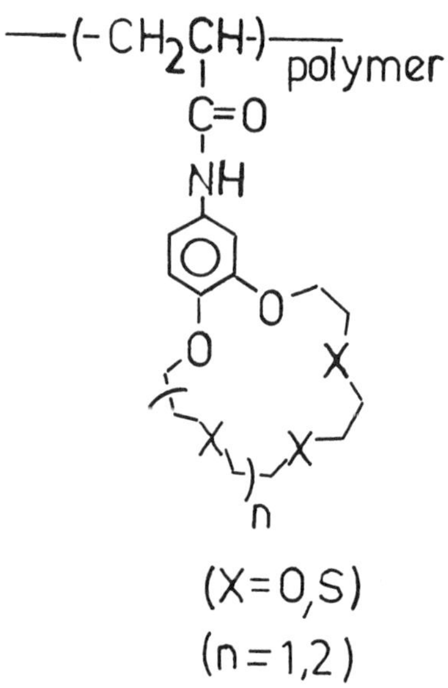

(74)

have proved to be more effective extraction reagents for metal picrates than the corresponding macrocycles. The high extractibilities and marked extraction selectivities can be explained by the ready formation of 2:1-sandwich complexes and their increased lipophilic nature as compared with the corresponding monocyclic structures [264].

Another interesting development in ligand design offers secondary donor groups by means of a flexible side arm attached to a basic crown ring. Carbon

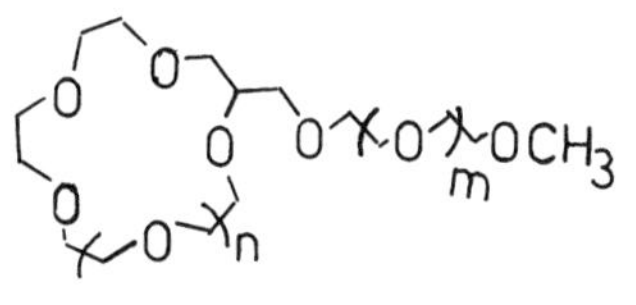

(m=0,1,2)
(n=0,1,2)

(75)

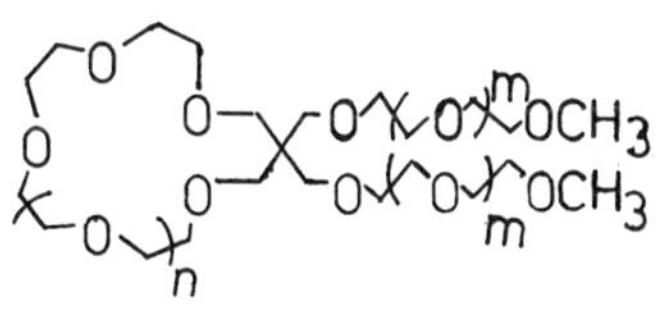

(m=0,1)
(n=1,2)

(76)

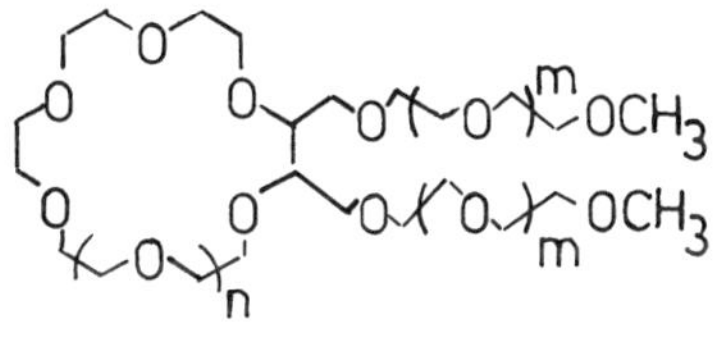

(m=0,1)
(n=0,1)

(77)

and nitrogen centers

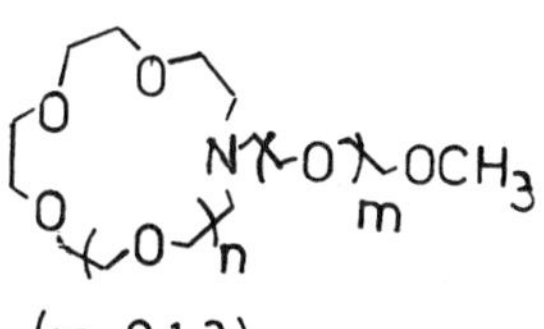

(m=0,1,2)
(n=1,2)

(78)

have been introduced as the pivot atom. This new type of crown compound, called a lariat ether, is reported to show enhanced cation binding and extraction as compared with the ring analogues without side arms [265].

A crown ether hexacarboxylic acid,

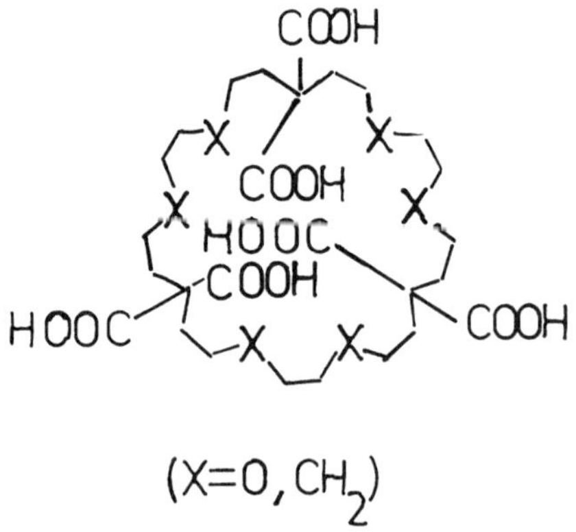

(79)

was designed and was found to bond uranyl ions highly selectively [265].

The complexing affinity of crown ethers opens up the prospect of the molecular design of chromogenic reagents sensitive to alkali metal/alkaline-earth metal ions [266–268]. A simple design for such a reagent molecule would be the introduction of a monobasic and chromophor into a crown ether, such as in

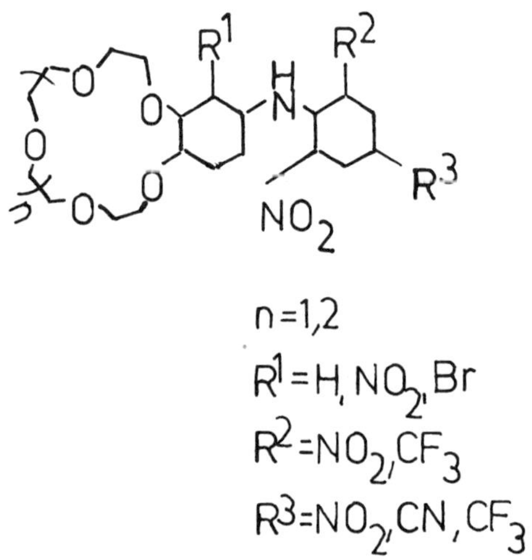

(80)

or by the design of new crown derivatives, such as

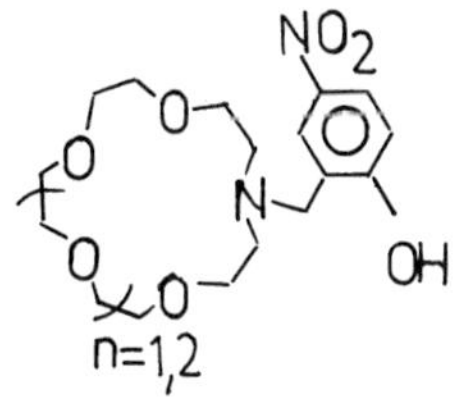
NO2
O
O
N
OH
O
O
n=1,2

(81)

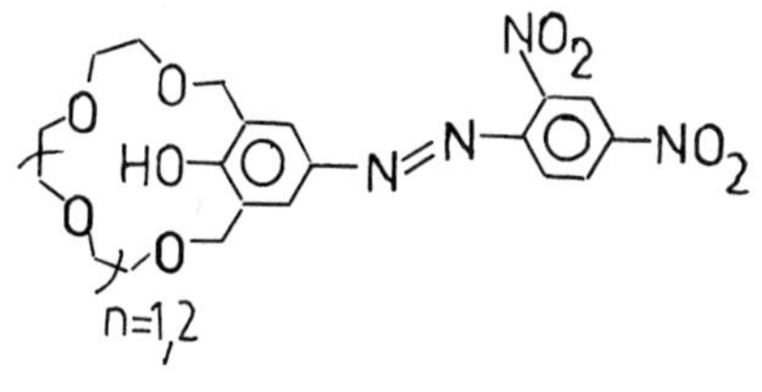
NO2
O
O
HO
N=N
NO2
O
O
n=1,2

(82)

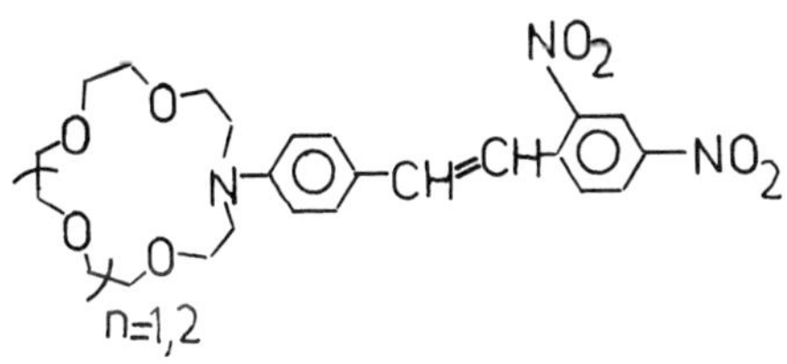
NO2
O
O
N
CH=CH
NO2
O
O
n=1,2

(83)

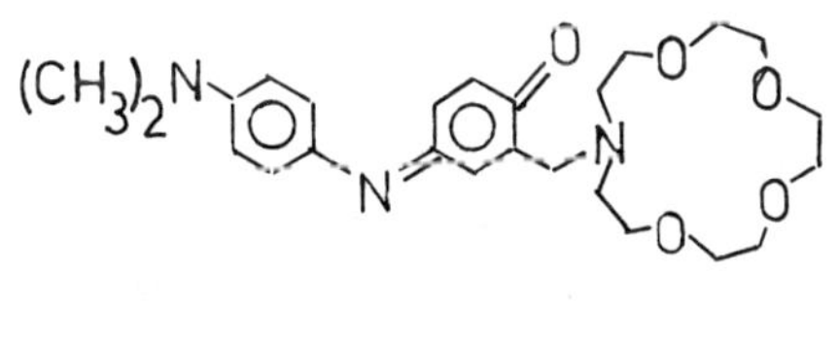
(CH3)2N
N
O
O
O
N
O
O

(84)

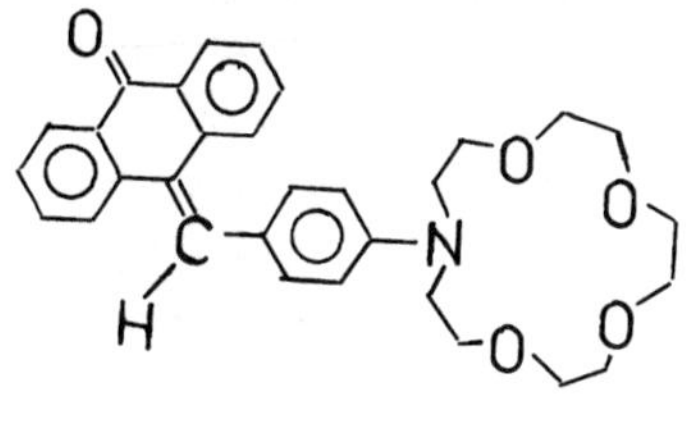
O
C
H
O
O
N
O
O

(85)

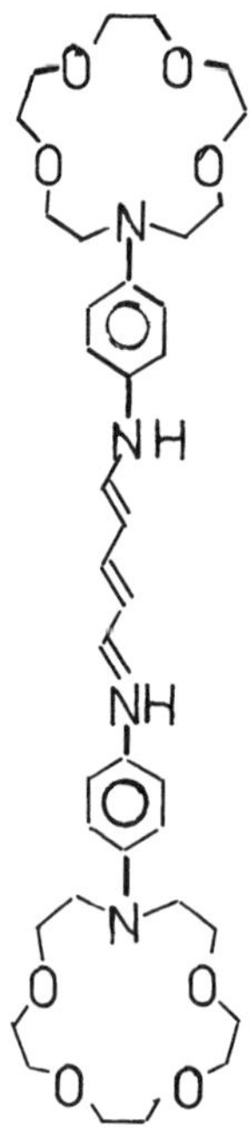

(86)

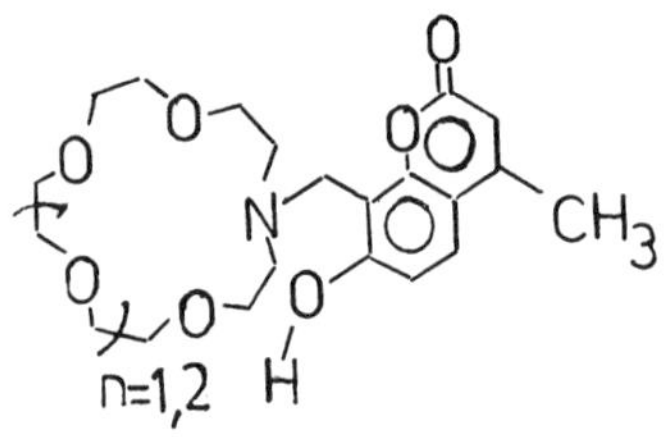

(87)

Based on the compound with structural formula (82), a new method for the spectrophotometric determination of Li^+ has been proposed [269].

Crown ethers are widely used in chromatography as well as in potentiometric, conductometric, polarographic, and volatametric determinations [269–273].

Together with the different titration methods, the ion-selective electrode technique has seen the widest application by far. Based on the pioneering work of the Simon group [274], and also on quantum chemical calculations [275], ion-selective electrodes have found wide clinical and biological application because of the possibility of performing fast and reliable local measurements of ion activities relevant for biological systems. The calcium-selective electrode is among the most widely used, as a direct consequence of the desire and need for determining free calcium ions in biological and cell fluids—also in water, soil extracts, industrial operations, and so forth. Since the neutral carrier Ca^{2+}-selective electrode with the compound

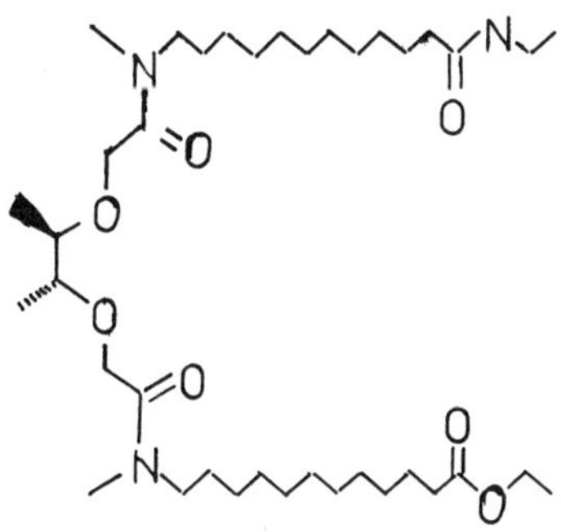

(88)

as the carrier has a far superior selectivity for Ca^{2+} versus H^+ and Zn^{2+} than the classical ion exchanger electrodes, it becomes possible to measure total Ca^{2+} in blood serum [276].

At present, little is known about the concentrations of ionized Mg^{2+} in living cells. A Mg^{2+}-selective microelectrode based on the newly developed compound

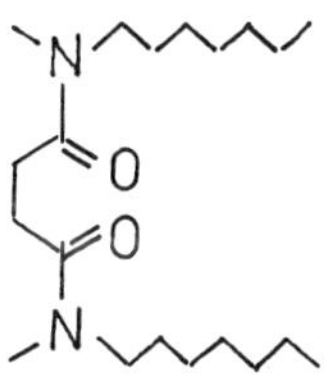

(89)

allows easy intracellular Mg^{2+}-activity studies [277].

The outlook for the application of sodium- and potassium-selective electrodes, carriers with structural formulas

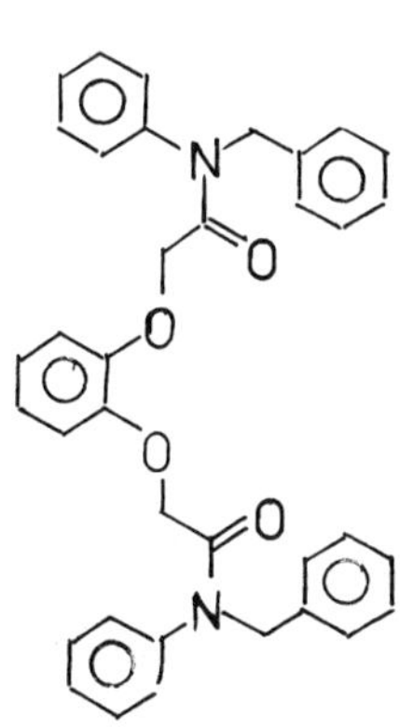

(90)

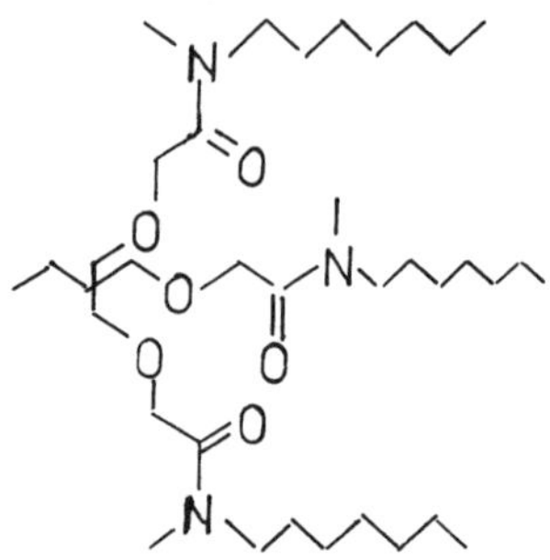

(91)

is most promising in clinical chemistry. Intracellular studies of sodium concentrations via microelectrodes [278] and online measurement of potassium on the surface of the beating hearts of dogs have been described [279]. At the same time, success was achieved in the continuous measurement of blood potassium ion concentrations during human open heart surgery using extracorporeal blood circulation coupled to a flow through a novel electrode system that is amenable to automation [280]. In view of routine clinical analysis and bedside monitoring, for example, in intensive care units, a special microelectrode measuring instrument has also been developed for the continuous (but separate) determination of Na^+ and K^+ activities in undiluted urine samples. A trend in electrode technology is the movement toward multielectrode systems that allow continuous intraoperational determination of different cations at the same time.

A crucial factor affecting analytical measurements, especially in intensive care situations, is speed—clearly an advantage of these electrode systems. The development of the ion-selective electrode technique has reached such a state that flame photometry is likely to disappear from clinical laboratories within a few years. There is no doubt that carrier membrane electrodes based on crown compounds will also become an applicable tool in other fields of chemical analysis.

B. Crown Ethers as Drug Carriers

The constant theme of organic chemists, which is to initiate the tricks of biology, is being approached by learning how molecules recognize each other. Mimicking the tricks of nature has long been the goal of the organic chemistry, with molecular recognition as its focus. Through their work, Pedersen [281], Lehn [282], and Cram [283] have shown the way.

Pedersen kicked off the initiative in 1967 when he published several papers that described the bonding of alkali metal ions (i.e., Li^+, K^+, Na^+, Ru^+, and Ce^+) by compounds called crown ethers, which collectively are made up of a large circle of carbon interrupted at regular intervals by oxygen atoms. In the absence of metal ions, the crown ether ring is "floppy." Introduce a metal ion, however, and it may be bound at the center of the circle, which now assumes a more organized, plate-like shape

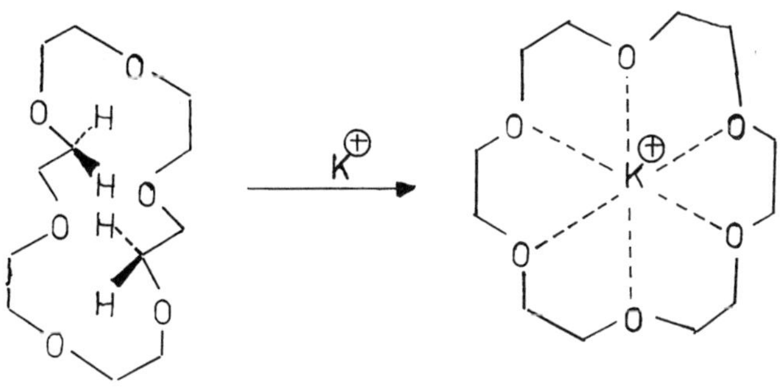

(92)

Soon afterward, in 1969, Lehn took Pedersen's principle into three dimensions by adding more layers, creating polycyclic compounds that he called cryptands because they contained molecular clifts or crypts. The effect of additional layers was to make the structure more rigid and to expand the range of substrates that could be found [282]. Nevertheless, whenever the cryptand bound something in its crypt, its overall structure was recognized to some degree:

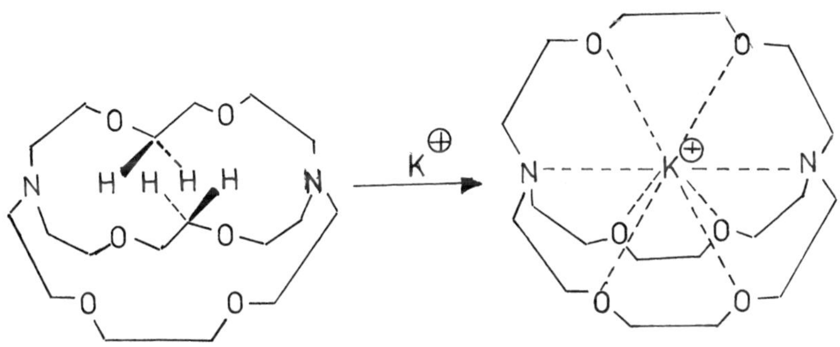

(93)

Cram, meanwhile, was also following Pedersen's inspiration, but by creating cyclic compounds that would remain rigid whether or not they were binding a substrate [283]. This principle of "preorganization" has thus been built into the more than 500 binding molecules, or *cavitands*, so far synthesized by Cram and his colleagues. Looking for some Anglo-Saxon terms with which to describe the system, Cram called it *host–guest chemistry*, the host being the receptor molecule, the guest being the substrate received:

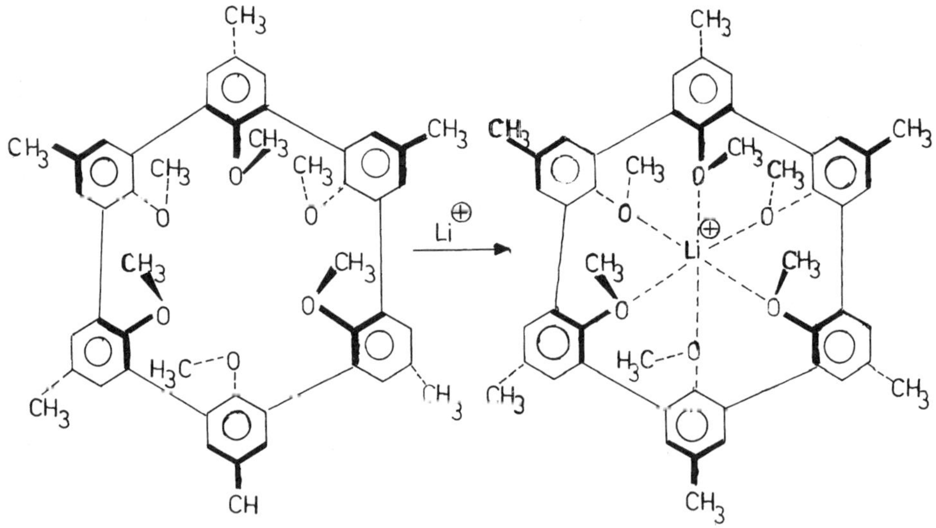

(94)

So, although Lehn and Cram employed different approaches, they were united in using synthetic organic chemistry to attempt to build compounds that were sufficiently rigid and contained cavities of a size and shape that would accommodate a desired substrate. The challenge was that, not only is the average-sized organic molecule tremendously flexible, it also typically lacks suitable cavities. Because chemical bonds radiate outward from nuclei, most organic compounds have convex surfaces of hydrogen bonds bound to carbon, oxygen, or nitrogen. To create concave surfaces–cavities it is necessary to have many more atoms.

Nature gets around this by having very large molecules, principally proteins and nucleic acids with molecular weights of 20,000 Da and more. However, organic chemists much prefer to handle compounds of about 1/10 that size. But when Cram and his colleagues embarked on their venture, only about a dozen of the existing 7 million synthetic organic compounds had significant concave surfaces. They could have chosen to work with cyclodextrine, torus-shaped carbohydrates manufactured by bacteria, but decided instead to ply the synthetic organic chemistry route because ultimately our options are limitless. Some researchers, incidentally, have brilliantly exploited the framework of cyclodextrins in creating synthetic enzymes [285].

In modifying and scrutinizing the efficiency of synthetic hosts, organic chemists have been able to dissect key components in molecular recognition processes, and in so doing have converged with biological chemists, who study the natural systems. As a result, organic chemists now can tailor host molecules so as to determine the rates at which the guest enters and leaves. Chemistry has a very long way to go before it can match natural systems, but it is on the road.

Cram made so-called molecular cells by bonding two hemispherical compounds together at their rims, thus forming prisons for whatever is inside. The contents are determined by whatever is in the reaction mixture when the hemispheres are bonded. These "cells," or *carcerands*, will be the potential vehicles for the slow-release delivery of drugs or pesticides. Carcerands can also be little laboratories, where we can match chemical reactions going on under extraordinary conditions.

Alkali metal ions are involved in numerous important biological processes, such as transmission of nerve impulses, nervous control of secretion and muscle functions, protein synthesis, and enzymatic regeneration of metabolism. Crown compounds, being complex formers, are naturally apt to interfere in such processes. For instance, it was found that lipophilic cryptands of structural formula

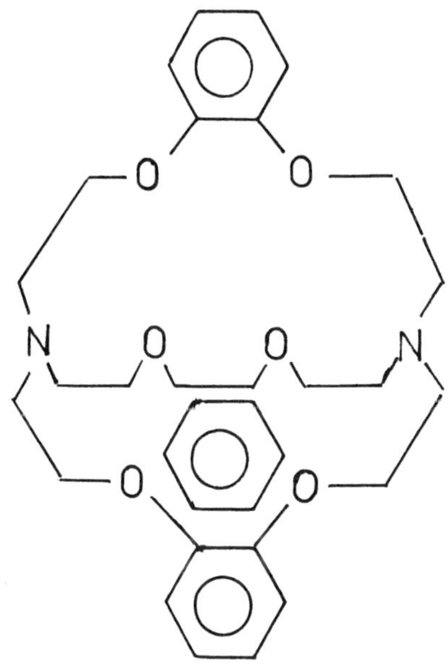

(95)

and also a pyridinoamide analogue show pronounced electrophysiological effects on sheep cardiac Purkinje fibers [286]. Guinea pig tracheal smooth muscle demonstrated a variety of in vitro positive and negative ionotropic response to concentrations of crown ethers of podando-coronand type in the mol/L to μmol/L range, suggesting that these ionophoric compounds have potential as therapeutic agents.

This is also true for the so-called drug-analogue crown compounds made of classical drug and crown ether building blocks, for example isoprenaline

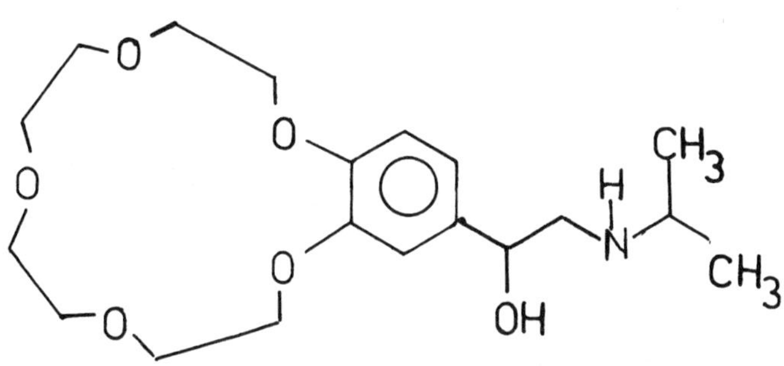

(96)

and eupaverine

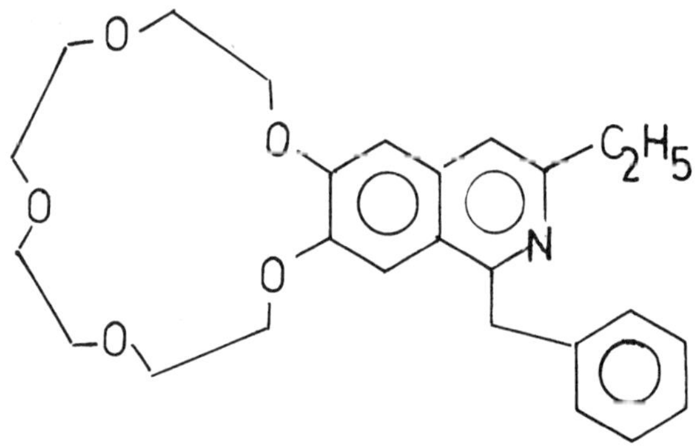

(97)

Great combination effects of such compounds had been hoped for, but this seems only to be the case for drug-analogous crown compounds with strong antiulcerogenic effects against histamine-induced ulcers [287,288]. Decontaminating effects of crown compounds in heavy metal intoxication have also been reported [289].

As early as 1975, Ringsdorf [290] established some peculiarities specific to polymeric drugs. These refer in the main to the "depot effect" (retarded absorption, retarded excretion), toxicity, pharmacokinetic variation (variable release of the active components, different metabolic pathways, and influence of the structure of the polymer, of the molecular weight, and of comonomer units incorporated), differential body distribution, polymeric specific effects, and drug combination along the polymer chain. All these ideas were demonstrated by the author and show the range of problems related to introducing polymeric drugs into medical practice.

Yet it is necessary to interpret some aspects from more comprehensive points of view. We consider that macromolecules represent, as compared with low molecular weight substances, not only a quantitatively modified material form but a qualitative development, too, with a higher information content. On the other hand, the process of drug action should be considered as a phenomenon of mutual recognition in which the two partners, the cell and the drug, play specific and equivalent roles. Thus, the necessity that appears for us is to ask whether the presence of macromolecules as a support for the drug obliges the polymer to be only an inert carrier in view of prolonged action of the drug, or whether it can be used for targeting the active principle.

An interesting example comes from the work of Schlipköter [291], Holt [292], and Allison [293], who found that poly(vinylpyridine)-*N*-oxides are active against silicosis, although the low molecular weight models such as isopropyl-*N*-oxides showed no activity. The polymeric specific activity of these *N*-oxides in protecting macrophages against silica can be explained by the fact that the polymer coats silica particles and thus prevents their rupture. Recently, Uglea et al. [294] found that many polyanionic polymers have antitumor activity although some low molecular weight models such as succinic or adipic acids showed no activity. However, there are various examples showing that the alteration of a low molecular weight drug to enhance its biological activity or to incorporate a specific property may be difficult or even impossible [295].

On the other hand, the overall molecular weight of the polymer used and their normal body distribution should be of significance. Excretion via the highly porous glomerular membrane of the kidney is limited for most synthetic polymers, since those with molecular weight above 80,000–100,000 cannot pass the tubular epithelium. On the other hand, such polymers can slowly excreted via the liver and its biliary system into intestine. Synthetic polymers as well as oligomers with molecular weights even lower than 1000 cannot pass the so-called "blood–brain barrier" and thus are not able to enter the brain and the cerebrospinal fluid. Clearly, molecular weight and its influence on total excretion is a crucial problem in the use of undegradable polymeric or oligomeric drugs. Many examples could be given showing that molecular weight is one of the important parameters influencing the biological activity of polymers [296].

It should also be pointed out that pharmacologically active polymers, simply because of their high molecular weights, may induce some cell-specific uptake without regard to other factors. As will be discussed later, the intracellular penetration of polymer is restricted to the endocytotic route (Fig. 3) [297], and one can therefore use the high endocytotic activity of certain cells to increase preferentially their polymer or oligomer drug load. This concept was the basis of first attempt to apply polymers in cancer chemotherapy [298], by taking advantage of the high endocytotic activity of many tumor cells. This is also the case for all types of diseases of the reticuloendothelial system, which is in fact the main target for polymeric and oligomeric drugs.

Synthetic polymers and oligomers normally cannot enter cells by diffusion through a membrane or by active processes via membrane proteins. The normal mechanism whereby such a polymer passes the cell membrane is the process of endocytosis [299]. This represents engulfment by an infolding of the plasma membrane with formation of a cytoplasmatic vacuole (phagosome, Fig. 3). Following uptake, fusion with the enzyme-containing lysosomes yields the digestive vacuole.

The endocytosis of polymers or oligomers is initiated by their adsorption at the cell membrane. This adsorption process plays an important role in the cell uptake of large particles and is influenced by various factors such as molecular weights and charge effects [300]. The higher the molecular weight, the higher the rate of endocytosis. However, the molecular weight–endocytosis rate relationship of drug–oligomer (or polymer) complexes must be considered taking into account other factors such as toxicity and biocompatibility. Compounds that are taken up selectively into lysosomes following formation of the phagosome are called *lysosomotropic* [301]. The importance of lysosomotropic agents can be explained in connection with the process of piggyback endocytosis, which is also depicted in Fig. 3. Piggyback endocytosis is the cell uptake of a compound, for example, a drug complexed with or fixed to a carrier, that is itself subject to endocytosis. Such drugs are released within the cell only after digestion of the carrier or cleavage of the detachable unit by lysosomal enzymes. After their attachment to the carrier, the drug molecules can no longer diffuse freely, with the consequence that their intracellular penetration is restricted to the endocytotic route, thus giving rise to specific uptake into those cells that display high endocytotic activity.

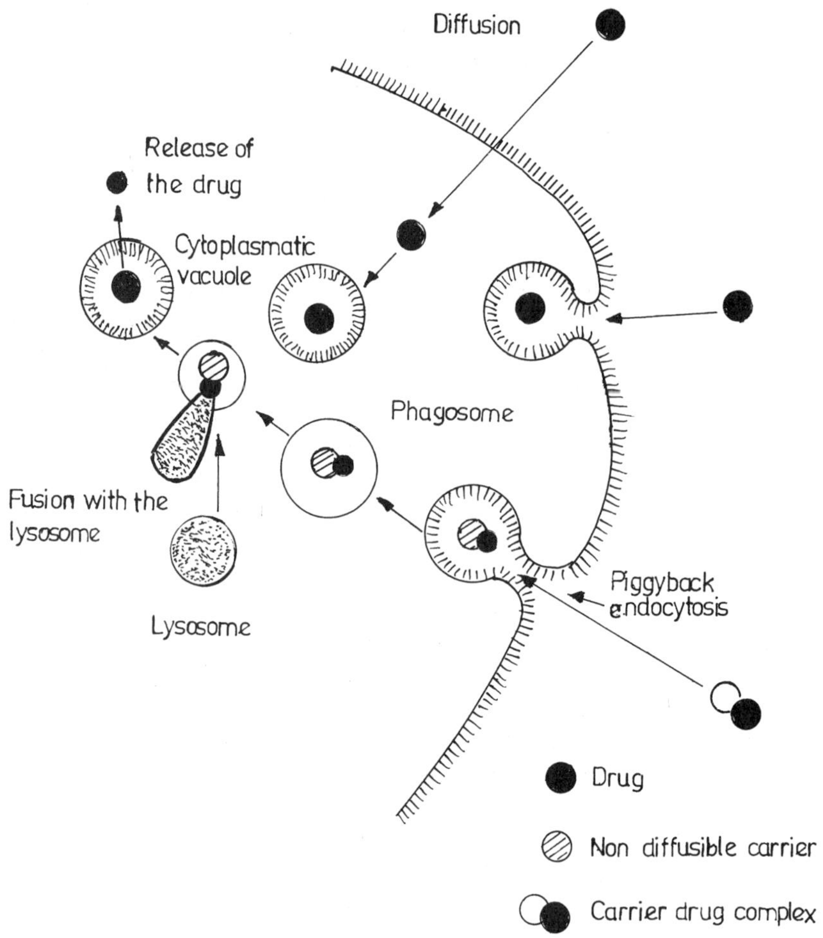

Figure 3 Mechanism of endocytosis and piggyback endocytosis.

C. Cyclic Ethers as "Physical Catalysts" in Biological Processes

Since times immemorial, humankind wished for a competition with nature. As people often felt powerless before the overwhelming forces of nature, they have tried, at the beginning rather shyly, then quite perseveringly, to appropriate the means of nature and then to achieve its performances. Prometheus and Icarus are the first examples.

Learning from their own failures, humans realized that their only chance in this encounter is a profound knowledge of the mechanisms of which natural phenomena are based. Their reproduction was the immediate purpose of such endeavors.

This is how biomimetics was born, a science the main aim of which is to reproduce phenomena and natural products by means of some systems able to decompose complex natural processes into unsophisticated and measurable steps.

The living cell has been considered since the very beginning to be a miniature factory. Capable of selectively providing high-purity raw materials, it possesses its own highly efficacious "technological processes." Obtaining and storing of useful products as well as elimination of residues from the living body are the main actions that ensure the survival of the living body. Liquid membranes have represented since the beginning a favorite model for the living cells.

We now turn to biological membranes, which are organized assemblies consisting mainly of proteins and lipids. The functions carried out by membranes are indispensable for life. Membranes are highly selective permeability barriers rather than impervious walls; they contain specific molecular pumps and gates. These transport systems regulate the molecular and ionic composition of the intracellular medium.

Membranes also control the flow of information between cells and their environment. They contain *specific receptors for external stimuli.* The movement of bacteria toward food, the response of target cells to hormones such as insulin, and the perception of light are examples of processes in which the primary event is the detection of a signal by a specific receptor in a membrane. In turn, some membranes generate signals, which can be chemical or electrical. Thus, membranes play a central role in biological communications.

Biological membranes have a thickness of about 100 Å and they are made up of lipids and proteins in a more or less ordered arrangement. The lipid molecules are oriented in such a way that their polar head groups are in contact with the aqueous phases, whereas the hydrocarbon chains form the interior of the membrane. Because hydrocarbons have a low dielectric constant, the energy required to bring a small ion, such as sodium or potassium, from the aqueous medium into the membrane is many times the mean thermal energy. This means that the lipid portions of the membrane represent an extremely high barrier for the passage of these ions.

Across some biological membranes there exist concentration differences in ions, such as potassium and sodium. There are times when these ions diffuse across the membrane in a normal manner, that is, from a point of higher concentration or chemical potential to a point of lower chemical potential. There are other times, though, during which the sodium ions move in the opposite direction, from lower to higher chemical potential. This phenomenon is known as the sodium pump in the biochemical literature. At first glance, it would appear that this phenomenon violates the second law of thermodynamics, and the operation of the pump is usually explained in biological texts as follows. Energy is usually supplied for biological systems by converting ATP and ADP using the enzyme ATPase. Further, the ATPase acts as a carrier for sodium ions while also providing energy by decomposing ATP [302,303].

Nevertheless, there are three ways of widely differing physiological significance in which metabolites can pass through biological membranes: free diffusion, facilitated diffusion, and active transport. It is obvious that free diffusion and active transport are energetically dependent, whereas facilitated diffusion is independent of energy. Free diffusion is approximately realized only for low molecular weight metabolites [304].

Facilitated diffusion is also a "passive" flux, the driving force of which can be entirely derived from the concentration gradient (or electrochemical gradient) of the particles between the two sides of the membranes. Facilitated diffusion, unlike free diffusion, is characterized by high specificity. The particle flux can be saturated with increasing particle concentration in front of the membrane. The difference between facilitated diffusion and that form of passive permeation known as free diffusion can be very satisfactorily explained by assuming the participation of a *carrier*. In this view, a specific carrier molecule, which may itself be unable to permeate the membrane, forms a complex with its transport substrate on or in the membrane. The complex dissolves in the membrane, through which it then can pass by diffusion, rotation, oscillation, contraction and expansion, or the like, and dissociate on the other side [305].

An important characteristic of facilitated diffusion is that it proceeds with no perceptible chemical change in the partner involved; the only chemical reactions are the association and dissociation of the carrier and the substrate. This kind of diffusion may be discontinuous or continuous; in the former case, association and dissociation take place only on the outside of the membrane, whereas in the latter, as exemplified by the transport of O_2 by hemoglobin, these processes also occur during passage through the membrane [305]. Heinz and Walsh [306] have shown that though the accumulation of glycine through isolated membrane preparations from *Escherichia coli* is energy dependent as an overall process, the actual transport through the osmotic barrier occurs by facilitated diffusion, that is, independently of energy.

Active or uphill transport exhibits the same kinetic characteristics as facilitated diffusion. A carrier mechanism is generally assumed to operate here also [306].

The concept of a carrier that facilitated the transport of ions and small hydrophilic molecules such as sugars and amino acids across the cell membrane dates back to the experiments performed by Osterhout [307]. This concept has since been worked out in great detail [305], but it was hypothetical until such compounds as valinomycin, monactin, and enniatin B were isolated and characterized [308]. These compounds are produced by certain microorganisms and possess antibiotic activity. Valinomycin and enniatin B are depsipeptides, that is, they are built by α-amino acids and α-hydroxy acids in alternating sequences:

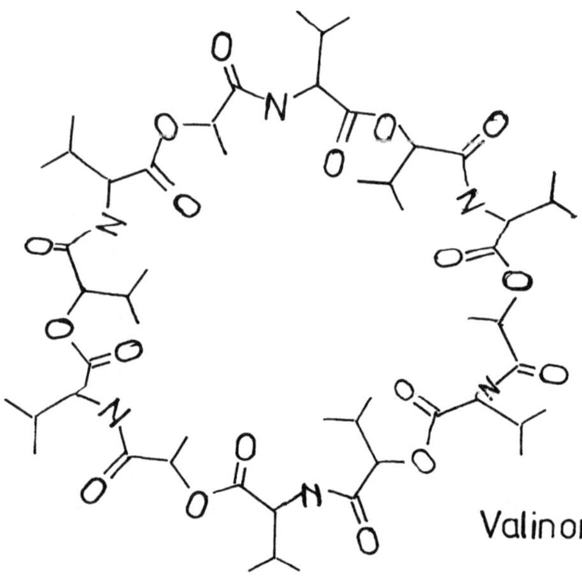

Valinomycin

(98)

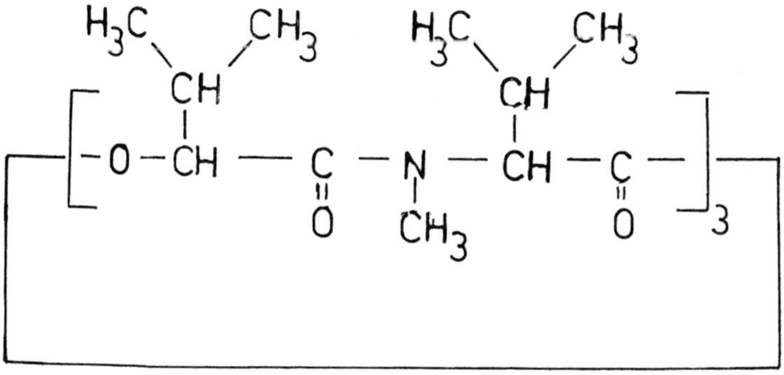

Enniatin B (99)

Taking into account the molecular dimensions and shapes, these compounds are cyclic oligomers, more precisely, cyclic trimers.

Monactin and other macrotetrolides (nonactin, dinactin, trinactin),

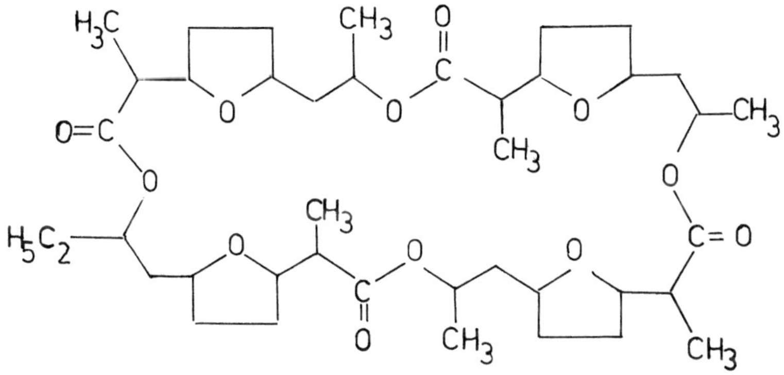

Monactin

a

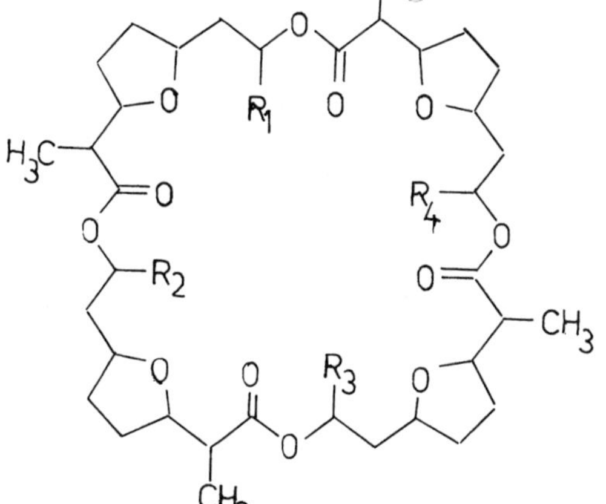

b

(100)

are cyclic compounds that contain four ether and four ester bonds. All these substances share a common property: they are macrocyclic molecules in which one side of the ring is hydrophilic, the other strongly hydrophobic. These compounds form complexes with alkali ions in organic solvents with a high degree of specificity [309]. The structure of the valinomycin–K^+ complex has been studied by spectroscopic methods as well as by X-ray diffraction [310]. The oxygen atoms of the six ester carbonyl groups form an octahedral cage around the potassium ion. In this way, the interior of the complex offers to the cation an environment that is similar to the hydration shell of the ion in aqueous solution. On the other hand, the exterior of the complex is strongly hydrophobic. The conformation of the molecule is stabilized by hydrogen bonds that are formed between neighboring amide groups.

Metal ion complexes of ionophores can be considered as host–guest complexes in which the guest entity is of spherical shape and is entrapped in a cavity-like structure formed by the cyclic or open-chain host molecule. This cavity site can either be preformed to accept the metal ion without major conformational change, or it can adopt its final shape upon complexation of the cation, which is associated with structural rearrangements. In all cases, a mutual geometrical and topological fit between host and guest molecules is essential for adduct stabilization, the adduct being in general a 1:1 complex for the ionophores. For an alkali metal ion as a spherical guest, the optimum complementary structural feature is a cavity of corresponding size, linked with polar groups in order to provide maximum interaction through ion–dipole forces. The polar ligand groups, which usually contain atoms such as oxygen, nitrogen, or more rarely, sulfur, should be situated in such a way that they can replace step by step the solvation shell of the cation during complex formation. The exterior of the ligand molecules, however, should be lipophilic to provide an appropriate surface for the nonplanar medium into which the metal ion is being transferred.

These structural features are maintained more or less by all the ionophores in their complexed forms.

Until recently, valinomycin has been regarded as a classic monocarrier, which only forms complexes of 1:1 stoichiometry with alkali metal ions. However, in 1974 a Soviet team reported evidence for the formation of adducts with 2:1 valinomycin:cation ratio, and they proposed a sandwich-type structure for the latter [311].

In the case of valinomycin and enniatin depsipeptides, the explanation for the structural origins of the more or less pronounced ion selectivities exhibited by these antibiotics was somewhat tentative. However, a detailed discussion of the structural features that lead to metal ion selectivities should be based on a whole set of comparable data on complex structures with various metal ions of different sizes. Fortunately, information of this kind has been provided for the macrotetrolide antibiotics [312–314]. The macrotetrolide antibiotics are 32-membered cyclic tetralactones that can be isolated from various *Actinomyces* species. Five homologues of the general formula,

Nonactin ($R_1=R_2=R_3=R_4=CH_3$);
Dinactin ($R_1=R_3=CH_3$, $R_2=R_4=C_2H_5$);
Trinactin ($R_1=CH_3$, $R_2=R_3=R_4=C_2H_5$). (101)

are known, which are referred to as nonactin, neononactin, and so on, depending on the number of methyls replaced by ethyl groups. The compounds are built up by four ω-hydroxycarboxylic acid subunits of alternating enantiomers condensed to each other by esterification.

These ionophores, which are frequently also called *nactins*, exhibit high selectivity in complex formation with alkali metal ions [315] as well as in ion transport through biological and artificial membranes [316].

Nigericin antibiotics are another category of compounds involved in transmembrane transport and other biological processes. Other common designations are *carboxylic acid ionophores* and *polyether antibiotics*, describing essential structural features of these biomolecules, which are unique among the ionophores described so far because they are linear and contain a terminal carboxy group, one or two hydroxy groups at the other end of the chain, and several ether oxygen atoms provided by tetrahydrofuran and tetrahydropyran rings, which may or may not be connected to each other by spiro-type junctions.

In contrast to the above-described carriers, which form positively charged complexes with cations, polyether antibiotics form neutral salts L^-M^+ with monobasic metal ions, because their carboxy group is dissociated at physiological pH [317].

Some of the carboxylic acid ionophores can complex divalent cations. Whether or not they have this ability can be used to classify the various members of the nigericin family; those ionophores not able to transport divalent cations are called monovalent polyethers, some of which are

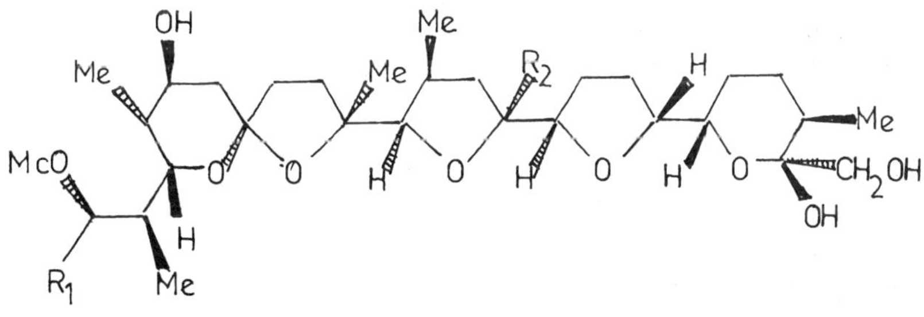

Mononxin: $R_1=\underset{\displaystyle CH_3}{CH}-COOH$; $R_2=C_2H_5$ (102)

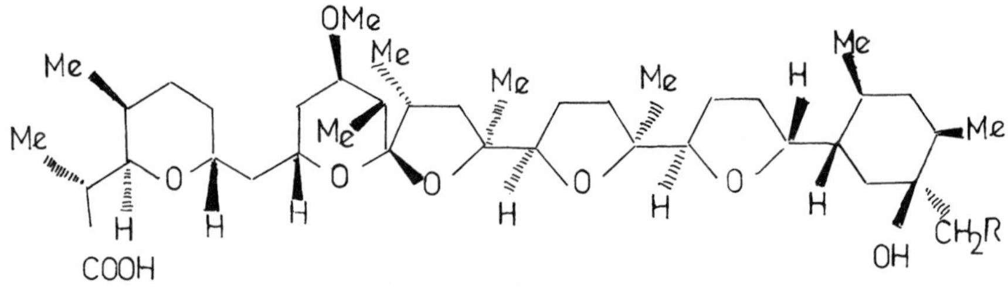

Nigericin: R=OH

Grisorixin: R=H (103)

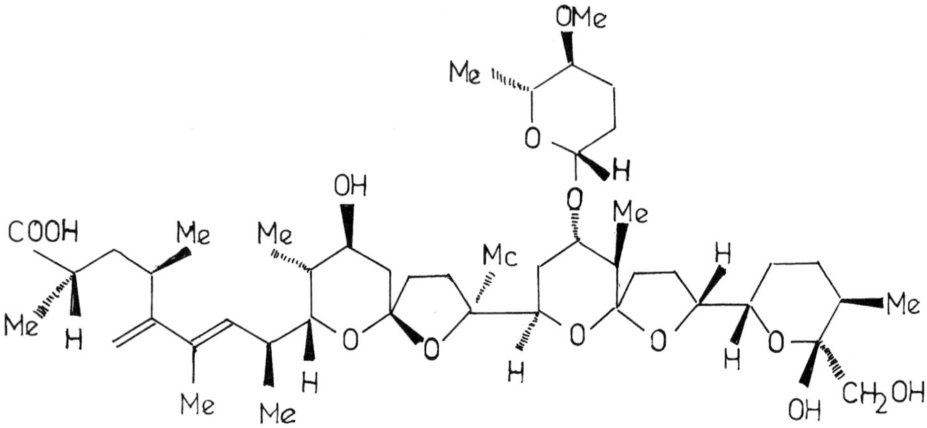

(104)

These may be further divided into two subgroups, the distinction between which rests on whether the ionophores contain a hexapyranose moiety attached to the polycyclic ligand or not. The former antibiotics, such as dianemycin, lenoremicyn, and A204A, are called *monovalent monoglycoside polyether antibiotics*. The sugar unit is bound as an α-glycoside in antibiotic A204A, and as a β-glycoside in the other members of this class.

The *divalent polyether antibiotics* are much fewer in number than those of the first group. The most popular representatives are lasalocid, formerly known as antibiotic X-537A, and A-2317. More recently discovered examples are lysocellin [318], ionomycin [319], and antibiotic X-14547 [320]:

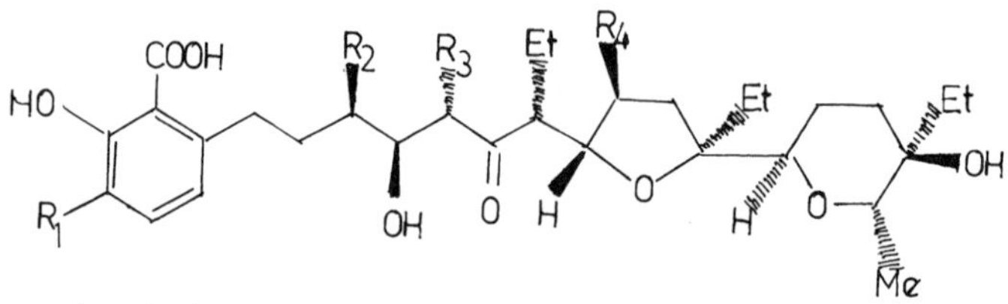

Lasalocid A: $R_1 = R_2 = R_3 = R_4 = Me$ (105)

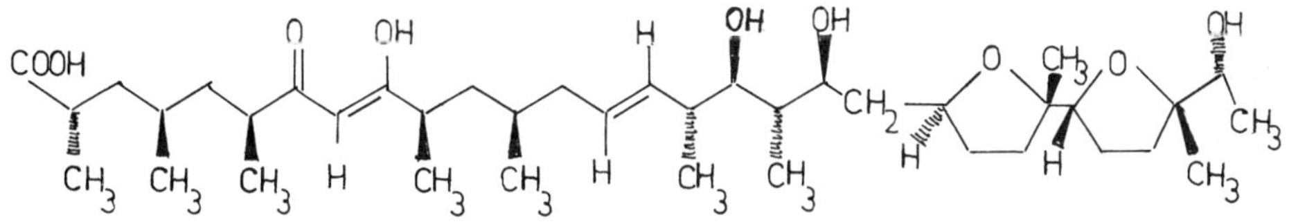

Ionomycin (106)

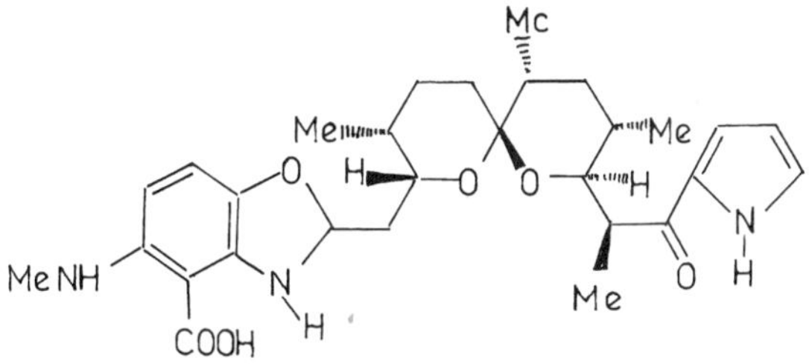

Antibiotic A 23187 (107)

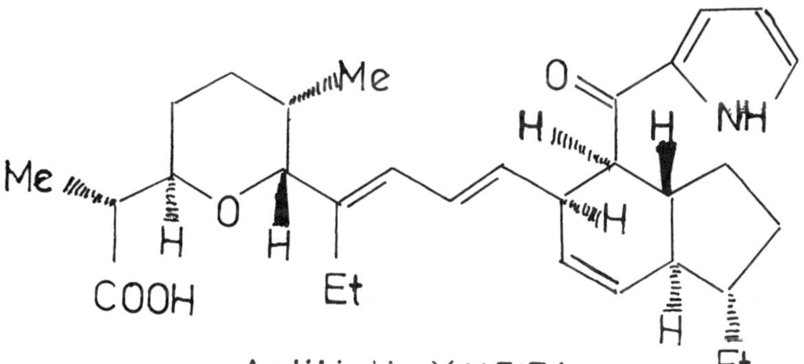

Antibiotic X14547A (108)

The discovery of the calcium-transporting properties of lasalocid and A-23187 has prompted numerous studies concerning their physiological activities; they were shown to be potential cardiovascular agents. Although clinical applications have been hampered by their potential toxicity, almost all of these antibiotics have become particularly important as coccidiostatics for poultry industries.

The ability of valinomycin and of macrotetrolides to increase the potassium permeability of biological membranes was interpreted from the viewpoint that these substances act as carriers. Because of the great complexity of biological membranes, however, detailed information on the transport mechanism could not be obtained from these experiments. For this reason, artificial or synthetic model membranes are now used in many studies, with macrocyclic antibiotics or with other synthetic carriers.

Artificial lipid membranes of macroscopic area and with a thickness of less than 100 Å may be formed by a technique originally developed by Mueller and his colleagues [321,322]. A small amount of a lipid solution, such as lecithin in *n*-decane, is put over the hole in a Teflon diaphragm, which is completely immersed in an aqueous solution. In this way, a lamella forms, which gradually becomes thinner and, in an intermediate stage, shows bright interference colors, corresponding to a thickness of less than 1 μm. After some time, a discontinuous transition to a much thinner structure take place at localized spots in the lamella where the light reflection drops to almost zero. These "black" spots expand until nearly the whole area of the hole is covered with a uniform membrane. The thickness d of this membrane may be measured.

For a lecithin with chain containing 18 carbon atoms, a value of d = 70 Å is found, which is nearly equal to two times the length of a fully extended lecithin molecule. This and other findings support the conclusion that the membrane consists essentially of a bimolecular layer (bilayer) of oriented lipid molecules. The polar head groups of the lipid molecules point toward the aqueous medium, whereas the fatty acid chain forms the interior of the membrane. Some hydrocarbon solvent remains dissolved in the film, but otherwise the structure of the artificial bilayer closely resembles the arrangement of the lipid molecules in biological membranes. Experiments with artificial lipid membranes have indicated some of the basic mechanisms by which ions may cross biological membranes [323].

In a solution of KCl, the electrical resistance of an unmodified bilayer membrane is very much higher, a typical value being 10^8 ohms/cm^2. If small amount of nonactin or valinomycin is added either to the aqueous phase or to the film-forming solution, the resistance of the bilayer drops by several orders of magnitude. This increase in the electrical conductivity comes about by the selective increase in the K^+ transport ability of the membrane.

Valinomycin is highly specific for certain alkali ions, especially for K^+. The experimental data show, for example, that K^+ gives a conductivity that is higher by a factor of 10^3 than the conductivity in the presence of Na^+. The reason for this high specificity is not fully understood. One possible explanation may lie in the steric constraints of the macrocyclic ring, which preclude an optimal electrostatic interaction between the relatively small sodium ion and the carbonyl oxygen atoms of the ring [324]. Alternatively, the selectivity may reflect differences in ion–ligand interactions of the type found in glass electrodes and certain organic solvents [325].

The increase of membrane selectivity λ_0 as a function of the increase of K^+ concentration and in the presence of valinomycin, monactin, and macrotetrolides support the hypothesis that these substances act as mobile carriers. In particular, the finding that the membrane conductivity is proportional to the carrier concentration over a large range of concentrations excludes the possibility that several antibiotic molecules together form an aggregate that bridges the membrane and may act as a pore. However, the assumption that a single valinomycin molecule forms a fixed pore in the membrane is extremely unlikely for geometrical reasons (the bilayer membrane has a thickness of about 70 Å, whereas the valinomycin molecule is a flat cylinder of 12 to 16 Å).

If valinomycin or other aforementioned natural compounds act as carriers in the classical sense, then the ion transport mediated by those compounds occurs in four distinct steps:

1. Formation of the carrier–substrate complex at one interface
2. Diffusion of the complex through the membrane phase
3. Release of the substrate at the other interface
4. Back diffusion of the free carrier (Fig. 4)

In the first step, the carrier S in the membrane combines with a metal ion M^+ from the aqueous solution. This association reaction takes place at the membrane–solution interface and may be described by the rate constant K_R. If c_M is the aqueous concentration of M^+ and N_S is the concentration at the interface of the carrier S (expressed in mol/cm^2), then the number of complexes MS^+ that are formed per square centimeter of the interface per second is equal to $K_R c_M N_S$.

In the second step, the complex MS^+ migrates to the opposite interface. The transport of MS^+ through the membrane is not a simple diffusion process; it may be more adequately described as a jump over a barrier of high activation energy. This is because of the peculiar shape of the potential energy profile of large hydrophobic ions such as that represented by the complex MS^+. It may be shown [326] that the interaction of a hydrophobic ion with the membrane leads to an energy minimum at the membrane–solution interface. The two minima are separated by a broad energy barrier, which is determined by the electrical image forces acting on the ion near the boundary between media of different electric constants [327]. The jump of MS^+ over the barrier may be characterized by a translocation rate constant K_{MS}; K_{MS} then gives directly the jump frequency of the complex.

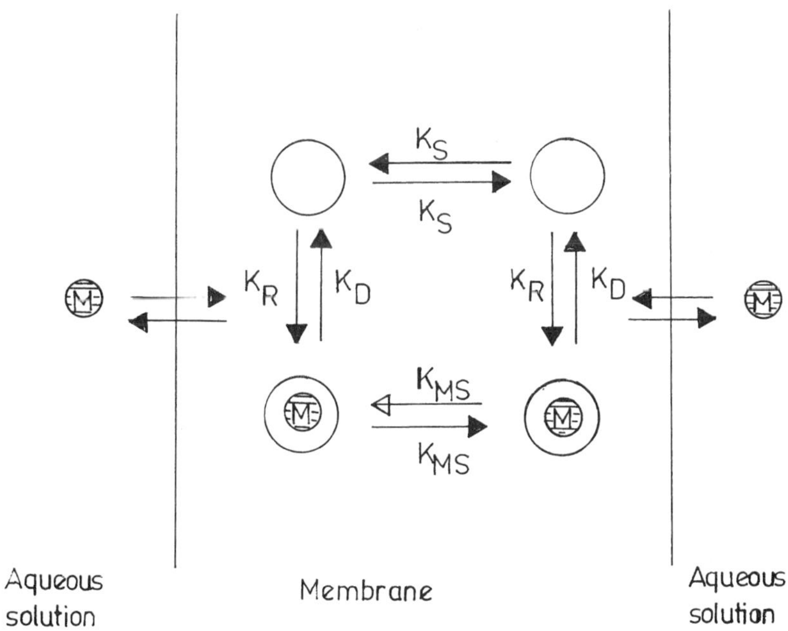

Figure 4 Transport of cation M^+ by the carrier S, where K_S and K_{MS} are the translocation rate constants, K_R is the rate constant of the association reaction, and K_D is the dissociation rate constant.

In the third step, the complex dissociates and releases the ion into the aqueous phase. This reaction is described by a dissociation rate constant K_D.

In the fourth step, the cycle is closed by the back transport of the carrier S.

To simplify the mathematical analysis of the system, the transport of S, too, is treated as a rate process that may be characterized by a translocation rate constant, K_S. It may be shown that both descriptions, diffusion or rate process, are correlated by the relation

$$K_S \approx \frac{D_S}{d^2} \tag{109}$$

where D_S is the diffusion coefficient of S in the membrane and d is the membrane thickness.

From this description of ion transport, several interesting questions arise. Is there a rate-limiting step in the overall reaction sequence, or do all reactions take place at comparable rates? Is the ion specificity of the carrier determined by thermodynamic factors alone (stability constant of the complex MS^+), or also by kinetic parameters (rate constants)? To answer these questions, a detailed kinetic analysis of the carrier system must be made. Such an analysis appears difficult at first because of the need to determine not only the four rate constants, K_R, K_D, K_S, and K_{MS}, but also the concentration of the carrier in the bilayer. The analysis becomes possible, however, by combining measurement of steady-state conductance with results obtained from electrical relaxation experiments [328].

Useful information on the transport kinetics may be obtained simply by measuring the current–voltage characteristics of the bilayer membrane in the presence of the carrier. This may seem surprising, because in most systems in which the conductance is ionic, the current–voltage characteristic is linear and therefore does not contain much information. In the case of carrier-facilitated transport, the current–voltage curve is influenced by the relative rate of the individual steps in the transport process.

For instance, if the rates of association and dissociation of the complex MS^+ are very high, the chemical reaction at the interface is always very close to equilibrium even if a current flows through the membrane. The overall transport is then limited by the translocation of MS^+ across the membrane. As was mentioned, the translocation of the complex may be treated as a jump over an activation barrier. In the presence of an external voltage V, the shape of the barrier is modified by the electrostatic energy; as a consequence, the rate constants for the jump from left to right (K'_{MS}) and from right to left (K''_{MS}) become unequal. This may be expressed to a first approximation by the following relations [325]:

$$K'_{MS} = K_{MS} e^{u/2} \tag{110}$$

$$K''_{MS} = K_{MS} e^{-u/2} \tag{111}$$

where K_{MS} is the jump frequency in the absence of an external voltage, and u denotes the reduced voltage:

$$u = \frac{V}{RT/F} \tag{112}$$

where R is the gas constant, T the absolute temperature, V the voltage, and F the

Faraday constant. Equations (110) and (111) predict an exponential increase of the current with the voltage at higher values of V. In the limiting case where the overall transport rate is determined by the translocation of the complex, the current–voltage curve is bent toward the current axis.

Conversely, if K_D is small, that is, if the dissociation step is rate limiting, the complex MS^+ will accumulate at one interface at large values of the voltage. The current J is then determined by the dissociation rate constant K_D and becomes independent of voltage. In this case, a saturating current–voltage characteristic is expected, that is, $J(V)$ is bent toward the voltage axis. It should be mentioned that a saturating $J(V)$ curve is only observed at $K_R \approx K_{MS}$, but it also occurs at large concentrations of the transport ion $K_R \neq K_{MS}$. Then the concentration of S in the membrane becomes small compared with the concentration of MS^+, with the consequence that the current is limited by the back transport of free carrier molecules.

It is useful to summarize briefly the results of the mathematical analysis of the carrier system in the steady state [329]. The specific membrane conductance λ is defined as the ratio of current density and voltage:

$$\lambda = \frac{J}{V} \tag{113}$$

In the limit of small voltages, λ reduces to the dynamic conductance λ_0:

$$\lambda_0 \frac{J}{V} V \approx 0 \tag{114}$$

The formal treatment of the carrier model then yields a rather simple relation for the ratio λ/λ_0:

$$\frac{\lambda}{\lambda_0} = \frac{2}{u}(1 + A)\frac{\sin(u/2)}{1 + A\cos(u/2)} \tag{115}$$

The ratio λ/λ_0 depends on the reduced voltage u and on the parameter A, which is given by a combination of the ion concentration c_M and the rate constants:

$$A = \frac{2K_{MS}}{K_D} + c_M \frac{K_R K_{MS}}{K_D K_S} \tag{116}$$

The parameter A can be determined from the experimental values of λ/λ_0; if λ/λ_0 is measured as a function of ion concentration c_M, the two terms in A may be obtained separately. A third combination of rate constants follows from the theoretical expression for λ_0:

$$\lambda_0 = \frac{F}{RT} N_S K_{MS} \frac{c_M K_R/K_D}{V + A} \tag{117}$$

where N_S is the interface concentration of free carrier molecules in the equilibrium state ($J = 0$). Thus, steady-state conductance measurements can yield only three independent combinations of the five unknown parameters K_R, K_{MS}, K_S, K_D, and N_S. The additional information needed for a complete kinetic analysis of the carrier system is obtained by electrical relaxation experiments.

Relaxation techniques have been widely used in chemical kinetics for the evaluation of the rate constants. This method is not restricted to homogeneous chemical systems; it may also be used for the kinetic analysis of transport processes in membranes [329]. The principle of the method is simple. The system is disturbed by a sudden displacement of an external parameter such as temperature or pressure, and the time required by the system to reach a new state is measured. In experiments with bilayer membranes, a suitable external parameter that may be suddenly changed is the electrical field strength within membrane (Fig. 5). For this purpose, the cell with the membrane is connected to a voltage source and an electronic switch permits an increase in the field strength to be made in less than 1 μs. The current through the membrane is measured with an oscilloscope as a voltage drop across the external resistor R_e. Immediately after the voltage jump, a current spike is observed, which results from the charging of the membrane capacitance. This capacitive spike limits the time resolution of the method; under favorable conditions, the resolution is about 1 μs. After the disappearance of the capacitive transient, the current in general does not immediately become constant, but approaches the stationary value J_∞ with a characteristic time course. The origin of this relaxation process may be explained as follows: At the first moment after the voltage jump, the concentrations of the complex MS^+ in the two interfaces retain their equilibrium values and are the same on both sides. Under the influence of the electrical field, charged complexes jump across the membrane, resulting in a certain initial current J_0. In the steady state, however, the concentrations of MS^+

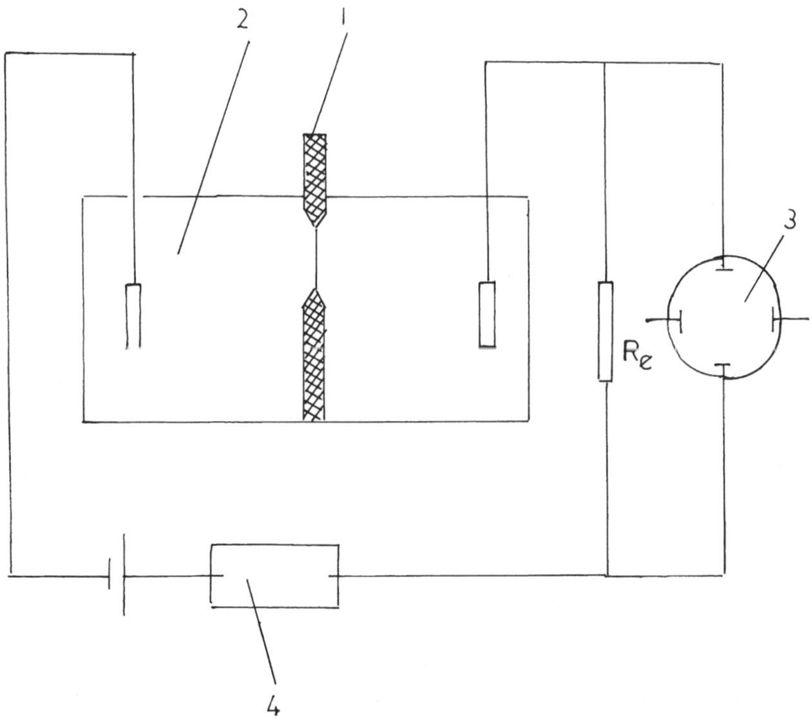

Figure 5 Principle of electrical relaxation measurements with lipid bilayer membranes. 1: Lipid bilayer membrane; 2: aqueous phases; 3: oscilloscope; 4: electronic switch.

at the two interfaces have become unequal, and, accordingly, the stationary current J_∞ is different from J_0.

A measurement of the membrane current J as a function of time t contains useful information about the rate constants of the system. An analytical expression for the function $J(t)$ may be obtained by a mathematical treatment of the carrier model in the nonstationary state [329]. In general, the time course of J is governed by two relaxation times τ_1 and τ_2:

$$J(t) = J_\infty(1 + \alpha_1 e^{-1/\tau_1} + \alpha_2 e^{-1/\tau_2}) \tag{118}$$

The "amplitudes" α_1 and α_2 as well as the relaxation times τ_1 and τ_2 are complicated functions of the four rate constants, the ion concentrations, and the voltage, and these are described elsewhere [329]. The essential point is that if any two of the four parameters, α_1, α_2, τ_1, τ_2 can be measured in addition to the steady-state conductance parameters, a complete kinetic analysis of the carrier system can be performed.

From the numerical values of the rate constants, a number of interesting conclusions may be drawn. First, the data show that K_D, K_S, and K_{MS} are of the same order of magnitude. This means that the translocations of free carrier and complex across the membrane as well as the dissociation of the complex at the interface occur at comparable rates (10^4 to 10^5 per second).

The finding that K_S and K_{MS} are equal within the limits of error is surprising. Of course, if the carrier transport mechanism were simple diffusion process in a viscous medium, we would expect the two translocation rates to be similar because the free carrier and the complex are of similar size. But as shown previously, because the charged complex has to overcome a high electrostatic energy barrier, however, which does not exist for the transport of the neutral carrier molecule, we would expect therefore K_S to be considerably larger than K_{MS}. A possible explanation may lie in the fact that valinomycin is highly surface active. It is therefore probable that valinomycin is adsorbed at the membrane–solution interface. In the adsorbed state, some of the carbonyl groups of the molecule may point toward the aqueous phase, whereas the apolar side chains may be oriented toward the hydrocarbon interior membrane. This means that energy is required when the molecule is released from the interface and enters into the hydrocarbon core of the membrane. In contrast, when the carrier molecule binds the alkali ion, the carbonyl groups are turned toward the interior of the complex, and the surface activity of the molecule is lost. If this picture is correct, then one may conclude from $K_S \approx K_{MS}$ that the activation energy for the adsorption of the free carrier into the interior of the membrane is determined by the height of the electrostatic barrier for the charged complex.

The rate of formation of the complex at the interface is described by the rate constant K_R. It is important to note that K_R depends on the surface charge of the membrane. This is simply a consequence of the definition of K_R: the number of associations per square centimeter per second is set equal to $c_M K_R$, where c_M is the bulk concentration of the cation in the aqueous phase. Now, as a consequence of the negative charge of, for example, the phosphatidylinositol membrane, the cation concentration $\bar{c}_M$ at the membrane surface is larger than the bulk concentration by a Boltzmann factor exponent $(-\phi)$, where ϕ is the electric potential (expressed in units RT/F) at the membrane surface [330].

Thus, we may introduce a "true" association rate constant $\bar{K}_R$:

$$c_M K_R = \bar{c}_M \bar{K}_R = c_M e^{-\phi} \bar{K}_R \quad (119)$$

and

$$K_R = \bar{K}_R e^{-\phi} \quad (120)$$

The Boltzman factor $e^{-\phi}$ is of the order of 20 for a phosphatidylinositol membrane under the conditions of these experiments. The values of K_R and $\bar{K}_R$ indicate that the conditions for the reaction in solution and at the interface are rather different. In solution, the association takes place in a homogeneous medium; on the other hand, the complex formation at the interface is a heterogeneous reaction in which the ion comes from the aqueous phase and combines with a carrier molecule that is bound to the membrane. The detailed mechanism of this heterogeneous reaction is not clear, however, and therefore we cannot explain why the reaction at the membrane is so much slower than in solution. It is possible that the carrier molecule at the interface is stabilized in a conformation that is less favorable for complex formation.

Because of its cyclic structure, the role of the carrier presents analogies with molecular catalysts; it may be considered as a *physical catalyst* operating a translocation in the substrate, as a chemical catalyst operates a chemical transformation. The carrier is the transport catalyst, which strongly increases the rate of the transport of the substrate with respect to free diffusion and shows enzymelike features (saturation, kinetics, and competition and inhibition phenomena). The transport of substrate may be coupled to the flow of a second species in the same (*symport*) or in the opposite (*antiport*) direction.

To conclude, it is interesting, as just mentioned, to compare the action of a carrier molecule to the function of an enzyme. An enzyme reduces the energy barrier that separates the reactants from the products of a chemical reaction. In an analogous way, the function of a carrier molecule is to reduce the extremely high activation barrier for the transport of an alkali ion across the hydrocarbon interior of a lipid membrane. The activity of an enzyme may be characterized by two parameters: the half-saturation concentration and the turnover number. In the case of the carrier molecule, the half-saturation concentration corresponds to that ion concentration in the aqueous phase at which half the membrane-bound carrier molecules are in the complexed form. This concentration $\bar{c}_M$ is equal to $K_D/\bar{K}_R \approx$ 20 M. Thus, under most conditions, the membrane-bound carrier is far from saturation.

In analogy with an enzyme, the rate of turnover, or the turnover number of a carrier, may be defined by the following fictitious experiment. Consider a lipid membrane that separates two aqueous phases and that contains a fixed number, N, of carrier molecules. The solution on the left side contains ions of concentration c_M, which are transported by the carrier; the ion concentration in the solution on the right side is assumed to be zero. If the aqueous phases are electrically short-circuited, a carrier-mediated ion flux of magnitude ϕ appears through the membrane. At low ion concentration c_M, the flux increases linearly with c_M. At high concentration, however, the carrier becomes gradually saturated, and ϕ finally approaches a maximal values ϕ_{max} (in a similar manner as an enzyme-catalyzed reaction approaches a maximal rate in the limit of high substrate concentration).

The turnover number f is then defined as the maximum number of ions that may be transported per second by a single carrier molecule:

$$f = \frac{\phi_{max}}{N} \tag{121}$$

This theory lends to the following simple expression for the turnover number:

$$f = \left(\frac{1}{K_S} + \frac{1}{K_{MS}} + \frac{1}{K_D}\right)^{-1} \tag{122}$$

In the case of valinomycin–phosphatidylinositol membrane–K^+ system [330], one finds $f \approx 10^{-4} s^{-1}$, which means that a single valinomycin molecule is able to transport 10^4 K^+ ions per second across the membrane. This number is much higher than the turnover number of most enzymes.

D. Other Considerations in Liquid Membranes

Spherical liquid membranes consist, in simplest terms, of an emulsion suspended in a liquid that does not destroy the emulsion. In a typical application, small droplets of aqueous solution are encapsulated in a thin-film oil; this emulsion is then suspended in another aqueous solution. Alternatively, small droplets of oil can be emulsified with water and the emulsion suspended in oil. In the first case, the oil phase is the liquid membrane; in the second case, the water is the liquid membrane. A typical droplet might be about 100 μm in diameter. These spherical liquid membrane systems have many potential medical applications in the emergency treatment of drug overdoses and for oxygenating the blood system. Spherical liquid membranes may be applied in resource recovery and water purification, as encapsulated cells as well as liquid membrane encapsulated enzymes [331].

Synthetic bulk liquid membranes consist of a cylindrical glass vessel and a coaxially arranged glass cylinder, which separates the two aqueous phases (Fig. 6a) [332,333]. In the U-shaped cell (Fig. 6b), the separation of the receiving and the source aqueous phase is accomplished by means of the two side arms [334,335]. H-type cells (Fig. 6c) represent a modification of the U-shaped cell and also work with a less dense solvent (e.g., 1-hexanol) as the liquid membrane component [336].

Specific binding of small molecules or ions with natural oligomeric compounds plays, as already mentioned, an important role in biochemical processes. Structural information recently accumulated for naturally occurring biologically active oligomers has allowed organic chemists to design highly structured host molecules. A variety of synthetic oligomers, linear or cyclic, for metal ions and organic ion binding have been investigated.

Using a transport system composed of an aqueous salt solution, a chloroform membrane, and a receiving water phase, various crown ethers were found to act as highly selective and very efficient Pb^{2+} carriers, even in the presence of large concentrations of other competing metal ions, such as the biologically important cations Na^+, K^+, Ca^{2+}, Fe^{2+}, Cu^{2+}, and Zn^{2+} [337]. Among the cations chemically most similar to Pb^{2+}, for example, Sr^{2+} and Ba^{2+}, good selectivity for Pb^{2+} is still obtained. A remarkable result of this is that the crown ethers

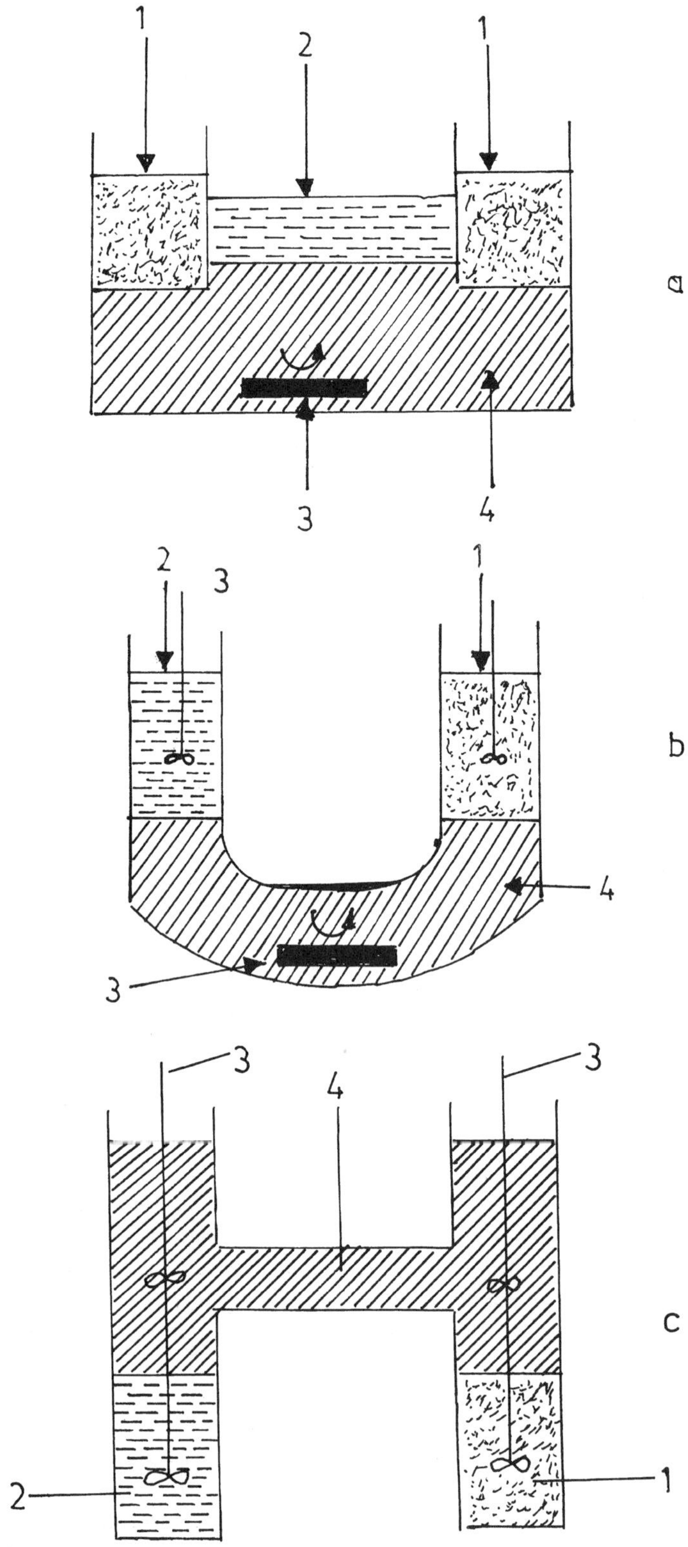

Figure 6 Cell arrangement for ion transportation: (a) cylinder cell; (b) U-shaped cell; (c) H-type cell. 1: Water phase, receiving phase; 2: salt solution, source phase; 3: stirring; 4: carrier containing membrane.

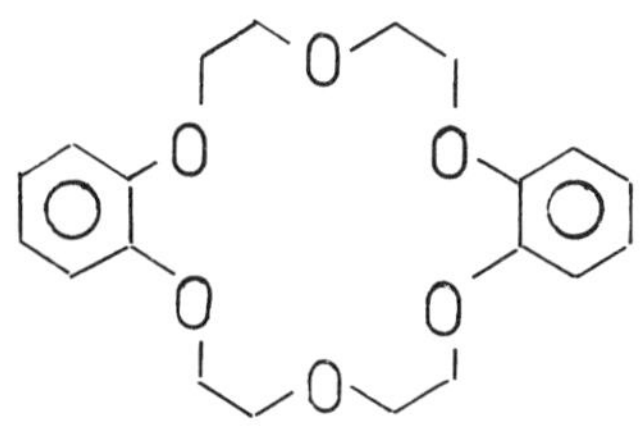

(123)

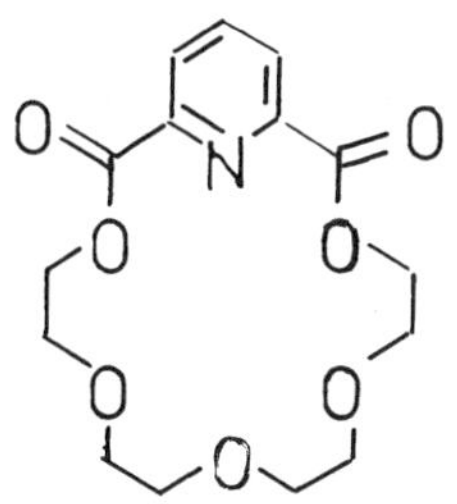

(124)

have greater selectivity for Pb[2b] over the other cations when they are together than when separate. This behavior is rather unexpected and led to the conclusion that the efficiency and selectivity of a given ligand in membrane transport may be also controlled by the presence of a co-cation.

Measurements were made of the relative rate of the transport of alkali metal cations through a chloroform liquid membrane containing crown ether with the structural formula [338]

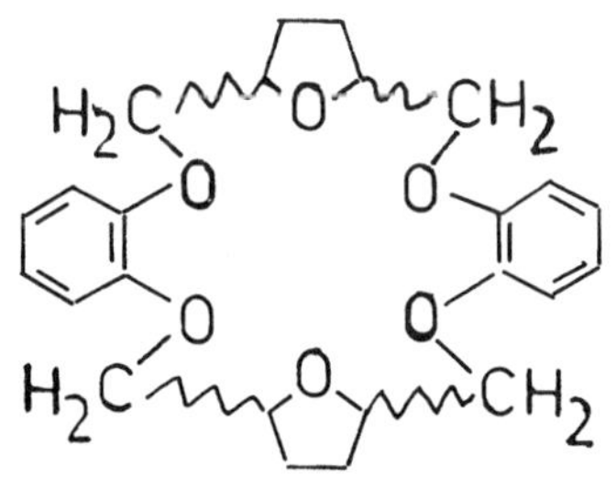

(125)

Although K^+ ion transport is differentially preferred in each case of stereoisomers, the stereoisomers differ considerably in their transport ability; for example, *meso*-C_{2b}-*trans-trans*-(125) has an outstanding transport rate for K^+ ions, which is higher than that of the flexible parent compound (125). These observations seem to be a result of its unique structure, having two chiral components with opposite chirality, somewhat reminiscent of nonactin (S_4 symmetry).

The compound with structural formula (31) showed excellent transport selectivity (liquid membrane = chloroform) for some organic ammonium salts, biogenic amines, and drugs (such as the hydrochlorides), these selectivities being superior to those for alkali metal cations [339]. In particular, this carrier can discriminate between organic ammonium cations and K^+ in the transport process; crown ether (123), which transports both K^+ and ammonium, was surprisingly nonselective in this experiment. Quinoline nitrogen terminal groups are, therefore, assumed to play an essential role in transport processes of organic ammonium ions, possibly via hydrogen bonding.

Phenylethylammonium chlorides with different substitution patterns are transported through a bulk chloroform layer [340,341] with the aid of the compound (123). The transport specificity is affected by several structural factors, depending on steric hindrance around the complexation side and decreasing number of hydrogen bonds between crown ether and substrate. As a result, primary ammonium salts are preferentially transported with respect to more substituted ones (the selectivity ratio RNH_3^+ to RNH_2^+Me is about 15:1). α-Methyl or α-*gem*-dimethyl substitution decreases the transport rate by a factor of about 2 or 7, respectively, in comparison with the unsubstituted case. This specific behavior may be envisaged as a simple separation method for organic ammonium cations.

Enantiomer separation (chiral recognition) occurred when designed chiral host compounds of binaphthyl type (formula 61) were used to carry racemic ammonium salts (α-methylbenzyl-tyrosine methyl ester hydrochloride) from an aqueous solution through chloroform to a second aqueous phase, where the guest is released [342]. Rate constants for transport were measured separately for the faster moving A-enantiomer (K_A) and the slower moving B-enantiomer (K_B). Values of K_A/K_B were found to vary from 1.45 to 1.15, depending on both the host and the guest structure. Thus, chiral recognition in transport has been realized by a rational design of host–guest structural relationship. In addition to these findings, a W-tube (combination of two U-tubes, cf. Fig. 6b) were designed that allows the continuous and simultaneous removal of each enantiomer of racemic ammonium salts from a renewable central aqueous reservoir (central arm [342], which is in contact with two separate chloroform pools containing carriers of inverse chirality. The enantiometric guests were delivered to separate aqueous solutions, one in the left and one in the right arm of the W-tube. Depending on experimental details, the optical purities of each enantiomer, obtained in the different receiving aqueous pools, range from 70 to 90%. These experiments show a realistic possibility for the construction of a simple resolving machine for racemic compounds based on membrane transport.

The rate of transport of a given cation through a liquid membrane can be varied by several orders of magnitude when the anion accompanying the cationic carrier is altered [343]. The ordering of anions according to transfer is consistent with size/charge considerations; for example, large monopositive anions like picrate, PF_6^-, ClO_4^-, and BF_4^- transfer more rapidly than the smaller ones in chloroform. The influence of the anion in membrane transport holds significant implications, one pointing to the possibility of separating or detecting anions themselves. Further, a polymeric crown ether polyamide membrane was found to strongly absorb sodium salts and to remain permeable to water [344,345]. For a cellulose nitrate membrane, the proportion of PEG blend (PEG = polyethylene glycol) was found to be responsible for the increased permeoselectivity for CO_2 [346]. These principles may be in the future allow the construction of highly selective hyperfiltration membranes for certain neutral small molecules.

A new type of model membrane for ion pumping exists as a membrane medium of a polyacrylamid derivate of ligand type,

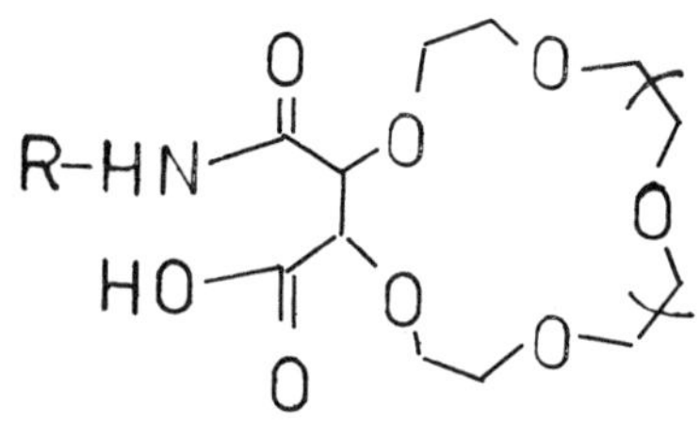

(126)

soaked into microporous Teflon filter [347]. The principles of coupled counter-transport (potassium versus proton) and the close analogies with mobile carrier membranes are retained. The physical resistance and multiple uses of these polymeric ionic pumps will be of additional benefit in the practical application of this separation strategy.

In some biological transport systems, certain monovalent cations such as Na^+, K^+ are believed to regulate amino acid transport [348,349]. Simple macrocycle carriers (123) and

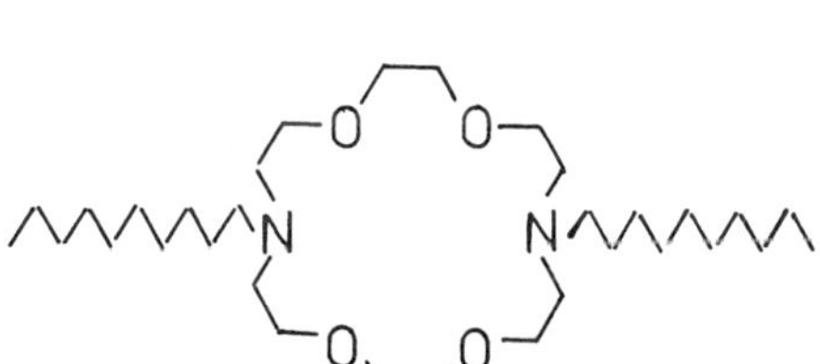

(127)

provide an artificial analogue of cation-dependent amino acid transport [350]. The regulation of the active transport of amino acid anions is clearly demonstrated by appropriate choice of cotransported cations. A K^+-ion gradient was used to pump the amino acid anions through a chloroform membrane; transport with Li^+, Na^+, and Cs^+ hardly occurred due to the cation selectivity of the carrier. This active transport system showed high substrate specificity for a series of amino acid and peptide anions, for example, (123), (126), and (127), giving the order Gly > Ala > His > Met > Leu > Phe (all as the N-benzoyl derivatives). Cryptand (128) gave a reversed sequence in the transport selectively and also was found to transport amino acid anions actively in the opposite direction, suggesting that coupling to the Cl^--anion gradient was used to pump amino acid anions up.

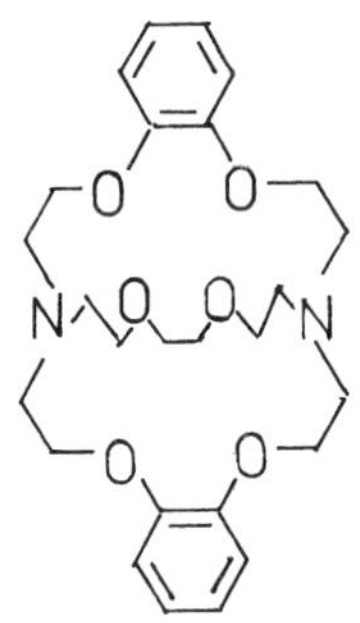

(128)

A new group of "metallo-carriers," employing transition metal complexes of a macrocyclic polyamine ligand,

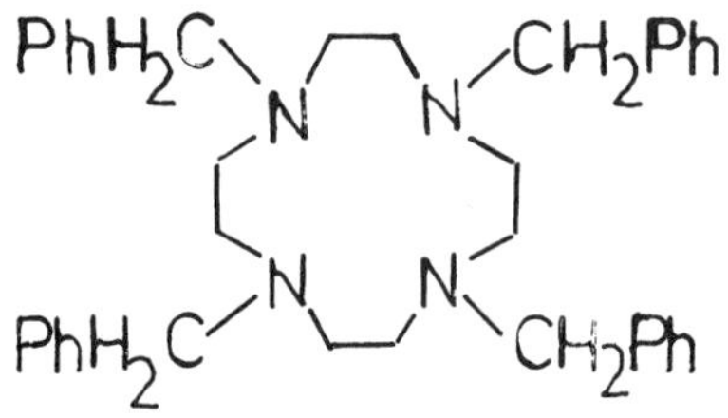

(129)

were demonstrated as a new type of powerful carrier for both active and passive transport of amino acid and oligopeptide anions across a chloroform membrane [351]. Transport properties are generally different from the system previously mentioned, and they are essentially controlled by factors such as the antiport anion and the nature of the central metal ion (Cu^{2+}, Ni^{2+}, Co^{2+}) in particular. For instance, the nickel complex of (129) more effectively mediates transport of Gly- and Ala-containing substrates, while the copper complex of (129) resulted in relatively fast transport of Leu- and Phe-containing substrate; the cobalt complex carrier gives higher transport rates for some dipeptides.

The present transport systems allow an artificial mimicking of the biological symport of amino acids, and they can be considered as a prototype for the design of specific anion transport membranes. A number of variations and extensions may be envisaged, either as biological models or potential applications [352,353].

REFERENCES

1. Lebedev, V. S., Shtykh, V. S., Reusov, A. V., and Kiselev, A. K., Current processes for production of oligoethers, *Plast. Massy*, No. 8, p. 48, 1979 (Russian).
2. *Technical Data* Hoechst, 1977.
3. Smyth, H. F., Carpenter, Ch. P., and Weil, C. S., Reaktion von Ratter auf wiederhalte inhalation von alkylene glycol dampfen, *J. Am. Pharm. Assoc. Sci. Ed.*, *39*, 349, 1950.

4. Löser, A., and Störmer, E., Adhesion von flussigkeiten an fasten substanzen. Molekulare wechsewirkung zwischen metalloxiden und adsorbieren dampfen, *J. Phys. Chem.*, *57*, 251, 1953.
5. Rowe, V. K., *Industrial Hygiene and Toxicology*, Patty, F. S., Ed., Interscience, New York, 1963, 1510–1515.
6. Smyth, H. P., Carpenter, Ch. P., and Weil, C. S., Toxicologische Angsben-Informations quellen und zucunftige Natwendigkeiten, *Ann. Ind. Hyg. Assoc. Quarterly, 15*, 203, 1954; *Chem. Z.*, 1955, 9385.
7. Smyth, H. P., Carpenter, Ch. P., and Shaffer, C. B., Akute schadigung von Meerschweinischen und ratten durcheinatmen der dampfen von PEG, *J. Ind. Hyg. Toxicol.*, 22, 477, 1940; *Chem. Z.*, 1945, 1227.
8. Tusing, T. W., Answerbang von einwirkungen auf die hant, *Drug Cosmet. Ind.*, *76*, 176, 1955; *Chem. Z.*, 1957, 11452.
9. Smyth, H. P., Carpenter, Ch. P., and Weil, C. S., Die nische giftwirkung von pereralen PEG, *J. Am. Pharmac. Assoc. Sci. Ed.*, *44*, 27, 1955; *Chem. Z.*, 1957, 4578.
10. Pfordte, K., Toxicologische angaben der poly(ethylene glycol) *Zentralhl. f. Pharmazie, 110*, 449, 1971.
11. Shaffer, C. B., and Critchfield, F. H., Renale Ausschindung und volumen-verteilung einiger PEG beim bund, *Amer. J. Physiol.*, *152*, 93, 1948; *Chem. Z.*, 1949, 558.
12. Carpenter, Ch. P., Critchfield, F. H., Nair, J. H., and Shaffer, C. B., Giffvirkung von zwei der flieggenatwehr dienenden butoxypolypropyllenglykolen, *Arch. Ind. Hyg. Occupat.*, *4*, 261, 1951; *Chem. Z.*, 1950–1954, 6304.
13. Pitter, P., Uber die biologische abbaufahigkeit der polyethylenoxide, *Coll. Czech. Chem. Commun.*, *38*, 2665, 1973.
14. Swischer, R. D., *Surfactant Biodegradation*, Academic Press, New York, 1970, 327–405.
15. Iwanow, N., and Schneider, R., Untersuchungen und diffusion und absorption von polyglycolen, *Fette, Seifen, Abstrichmittel, 58*, 549, 1956.
16. Scheller, H., Die bedentung des hydrophyl-lipophilen gleichgewischts fur die herstellung von emulsionen, *Parfumerie und Kosmetik, 41*, 85, 1960.
17. Armitage, F., and Trace, L. G., BRIT Patent 744, 519, 1952.
18. Vieweg, R., and Höchtlen, A., *Polyurethane, Kunstoff-Handbuch*, vol. VII, Carl Hanser Verlag, Munich, 1966.
19. Nguyen, H. A., and Maréchal, E., Synthesis of reactive oligomers and their use in block polycondensation, *J. Macromol. Sci., Rev. Macromol. Chem. Phys, C28*, 187, 1988.
20. Colwell, Ch. E., U.S. Patent 2,866,681, 1959.
21. Wurzschmitt, B., Beitrage zur mikro-elentaranalyse, *Dochman Monogr.*, *17*, 20, 1951.
22. Schuz, E., Toxicität von polyethylenglykole, *Berufadermatose* 7, 266, 1959.
23. Springer, E., and Isak, H., Die kapazitat von aluminium oxid gegen uber salbengrund stoffen. 5., *Z. Lebensmittel Unters u. Forschung, 125*, 428, 1964.
24. Springer, E., Isak, H., and Weigand, H., Die kapazitat von aluminium oxid gegen uber salbengrand stoffen, *Arch. Pharmaz. Ber. dtsch. Pharmaz. Ges.*, *291*, 339, 1958.
25. Feist, W. C., and Tarkow, H., A new procedure for measuring fiber saturation point, *J. Appl. Polymer Sci.*, *11*, 149, 1963.
26. Heitzpfal, W., Fractionation of poly(ethylen glycol) by column chromatography, *Makromol. Chem.*, *121*, 102, 1969.
27. Uglea, C. V., *Polymer Characterization. Fractionation*, Academic Editorial House, Bucharest, 1983.
28. *Manualul Ingrinerului Chimist*, Editura Technica, Bucharest, 1973, Vol. 2.
29. Blumberg, A. A., and Pollack, S. S., Polyethylene glycol derivatives as phase-transfer catalysts, *J. Polym. Sci.*, *A-1*, *2*, 2499, 1964.

30. Lundberg, R. B., and Bailey, F. R., Poly(ethylene glycol) monomethyl ether and their role in the phase transfer reactions, *J. Polym. Sci., A-1, 4*, 1563, 1966.
31. Chan, L. L., and Smid, J., Effect of the polyethylene glycol derivatives upon the reaction rate in phase-transfer processes, *J. Am. Chem. Soc., 89*, 4547, 1967.
32. Panayotov, I. M., and Tsvetanov, C. B., The role of poly(ethylene glycol) in phase transfer catalysis, *Makromol. Chem., 134*, 313, 1970.
33. Panayotov, M., Bobrinka, T., Petrova, T., and Tsvetanov, C. B., On the nature of "Red" potassium solution in benzene, obtained in the presence of poly(ethylene oxide), *Makromol. Chem., 176*, 815, 1975.
34. Swark, M., Ed., *Ion and Ion Pairs in Organic Chemistry*, vol. 2, Wiley Interscience, New York, 1972.
35. Hirao, A., Nakahama, S., Takahashi, M., and Yamakaki, N., Additive effect of poly(ethylene oxide). 1. Acceleration effect of poly(ethylene oxide) in several nucleophilic reactions, *Makromol. Chem., 179*, 915, 1978.
36. Hirao, A., Nakahama, S., Takahashi, M., and Yamakaki, N., Additive effect of poly(ethylene oxide). 2. An acceleration effect of poly(ethylene oxide) in a Williamson reaction, *Makromol. Chem., 179*, 1735, 1978.
37. Vögtle, F., and Weber, E., Naturale organische komplex-liganden und ihre alkalikomplexe, I. Kronen ether, cryptanden, podanden, *Kontakte*, 1977, 11.
38. Lee, D. D., and Chang, V. S., Oxidation of hydrocarbons. 8. Use of dimethyl polyethylene glycol as a phase transfer agent for the oxidation of alkenes by potassium permanganate, *J. Org. Chem., 43*, 1532, 1978.
39. Toke, L., Szabo, G. T., and Arangosi, K., Polyethylene glycol derivatives as complexing agents and phase-transfer catalysts, II., *Acta Chim. Acad. Sci. Hung., 100*, 257, 1979.
40. Toke, L., Szabo, G. T., and Somogyi-Werner, K., Polyethylene glycol derivatives as complexing agents and phase-transfer catalysts III. Behaviour of polyethylene derivatives in liquid-liquid phase equilibria, *Acta Chim. Acad. Sci. Hung. 101*, 47, 1979.
41. Panayotov, I. M., Dimov, D. K., and Tsvetkov, Ch. B., *J. Polym. Sci., Polym. Chem. Ed., 18*, 3059, 1980.
42. Yanagita, S., Takahashi, K., and Okahasa, M., Metal-ion complexation of noncyclic poly(oxyethylene) derivatives. I. Solvent extraction of alkali and alkaline earth metal thiocyanates and iodides, *Bull. Chem. Soc. Japan, 50*, 1386, 1977.
43. Sam, D. J., and Simmons, H. E., Phase-transfer of potassium fluoride in the presence of oligoethylene glycol derivatives, *J. Am. Chem. Soc., 96*, 2252, 1974.
44. Brown, H. C., and Krishnamurthy, S., Poly(ethylene glycol) as catalyst in the reduction reaction of esters, *Tetrahedron, 35*, 567, 1979.
45. Santanielle, O. E., and Ferraboschi, P., Poly(ethylene glycol)-derivatives as catalyst in certain organic reactions, *J. Org. Chem., 46*, 4584, 1981.
46. Balasabramanian, D., and Sukumar, P., The influence of molecular weight upon the PEG activity in the acceleration effect on organic reactions, *Tetrahedron Lett.*, 1979, 3545.
47. Ringsdorf, H., Structure and properties of pharmacologically active polymers, *J. Polym. Sci., Symposia, 51*, 135, 1975.
48. Ringsdorf, H., Synthetic polymeric drugs, *Med. Macromolecular Monogr., 5*, 197, 1976.
49. Ferruti, P., Macromolecular drugs acting as precursors in macromolecular active substances. Preliminary considerations, *Pharmacol. Res. Commun., 7*, 1975.
50. Ferruti, P., Macromolecular drugs, *Il Farmaco, Ed. Sci., 3*, 220, 1977.
51. Batz, H. G., Polymeric drugs, *Adv. Polymer Sci., 23*, 25, 1977.
52. Pitha, J., Polymer-cell surface interactions and drug targeting, *Targeted Drugs*, Goldberg, E., Ed., New York, Wiley and Sons, Ch. 6, 113, 1977.

53. Drobnik, J., and Rypacek, F., Soluble synthetic polymer as potential drug systems, *Adv. Polymer Sci., 57*, 1, 1984.
54. Duncan, R., and Kopacek, J., Soluble synthetic polymers as potential drug carriers, *Adv. Polymer Sci., 57*, 43, 1984.
55. Ferruti, P., New polymeric and oligomeric matrices as drug carriers, *CRC Crytical Reviews in Therapeutical Drug Carriers Systems, 2*, 175, 1986.
56. Flory, P. J., *Principles of Polymer Chemistry*, Cornell University Press, Ithaca, New York, 1956.
57. Ferruti, P., Betelli, A., and Fere, A., High polymers of acrylic and methacrylic esters of N-hydroxysuccinimide as polyacrylamides precursors, *Polymer, 13*, 462, 1962.
58. Middendorf, L., Use of poly(ethylene glycol) as solvent in pharmaceutical industry, *Pharm. Ind., 16*, 44, 1954.
59. Schütz, E., Auxiliary based on PEG in pharmaceutical industry, *Arzneimittel-Forsch., 3*, 451, 1953.
60. Beuttner, W., and Steiger-Trippi, K., Galenic preparations, *Schw. Apoth.-Ztg., 96*, 293, 1956.
61. Albertson, A. G., Donaruma, L. G., and Vogl, O., Synthetic polymers as drugs, *Ann. N.Y. Acad. Sci., 446*, 205, 1985.
62. Cecchi, R., Rusconi, L., Tenzi, M. C., Danusso, T., and Ferruti, P., Synthesis and pharmacological evaluation of poly(oxyethylene) derivatives of 4-isobutylphenyl-2-propionic acid (Ibuprofen), *J. Med. Chem., 24*, 622, 1981.
63. Dal Pozzo, A., Acquasaliente, M., Donzalli, G., Delir, F., and Ferruti, P., Synthesis and antitumoral properties of new aroylacrylic esters of PEG, *Il Farmaco, Ed. Sci., 41*, 622, 1986.
64. Ferruti, P., Tenzi, M. C., Rasconi, L., and Cecchi, R., Succinic and glutaric half-esters of PEG's and their benzotriazole and imidazole derivatives as oligomeric drug-binding molecules, *Makromol. Chem., 182*, 2183, 1981.
65. Rusconi, L., Tenzi, M. C., Zambelli, V., and Ferruti, P., Activated derivatives of succinic and glutaric half-esters of PEG and their exchange reactions with hydroxy and amino-compounds, *Polymer, 23*, 1177, 1983.
66. Zalpiscky, S., Gilin, G., and Zilka, A., Attachment of drugs to PEG, *Eur. Polymer J., 19*, 1177, 1983.
67. *Technical Bulletin* Hoechst, 1977.
68. Singh, P., PEG's action upon the barbituric acid, *J. Phar. Sci., 55*, 63, 1966.
69. Kaiser, H., *Pharmazeutisches Tacchenbuch*, Wissenschaftliche Verlaggesellschaft, Stuttgart, 1962, 226.
70. Czetsch-Lindewald, H., *Supositoren*, Editio Cantor, Aulendorf, Würtenberg, 1958.
71. Kedvessy, G., and Redgon, G., Use of PEG as antiseptic suppositories, *Phar. Ind., 25*, 445, 1963.
72. Collins, A. P., Molecular weight and physical characteristics of PEG. The influence upon the pharmacological applications, *Am. Prof. Pharmacist, 23*, 231, 1957.
73. Gastirner, F., and Said, G., Application of PEG mixtures in formulation of drugs, *Pharm. Ind., 32*, 757, 1970.
74. Spiegel, A. J., and Woseworthy, M., Application of PEG in coating freeze-dried suppositories, *J. Pharm. Sci., 52*, 917, 1963.
75. Anschel, J., The dependence of PEG molecular weight and the application capacity in pharmacy, *Pharm. Ind., 35*, 273, 1973.
76. Ritschel, J., Use of PEG as solvent in injections, *Pharm. Ind., 27*, 781, 1965.
77. Carpenter, C. P., and Shaffer, C. B., The behaviour of PEG 300 in intravenous injection, *J. Am. Pharm. Assoc. Sci., Ed., 41*, 27, 1952.
78. Eli Lilly Comp., U.S. Patent 2,605,209, 1952.
79. Ritschel, W. A., *Die Tablette*, Editio Cantor KG, Aulendorf, 1966.

80. Miller, B., and Chavkin, L., Application of PEG in tablet manufacture, *J. Am. Pharm. Assoc. Sci. Ed., 43*, 486, 1954.
81. Maly, J., and Jarca, A., Use of PEG as anti-stick agent, *Pharm. Ind., 29*, 399, 1967.
82. Gans, E. H., and Chavkin, L., PEG's in coating tablet technology, *J. Am. Pharm. Assoc. Sci. Ed., 48*, 483, 1954.
83. Kedvessy, G., and Selmeczi, A., Advantages of PEG 6000 in coating technology, *D. Apoth.-Ztg., 102*, 635, 1962.
84. Seiller, M., and Duchène, D., L'Utilisation des dérivés de poly(ethylene glycol) pour l'injection, *Ann. Pharm. Fr., 26*, 291, 1968.
85. Kedvessy, G., and Selmeczi, A., The behaviour of PEG 6000 in the drug industry, *Pharm. Ind., 31*, 412, 1969.
86. Abbott Laboratories, U.S. Patent 2,881,085, 1953.
87. Wurster, D. W., Formulation in coating fluidized bed, *J. Am. Pharm. Assoc. Sci. Ed., 48*, 451, 1951.
88. Wisconsin Alumni Research Foundation, U.S. Patent 2,799,241, 1957.
89. Cordes, G., Additive effect of PEG derivative in coating compounds, *Pharm. Ind., 31*, 566, 1969.
90. Lelman, K., and Breher, D., Film coating with aqueous acrylic resins. The influence of PEGs additives, *Pharm. Ind., 34*, 894, 1972.
91. Rothe, W., and Groppenbächer, G., Permeability of film formulated with acrylic resins and PEG derivatives, *Pharm. Ind., 35*, 723, 1973.
92. Boeringer GmbH., GER 1,184,459, 1962.
93. Riekman, P., High-speed coating methods in pharmacology, *Pharm. Ind., 25*, 172, 1963.
94. Kunze, K. H., Technology of film coating in drug industry, *Pharm. Ind., 28*, 75, 1966.
95. Nowak, G. A., *Die kosmetischen präparate*, Verlag für Chem., Industrie, Zielkowski, H., Ed., Augsburg, 1969.
96. Jousef, R. T., Uses of PEG derivatives in cosmetic formulations, *Pharm. Ind., 35*, 154, 1973.
97. Wallhäuser, K. H., PEG based formulation in cosmetic additives, *Seifen, Ole, Fette, Wachse, 100*, 11, 1974.
98. Continental-Gummiwerke, GER Patent 1,248,286, 1963.
99. Doulon and Co., GER Patent 1,232,866, 1964.
100. Koch, O., Dust-free enzyme concentrates, *Fette, Seifen, Anstrichmittel, 75*, 331, 1973.
101. Roth, K., Stabilizing effect of PEG upon the enzyme formulations, *Fette, Seifel, Anstrichmittel, 76*, 28, 1974.
102. Tschakert, E., Storage stability of enzymatic detergents, *Seifen, Ole, Fette, Wachse, 97*, 633, 1971.
103. Gundersen, R. C., and Hart, A. W., *Synthetic Lubricants*, Reinhold Publishing Corp., New York, 1962.
104. *Ullmann Enzyklopädie der Technische Chemie*, 3rd ed., vol. 15, p. 301.
105. Mueller, E. R., and Martin, W. H., Polyalkylene glycol lubricants uniquely water soluble, *Lubrication Engineering, 31*, 348, 1978.
106. Mueller, E. R., Synthetics: Rx for lubrication problems, *Chemical Engineering*, 1971, June 28, p. 91.
107. Mueller, E. R., Polyalkylene glycol lubricants established industrial products, *ACS Symposium*, Atlantic City, New Jersey, 1974.
108. Hoechst AG, Applications of polyglycols, *Technical Bulletin*, 1985.
109. Kiefer, W. C., and Mueller, D., Lubricating earth moving equipment, *Automotive Engineering, 83*, 26, 1975.
110. Lee, C., Engine clinic, *R/C Modular Magazine, 11*, 8, 1974.

111. O'Connor, J. J., Synthetic liquid lubricants, *Standard Handbook of Lubricating Engineering*, McGraw Hill, New York, 1968, Ch. 11.
112. Braithwaite, E. R., *Lubrication and Lubricants*, Elsevier, Amsterdam, 1967, Ch. 4.
113. Morán, R., Dimensionally stable wood. The role of PEG additives, *Holz als Roh und Werstoff, 23*, 142, 1965.
114. Schneider, A., Incorporation of PEG in cell walls, *Holz als Roh und Werkstoff, 27*, 209, 1969.
115. Mitchell, H. L., and Wahlgren, H. R., Dimensional stabilization of wood, *Forest Prod. J., 9*, 437, 1959.
116. Schneider, A., Technology of wood treatment with aqueous solution of PEG, *Holz als Roh und Werkstoff, 28*, 20, 1970.
117. Mitchell, H. L., and Iverson, E. S., PEG formulation for wood treatment, *Forest Prod. J., 11*, 6, 1961.
118. Mitchell, H. L., and Fober, E. W., Effects of PEG application in wood storage, *Forest Prod. J., 12*, 476, 1962.
119. Englerth, G. H., and Mitchell, H. L., Processing of wood in the presence of PEG solutions, *Forest Prod. J., 13*, 48, 1963.
120. Morán, R., and Johansson, F., Softening effect of PEG in normal surface coating, *Schwed. Tischrzeitung*, 1971, No. 1.
121. Gaudel, P., Der Preparator, *Z. f. Museumstechnik, 9*, 202, 1963.
122. Kopeke, G., Preservation of ancient moist wood by PEG solutions, *Archäologischer Anssiger*, 1967, No. 2, 165.
123. Noack, D., Preservation of ancient ships by PEG solution, *Kunst u. Denksalppflage*, January 1969, 130.
124. Barkman, L., and Fransan, A., *Untarwasser Archheologie*, Hans-Putty-Verlag, Berlin, 1973, 241.
125. Daul, G. C., and Muller, T. E., Production of high-tenacity regenerated cellulose, *J. Appl. Polymer Sci., 12*, 487, 1968.
126. Rose, K. P., Mixture of PEG and fatty amine ethoxylates as viscose modifiers, *Faserforsch. u. Textiltech., 19*, 499, 1968.
127. Kelly, G. B., and Buttrick, G. W., *Tappi Report, 21st Coating Conference*, Houston, 1969, 563.
128. Schmid, A. F., Study of the PEG application in paper industry, *Adhäsion*, 1959, 568.
129. Apps, E. A., PEG derivatives as carriers of inks and stamping inks, *Paint Manufacture*, 1962, No. 1, 5.
130. *Geburthilfe und Frauenheilkunde, 25*, 1037, 1965.
131. Hirsch, Th., and Boellaard, J. W., Uses of PEG's in biological science, *Z. f. wiss. Mikroscopie, 64*, 24, 1958.
132. Boellaard, J. W., and Hirsch, Th., PEG's based blends in preparing animal and human organ preparations, *Mikroscopie, 13*, 368, 1969.
133. *Der Preparator, 20*, No. 1/2, 1974.
134. Hoechst, A. G., GER Patent 2,109,199, 1971.
135. Hoechst, A. G., GER Patent 2,300,844, 1973.
136. *Adhesive Age*, December 1967, 23.
137. Hoechst, A. G., GER Patent 2,204,009, 1972.
138. *Zinn und seine Verwendung, 51*, 13, 1961.
139. Thwaites, C. J., *Sheet Metal Industries, 38*, 583, 1961.
140. *Zinn und seine Verwendung, 70*, 12, 1966.
141. Du Pont, U.S. Patent 2,531,832, 1950.
142. Merriot, A. W., and Picker, D., PEG derivatives as phase transfer catalysts, *J. Am. Chem. Soc., 97*, 2345, 1975.
143. Brandstrem, A., Synthesis of triaryl phosphate, *Pure Appl. Chem., 54*, 1769, 1982.

144. Tolton, G. E., and Clinton, N. A., PEG derivatives as phase transfer catalysts and solvents for organic reactions, *J. Macromol. Sci., Rev. Macromol. Chem. Phys., C 28*, 293, 1988.
145. Peupaert, J. A., De Keyser, J. L., Vanderverst, D., and Dument, P., Phenytoin synthesis in the presence of phase-transfer catalysts, *Bull. Soc. Chem. Belg., 93*, 493, 1984.
146. Krisnakumar, V. K., Implications of poly(ethylene glycol) as phase transfer catalyst of organic reactions, *Synth. Commun., 14*, 189, 1984.
147. Sukata, K., Synthesis of nitrogen heterocycles by phase transfer catalysis, *Bull. Chem. Soc. Japan, 56*, 289, 1983.
148. Regen, S. L., Dreiphasen-Katalyse, *Angew. Chem., 91*, 464, 1979.
149. Davidson, R. S., Patel, A. M., and Sufdar, A., Alkylation of aromatic amines using PEG as phase transfer catalyst, *J. Chem. Res.*, 1984, 88.
150. Zupancic, B. G., and Sopcic, M., Synthesis of N-alkoxyalkylacylanilines using oligo(ethylene glycol) as phase transfer catalyst, *Synthesis*, 1982, 942.
151. Kimura, Y., and Ragen, S. L., Dehydrohalogenation reactions catalysed by poly(alkylene glycol)'s, *J. Org. Chem., 48*, 198, 1983.
152. Angeletti, E., Tundo, P., and Venturello, P., Ether synthesis using poly (alkylene glycol) as phase transfer catalysts, *J. Chem. Soc., Perkin Trans., 1*, 1137, 1982.
153. Slaoui, S., Le Gouller, R., Pierre, J. L., and Luche, J. L., Application of PEG's as phase transfer catalysts in Williamson synthesis, *Tetrahedron Lett., 23*, 1681, 1982.
154. Sukata, K., *Yuki Cozei Kagaku Koikaiashi, 39*, 443, 1981; *Chem. Abstr., 95*, 800375, 1981.
155. Evans, P. R., and Berenbaum, M. B., U.S. Patent 4,174,1979.
156. Banthia, A. K., Lunsford, D., Webster, D., and McGrath, J. E. Polyether synthesis using PEG's as phase transfer catalysts, *J. Macromol. Sci.-Chem., A15*, 943, 1981.
157. Imai, Y., Ueda, M., and Li, M., Synthesis of aromatic oligoethers, *J. Polymer Sci., Polymer Lett. Ed., 17*, 85, 1979.
158. Bartach, R. A., and Yang, D. W., Aryldiazonium salt reactions in the presence of PEG's as phase transfer catalysts, *Tetrahedron Lett.*, 1979, 2503.
159. Starcks, C. M., and Lietta, C., *Phase Transfer Catalysis, Principles and Techniques*, Academic Press, New York, 1978.
160. Pieser, L. F., and Pieser, M., Eds., *Reagents for Organic Synthesis*, Wiley, New York, 1967, 333.
161. Clark, L. V., in *The Chemistry of Carboxylic Acids and Esters*, Wiley, Toronto, Ch. 12.
162. Ugelstad, J., Berge, A., and Listcu, H., Study on non-supported oligoethylene glycol dimethylesters, *Acta Chim. Scand., 19*, 208, 1965.
163. Toke, L., and Szabo, G. T., Polyethylene glycol derivatives as complexing agents and phase transfer catalysts, *Acta Chem. Acad. Sci. Hung., 93*, 421, 1977.
164. Balasubramanian, D., Sukamar, P., and Chandoni, B., Synthesis of benzaldehyde by partial oxidation of benzyl alcohol, *Tetrahedron Lett.*, 1979, 3543.
165. Yamazaky, N., Hirav, A., and Nakahama, S., *Trans*-stylbene oxidation under the influence of phase-transfer catalysts, *J. Macromol. Sci., Chem., A13*, 321, 1979.
166. Neuman, R., and Sasson, Y., Autooxidation of weak carboxylic acids. The effect of PEG's and the mechanism of phase transfer catalysis, *J. Org. Chem., 49*, 1282, 1984.
167. Branelle, D. J., U.S. Patent 4,410,422, 1983.
168. Keller, W. E., *Compendium of Phase-Transfer Reaction and Related Synthetic Methods*, Fluka AG Buchs, Berna, Switzerland, 1979.
169. Dehmlow, E. V., and Dehmlow, S. S., Eds., *Phase Transfer Catalysis. Monographs in Modern Chemistry 11*, Verlag Chemie, Weinheim, 1980.
170. Weber, W. P., and Gokel, G. W., *Reactivity and Structure Concepts in Organic Chemistry, 4*, Springer Verlag, Berlin, 1977.

171. Sukata, K., Synthesis of isoprene by decomposition of dioxane derivatives using PEG's as phase transfer catalysts, *J. Chem. Soc. Japan*, *57*, 613, 1984.
172. Pedersen, C. J., U.S. Patent 3,689,225, 1972.
173. Santanielle, E., Manzocchi, A., and Ferraboschi, P., *Polymer Prepr., Am. Chem. Soc., Div. Polym. Chem.*, *23*, 192, 1982.
174. Sawicki, R. A., U.S. Patent 4,421,675, 1983.
175. Kimura, Y., and Regen, S. L., Supported PEG's and their application in organic synthesis, *J. Org. Chem.*, *48*, 195, 1983.
176. MacKenzie, W. M., and Sherington, D. C., Supported polyethylene glycol and phase transfer catalysis of dehydration reactions, *Polymer*, *22*, 431, 1981.
177. Hradill, J., and Svec, F., Macroporous glycidyl methacrylate copolymer as support PEG's in the phase transfer catalysis, *Polymer Bull.*, *11*, 159, 1984.
178. Itatt, R. M., and Christensen, J. J., *Synthetic Multidentate Macrocyclic Compounds*, Academic Press, New York, 1978.
179. Hiracka, M., *Crown Compounds. Their Characteristics and Applications. Studies of Organic Chemistry 12*, Elsevier, Amsterdam, New York, 1982.
180. Weber, H., *Cyclic Oligoethers, Top. Curr. Chem.*, *98*, 1, 1981.
181. Uglea, C. V., and Neguleseu, I. I., *Syntheses and Characterization of Oligomers*, CRC Press, Boca Raton, Florida, 1991, Ch. 3.
182. Pedersen, C. J., and Frensdorf, I. G., Halogenation and nitration reactions of aromatic crown ethers, *Angew. Chem.*, *84*, 16, 1972.
183. Agai, B., Bitter, L., Hell, E., and Szöllosy, A., Hydrolysis of crown ethers, *Tetrahedron Lett.*, 1985, 2705.
184. Pedersen, C. J., U.S. Patent 3,678,978, 1972.
185. Hee, G. S., and Bartach, R. A., Formation of oxonium compounds under the influence of crown ether and Lewis acids, *J. Org. Chem.*, *47*, 3557, 1982.
186. Batt, S. G., Kynst, U., and Atwood, J. C., Reaction of crown ethers in the presence of Lewis acids, *J. Incl. Phenom.*, *4*, 241, 1986.
187. Weber, W., in *Progress in Macrcyclic Chemistry*, Izatt, R. M., and Christensen, J. J., Eds., Wiley, New York, 1987, vol. 3, p. 137.
188. Truter, R. M., and Pedersen, C. J., *Kryptate*, Endeavour, Amsterdam, 1971, 142.
189. Weber, E., Jozel, M. F., Puff, H., and Franken, S., Solid-state inclusion compounds of new host macrocycles with unchanged organic molecules. Host synthesis, inclusion properties and X-ray crystal structure of an inclusion compound with 1-propanol, *J. Org. Chem.*, *50*, 3125, 1985.
190. Vögtle, F., and Weber, E., in *The Chemistry of Ether Linkage*, Patai, E., Ed., Suppl. E, Part 1, Wiley, London, 1981, 59.
191. Weber, E., Mehgfach benzokondensierte kronenether-synthese, ionenselektivität in membranelektroden und wasere inschluss, *Chem. Ber.*, *118*, 4439, 1985.
192. Weber, E., Polytropic cation receptors. 2. Synthesis and selective complex formation of spiro-linked "multitrop" crown compounds, *J. Org. Chem.*, *47*, 3478, 1982.
193. Goldberg, I., Geometry of the ether, sulphide and hydroxyl groups and structural chemistry of macrocyclic and noncyclic polyether compounds, in *The Chemistry of the Ether Linkage*, Patai, E., Ed., Suppl. E, Part 1, Wiley, London, 1981, 175.
194. Ouchi, M., Inone, Y., Wada, K., Iketani, S., Hakushi, T., and Weber, E., Molecular design of crown ethers. 4. Synthesis and selective cation binding of 16-crown-5 and 19-crown-6 lariats, *J. Org. Chem.*, *52*, 2420, 1987.
195. Lehn, J. M., Dimetallic macrocyclic inclusion complexes. Concepts-design-prospects, *Pure Appl. Chem.*, *52*, 2441, 1980.
196. Weber, E., Progress in crown ether chemistry, *Kontakte*, 1984, 26.
197. Cameron, G. G., Buchan, G. M., and Law, K. S., Polyester synthesis by phase transfer catalysis, *Polymer*, *22*, 558, 1981.

198. Weber, E., Progress in crown ether chemistry, Part IV.C, *Kontakte*, 1982, No. 1, 24.
199. Weber, E., Progress in crown ether chemistry, Part IV.D, *Kontakte*, 1983, No. 1, 38.
200. Grokel, G. W., and Durst, H. D., *Crown Ether Chemistry; Principles and Applications, Aldrichimica Acta, 9*, 3, 1976.
201. Grokel, G. W., and Durst, H. D., Principles and applications of crown ethers, *Synthesis*, 1976, 168.
202. Knipe, A. G., Crown ethers, *J. Chem. Ed., 53*, 618, 1976.
203. Weber, E., and Vögtle, F., Neutrale organische complex-liganden und ihre alkalicomplexe. II. Kronenäther, cryptanden als reagensien und katalysateren (Teil B), *Kontakte*, 1977, No. 3, 36.
204. Grokel, G. W., and Weber, G. W., Phase transfer catalysis. Part 1. General principles, *J. Chem. Ed., 55*, 350, 1976.
205. Grokel, G. W., and Weber, G. W., Phase transfer catalysis. Part 2. Synthetic applications, *J. Chem. Ed., 55*, 429, 1978.
206. Kage, K., Macrocyclic polyethers—applications to organic reactions, *Yuki Cosei Kagaku Shi, 33*, 163, 1975.
207. Dehmlow, E. W., Fortschritte der phasentransfer-katalyse, *Angew. Chem., 89*, 521, 1977.
208. Lietta, C. L., Applications of macrocyclic polydentate ligands to synthetic mediated transformations, in *Synthetic Multidentate Macrocyclic Compounds*, Izatt, R., and Christensen, J. J., Eds., Academic Press, London, 1978, 111.
209. Lietta, C. L., Organic transformations mediated by macrocyclic multidentate ligands, in *The Chemistry of Functional Groups*, Patai, E., Ed., Suppl. E, Part 1, Wiley, 1980, 157.
210. Montanari, F., Landoni, D., Ralla, F., Phase-transfer catalysed reactions, in *Host-Guest Complex Chemistry II*, Vögtle, F., Ed., *Top. Curr. Chem., 101*, 1982.
211. Cocagne, P., Galle, R., and Wiguerre, J., The present use and the possibilities of phase transfer catalysis in drug synthesis, *Heterocycles, 20*, 1379, 1973.
212. Vögtle, F., and Weber, E., Neutrale organische komplex-liganden und ihre alkalikomplexe. II. Kronenäther, cryptanden und podanden als reagensien und katalysatoren (Teil A), *Kontakte*, 1977, No. 2, 16.
213. Regen, S. L., Dreiphasen-katalyse, *Angew. Chem., 91*, 464, 1979; *Angew. Chem. Int. Ed. Engl., 18*, 421, 1979.
214. Yanagida, S., and Takahashi, K., Losid-solid-liquid three phase transfer catalysis of polymer bound acyclic poly(oxyethylene) derivatives. Application to organic synthesis, *J. Org. Chem., 44*, 1099, 1979.
215. Mac Kenzie, W. M., and Sherrington, D. C., Nucleophilic substitution at determined carbon atom catalysed by supported podands, *Polymer, 21*, 791, 1980.
216. Neffermann, J. G., Mac Kenzie, W. M., and Sherrington, D. C., Non-supported and resin-supported oligo(oxyethylene)s as solid-liquid phase transfer catalysts. Effect of chain length and head groups, *J. Chem. Soc., Perkin-Trans.* 2, 514, 1981.
217. Markel, G., and Beier, M., Halogen exchange reactions at the silicon atom catalysed by crown compounds, *J. Organomet. Chem., 173*, 129, 1979.
218. Nordlander, J. E., and Catalano, D., Crown ether as catalyst of N-alkylation reactions, *Tetrahedron Lett.*, 1978, 4987.
219. Guido, W., and Mathre, D. J., Phase-transfer alkylation of heterocycles in the presence of 18-crown-6 and potassium *tert*-butoxide, *J. Org. Chem., 45*, 3172, 1980.
220. Mariani, G., Modena, G., Pizzo, G. P., and Scorrano, G., The effect of crown-ethers on the reactivity of alkoxides. Part 2. The reaction of potassium isopropoxide and 2,4-dinitrohalogenobutanes in propan-2-ol-benzene, *J. Chem. Soc., Perkin Trans.* 2, 1187, 1979.

221. Pettit, G. R., Blazer, R. M., Eirek, J. J., and Yamauki, K., Structural biochemistry. 20. Methylation of purine nucleosides, *J. Org. Chem.*, *15*, 1073, 1980.
222. Belsky, L., "Naked" fluoride catalysed Michael-additions, *J. Chem. Soc., Chem. Commun.*, 1977, 237.
223. Soga, K., Hosada, N., and Ikeda, S., A new synthetic route to polycarbonate, *J. Polym. Sci., Polym. Lett. Ed.*, *15*, 611, 1977.
224. Sega, K., and Hosada, S., New condensation catalyst for polymer synthesis, *J. Polym. Sci., Polym. Lett. Ed.*, *17*, 517, 1979.
225. Cuomo, J., and Olefson, R. A., An efficient and convenient synthesis of fluoroformates and carbonyl fluorides, *J. Org. Chem.*, *44*, 1016, 1979.
226. Bianchi, T. A., and Gate, I. A., Fluoride assisted acid derivatives transformations, *J. Org. Chem.*, *42*, 2031, 1977.
227. Effenberger, F., and Konig, G., Synthesis of phosphoryl fluorides assisted by crown compounds, *Synthesis*, 1981, 70.
228. Vögtle, F., New ligand systems for ions and molecules and electronic effect upon the reactions, *Pure Appl. Chem.*, *52*, 2405, 1980.
229. Susuki, H., and Koga, K., *Heterocycles*, *12*, 1305, 1979.
230. Gram, D. J., and Gogan, G. B. T., Chiral crown complexes catalyse Michael addition reaction to give adducts in high optical yields, *J. Chem. Soc., Chem. Commun.*, 1981, 624.
231. Akabari, S., Ohtomi, N., and Yatabe, S., Two-phase Darzene condensation reaction with octopus compounds as catalysts, *Bull. Soc. Chem. Japan*, *53*, 1463, 1980.
232. Dehmlow, E. W., and Lissel, H., Alkyne synthesis using *tert*-BuOH-18C6 crown ether-petroleum ether as catalytic system, *Liebigs Ann.*, 1980, 1.
233. Dehmlow, E. W., and Lissel, H., Reactions conditions of alkyne synthesis under the influence of crown ether, *Tetrahedron*, *37*, 1563, 1981.
234. Ome, S., and Takata, T., Potassium superoxide-18C6 system catalyse the synthesis of organic sulfur compounds, *Tetrahedron*, *37*, 37, 1981.
235. Lehn, J. M., Cryptate inclusion complexes: Effects on solute-solute and solute-solvent interactions and on ionic reactivity, *Pure Appl. Chem.*, *52*, 2303, 1980.
236. Lacoste, J., Schue, F., Bywater, S., and Kaempf, B., Use of cryptates in anionic polymerization: Polymerization of styrene, *J. Polym. Sci., Polym. Lett. Ed.*, *14*, 201, 1976.
237. Izatt, R. N., and Christensen, J. J., Eds., *Progress in Macrocyclic Chemistry*, vol. 1, Wiley, New York, 1979.
238. Goldberg, I., Complexes of crown ethers with molecular guests, in *Inclusion Compounds*, Atwood, J. L., Davies, J. E. D., and MacNicol, D. D., Eds., vol. 2, Academic Press, London, 1984, 261.
239. Vignier, M., Abadie, M., Schué, F., and Kaempf, B., Preparation and UV characterization of crown-ether complexes with alkaline metals, *Eur. Polymer J.*, *13*, 213, 1977.
240. Vignier, M., Collet, A., Schué, F., and Kaempf, B., Preparation and characterization of alkaline radical anions in THF, *J. Phys. Chem.*, *82*, 1578, 1978.
241. Deffieux, A., Oligomerization of propylene sulfide catalysed by crown compounds, *Polymer*, *18*, 1047, 1977.
242. Boileau, S., Polymerization of isobutylene sulfide under the influence of crown complexes, *Tetrahedron Lett.*, 1978, 1767.
243. Sigwalt, P., Polymerizations catalysed by crown ether complexes, *J. Polym. Sci., Polymer Symposia*, *62*, 51, 1978.
244. Weber, E., and Vögtle, F., Crown-type compounds—an introductory overview, in *Host-Guest Complex Chemistry*, Vögtle, F., Ed., Springer Verlag, Berlin, 1981, 11.
245. Hemery, P., Warzelhan, V., and Boileau, S., Kinetics of ring opening of propylene sulfide. 1. Alkali metal and cryptated metal carbonyl, *Polymer*, *21*, 77, 1980.

246. Suparne, S., Crown-ethers-alkali metal complexes as catalyst of ε-caprolactam polymerization, *Polymer J., 13*, 313, 1981.
247. Tada, M., and Hirane, N., Photochemical behaviour of crown-ether type compounds, *Tetrahedron Lett.*, 1978, 5111.
248. Turre, N. J., Von Nicholas, J., Turre, J., Gratzil, M., and Braun, A. M., Photopphisikalische und photochemische processes in micellaren systemen, *Angew. Chem., 92*, 712, 1980.
249. Weber, E., Progress in crown ether chemistry, Part IV.E, New applications of crown compounds in chemical analysis, *Kontakte*, 1978, No. 2, 16.
250. Sekine, T., and Hasegawa, Y., Solvent extraction by crown ethers, *Kagakuna Ryoiki, 33*, 464, 1979.
251. Takeda, Y., The solvent extraction of metal ions by crown compounds, in *Host-Guest Complex Chemistry*, Vögtle, F., and Weber, H., Eds., Springer Verlag, Berlin, New York, 1984.
252. Blasius, E., and Jansen, K. P., Analytical applications of crown compounds and cryptands, in *Host-Guest Complex Chemistry*, Vögtle, F., Ed., Springer-Verlag, Berlin, Heidelberg, New York, Tokyo, 1985, 189.
253. Moody, G. J., and Thomas, J. D. E., Eds., *Selective Ion Sensitive Electrodes*, Morrow Publishing, Watford, 1971.
254. Cammann, K., Ed., *Das arbeiten mit ionenselektiven elektroden*, Springer-Verlag, Berlin, 1973.
255. Kessler, M., Clark, L. C., Lubbers, D. W., Silver, I. A., and Simon, W., Eds., *Ions and Enzyme Electrodes in Biology and Medicine*, Urban and Schwartzenberg, Munchen, 1976.
256. Koryta, J., Ed., *Ion-Selective Electrodes*, Wiley, Chichester, New York, Brisbane, Toronto, 1980.
257. Takagi, M., and Ueno, K., in *Host-Guest Complex Chemistry*, Vögtle, F., and Weber, E., Eds., Springer Verlag, Heidelberg, New York, 1984, 39.
258. Takagi, M., and Uene, K., in *Host-Guest Complex Chemistry, Macrocycles Synthesis, Structure, Applications*, Vögtle, F., and Weber, E., Eds., Springer Verlag, Berlin, Heidelberg, New York, Tokyo, 1985, 217.
259. Löhr, E. C., and Vögtle, F., *Chromo- and Fluoro-Ionophores. A New Class of Dye Reagents, Acc. Chem. Res., 18*, 65, 1985.
260. Weber, E., and Vögtle, F., Neutrale organische complex-liganden und ihre alkalicomplexe, II. Kronenäther und kryptanden als reagenzien und katalysateren (Teil C), *Kontakte*, 1978, No. 2, 16.
261. Vögtle, F., Weber, E., and Elben, U., Neutrale organische complexliganden und ihre alkalimetalcomplexe, III. Biologische wirkungen synthetischer und naturlicher ionophore (Teil A), *Kontakte*, 1978, No. 3, 32.
262. Izatt, R. M., Lamb, J. D., Katough, D. J., Christensen, J. J., and Rytting, J. H., Design of selective ion binding macrocyclic compounds and their biological applications, *Drug Design, 8*, 356, 1979.
263. Lindenbaun, S., Rytting, J. H., and Sterason, L. A., Ionophores—biological transport mediations, in *Progress in Macrocyclic Chemistry*, Izatt, R. M., and Christensen, J. J., Eds., vol. 1, Wiley, New York, 1979, 225.
264. Tobushi, I., and Shimokawa, K., Model approach to retinal pigments; remarkable red shift to proximal ammonium ion, *J. Am. Chem. Soc., 102*, 5400, 1980.
265. Dietrich, B., in *Cryptate Complexes in Inclusion Compounds*, Atwood, J. L., Davies, J. E. D., and Mac Nichol, D. D., Eds., vol. 2, Academic Press, London, 1984, 337.
266. Vögtle, F., New ligand systems for ions and molecules and electronic effects upon complexation, *Pure Appl. Chem., 52*, 2405, 1980.
267. Dix, J. F., and Vögtle, F., Neue chromoionophere, *Chem. Ber., 114*, 638, 1981.

268. Merck, GmbH, Eur. Pat. Appl., 83,100,281, 1983.
269. Nakashima, K., and Nakataugi, S., A new colorimetric determination of Li^+ using crown type compounds, *Chem. Lett.*, 1982, 1781.
270. Kolthoff, I. E., Application of macrocyclic compounds in chemical analysis, *Analyt. Chem., 51*, 1R, 1979.
271. Pedersen, C. J., Macrocyclische polyäther und ihre complexe, *Angew. Chem.*, 84, 16, 1972.
272. Cox, B. G., Schneider, H., and Stroka, J., Kinetics of alkali metal complex formation with cryptand in methanol, *J. Am. Chem. Soc., 100*, 4746, 1978.
273. Cox, B. G., Knop, D., and Scheider, H., Kinetics of the protolysis of cryptande in basic aqueous solution, *J. Am. Chem. Soc., 100*, 6002, 1978.
274. Simon, W., in *Molecular Movements and Chemical Reactivity as Conditioned by Membranes, Enzymes and Other Macromolecules*, Lefever, E., and Goldbeter, A., Eds., Wiley, 1978, 282.
275. Protech, E., and Banda, J., *Proc. Symp. Steric Effects in Biomolecules*, Eger, Hungary, 1981.
276. Anker, F., Wieland, E., Ammann, D., Bohner, R. E., Asper, R., and Simon, W., Neutral carrier based ion-selective electrode for the determination of total calcium in blood serum, *Anal. Chem., 52*, 2400, 1980.
277. Lanter, F., Erne, D., Ammann, D., and Simon, W., Neutral carrier based ion-selective electrode for intracellular magnesium activity studies, *Analyt. Chem., 52*, 2400, 1980.
278. Steiner, N. A., and Oehme, M., Neutral carrier based ion-selective electrode for intracellular determination of sodium concentration, *Analyt. Chem., 51*, 351, 1979.
279. Guggi, M., and Kessler, M., in *Frontiers of Biological Energetics*, Dalton, P. L., Ed., vol. 2, Academic Press, New York, 1978.
280. Osswald, E. F., Continuous determination of potassium in blood, *Clin. Chem., 25*, 32, 1979.
281. Pedersen, C. J., Cycloethers, *J. Am. Chem. Soc., 89*, 2495, 1967.
282. Lehn, J. M., Cryptate: Macrocyclic inclusion complexes, *Pure Appl. Chem., 49*, 857, 1977.
283. Cram, D. J., and Trueblood, E. N., Concept, structure and bonding in complexation, in *Host-Guest Complex Chemistry-Macrocycles*, Vögtle, F., and Weber, E., Eds., Springer-Verlag, Berlin, 1985, 125.
284. Lehn, J. M., Chemistry of transport processes—design of synthetic carrier molecules, in *Physical Chemistry of Transmembrane Ion Motion*, Spach, G., Ed., Elsevier, Amsterdam, 1983, 181.
285. Breslow, D. S., Cyclodextrine as carrier of enzymes, *Pure Appl. Chem., 46*, 103, 1976.
286. Achenbach, G., Hauswirth, G., Kossman, J., and Ziskoven, R., Cyclic compounds effects upon the sheep cardiac, *Physiol. Chem. Phys., 12*, 277, 1980.
287. Elban, U., Drug analogous crown compounds of isoprenaline and eupaverine, *Liebigs Ann. Chem.*, 1979, 1102.
288. Voronkov, M. G., Antiulcerogenic effect of analogous crown compounds, *Khim. Farm. Zhur., 19*, 819, 1985 (Russian).
289. Izatt, R. M., Lamb, D. J., Eatough, D. J., Christensen, J. J., and Rytting, J. H., Design of selective ion binding macrocyclic compounds and their biological applications, *Drug Design, 6*, 356, 1979.
290. Ringsdorf, H., Synthetic polymeric drugs, *Med. Macromol. Monogr., 5*, 197, 1975.
291. Schilpköter, H. W., and Brohaus, A., Poly(vinylpyridine)-N-oxide activity against silicosis, *Fortschr. Staubhangenforsch.*, 1963, 397.
292. Holt, P. F., and Lindsay, H., Medical applications of isopropylpyridine-N-oxides, *J. Chem. Soc., B*, 1969, 54.

293. Allison, A. C., Medical application of vinylic polymers, *Arch. Int. Med. Dig., 128*, 131, 1971.
294. Uglea, C. V., Ottenbrite, R. M., Offenberg, H., Grecianu, A., and Negulescu, I. I., Anesthesine modified maleic anhydride cyclohexyl-l,s-dioxepine copolymer. Preparation and potential medical application, *12th Annual Int. Conf. Eng. Med. Biol. Soc.*, Philadelphia, 1990.
295. Abel, G., Connors, T. A., Hofman, V., and Ringsdorf, H., Pharmacologically active polymers, 13. Elucidation of the tumor affinity of poly(sulfadiamine acrylamide), *Makromol. Chem., 177*, 2669, 1976.
296. Ringsdorf, H., *Biological Aspects of Radiation Protection*, Brow, Boston, 1963, 1.
297. DeDuve, G., *Lysosomes*, Brow, Boston, 1963, 63.
298. Poole, A. E., in *Lysosomes in biology and Pathology*, vol. 3, Dingle, J. T., Ed., North-Holland Publ. Co., Amsterdam, 1973, 303.
299. Jaques, P. J., in *Lysosomes in Biology and Pathology*, vol. 1, Dingle, J. T., Ed., North-Holland Publ. Co., Amsterdam, 1969, 365.
300. Brebnik, J., and Rypacek, F., Soluble synthetic polymers in biological systems, *Adv. Polym. Sci., 57*, 1, 1984.
301. DeDuve, G., in *Biological Approaches to Cancer Chemotherapy*, Harris, E. J. C., Ed., Academic Press, London, 1961, 101.
302. Stanford, A. L., in *Foundation of Biophysics*, Academic Press, New York, 1974, 113.
303. Lehninger, A. L., *Biochemistry*, 2nd ed., Worth Publ., New York, 1975.
304. Heckman, K., Zur Theorie der "single file" Diffusion, *Z. Physik. Chem., 44*, 184, 1965.
305. Blumenthal, R., and Katchalsky, A., The effect of carrier association-dissociation rate on membrane permeation, *Biochim. Biophys. Acta, 173*, 357, 1969.
306. Heinz, K., and Walsh, F. M., Exchange diffusion, transport, and intracellular level of glycine and related compounds, *J. Biol. Chem., 233*, 1488, 1958.
307. Osterhout, W. J. V., Some aspects of transmembranar transport, *Ergeb. Physiol. 35*, 967, 1933.
308. Simon, W., and Morf, W. S., in *Membrane—A Series of Advances*, vol. 2, Riseman, G., Ed., Dekker, New York, 1972.
309. Pioda, I. A. R., Dohner, M. A., Wachter, R. B., and Simon, W., Complexe von nonactin und monactin mit natrium, potassium und amonium ionen, *Helv. Chim. Acta, 50*, 1373, 1967.
310. Onishi, M., and Urry, D. W., Spectroscopic determination of the Valinomycin-K^+ complex structure, *Science, 168*, 1091, 1970.
311. Ovchinikov, Yu. A., Ivanov, V. T., and A. M. Shkrob, A. M., *Membrane-Active Complexes*, Elsevier, Amsterdam, 1974.
312. Debler, M., Dunitz, J. D., and Silbaum, B. T., Die struktur des KNCS-Complexes von nonactin, *Helv. Chim. Acta, 52*, 2473, 1969.
313. Debler, M., Phizachkerley, R. F., *Helv. Chim. Acta, 57*, 664, 1974.
314. Neupert-Laves, K., and Debler, M., The structure of tetrolides-cations complex, *Helv. Chim. Acta, 59*, 614, 1976.
315. Morf, W. E., and Simon, W., Berechung von freien hydration-enthalpien und koordinationszahlen für kationen aus leight zugglichen parameters, *Helv. Chim. Acta, 54*, 2683, 1971.
316. Szabo, G., Eiseman, G., and Ciani, S., The investigation with spectroscopic methods of nactin-cation complexes, *J. Membr. Biol., 1*, 2683, 1971.
317. Gertenbach, P. G., and Popov, A. T., Solution chemistry of monensin and its alkali metal complexes. Potentiometric and spectroscopic studies, *J. Am. Chem. Soc., 97*, 4738, 1975.

318. Ebata, W., Kasahara, H., Sekine, M., and Inine, V., Lysocellin, a new polyether antibiotic. I. Isolation, purification, physico-chemical and biological properties, *J. Antibiot., 28*, 118, 1975.
319. Liu, C., and Hermann, T. E., Ionomycin—a new polyether compound with biological activity, *J. Biol. Chem., 253*, 5892, 1978.
320. Liu, C., X-14547A, a new ionophore antibiotic produced by *Streptomyces antibioticus*. Discovery, fermentation, biological properties of the producing culture, *J. Antibiot., 32*, 95, 1979.
321. Mueller, P., Rudin, P. O., Tion, N. T., and Wescott, W. C., Method for the formation of artificial membranes, *Nature, 194*, 979, 1962.
322. Mueller, P., Rudin, P. O., Tion, N. T., and Wescott, W. C., Methods for the formation of single bimolecular lipid membranes in aqueous solution, *J. Phys. Chem., 67*, 534, 1963.
323. Liberman, E. A., and Topaly, V. T., Electrical resistance of bilayer membranes, *Biochim. Biophys. Acta, 163*, 125, 1968.
324. Morf, W. E., and Simon, W., Thermodynamic parameters of the ring compounds-alkali complexes in aqueous solvent, *Helv. Chim. Acta, 54*, 794, 1971.
325. Kraane, S., and Eismann, G., in *Membrane—A Series of Advances*, vol. 2, Eisemann, G., Ed., Dekker, New York, 1972.
326. Katterer, B., Neumcke, P., and Länger, P., Potential energy profile of hydrophobic ions, *J. Membrane Biol., 5*, 225, 1971.
327. Nemcke, P., and Länger, P., Electrical image forces of the membrane-aqueous phase interphase, *Biophys. J., 9*, 1160, 1969.
328. Länger, P., and Stark, G., The kinetic analysis of transmembranar transport, *Biochim. Biophys. Acta, 211*, 458, 1970.
329. Stark, G., Katterer, B., Benz, R., and Länger, P., Mathematical analysis of the facilitated diffusion, *Biophys. J., 11*, 981, 1971.
330. Lesslauer, W., Richter, P., and Länger, P., Thermodynamic and some electrical properties of bimolecular phosphatidylinositol membrane, *Nature, 214*, 1224, 1964.
331. Maugh, W. W. H., Liquid membrane: New techniques for separation and purification, *Science, 193*, 334, 1976.
332. Newcomb, M., Termar, W. L., Regelson, N. C., and Cram, D. J., Host-guest complexation. 20. Chiral recognition in transport as a molecular basis for a catalytic resolving machine, *J. Am. Chem. Soc., 101*, 4941, 1979.
333. Lamb, J. D., Christensen, J. J., Izatt, S. R., Bedke, K., Austin, M. S., and Izatt, R. M., Effects of salt concentration and anion on the rate of carrier-facilitated transport of metal cations through bulk liquid membrane containing crown ether, *J. Am. Chem. Soc., 102*, 3399, 1980.
334. Shiube, T, Kurihara, K., Kabataki, T., and Kamo, M., Active transport of picrate anion through organic liquid membrane, *Nature, 270*, 277, 1977.
335. Merebaka, A., Wodzki, R., and Wyozynska, A., Liquid membrane with organic polyphosphates as ionic carriers, *Makromol. Chem., 190*, 1901, 1989.
336. Conolly, J. S., *Photochemical Conversion and Storage of Solar Energy*, Academic Press, New York, 1981, 444.
337. Lamb, J. D., Izatt, S. R., Robertson, P. A., and Christensen, J. J., Highly selective membrane transport of Pb^{2+} from aqueous metal ion mixtures using macrocyclic carriers, *J. Am. Chem. Soc., 102*, 2452, 1980.
338. Nagasaki, M., Highly selective membrane transport of heavy metal ions, *J. Org. Chem., 47*, 2429, 1982.
339. Tsukube, H., Maruyama, K., and Araki, T., Active and passive transport of amino acid derivatives via metal complex carriers, *Tetrahedron Lett.*, 1981, 2001.
340. Bacon, E., and Jung, L., *J. Chem. Res.*, 1980, 136.

341. Maruyama, K., Tsukube, H., and Araki, T., Carrier-mediated transport of amino-acid and simple organic amines by lipophilic metal complexes, *J. Am. Chem. Soc., 104*, 9197, 1982.
342. Newcomb, E., Toner, J. C., Regelson, G., and Cram, D. J., Host-guest complexation. 21. *J. Chem. Soc., Perkin Trans. 2*, 1485, 1983.
343. Christensen, J. J., Lamb, J. D., Izatt, J. R., Starr, S. E., Weed, G. C., Astin, M. E., Stitt, B. D., and Izatt, R. M., Effect of anion type on rate of facilitated transport of cations across liquid membranes via neutral macrocyclic carriers, *J. Am. Chem. Soc., 100*, 3219, 1978.
344. Shehari, E., and Jagur-Grodzinski, M., *J. Appl. Polym. Sci., 20*, 773, 1976.
345. Sheheri, E., and Jagur-Grodzinski, J., *J. Appl. Polym. Sci., 20*, 1665, 1976.
346. Kawakami, M., and Iwanaga, H., *Chem. Lett.*, 1980, 1445.
347. Araki, T., and Tsukube, H., Oligomeric N-(phenylcarbonyl)-ethylenimine for highly specific absorption of mercury(II) and copper(II) ions, *Macromolecules, 11*, 250, 1978.
348. Tsukube, H., Araki, T., Inoue, H., and Nakamura, A., Nonclassical urea oligomers. VI. Oligomeric N-(phenylthiocarbamoyl)-ethylenimine: A new selective absorbant of Cu^{2+} and Hg^{2+} ions, *J. Polym. Sci., Polym. Lett. Ed., 17*, 437, 1979.
349. Tsukube, H., Nakamura, A., and Maruyama, K. Synthesis and metal binding properties of novel sulfur-containing functional oligomer, *J. Polymer Sci., Polym. Chem. Ed., 18*, 1359, 1980.
350. Tsukube, H., *J. Chem. Soc., Perkin Trans. 1*, 2359, 1982.
351. Tsukube, H., *J. Chem. Soc., Perkin Trans. 1*, 29, 1983.
352. Tabushi, I., and Kobuke, M. *J. Am. Chem. Soc., 103*, 6152, 1981.
353. Behr, J. P., and Lehn, J. M., Transport of amino acids through organic liquid membranes, *J. Am. Chem. Soc., 95*, 6108, 197.

5

Oligoesters

I. OLIGO(ETHYLENE TEREPHTHALATE)

Poly(ethylene terephthalate) is produced by polycondensation of bis(hydroxyethyl)terephthalate (BHET) or its oligomers. BHET may be synthesized both by the reaction of dimethyl terephthalte (DMT) and ethylene glycol (EG) and by the direct esterification of EG with terephthalic acid. Although direct esterification has recently gained importance, the DMT method remains the main process for obtaining BHET. This latter process provides the formation of DMT solution in EG, the transesterification of DMT with EG and distillation of methanol with formation of BHET, and finally, BHET polycondensation. The DMT:EG molar ratio is 1:2 with an excess of EG (0.2 to 0.5 mol). Preheating of EG to 120 to 160°C and introduction of DMT in the molten state shortens the DMT dissolving time.

The DMT transesterification with EG takes place in the presence of catalyst introduced in amounts that represent 0.02 to 0.08% vs. DMT mass and at 165 to 200°C [1]. The catalyst is presolved in EG, 2 to 3 parts catalyst to 100 parts EG. Bivalent metal (Zn, Mn, Co) acetates are used in the main. Catalyst solution is introduced into the system under powerful stirring with a view to prevent large concentration areas of the catalyst; otherwise, violent degassing of methanol can occur.

The transesterification step can last 3 to 4 h, so that 90 to 95% of the theoretical methanol amount is liberated (about 330 kg methanol per 1000 kg DMT). The reaction product formed contains 40% BHET, 30% dimer, and 30% oligomers with low molecular weight [2]. When the transesterification step is over, excess EG is distilled at 220 to 245°C and then polycondensation catalyst is introduced (in the form of EG solution) along with other additives. Some metal oxides (TiO_2, SbO_3, or GeO_2) are used as polycondensation catalysts. The use of a Bronsted acid (H_2SO_4) as the catalyst diminishes the oligomer content in the polymer [3].

In what follows, we shall deal particularly with the analysis of the transesterification reaction, considered as a source of oligomer yield. At the same time, one should not overlook degradation processes that take place concomitantly with the polycondensations. The main result of the latter is the formation of some species with carboxylic end groups [4]. These degradation processes can be intensified by certain additives, for example, dyestuff [5].

Let us now examine the intrinsic causes of oligomer formation during the transesterification reaction. In the transesterification process, the extent of reaction has traditionally been followed by measuring the methanol withdrawn from the reactor. It has been shown [6] that a model based on the ester interchange reaction accounts for the methanol data obtained under a wide range of experimental conditions. Additionally, some attempts at using methanol data to elucidate the existence of oligomerization has been reported [7,8]. Oligomerization is produced through the following reactions:

$$\underset{E_m}{\sim\!\!-C_6H_4-COOCH_3} + HOCH_2CH_2OH \overset{k_1}{\longleftrightarrow}$$

$$\underset{E_g}{\sim\!\!-C_6H_4-COOCH_2CH_2OH} + \underset{M}{CH_3OH} \tag{1}$$

$$\underset{E_m}{C_6H_4-COOCH_3} + \underset{E_g}{HOCH_2CH_2OOC-C_6H_4-\sim} \overset{k_2}{\longrightarrow}$$

$$\underset{Z}{\sim\!\!-C_6H_4-COOCH_2CH_2OOC-C_6H_4-} + \underset{M}{CH_3OH} \tag{2}$$

$$2\ \underset{E_g}{\sim\!\!-C_6H_4-COOCH_2CH_2OH} \overset{k_3}{\longrightarrow}$$

$$\underset{Z}{\sim\!\!-C_6H_4-COOCH_2CH_2OOC-C_6H_4-\sim} + HOCH_2CH_2OH \tag{3}$$

To fit his methanol data, Fontana [7] developed a model based on reactions (1) and (2). He considered that the reactivity of OH groups of EG was twice that of OH end groups of E_g and assumed that this ratio was independent of temperature. This allowed him to define an average reactivity of OH end groups, which he used in parameter estimation. Yamanis and Adelman [8] pointed out that Fontana [7] forced the system to have a given degree of oligomerization by assuming a reactivity ratio $k_2/k_1 = 0.5$. Therefore, they extended Fontana's approach to the case with reactivity ratio an adjustable parameter. Yamanis and Adelman fitted the nonisothermal data of Fontana by using the differential method of data analysis. They found that the data were well fitted by reactivity between 0.125 and 1.0. Since higher values of the reactivity ratio meant significant oligomerization and low values did not, the previously mentioned results did not allow any conclusion to be reached about the importance of oligomerization. Also, Yamanis and Adelman concluded that oligomerization was negligible.

To use the elegant approach proposed by Fontana [7] and Yamanis and Adelman [8], isothermal data are needed because the definition of the average reactivity of OH end groups implies that k_2/k_1 should be independent of temperature. However, because of the experimental method used in these experiments, the reactions were realized in a highly nonisothermal way. Contrary to what previous authors have reported, Baradiaran and Asua [9] concluded that more observed variables are needed to elucidate the significance of oligomerization reactions. To do this, more information, namely, the number of groups Z, needs to be available. The degree of oligomerization during the reaction of DMT and EG was determined experimentally by measuring the number of ethylene diester groups (Z) by Fourier transform infrared (FTIR) spectroscopy. It was found that [10], contrary to what was previously been reported based on methanol data [6–8], significant oligomerization did occur during the reaction of DMT with EG. In addition, the degree of oligomerization was found to increase as the reaction temperature increased and the ratio DMT/EG decreased.

Other experimental factors have been identified as influencing the formation of oligomers in the process of poly(ethylene terephthalate) synthesis. Thus, Meyer [11] and Ramm [12] consider a model of the transesterification reaction in a continuous process. Data obtained have led to the elaboration of best reaction conditions. Among these, Meyer noticed the reactor geometry. Figure 1 shows the principal diagram of the plant for oligo(ethylene terephthalate) production in a continuous process.

The reactants (DMT and EG) together with the transesterification catalyst are introduced continuously in the horizontal reactor 1, the geometry of which is determined by an *L/D* ratio ≥ 4 with a view to providing enough contact time. Reaction mixture in the form of melt has sufficiently low depth so that gaseous methanol may liberate as soon as it formed. The degassing process also provides for reaction mass stirring.

The reactor is separated by punched metal walls (2) in four reaction areas built in such a way that they allow melt circulation but prevent mixing of the valors that liberate from the melt. Above each reaction area there is a filler column that provides for the purification of methanol vapors. These are then passed through condenser 3 and collected into vessel 4.

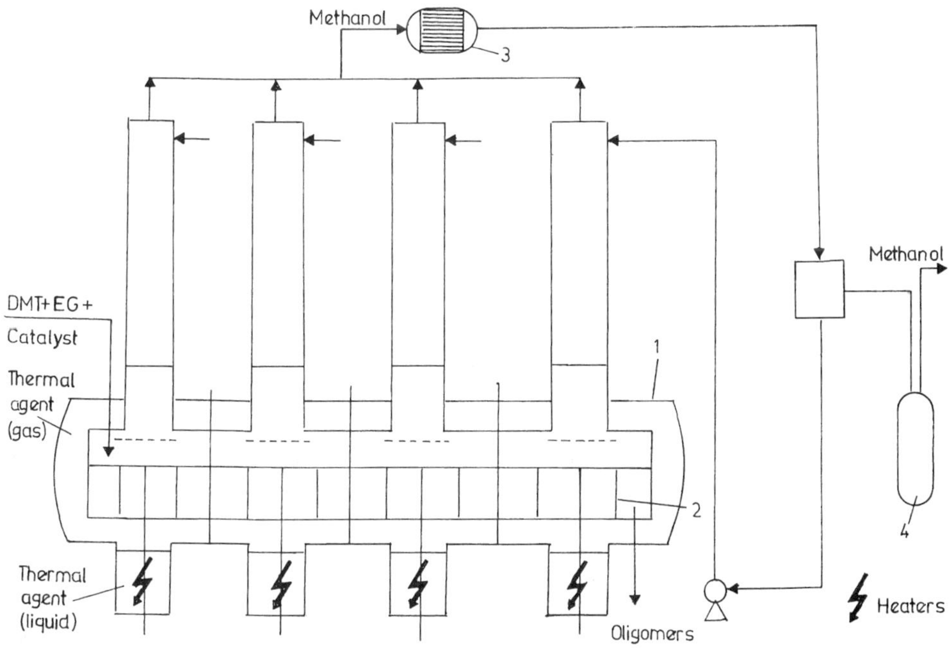

Figure 1 Basic layout of a pilot plant for production of oligo(ethylene terephthalate)s in a continuous process. 1: Polycondensation reactor; 2: punched metal walls; 3: condenser; 4: collector. (Adapted from Ref. 11.)

In polymer dope dyeing, the dyestuff is introduced into the polymerization system during the synthesis. For this purpose, either reactive or inert dyes may be used. In the first case, the dyestuff becomes a part of the main chain by combination at some suitable functional groups. This process is not used industrially. In the second case, the dyestuff is dissolved in polymer molten mass and a physical blend of macromolecular product and dyestuff is obtained. This last process is applied in the industry for poly(ethylene terephthalate) (PET) dope dyeing. The difference between dope-dyed polyester and colorless polyester obtained in the same experimental conditions is evidenced by a lower intrinsic viscosity and a higher number of COOH end groups and diethylene glycol (DEG) units. These results have been interpreted by Uglea and Stan [5] as being a result of higher content of oligomers produced due to the influence of dyestuff. Since the coloring agents used have a 1,4-aminoanthraquinone structure in which hydrogen atoms of the amine groups are partially substituted for by different radicals, the higher content of COOH end groups in colored PET may be explained easily. The basic character of the aromatic amines is decreased by $p-\pi$ conjugation that involves the unpaired electrons of nitrogen and the electrons of the benzene ring [13]. This process gives an extra positive charge at the nitrogen atom, which induces polarization of the ester link. The following reaction may be written:

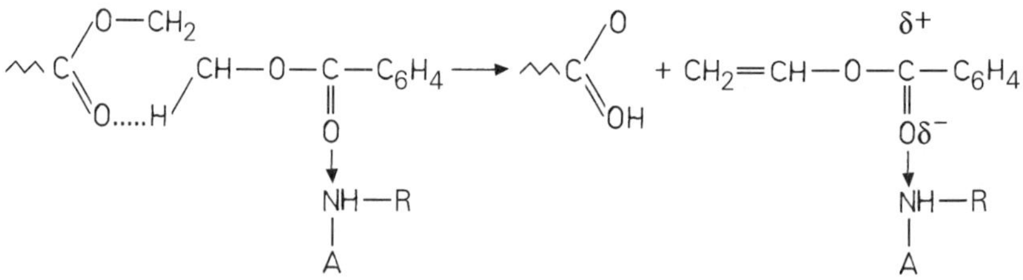

(4)

A = $—C_{10}H_4NH—R$; R = 1,3,5-trimethylphenyl or $—CO—C_6H_5$.

The gel permeation analysis of the polycondensation process reveals some interesting facts concerning the occurrence of oligomers [5]. These data show the systematic influence of dyestuff on the value and distribution of molecular weight in dope-dyed PET. The analysis of product characteristics shows that in the first stages of the polycondensation process, the dyed polymer has its molecular weight distribution shifted to the range of higher molecular weights. The same results are obtained from the values of M_n and M_w. In the subsequent stages of polycondensation, the presence of dyestuff is accompanied by a systematic shift of the distribution curves of colored PET to the range characteristic of the low molecular weight species. This is shown by the position of the gel chromatograms. The change caused by the dyestuff in the direction of the value of the molecular weight and molecular weight distribution of colored PET is the result of the mechanism that dominates the interaction between polymer and dyestuff. In the first stage of polycondensation, the main influence of the dyestuff is to decrease the melt viscosity (in industrial autoclaves, this influence is seen by the decrease of the power required to drive the stirrer). The decrease of melt viscosity is not accompanied by a reduction of polymer molecular weight, but it is more a consequence of PET melt plasticization caused by the presence of the dyestuff in the system. As a result of the easier removal of EG from the system due to the decrease of melt viscosity, the molecular weight of the colored polymer is higher than that of the colorless one. In the further stages of the polycondensation, the presence of the dyestuff is accompanied especially by a supplementary degradation effect evidenced in the dyed samples by the diminished value of the molecular weight, and by the higher amount of COOH end groups and DEG. The presence of the dyestuff in PET results in the appearance in the gel chromatograms of some specific peaks in the domain of low molecular weight species. These peaks correspond to M_n values of 1100, 2560, and 1600. The possibility of separating the dyestuff from the polymer by gel permeation chromatography supports the conclusion that there are no covalent bonds between polymer and dyestuff.

The introduction of some "colored structures" in the PET backbone by using the corresponding dyestuffs affords the synthesis of some colored products without altering the characteristics of the white polymer [13–23].

By direct esterification or transesterification of dicarboxylic acids with various glycols in the presence of monocarboxylic acids and alcohols, oligoesters are produced. As catalysts, benzene- and toluenesulfonic acids are used as well as carboxylic acids salts or tetrabutoxytitane. The main use of these is in the plasticizing of various plastics [24,25].

II. OLIGO(BUTYLENE TEREPHTHALATE) AND RELATED OLIGOESTERS

Oligo(butylene terephthalate) is obtained by DMT transesterification with butanediol or by direct esterification of terephthalic acid with same diol. The first process is catalytic and comprises two steps: formation of bis(β-oxybutylterephthalate), and its condensation in order to obtain poly- or oligo(butylene terephthalate). As catalysts, the organotitanium compounds used are introduced into the process at the beginning or during the reaction. The working temperature for the transesterification reaction is 130 to 250°C, whereas the polycondensation requires higher temperature (250 to 300°C) and low pressure.

Zimmer Comp. (Düsseldorf, Germany) and Mitsubishi Rayon Corp. (Osaka, Japan) claimed a continuous process for the production of oligo(butylene terephthalate) using catalysts based on organotitanium compounds [26]. The key point in these technologies is to produce the polycondensation at a low pressure with a view to diminishing the tetrahydrofuran content of the reaction product.

By introducing some glycols or polyols with various structures or molecular weights into the system [27–34], numerous products with various and important applications have been obtained. These include protective coatings [35–37], materials with improved mechanical properties [38,39], hot melt compositons [40], and paints [41].

III. UNSATURATED OLIGOESTERS

The introduction into the process of esterification or transesterification of some unsaturated dicarboxylic acids (or their anhydrides) leads to the production of unsaturated oligoesters, also called *polyester resins*. Mixtures of saturated and unsaturated dicarboxylic acids are equally commonly used. The reaction runs in an inert atmosphere at 100 to 250°C, under intensive stirring with continuous elimination of water. The reaction is catalyzed by organic acids, cationites, organotitanium compounds, and nitrogen- or phosphorus-containing organic compounds [42, 43].

Berzelius [44] prepared glyceryl tartrate, Smith prepared glyceryl phthalate resins in 1901, and these glyptal resins were patented in United States in 1914 [45]. Glycol maleate was prepared as early as 1894 [46], but the commercial development of unsaturated polyesters and fiberglass-reinforced composites was carried out at Pittsburgh Plate Glass and United States Rubber Co., beginning in the mid 1930s. The impetus for large-scale production come from the military and marine needs in World War II. More than 500 thousand tons of unsaturated polyester resins were produced in United States in 1978.

Unsaturated polyesters are usually liquid oligomers of phthalic anhydride, maleic anhydride, and a molar excess of propylene glycol, crosslinked by copolymerization with styrene. A general-purpose unsaturated polyester is composed mainly of phthalic acid or isophthalic acid as the saturated acid component, fumaric acid or maleic acid as the unsaturated acid component, and propylene glycol or a mixture of this glycol with other glycols, such as diethylene glycol, as the glycol component. Within these basic groups, there exist theoretically a broad range of compounds that can be employed to modify the polymer structure. The unsaturated dibasic acid amide was used for preparing three novel unsaturated polyesters with ethylene glycol, diethylene glycol, and tetraethylene glycol, in turn. These unsaturated polyesters were diluted with a styrene/acrylonitrile mixture to prepare curable resins with inhibited premature gelation [47].

Also interesting is the synthesis of unsaturated polyesters by copolycondensation of oligo(ethylene terephthalate) with maleic acid. The raw material oligo(ethylene terephthalate) for this process is obtained by degradation of residues coming from the textile industry [48]. Thus, high-flexibility lacquers [49], adhesives for the footwear industry [50], textile additives, and ingredients for lubricants and rubber have been obtained [51–55].

A group of acid-terminated unsaturated oligoesters having molecular weights of 500 to 1400 and incorporating a lactone as co-reactant with an aromatic, or partially or fully hydrogenated aromatic, dicarboxylic acid and an aliphatic diol are valuable plasticizers for poly(vinyl chloride) [56–59].

Polyfunctional oxazoline-terminated unsaturated oligoesters are useful in preparing coatings, caulks, impregnants, adhesives for textiles, leather, plastics, wood, and metals, and binding agents for pigments, fibers, and nonwoven fabrics [60–67].

The vinyl monomers used to effect crosslinking include styrene [68–70], or a combination of styrene with other copolymerizable monomers, such as α-methylstyrene, methyl acrylate, methyl methacrylate [71], diallyl phthalate [72], triallylcyanurate, and so forth.

Peroxide catalysts and ozonides are generally employed as initiators for crosslinking unsaturated polyesters. The formation of active radicals that can initiate the hardening copolymerization reaction may be begun also by ferrocene and some of its derivatives (acetyl ferrocene, benzyl ferrocene, etc.) [73]. Note that styrene-containing unsaturated polyesters will polymerize at room temperature without a catalyst. Such premature reaction can be effectively prevented by the use of inhibitors (hydroquinone, phenolic resins, aromatic amines, pyrogallol, picric acid, quinones, and quaternary ammonium compounds).

Maleated polyisobutylene was used as a toughener for unsaturated polyester resins by Abbati et al. [74]. The effect was found to depend on the grafting degree of the rubber and on the time period during which the two-component mixture was allowed to react prior to the curing process.

The properties of the cured polyester are largely imparted by the raw materials added, the reaction conditions, the type and concentration of the copolymerizable monomer or the method used for curing, and the curing conditions.

Normal unsaturated polyester resins have a poor adhesion on metals, especially on steel substrates, which is mainly due to styrene or vinylic monomers. Moreover, these monomers have some toxic and nauseating effects and, due to limited supplies, are becoming more expensive. The unsaturated polyester resin films cured by electron beams have mechanical properties that even exceed those of unsaturated polyester films that contain styrene or other vinylic monomers. It is supposed that this improvement, especially the higher elongation at break in combination with a high modulus, is essentially due to a much more uniform distribution of the crosslinks within the network [75,76].

Crosslinking occurs via free-radical polymerization. In the case of styrene-cured polyester resins, three types of radicals are formed: those of fumarate, of styrene, and of the catalyst. The catalyst is added in concentrations of 0.5–2.0%, depending on the type of catalyst and the temperature of the medium. It has been shown that styrene does not homopolymerize under the reaction conditions, and that no measurable amount of polystyrene is present in the crosslinked resin [77].

Beyond a molar ratio of styrene to fumarate of 2:1, more unreacted styrene is present, whereas the reacted fumarate portion remains relatively constant over a broader range of styrene concentrations.

An important factor in the conversion of unsaturated groups into crosslinked units is the degree of conversion of maleate into fumarate isomer. As a general rule, the higher the degree of isomerization of the maleate, the greater the proportion of reacted double bonds in the crosslinked structure [78].

The structure of a cured polyester can be described as follows: The network is composed of two polymolecular species, namely, the polycondensation chains and the copolymer chains. Both are connected with each other through covalent bonds with the fumarate groups; both types of chain polymers have the fumarate groups in common, thus forming a network.

Structural elements of unsaturated polyester resins that affect the properties of the cured products include molecular weight and molecular weight distribution, distribution of various repeating units, type of unsaturation, and ratio of acid to glycol components. An overall concept regarding the structural features of polyester resin and the performance properties of cured product is not available. However, it is known that the structural features determine the values of hardness, impact strength, heat-distortion temperature, water absorption, chemical resistance, and heat resistance.

The unsaturated polyester resins are used as fire retardants, binders, and powder coatings [79], and in the reinforced plastics industry. Fire-retardant polyester resins generally contain halogen in the acid or glycol component, or other groups imparting a degree of resistance to burning. In addition, to meet the stringent test requirements of fire retardancy a polyester must be appropriately compounded. To increase flame retardance, various compounding procedures are employed. These include the addition to the polyester resin of halogenated aliphatic hydrocarbons, inert fillers, antimony oxide, as well as partial substitution of styrene by halide-containing monomers. Unsaturated polyester resins containing SO or SO_2 groups in their chain also retard burning.

Unsaturated polyesters are the main binders for the 2-billion-pound reinforced plastics industry, and they are of importance in automobile bodies, boat hulls, building panels, chemical piping, and so forth. Indeed, if binders for reinforced plastics were classified as adhesives, the unsaturated polyesters would top the list.

REFERENCES

1. Petuhov, B. V., *Polyester Fibers*, Khimia, Moscow, 1976, 271 (Russian).
2. Korshak, V. V., *Technology of Plastic Materials*, Khimia, Moscow, 1985, 607 (Russian).
3. Smith, R. R., and Sartorelli, D. W., U.S. Patent 4,421,337, 1984.
4. Uglea, C. V., and Stan, V., The characterization of dope dyed poly(ethylene terephthalate), *Br. Polymer J., 14,* 39, 1982.
5. Uglea, C. V., Corduneanu, I., and Negulescu, I. I., Characterization of polyesters: Thermo-oxidative degradation, *Monatsch. Chem., 115,* 349, 1984.
6. Barandiaran, M. J., and Asua, J. M., Low molecular species in the reaction of dimethyl terephthalate with ethylene glycol, *J. Polym. Sci., Polymer Chem. Ed., 27,* 4241, 1989.
7. Fonatana, C. M., Polycondensation equilibrium and the kinetics of the catalyzed transesterification in the formation of polyethylene terephthalate, *J. Polymer Sci., A-1, 6,* 2343, 1968.
8. Yamanis, J., and Adelman, M., Condensation of dimethyl terephthalate with ethylene glycol. I. Low molecular weight species, *J. Polymer Sci., Polymer Chem. Ed., 14,* 1495, 1976.
9. Barandiaran, M. J., and Asua, J. M., Study of the significance of oligomerization in the reaction of DMT with EG by using methanol data, *Polymer, 31,* 1347, 1990.
10. Barandiaran, M. J., and Asua, J. M., Analysis of oligomerization in the reaction of DMT with EG, *Polymer, 31,* 1352, 1990.
11. Meyer, H., Zur reaktionsführung bei der kontinuerlichen herstelung der oligomeren des bis(β)-hydryäthyl)-terephthalates aus dimethylterephthalat, *Faserforsch. u. Textiltechnik, 22,* 341, 1971.
12. Ramm, H., GER Patent 2,200,832, 1973.
13. Nenițescu, C. D., *Organic Chemistry*, 7th ed., vol. 1, Editura Didactică şi Pedagocică, Bucharest, 1973, 567.
14. Maréchal, E., Macromolecular dyes. Oligomeric and unsaturated dyes for UV curing, *Pure Appl. Chem., 52,* 1932, 1980.
15. Uglea, C. V., Offenberg, H., and Cascaval, A., Synthesis and characterization of self-colored poly(ethylene terephthalate), *Eur. Polymer J., 21,* 681, 1985.
16. Uglea, C. V., and Mihăescu, A., Unperturbed dimensions and intrinsic viscosity-molecular weight relationship of copolyesters poly(ethylene terephthalate-co-ethylene isophthalate), *J. Chem. Tech. Biotech., 35A,* 1, 1985.
17. Radovici, A., Stan, V., and Uglea, C. V., Gas-chromatographic method for the characterization of transesterification process, *Materiale Plastice (Bucharest), 18,* 186, 1981.

18. Uglea, C. V., Characterization of poly(ethylene terephthalate). I. Intrinsic viscosity-molecular weight relationship, *Materiale Plastice (Bucharest), 18,* 282, 1981.
19. Uglea, C. V., Heinisch, P., Mihăescu, A., and Negulescu, I. I., Characterization of poly(ethylene terephthalate). II. Fractionation by coacervate extraction, *Materiale Plastice (Bucharest), 19,* 90, 1982.
20. Uglea, C. V., Heinisch, P., and Mihăescu, A., Characterization of poly(ethylene terephthalate). III. Determination of unperturbed dimensions, *Materiale Plastice (Bucharest), 19,* 218, 1982.
21. Uglea, C. V., Matei, E., and Mihăescu, A., Poly(ethylene terephthalate) characterization. V. Fractionation of poly(ethylene terephthalate) modified with dimethyl isophthalate, *Bull. Inst. Polit. Iassy, 33,* 99, 1987.
22. Uglea, C. V., Mihăescu, A., and Matei, E., Characterization of poly(ethylene terephthalate). VI. Intrinsic viscosity-molecular weight relationship and unperturbed dimensions of PET modified with dimethyisophthalate, *Bull. Inst. Polyt. Iassy, 33,* 107, 1987.
23. Uglea, C. V., Aizicovici, S., and Mihăescu, A., Gel permeation chromatography of poly(ethylene terephthalate)-coethylene isophthalate, *Eur. Polymer J., 21,* 677, 1985.
24. Barstein, R. S., and Sorokina, I. A., *Catalysed Polycondensation*, Khimia, Moscow, 1988, Ch. 3.
25. Barstein, R. S., Kirilovitsch, V. I., and Mosovskii, Iu. E., *Plastifiants of Polymers*, Khimia, Moscow, 1982, 197.
26. *Poly(Butylene Terephthalate). Scientific Informations*, NIITEHIM, Moscow, 1977, 29 (Russian).
27. Hann, M., GER Patent 2,259,712, 1973.
28. Kimura, T., Kobayashi, J., and Nakamoto, H., BRIT Patent 1,353,476, 1974.
29. Lindner, G., GER Patent 2,261,639, 1978.
30. Hintermayer, K., GER Patent 2,438,379, 1976.
31. Trifonov, T. G., BULG Patent 25,307, 1978.
32. Fichtner, W., GER(EAST) Patent 112,656, 1973.
33. Kreutzer, H., GER Patent 2,443,451, 1976.
34. Jelev, J. I., BULG Patent 14,039, 1975.
35. Wolfe, J. R., U.S. Patent 3,891,604, 1975.
36. Laganis, D., and Begley, F. M., U.S. Patent 4,133,787, 1979.
37. Tobias, M. A., and Lynch, C. L., U.S. Patent 4,140,729, 1979.
38. Bopp, R. C., U.S. Patent 4,161,498, 1979.
39. Georgoudis, P. C., U.S. Patent 3,975,323, 1976.
40. Hoh, G. L. K., and Tsukamoto, A., U.S. Patent 3,832,314, 1975.
41. Fujiyashi, K., Mizumura, Y., and Sono, J., U.S. Patent 3,867,480, 1975.
42. Sedov, L. N., *Unsaturated Polyesters*, Khimia, Moscow, 1977, 231 (Russian).
43. Mărculescu, B., RO Patent 73,605, 1980.
44. Berzelius, J., *Rapp. Annu. Inst. Geol. Congr., 26,* 1847; cited in Seymour, R. B., *J. Macromol. Sci.-Chem., A15,* 1165, 1981.
45. Callahan, M., U.S. Patent 1,191,732, 1914.
46. Vorlander, D., *Ann.*, 1894, p. 280.
47. Attia, I. A., Abdilhalia, M. S., and Abdelazim, A. A., Synthesis and characterization of some novel unsaturated polyester resins containing amide groups, *Polymer Bull., 34,* 377, 1995.
48. Mărculescu, B., RO Patent 73,606, 1980.
49. Nicola, G., RO Patent 71,014, 1981.
50. Donescu, D., RO Patent 67,032, 1979.
51. Stere, E. A., RO Patent 62,318, 1977.
52. Mancini, G., RO Patent 70,591, 1980.

53. Warner, K. N., FR Patent 2,458,563, 1981.
54. Miyamoto, A., FR Patent 2,457,305, 1980.
55. Maruzen Oil Corp., FR Patent 2,217,368, 1974.
56. Lamb, F., BRIT Patent 1,454,920, 1976.
57. Lamb, F., BRIT Patent 1,455,196, 1976.
58. Lamb, F., BRIT Patent 1,458,433, 1976.
59. Lamb, F., BRIT Patent 1,455,390, 1976.
60. Malek, J., CZECH Patent 156,093, 1973.
61. Ludina, V. S., and Tarasov, A. I., SU Patent 417,448, 1974.
62. Enikopolov, N. S., SU Patent 615,091, 1978.
63. Maréchal, E., Polymeric dyes—synthesis, properties and uses, *Progr. Org. Coating, 10,* 251, 1982.
64. Emmons, W. D., and Stevens, T. E., U.S. Patent 4,138,545, 1979.
65. Chatta, M. S., and Cassatta, J. C., High solid coatings from new oligomers, *J. Coat. Tech., 55,* 39, 1983.
66. McCallum, A. W., and Hull, D. C., U.S. Patent 3,929,867, 1975.
67. Lemper, A. L., and Rosenfeld, J. C., U.S. Patent 4,137,278, 1979.
68. Liu, S. B., and Yu, T. L., Study of the microgelation of unsaturated polyester resins by dynamic light scattering, *Macromol. Chem. Phys., 196,* 1307, 1995.
69. Abdelazim, A. A, Mechanical properties and curing characteristics of unsaturated polyesters synthesized for long casting, *Polymer Bull., 35,* 229, 1995.
70. Anisimov, Y. M., and Grekhova, O. B., Initiation of curing and copolymerization of unsaturated oligoesters resins with styrene, *Russian J. Appl. Chem., 67,* 1034, 1994.
71. Nisimov, G. B., and Grekhova, G. B., Hardening and copolymerization of unsaturated oligoester resins with methyl methacrylate and copolymer composition, *Russian J. Appl. Chem., 67,* 1719, 1994.
72. Jang, J. S., and Yi, J., Curing kinetics of allyl ester resins from differential scanning calorimetry, *Polymer J., 27,* 404, 1995.
73. Kalenda, P., Ferrocene and some of its derivatives used as accelerators of curing reactions in unsaturated polyester resins, *Eur. Polymer J., 31,* 1099, 1995.
74. Abbati, M., Martuscelli, E., Musto, P., Ragosta, G., and Scarinzi, G., Maleated polyisobutylene: A novel toughener for unsaturated polyester resins, *J. Appl. Polymer Sci., 58,* 1825, 1995.
75. Ahmed, T., and Funke, W., Structure and properties of radiation cured unsaturated polyesters in absence of vinyl monomers, *ACS Polymer Prepr., Vol. 1, 34,* 725, 1974.
76. Pucic, I., and Ranogajec, F., DC electrical conductivity as a method for monitoring radiation curing of unsaturated polyester resins. I. Measurement conditions and comparison with extraction analysis data, *Radiat. Phys. Chem., 46,* 365, 1995.
77. Hammann, D. K., Funke, W., and Gilch, H., *Angew. Chem., 71,* 19, 1959.
78. Boenig, H. V., in *Encyclopedia of Polymer Science and Technology*, Vol. 11, 1969, 135.
79. Homma, M., Shoji, A., and Nakamura, H., U.S. Patent 3,989,767, 1976.

6

Alkyd Resins

I. INTRODUCTION

The performance of glyptal polyester coatings was upgraded by Kienle [1], who in 1927 incorporated unsaturable vegetable oils in these polyesters and coined the acronym *alcid* from the first and last syllables in the reactants, viz. *al*cohol and a*cid*, and then changed this name to the more euphonious alkyd.

Alkyd resins have been defined as any condensation product involving a polybasic acid, like phthalic, maleic, or succinic acid, and a polyhydric alcohol, like glycerine and the glycols, almost always with addition of modifying agents such as higher fatty acids. This definition includes polyester resins, of which alkyds are a particular type. Alkyd resins are used chiefly in the coating industry. In this industry, the terms alkyd, alkyd resin, and alkyd solution are used interchangeably, even though most alkyds are handled as solutions in hydrocarbon solvents (30–70% by weight of alkyd resin).

The term *modified alkyd*, which formerly was used to describe these products, now is associated with chemical modifications that are carried out during alkyd preparation, and that incorporate chemical agents of types other than those included in the definition. For example, the term *rosin-based alkyd resins* or *rosin-modified alkyd* refers to alkyd resins in which all or a portion of the monobasic fatty acid is replaced by rosin. Unmodified alkyd resins are polyester products composed of polyhydric alcohol, polybasic acid, and monobasic fatty acid. When no fatty acids are used, or when they are completely replaced by other types of acid, the products can be considered as "oil-free" alkyds.

The formation of polyesters with resinous characteristics has long been known. Berzelius [2] in 1847 reported a resinous product from the reaction of tartric acid with glycerol. Berthelott [3] in 1853 prepared the glyceryl ester of camphoric acid, and Van Bremmelen [4] in 1856 prepared glyceryl succinate and glyceryl citrate. In 1901, Watson Smith [5] in England obtained a hard brittle polymer by the reaction of glycerol with phthalic anhydride. Baekeland's early work on phenol formaldehyde resins stimulated the interest of the electrical industry in new insulating materials, and in 1910, extensive investigations were begun at the General Electric laboratories on the glycerol–phthalic anhydride reaction. The work of Callahan [6], Friedberg [7], and Howell [8] showed that when part of the phthalic anhydride is replaced with monobasic acids, such as butiric or oleic acid, more flexible and more soluble resins result.

II. SYNTHESIS

The chemical reactions that occur during preparation of alkyd resins are polycondensation, polyaddition, and side reactions such as decarboxylation.

The most common alkyd raw materials are *polybasic acids* (phthalic anhydride, isophthalic acid, maleic anhydride, fumaric, azelaic, succinic, adipic, and sebacic acids), *oils* (linseed, soya, dehydrated castor, tung, fish, safflower, oiticica, cotton seed, and coconut), *polyhydric alcohols* (glycerol, pentaerythritol, dipentaerythritol, trimethylolethane, sorbitol, trimethylolpropane, ethyleneglycol, propylene glycol, neopentyleneglycol, and dipropylene glycol), and *monobasic acids* (tall-oil fatty acids and synthetic saturated fatty acids).

The most important polybasic acid for alkyd resins is phthalic acid, which is produced and used in the form of its anhydride. The great demand for terephthalic acid for polyester fiber has spurred the development of methods for the separation of the isomers of mixed xylenes. This has resulted in the availability of large quantities of *m*-xylene, which is oxidized to isophthalic acid. The first use of isophthalic acid in drying oil-modified alkyd resins was disclosed in a patent [9], which claimed that alkyd resins made from isophthalic and terephthalic acids had better drying characteristics than the corresponding orthophthalic resins. On the other hand, Hovey and Hodgins [10] discussed the use of isophthalic and terephthalic acid in oil-modified alkyd resins and concluded that the difference in drying time claimed could be due to the presence of antioxidant impurities in phthalic anhydride made from naphthalene. The other advantages of making alkyd resins from isophthalic and terephthalic acids are the low acid losses by sublimation and the greater thermal stability at oil bonding temperatures. Superior film properties of alkyd resins made from isophthalic and terephthalic acids were also observed [11].

Favorable price considerations have contributed to increasing interest in isophthalic acid. Phthalic anhydride is less reactive than isophthalic acid, even though phthalic anhydride has the advantage of nearly instantaneous formation of the monoester. Since isophthalic acid has a very high melting point (347°C) and a low solubility in other alkyd ingredients, such as monoglyceride, the processing of its alkyds has presented some difficulties. It does not sublime like phthalic anhydride; it evolves twice as much water during the processing, and this can lead to some loss of polyols and acids by steam distillation and/or mechanical entrainment. To avoid such losses, producers of isophthalic acid have recommended a slow upheat of at least 2 h and the use of a low inert gas rate during this period in the alcoholysis and fatty acid process.

Maleic acid, or more commonly the anhydride, and fumaric acid possess the basic requirement (difunctionality) for polyester formation. They also possess an additional functionality from their double bond, which enables them to form Diels–Alder and other adducts with unsaturated acids in drying oils [12].

Maleic and fumaric acids often are combined with rosin by the Diels–Alder reaction, and tribasic adducts such as the one shown in the following reaction are formed:

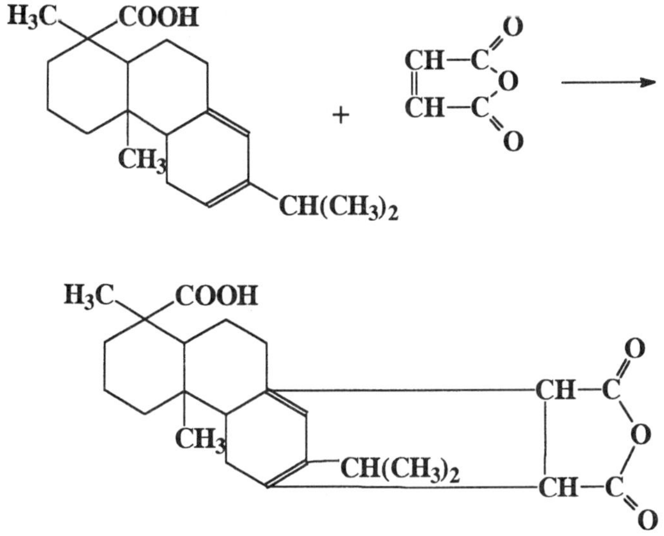

(1)

Alkyds with either rosin–maleic acid, maleic anhydride, or fumaric acid adducts as the polybasic acids are used in many low-cost coating applications, but the films from these resins do not have the color retention, toughness, adhesion, gloss retention, or exterior durability that the phthalic anhydride alkyds have.

Other dibasic acids used in alkyds to impart special properties are adipic acid, azelaic acid, sebacic acid, tetrachlorophthalic anhydride, chlorendic anhydride, dimerized fatty acids, and trimellitic anhydride. Adipic, azelaic, and sebacic acids impart flexibility in the alkyd structure and are primarily used in alkyds designed for application as plasticizers. Tetrachlorophthalic anhydride and chlorendic anhydride are used to impart fire-retardant properties to the resin system.

The monobasic acid modifies the properties of the resin in two ways: by its capacity to control functionality and so allow control of polymer growth; and by the nature of its inherent chemical and physical properties. The extent and kind of unsaturation in the drying fatty acids has a strong bearing on the properties of the finished alkyd. In general, triene unsaturation contributes more to improve drying rate and other physicomechanical properties of the alkyd resin. Conjugated systems are slightly better than nonconjugated systems in the development of initial drying. The monounsaturated portions have little drying tendency. Saturated acids appear to prevent drying.

Rosin is used to replace the fatty acids in a wide variety of commercial products, which may also include glycerol terephthalates.

The term *rosin* refers to the resinous materials that occur naturally in the oleoresin of pine trees, as well as derivatives thereof including rosin esters, modified rosins such as fractionated, halogenated, and polymerized rosins, modified rosin esters, and the like.

Rosin is a friable solid at room temperature. It is insoluble in water but soluble in organic solvents. Rosin may be characterized by softening point and by the value of the acid number. Rosin is a complex mixture of mutually soluble, naturally occurring high molecular weight organic acids and related neutral materials. The nonacidic constituents are a mixture of high molecular weight esters, alcohols, aldehydes, and hydrocarbons that vary in nature, but which for the most part have structures related to the rosin acids. They may vary in relative amounts according to the source of the rosin and extent to which it is refined.

The reactions of rosin that yield derivatives that are used in the production of synthetic polymers, and as additives or modifiers for natural and synthetic polymers, involve its carboxyl group and double bonds, singularly or together. Oxides, bases, salts, and other metal compounds react with the carboxyl group of the rosin acids.

Rosin-derived "semielastomeric" resins are used with vinyl polymers and copolymers in certain types of grease- and oil-resistant specialty inks, as well as in some forms of overprint coatings. These synthetic resins have an alkyd structure. Since they are rosin derived, they exhibit the expected good pigment-wetting properties, outstanding gloss, and excellent specific adhesion to a variety of surfaces ranging from coated paper and glassine to metal foil. As alkyd-type materials, they possess flexibility and strength characteristics approaching film-forming polymers. They also permit formulation, difficult to achieve with vinyl resins alone, of full-bodied vehicles for vinyl-based inks.

Rosin, glycerol, and pentaerythritol esters of rosin, and dibasic acid-modified rosin esters are common ingredients in alkyd resins used as vehicles for various protective coatings. These rosin constituents are either added during the manufacturing process, or they are blended by "cold-cutting" with the finished alkyd. In an alkyd preparation where rosin (unmodified) is used as part of the monobasic acid ingredient, the finished product is known to the industry as a *rosin-modified alkyd*. The presence of rosin or rosin derivatives in the polymer molecule imparts the following improvements: better brushability, faster drying, better gloss characteristics and retention, greater hardness, better chemical stability, and improved adhesion to substrates.

The use of rosin-derived alcohol, hydroabietyl alcohol, as a portion of the polyol ingredient yields a higher molecular weight modified alkyd, with improved solubility, better brushability, flow, hold-out, and greater hardness.

Unmodified rosins combine with fumaric acid and maleic anhydride in accordance with the Diels–Alder reaction to form tribasic adducts. The resulting polybasic acids are often used instead of phthalic anhydride in the production of alkyd resins for low-cost coatings. These rosin-derived, nonphthalate alkyds are used to improve gloss and other properties and are suitable mainly for interior coatings. They do not possess the high degree of outdoor durability characteristic of alkyd resins derived solely from phthalic anhydride.

The rate of drying is a function of polyunsaturated or polyenoic acid content. This rate increases rapidly up to a polyenoic acid content of about 50% [13]. Above that figure, a limiting value is gradually approached. The limiting value in the particular alkyd used represents that point at which intermolecular polymerization is largely completed and the polymer is of such a size that any additional polyunsaturation reacts intramolecularly.

The hardness of the alkyd resins was increased by Bhow and Payne [14], who styrenated the fatty acids and then reacted the product with phthalic anhydride and glycerol. According to these investigators, styrenated dehydrated castor oils were liquid when less than 70% styrene was added, and solid at higher concentrations of styrene.

Peterson [15] described a styrenated oil with 50–55% styrene, which corresponded to conventional alkyds with 12.5–15.2 gal of oil to 100 lb of resin. According to Payne [16], there are several competing reactions in this polymerization, and the dominant, that is, faster, reaction determines the type of product produced. Hewitt and Armitage [17] proposed that the copolymerization with conjugated oils, such as tung and oiticia oils, is comparable to the copolymerization of styrene and butadiene in which 1,4-addition is predominant.

Styrenated alkyds may also be produced by the addition of glycerol and phthalic anhydride sequentially to the styrenated oil by styrenation of a monoglyceride, followed by reaction with phthalic anhydride and by styrenation of the alkyd resin. According to the literature [18], styrenated alkyds dry rapidly and are insoluble in aliphatic solvents but are soluble in xylene. However, styrenated alkyds produced from vinyl toluene are soluble in aliphatic solvents such as white spirit. It is generally agreed that styrenated alkyds are faster drying, harder, and more resistant to water and alkaline solution than conventional alkyds.

The resins of conventional paints were classified into two main groups from the point of their curing mechanism at room temperature. The first one did not need any special curing agent, and the second was cured with a reactive hardener added just before painting because of the short pot life. Alkyd paint modified with drying oil was most popular and convenient among resins in the first group because of easy painting. The curing mechanism, namely, the formation of three-dimensional crosslinked structure, was based on the chemical behavior of the modified drying oil. The carbon–carbon double bond was produced by absorbing an oxygen molecule, and the decomposed radical initiated the crosslinking polymer reactions. A so-called dryer composed of some metal naphthenates was added to the paint in the manufacturing process to accelerate the curing rate at air drying time.

Most paints are used with some organic solvents to facilitate their application by lowering the viscosity. The solvents, however, dissipate into the atmosphere after painting. This conventional painting method not only contributes to air pollution, it is also a waste of valuable chemical petroleum products. An interesting method for the curing of alkyd resins was proposed by Satone et al. [19]. In this method a reactive solvent was investigated with a view to reducing the content of the organic solvent. 1,1-Bis-(1′-methyl-2′-vinyl-4′,6′-heptadienoxy)-ethane was found to be suitable as the reactive diluent. A small addition of this compound to alkyd paint reduced the content of solvent to less than 15% with a conventional alkyd resin of high molecular weight (high-solid type) and, further, a super high paint (5% solvent) could be manufactured when this compound was added to a superlong-oil-length alkyd resin of low molecular weight.

The film-forming ability of alkyd resins, that is, fatty acid–modified polyesters, depends on the functional groups present, namely hydroxyl, carboxyl, and in some cases, carbon–carbon unsaturation. Of these, the hydroxyl group serves a number of crucial functions. It is involved in the esterification reaction of the monomeric and low molecular weight entities, which leads to polymer formation.

Glycerol is the workhorse polyol of alkyds, closely followed by pentaerythritol. Mixtures of pentaerythritol and ethylene glycol are used quite extensively in the preparation of medium- and short-oil alkyds containing 30–50% fatty acids. These resins exhibit better compatibility properties, gloss retention, and durability than alkyds based on glycerol as the sole polyol. However, the volatility of ethylene glycol presents a problem in the processing of alkyds.

A number of other polyols are used in alkyd resins in much smaller quantities than glycerol, pentaerythritol, and ethylene glycol. In general, the greater the distance between hydroxyl groups, the softer and more flexible will be the resultant resin. The nonhydroxyl portion of the polyol also affects the properties of the resin. The hydrocarbon chain in trimethylolpropane leads to alkyds that, in comparison with glycerol resin, are more readily soluble in hydrocarbon solvents. They are also more flexible and are more water and alkali resistant. The latter properties are probably due to a shielding effect of hydrocarbon chain that protects the ester groups from attack.

The function of the hydroxyl group in film formation depends on the drying conditions and/or the presence of film formers other than the alkyd. Where drying takes place by autooxidation process involving the formation and decomposition of hydroperoxides, the hydroxyl group can adversely affect the drying rate by complex formation and the consequent stabilization of the hydroperoxide [20]:

$$\mathrm{R{-}O{-}O{-}H + R'{-}O{-}H \rightleftarrows \begin{array}{l} R{-}O{-}O{-}H \\ \quad\vdots \\ \quad H{-}O{-}R' \end{array}} \tag{2}$$

Examination of a number of alkyds, with formulations typical of the range in common use, showed that the measured hydroxyl value corresponded to the theoretical figure only when the system had a functionality of two. In connection with the availability of the hydroxyl groups, Kienle et al. [21] could not explain the glycerol–phthalic anhydride reaction by simple second- or third-order equations, whereas in fatty acid–modified polyester Wekna and Klausch [22] and Bergmann [23] claim second-order kinetics apply.

Solomon and Hopwood [24] find that, in the case of alkyd prepared from fatty acid, glycerol, and phthalic anhydride in the molar ratio 1.0:1.26:1.11, the measured hydroxyl values are considerably less than the calculated figures; and the difference between these values increases with the molecular weight and the complexity of the polymer molecule. The difficulties encountered in the kinetic analysis of the formation of alkyd resins could be related, at least in part, to the availability of the hydroxyl group.

The formation of the 52%-oil-length alkyd described by Solomon and Hopwood [24] does not follow either second- or third-order kinetics on the basis of either the measured or theoretical hydroxyl values. Of the possible explanations for this observation, the three most likely would appear to be the following: (1) fractional kinetics apply, possibly because of the difficulty associated with the removal of the water of reaction at the high viscosities; (2) under esterification conditions, more hydroxyl groups are available than is indicated by acetylation; (3) some of the carboxyl groups also are not available for reaction.

We do not yet have sufficient evidence to decide among these alternatives, but it is worth noting that phthalylation, which should closely parallel the conditions operating during polyesterification, gave substantially the same hydroxyl values as acetylation.

The lack of availability of the hydroxyl groups in these polyesters could also partially explain the divergence between calculated and observed gel points.

The significance for film-forming reactions of the available, rather than the theoretical, hydroxyl values, is demonstrated by the studies on typical crosslinking reactions. When the alkyd, a malamine–formaldehyde condensate, and an acid catalyst were heated together, free hydroxyl groups remained even after 75 min at 150°C, as shown by absorption at 3500 cm^{-1} in the infrared spectrum [24]. In a similar experiment with the acetylated alkyd, no change in the infrared spectrum occurred, indicating that the hydroxyls involved in the condensation with the malamine–formaldehyde resins are those that react with acetic anhydride. Likewise, reaction of the alkyd with excess phenyl isocyanate indicated that some hydroxyl groups were not attacked and, since the acetylated resin did not show any change in hydroxyl intensities when treated with isocyanate, it is concluded that the same hydroxyl groups react with both acetic anhydride and the isocyanate. Therefore, measurement of the available hydroxyl groups by acetylation offers a method for predicting the potential reactivity of the alkyd resin in thermosetting film-forming reactions.

It is much more difficult to demonstrate the effect of available hydroxyl groups on the rate of air drying of the alkyd and in the solubility of a similar polymer in water, since in removing the active or available hydroxyl groups by acetylation, the total hydroxyl value is naturally decreased. However, the acetylated alkyd formed a dry film in one-quarter of the time required by the alkyd [24], and the magnitude of the improvement warrants further studies to determine the relative importance of available and total hydroxyl contents on the rates of autooxidation and film formation. An alkyd of a lower degree of condensation and soluble in a butyl–cellosolve–amine–water mixture had its water tolerance reduced from infinity to zero by the acetylation of one-third of its total hydroxyl groups; once again, the magnitude of this change suggests that available hydroxyl groups could be of vital importance in conferring water solubility on the polymer.

As discussed earlier, the preparation of alkyd resins is essentially an esterification process in which the polybasic acids and polyhydric alcohols are reacted with various oils or fatty acids and modifying agents. When a fatty acid is used, the process chiefly involves direct esterification. When an oil is the fatty acid source, an alcoholysis or ester-exchange reaction between the oil and the polyhydric alcohol is usually carried out before the esterification step.

Four basic methods for manufacturing most alkyd resins are recognized: the fatty acid method, the fatty acid–oil method, the alcoholysis method, and the acidolysis method. The fatty acid and alcoholysis methods are the most important.

The *fatty acid method* is the simultaneous direct esterification of all ingredients. This method allows greater freedom in formulation since any polyhydric alcohol or polyhydric alcohol blend can be used, and fatty acids not available as glycerides, such as fatty acid from tall oil, pelargonic acid, 2-ethylhexanoic acid, and so forth, can be used as well as special fatty acids that have been segregated and refined for specific alkyd performance. The entire mixture of fatty acids, polyols, and dibasic acids is heated to a reaction temperature of 220–260°C until the desired alkyd specifications are obtained.

The *fatty acid–oil method* is a process in which a mixture of fatty acid, triglyceride oil, glycerol, and phthalic anhydride is processed at normal reaction temperature, and a homogeneous resin results.

The *alcoholysis or monoglyceride method* starts with an oil, a polyol, and a dibasic acid or anhydride to produce alkyds. If such a mixture is heated, the polyol reacts solely with the dibasic component, and a useless heterogeneous mixture is obtained. The polyester is insoluble in the oil phase, and it rapidly forms an unmodified polyester, which gels at low degree of reaction. This incompatibility is overcome by the reaction of triglyceride oil with glycerol or another polyol in the presence of a catalyst at a temperature of 225–250°C. When glycerol is used, the following alcoholysis reaction occurs:

$$\begin{array}{lcl} CH_2\text{—O—CO—R} & & CH_2OH & & CH_2\text{—O—CO—R} \\ | & & | & & | \\ CH\text{—O—CO—R} & + 2 & CHOH & \xrightarrow[\text{Catalyst}]{\Delta} 3 & CH\text{—OH} \\ | & & | & & | \\ CH_2\text{—O—CO—R} & & CH_2OH & & CH_2\text{—OH} \\ \text{Glyceride oil} & & \text{Glycerol} & & \text{Monoglyceride} \end{array} \qquad (3)$$

In the alcoholysis reaction, the catalysts most frequently used are litharge (PbO), calcium hydroxide, lithium carbonate, or the soaps derived therefrom.

In most commercial alkyd preparations, sublimed litharge has been the preferred catalyst, in quantities ranging from 0.02 to 0.05% based on the oil. With recent restrictions on the use of lead compounds, the use of calcium and lithium hydroxides and their soaps has increased. With these catalysts, about 0.008–0.01% as metal, based on the oil, is normally used. The oil is heated to 230–250°C with good agitation and inert gas blanket, the catalyst and the polyol are added, and the mixture is reheated to the reaction temperature of 230–250°C. Since the oil is insoluble and the monoesters are soluble in anhydrous methanol, the course of the alcoholysis reaction is followed when 1 volume of reaction mixture gives a clear solution in 2–3 volumes of anhydrous methanol, the dibasic acid is added, and the polyesterification is conducted at 210–260°C to the desired alkyd specifications.

The alcoholysis reaction is further complicated by the tendency of the polyols, under the influence of heat and the alkaline catalysts used for alcoholysis, to undergo interetherification with the resultant loss of available hydroxyl groups and the formation of higher functional polyols.

Certain significant differences are also observed for alkyds that are identical from a chemical composition standpoint but differ in properties and performance depending on whether they are made by alcoholysis or the fatty acid method: rate of esterification slows down at somewhat higher acid number; bodying and gelation occurs at slightly higher acid numbers; air-dry set time is somewhat slower, and the resin tolerates more aliphatic thinner.

The application of isophthalic acid as a raw material in the synthesis of alkyd resins involves the alcoholysis of the oil with the pentaerythritol in long-oil alkyds, and with a 65:35 pentaerythritol–ethylene glycol–equivalent blend in the medium-oil-length alkyd using a 15–20% excess of hydroxyl content. A water-cooled reflux condenser should be used during the alcoholysis. On completion of the alcoholysis, the reaction mixture is cooled to 200°C, and isophthalic acid added. With equipment set up to prevent loss of glycol, the temperature is slowly raised to 235–245°C with a slow inert gas sparge and held for clarity. Then, with increased inert gas sparge, the esterification is completed to the desired endpoint.

The *acidolysis process*, another less commonly used method of forming alkyd resins, is based on the reactions

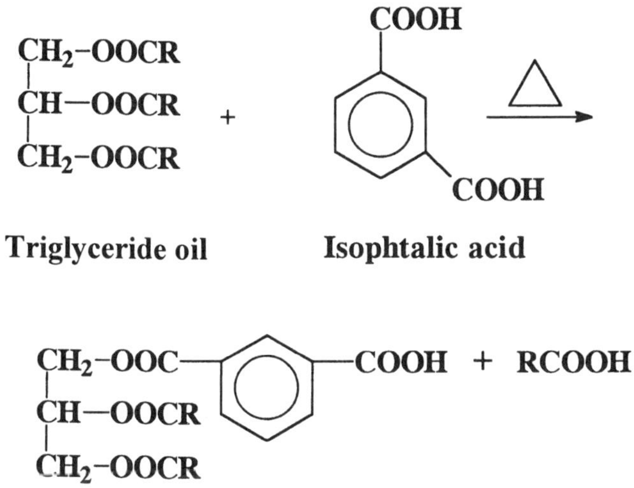

(4)

This reaction occurs at higher temperature without catalyst and its use is limited to polybasic acids, such as isophthalic and terephthalic acids, which unlike phthalic acid or anhydride do not sublime and are quite insoluble in the monoglyceride until considerable esterification has occurred.

III. MODIFICATIONS AND APPLICATIONS

Alkyds are better for the purpose of film formation and protection against corrosion than are the most of the resins. Consequently, they are used as coatings. A coating is a layer of a substance used to protect and/or decorate a surface. It must adhere to its substrate, as opposed to a covering, which may simply surround or lie over the substrate. At the time of application, most coatings are composed of three principal ingredients: the polymeric binder, the pigment, and the volatile thinner or solvent. Each ingredient, in itself, may be a complicated mixture.

Alkyd resins are characterized by rapid drying, good adhesion, flexibility, and durability. Their principal disadvantage is the ease with which the ester groups, which form so large a part of the molecules, are hydrolyzed under alkaline conditions. It is possible to produce alkyds with greatly improved resistance to hydrolysis by the use of special raw materials or by modification with a wide variety of reactive chemicals and other polymeric materials. The typical examples are long-oil (based on 1:1:1 mol ratio of glycerol–phthalic anhydride–fatty acids and 60.5% oil) and short-oil (based on 6:6:2 mol ratio of glycerol–phthalic anhydride–fatty acids and 31.2% oil) with the structural formulas

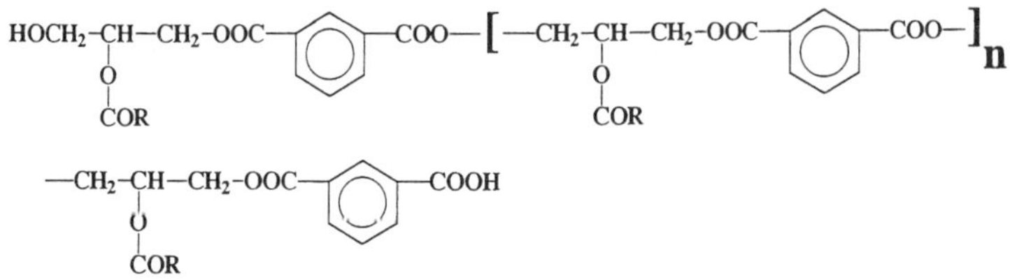

(5)

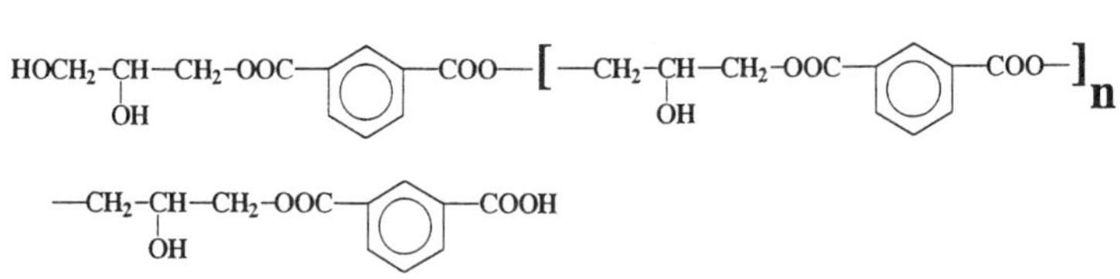

(6)

In a long-oil resin, the large number of long-chain fatty acid groups imparts a nonpolar character to the molecule. In a short-oil alkyd, the high proportion of hydroxyl groups imparts not only polarity but centers of potential reactivity with a host of hydroxyl-reactive materials. Moreover, the unsaturation in the fatty acid groups allows interpolymerization with many vinyl monomers, epoxidation, and the other reactions of double bonds. There are many polymeric materials and reactive functional materials with which suitably designed alkyds are modified in order to impart improved and/or special film-forming properties [25].

Castor oil fatty acids have one double bond per molecule, whereas hydrogenated castor oil fatty acids are saturated; consequently, the oils, or alkyds derived from them, are not film forming under autooxidation conditions. However, alkyds prepared from these oils are widely used as polymeric plasticizers for other film-forming resins; the two most important types are cellulose nitrate and melamine–formaldehyde condensates.

The use of alkydes with cellulose nitrate in lacquers resulted in the remarkable growth of cellulose nitrate lacquers. Compatibility of alkyds with cellulose nitrate extends up to 55% oil modification. However, the best compatibility is with short-oil alkyds that have a high degree of polarity from ester and excess hydroxyl groups.

Lacquers used for finishing wood furniture are classified as sealers and topcoats. The sealers are intended to overcome the irregularities from the porosity of wood and filler, and they should sand easily so that a smooth, uniform surface may be obtained on which to apply the final topcoat lacquers.

A high-quality alkyd–cellulose nitrate blend for lacquers used as the topcoat for wood furniture is given in Table 1.

Blends of alkyd resins, cellulose nitrate, and catalyst-convertible butylated urea–formaldehyde resins are another development in wood-furniture lacquers. In the presence of an appropriate catalyst such as an alkyl acid phosphate, these components react under air-drying conditions, as well as at the higher temperature (150°F), to form an insoluble film. These lacquers show improvements over the conventional alkyd–cellulose nitrate blends in the alcohol resistance of the film and in having higher solids at spraying viscosities.

Table 1 Formulation for a Topcoat Lacquer for Furniture Using an Alkyd–Cellulose Nitrate Blend

Component	Percent by weight
RS Nitrocellulose[a]	11.4
Short-oil nondrying alkyd	17.0
Dioctyl phthalate	2.0
Toluene	33.0
Ethanol	1.6
Butanol	5.0
Butyl acetate	30.0

[a]Registered trademark of the Hercules Powder Comp.
Source: Technical Bulletin, Hercules Powder Comp.

Exterior-grade alkyd–cellulose nitrate blends are used in large volume for metal finishing and for automotive refinishing applications. The durability and gloss retention of exterior lacquers are the most important factors contributing to their marketability.

In the field of automotive finish coats, between about 1930 and 1955 two general systems were common: those based on alkyd–cellulose nitrate plasticizer combinations as binder, and those based on combinations of alkyd and amino–formaldehyde resins. The lacquer (cellulose nitrate) system required lower baking temperature or shorter schedules at higher temperature. No extensive chemical reaction occurred during baking, and film formation, followed by hardening, resulted from solvent evaporation. Such finishes had the advantage that they could be polished or rubbed to a very attractive luster. This advantage is not easily attained in reactive curing enamels of the alkyd–aminoplast type, or in more modern vinylic or acrylic lacquers. Cellulose nitrate lacquers are thus still used where custom finishing involving much labor can be justified. Touch-up and repair is easy with such air-drying lacquer systems, and films of considerable thickness can be developed, with or without intermediate sanding between coats. These old lacquers, however, at their best, lacked the exterior durability of chemically reactive or curing enamels, nor do they compare with more modern lacquers in gloss retention under exposure to sunlight and water.

Pigmented industrial enamels prepared from the castor oil alkyds were different, particularly for gloss and humidity resistance. The alkyds prepared directly from oil had better properties.

Alkyd resins, particularly of the long-oil class, are used in volume in letterpress and offset printing inks. They are normally used at 75–100% solids and are improvements over bodied linseed oil or oleoresinous varnishes. In letterpress and offset printing, alkyd resins improve rub resistance. This is accomplished without losing pigment dispersability and other essential properties. Alkyd resins usually contain 5–20% isophthalic or *o*-phthalic constituent. Isophthalic alkyd resins are used in printing ink, principally in air-drying compositions for offset and letterpress printing. The advantages of isophthalic alkyd resins are better resistance to water and harder drying (tougher films are produced). *o*-Phthalic compositions are generally employed only in metal-decorating printing inks where the major property under consideration is color retention. The drying-oil segment of both *iso*- and *o*-phthalic alkyd resins of polyesters is essentially the same, depending on the end use. For example, if color retention is important, soybean or tall-oil fatty acids, both high in linoleic acid, are used. These vehicles will normally be used for backing purposes. For drying applications where the color retention is not a factor, linseed oil and linseed oil–China wood oil blends are employed. These oils contain substantial quantities of conjugated unsaturation, such as linolenic acid, and give faster drying speeds, harder films, and films with improved water and chemical resistance. The fatty acid portion is soya, or refined fatty acids from tall oil or linseed oil. In letterpress, the differences between phthalic and isophthalic alkyd resins are slight; however, in offset, the isophthalic compositions seem to satisfy the demands of the process more effectively. Pigment wetting also favors isophthalic alkyd resins. The use of alkyd resin in offset ink must be carefully controlled. Ester polymers terminating in acid and hydroxy groups can cause problems in lithography. The hydroxyl group is the most vulnerable to attack by the aqueous fountain solution. It is, therefore, important that alkyd resins for offset printing ink have low hydroxyl values. It is common practice to use oleoresinous vehicles in combination with alkyd resins. In this manner, a balanced system is reached that satisfies almost all letterpress and offset needs.

The metal-decorating industry utilizes lithography as its standard printing process. Oleoresinous varnishes and alkyd resins play important roles.

White inks for can decorating are composed of soya oil, dehydrated castor oil, or safflower alkyd resins with or without maleic oleoresinous varnishes.

The short-oil alkyd resins containing 38–45% phthalic anhydride contain a higher proportion of hydroxyl groups, which provide compatibility and reactive sites with alkylated urea–formaldehyde and melamine–formaldehyde resins. This combination is widely used in industrial baking enamels for metal cabinets, appliances, venetian blinds, toys, and so forth. These alkyds are usually based on tall-oil fatty acids and soybean oil or fatty acids, although coconut oil or short-chain saturated fatty acids provide the best color retention on baking, and improved gloss retention. Where superior adhesion and impact resistance are required, the alkyd is based on dehydrated castor oil. The short-oil-drying alkyds combined with urea resin and an acidic catalyst are used in force-dry wood furniture finishes with excellent resistance properties.

Chlorinated rubber is compatible with alkyds of similar linearity and low polarity. Most of the highway marking points in the United States are based on a combination of chlorinated rubber and a compatible alkyd resin based on soybean oil.

Phenolic resins and drying-oil alkyds combine to form a chroman-oil type structure in which the excellent gloss retention and durability of alkyds is combined with the water and alkali resistance of phenolics.

The free hydroxyl groups of alkyds react with polyisocyanates. The products dry faster and have improved chemical and abrasion resistance. Alkyds with less than 45% fatty acids can be designed to be compatible with low molecular weight epoxy resins and certain melamine–formaldehyde resins. Pigmented systems based on this combination have excellent adhesion to metal, improved gloss and color retention, and excellent water and chemical resistance [26].

Reactive silicone intermediates react with the hydroxyl groups in long-oil air-drying alkyds to give copolymers with greatly improved durability and gloss retention [27].

The unsaturation in the fatty acid groups of alkyds allows interpolymerization with a variety of reactive vinyl or acrylic monomers. Vinyl- or acryl-modified alkyds can be prepared by blending or by copolymerization involving the unsaturation of the fatty acids or combination of the polymers via other functional groups [28].

Other methods of combining vinyl or acrylic polymers with alkyds are based on a vinyl or acrylic copolymer that contains groups capable of being esterified by either the hydroxyl or carboxyl of the alkyd.

Certain alkyd resins can be combined with modified polyamide resins to produce materials exhibiting a marked degree of true thixotropy when dissolved in suitable solvents. This valuable property is maintained when such products are used in the manufacture of paints. When the alkyd and polyamide are heated at elevated temperature, amide and ester interchange reactions occur that split the polyamide chains and affix the fragments to the alkyd molecule. In further reactions, the amide groups are scattered throughout the molecule. The reactions probably proceed in the following way:

$$RCOOH + R'\text{—}CO\text{—}NH\text{—}R'' \rightarrow R'COOH + R\text{—}CO\text{—}NH\text{—}R'' \quad (7)$$

$$R\text{—}OH + R'\text{—}CO\text{—}NH\text{—}R'' \rightarrow R'COOR + H_2N\text{—}R'' \quad (8)$$

$$R''\text{—}NH_2 + RCOOH \rightarrow R\text{—}CO\text{—}NH\text{—}R'' + H_2O \quad (9)$$

where R, R′, and R″ are alkyd or polyamide molecule residues.

The thixotropic behavior of these resins has been attributed to hydrogen bonding of carboxyl and the amide groups throughout the alkyd–polyamide complex [29]. These thixotropic alkyds are particularly valuable in architectural and maintenance paints, where they impart nondrip properties, freedom from pigment settling, and ease of brushing and allow the application of thicker coats without danger of sagging.

In automotive finish coats, the alkyd–aminoplast resin enamels had better initial luster than cellulose nitrate lacquers, and greater resistance to road-tar stain and to chalking or exterior exposure, particularly in the areas receiving much ultraviolet light. The relative advantages, however, varied greatly from one pigment to another.

New developments in upgrading color quality and stability of organic pigments since about 1950 made a reappraisal of binder formulations imperative. This research had the goal of upgrading enamels through better design of alkyds and amino resins.

Oil-free alkyds refers to the alkyd resins obtained when no fatty acids are used or when they are completely replaced by other types of acids. Rosin is used to replace the fatty acids in a wide variety of commercial products.

Rosin is a clear, pale yellow to dark amber, thermoplastic resinous solid derived from naturally occurring constituents found in the ducts and cells of dead and living pine trees of the species indigenous to the temperate regions of the world. It is obtained by three widely different procedures, each of which yields a commercial form of rosin named for the raw material processed. These forms are as follows: *gum rosin*, obtained by diluting the oleoresin (from the living pine tree) with turpentine, filtering out trash, and steam distilling away the spirits of turpentine; *wood rosin*, from solvent extracts of aged pine stump wood; and *tall-oil rosin*, obtained by depitching and fractional distillation of crude tall oil, a byproduct of the softwood (pine species) pulp industry [30].

Oil-free alkyds containing glycol, maleic anhydride, and hydroabietyl alcohol have low viscosity and pale color. Their coatings have good flexibility, but the color retention is inferior to that of coatings from alkyds containing saturated fatty acids such as lauric and pelargonic.

A different type of oil-free alkyd, which has found considerable use in paper and cellophane lacquer, is based on the ester of glycol and a terpene–maleic anhydride adduct.

A large volume of alkyds is used in pigmented undercoat and topcoat enamels for mass-produced metal products. Enamels that require forced drying by heat (baking) are preferred. Alkyd–amino resin blends are used mainly for the quality finishes required in the appliance, metal furniture, and automotive fields.

Washing machines, dryers, stoves, refrigerators, and similar appliances pose rather demanding finishing problems. To be successfully marketed, these appliances must have finishes that are not only mechanically adequate under all service conditions, but also aesthetically pleasing. Thus, these finishes must have features such as hardness, toughness, mar resistance, and good adhesion to the metal so that the finish does not separate if the sheet metal is accidentally flexed. At the same time, excellent initial color and gloss and resistance to aging, temperature extremes, humidity, exposure to sunlight, fruit acids, greases, and soaps are necessary so that yellowing or film degradation does not occur.

The coating industry has met these problems with blends of alkyds and heat-reactive amino resins. With such vehicles, adequate performance can be achieved through baking cycles of 30–40 min at temperatures of 250–300°F. In general, where initial color and color retention are the prime requisites, nondrying alkyds are most widely used.

Nondrying oil alkyd–amino resin blends give baking enamels with excellent resistance to soap and alkali. A replacement of 25% of the short-oil, nondrying alkyd with a short-oil, drying alkyd will give slightly better adhesion with a sacrifice in soap and alkali resistance.

The latter type of formulation permits one-coat application. Replacement of the butylated melamine–formaldehyde resin with a butylated urea–formaldehyde resin gives lower cost with a slight sacrifice in properties.

The automotive finishes require exterior durability, gloss, gloss retention, low dirt pickup, color quality and retention, combined with the requisite hardness and toughness. To meet these specifications, many automotive enamels are currently being made with blends of alkyd–amine resin vehicles having a relatively high amino resin content (25–35%), but the ultimate development of these coating properties is, in general, influenced mainly by the alkyd resin used. To meet these requirements, especially those of durability and gloss retention, short-oil, nondrying alkyds are used in preference to short-oil, drying alkyd resins in the enamels.

The availability of a wide variety of glycols, triols, fatty acids, and polybasic acids used with trimellitic anhydride provides resin chemists with the opportunity to develop water-borne alkyds with various properties [31]. Both air-drying and baking types having varying degrees of fatty acid modification can be prepared.

A résumé of the formulation possibilities of alkyds for coating is given in Table 2. Water-soluble oil-free alkyd or reactive saturated polyester coating resins can be used to formulate gloss enamels that have excellent performance properties. The water-soluble oil-free alkyd resins have carboxyl and hydroxyl groups and are based on glycols, triols, and polybasic acids [32]. As in the fatty acid–modified types, water solubility is imparted by the residual carboxyl groups, which are neutralized with organic amines. Although these resins are generally more expensive than conventional solvent-based coatings, restrictions on the use of solvents will result in their increasing use in the industrial equipment, automotive, and appliance markets.

High-solids coatings offer an alternative solution to the restrictions on the use of organic solvents and the soaring cost of energy. Compared with a conventional industrial enamel applied at 30 vol % solids, an 80 vol % solid system shows an 89% reduction in total solvent required. Similarly, a 70 vol % solid system yields an 81% reduction in total solvent required. These considerations have been an impetus to the development of high-solids alkyds and hydroxylated saturated polyesters. The three criteria for resins for high-solids coating resins have been summarized by Antonelli [33]: controlled reactivity, low viscosity, and low volatility.

Considerable developmental work is currently being conducted on special alkyds and reactive polyesters for coatings applied as solvent-free powders.

The noncoating uses of alkyds represent about 5% of the alkyd market. Alkyd resins can be components of the following: plastics, caulking and sealing compounds, adhesives, nonwoven fabric binders, textile, printing paste, coating for printing plates, printing ink, phonograph records, drawing and lubricating oils and greases, polyurethane foams, finely divided metal dispersions, shoe-filler com-

Table 2 Modified Alkyd Resins

Resin	Uses in paints
Oxidizing alkyd resins	Architectural enamels, house paints, baking and air-drying undercoats and enamels for machinery, interior paints, flat wall paints, prefab housing structural units, and other factory products
Alkyd and phenoplast Alkyd and cellulose nitrate Alkyd and chlorinated rubber Alkyd and diisocyanate Alkyd and vinyl + epoxy	Air-drying or low-temperature-baking undercoats and enamels (for metal products) that have more plasticlike film properties than can be attained with alkyds alone
Alkyd and aminoplast Alkyd and aminoplast + epoxy Alkyd and silicone	Similar uses as above, but where a high premium is placed on color retention, and superior chemical and heat resistance

positions, and impregnating agents for seals and shoe-filler compositions. In addition to these uses, alkyds find application as binders in linoleum floor coverings and for sand in the manufacture of cores for foundry operations (core oils).

In general, the alkyd resins for these uses are very low cost products of very high oil content.

In adhesives, alkyds can play a role as binders and/or plasticizers.

The powerful forces of ecology, energy shortage, and inflation have presented technical problems of great magnitude. The increasing ecological restrictions on the use of photoreactive solvents and the dependence on limited resources of oil presents a bleak picture for the long-term future availability of petroleum-derived solvents. This situation has stimulated intensive research on water-soluble alkyds and polyesters, and although considerable progress has been made, this effort will continue at an increased pace with the availability of new polyfunctional building blocks.

The high-solids coatings offer great promise in the field of industrial and maintenance coatings, and progress has been made in developing viscosity reactive alkyds for high-solids coatings that can be crosslinked at ambient temperature. Fast-curing high-solids coatings based on acrylated alkyds and multifunctional acrylates can be formulated for a wide range of industrial applications [34]. The use of small proportions of multifunctional acrylates in the preparation of water-soluble air-drying alkyds greatly decreases drying time.

Another major line of investigation now being carried out in the alkyd field includes improvement in corrosion prevention and solvent resistance of coatings by combining alkyds with epoxies and other crosslinking polymers.

IV. PRODUCTION AND CONSUMPTION

Asphalt, shellac, and other naturally occurring resins have been used as protective and decorative coatings for centuries and were the major products of the polymer industry in the 1880s. Ready-mixed oleoresinous paints were available prior to the 1880s and accounted for 50% of commercial paints in the 1930s.

Elastomeric vulcanized natural rubber and hard rubber and oil-resistant polyethylene sulfide (Thiokol) and polychloroprene (Duprene, Neoprene) were all available commercially in the 1930s.

Viscose rayon and cellulose acetate rayon, as well as cotton, linen, silk and wool fibers, and nylon-6,6 bristles were also available in the 1930s. Shellac, casein, gutta percha, asphalt, cellulose acetate, and cellulose acetate butyrate as well as

Table 3 U.K. Production of Plastic Materials, 1951–1968 (1000 tons)

	1951	1953	1956	1959	1965	1968
Thermosetting plastics	130	110	158	235	312	330
Thermoplastics	65	100	155	423	630	855
Total	195	210	313	658	742	1185

Source: Ref. 35.

Table 4 U.K. Production of Thermosetting Plastics, 1968–1972 (1000 tons)

	1968	1969	1970	1971	1972
Aminoplasts	130.1	136.3	137.1	137.8	143.0
Phenolic and cresylics	68.4	73.5	59.9	55.3	54.5
Alkyds	65.0	64.3	66.4	71.3	60.6
Unsaturated polyester resins	34.0	38.5	40.8	39.4	47.9
Polyurethanes	35.3	30.9	33.4	36.4	41.8
Other thermosets	23.5	33.1	38.8	37.4	38.2
Total	356.3	376.6	376.4	377.6	386.0

Source: Ref. 35.

synthetic moldable polymers from phenol and formaldehyde (Bakelite), urea and formaldehyde, acrylic esters (Plexiglas and Lucite), polyvinyl acetals, vinylidene chloride (Saran), styrene, vinyl chloride, and vinyl acetate (Vinylite), and phthalic anhydride, glycerol, and unsaturated acids (Alkyds) were produced commercially in the 1930s. However, the annual worldwide production of resinous products at that time was less than 200 thousand tons; it increased to 1.5 million tons in 1950.

While the early 1930s and World War II saw the development of the newer family of thermoplastics, thermosetting plastics were still the most important until the early 1950s, as Table 3 shows [35]. Of the total production of 330,000 tonnes in 1968, 73% was accounted for by the older materials phenol–formaldehydes (PFs), urea–formaldehydes (UFs), and alkyds. The remaining 27% was accounted for by the newer materials such as unsaturated polyesters (UPs), epoxides (EPs), and polyurethanes (PUs).

The data regarding U.K. production of thermosetting plastics (Table 4) over 1968–1972 reveal an annual growth rate of 2.4% for aminoplasts, and decline of 5.5% a year for phenolics and cresylics, and a decline of 1.7% a year for alkyd resins.

In contrast, the output of polyester resins rose by an average of 8.9%, that of polyurethanes by 4.3%, and that of the other thermosetting plastics by no less than 12.9%. Overall, the growth of thermosetting plastics output was only 1.9% a year. During this time, the trend in consumption was as given in Table 5.

Table 5 U.K. Consumption of Thermosetting Plastics, 1968–1980 (1000 tons)

	1968	1969	1970	1971	1972	1980
Aminoplasts	110.8	116.8	111.3	110.7	110.9	—
Phenolic and cresylics	54.2	51.7	39.8	33.9	42.4	100
Alkyd resins	55.9	53.8	57.6	63.1	54.5	90
Unsaturated polyesters	33.4	37.2	40.9	38.5	48.0	100
Polyurethane foams	35.7	30.5	33.1	36.2	41.3	140
Other thermosets	—	—	40.0	37.5	34.8	—

Source: Ref. 35.

Table 6 U.S. Production of Alkyd Resins (1000 tons)

Year	Total alkyds	Phthalic alkyds	Nonphthalic alkyds
1960	257	257	—
1965	291	256	35
1970	282	266	16
1971	322	288	34
1973	311	280	31
1975	315	293	22
1980	340	310	30
1985	360	315	45
1990	380	335	45

Source: Ref. 36.

Tables 6 and 7 list, respectively, production figures and the consumption of alkyd resins in United States during 1960–1975 [36,37].

As is generally known, 1973 was a year of considerable expansion in demand and output. According to *Europlastic Monthly*, for alkyd resins decorative and industrial paints are of equal importance—amounting together to nearly 90% of usage. The remainder is accounted for by marine paints.

Table 7 Consumption in United States of Alkyd Resins in Paints and Coatings by Type of Alkyd, in 1975 and 1990 (1000 tons)

	Consumption	
Resin	1975	1990
Phthalic anhydride, unmodified	173	190
Drying and semidrying oil type	148[a]	160[a]
Nondrying oil type	25	30
Phthalic anhydride, modified	80	120
Resin or phenolic resin, modified	32	40
Styrene or vinyl toluene, modified	23	35
Acrylic, modified	9	10
Benzoic acid, modified	9	15
Other modified resins[b]	7	20
Isophthalic acid[c]	41	50
Polybasic acid (nonphthalic)	22	20
Total	316	380

[a]Includes some water-based resins and some resins modified to obtain thixotropic properties(typically, these alkyds contain 1–2% polyamide resin).
[b]Includes alkyds modified with silicone resins, vinyl monomers, and miscellaneous modifiers.
[c]Includes unmodified and modified resins.
Source: Ref. 37.

For building, the main application areas of alkyd resins are for binders and adhesives, surface coatings, and insulation [38].

REFERENCES

1. Kienle, R. H., U.S. Patent 1,893,873, 1933.
2. Berzelius, J., *Rappt. Ann. Inst. Geol. Hongrie*, 1847, 26.
3. Berthelott, M. M., *Compt. Rend. H., 37,* 398, 1953.
4. Bremmelen, J. Van, *J. Prakt. Chem., 69,* 84, 1956.
5. Smith, W., *J. Soc. Chem. Ind., London, 20,* 1099, 1961.
6. Callahan, M. J., U.S. Patent 1,108,330, 1914.
7. Friedberg, L. H., U.S. Patent 1,119,592, 1914.
8. Howell, K. B., U.S. Patent 1,098,728, 1914.
9. I. G. Farbenindustrie A. G., GB Patent 414,665, 1934.
10. Hovey, A. G., and Hodgins, T. S., *Paint, Oil, Chem. Rev., 102,* 2, 1940.
11. Lum, F. G., and Carlson, E. F., *Ind. Eng. Chem., 44,* 1595, 1952.
12. Cowan, J. C., in *Applied Polymer Science*, Craver, J. K., and Tess, R. W., Eds., ACS, Washington, D.C., 1975, Ch. 26.
13. Moore, D. T., *Ind. Eng. Chem., 43,* 2348, 1951.
14. Bhow, N. R., and Payne, H. F., *Ind. Eng. Chem., 42,* 700, 1950.
15. Peterson, J. N. R., *Am. Paint. J.*, 1948, 32.
16. Payne, H. H., *Organic Coatings Technology*, vol. 1, John Wiley, New York, 1954.
17. Hewitt, D. H., and Armitage, F., *J. Oil & Colour Chemists Assoc., 29,* 109, 1946.
18. Nylen, P., and Sunderland, E., *Modern Surface Coatings*, Wiley-Interscience, New York, 1965.
19. Satone, E., Todao, N., Wada, H., Makainda, Y., and Yanaka, M., Study of reactive diluent for air-dried alkyd paint, *J. Appl. Polymer Sci., 22,* 253, 1978.
20. Khan, N. A., *Pakistan J. Sci., 12,* 95, 1960.
21. Kienle, R. H., Van der Meulen, R. H., and Petke, F. E., *J. Am. Chem. Soc., 61,* 2258, 1939.
22. Wekna, K., and Klausch, W., *Farbe Lack, 59,* 85, 1953.
23. Bergmann, D. W., *J. Oil-Colour Chemists Assoc., 42,* 393, 1959.
24. Solomon, D. H., and Hopwood, J. J., Reactivity of functional groups in surface coating polymers. Part I. Hydroxyl groups in alkyd resins, *J. Appl. Polymer Sci., 10,* 981, 1966.
25. Lanson, H. J., *Paint Vern. Prod., 49,* 25, 1959.
26. Somerville, G., in *Epoxy Resins*, Skeist, I., Ed., Reinhold Publ. Corp., New York, 1955, 221.
27. Bulletin No. 03-003, Dow Corning Corp., Midland, Michigan.
28. Technical Bulletin, The Dow Chemical Co.
29. North, A. G., *J. Oil-Colour Chem. Assoc., 39,* 695, 1956.
30. Zachary, L. G., Bajak, H. W., and Eveline, F. J., *Tall Oil and Its Uses*, McGraw-Hill, New York, 1965, 5–21.
31. Technical Bulletin, Amoco Chemical Corp., St. Louis, Missouri.
32. Technical Bulletin, Eastman Chemical Products Inc., Baton Rouge, Louisiana.
33. Antonelli, J., *Am. Paint Coating J., 59,* 44, 1975.
34. Levine, E., Kuzma, E. J., and Nowak, M. T., *Mod. Paint Coatings, 66,* 8, 1976.
35. Whelan, A., and Bryson, J. A., *Developments with Thermosetting Plastics*, Applied Science Publ., London, England, 1975, 1–12.
36. *Chemical Economic Handbook*, Stanford Research Institute, Menlo Park, California, 1992.
37. NPCA Data Bank Program, Stanford Research Institute, Menlo Park, California.
38. Layman, R. E., U.S. Patent 3,893,959, 1975.

7

Epoxy Resins

I. DEFINITION AND HISTORY

Epoxy resins are those oligomers containing in their molecule at least two epoxy groups or two glycidylic groups able to participate in further crosslinking reactions. On the other hand, in commercial nomenclature there is one broad exception to this definition [1]. Certain aromatic-based polyalcohols are by convention called epoxy resins, although the epoxy group content may be less than two groups per molecule—or even zero.

The use of resins dates back to ancient times. Abundant evidence exists in the literature of Egypt, Babylon, Persia, India, Greece, and Rome to indicate that naturally occurring resins were used in incense, medicine, coating compositions, paints, and dyes. Franciscus Teophilus, an 11th-century monk, made an oil-modified varnish from a resin (probably amber) and linseed oil. The Renaissance painters used amber varnishes, and the notes of Da Vinci bear witness to his experiments with oils, varnishes, and pigments. The art of lacquering, based on the use of native natural resins, originated in China and reached its highest development in Japan, where the records of its use there go back to the fourth century [2,3].

It is hard to believe that the Swiss chemist Castan [4], in the 1930s when he was investigating glycidylic products with a view to obtaining a new dental material, even guessed that epoxy oligomers discovered by him would become so significant in 20 years, so that many industrial branches would be potential fields of application of these product properties. For glass fiber–reinforced plastics, a wide variety of adhesives, and numerous electroinsulating materials and in fields such as the aeronautics and aerospace industries, electronics, and civil engineering, even the existence of these new products and field's of activity cannot be imagined without epoxy resins.

These prepolymers, which have been called ethoxylenes, are now called epoxy resins. Castan [4] crosslinked these polyethers by heating them with phthalic anhydride. Dicyanodiamide and ditolylguanidine were also used as crosslinking agents, and patents were granted for the use of amines and reaction products of dimerized fatty acids and aliphatic polyamines (Versamids) [5].

Cross-licensing agreements for the many patents issued have been arranged among Devoe (Memphis, TN) and Reynolds (St. Louis, MO), Ciba-Geigy (New Orleans, LA), Shell (Atlanta, GA), American Marietta (Basel), Dow (Cleveland, OH), Reichold (Boston, MA), and Union Carbide. The trade names of Epon and Araldite are used to describe the epoxy resins sold by Shell and Ciba-Geigy, respectively. More than 140 thousand tons of these resins were sold in the United States in 1978.

Cycloaliphatic epoxy resins are produced by peracid epoxidation of unsaturated cyclic hydrocarbons by Union Carbide [6], and epoxy acrylates, called vinyl esters, are sold by Dow under the trade name of Derakane [7].

II. SYNTHESIS

The basic chemistry of the epoxy group and the application of this chemistry to the manufacture of epoxy resins is well covered in the literature [8,9].

The most common hydroxyl-containing compound used to prepare epoxy resins is bisphenol A (BPA), which reacts with epichlorohydrin to produce, via the following reactions, diglycidyl ethers:

$$R{-}H + ClCH_2{-}\overset{O}{\overbrace{CH{-}CH_2}} \longrightarrow R{-}CH_2{-}\overset{OH}{\overset{|}{CH}}{-}CH_2{-}Cl \qquad (1)$$

$$R{-}CH_2{-}\overset{OH}{\overset{|}{CH}}{-}CH_2{-}Cl + NaOH \longrightarrow R{-}CH_2{-}\overset{O}{\overbrace{CH{-}CH_2}} + NaCl + H_2O \qquad (2)$$

The preparation of the lowest member of the series, the diglycidyl ether of BPA, from epichlorophydrin (ECH), BPA, and sodium hydroxide can be represented as follows:

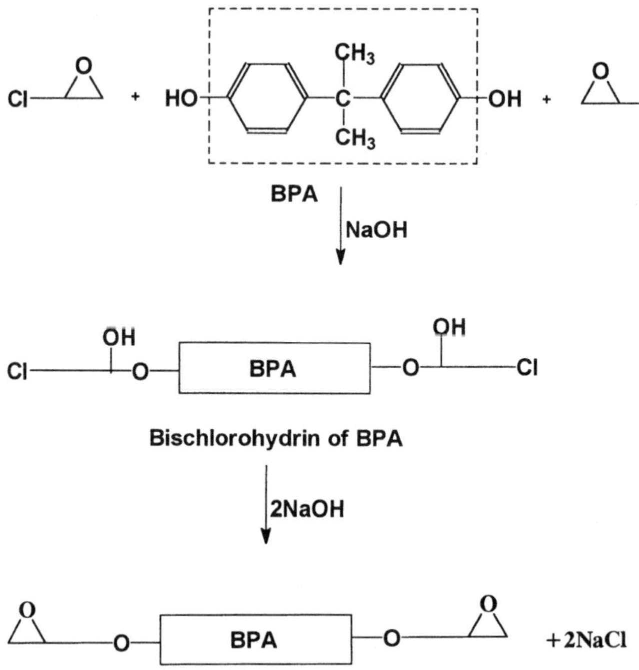

(3)

(4)

Diglycidyl ether of BPA (5)

From a mechanistic point of view, the reaction has been clearly demonstrated to consist of two separate sequential steps, the first catalyzed by sodium hydroxide and the second involving sodium hydroxide as a reactant. In commercial practice, sodium hydroxide in excess of catalytic amounts is present from the start, and the preparation is conducted as though it were a one-step process. Under these conditions, the ECH and the diglycidyl ether of BPA are in competition for the as yet unreacted BPA. As a consequence, some higher homologues are always produced:

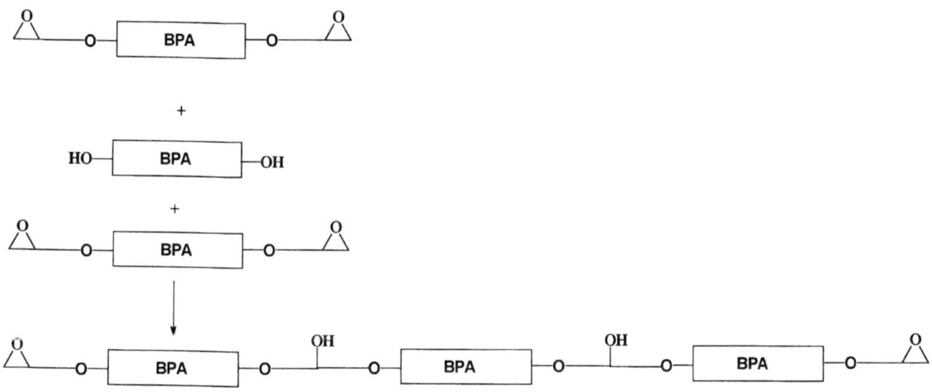

(6)

In the commercial manufacture of the low molecular weight liquid epoxy resins, the production of the higher molecular weight homologues is suppressed by providing a very high excess of ECH. Even so, a typical liquid epoxy resin contains about 88% diglycidyl ether of BPA and 12% higher homologues.

The manufacture of the higher molecular weight solid grades, which are of primary interest in the derivation of epoxy-based protective coatings, can be accomplished by two routes. In the first route, the so-called "tuffy" process, the scheme just outlined is followed, but the ECH/BPA ratio is kept low to encourage the formation of the higher homologues. By proper control it is possible to make numerous grades ranging up to molecular weights of 3500–4000 and corresponding epoxy equivalent weights of 2500–4000.

It is also possible to make solid epoxy resins by leaving out the ECH altogether and using the low molecular weight liquid epoxy resin as the only source of epoxy groups ("fusion" or "advancement" process). By using the proper amount of BPA and forcing the reaction to the point of essential exhaustion of phenolic hydroxyl groups (by the use of basic catalysts), the same grades of resin may be produced.

The "tuffy" process leads to products containing both odd- and even-membered diepoxide oligomers, whereas the "fusion" process leads to products that are composed of molecular species containing even-membered oligomers as the major component and smaller amounts of odd-membered diepoxide oligomers.

The reactions leading to the tuffy resins are schematically illustrated in the following:

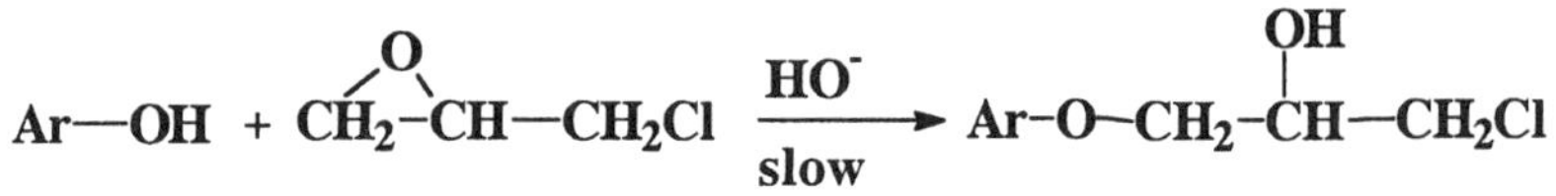

(7)

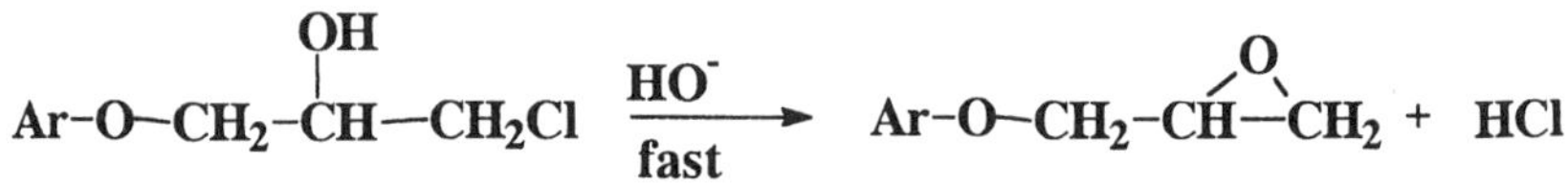

(8)

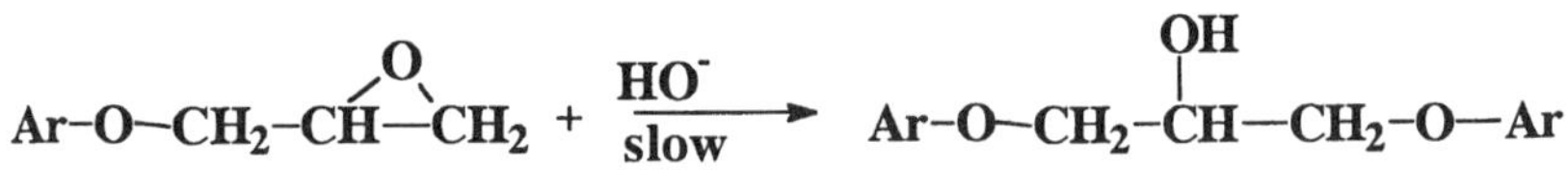

(9)

where ArOH represents a phenolic functional group.

The reactions shown above have been presented as normal additions of the phenolic group to the epoxide. Brode and Wynstra [10] have shown by studies on model reactions between phenol and ECH that the abnormal addition of the phenolate ion to the epoxide group,

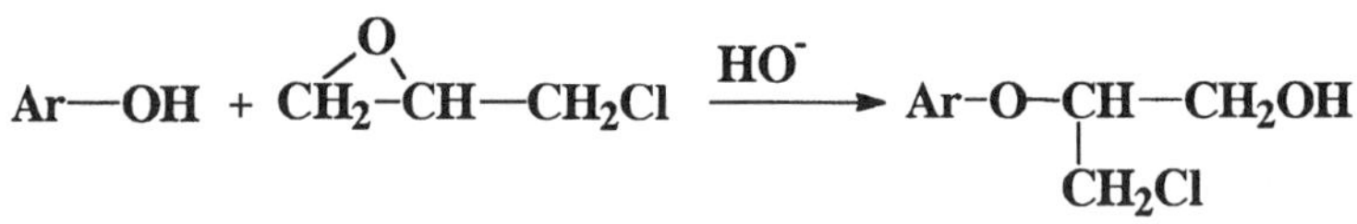

(10)

occurs only to a very small extent.

The dehydrohalogenation (reaction 8) is normally very rapid, as compared with reactions (7) and (9). In practice, tuffy resins are made under heterogeneous conditions. Under these conditions, it is likely that complete dehydrohalogenation does not occur. Furthermore, the possibility of hydrolysis of a small percentage of the epoxide groups leading to α-glycol end groups, cannot altogether be excluded.

A side reaction that can lead to chain branching is the base-catalyzed addition of epoxide to the aliphatic hydroxyl groups (mainly secondary OH) present in the mixture:

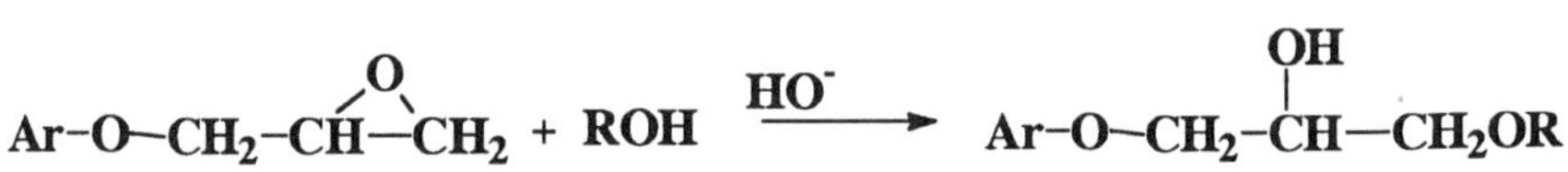

(11)

where R is an aliphatic group, such as $Ar{-}O{-}CH_2{-}\overset{|}{C}H{-}CH_2{-}O{-}Ar$.

At the beginning of the reaction of phenolic groups with epoxide, the concentration of aliphatic hydroxyl groups is zero. As the reaction progresses, the concentration of secondary alcoholic groups increases and the possibility of side chain branching increases.

The occurrence of such side reactions can partially account for the fact that the tuffy resin has a lower epoxide value than is implied by the measured molecular weight distribution [11]. The fact that the stoichiometric balance cannot be maintained due to incomplete dehydrohalogenation and the presence of monofunctional epoxides in the reacting mixture can appreciably limit the molecular weight attainable.

In the case of resins made by the advancement or fusion process, the situation is quite different. Side reactions such as cyclization during chain growth can also lead to deviation from the theoretically predicted structure. It is known that the reactions of BPA and ECH in ethanol or methanol lead to small amounts of cyclic dimer [12]. In practice, most epoxy resins obtained by tuffy or advancement processes contain small amounts of functional groups other than epoxide groups either as end groups or as side chains pendent to the main polymeric backbone, for example,

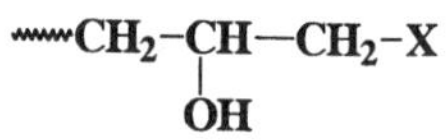

(12)

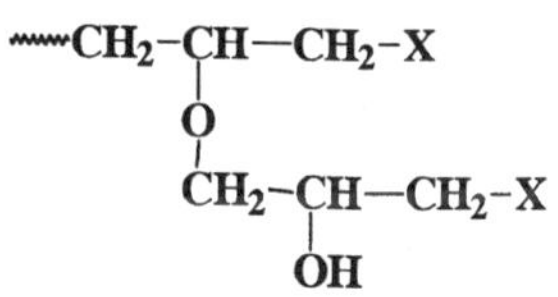

(13)

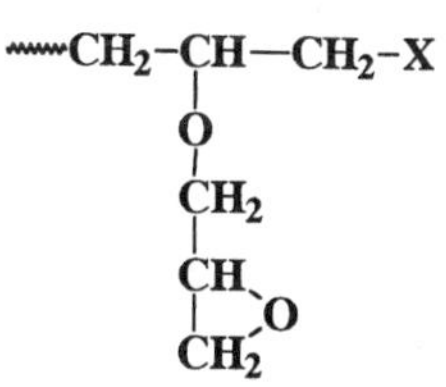

(14)

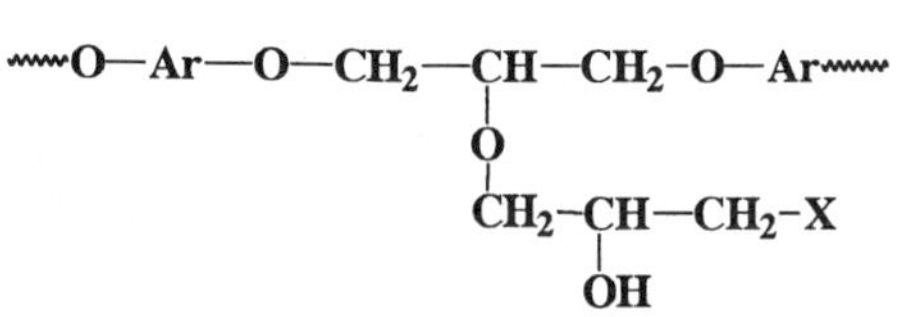

(15)

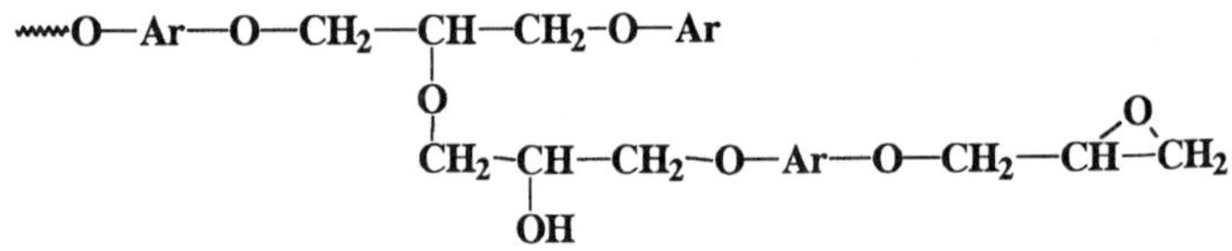

where X = Cl, OH and Ar = 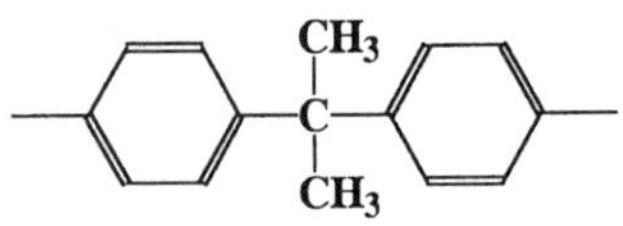

(16)

However, the results obtained by Batzer and Zahir [13] indicate that the content of such branching in commercial epoxy resins is very small.

The fusion process has many desirable features. For example, it is not necessary to handle the relatively volatile and toxic ECH. Nor are there volatiles or inorganic salts produced that must be removed, and all materials charged to the kettle end up as final product. There have, however, been serious drawbacks to the use of the fusion process, arising primarily from the catalysts heretofore available. These basic catalysts, while very effective in promoting the desired epoxy/phenolic hydroxyl condensation reaction, are also effective to a varying degree in promoting one or both of the known side reactions, namely, the epoxy/epoxy homopolymerization and the epoxy/aliphatic hydroxyl condensation.

Either of these side reactions causes the exothermic heat output of the process. Consequently, there is always the real possibility of an uncontrolled reaction leading quickly to gelation.

Most of these basic catalysts are susceptible to partial deactivation by the chlorohydrin content of the liquid epoxy resin feedstock. Therefore, the proper amount of base to be added must be redetermined for each new lot of feed resin.

At the completion of the fusion reaction the residual basic catalyst must be determined, or removed if a self-stable product is desired.

Fortunately, new catalysts have been developed for the fusion process that have none of the drawbacks of the conventional catalysts. These new catalysts, while very effective for the desired epoxy/phenolic condensation, are totally ineffective in promoting the competing epoxy/epoxy and epoxy/aliphatic hydroxyl reactions. With these catalysts the process is essentially self-controlling, proceeding to the exhaustion of the phenolic groups and then stopping. Quick quenching of the reaction is not important, as extended reaction time, within reason, does not affect product quality. Last, but not least, the catalyst essentially deactivates itself during reaction, remains completely soluble, and may be left in the product without detriment to subsequent product performance. Certain catalysts such as lithium salts and 1,1,3,3-tetramethylguanidine [14] are known to promote extensive etherification during the synthesis of epoxide resins.

Aromatic diglycidylethers of the bisphenol type form an important group of epoxy resins. Epoxy resins based on BPA have achieved the greatest significance. Efforts to expand the application areas of epoxides led to the investigation of other bisphenols as raw materials for the production of epoxy resins. First of all, 4,4′-dihydroxydiphenylmethane (bisphenol F) and bisphenol S were used. Epoxy resin based on bisphenol S,

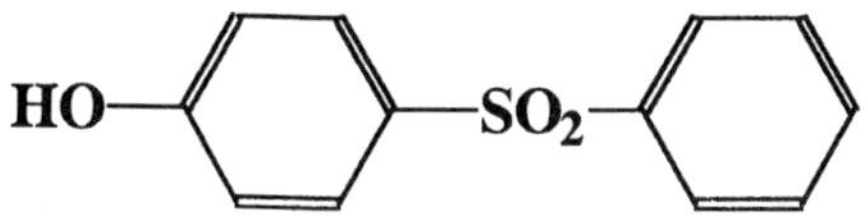

Bisphenol S (16a)

prepared by condensation of bisphenol S and epichlorohydrin, was described for the first time by Beavers [15]. It was obtained in the form of white crystalline powder with melting point 162–163°C.

At present, epoxy resins based on bisphenol S are used, for instance, in castings with good flexural and compressive strength [16], in production of laminates with greater toughness [17], for laminates and two-component adhesives with good thermal stability and chemical resistance [18], for coatings with higher thermal resistance than that of epoxy resins based on BPA [19–21], and for thermoplastic products characterized by relatively high strength at elevated temperatures and high resistance to mechanical shock [22–24].

Epoxy resins containing bromine atoms are particularly useful in circuit boards, electrical laminates, and potting and encapsulation applications in which flame-retardant properties are desired. Brominated resins that contain bromine atoms *ortho* or *para* with respect to a glycidyl ether group or a hydroxyl group have been proved to be the cause of inferior thermal stability and wire bond failure [25].

Brominated resins in which the bromine atom is in *meta* position to the phenolic hydroxyl or glycidyl ether group have been shown to be more hydrolytically and thermally stable than their *ortho*- or *para*-brominated counterparts [26,27].

Novel 2,2′,6,6′-tetrabromo-3,3′,5,5′-tetramethyl-4,4′-biphenol (TBTMBP) and its epoxy derivatives [28] were synthesized to incorporate the stable *meta*-terminated phenol moiety into epoxy resin systems. In electronic encapsulation and laminate applications, epoxy systems derived from TBTMBP exhibit superior hydrolytic and thermal stability as compared with the conventional *ortho*-brominated epoxy resins. These properties have resulted in an extended device life for semiconductors and a high T_g with excellent blister resistance for the printed circuit board, while meeting flame retardancy requirements as well.

Tetrafunctional epoxydic resins are prepared starting from aromatic diamines and epichlorohydrine, as shown in the following reactions [29]:

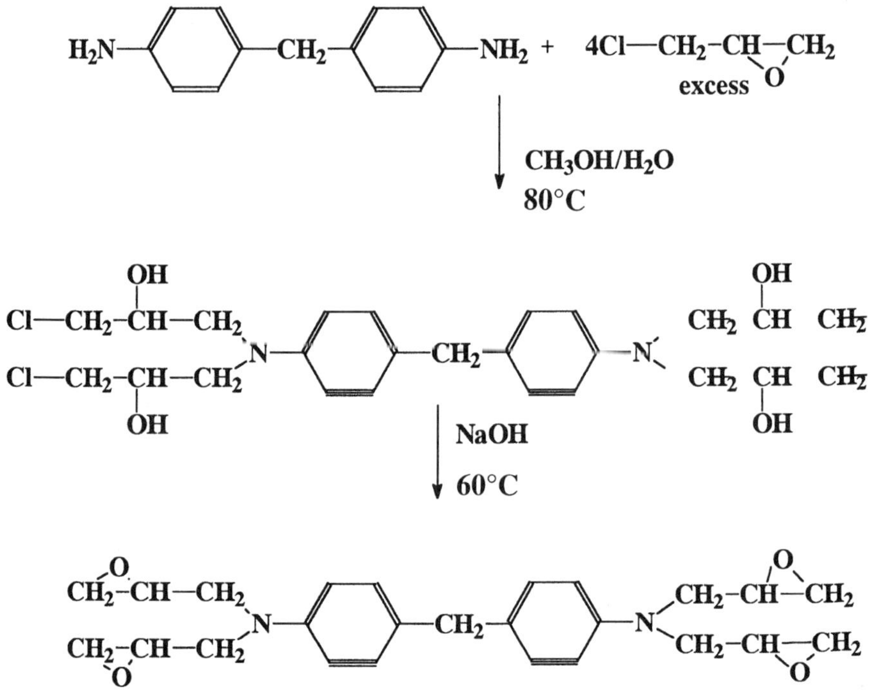

(17)

(18)

(19)

In these reactions, diamines presenting variable basicity (diamino-4,4′-diphenyl methane, diamino-4,4′-diphenyl sulfone, and 4-amino phenyl ether) were used. It was found that diamino-4,4′-diphenyl sulfone is less reactive than its homologue 4,4′-diaminodiphenyl methane [29].

The development of photonics and microoptoelectronics has created a strong need for transparent organic materials. These materials are suitable for molding lenses and prisms and for joining optical components to other components. Epoxy resin compositions are widely used for molding resins and adhesion. It is particularly important to be able to control the refractive index of optical resins in order to match them with optical components. Optical adhesive compositions containing fluorinated epoxy resins are a practical method of controlling refractive index [30–32]. However, conventional epoxy resins still have high refractive indices (1.51–1.58 at 0.59 μm wavelength), and so they cannot be matched with low refractive index optical components such as quartz and optical fibers (refractive indexes are ~1.46).

A novel epoxy resin with low refractive index, high optical transparency from UV to near infrared, high adhesive strength, and high resistance was obtained by Maruno et al. [32]. This resin is synthesized from benzene, hexafluoroacetone, and epichlorohydrin as shown in Schemes 1 and 2. Uncured epoxy resins IV and V (Scheme 2) show high UV transparency because they consist of nonaromatic hydrocarbons. Therefore, these resins should be advantageous in formulating UV curable epoxy compounds.

Using as starting material five fluorinated diols, Maruno et al. [33] reported the synthesis of fluorine-containing epoxy acrylate resins and epoxy methacrylate resins by the two reactions paths shown in Scheme 3. These resins are suitable as

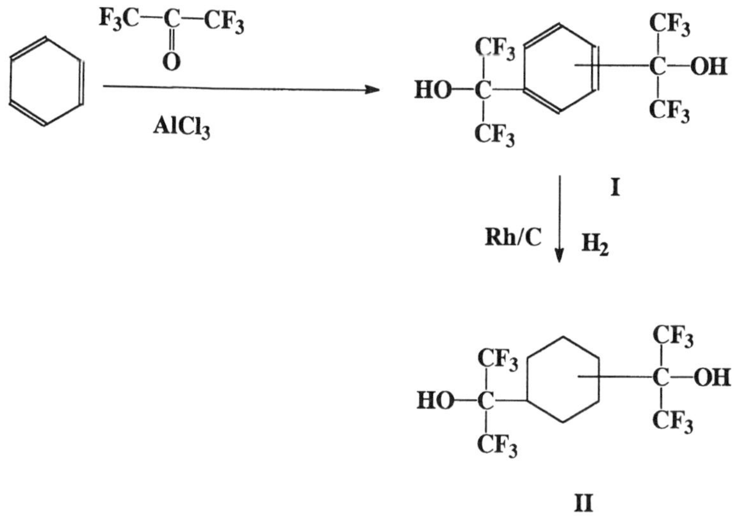

1

Scheme 1 Synthesis of fluorinated diol.

adhesives for fabricating optical communication devices, since they show high adhesive tensile shear strength and can match the refractive indices of optical glasses.

Manaka [34] prepared, by reaction of several diamines with opoxy-2,3-propoxymethacrylate (or glycidylmethacrylate, GMA), several methacrylated mono-

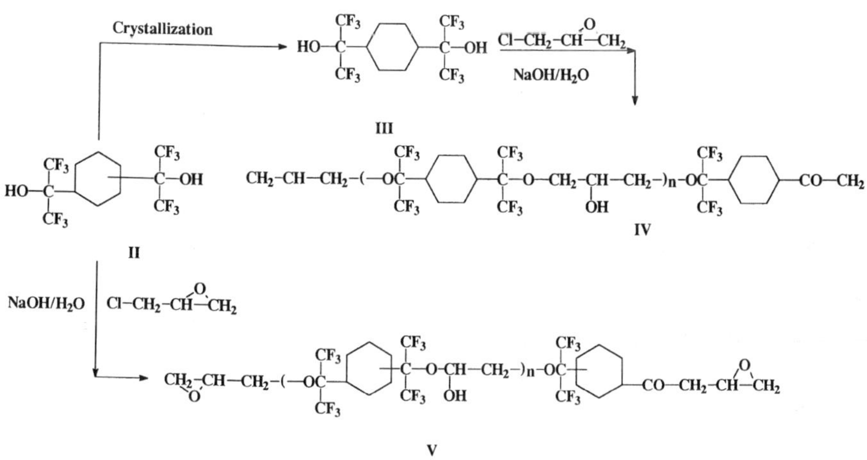

2

Scheme 2 Synthesis of fluorinated epoxy resins.

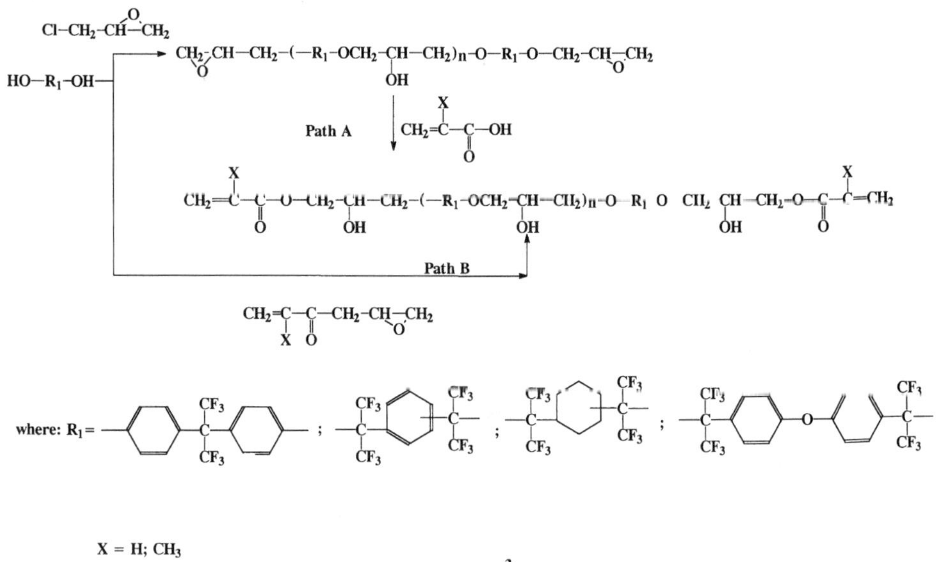

3

Scheme 3 Synthesis of fluorine-containing epoxy acrylated resins.

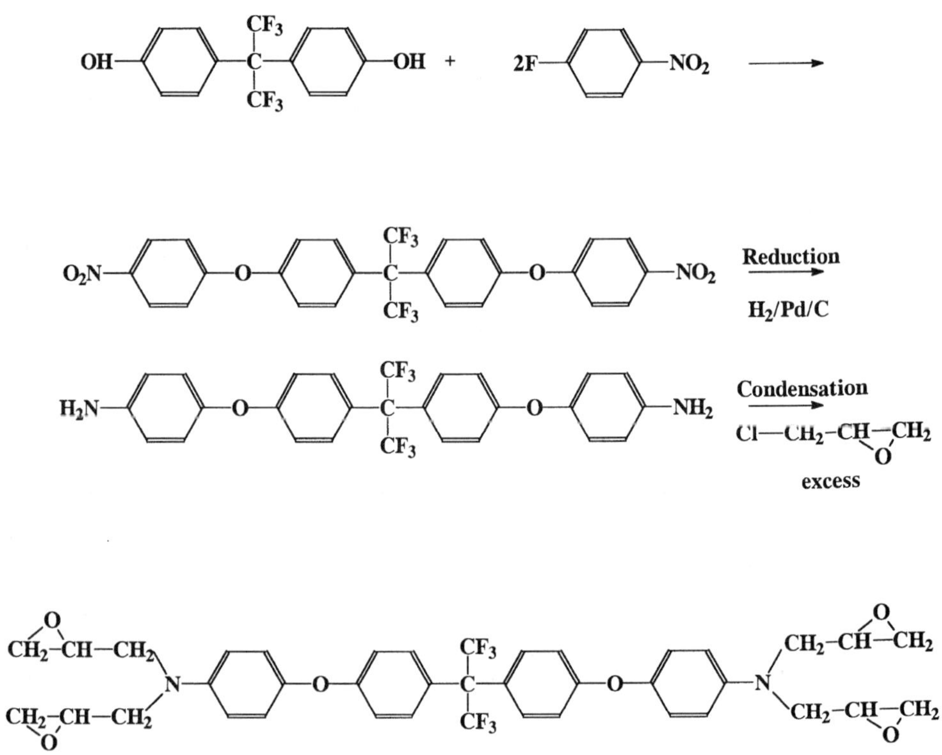

4

Scheme 4 Synthesis of fluorine-containing tetraepoxide.

mers. By radical polymerization of these monomers, he obtained very brittle compounds, due to the high crosslinking density of the networks formed. Klee et al. [35,36] prepared similar prepolymer by reaction of GMA with a BPA–diglycidyl ether diamine adduct, and Titier et al. [37,38] applied this reaction using a reactive extrusion procedure. The utilization of the extruder allows the preparation of these compounds ecologically and cost-effectively.

Based on the results obtained by Smolinski and Halaza [39], Boutevin et al. [29] reported the synthesis of fluorine-containing tetraepoxide with the chemical structure shown in Scheme 4.

Epoxy resins may be also obtained through the reaction of novolac resins or alcohols with epichlorohydrin.

The epoxy resins based on aliphatic alcohols are interesting because they impart flexibility to the final product, and the glass transition temperature, T_g, is lower than for aromatic epoxy resins, which can be an important feature when the cure is close to room temperature. The aliphatic resins should be useful products in civil engineering, where cure at room temperature is necessary.

The chemical structure of aliphatic epoxy resins was studied by Simoe Gandaro et al. [40] via gel permeation chromatography, Fourier transform infrared (FTIR) spectrometry, mass spectrometry, and wet analysis. The experimental data obtained reveal that when trimethylolpropane is used as starting monomer, only two OH groups are epoxidized. The resin and its different fractions have a functionality of two. Under the condition of synthesis used by the supplier (Triepox GA, Gairesa, Spain), three main products were found with molecular weights of 246, 339, and 430 and two other compounds with molecular weights of 282 and 375. No trifunctional product was detected (Schemes 5a–c).

When the reaction is carried out in an aqueous medium, such as a water solution of NaOH, other secondary reactions such as the following can occur:

$$R{-}O{-}CH_2{-}\underset{\diagdown\ \diagup \atop O}{C{-}CH_2} + NaOH \rightarrow R{-}O{-}CH_2{-}\underset{|\atop O^-}{CH}{-}CH_2{-}OH \qquad (20)$$

This reaction becomes significant without adequate control of all parameters. When the chemical structure of the resin is aliphatic, its water solubility (higher than aromatics) obliges us to protect the resin to avoid its destruction under alkali conditions. This is one of the most important reasons for producing aliphatic epoxy resins in a nonaqueous medium [41]. If alkali water solutions are used, a water-immiscible solvent will be necessary, which extracts the glycidyl ether as it is formed.

Alkyl acid phosphates react very rapidly with aliphatic epoxy resins to produce hydroxy oligomers. This reaction can be carried out in the presence of alkylated melamine to obtain crosslinked polymers suitable as coatings.

The reactions of phosphoric acid esters with epoxy resins have been employed to obtain fast-curing adhesives [42], corrosion-inhibiting coatings [43], fire-resistant materials [44], and high-solids coatings [45–48].

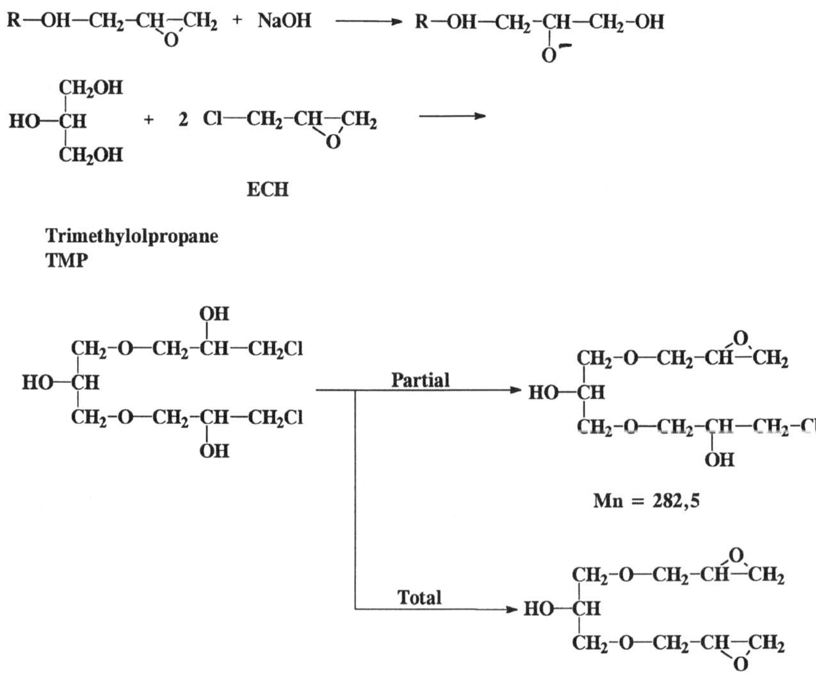

5a

Mn = 246

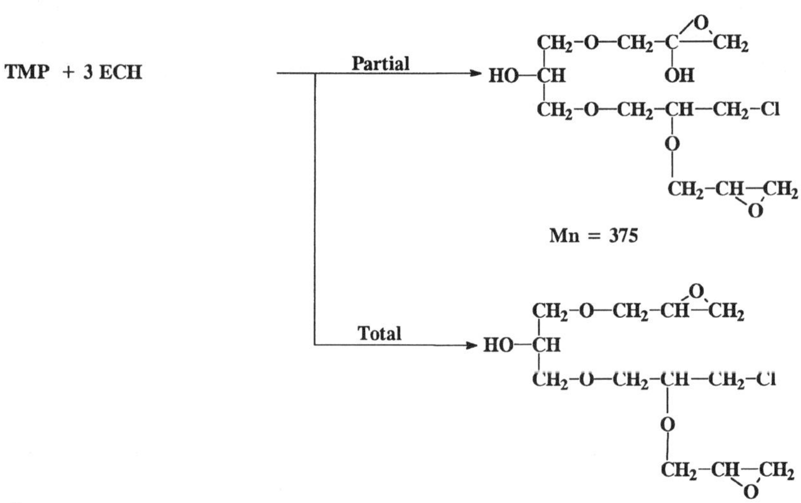

5b

Mn = 338,5

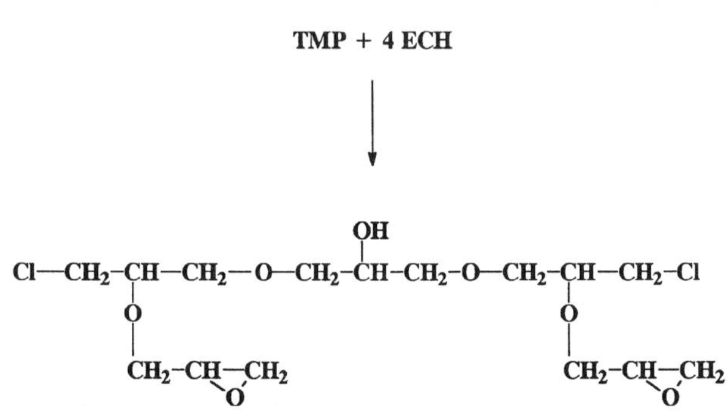

5c

Mn = 430

Coatings with high solids content (65% solids) are attractive for meeting emission control regulations and because of their lower energy consumption. To obtain high-solids coatings with adequate spray viscosity, one must start with low molecular weight resins, which on curing produce a desirable network only if the crosslinking reactions employed are fast and proceed without undesirable side reactions. The reaction of epoxy resins with partial esters of phosphoric acid impart these features. This reaction can be carried out in the presence of alkylated melamine to obtain crosslinked polymers suitable as coatings, as shown in the following reaction:

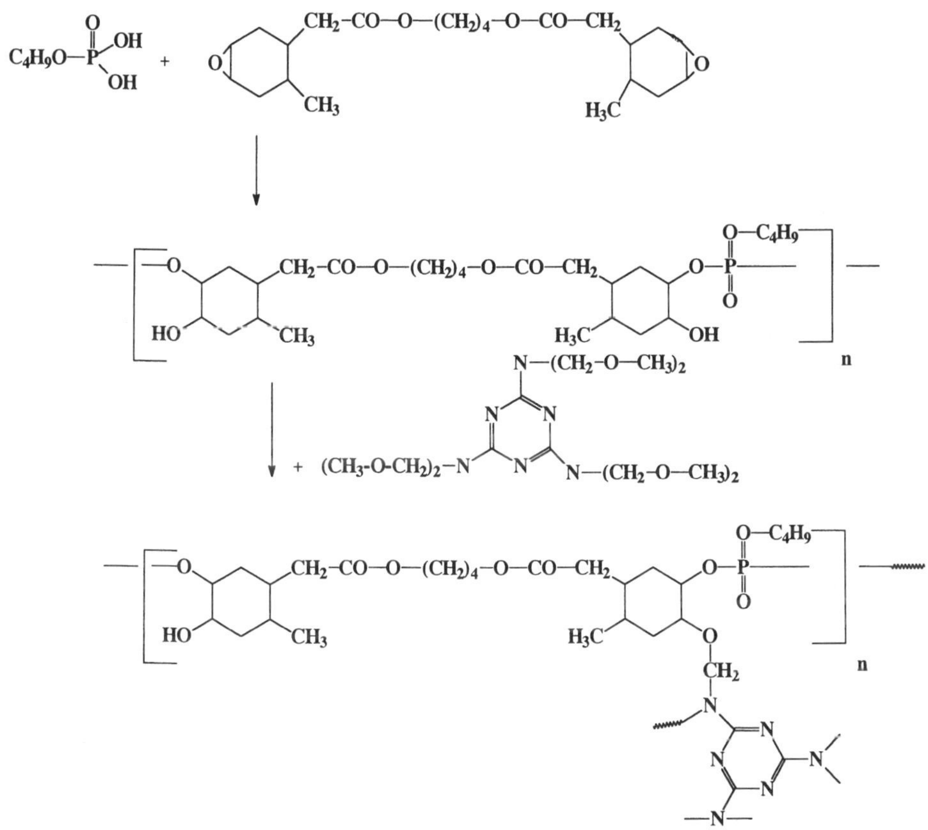

(21)

The hydroxy oligomers with the structural formula given in reaction (21) (compound A) are compatible with hydroxy acrylic copolymers, with which they can be combined to obtain paints with excellent physical properties. Because of the fast reaction between epoxy resins and phosphate ester, leading to an increase

---

Scheme 5 The chemical structure of the main constituents of epoxy resin TRIEPOX GA (Gainesa, Spain). (a) Molar ratio TMP/ECH = 0.5 (TMP = trimethylol propane; ECH = epichlorohydrine); (b) molar ratio TMP/ECH = 0.33; (c) molar ratio TMP/ECH = 0.25.

in viscosity, the paint should be formulated as a two-package system. These coatings have excellent physical properties and are very promising for automotive topcoats. The bake temperature (130°C/20 min) of these coatings is higher than that of the urethane coatings.

In epoxy resin technology, the chlorohydrin group is usually produced by reaction of ECH with an active hydrogen–containing compound. However, chlorohydrin groups may be generated in other ways, for example, by the reaction of hypochlorous acid with olefins, or by reduction of α-chloroketones. Epichlorohydrin, in addition to being used to produce resins as described previously, may also be used to produce epoxy resins from organic diacids, diamines, dithiols, and other active-hydrogen compounds.

III. CURING REACTIONS

The term *cure* in the chemical sense means complete or total cure. That is to say, all of the reactive groups, such as methylol and epoxy, or all of the reactive sites, such as unsaturation and hydroxyl, in a resin molecule are consumed during the hardening step. However, in the plastics industry, "cure" has come to mean "optimum cure," which is that degree of cure necessary to give the best physical properties for the application at hand.

The curing of epoxy resins may be divided into two classes: curing with hardeners, and curing with catalysts. Hardeners are defined as polyfunctional compounds that are used with an epoxy resin in a stoichiometric or near-stoichiometric ratio. Catalysts, on the other hand, are compounds that cause the epoxy resin to self-polymerize. They may be monofunctional and are always used in much lower amounts than the stoichiometric.

Cure of thermosetting resin is one of the key processes for thermosetting adhesives, and also for the process of resin-based composites. The strength of adhesive join and composite are directly related to the curing process. In industry, the cure of the resin system must be judged in order to determine the optimum conditions, including cure system, curing temperature, time, and so forth, and guarantee the quantity of cured resin or articles.

The cure of thermosets is the crosslinking of a linear macromolecule via a very complicated mechanism. As soon as the crosslinkings form, the resin cannot be solved and melted, which leads to quite a bit of difficulty in studying it. Traditionally, chemical analysis, IR, and calorimetry detecting the degree of conversion of reactive groups have been used to study the cure process. However, increased consumption of reactive groups is already not obvious at the last stage of cure. Thus, the sensitivity and function of these analysis techniques will be greatly reduced. But it is the last stage of cure that has a large effect on the physical and mechanical properties of the cured resin. To a large extent, it will determine the optimum properties of the resulting resin.

Certainly, the cure is related to mechanical strength. The cure is the process during which the modulus of the resin becomes large. In the last stage of cure, in which the sensitivity of some analysis techniques decreases rapidly, cure degree can be reflected in the mechanical strength in an obvious way. Therefore, the cure process can be studied successfully with mechanical methods [49,50]. Torsional braid, the dynamic spring method, and the dynamic torsional vibration method are some successful examples.

Hsich's nonequilibrium thermodynamic fluctuation theory [51,52] directly describes the changes in physical or mechanical properties of the curing system during cure. According to this theory, the physical or mechanical properties of the resin system during cure can be expressed as

$$\frac{G_\infty - G_{(t)}}{G_\infty - G_0} = \exp\left(-\frac{t}{\tau}\right)^{\beta} \qquad (22)$$

where G_0 and G_∞ are the initial and final physical or mechanical properties during cure, respectively, $G_{(t)}$ is the property at time t, τ is time parameter (relaxation time) of the reaction system, and β is the constant describing the width of the relaxation spectrum.

Using Eq. (22), He and Li [50] determine the theoretical and experimental cure curves and find a good agreement between them.

When a thermoset system cures, two principal structural transitions may occur: gelation and vitrification. Gelation marks the transition from a liquid to a rubbery state, since the crosslinking network has elastic properties not present in the low molecular weight, linear or branched prepolymers. The gel point is defined as the conversion at which the weight-average molecular weight, $\bar{M}_w$, diverges and an infinite network begins to form, which is insoluble in any solvent.

Vitrification involves a transition from the liquid or rubbery state into the glassy state as a consequence of an increase in molecular weight before gelation, or an increase in crosslinking density after gelation.

The epoxy resins were first introduced commercially in 1947. Like the polyesters, this type of resin does not liberate any volatiles on curing. For polymerization and crosslinking, it depends on the opening of the oxirane or epoxy ring with active hydrogen atoms. This type of resin is a fundamental improvement on phenolics, and it brings to industry some extraordinarily fine properties of chemical resistance, flexibility, adhesiveness, castability, and dimensional stability.

Epoxy resins may be converted to the thermoset state through the epoxy groups by reaction with well over 50 different classes of chemical compounds. Amines, polyamides, anhydrides, Lewis acids, ureas, melamine, imidazoles, BF_3: amine complexes, imides, and so forth, have been commonly used as cure agents for epoxy resins [53].

A. Amines as Curing Agents

The mechanism and kinetics of the epoxy–amine curing reaction have been widely analyzed [54–62]. Three principal reactions take place:

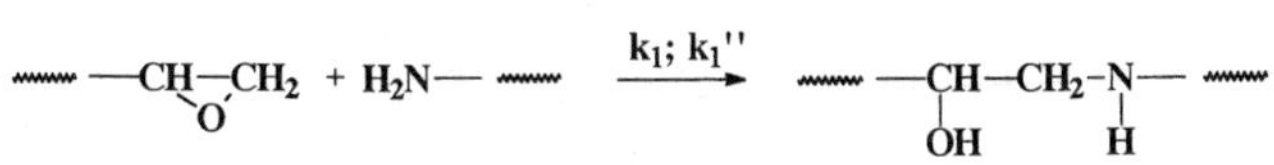

(23)

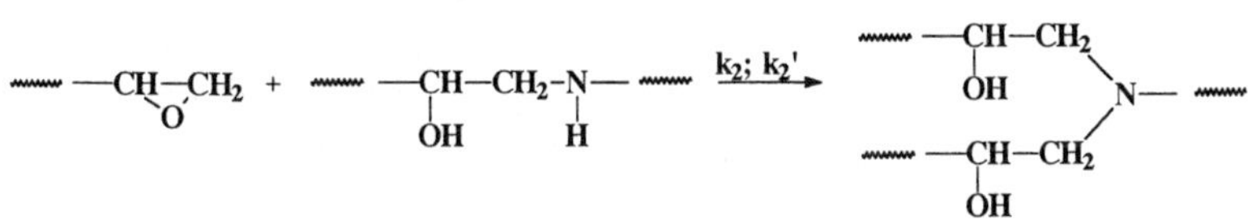

(24)

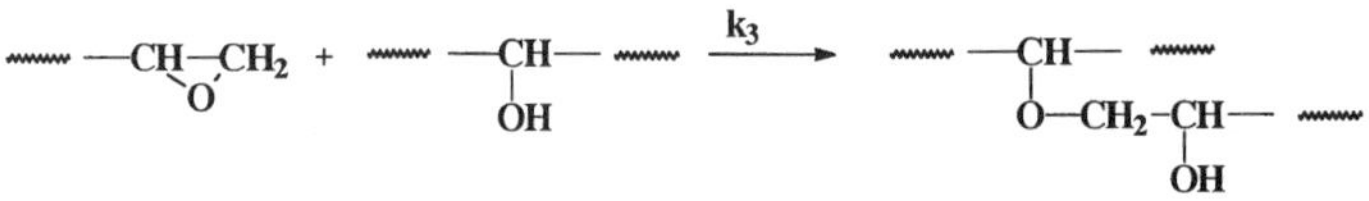

(25)

where k_1, k_2 and k'_1, k'_2 are catalytic and noncatalytic rate constants, respectively.

Reaction (23) illustrates the reaction of the primary amine hydrogen with the epoxy group. This is followed by the secondary amine hydrogen reacting with another epoxy group (reaction 24). Evaluation of the ratio of the rate constants $r = k_2/k_1$ was the subject of many studies [63–65]. This ratio is indeed supposed to have a strong effect on the cure process. Unfortunately, there is general disagreement in the literature concerning the values obtained. Moreover, though the ratio is generally reported to be temperature independent [54–57], some authors [66,67] found an increasing trend with cure temperature.

The reaction rate for an epoxy system can be expressed by the sum of all partial reaction rates.

The tendency of the etherification reaction (reaction 25) to take place depends on the temperature [63] and the basicity of amine [68], and it increases with the initial ratio of epoxy/amine [56].

It is generally accepted that the curing reaction occurs by two competitive mechanisms [54,58]: one is catalyzed by the hydroxyl groups initially present in the epoxy prepolymer and those generated during the reaction (rate constants k_1, k_2), and the other is a noncatalytic mechanism (rate constants k'_1, k'_2) identified by a second-order reaction. The catalytic mechanism vanishes at high temperatures due to the difficulty of forming the ternary transition complex, whereas the noncatalytic reaction takes place over the whole temperature range. Rosenberg [69] showed, in fact, using highly pure reactants, that the noncatalyzed reactions do not really occur and that nucleophilic addition on the epoxy ring proceeds with the preliminary activation of the epoxy ring by amines, which act both as electrophilic and as nucleophilic agents. However, both kinetic treatments can be used for the description of epoxy–amine cure reactions, because they lead to the same kinetic constants and activation energies.

In the early stage of curing, that is, when T_g is well below the cure temperature T_{cure}, the kinetics are mainly controlled by the rate of the chemical reactions. At higher conversion and after vitrification, the reactions become diffusion controlled.

In a recent study, Girard-Reydet et al. [62] found that the reactivities of different aromatic diamines, namely, 4,4′-diamine-diphenyl sulfone (DDS), 4,4′-methylenedianiline (MDA), 4,4′-methylene-bis-2,6-diethylaniline (MDEA), and 4,4′-methylene-bis-3-chloro-2,6-diethylaniline (MCDEA), cured with an epoxy prepolymer (diglycidyl ether of bisphenol A), are in the following order: MDA > MDEA > DDS > MCDEA. Although MDA, MDEA, and MCDEA exhibit very different reactivities, the ratio of the reactivity of secondary to primary amine hydrogens (k_2/k_1) is found to be the same ($r = 0.5$). On the other hand, the reactivity ratio of DDS is found to be $r = 0.45$. This lower value may be the result of the lower number of stable conformations afforded by the —SO_2— group as compared with the —CH_2— group. In this study no temperature dependence was observed.

The use of substituted ureas as curing agents for epoxy resins has been the subject of few studies. Ywakura and Izawa [70] analyzed the reactions between phenylglycidyl ether and substituted ureas. In every case, the reaction, carried out under mild conditions (T = 90°C, t = 1 h, mol ratio of epoxy/urea = 1–3, with or without the use of $N(C_2H_5)_3$ as catalyst), led to the formation of a 2-oxazolidone and a secondary amine:

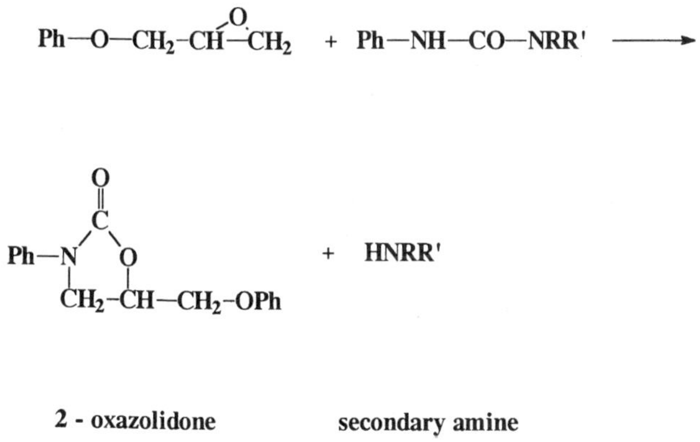

(26)

In excess of epoxides, the generated secondary amine may react further to give a tertiary amine:

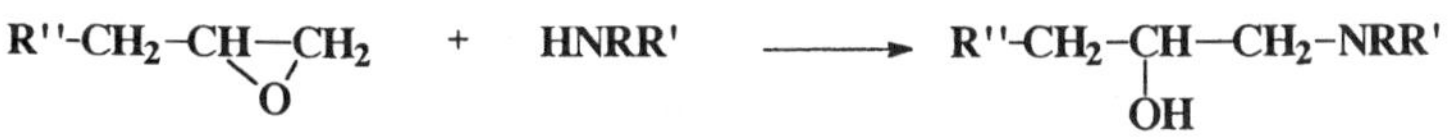

(27)

When the nature of R and R′ is such that the basicity of the tertiary amine is enhanced, an anionic homopolymerization of epoxides takes place through the following mechanism [71]:

(28)

(29)

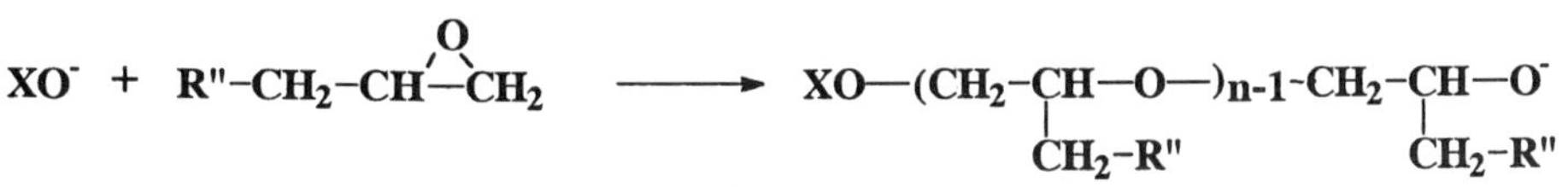

(30)

The need to have OH groups in the formulation in order to generate the initiator through reaction (29) has been experimentally confirmed [72].

Therefore, substituted ureas may be used as latent initiators for anionic homopolymerization of epoxy resins. Pearce and Morris [73] reported the use of 1,1-pentamethylene-3-phenyl urea, prepared by the reaction of phenyl isocyanate with piperidine in dry benzene, to cure a tetraglycidyl methylenedianiline resin toughened with carboxyl-terminated butyl rubber. These formulations showed an excellent stability at 23°C and led to a high T_g product when cured at 170°C.

As substituted ureas result from the reaction of a secondary amine with an isocyanate, a great variety of latent initiators may be generated by dissolving a convenient amount of a secondary amine in the epoxy resin and adding a stoichiometric amount of an isocyanate. By adjusting the addition rate of the isocyanate, the evolution of reaction heat may be kept under control. Moreover, as the isocyanate reacts very much faster with secondary amines than with any secondary OH present in the epoxy molecule, and no reaction is observed between NCO and epoxides at moderate temperature [74], the substituted ureas are the only products generated in the reaction medium.

In situ generation of the substituted urea may be used with advantage to replace the use of a volatile secondary amine, such as piperidine, in the cure of epoxy resin. In this case, it has been reported that an increase in the curing temperature leads to a decrease in the glass transition temperature T_g [75]. Losses of piperidine (b.p. = 106°C) during the cure lead to a decrease in the maximum epoxy conversion and, consequently, to a reduction in the T_g value. To avoid both the volatilization of piperidine and the reaction propagation in the fresh mixture, it is possible to convert the piperidine into a substituted urea by adding a stoichiometric amount of an isocyanate. This will increase the latency of the initial formulation and give reproducible curing schedules [76].

B. Carboxylic Acids and Anhydrides as Curing Agents

The useful Bronsted acid type curing agents for the epoxy resins are polyfunctional phenols, polyfunctional organic acids, and anhydrides of mono- and difunctional organic acids.

The reactions of epoxy resins with the acid curing agents are more complex than those with the amine curing agents and are somewhat less well understood [77,78]. The reactions may be brought about with or without catalysts. The uncatalyzed reaction, in simplified form, may be represented by

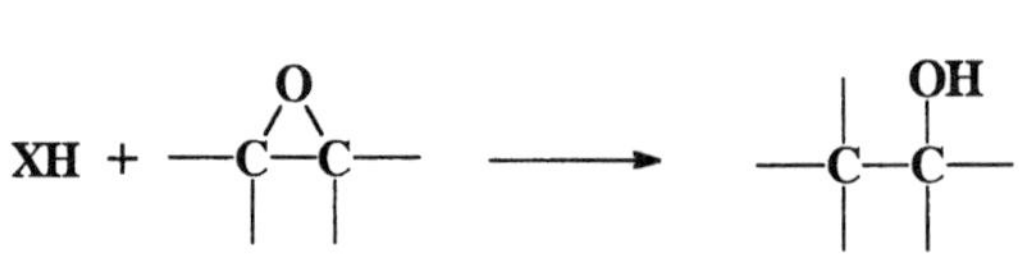

Esterification (31)

Independently, under acid conditions, a second reaction occurs, which is catalyzed by the acid:

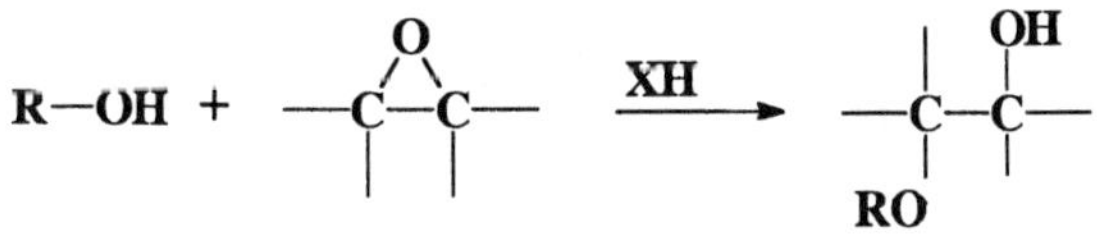

Etherification (32)

Uncatalyzed, the alcohol–glycidyl ether reaction was found [77] to be rather sluggish; a temperature of 200°C or higher was required to realize a conveniently rapid rate.

Several considerations regarding uncatalyzed system apply to the reactions of acids with epoxy groups:

In the reaction of acids with diepoxides, the independent etherification reaction is required to obtain the necessary system functionality for a crosslinked system.
The ratio of the etherification reaction to the esterification reaction will determine the cured properties.
The epoxy groups with higher electron densities will be more reactive.
The speed of the reaction will be a function of the value of system acidity.

Although acid-catalyzed and uncatalyzed alcohol–epoxide reactions are known to lead to a mixture of isomeric species, the base-catalyzed reactions give a product that is almost exclusively a primary ether and a secondary alcohol. If, then, a secondary alcohol is reacted with a glycidyl ether under conditions of base catalysis (to yield a new secondary alcohol), the almost exclusive reaction will be that of glycidyl ether with secondary alcohol. The reaction would still be complicated by the competition of the new hydroxyls with the starting alcohol for epoxide.

The polymerization character and the kinetics of the base-catalyzed alcohol–glycidyl ether reaction are consistent with the following mechanism: First, some concentration of alkoxide ion is generated from the catalyst and either the epoxide or the alcohol. This anion then reacts with epoxide to form a new alkoxide ion, which can continue to add epoxide:

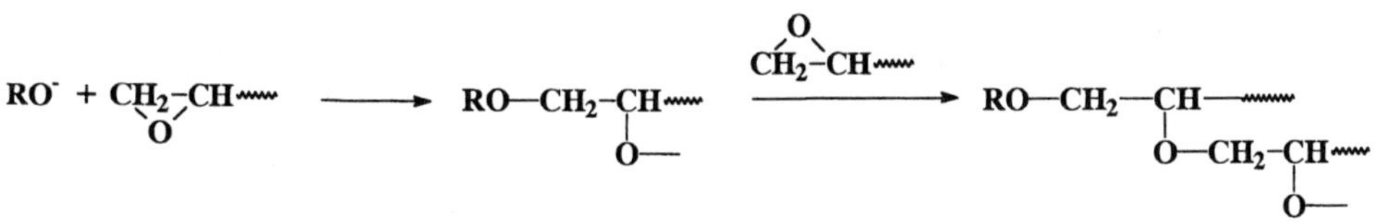

(33)

Contact with another molecule of alcohol results in termination of this polymer chain and generation of a new alkoxide ion to start the polymerization process over again:

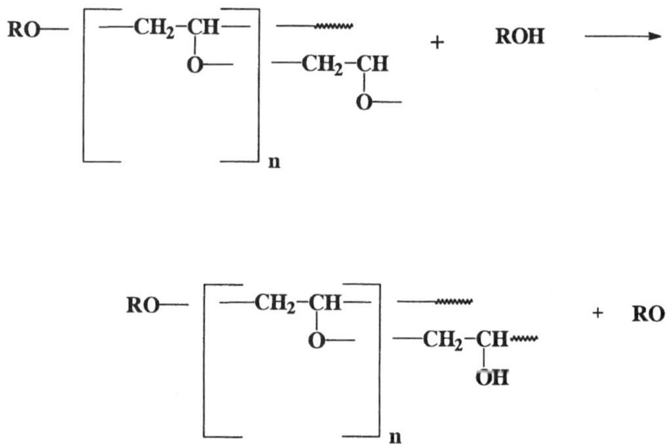

(34)

Tertiary amines are believed to initiate epoxide polymerization by their addition to epoxide, followed by reaction with alcohol,

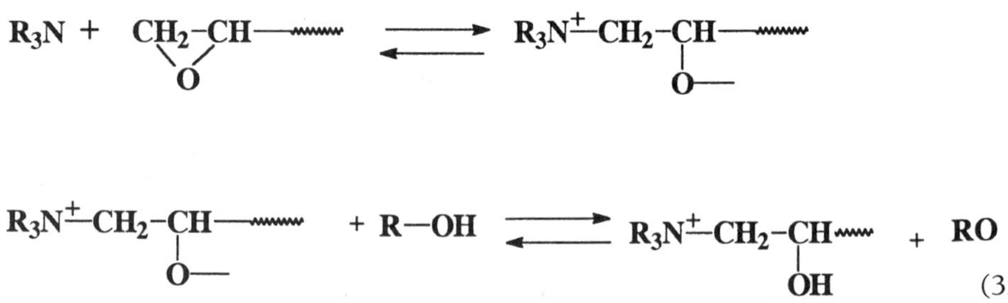

(35)

The base-catalyzed reaction of epoxide with phenol, in equimolar amounts, is somewhat slower than that of the alcohol–epoxide with the same catalyst. The mechanism of reaction is given in the following reactions:

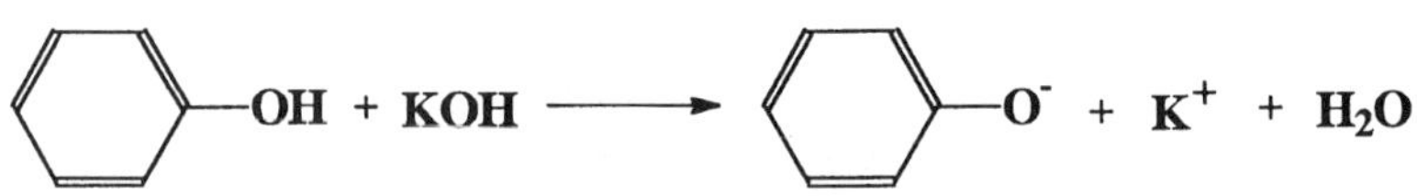

(36)

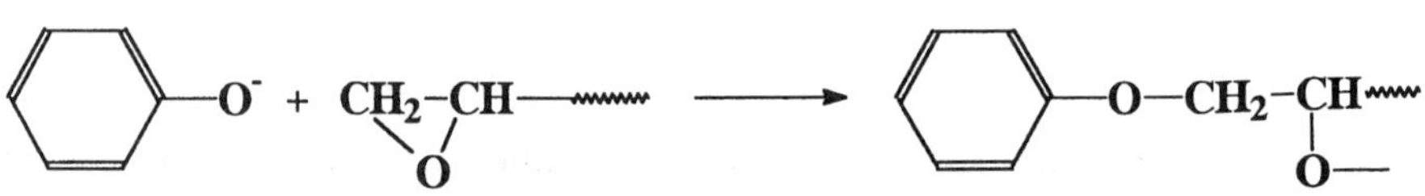

(37)

This alkoxide ion, because of its high order of basicity, reacts immediately with the phenol present in excess to generate phenoxide ion to repeat the cycle and at the same time exclude the possibility of the following side reaction:

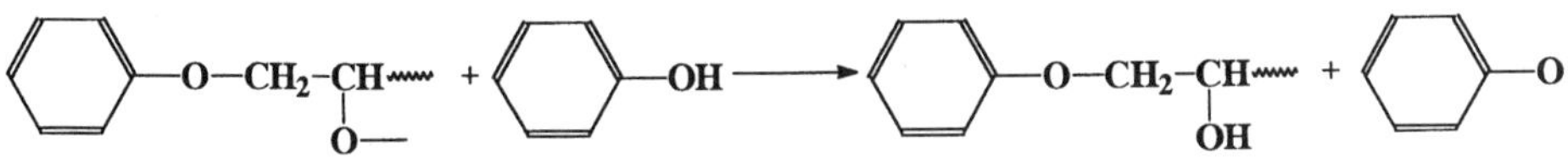

(38)

In the case of cure reaction with carboxylic acids, four types of possible reactions were anticipated [79]:

(39)

$$\text{R—COOH} + \text{—CH(OH)—} \longrightarrow \text{—CH(OOC—)—} + H_2O \qquad (40)$$

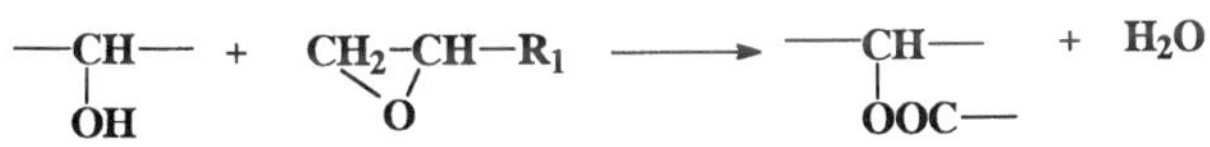

(41)

(42)

When base catalyzed, the following sequence of reactions should result in a high order of selectivity:

$$\text{R—COOH} + \text{Base} \longrightarrow \text{R—COO}^-$$

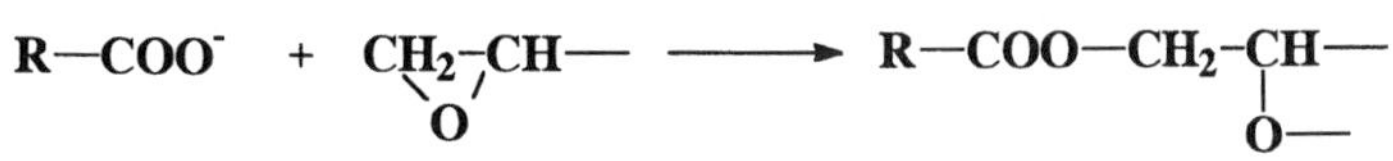

(43)

In practice, difunctional acids and diphenols are little used as curing agents for epoxy resins. They are, however, often used in combination with a second, more highly functional or efficient curing agent (novolac resins or fatty trimer acids).

The curing reaction between an epoxy resin and an acid anhydride with or without catalyst has been studied by Tanaka and Kazinschi [78], Fisch and coworkers [80,81], Wegler and coworkers [82], Dearborn and coworkers [83,84], Fischer [85], and Schechter and Wynstra [79]. Anhydride curing agents are widely used with epoxy resins. The majority of these are cyclic anhydrides.

Inspection of the formulas of an epoxide and cyclic anhydride side by side reveals that an equimolecular mixture has the elemental composition of a polyester. Only a catalyst is needed to isomerize, in effect, such a mixture to a linear polyester. In such a plymerization, both epoxide and anhydride must be strictly bifunctional toward each other. Careful studies of the uncatalyzed reaction have been made by Fisch and Hoffman [80,81], who showed that some epoxide homopolymerization occurs. This gives systems in which the anhydride is only 0.5 to 0.85 mol per mol of epoxide.

It has been found that tertiary amines are catalysts that not only affect the rate of the anhydride–epoxide reaction, but its course as well [82,83]. Dearborn and coworkers [84], for example, report that the highest crosslinking in amine-catalyzed anhydride–epoxide resin cures was obtained at a 1:1 anhydride:expoxide ratio. This implies that the amine largely or wholly suppresses the undesirable side reaction of epoxide polymerization. This method of curing is potentially a powerful tool in polymer chemistry, since in its broadest sense it enables one to prepare a linear polyester containing at regular intervals along the chain various substituents available for further reactions.

The following mechanism is consistent with the experimental data described in the literature [82,83] for the cure of epoxy resins with anhydrides in the presence of tertiary amines. In the first step, an activation of anhydride by the amine takes place:

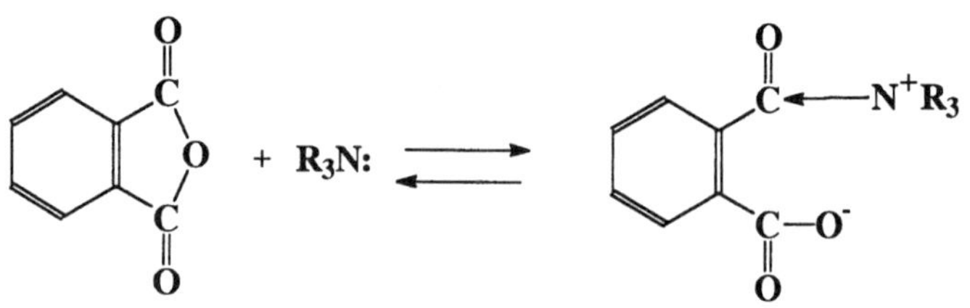

(44)

Note that the activation of the anhydride causes deactivation of the amine. This helps to explain the actual suppression of epoxide homopolymerization by the anhydride–amine combination.

In the second step, the carboxyl anion reacts with the epoxide:

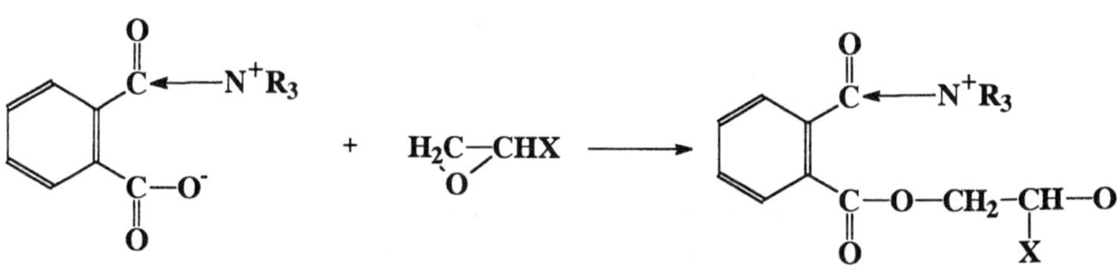

(45)

Further, the alkoxide anion reacts with the anhydride:

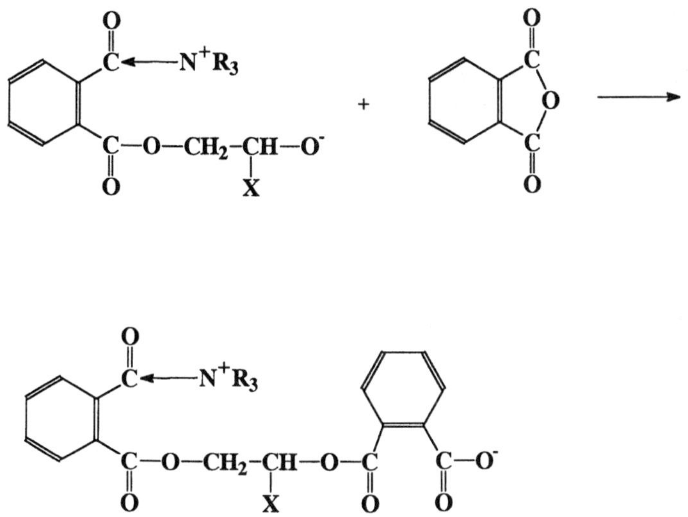

(46)

Continuation of these alternating steps would give rise to the polyester.

In a further step, a displacement of the amine by an alkoxide ion would also continue polyester formation:

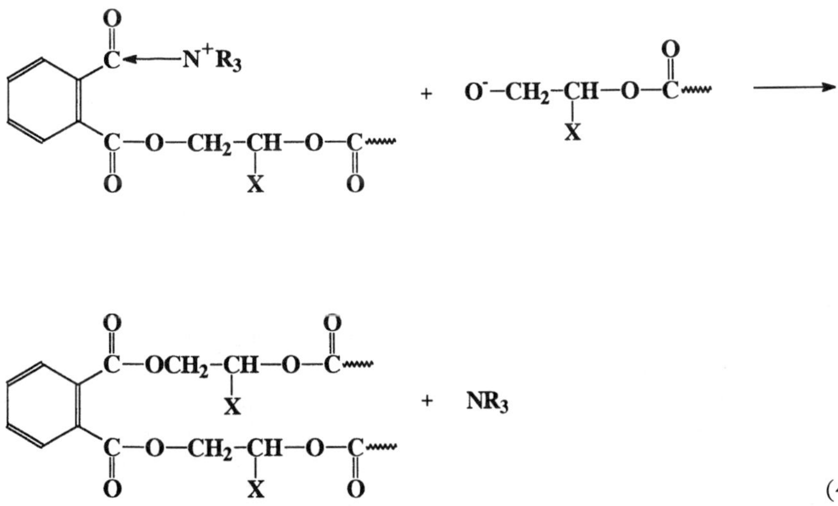

(47)

Other workers, notably Schechter and Wynstra [79], have proposed a stepwise mechanism involving initiation by small quantities of alcohols, which react with anhydrides to form carboxy esters, which in turn react with the epoxides giving hydroxy esters. In this case, the hydroxyl impurities control the molecular weight of the product by chain initiation as well as termination. The rate, moreover, should also depend on hydroxyl concentration.

Among the anhydrides, phthalic anhydride has been most used because of its low cost, high commercial purity, satisfactory reaction rate, and light-color products. Hexa- and tetrahydrophthalic anhydrides have also been very satisfactory. Aliphatic anhydrides, especially succinic, have been slower reacting and have given dark products. A number of polymeric monocyclic anhydrides, notably adipic, isophthalic, and terephthalic acid anhydride, have been used successfully.

Maleic anhydride turns very dark with tertiary amines. In the epoxide system, it also gels when very little anhydride has reacted. Gelling probably occurs when an anion is added to one of the resonance forms of maleic anhydride:

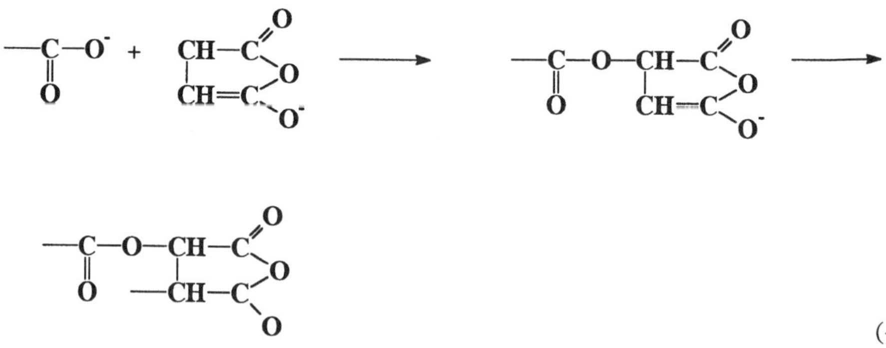

(48)

This product can then further react bifunctionally, giving crosslinked materials.

C. Lewis Acids as Curing Agents

Although a variety of boron trifluoride complexes have been described in the literature [84] as catalytic curing agents for epoxy resins, only one of these, the boron trifluoride monoethylamine complex (BF_3–MEA), has been used to a great extent in epoxy technologies: coatings, encapsulation, preparation of glass fiber–reinforced materials, and so forth. The BF_3–MEA-cured epoxy resins have good strength, heat resistance, and electrical properties. This type of initiator reacts as a semilatent initiator and provides long pot life at room temperature. It is conveniently applied in the reaction-injection molding (RIM) process and can also act as an accelerator for diamino-diphenyl sulfone (DDS)-cured epoxy resins [85,86].

The mechanism of epoxy–prepolymer curing with amine–boron trifluoride complexes is controversially discussed in the literature. It is often postulated that the BF_3–amine complex reacts at its dissociation temperature [85,86], but this dissociation temperature has not been determined experimentally. Arnold [87] postulated that a proton is released in the dissociation of the amine complex,

$$F_3B\cdot H_2NR' \rightarrow [F_3B\cdot HNR']^- + H^+ \quad (49)$$

followed by an addition reaction of this proton to the oxygen atom of the oxirane ring to yield an oxonium ion,

(50)

the counterion being $[F_3B\cdot HNR']^-$ = X produced in reaction (49).

Chain propagation occurs by reaction with an additional epoxide:

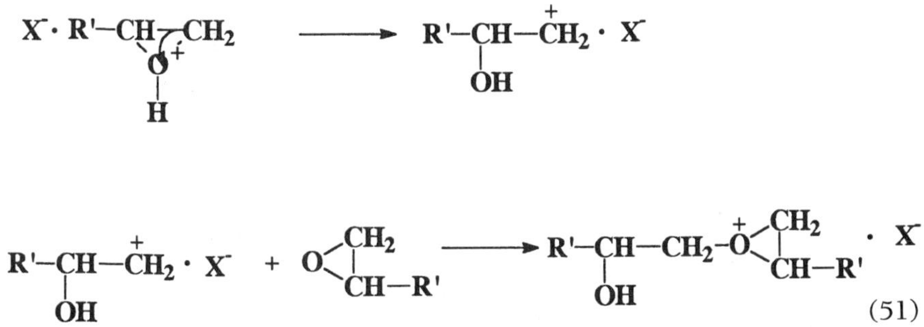

(51)

In the meanwhile, Smith et al. [88] proposed another path of BF_3–amine complex dissociation:

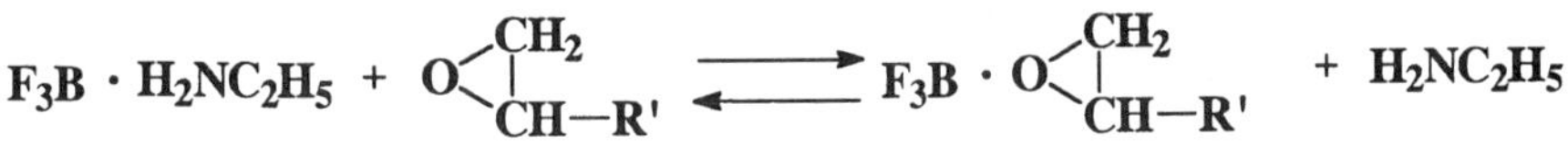

(52)

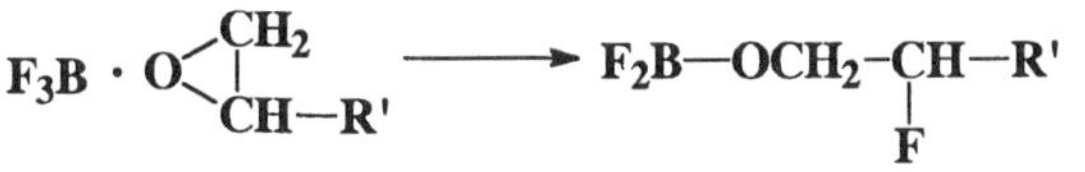

(53)

(54)

However, the primary amine complexes of BF_3 have not yet been reported in the literature to dissociate reversibly. Therefore, it seems clear that dissociation cannot be the primary reaction proceeding the cure of epoxy oligomers by BF_3 complexes.

On the other hand, following Smith and Smith [89], BF_3–MEA is useful because it is a latent catalyst, that is, it is inactive at room temperature and requires elevated temperature (120°C or higher) to be activated. The experimental data obtained by Smith and Smith [89] indicate that BF_3–MEA is not a true catalyst at all. At epoxy resin cure temperature (120–150°C), it rapidly converts to fluoroboric acid, and it is the fluoroboric acid that catalyzes the epoxy cure [89]. Perhaps, the fluoroboric acid is produced as follows:

$$2BF_3 \cdot NHR_2 \longrightarrow BF_4^- \cdot N^+H_2R_2 + BF_2^+ \cdot N^-R_2$$

$$BF_4^- \cdot N^+H_2R_2 \longrightarrow HBF_4 + NHR_2 \qquad (55)$$

with the fate of the $BF_2^+NHR_2^-$ not yet determined.

Harris and Temin [90] have proposed that the complexes react directly with the oxirane ring:

(56)

With respect to this reaction, weak bonds are formed between the oxygen of the oxirane ring and one hydrogen atom of the amine. This explanation is also supported by Ito and Okahashi [91]. In addition, they have suggested a termination in which boron trifluoride complexes with tertiary amine are produced:

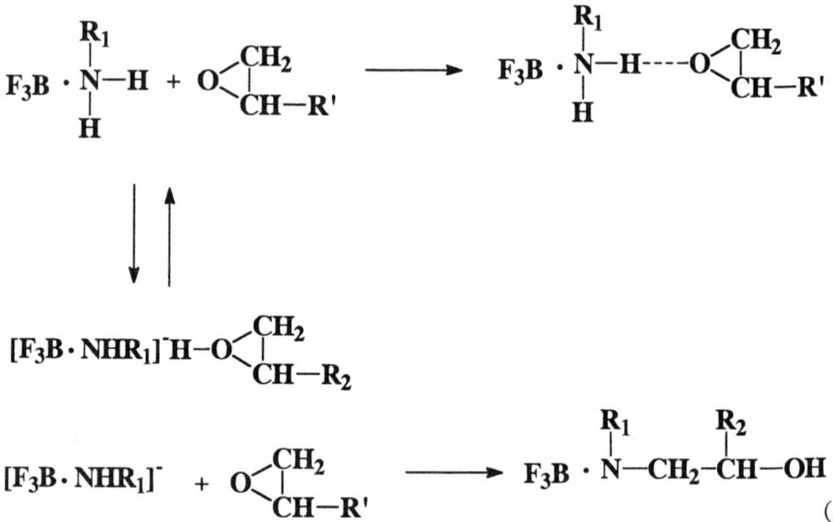

(57)

Okamoto [92] has studied the cationic polymerization of epichlorohydrin in the presence of ethylene glycol, using as initiator triethyloxonium hexafluorophosphate, and proposed a propagation mechanism similar to living polymerization.

On the other hand, Penczek et al. [93], using similar conditions, proposed two possible paths of propagation:

1. An active chain end:

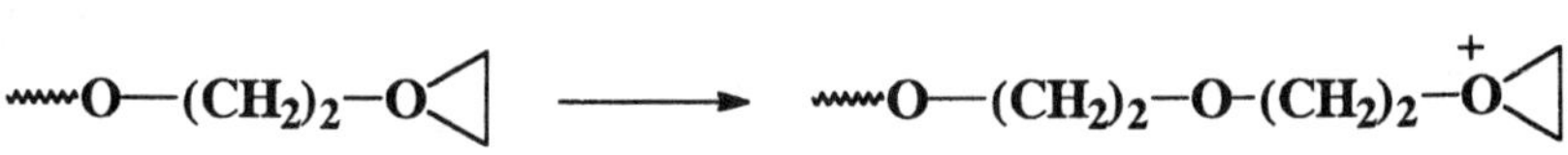

(58)

2. An activated monomer:

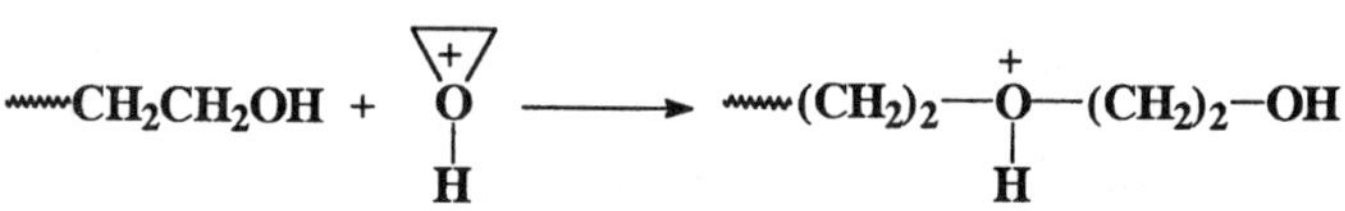

(59)

Bouillon et al. [94] compare these results with the experimental data obtained in the case of homopolymerization of epoxy–prepolymers and monomers initiated by a BF_3–amine complex in the presence of low molecular weight poly(ethylene oxide). Based on these data, they proposed the following mechanism of polymerization of the system epoxy–prepolymer–monomer–$BF_3 \cdot$amine complex.

1. Initiation and propagation by the activated chain end
   Initiation:

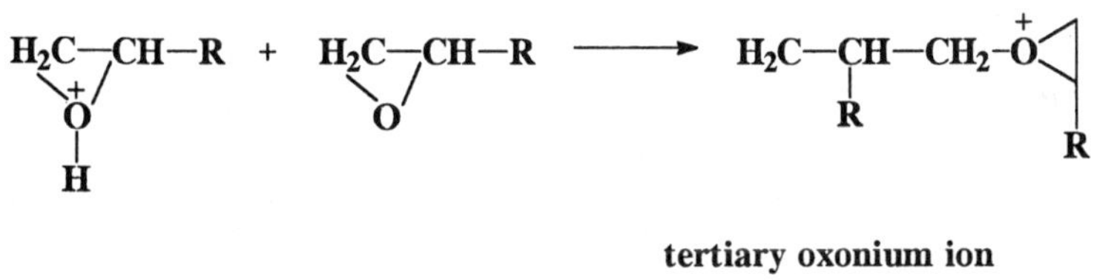

(60)

Propagation:

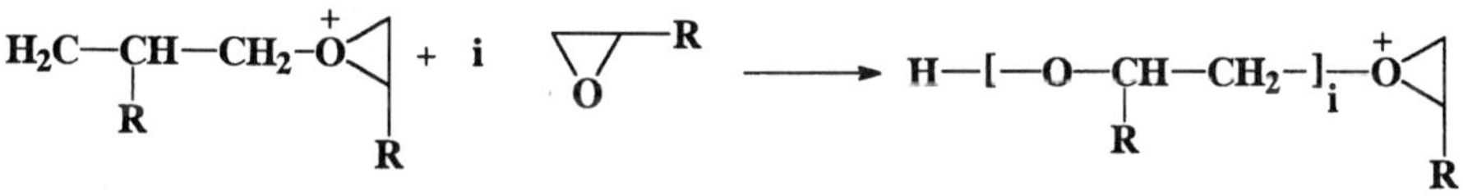

(61)

2. Initiation and propagation by the activated monomer mechanism
   Initiation:

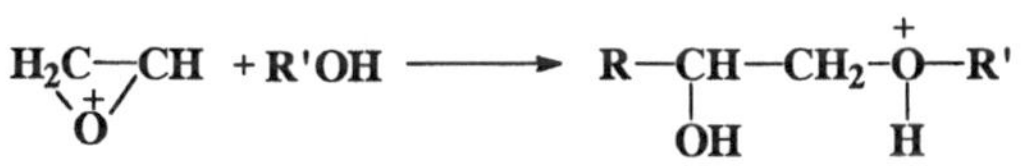

(62)

Propagation:

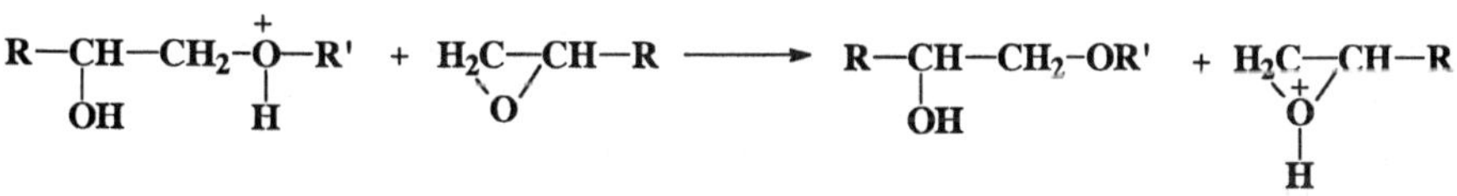

(63)

D. Imidazoles as Curing Agents

Azole compounds as imidazole, benzimidazole, benzotiazole, and 2-mercaptobenzimidazole are of interest as curing agents for epoxy resins [95–97].

Imidazole-cured resins are widely used in the electronics industry as molding and sealing compounds because they exhibit better heat resistance, less tensile elongation, higher modulus, and a wider range of cure temperatures than amine-cured epoxy [98–100].

There have been few attempts to investigate more profoundly the process occurring in the curing of epoxy resins with imidazole compounds, even though the use of imidazoles as curing agents for epoxies has been known for about 20 years.

Farkas and Ströhm [97] were the first to study this problem seriously. They found the curing of epoxies with imidazoles to be a complex process beginning with addition of the carbon of the epoxy group to the nitrogen of the imidazole. Barton and Stepherd [101] postulated that the reaction involved the attack of the basic pyridine-type nitrogen, but no secondary pyrrole-type nitrogen as suggested by the first study. They supported this theory by evidence of the same rates of reaction of the epoxides both with the original imidazole and with the 1:1 adduct of this imidazole and the epoxide. Both authors [97,101] noted the rate of the adduct formation was higher than the rate of polymerization and, therefore, not overall rate determining. They suggested that the adducts of imidazoles with epoxides are the proper catalysts of polymerization of epoxides.

Active imidazoles are present throughout the course of curing, and overall curing rates (formation of polyethers) vary with imidazole structure, even though all imidazoles form adducts rapidly. Riccardi et al. [102] explain this discrepancy as the instability of adducts of imidazoles with epoxides. Parallel to the polymerization, there occur cleavage of N–C bonds and the deprotonation of the carbon at the 2-position of imidazole, and thus the imidazoles are regenerated.

A major pathway for the regeneration of the catalyst during the curing process is N-dealkylation of the imidazoles via a substitution process and concomitant formation of ethers. The second route is the re-formation of imidazole or adduct with only one substituted nitrogen by β-elimination of an N-substituent from either the adduct or the polymer accompanied by production of ketones and, eventually, double bonds. The elimination can occur at any stage of the curing, and proton abstraction may take place through an inter- or intramolecular mechanism.

For the mechanism of the curing of the epoxides with the imidazoles, as seen from previous studies and the experimental data obtained by Jisova [103], the reaction scheme shown in Schemes 6–9 is suggested. The initiation occurs with an induction period. The 1:1 (resp., 1:2) adducts of the imidazole and the epoxide are formed quickly (Scheme 6). In the presence of the excess epoxides, growth centers appear after the dissociation or the rearrangement of the formed adduct (Scheme 7). The addition of a further molecule of epoxide to the growth center starts the propagation of polymer (Scheme 8).

The termination results from N-dealkylation or β-elimination of an N-substituent from the growing polymer (Scheme 9). The formation of ethers or ketones and the regeneration of 1:1 adduct and, eventually, the imidazole are concomitant with the termination reactions and can occur at any stage of the curing.

In a simpler form, the mechanism of curing the epoxy resins by imidazole has been formulated recently by Heise and Martin [104–106].

Ding and coworkers [107,108] found that azole compounds could react with copper and other transition metals at zero oxidation state to form metal(+1) or metal(+2) "inner" complexes, which are usually insoluble in organic solvents and covered the metal surface in the form of a polymeric layer [109–111]. This film could change the physical and chemical properties of a metal surface, and it might provide active sites for coordination or covalent bonding by which strong adhesion of polymers to metals could be promoted. With a suitable combination of chemical

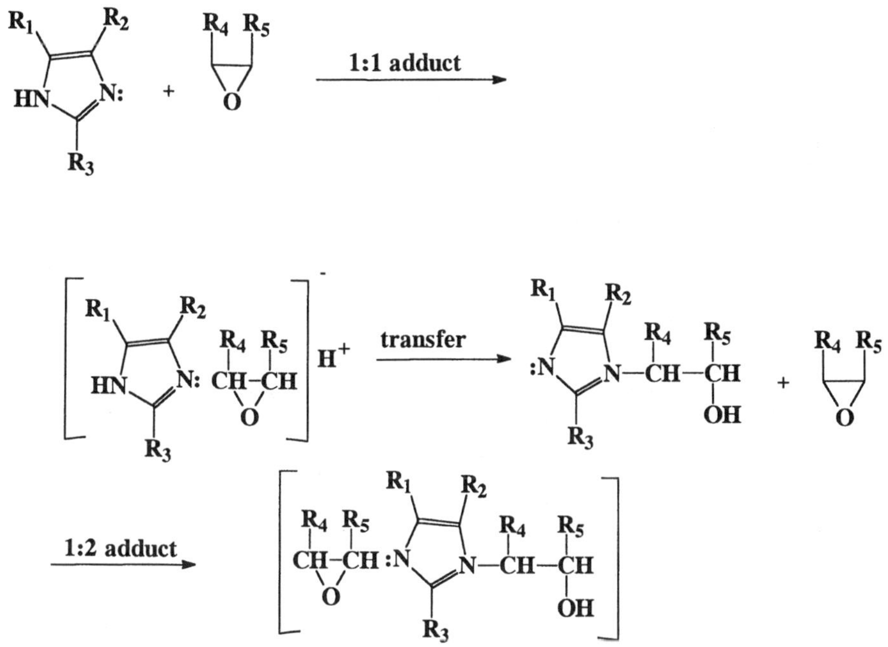

6

Scheme 6 Curing mechanism of epoxides by imidazoles: forming of adducts.

reactivities of the azole compounds toward metals and epoxy resins, it was thought that a new kind of agent could be designed for coupling epoxy resins to metals.

Imidazoles are known to display poor storage stability and, in the past, work has been directed at overcoming this problem [112]. Barton and coworkers [113–116] reported the use of copper complexes of the adducts of phenyl glycidyl ether (PGE) and 2-ethyl-4-methylimidazole (EMI),

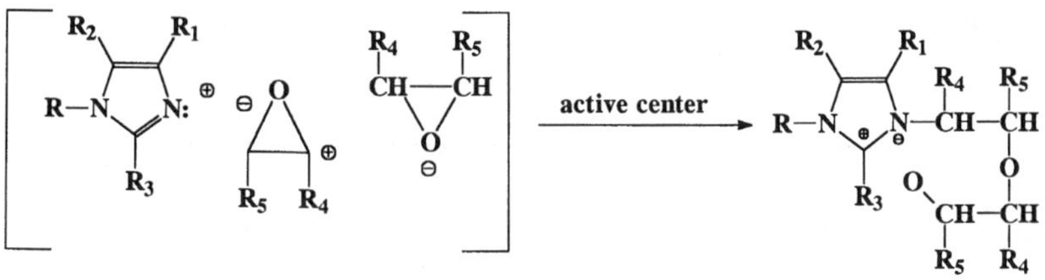

7

Scheme 7 Curing mechanism of epoxides by imidazoles: initiation.

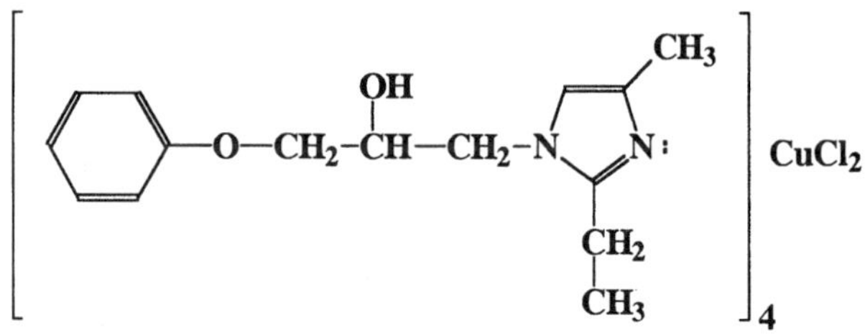

(64)

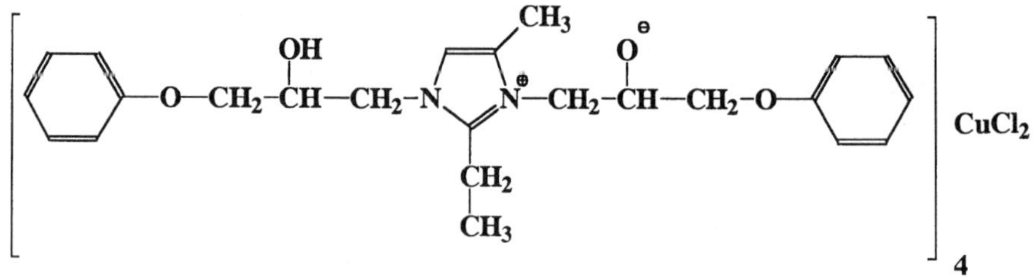

(65)

as curing agents for epoxy resins, which display improved solubility and storage stability, enabling their formulation in one-pot compositions. While remaining stable at ambient and low temperature, the complex given in (64) dissociates at ca. 120°C to produce the adduct with structural formula

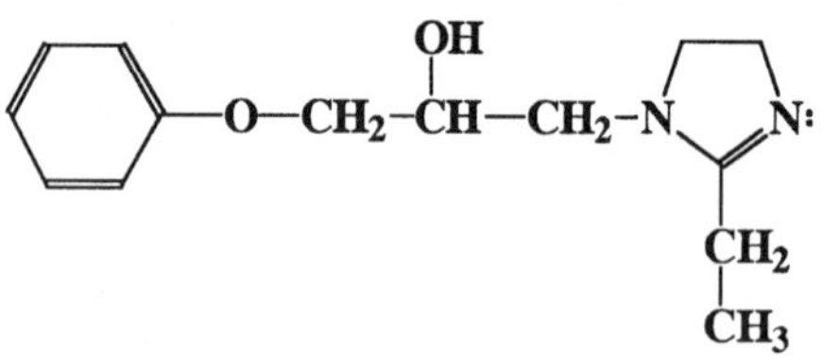

(66)

Imidazoles are added to epoxy systems to catalyze the homopolymerization of epoxide groups (polyetherification), but unmodified imidazoles begin to react when mixed with epoxides (cure occurs slowly at room temperature), making them unsuitable for use in one-pot compositions. Transition metals have been used to prepare complexes of imidazoles that react very slowly at room temperature,

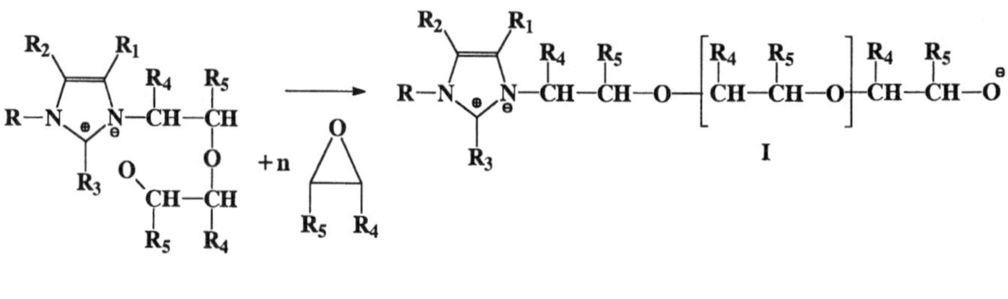

8

Scheme 8 Curing mechanism of epoxides by imidazoles: propagation.

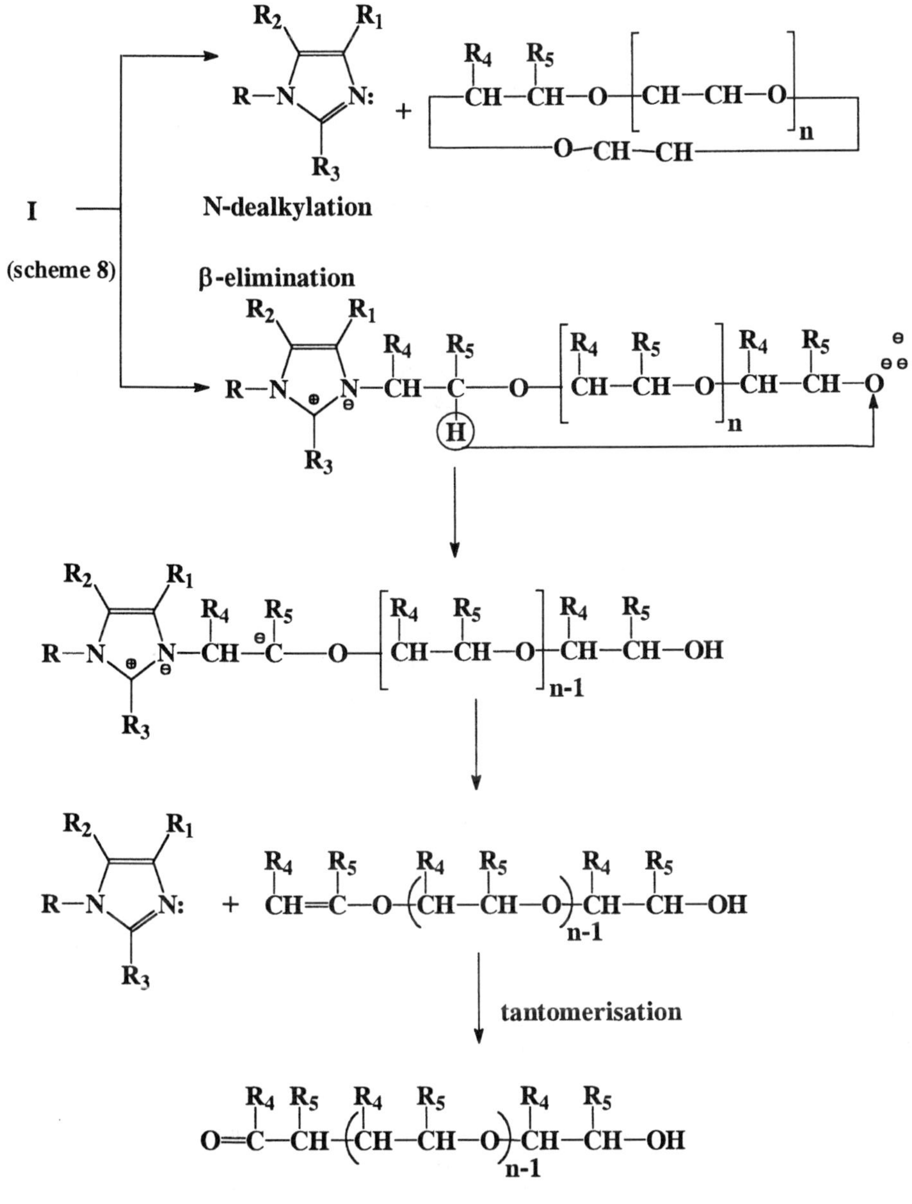

9

Scheme 9 Curing mechanism of epoxides by imidazoles: termination.

but they have exhibited a rapid cure at elevated temperature. The complexation of a transition metal (in this case, copper) effectively arrests the polyetherification reaction at ambient temperature until the temperature is elevated to effect cure. A further advantage over existing imidazole curing agents is the efficiency of complexation. After a period of time at the cure temperature (e.g., 120 or 140°C), the epoxy complex mixture can be quenched at room temperature with no further advancement of cure. This development makes the tailoring of a commercial epoxy system possible, thus facilitating more complex cure schedules in a more controllable manner. At present, the exact mechanism by which the reaction is quenched is unclear. Reassociation is a distinct possibility, but it must be borne in mind that the copper plays no further part in the reaction once the cure has been initiated. Hence, the possibility of association with the alkoxide (RO^-) groups formed during the reaction cannot be discounted at this stage.

E. Imides as Curing Agents

In recent years, attention has been focused on the development of novel cure agents containing imide groups for epoxy resins with a view to improving the performance at elevated temperature. A patent by Lee [117] describes the synthesis of anhydride-terminated polyimide through the reaction of 2,4-toluidinediamine and 3,3′,4,4′-benzophenone tetracarboxylic dianhydride and the use of this compound as a cure agent for epoxy resins for making composites. Another patent, by Serafini et al. [118] deals with the preparation of composites from epoxy resins using bis-aminoimides as cure agents. Ichino and Hasuda [119] studied the curing of epoxy resins with bis(hydroxyphthalimide)s obtained through the reaction of aromatic/aliphatic diamines and 4-hydroxyphthalic anhydride and evaluated the adhesive properties.

Adhinarayanan [120] describes the adhesive properties of epoxy–imide resins obtained through the reaction of epoxy resins, namely, Araldite GY250 (diglycidylether of bisphenol A and epichlorohydrin; difunctional) and Araldite EPN1138 (novolac epoxy resin; polyfunctional), both manufactured by Hidustan Ciba-Geigy India Ltd (Calcutta), with bis(carboxyphthalimide)s. The chemical structures of these epoxy resins and bis(carboxyphthalimide) are

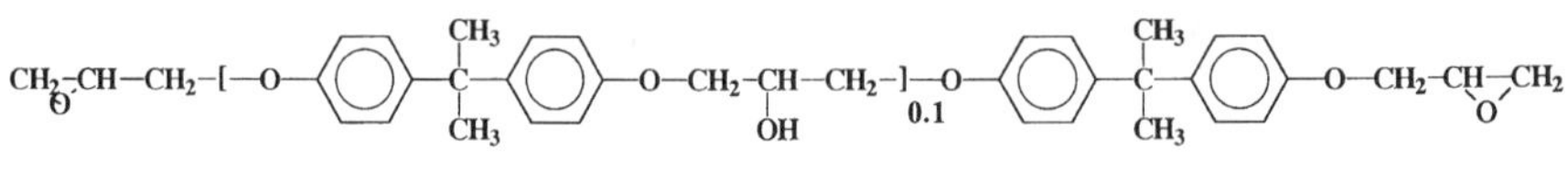

Araldite GY 250 (67)

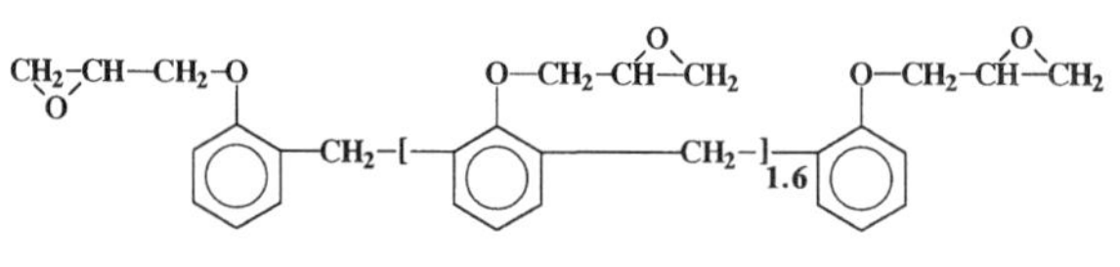

Araldite EPN 1138 (68)

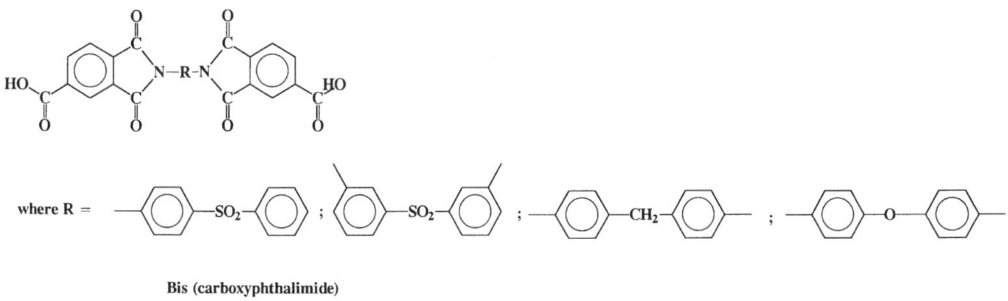

(69)

The curing reaction of the epoxy resins with the bis(carboxyphthalimide) is

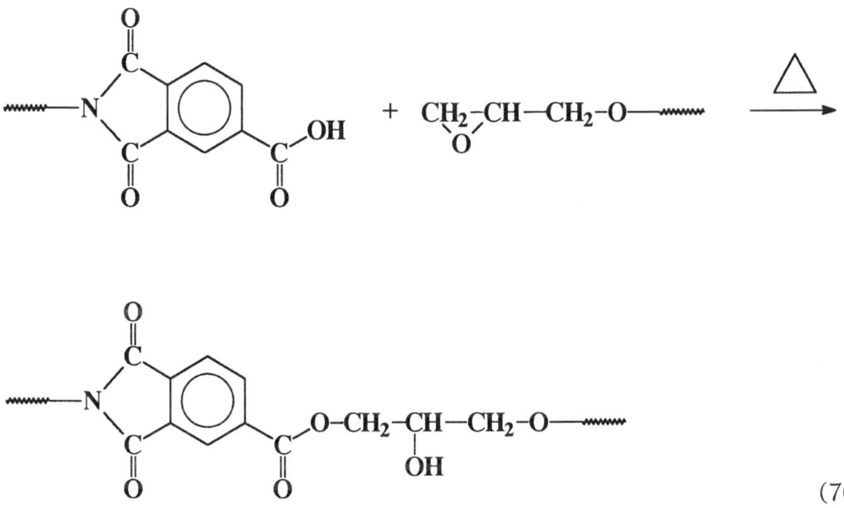

(70)

The experimental data obtained by Adhinarayanan reveal that the epoxy resins cured with bis(carboxyphthalimide) are stable up to 370–380°C and have good adhesive properties.

On the other hand, Jun-Lin Fan et al. [121] report the curing of EPN1139 with 1-cyanoethyl-2-phenyl-4,5-(dicyanoethoxyl methyl)imidazole, with the following structural formula:

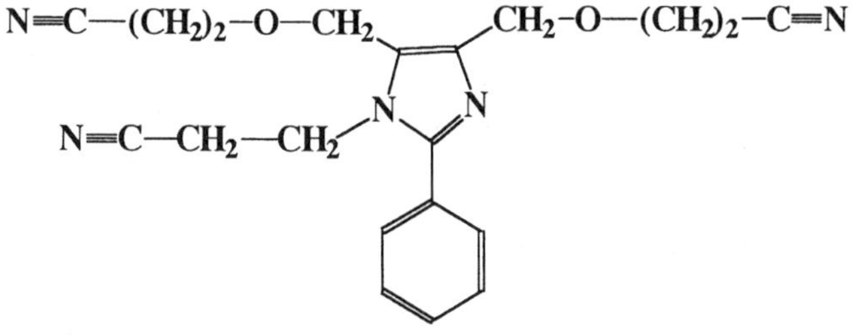

(71)

The properties of the cured resins and the mechanism of cure reaction are not yet well defined.

Network formulation during cure of epoxy resins is an important subject; it affects the ultimate structure and properties of cured epoxy resins. It is frequently assumed that gelation occurs during network formation of branching polymerization when the first infinite molecular weight network is formed. Mean-field theories of gelation of polymers [122–124] have been developed based on the classic Flory–Stockmayer theories [125,126]. In these theories, the probability of forming an infinite molecular weight network is derived from the probability of finding a branch point during the polymerization. For a reaction of a difunctional epoxy and an amine of functionality f, the extent of reaction at the gel point, p_G, can be calculated using

$$p_G = \frac{1}{[1 + r\rho(f - 2)^{0.5}} \tag{72}$$

where r is the ratio of the epoxide to the amine hydrogen and ρ is the fraction of the amine hydrogen in the multifunctional reactants ($f > 2$). For stoichiometric diepoxy–diamine reactions, $f = 4$, $r = \rho = 1$, and $p_G = 0.577$.

Percolation theory [127–129] and dynamic scaling theory [130] have been also applied to the gelation of polymers. These theories assume that the macromolecule has unrestricted conformational freedom during the reaction. Another mechanism for the solidification of a curing network system is progressive phase separation as suggested by Bobalek et al. [131] and Solomon [132]. This mechanism suggests that, as the solidification proceeds, insoluble material separates as a second phase from the residual material in the reaction bath. However, as has been described in detail by Dusek [133], if the reacting system separates into multiple phases, chemical kinetics should show major changes as phase separation occurs. Very many kinetic studies of a wide variety of curing epoxy systems have not shown evidence of any such deviations up to and for at least a short way past the observed onset of solidification.

In a critical survey on the mechanisms of solidification of epoxy–amine resins during cure, Huang and Williams [134] reveal that for most epoxy systems, solidification does not occur near the predicted degree of reaction; it may occur before or after the predicted point. Analysis of the scaling properties of systems studied approaching the critical degree of reaction have predicted that certain scaling laws should be observed near the critical degree of reaction. These authors pointed out that the systems may be divided into two categories: those that exhibit scaling as predicted, and those that do not exhibit scaling. In a system for which the scaling predictions are observed, solidification occurs at the degree of reaction predicted by the simple statistical theory.

On the other hand, it is well known that the crosslinking density of the formed network has a pronounced effect on the glass transition temperature [135] and also on the lower-temperature processes, especially at low crosslinking extents, although the influence of other factors—such as free volume at high crosslinking densities—remains to be clearly demonstrated.

Usually, these transitions have not been related to thermoset morphology, although some works about the existence of inhomogeneities in the morphology of highly crosslinked epoxy resins have been published [136,137]. These inhomogeneities should be highly crosslinked nodules immersed in an intermolecular matrix of lower crosslinking density; indeed, the intranodular crosslinking density should be responsible for the onset of the molecular motion corresponding to the glass transition and also for some variations in low-temperature relaxations, the last one especially at stoichiometric compositions [137]. Many investigations have been conducted into variations in the low-temperature transitions as a function of cure conditions and/or chemistry and concentration of resins and curing agent. These relaxation processes at temperatures far below the glass transition temperature have been denoted as β, between -30 and $-60°C$, and γ, occurring near $-140°C$; relaxations that result from the motion of glyceryl amine groups and other parts of the network structure and the movement of methylene units, respectively. Another relaxation, denoted as ω or β', at 50 to 90°C (also called ω-transition by Sanz et al. [138] has been also reported [139], although its meaning is not very clear.

These aforementioned high crosslinking density regions are responsible for the ω relaxation and are produced by a heterogeneous crosslinking reaction. Based on these data, Grof et al. [140] have recently suggested that the growth of polymers near the gelation threshold is a connectivity phenomenon rather than a statistical process.

Industrial-scale production of certain composites based on an epoxy matrix involves employing a solvent, and it is often assumed that the solvent, after being removed, does not influence the behavior of the final products. Nevertheless, as some investigations have shown, the solvent employed can affect the physicochemical behavior of epoxy resins as well as that of their composites [141,142]. The properties of thermoset networks depend on the conditions under which the cure reactions occur, because the viscosity, the crosslinking density, and the number of accessible conformations all change with time. Therefore, variations of the physical properties of crosslinked epoxy resins have been attributed to a lower crosslinking density of the formed network structure because of the solvent employed [143]. It has also been shown, however, that for a complete cure schedule the epoxy conversion remains similar for mixture cast with or without solvent [143], and therefore the crosslinking density could also be similar. These contradicting observations could be related to differences in the method of solvent removal used in making these investigations.

The influence of residual solvent and that of thermal history during curing on the mechanical properties of diglycidyl ether of bisphenol A cured with diaminodiphenyl methane epoxy resins can be summarized as follows [138]:

The existence of residual solvent at the stages of curing clearly affects the relaxational behavior of epoxy resins. This influence seems to be related to differences in local density inside cured networks.

The local motions of chains determine the variations in the ω relaxation.

Changes in cure temperature at the earlier curing stages can modify the dynamic mechanical behavior of epoxy systems.

Crosslinking densities of fully cured networks, and thus the glass transition temperatures, are high for stoichiometric mixtures than for any other composition. However, epoxide resins usually shrink during the curing and cooling processes. If the shrinkage is constrained by adhesion with other materials, the constrained shrinkage is converted to internal stress [144]. The internal stress causes various defects, such as cracks, delaminations, and voids. Therefore, it is of great importance to prevent or reduce shrinkage and internal stress of cured epoxide resins. The magnitude of internal stress in the two- and four-functional resin systems depends on the chemical structure of aromatic diamines used as cure agents. The internal stress of four-functional resin systems may be reduced by the introduction of larger substituent groups into curing agents, even though the T_g's of these systems are similar. The reduction of internal stress is explained as a result of the decreasing of the shrinkage after the vitrification point in the curing process.

IV. MODIFICATION OF EPOXY RESINS

Essentially, the methods of epoxy resin modification are based on attaining a dispersion of second-phase particulates in the epoxy matrix; the second phase may be either rubbery or rigid in nature, or both. Substantial enhancement of fracture toughness has been achieved in the hybrid systems.

Since cured epoxy resins are inherently brittle materials, modification is necessary to make them usable in many applications. Hence, various methods have been employed to overcome the tyranny of the Griffith equation in this class of materials.

Improved toughness can be achieved by additions to the epoxy/hardener formulation [145,146]. However, if these modifiers are fully compatible with the cured epoxy, they act as plasticizers and cause a considerable reduction in the glass transition temperature (T_g) of the system. This results in toughened systems at an appreciable sacrifice in heat distortion temperature and, therefore, elevated-temperature properties.

An early theory to explain rubber toughening of brittle glassy polymers is due to Bucknall and Smith [147]. The basis of this theory is that rubber particles initiate the formation of crazes and control their growth [148]. Under tensile stress, crazes are formed at the points where the principal strain is maximal, namely, near the equator of the rubber particles, and propagate outward. The growth of crazes stops as a result of stress relaxations when they meet energy-absorbing obstacles, such as smaller rubber particles, preventing them from developing into large cracks. A high level of adhesion between the matrix and the rubber particles is necessary for the craze-arresting mechanism to operate. A weakly adhering rubber particle would be pulled away from the matrix, leaving a hole that would intensify the stresses locally and induce fracture.

This mechanism is unlikely to operate, however, in the case of highly crosslinked systems, since the chain length between crosslinks is too short to effectively produce the fibrils of oriented chains acting as bridges between the two surfaces of the craze [149].

More recent theories [150–152] for the toughening of highly crosslinked polymers by rubber inclusions involve the concept of shear yielding of the matrix around the particles of a mechanism for increasing the level of strain that the matrix will reach before fracture. Although the adhesion between matrix and rubber particles is a crucial factor in improving toughness by the shear yielding mechanism, it is the higher compliance of the particles relative to the matrix that is responsible for the onset of shear deformation in the matrix. Wu [153] has put forward the concept of critical interparticle surface distance as a condition for inducing yielding in a matrix containing rubber particles. It is possible that the random distribution of particles in the matrix, together with an optimized packing arrangement of the particles, plays an important role in developing shear deformations within the matrix.

This mechanism, however, still relies on the ability of the matrix around the particles to undergo a certain amount of yielding, and therefore, it is unlikely that it will operate when the crosslinking density of the matrix is very high. For the latter case, the only possibility to toughen the resin is by energy absorption through deformations within the rubber particle itself, particularly through crack bridging or volumetric dilatation [154].

Obtaining a good bond between rubber particles and matrix is primarily a matter of "compatibility" and chemical reactions between the two phases. Full miscibility between a rubber and a resin does not produce effective toughening, but merely serves to plasticize the matrix. Total immiscibility is equally undesirable, since a completely immiscible rubber of reasonably high molecular weight, which would be required to achieve adequate mechanical properties, will not form a fine dispersion in the resin due to inevitable large differences in viscosity between dispersed particles and the matrix, nor will it produce a strong bond at the rubber–matrix interface. The ideal rubber for toughening purposes is one that is semimiscible, that is, the mixture is near the critical solution conditions around the cure temperature.

It is then possible to improve the adhesion between the epoxy matrix and the toughening phase by allowing the particle to precipitate from a homogeneous matrix, so that the particle will contain not only the rubbery material but also the epoxy resin and possibly the catalyst to promote reactions between the two. Demixing via spinodal decomposition is important for designing the morphology of polymer blends, since the characteristic morphology can be fixed or frozen at various stages during the evolution of the phase separation process by quenching the demixed system below T_g. For thermosetting systems, a spinodal decomposition induced by chemical reaction is even more effective than one by thermal means [155]. It has been found that various reactive end groups in a liquid rubber modifier, for example, carboxyl, hydroxyl, epoxy, mercaptan, phenol, methylol, and amine, can be effective in inducing particle precipitation through reactions with an epoxy resin and in achieving a good interfacial bond [156,157].

Another weakness of conventional epoxy resins is their relatively large level of water absorption, which causes a large reduction in glass transition temperature [158,159] and a deterioration in mechanical properties [160,161]. The high affinity of epoxy resins toward water is explained by the presence of high number of polar groups in the epoxy network. It has been proposed, in fact, that the diffusion of penetrant molecules into polymers depends on two factors [162], namely, the zones of attraction between the polymer and the pendent molecules, and the availability of appropriate molecular size holes in the polymer network. The first factor concerns the chemical nature of the penetrant in relation to that of the polymer. The relatively high water absorption capacity of epoxy resin derives from the presence of hydroxyl groups in the epoxy chains, which attract the polar water molecules through hydrogen-bond formation [163,164]. The second factor, on the other hand, involves the presence of holes determined by the polymer structure and morphology, which in turn depends on the fluctuations in the crosslinking density, and molecular chain stiffness, hence their ability to closely pack in an amorphous system. The formation of an appropriate hole also depends on the cohesive energy density of the polymer and on the size of the penetrating molecules. Water molecules, for example, are hydrogen bonded and can form clusters within the polymer. Thus, this particular factor affecting diffusion is essentially a geometrical one: the free volumes available within the polymer for occupation by the penetrant molecules.

The efficiency of modifiers depends on their ability to form diffusion barriers consisting of co-continuous (interpenetrating) phases within the intrinsically hydrophobic matrix.

A. Modification of Epoxy Resins by Functionalized Butadiene Oligomers

In the mid 1960s, carboxyl-terminated polybutadiene–acrylonitrile (CTBN) liquid polymers were introduced for the purpose of epoxy resin modification. These telechelic polymers are essentially macromolecular diacids. They offer processing ease (and therefore advantage) over the solid carboxylic nitrile elastomers. It is no surprise that the epoxy prepreg industromer (adhesive and nonadhesive grades) found the liquid and solid carboxylic nitrile elastomers species useful together in processing liquid and lower molecular weight epoxy resins where elastomer modification was needed.

Later, in 1974, amine reactive versions of the liquid nitrile polymers (ATBN) were issued, thereby offering another way to introduce rubbery segments into a cured epoxy resin network. Detailed discussion of nitrile rubber, carboxylic nitrile rubber, and both carboxyl- and amine-terminated nitrile liquid polymers is available elsewhere [165–169]. Wide-ranging documentation exists that covers modification of epoxy resins using carboxyl-terminated polybutadiene/acrylonitrile liquid polymers in which addition esterification (alkylhydroxyl esterification) reactions are employed to prepare the epoxy base [170–173]. While such reactions proceed uncatalyzed at excess epoxy equivalents normally greater than 3:1 epoxy: carboxyl, instructive examples are given [168] for *t*-phosphine- or quaternary phosphonium salt–catalyzed systems. To a lesser extent, amine-terminated polybutadiene/acrylonitrile liquids are also used by epoxy formulators. This polymeric disecondary amine is employed in admixture with amines, amidoamines, or fatty polyamides of choice [174]. Thus, one admixes low molecular weight or polymeric amines to create a novel toughening or flexibilizing hardener package.

In each instance of nitrile elastomer modification—whether rubber is added to the epoxy portion or to the hardener portion—the level of rubber largely determines whether a toughened or a flexibilized epoxy results. The former is characterized by little loss in thermal/mechanical properties. The latter shows a dominant influence of the added rubber.

Consequent to documentation surrounding methods of employing reactive nitrile elastomers to modify epoxy resins is a growing body of literature that characterizes and elucidates these systems. Such topics as morphology in the cured and uncured state, transitions from toughening to flexibilization, viscoelastic ef-

Table 1 Carboxylic Butadiene–Acrylonitrile Elastomers

Grade	Physical form	Acrylonitrile (%)
Hycar 1072 CG	slabs	26–28
Hycar 1472[a]	3/8-in. chips	28

[a]"Crum" version of Hycar 1072 CG.
Source: Ref. 176.

fects, equilibrium physical properties, and phase structure are available to the investigator [167,168].

McGarry and Willner [175] showed that low molecular weight carboxyl-terminated butadiene–acrylonitrile copolymers are the most effective agents for improving the fracture toughness of epoxy resins.

Carboxyl-containing butadiene–acrylonitrile copolymers epoxy resins are available in two physical forms. Table 1 lists and characterizes two of the high molecular weight (solid) copolymers of butadiene–acrylonitrile having acid groups distributed randomly in the polymer backbone. Both of these polymers are directly soluble in polar solvents such as methyl ethyl ketone.

The carboxyl-terminated liquid polymers (CTBN) are telechelic polymers ranging in comonomer acrylonitrile content from 0 to 27 percent. Table 2 gives the characterization of these polymers. They are solvent-free and range in M_n from 3400 to 4000. The higher acrylonitrile-containing CTBN type possesses the best balance of oil- and solvent-resistant properties.

The solids Hycar 1072 and Hycar 1472 are employed with epoxy resins using processing techniques that employ solvent solutions and rubber milling operations (or some type of shear mixing). The carboxyl-terminated liquid polymers offer the capability of solventless preparation and formulation of the base epoxy resin(s). The two forms (solid and liquid) may be used together advantageously to strike a balance between high and low molecular weight elastomer addition providing (1) a bimodal particle-size dispersed phase, (2) an edge in film-forming properties, and (3) a higher rubber level.

The chemistry of carboxylic elastomer crosslinking reactions has been thoroughly discussed by Brown [177] in a review article. The experimental data reveal

Table 2 Typical CTBN Reactive Liquid Polymers

Polymer properties	CTB	CTBN	CTBN	CTBN	CTBNX
Acrylonitrile content (%)	0	10	10	27	18
M_n	4000	3600	3500	3400	3400

Source: Ref. 176.

the presence of crosslinked rubber as a discrete, well-dispersed, discontinuous phase in epoxy resin models that utilized CTBN and are amine catalyzed. It was determined that low levels of rubber could significantly improve the epoxy resin's crack resistance and impact strength.

Siebert and Riew [178] first described the chemistry of rubber particle formation in an admixed model involving CTBN, a DGEBA (diglycidyl ether of bisphenol A) liquid epoxy resin, and a selective catalyst. They proposed that the composition of the rubber particles in the dispersed phase critically depended on the epoxy–CTBN–epoxy adduct being formed *in situ* and then further chain-extended and crosslinked with additional resin.

In structural applications for epoxy resins, there are useful nitrile–epoxy systems that are prepared from (1) noncarboxylic nitrile rubber, which has been milled to reduce molecular weight and then transferred to solvents with the epoxy resins, and (2) specialty nitrile latexes, which are added directly to the epoxy resin and then vacuum-stripped to remove water. These approaches for preparing a nitrile–epoxy formulating base are not widely used and are confined primarily to adhesive applications.

Methods that are gaining considerably wider acceptance involve the CTBN and/or solid carboxylic nitrile elastomers. The epoxide–carboxyl reaction offers a simple, reproducible method for forming epoxy–rubber adducts that are acid-free. This permits formation of the proposed major ingredient of the rubber dispersed phase prior to formulating and curing. This approach to preparing the nitrile–epoxy formulating base (1) attaches or adducts the rubber and resin, (2) retains linearity, (3) rids the system of carboxyl functions, which plague certain epoxy resin cure catalysts, and (4) provides an increase in alcoholic hydroxyl groups. Since all species after esterification are essentially epoxy species, latitude in formulation is achieved with regard to latency, use of Lewis acid type catalysts, maintenance of relatively low viscosities, and so forth.

Table 3 lists some generally known catalysts that are useful in preparing the rubber/epoxy adducts. Catalysts specific to alkyl-hydroxy esterification reactions are advised since adducting may be done successfully at lower temperatures in a reasonable time. Prolonged reacting at elevated temperatures (>150°C) could effect some homopolymerization of the base epoxy resin employed.

Epoxy–methacrylates containing rubber are possible using CTBN. Again, esterification reactions are useful in producing not only this well-known resin type for structural applications, but also for incorporating rubber into the structure. It is possible that rubber-modified epoxy–acrylates would be similarly useful.

Table 3 Alkylhydroxy Esterification Catalysts

Triphenyl phosphine
N-Methyl morpholine
Potassium acetate
Methyl triphenyl phosphonium iodide
Triethanolamine
Lithium benzoate

The use of ATBN (amine-terminated butadiene–acrylonitrile rubber) modifiers is attractive because of the high reactivity of the end groups toward the epoxide ring, which will result in shorter gel times.

The influence of the curing temperature on the morphology and mechanical properties of blends obtained by adding different amounts of ATBN to a bisphenol A type epoxy resin has been studied by Butta et al. [179], who found that the final morphology of the toughened systems depends strongly on cure temperature and rubber content. Curing at low temperature promoted the formation of optically clear material. The temperature at which clear materials were obtained was found to increase with increasing rubber content. Curing at high temperatures, on the other hand, gave rise to the formation of large spherical domains. With increasing ATBN content these domains became bigger and more closely packed, reaching phase inversion conditions above 20% rubber.

Although CTBN and ATBN oligomers are very efficient for improving the fracture properties of epoxy resins without sacrificing excessively the modulus and strength, these two elastomeric modifiers have some drawbacks. The main deficiency of these oligomers is the high level of unsaturation in their structure, which provides sites for degradation reactions in oxidative and high-temperature environments [180]. The presence of double bonds in the chains can cause oxidation reactions and/or further crosslinking, with the loss of elastomeric properties and ductility of precipitated particles.

Several attempts have been made over the last two decades to find alternative modifiers to CTBN and ATBN for the toughening of epoxy resins.

B. Modification of Epoxy Resins with Glassy Polymeric Agents

Engineering thermoplastics are interesting materials as modifiers for epoxy resins, from the viewpoint of maintaining mechanical and thermal properties of the matrix resins. Modifications of epoxy resins with various types of ductile thermoplastics have been studied as alternatives to reactive rubber for improving the toughness of epoxy resins. At first, commercial poly(ether sulfone)s (PESs) such as Victrex were used in the modification of polyfunctional epoxies, but these were less effective because of the high crosslinking structure and the poor interfacial bonding between two incompatibilized phases in the cured resin [181].

Raghara [182] has blended a low molecular weight PES containing hydroxyl end groups with a tetrafunctional epoxy resin, expecting the reactive end groups of PES to participate in the chemical reactions during curing of the epoxy resin and to achieve in this way a high level of interfacial adhesion. The increase in toughness that has been achieved with these epoxy–PES systems, however, is only marginal.

The very high crosslinking density of the cured tetrafunctional epoxy add to the possibility that the inclusion of low concentrations of PES is not sufficient to produce formation of shear bands. The Young's modulus of the modified system was found to be slightly lower than the values for the neat epoxy resin, possibly due to a reduction in crosslinking density as a result of the dilution effect of the uncrosslinked PES chain.

Bucknall and Partridge [183] have considered the results reported by Raghara [182] and have also found only a small increase in fracture toughness for mixtures of PES with tetrafunctional and trifunctional epoxy resins, separately cured with diamino-diphenyl sulfone (DDS) and dicyandiamide (DICY). They also reported that the PES-toughened tetrafunctional epoxy resin cured with both types of hardeners did not show phase separation, whereas phase separation was observed in the case of a trifunctional resin.

Diamant and Moulton [184] have also investigated the toughening of a tetrafunctional epoxy resin using various ductile thermoplastic polymers possessing a high glass transition temperature. They observed that a mixture of epoxy resin and PES did not possess a two-phase structure, and that the fracture toughness of the modified system was not markedly different from the values measured for the unmodified epoxy.

Hedrick et al. [185,186] used phenolic hydroxyl and amine end functionalities in poly(arylene ether sulfone) oligomers to chemical modify the network of a difunctional DGEBA resin. The oligomers reacted with a large molar excess of epoxy resin and then cured into crosslinked networks with a stoichiometric quantity of 4,4′-diamino-diphenyl sulfone (DDS). The aryl ether sulfone was found to be molecularly miscible with the epoxy precursor over the entire range of compositions and molecular weight investigated, developing a two-phase structure during curing, in which polysulfone formed discrete particles evenly dispersed in the epoxy matrix. Despite the existence of a two-phase structure, the crosslinked systems were nearly transparent, due to a similarity in the refractive indices of the two components. The fracture toughness of these resins was improved significantly with a minimal deterioration in the flexural modulus. The authors attributed this result to the presence of strongly adhering polysulfone particles in the epoxy matrix, which deformed plastically during fracture and induced shear yielding in the epoxy matrix. Similar improvements have been reported recently by Yoon and coworkers [187] using aminophenyl-terminated polyethersulfone in a difunctional bisphenol A epoxy resin. The latter studies have provided evidence for the hypothesis relating the brittleness of trifunctional epoxy resins to their higher crosslinking density, which prevents the occurrence of yielding deformations in the matrix.

Martuscelli and coworkers [188] have used a bisphenol A–based polycarbonate (PC) as the glassy thermoplastic polymer modifier. The PC was dissolved directly in the uncured epoxy resin at high temperature (200°C). After the addition of the curing agent and the accelerator at low temperatures (80°C), the temperature was increased again for curing and subsequently for post-curing. FTIR analysis of the uncured epoxy/PC mixtures revealed the occurrence of chemical interactions between the two components producing PC chains with epoxide end groups. These functional groups take part in the subsequent crosslinking reactions, resulting in the incorporation of PC segments within the epoxy network. Dynamic mechanical measurements and scanning electron microscopy analysis of the epoxy/PC blends did not show any evidence of phase separation of the minor component during the curing process. Tests at both low and high strain rate showed a marked increase in toughness with increasing amount of PC in the blend. This toughening effect with the use of PC was achieved without significantly reducing other desirable properties of the matrix such as the elastic modulus. From scanning electron microscopy analysis, it was found that localized yielding occurs at the crack tip and that the mechanism is probably similar to what is normally observed with interpenetrating network modifications.

It is well established that the addition of rubber or thermoplastic to the resin or rubber coating of the glass fabric prior to lamination improves the impact resistance in glass/epoxy laminates [189,190]. Thermoplastic additions are more lucrative when considering their high-temperature applications and the compromises that are made in mechanical properties other than toughness in rubber-modified glass/epoxy composites.

In a recent study, Matheswaran and Padmanabhan [191] examined the influence of thermoplastic additions of plastisol (polyvinyl chloride plasticized with dioctyl phthalate) and PC, added independently in varying amounts, on the Izod impact energy and static properties such as flexural strength and interlaminar strength (ILSS). Plastisol forms a stronger interface with the epoxy resins than does PC. Polycarbonate forms a weaker interface, especially with a difunctional resin [192]. It was observed that plastisol in epoxy initially forms a homogeneous blend. Although plastisol separates out of epoxy, the interfacial bond between this phase and the epoxy matrix is quite good. At 10% addition of plastisol, an improvement in ILSS and flexural strength was observed; the glass–epoxy interface is virtually unaffected by this addition, but the epoxy is modified, with crack-blunting plastisol sites that tend to influence the nature of fracture from brittle to ductile.

However, 30 wt % addition of plastisol causes agglomeration of the phase. The adhesion of plastisol to epoxy become appreciable, and properties like the flexural strength and ILSS are affected. Furthermore, the matrix strength is also considerably lowered as a result of this plastisol addition.

The PC–epoxy blend formed a heterogeneous phase and exhibited a weak interface with both epoxy and glass fibers. Polycarbonate additions always lowered the flexural strength and the ILSS of the composite. The experimental data obtained by these authors [192] reveal that the energy absorbed in Izod impact is maximal for samples containing 30 wt % of plastisol or PC. Scanning electron microscopy observation revealed the following fracture mechanism. First, the toughening mechanisms are different: plastisol toughens by undergoing tearing and blunting of the crack tip, whereas PC increases the fracture energy by crack detour and blunting involving significant resin rippling. Shearing of the PC phase was also observed at high wt % additions due to the depletion of the epoxy matrix layer and the load being borne more by the polycarbonate phase. Second, toughening occurs in epoxies with any crosslinking density but is greater in epoxy resins with low crosslinking density [193], and toughness increases in the present case because the fillers are thermoplastics. This is a supplementary effect. Third, maximum tensile radial stresses at the poles of the spherical polycarbonate cause debonding of the PC phase from the matrix. Partial or total debonding of the PC phase will cause stress concentration at the PC–epoxy interface. Matrix plastic deformation, yielding, and rippling at the equator of the PC particle enhance the impact resistance of the composite.

Jijima and coworkers [194–197] have used *N*-phenylmaleimide–styrene copolymers (PMS), *N*-phenylmaleimide–styrene–*p*-hydroxystyrene (PMSH), and poly(aryl eter ketone)s as hybrid modifiers to improve the toughness of bisphenol A diglycidyl ether epoxy resin cured with *p,p'*-diamino-diphenyl sulfone. These hybrid modifiers were effective in toughening the epoxy resin. When using the modifier composed of 10% PMS (M_w = 313,000) and 2–5% PMSH (M_w = 316,000), the fracture toughness for the modified resins increased 100% with no deterioration in the flexural properties or the glass transition temperature. The improvement in toughness of the epoxy resins was attained because of the co-continuous phase structure and the improvement in interfacial adhesion.

Albert and coworkers [198,199] have used bis(4-hydroxybenzoate)-terminated poly(tetrahydrofuran) liquid rubbers in order to improve toughness of anhydride-cured epoxy resins.

C. Modification of Epoxy Resins with Rigid Crystalline Polymers

In the aforementioned examples of modification, it was pointed out that the addition of glassy rigid polymers to epoxy resins has been found to be successful only in a limited number of cases. A relatively high fracture toughness was obtained, in fact, only when the morphology became a two-phase co-continuous microstructure or when the thermoplastic component became the continuous phase with dispersed epoxy domains. This type of distribution of the glassy polymer–rich phase can lead to poor high-temperature creep resistance and/or reduced solvent resistance. On the other hand, if a rigid crystalline phase forms a co-continuous dispersed phase in the epoxy network, these deficiencies could be eliminated, but it is difficult to visualize how such rigid crystalline domains can provide an efficient toughening mechanism.

One possible toughening mechanism in such systems could operate by phase transformation, which is well known for ceramic materials [200]. An example is represented by zirconium-containing ceramics [201]. The metastable tetragonal phase of zirconium is incorporated into the ceramic, and under the influence of the stress field ahead of a crack tip, this phase transforms to the stable monoclinic phase. Because the monoclinic phase is less dense than the tetragonal phase, compressive stresses are set up on one of the phases, which superposes on the tensile stress field ahead of the crack tip producing shear deformations, with the effect of increasing the critical fracture energy.

Such a toughening mechanism might be applicable to brittle polymers if a stress-transformable crystalline polymer is used as the rigid second phase, and it might well provide an effective solution to the toughening of highly crosslinked thermosets. However, only a few studies have been carried out to verify this hypothesis for polymeric materials [202]. For dispersed crystalline polymers to be effective for toughening by phase transformation, several properties similar to those of the metastable tetragonal phase of zirconia are required. First, the polymers should be able to exist in a variety of crystalline states and transformations from one state to another should take place under the influence of an applied stress. The desired phase transformation is one in which volume dilution and distortion both occur in response to the stress field ahead of a crack tip. Second, the desired stress-free crystalline phases should be stable in the temperature ranges under which the brittle matrix resins are processed. Third, these polymers should form strong interfacial bonds with the matrix.

Kim and Robertson [203] studied the toughening of an aromatic amine-cured diglycidyl ether of bisphenol A epoxy with particles of crystalline polymers. The crystalline polymers chosen were poly(butylene terephthalate) (PBT), nylon-6, and poly(vinylidene fluoride) (PVDF). Each of these was found to be capable of undergoing phase transformation as a result of the application of a stress field, and each of these polymers could be bonded to epoxies. In the case of PVDF, however, an amine curing agent is necessary to achieve a good bond with the epoxy resin. The authors reported that nylon-6 and PVDF were found to toughen epoxy resins to an extent similar to what is achievable with an equivalent amount of CTBN rubber. Fracture toughness, on the other hand, was increased twofold with nylon-6 and PVDF.

In a recent work, Kubotera and Yee [204] investigated the possibility of using a crystalline block copolymer, containing amorphous chains, as a modifier to improve the fracture toughness of highly crosslinked epoxy resins. They reported that low molecular weight oligomers of PBT and poly(ether ketone) can form crystals in the form of triblock copolymers with amorphous poly(ether sulfone). The main reason for using copolymers rather than crystalline homopolymers was that the copolymer particles were expected to exhibit better interfacial adhesion, owing to the solubility of the amorphous blocks. All copolymers, in fact, gave homogeneous resin mixtures with loadings up to 15–20 wt %, resulting in multiphase structures after curing. The scanning electron micrographs revealed the presence of a second phase dispersed in the epoxy matrix, which did not result from spontaneous liquid–liquid phase separation but from crystallization growths.

D. Modification of Epoxy Resins with Acrylic Oligomers

Banthia et al. [205] found that acrylate oligomers exhibit extremely good miscibility with conventional epoxy resins and will precipitate as a distinct dispersed phase during curing. With the addition of 4–10 wt % of ethylhexyl acrylate oligomers, the cured castings were found to exhibit enhancements in impact strength comparable to traditional toughened epoxy systems. In essence, these authors have found that carboxy-terminated telechelic ethylhexyl acrylate oligomers are effective elastomeric toughening agents, exhibiting better oxidative and thermal stability with respect to CTBN toughening systems.

Wang et al. [206] used as modifiers novel polyfunctional acrylate elastomers with medium molecular weight to toughen epoxy resins. The polyfunctional poly(*n*-butylacrylate)s reported in their study were epoxy functionalized poly(*n*-butylacrylate) (ETP*n*BA) and carboxyl functionalized poly(*n*-butylacrylate) (CTP*n*BA), both obtained by photopolymerization. The effect of the level of functionality and type of functional group in the elastomers used as toughening agents was investigated by means of tensile and impact tests and electron microscopy. It was found that there is an optimum functionality of elastomers for maximum impact resistance in epoxidized (ETP*n*BA) and carboxylated (CTP*n*BA) copolymer-modified systems. Studies on morphology of the modified epoxy resin system indicated that the better toughening effects of the epoxy-functionalized modifiers ETP*n*BA were due to the presence of a multiple distribution of particle sizes. The aggregation of rubber particles occurring with the use of carboxyl-functionalized CTP*n*BA in the epoxy resin, on the other hand, was believed to be the cause of the observed reduction in toughness.

Lee and coworkers [207] obtained similar results by using a family of *n*-butyl acrylate/acrylic acid (*n*BA/AA) copolymers with a broad functionality range (1.62–9.93). They reported, in fact, that improvements in adhesion strength could be achieved by incorporating the *n*BA/AA copolymer in a diglycidyl ether bisphenol A epoxy matrix, and that there was an optimum functionality for achieving the highest interfacial adhesion.

A different approach was taken by Touhsaent et al. [208]. These authors synthesized two polymers, one of which formed a network, by simultaneous independent reactions in the same container. They have indicated that inter-crosslinking reactions are eliminated by combining free radical (acrylate) and condensation (epoxy) polymerization. By this method, they modified an epoxy resin with poly(*n*-butyl acrylate) polymer. They have found that a two-phase morphology developed, consisting of co-continuous rubber domains (about 0.1–0.5 μm) within the epoxy resin. The dimensions of the dispersed rubber phase domains and the extent of molecular mixing between the two components were found to depend on the relative reaction rates (or gel time) with respect to the rate of phase separation. Better mechanical properties resulted when the extent of molecular mixing was minimized and heterophase semi-IPNs were produced.

Similar observations were reported by Sperling and Sarge [209] in their work concerning the production of heterogeneous Independent Polymer Network (IPNs) of polymethylmethacrylate (PMMA) within a polydimethyl siloxane elasmer (PDMS). When polymerization of the network-forming methacrylate monomer and crosslinking of PDMS elastomer were carried out simultaneously, the resulting product was found to contain dispersed PMMA particles within the PDMS matrix, giving rise to a strain-hardening elastomer behavior.

E. Modification of Epoxy Resins with Hygrothermal Toughening Agents

When the methacrylate monomer was polymerized within an already crosslinked PDMS phase, the resulting product displayed a co-continuous two-phase morphology and exhibited a leathery type behavior.

The search for functionalized oligomers for the production of toughened epoxy resins that exhibit low water absorption characteristics and are suitable for high-temperature applications has been the focus of attention of several research workers. For example, Takahashi et al. [210] have examined several amine-terminated silicone oligomers as toughening agents for epoxidized novolac resins for use as encapsulants for semiconductor integrated circuit devices—low electrical stress application. In such cases, siloxane oligomers offer the following advantages: lower T_g values for the dispersed rubbery particles than those for conventional elastomers; and very good thermal stability. The miscibility of the silicone oligomers with epoxy resin was found to increase with increasing ratio of phenyl methyl siloxane units relative to dimethyl siloxane units in statistical copolymers. Total solubility in the epoxy resin was achieved, however, only in the case of phenyl methyl siloxane homopolymers, which produced transparent (monophase) cured products. For other siloxane oligomers, a two-phase morphology was observed in every case, but the dispersed particles became extremely small (i.e., about 0.01 μm) when the silicone oligomers were added to the resin as solutions in toluene.

Interestingly, lower thermal expansion coefficients were observed for these systems in compositions containing 70% fused silica filler. This has been attributed, however, to improved interfacial adhesion resulting from the reaction between the siloxane component and the silica particles, thereby improving the transfer efficiency of the thermal stresses.

Siloxane elastomers have been chosen also by other workers as an attractive alternative to traditional toughening systems, although some of these oligomers are quite expensive. Besides the advantages already mentioned, they exhibit good weatherability, oxidative stability, high flexibility, and moisture resistance [211]. Moreover, the nonpolar nature and low surface energy of siloxanes constitute a thermodynamic driving force for them to migrate to the air–polymer interface. This migration can occur with simple physical blends as well as with systems containing chemically linked microphase-separated segments.

During the early stages of curing of a siloxane-modified epoxy, that is, before extensive crosslinking begins to impair the diffusion characteristics, such migration is considered possible and is believed to lead to the formation of a very hydrophobic and chemically bound surface coating [212]. There is evidence to suggest that such a surface layer reduces friction and improves the wear properties [213].

Yorkgitis and coworkers [214] have chemically modified epoxy resins with functionally terminated poly(dimethyl siloxane), poly(dimethyl-co-methyltrifluoropropyl siloxane), and poly(dimethyl-co-diphenyl siloxane) oligomers and have analyzed the morphology, solid-state properties, and friction and wear properties of the system. They have found that the miscibility of siloxane modifiers in epoxy resins can be enhanced by increasing the percentage of methyltrifluoropropyl siloxane or diphenyl siloxane relative to dimethyl siloxane.

It is known that the solubility parameter is a good indicator of the miscibility of one substance with another and, together with considerations of molecular weight and temperature, can be used to predict the possible occurrence of phase separation of the elastomer from the resin during cure. Through copolymerization of dimethyl siloxane with partially aromatic diphenyl siloxane or polar methyltrifluoropropyl siloxane, it is possible to raise the solubility parameter of the siloxane elastomer from 3.7×10^{-3} $(J/m^3)^{1/2}$ to close to that of the epoxy resin, approx. 4.5×10^{-3} $(J/m^3)^{1/2}$ [156]. This is analogous to how the solubility parameters of ATBN and CTBN elastomers are controlled by adjusting the acrylonitrile content. In this way, the level of miscibility of the siloxane modifier in the epoxy resin controls the size and makeup of the phase-separated elastomeric domains, that is, the morphology and the resulting modulus and fracture toughness of the modified resins. The authors have reported that while unmodified polydimethylsiloxane, due to the large difference in solubility parameter, phase separates from the epoxy resin into large domains that do not increase the fracture strength, the fracture toughness of the epoxy resin can be improved by modification with siloxanes containing 40% or higher methyltrifluoropropyl content, or 20 and 40% diphenyl siloxane content.

Cecere et al. [215] have focused their work on the optimization of both molecular weight and diphenyl siloxane contents of the poly(diphenyl-dimethyl)siloxane oligomers in order to obtain the maximum impact strength without sacrificing the flexural modulus. They have found that siloxane copolymers when used as impact modifiers are not very effective in increasing fracture toughness over that of an unmodified system. It appears that a system incorporating 15 wt % of a 40% diphenyl/60% dimethyl copolymer with a molecular weight of approx. 5000 g/mol yields the highest impact strength with a small decrease in flexural modulus. This system phase separates into evenly dispersed particles with an average diameter of ca. 1 μm.

As epoxy and silicone rubber are completely immiscible, the addition of a compatibilizer is necessary to obtain a satisfactory dispersion of the rubber in the resin. The main objective of Kasemura and coworkers [216] was to find an appropriate surface-active agent to reduce the interfacial tension between the resin and the rubber, in order to compatibilize the two components. These authors achieved adequate compatibility in the epoxy resin with the use of a polyester-modified silicone oil to disperse an RTV (room temperature vulcanizing) silicone rubber or silicone diamine. The results showed that the impact fracture energy of the resin was increased by the addition of the RTV silicone rubber, up to two times that of the unmodified resin, whereas the addition of silicone diamine had almost no effect, possibly because the molecular weight was too low. Moreover, T-peel strengths of aluminum plates bonded by epoxy resin filled with RTV silicone rubber and with silicone diamine effectively increased with the silicone content, showing a maximum at 10–20 pph. By scanning electron microscopy, many particles of silicone rubber, 1–20 μm, were observed across the whole of the fracture surface.

More recently, other workers [217] have studied the rubber modification of bifunctional and tetrafunctional epoxy matrices by means of s block copolymer of polydimethylsiloxane and polyethylene elastomer or an anhydride-grafted polybutene. The choice of these types of liquid reactive elastomers was determined, and photooxidative resistance compared with that of classical unsaturated elastomers. The results showed that the mechanism of fracture of bifunctional resins can be positively influenced by the addition of the aforementioned rubbery systems, whereas for a tetrafunctional epoxy resin the same elastomers do not produce any improvement in impact properties. The authors have attempted to give an explanation on the basis of the different networks obtained in the two matrix systems [217]. In accordance with previous findings and interpretations, the authors concurred that in the case of tetrafunctional epoxy resins, the matrix has a very high crosslinking density and, therefore, its capacity to deform by shear yielding is highly reduced; hence, the contribution of the rubbery particles in enhancing fracture toughness by promoting localized shear yielding in the matrix is rather small.

The constant search for new elastomeric systems with specific properties to use as toughening agents for epoxy resins has induced some researchers to investigate the possible use of fluoro-elastomers.

Attempts to introduce fluorine atoms into the network of crosslinked epoxy formulations have been made through the addition of specially functionalized fluoro-elastomers with the aim of simultaneously enhancing the toughness and thermal stability characteristics [218]. The elastomer and the resin/hardener components were mixed from solutions to obtain an initially monophase system that would subsequently allow the precipitation of the elastomer into fine particles through post-curing heat treatment. Mijovic et al. [219] grafted functionalized groups onto fluorocarbon elastomers to enhance their miscibility and reactivity with resins. A large increase in fracture energy was observed with the addition of 15% elastomer to the resin, though this was accompanied by a decrease in T_g, due to the solubilization of some of the elastomer within the epoxy network.

By reacting in solution an acid fluoride-functionalized perfluorooligomer with a diglycidyl ether of bisphenol A, Rosser et al. [220] produced a prepolymer that was subsequently used to modify an epoxy/diamino-diphenyl sulfone resin matrix for a glass cloth composite. They have demonstrated that this immiscible elastomer prepolymer exhibits sufficient chemical reactivity with the epoxy resin to give rise to improvements in flexural ductility and impact resistance, without loss of strength and modulus or lowering of the glass transition temperature. The results suggest that a simultaneous interpenetrating polymer network (SIN) was formed, which gave rise to improvements in the mechanical properties of the composite.

The use of modified perfluoropolyether oligomers in an epoxy resin was found by Mascia et al. [221] to produce both co-continuous and particulate two-phase systems, the morphology depending on details of the procedure. Hydroxy-terminated fluoroalkane oxide oligomers were reacted with chlorendic anhydride and subsequently with ε-caprolactone to produce carboxyl-terminated perfluoroether prepolymers that were totally miscible with diglycidylether of bisphenol A. Curing the epoxy resin mixtures with hexahydrophthalic anhydride hardener and benzyl dimethylamine catalyst produced transparent products exhibiting a co-continuous two-phase (heterogeneous IPN) morphology. Pre-reacting the fluoroalkenoxide prepolymers with an excess of epoxy resin, prior to the addition of hardener and catalyst, resulted in opaque products displaying a two-phase dispersed particle morphology. Optical microscopy experiments on the resin mixture up to the gelation point have clearly revealed that particles form and grow only if "nuclei" are originally present. The nuclei can be considered as consisting of small swollen "gels" that grow into particles as a result of the inner diffusion of the hardener. Mechanical property measurements revealed remarkable improvements for both systems in fracture energy (450–600%), flexural strength (24–75%), and strain at break (300–500%), even with the use of only small amounts of perfluoropolyethers, namely, 3.5–5.0%. However, these were achieved at the expense of a small decrease in modulus (20–30%) and T_g (10–25°C). It must be noted that these effects were much more pronounced for products exhibiting a two-phase, dispersed particle morphology than for co-continuous phase (IPN) systems. The authors [222] also found that both IPN and particulate two-phase formulations showed a reduction in flexural strength after 21 days of aging at 200°C, but the measured values were always much greater than for the nonaged control samples. The strain at break for IPN systems, however, increased considerably with aging in proportion to the concentration of prepolymer used.

The main objective of Pascal et al. [223] was to find an effective toughening by introducing energy absorption processes that do not involve matrix ductility.

Few published works have been devoted to the toughening of thermostats with high-performance thermoplastic powders [224,225]. This approach allows a two-phase system to be obtained without passing through a phase-separation process. However, partial miscibility in the interphase area is possible depending on the thermoplastic chemical structure and blending conditions. In addition, the introduction of a dispersed powder into the uncured thermosetting resin should induce a limited viscosity increase, allowing a good processibility to be maintained. To evaluate the efficiency of the powder modification, Pascal et al. [223] investigated the effect of two semicrystalline polyamides PA6 and PA12, which are commercially available with a controlled powder size distribution. In their study, the authors used an epoxy resin that consists essentially of *N,N′,N′,N′*-tetraglycidyl-4,4′-diamino-diphenylmethane (TGMDA) cured with a mixture of 4,4′-diamino-diphenyl sulfone (DDS), dicyandiamide (DDA), and 3-(3,4-dichlorophenyl)-1,1-dimethylurea. The polyamide powders (PA6 and PA12) and the cardo polyimide (supplied by Atochem, Milan) were used in the powder modification method. The cardo polyamide was synthesized by reaction of benzophenone tetracarboxylic dianhydride (BTDA) and 4,4′-(9H-fluoren-9-ylidene)bisbenzeneamine (FBPA) in *N*-methylpyrrolidone (NMP) solution:

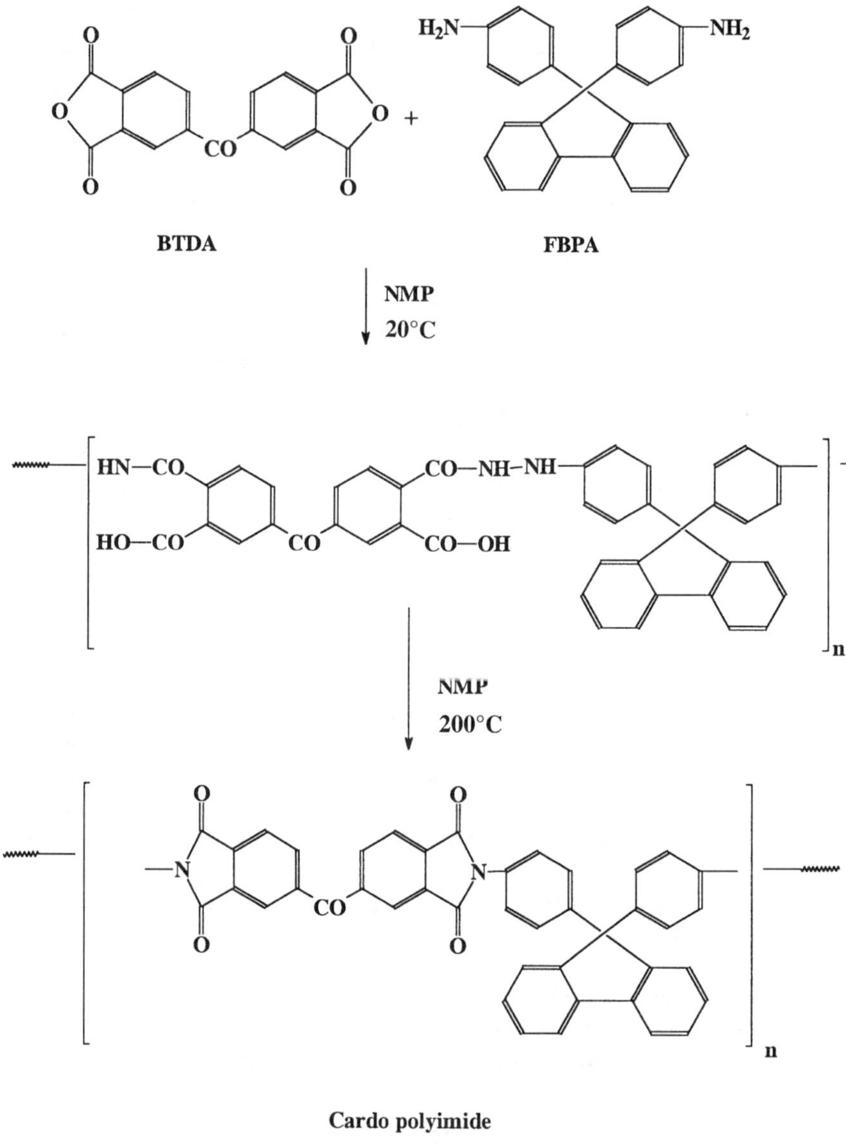

(73)

The powder-containing blends were prepared by mixing the basic epoxy resin with the cardo polyimide, and PA6 and PA12 polyamides. Stable fine dispersions containing about 8–16% of thermoplastic powder were readily obtained under usual mechanical stirring.

The versatility of the powder-toughening method appears particularly attractive for high-temperature thermosetting resins due to the possibility of combining relatively insoluble or infusible products. The initial problem lies, however, in preparing a polymer powder with a particle size distribution suitable for reinforced composite fabrication, namely, below interfiber spacing, which is typically in the μm range.

From analysis of the literature on modifiers for epoxy resins, one can arrive at the following conclusions regarding the toughening of epoxy resins [226]:

1. An efficient toughening mechanism can be set up if the crosslinking is sufficiently low. Soft or rubbery inclusions, the surface-to-surface distance of which is below a critical value, tend to induce shear deformations in the matrix and will, therefore, provide the required conditions for toughness enhancement.
2. When the crosslinking density is high, as in the case of tetrafunctional resins, toughness can be increased with the use of ductile glassy polymers capable of reacting with the epoxy resin in order to produce a strong interfacial adhesion between the two phases.

   The similar modulus of the two phases creates a uniform stress distribution across the boundaries. As a result, fracture is not likely to initiate near the boundaries, but within the matrix. The propagation of cracks will be hampered by plastic deformations or microcavitations in the dispersed ductile particles.

   Obviously, such a mechanism will operate more effectively if the two phases are co-continuous and/or the interfacial adhesion is sufficiently high to prevent fracture from propagating at the interphase boundaries.
3. The miscibilization of oligomers through telechelic end-of-chain extensions is likely to constitute the most efficient mechanism for the nucleation of particle precipitation during curing of the resin. These are also likely to minimize reductions of the T_g of the matrix by residual (non-precipitated) oligomer species, owing to their inability to form true solutions in the resin—only highly swollen molecular aggregates. Consequently, phase separation by nucleation and growth during curing is likely to occur very readily and completely.

V. APPLICATIONS OF EPOXY RESINS

It was hard to even guess that the glycidylic compounds discovered by the Swiss chemist P. Kastan in the 1930s would become so significant in 20 years. Many industrial branches would become potential fields of application of these products. Glass fiber–reinforced plastics, a wide variety of adhesives, and numerous electroinsulating materials; the aeronautics and aerospace industries, electronics, and civil engineering—all these new products and fields of activity could not exist without epoxy resins.

The epoxy resins are the most versatile class of contemporary plastics. When first introduced in the late 1940s, the epoxy resins were proclaimed to be a "miracle plastic." Throughout the years, disadvantages have become apparent. In fact, the epoxy resins are engineering materials rather than a philosopher's stone. They must be considered as such.

It is very difficult to estimate the volume usage in the various applications of epoxy resins. From the entire yield of epoxy resins, 50% is used to produce coverings, 20% to obtain reinforced materials used in building, and the rest to obtain oils.

Formulations designed for specific applications generally consist of a number of ingredients in combination, and in this manner the properties are selectively modified. Broadly speaking, the fields of coatings, adhesives, and casting compounds require the most sophisticated formulations. These systems are often quite complex, and many contain from 6 to 10 ingredients. In the composite field, the primary emphasis is on the selection of the proper epoxy resin and the proper curing agent. Other ingredients are seldom present.

Numerous types of epoxy oligomers as obtained in the laboratory or on an industrial scale are known. Such a diversity came to be by the requirements called for in their various applications. Epoxydianic oligomers occupy the most important volume of production. Besides these, also worth mentioning are the epoxy-novolac type oligomers, obtained from phenol–formaldehyde oligomers and epichlorohydrine (ECH). Another important group are *N*-glycidylic oligomers based on amines, amides, cyanuric acid, and melamine. The interaction of epoxy oligomers with isocyanates produces epoxyurethanes, whereas by the reaction with phosphorus or silicic acids, epoxy oligomers with P or Si content are obtained [227].

Another type of epoxy oligomers are the cyclo-aliphatic ones, obtained by oxidation of cyclic diolefins with various peroxidic compounds (organic or inorganic hydroperoxides), hypochlorous acid, halogenhydrines and their derivatives, and oxygen or ozone. At present, the main industrial procedure for their production uses peracetic acid. In keeping with initial cycloolefin structure, diepoxides with different structures are produced, with the peculiarity that peroxidic oxygen is bound to the aliphatic cycle.

Union Carbide produces a number of epoxyaliphatic resins under the name of ERLA, obtained by homo- or copolymerization of the mixture of esters formed between bis-2,3-epoxychloropentylol with ethylene glycol in the presence of benzyldimethylamine. The diepoxide formed is commercialized under the name ERLA-4205. Compared with other epoxy resins, ERLA epoxies show a higher resistance by 25 to 110% to bending, compression, and tension.

The properties of some industrial or semi-industrial oligomeric epoxies based on diphenylol propane produced by Ciba-Geigy (Basel) and Shell (Cleveland, OH) are listed in Table 4 [228]. Several characteristics of epoxy oligomers produced in Russia are listed in Table 5 [228]. Tables 6 and 7 list characteristic properties of epoxy oligomers with a nitrogen content and cycloaliphatic epoxy oligomers, respectively, produced in Russia [228].

Table 4 Characteristics of Some Epoxy Oligomers Based on Diphenylol Propane

Trade name	M_n	Epoxy group content (%)	Softening temperature (°C)
Araldit GY-250	—	30.0	—
Aradit GY-280	—	21.6	—
Araldit GY-6071	—	11.4–12.5	64–76
Araldit GY-7072	—	13.8–18.9	73–89
Araldit GY-6084	—	5.7–6.3	95–155
Araldit GY-6099	—	1.4–2.3	140–155
Epikot 815	—	28.5–31.4	—
Epikot 828	380	28.5–31.4	—
Epikot 834	470	19.9–25.0	2
Epikot 1001	900	11.4–12.9	64–76
Epikot 1004	400	5.7–63.0	95–100
Epikot 1007	2900	2.8–3.4	125–132
Epikot 1009	3750	1.4–2.3	140–155

Source: Ref. 228.

For solidification of epoxy oligomers, numerous substances are produced, such as aliphatic polyamines, aromatic amines, dicyanoamines and their derivatives, the anhydrides of di- and polycarboxylic acids, as well as a number of catalysts for solidification. These substances impart a large variety of technological properties to epoxy oligomers. Some properties of the main solidification agents produced worldwide are listed in Table 8 [229–231].

Table 5 Properties of Epoxy Oligomers Based on Diphenylol Propane Produced in Russia

Trade name	Epoxy group content (%)	M_n	Content (max.) (%)		
			Volatiles	Cl^-	Cl_2
ED-24	23	340–370	0.3	0.007	0.5
ED-5	18–23	360–470	2.0	—	1.5
ED-22	22.5–23.5	390	0.5	0.007	1.0
ED-20	19.9–22.0	390–430	1.0	0.007	1.0
ED-16	16.0–18.0	480–540	0.8	0.007	0.7
ED-6	13.0–18.0	480–600	1.0	—	0.7
ED-14	13.9–15.9	540–640	0.8	0.007	0.7
ED-10	10.0–13.0	660–860	0.8	0.007	0.6
ED-8	8.0–10.0	860–1100	0.8	0.007	0.6
E-40	16.0–21.0	600	—	—	—
E-44	6.0–8.5	1600	—	—	0.2
E-49	2.0–4.5	2500	—	—	0.2

Source: Ref. 228.

Table 6 Characteristics of Nitrogen-Containing Epoxy Oligomers Produced in Russia

Trade name	Viscosity (Höppler, 40°C, cP)	Content (%)		
		Epoxy groups	Volatiles	Cl_2
EA	120	30	1.5	2.5
EMDA	10×10^3–12×10^3	31–38	0.4–1.5	0
UP-610	1.1×10^3–1.8×10^3	36–38	1.0	1.16
UP-622F	10.0	26–29	2.0	—
UP-622	10.0	25–28	2.0	—
UP-622A	0.5×10^3–1.5×10^3	25–28	2.0	1.5
UP-622T	1.5×10^3	32	1.0	—
UP-631	—	9	0.5	1.5
UP-67	15[a]	23.5	1.0	1.5
UP-643	90[b]	22.0	0.5	1.0

[a]Dynamic viscosity, 25°C, $Pa \cdot s^{-1}$.
[b]Dynamic viscosity, 50°C, $Pa \cdot s^{-1}$.
Source: Ref. 228.

It is already known that the alpha epoxy cycle shows a large capacity of reaction toward various compounds through a polyaddition process implying opening of the cycle. Partners in this reaction can be various categories of substances, which have been partly listed in Table 8. As a result of this process, a tridimensional structure emerges, the characteristics of which depend on the molecular weight distributions and certain structural details of the initial oligomers.

Commercial applications for epoxy resins systems were at one time predominantly, indeed almost exclusively, held by the structural adhesives industry. These systems operate as some of the best available coatings in automotive and aerospace industries.

Though organic coatings have been long employed to protect metals from corrosion, there are still many unsolved problems in corrosion mechanisms, adhesion, and practical testing.

A widely used approach to reducing the corrosion of steel involves the application of a polymeric coating, which serves as a barrier between the steel and the corrosion-causing environment. However, organic coatings can undergo degradation that reduces their effectiveness under service conditions. An essential

Table 7 Properties of Cycloaliphatic Epoxy Oligomers Produced in Russia

Trade name	Dynamic viscosity (Pa)			Content (%)	
	25°C	40°C	80°C	Epoxy groups	Volatiles
UP-612	—	9	—	27.0	1.0
UP-632	0.2–0.6	—	—	27.0	1.5
UP-644	1.4–4.0	—	—	19.5	1.5
UP-647	—	—	1.5–7.0	24.0	1.5

Source: Ref. 228.

Table 8 Properties of Solidification Agents Used for Epoxy Resins

Trade name	Producer	Viscosity (25°C, cP)	Density (25°C)
L-18	Russia	50,000	—
S-19	Russia	15,000	—
T-19	Russia	15,000	—
I-5M	Russia	700	0.96
UP-0618	Russia	1,200	1.0
DEN-10SH60	Dow Chemical	950	—
ZZL-035	Union Carbide	375	—
NU-215	Ciba-Geigy	4,000	0.95
Ancamed-250	Anhor (England)	4,500	—
A-120	Ciba-Geigy	100,000	1.07
Beckopox EH-650	Hoechst	950	0.98
Versamid 100	Schering	850	0.98
G-MI-250	Schering	750	0.95
Rutapox H-155	Bakelite	10,000	—
Goodmide G-175	Tohto	60,000	—
Goodmide G-645	Tohto	3,000	—

Source: Refs. 229–231.

element for developing improved coatings and coating systems is understanding their degradation mechanisms, particularly at the coating–substrate interface. The exact loci and mechanisms of delamination and degradation of a coating on a steel substrate exposed to a corrosive environment are still very much in debate at present. Early works indicated that hydroxide ions generated by the corrosion reactions cause the debonding [232–234].

Water is regarded as the major ingredient causing corrosion. The consensus is that a layer of water builds up under an organic coating. When a cathodic corrosion takes place, water, with the aid of oxygen and electrons, can produce hydroxyde ions according to

$$H_2O + \frac{1}{2} O_2 + 2\ e^- = 2\ HO^- \qquad (74)$$

This reaction is facilitated by cation counterions and a catalytic surface of metal oxide. Via this equation, we can speculate on some of the measures that should help to prevent or determine the corrosion (cathodic) process and the subsequent loss of adhesion of the coating:

1. Modify the polymer structure and achieve lower water permeability.
2. Reduce oxygen permeability to the metal through the coating by applying a certain primer.
3. Reduce the conductivity of the oxide layer [235].

4. Incorporate a cation-exchange material at the interface. Leidheiser and Wang [236] demonstrated that the rate of the cation counterion diffusion through the organic coating can control the hydroxide formation. Thus, Leidheiser [237] proposed the use of a cation-exchange medium at the interface. However, in reality, this proposal is not easy to carry out.
5. Use a corrosion inhibitor to suppress the catalytic activity of the metal oxide. This is perhaps the most common way to prevent corrosion.

In marine environments, corrosion by salt water has always been a severe problem. Aerospace missiles on ships are equally susceptible to attach by salt and water. A finish system of a cathodic base coat with an epoxy topcoat was demonstrated to provide exceptional resistance for the motor cases of missiles in severe marine environments.

Water is not only a major ingredient for metal corrosion, it is also a key product in the corrosion and degradation of electronic and optical devices.

For electronic or microelectronic devices, corrosion can cause catastrophic failures. To achieve reliability, corrosion-resistant materials, such as protective coatings or encapsulants, and corrosion-preventing agents, such as inhibitors, are commonly used. The corrosion can be atmospheric, galvanic, or electrolytic. Atmospheric corrosion results from the interactions of moisture, oxygen, ionic impurities, and low molecular weight organic compounds, whereas galvanic corrosion is caused by the contact of dissimilar materials. Electrolytic corrosion of conductors is induced by the potentials normally applied to an electronic circuit.

In space environments, temperature extremes require exotic polymers as adhesives and sealants for spacecraft, and bombardment by particles, such as molecules, ions, and atoms, requires special coatings on the surface of spacecraft.

In the case of amine-cured epoxy resins, the delamination is determined both by corrosion and by oxidative reactions [238]. This process probably involves the loss of the secondary alcohol:

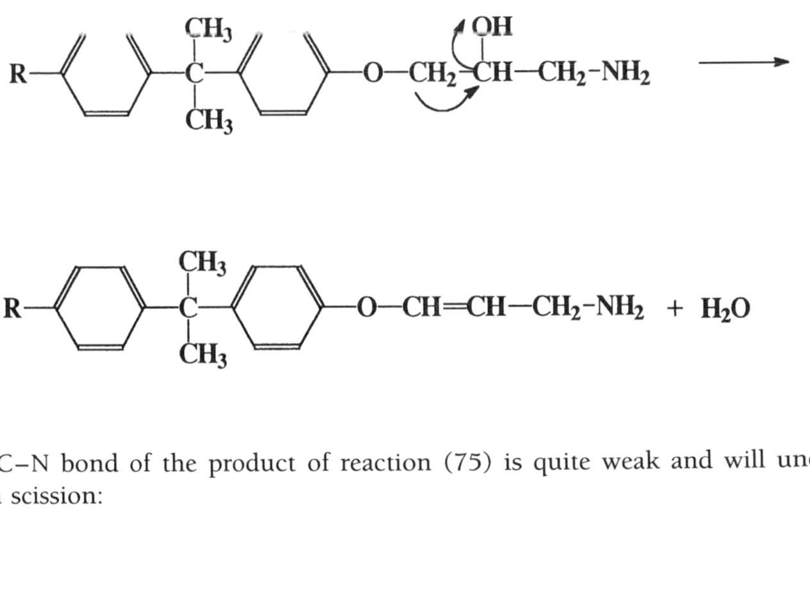

(75)

The C–N bond of the product of reaction (75) is quite weak and will undergo chain scission:

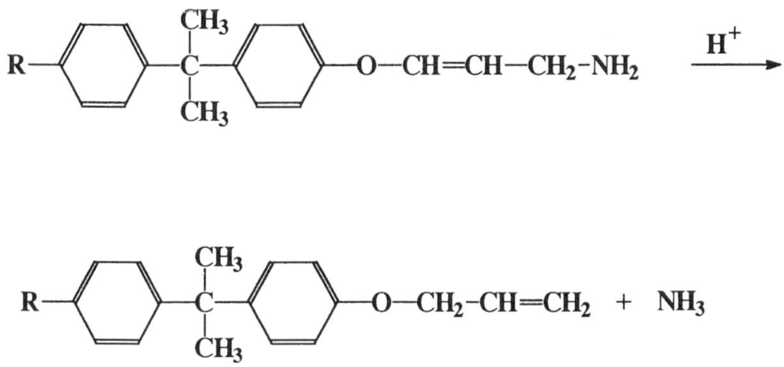

(76)

Epoxy resins may be converted to coatings with good resistance to mild acid and alkaline environments by esterification of the expoxide and hydroxyl groups with dehydrated castor oil fatty acids. Linseed or soybean oil acid may also be used.

Oil-free epoxy formulations are used for industrial maintenance and, in fact, have been introduced into the retail market for use on certain areas in dwellings. Curing of the epoxy resins to hard resistant films may be brought about by reacting them with amines, polyamines, or polyamides to produce crosslinking. These components (epoxy and amines) are mixed just before use, the type and amount of each part being adjusted to give eventual curing and also to provide several hours of pot life (i.e., retention of fluidity) at normal temperature after mixing.

A modified form of epoxy is known as coal-tar epoxy coating. This consists of a two package epoxy–amide combination with the addition of coal-tar pitch to either the amide or epoxy portion. It is possible to spray this material up to a thickness of 20 μm. Its chief use is on underground pipes or aboveground structures where high chemical resistance is needed and appearance is not important.

In the construction industry, epoxy resins find a variety of applications such as adhesive mortars, composite materials, and coatings. Epoxy polymers show excellent adhesion to the most common metals, glasses, ceramics, concretes, and other materials.

Their superior performance is, in essence, due to the basic chemical structure of epoxy. Epoxy resins have very high polarity. The presence of aliphatic hydroxyl and ether groups in the primary resin chain as well as in the cured polymer serve to create an electromagnetic bonding between the epoxy macromolecules and the surface being bonded [239].

In commercial applications, epoxy resins are rarely used without the incorporation of some other materials. Filling or polyblending are both used to enhance their performance by providing additional mechanical properties in the blends [240].

REFERENCES

1. Leeard, N., and Newille, K., in *Encyclopedia of Polymer Science and Technology*, vol. 6, Mark, H. F., Gaylord, N. G., and Bikales, M., Eds., Interscience, New York, 1967, 209.

2. Kirk, R. E., and Ozhmer, D. F., *Encyclopedia of Chemical Technology*, vol. II, Interscience Encyclopedia Inc., New York, 1977, 667.
3. Scheiber, J., and Sandig, K., *Artificial Resins*, Pitman, London, 1931, 1–7.
4. Castan, P., CH Patent 211,116, 1940.
5. Cowan, J. C., U.S. Patent 2,450,940, 1947.
6. Phillips, B., U.S. Patent 2,904,473, 1960.
7. May, C. A., *SPE J.*, *21*, 1105, 1965.
8. Lee, H., and Neville, K., *Handbook of Epoxy Resins*, McGraw-Hill, 1967, Chs. 2 and 5.
9. Myers, R. B., and Long, J. S., Eds., *Film Forming Compositions*, Marcel Dekker, New York, 1967, Ch. 7.
10. Brode, G. L., and Wynstra, J., *J. Polym. Sci.*, A-*1*, *4*, 1045, 1966.
11. Batzer, H., and Zahir, H. A., Studies in the molecular weight distribution of epoxide resins. I. Gel permeation chromatography of epoxide resins, *J. Appl. Polym. Sci.*, *19*, 575, 1975.
12. Tanaka, S., Yokoyama, K., and Takashima, M., *Polym. Lett.*, *6*, 385, 1968.
13. Batzer, H., and Zahir, H. A., Studies in the molecular weight distribution of epoxide resins. II. Chain branching in epoxide resins, *J. Appl. Polym. Sci.*, *19*, 601, 1975.
14. Batzer, H., and Zahir, H. A., Studies in the molecular weight distribution of epoxide resins. III. GPC of epoxide resins subject to postglycidylation, *J. Appl. Polym. Sci.*, *19*, 609, 1975.
15. Beavers, E. M., U.S. Patent 2,765,322, 1953.
16. Singley, J. E., and Whittle, G. P., U.S. Patent 3,060,151, 1962.
17. Stutz, H., Tesch, P., Neumann, P., and Schaefer, G., GER Patent 3,523,318, 1987.
18. Griebsch, E., and Hilgert, H., GER Patent 1,106,072, 1961.
19. Sorokin, M. F., and Ivanov, S. A., *Lakokras. Mater. Ikh. Primen.*, *5*, 11, 1984 (Russian).
20. Sorokin, M. F., and Kirillova, S. G., *Lakokras. Mater. Ikh. Primen.*, *4*, 16, 1982 (Russian).
21. Sorokin, M. F., and Volkova, L. I., *Lakokras. Mater. Ikh. Primen.*, *5*, 2, 1984 (Russian).
22. Kreps, R. W. F., and Goppel, J. M., U.S. Patent 3,364,178, 1968.
23. Sykova, V., Spacek, V., and Dobas, I., *J. Appl. Polym. Sci.*, *54*, 1463, 1991.
24. Podzimek, S., Sykova, V., and Svestka, S., *J. Appl. Polym. Sci.*, *58*, 1491, 1995.
25. Khan, M. M., and Fatemi, H., *Proc. Int. Symp. Microel.*, *9*, 420, 1986.
26. Factor, A., *J. Polym. Sci., Polym. Chem. Ed.*, *11*, 1691, 1973.
27. Zaks, Y., Lo, J., Paucher, D., and Pearce, E. M., *J. Appl. Polym. Sci.*, *27*, 913, 1982.
28. Chun-Shan Wang, Berman, J. R., Walker, L. L., and Mendoza, A., Meta-Brombiphenol epoxy resins: Applications in electronic packaging and printed circuit board, *J. Appl. Polym. Sci.*, *43*, 1315, 1991.
29. Boutevin, B., Robin, J. J., and Roume, C., Synthèse de résines époxydes tetrafonctionelle par l'intermediaire de diamines et de l'épichlorohydrine, *Eur. Polym. J.*, *31*, 313, 1995.
30. Maruno, T., and Nakamura, K., *J. Appl. Polym. Sci.*, *42*, 2141, 1991.
31. Nakamura, K., and Maruno, T., *Electr. Commun. Lab. Tech., J.*, *35*, 1227, 1986.
32. Maruno, T., Nakamura, K., and Murata, N., Synthesis and properties of a novel fluorine-containing alicyclic diepoxide, *Macromolecules*, *29*, 2006, 1996.
33. Maruno, T., Ishibashi, S., and Nakamura, K., Synthesis and properties of fluorine-containing epoxy(meth)acrylate resins, *J. Polym. Sci., Part A, Polym. Chem.*, *32*, 3211, 1994.
34. Manaka, K., U.S. Patent 3,975,340, 1976.
35. Klee, J., Clauben, H., Horhold, H., and Raddatz, J., *Polym. Bull.*, *27*, 511, 1992.
36. Klee, J., Horhold, H., and Schlz, H., *Acta Polymerica*, *42*, 17, 1991.

37. Titier, C., Pascault, J. P., Taha, M., and Rosenberg, B. J., Epoxy-amine multimethacrylic prepolymers, kinetic and structural studies, *J. Polym. Sci., Part A, Polym. Chem.*, *33*, 175, 1995.
38. Titier, C., Pascault, J. P., and Taha, M., Synthesis of epoxy-amine multiacrylic prepolymers by reactive extrusion, *J. Appl. Polym. Sci.*, *59*, 415, 1996.
39. Smolinski, S., and Halaza, E., *Rocz. Chem.*, *48*, 1459, 1974.
40. Simoe Gandara, J., Paseiro Lozada, P., Perez Mamela, C., and Paz Abuin, S., Epoxy resins based on trimethylol propane. I. Determination of chemical structure, *J. Appl. Polym. Sci.*, *55*, 225, 1995.
41. Quentela, A. L., and Abuin, S. P., Epoxidation reaction of trimethylol propane with epichlorohydrin: Kinetic study of chlorohydrin formation, *Polym. Engng. Sci.*, *36*, 568, 1996.
42. St Cyr, M. C., *Soc. Plast. Eng. Trans.*, *1*, 47, 1961.
43. Cupery, M. E., U.S. Patent 2,692,876, 1954.
44. Apice, P. J., U.S. Patent 3,433,854, 1969.
45. Chatta, M. S., *J. Coat. Technol*, *52*, 43, 1980.
46. Chatta, M. S., Beckwith, E. C. S., and Henk van Oene, U.S. Patent 4,184,785, 1980.
47. Chatta, M. S., U.S. Patent 4,181,784, 1980.
48. Chatta, M. S., and Henk van Oene, New approach to high solid coatings, *I & EC Prod. Res. Develop.*, *21*, 437, 1982.
49. He Pingsheng and Li Chune, *J. Mater. Sci.*, *24*, 2951, 1989.
50. He Pingsheng and Li Chune, Study on cure behaviour of epoxy resin-BF_3-MEA system by dynamic torsional vibration method, *J. Appl. Polym. Sci.*, *43*, 1011, 1991.
51. Hsich, H. S. Y., *J. Mater Sci.*, *13*, 2560, 1978.
52. Hsich, H. S. Y., *J. Appl. Polym. Sci.*, *27*, 3265, 1982.
53. Adhinarayanan, K., Epoxy-imide based on bis(carboxyphthalimide)s, *J. Appl. Polym. Sci.*, *49*, 759, 1991.
54. Horie, K., Hiura, H., Sauvada, M., Mika, I., and Kambe, H., *J. Polym. Sci., Polym. Chem. Ed.*, *8*, 1357, 1970.
55. Dusek, K., Ilavsky, M., and Lunak, S., *J. Polym. Sci., Polym. Symp. Ed.*, *53*, 29, 1975.
56. Lunak, S., and Dusek, K., *J. Polym. Sci., Polym. Symp. Ed.*, *53*, 45, 1975.
57. Charlesworth, J. J., *J. Polym. Sci., Polym. Symp. Ed.*, *18*, 621, 1980.
58. Riccardi, C. C., Adabbo, H. E., and Williams, J. J., *J. Appl. Polym. Sci.*, *29*, 2481, 1984.
59. Seung, C. S. P., Pymm, E., and Sun, H., *Macromolecules*, *19*, 2922, 1986.
60. Matejka, L., and Dusek, K., Curing of diglycidylamine-based epoxides with amines: Kinetic model and simulation of structure development, *J. Polym. Sci., Part A, Polym. Chem. Ed.*, *33*, 461, 1995.
61. Huang, M. L., and Williams, J. G., Mechanisms of solidification of epoxy-amine resins during cure, *Macromolecules*, *27*, 7423, 1994.
62. Girard-Reydet, Riccardi, C. C., Santerean, H., and Pascault, J. P., Expoxy-aromatic diamine kinetics. 1. Modeling and influence of the diamine structure, *Macromolecules*, *28*, 7599, 1995.
63. Riccardi, C. C., and Williams, R. J. J., *J. Appl. Polym. Sci.*, *38*, 8445, 1986.
64. Simon, S. L., and Gilham, J. K., *J. Appl. Polym. Sci.*, *46*, 1245, 1992.
65. Wisanrakkit, G., and Gilham, J. K., *J. Appl. Polym. Sci.*, *41*, 2885, 1990.
66. Wang, X., and Gilham, J. K., *J. Appl. Polym. Sci.*, *43*, 2267, 1981.
67. Min, B. G., Stachurski, Z. H., and Hodkin, J. H., *Polymer*, *34*, 4488, 1993.
68. Dusek, K., *Polym. Mater. Sci. Eng.*, *49*, 378, 1983.
69. Rosenberg, B. A., *Adv. Polymer Sci.*, *72*, 113, 1985.
70. Ywakura, Y., and Izawa, S., *J. Org. Chem.*, *29*, 379, 1964.
71. Schechter, L., and Wynstra, J., *Ind. Eng. Chem.*, *48*, 86, 1956.

72. Galy, J., Sabra, A., and Pascault, J. P., *Polym. Engrg. Sci.*, *26*, 1714, 1986.
73. Pearce, P. J., and Morris, C. E. M., *Polym. Commun.*, *29*, 93, 1988.
74. Kadurina, T. I., Prokopenko, V. A., and Omelchenko, S. I., *Eur. Polym. J.*, *22*, 865, 1986.
75. Manzione, L. T., Gilham, J. K., and McPherson, C. A., *J. Appl. Polym. Sci.*, *26*, 889, 1981.
76. Fasce, D. P., Galante, M. J., and Williams, R. J. J., Curing of epoxy resins with in situ generated substituted ureas, *J. Appl. Polym. Sci.*, *39*, 383, 1990.
77. Schechter, L., and Wynstra, J., *Ind. Eng. Chem.*, *48*, 86, 1956.
78. Tanaka, Y., and Kazinschi, H., Study of epoxy compounds. Part I. Curing reactions of epoxy resin and acid anhydride with amine and alcohol as catalyst, *J. Appl. Polym. Sci.*, *7*, 1063, 1963.
79. Schechter, L., and Wynstra, J., Glycidyl ether reactions with alcohols, phenols, carboxylic acids, and anhydrides, *Ind. Eng. Chem.*, *48*, 186, 1956.
80. Fisch, W., and Hoffmann, W., *J. Polym. Sci.*, *12*, 497, 1954.
81. Fisch, W., *J. Appl. Chem.*, *6*, 429, 1956.
82. Wegler, R., *Angew. Chem.*, *67*, 587, 1955.
83. Dearborn, E. C., Fuoss, R. M., and White, A. F., *J. Polym. Sci.*, *16*, 20, 1955.
84. Dearborn, E. C., *Ind. Eng. Chem.*, *45*, 2715, 1954.
85. Fischer, R. F., Polyesters from epoxides and anhydrides, *J. Polym. Sci.*, *44*, 155, 1960.
86. Bouillon, Nelly, Pascault, J. P., and Tighzert, L., Epoxy prepolymers cured with boron trifluoride amine complexes. 1. Influence of the amine on the curing, *Makromol. Chem.*, *191*, 1403, 1990.
87. Arnold, R. J., *Mod. Plastics*, *41*, 149, 1964.
88. Smith, R. E., Larsen, F. N., and Long, C. L., *J. Appl. Polym. Sci.*, *29*, 3697, 1984; *ibid.*, *29*, 3713, 1984.
89. Smith, R. E., and Smith, C. H., Epoxy resin cure. III. Boron trifluoride catalysts, *J. Appl. Polym. Sci.*, *31*, 929, 1986.
90. Harris, J. J., and Temin, S. C., *J. Appl. Polym. Sci.*, *10*, 523, 1966.
91. Ito, K., and Okahashi, K., *Mitsubishi Dinki Lab. Report*, *10*, 83, 1969.
92. Okamoto, Y., *Ring Opening Polymerization*, *ACS Symposium Series*, McGrath, J. E., Ed., 286, 1985.
93. Penczek, S., Kubisa, P., and Szymansky, R., *Macromol. Chem.*, *Symp.*, *6*, 201, 1986.
94. Bouillon, Nelly, Pascault, J. P., and Tighzert, L., *Makromol. Chem.*, *191*, 1417, 1990.
95. Vogt, J., *J. Adhesion*, *22*, 139, 1987.
96. Yashida, S., and Ishida, H., *J. Adhes.*, *16*, 217, 1984.
97. Farkas, A., and Ströhm, P. F., *J. Appl. Polym. Sci.*, *12*, 159, 1968.
98. Ito, M., Hata, H., and Kamagata, K., *J. Appl. Polym. Sci.*, *33*, 1843, 1987.
99. Jackson, R. J., Pigneri, A. M., and Gaigoci, E. C., *SAMPE J.*, *23*, 16, 1987.
100. Heise, M. S., and Marsin, G. C., Analysis of the cure kinetics of epoxy/imidazole resin systems, *J. Appl. Polym. Sci.*, *39*, 721, 1990.
101. Barton, H., and Stepherd, P. M., *Makromol. Chem.*, *176*, 919, 1975.
102. Ricciardi, F., Joullié, M. M., Romanchick, W. A., and Griscavage, A. A., *J. Polym. Sci.*, *Polym. Lett. Ed.*, *20*, 127, 1982.
103. Jisova, V., Curing mechanism of epoxides by imidazoles, *J. Appl. Polym. Sci.*, *34*, 2547, 1987.
104. Heise, M. S., and Martin, G. C., Analysis of the cure kinetics of imidazole resin systems, *J. Appl. Polym. Sci.*, *39*, 728, 1990.
105. Heise, M. S., Martin, G. C., and Gotro, J. T., Characterization of imidazole-cured epoxy phenol resins, *J. Appl. Polym. Sci.*, *42*, 1557, 1991.
106. Heise, M. S., and Martin, G. C., *Macromolecules*, 22, 99, 1989.

107. Jianfe Ding, Chinmin Chen, and Gi Hue, The dynamic mechanical analysis of epoxy-copper powder composites using azole compounds as coupling agents, *J. Appl. Polym. Sci.*, *42*, 1459, 1991.
108. Gi Hue, Juengfong Zhang, Guaquan Shi, and Peigi Wu, *J. Chem. Soc.*, *Perkin Trans. I*, 1989, 33.
109. Gi Hue, Quingping Dai, and Sanggen Jiang, *J. Am. Chem. Soc.*, *110*, 2393, 1988.
110. Gi Hue, Xueying Huang, Sanggen Jiang, and Gaoquanshi Ho, *J. Chem. Soc.*, *Dalton Trans.*, 1988, 1487.
111. Gi Hue, Jianfu Ding, Peigi Wu, and Gengding Ji, *J. Electroanal. Chem.*, *270*, 163, 1989.
112. Barton, J. M., BRIT Patent 2,135,316, 1984.
113. Barton, J. M., Buist, G. J., and Liu, S., *J. Mater. Chem.*, *4*, 379, 1994.
114. Barton, J. M., Howlin, B. J., and Liu, S., *Polym. Bull.*, *33*, 215, 1994.
115. Barton, J. M., Hamerton, I., and Liu, S., *Polym. Bull.*, *33*, 347, 1994.
116. Buist, G., Hamerton, I., and Barton, J. M., Comparative kinetic analyses for epoxy resins cured with imidazole-metal complexes, *J. Mater. Chem.*, *4*, 1793, 1994.
117. Lee, C. J., U.S. Patent 4,487,894, 1984.
118. Serafini, T. T., Delvigs, P., and Vannucci, R. D., U.S. Patent 4,244,857, 1981.
119. Ichino, T., and Hasuda, Y., *J. Appl. Polym. Sci.*, *34*, 1667, 1987.
120. Adhinarayanan, K., Epoxy-imide resin based on bis(carboxyphthalimide)s, *J. Appl. Polym. Sci.*, *43*, 783, 1991.
121. Jun-Liu Fan, Guan-Miu Wang, and Wen-Hsiung Ku, *J. Appl. Polym. Sci.*, *43*, 829, 1991.
122. Gordon, M., and Malcolm, G. N., *Proc. Royal Soc.*, *London*, *Ser. A*, 1966, 295.
123. Miller, D. R., and Macosko, G. W., *Macromolecules*, *11*, 656, 1978.
124. Dusek, K., *Adv. Polym. Sci.*, *78*, 1, 1986.
125. Flory, P. J., *Principles of Polymer Chemistry*, Cornell Univ. Press, Ithaca, New York, 1953.
126. Stockmayer, W. H., *J. Chem. Phys.*, *11*, 45, 1943.
127. Winter, H., and Chambon, F., *J. Rheol.*, *30*, 367, 1986.
128. Martin, J. E., and Wilcox, J. P., *Phys. Rev. Lett.*, *61*, 373, 1988.
129. Adolf, D., Martin, J. E., and Wilcox, J. P., *Macromolecules*, 23, 527, 1990.
130. Martin, J. E., and Wilcox, J. P., *Phys. Rev.*, *39*, 253, 1989.
131. Bobalek, E. G., Moore, E. R., Levy, S. S., and Lee, C. C., *J. Appl. Polym. Sci.*, *8*, 625, 1964.
132. Solomon, D. H., *Macromol. Sci. Rev.*, *C*, *1*, 179, 1967.
133. Dusek, K., *Adv. Polym. Sci.*, *78*, 1, 1986.
134. Huang, M. L., and Williams, J. G., Mechanisms of solidification of epoxy-amine resins during cure, *Macromolecules*, 27, 7423, 1994.
135. Gérard, J. F., Galy, J., Pascault, J. P., Cukierman, S., and Halary, J. L., *Polym. Engng. Sci.*, *3*, 615, 1991.
136. Bell, J. P., *J. Appl. Polym. Sci.*, 27, 3503, 1982.
137. Mijovic, J., and Tsay, L., *Polymer*, 22, 902, 1981.
138. Sanz, G., Garmendin, J., Andres, M. A., and Mondragon, I., Dependence of dynamic mechanical behaviour of GGBA/DDM stoichiometric epoxy systems on the conditions of curing process, *J. Appl. Polym. Sci.*, *55*, 75, 1995.
139. Sasuga, T., and Udagawa, A., *Polymer*, *32*, 402, 1991.
140. Grof, K., Markvickova, L., Konak, C., and Dusek, K., *Polymer*, *34*, 2816, 1993.
141. Cavaillé, J. Y., Johari, G. P., and Mikolajczak, G., *Polymer*, *28*, 1841, 1987.
142. Mikolajczak, G., Cavaillé, J. Y., and Johari, G. P., *Polymer*, *28*, 2023, 1987.
143. Hofer, K., and Johari, G. P., *Macromolecules*, 24, 4978, 1991.
144. Croll, S. G., *J. Coat. Technol.*, *51*, 49, 1979.

145. Lu, H., and Neville, K., *Handbook of Epoxy Resins*, McGraw-Hill, New York, 1967, 229.
146. Smith, S., U.S. Patent 3,644,567, 1972.
147. Bucknall, C. B., and Smith, R. R., *Polymer*, *6*, 437, 1965.
148. Bucknall, C. B., in *Toughened Plastics*, Applied Science Publishers, London, 1977, 77.
149. Donald, A. M., and Kramer, E. J., *J. Mater. Sci.*, *17*, 1871, 1982.
150. Sultan, J. M., and McGarry, F. J., *Polym. Engng. Sci.*, *13*, 29, 1973.
151. Rowe, E. H., and Riew, C. H., *Plast. Engng.*, *41*, 45, 1975.
152. Kinloch, A. J., Shaw, S. J., Tod, D. A., and Hunston, D. L., *Polymer*, *24*, 1341, 1983.
153. Wu, S., *Polymer*, *26*, 1855, 1985.
154. Kunz-Douglass, S., and Ashby, M., *J. Mater. Sci.*, *15*, 1109, 1980.
155. Yamanaka, K., and Inoue, T., *Polymer*, *30*, 1839, 1989.
156. Rowe, E. H., Siebert, A. R., and Drake, R. S., *Mod. Plast.*, *49*, 110, 1978.
157. Wang, H. B., Li, S. J., and Ye, J. Y., *J. Appl. Polym. Sci.*, *44*, 789, 1992.
158. Moy, P., and Karasz, F. G., *Polym. Engng. Sci.*, *20*, 315, 1980.
159. Carfagna, C., and Apicella, H., *J. Appl. Polym. Sci.*, *28*, 2881, 1983.
160. Apicella, H., Nicolais, L., Astarita, G., and Drioli, E., *Polymer*, *22*, 1064, 1981.
161. Browning, C. E., *Polym. Engng. Sci.*, *18*, 16, 1978.
162. Meares, P., in *Polymers, Structure and Bulk Properties*, Van Nostrand, London, 1965, 326.
163. Judd, N. C. W., *Br. Polym. J.*, *9*, 36, 1977.
164. Carfagua, C., Apicella, H., and Nicolais, I., *J. Appl. Polym. Sci.*, *27*, 105, 1982.
165. *Epoxy Resins and Epoxy Resins Based Materials Catalog*, OIHF, Moscow, NIITECHIM, 1977 (Russian).
166. Drake, R. S., Egan, D. R., and Murphy, W. T., Elastomer modified epoxy resins in coatings applications, *Rubber World*, 1968, October, 15.
167. Hunson, D. L., Bitner, J. L., Rushford, J. L., Ross, W. S., and Riew, G., *Adhesion and Adhesives: Science, Technology and Applications*, The Conf. of Plastics and Rubber Inst., (London), 1980, Paper 14.1.
168. Sohn, J. E., Morphology of solid uncured rubber-modified epoxy resins, *181st National Meeting, ACS, ORPL*, 14, 1981.
169. Moulon, R. J., Advanced technology in material engineering *SAMPE Int. Conf.* 1981, Cannes, France.
170. BF Goodrich Chem. Group, *Toughen Epoxy Resins with Hycar Reactive Polymers*, 1980, RLP-2.
171. Shelley, R. R., and Clarke, J. A., BRIT Patent 1,461,127, 1977.
172. Mendelsohn, M. A., U.S. Patent 4,298,656, 1981.
173. McPherson, C. A., U.S. Patent 4,121,015, 1978.
174. Tsuchyia, Y., U.S. Patent 4,253,930, 1981.
175. McGarry, F. J., and Willner, A. M., Toughening of an epoxy resin by an elastomeric second phase, R68-8, MIT, 1968; cited in Frigione, M. E., Mascia, L., and Acierno, D., *Eur. Polym. J.*, *31*, 1021, 1995.
176. Drake, R., and Siebert, A., Elastomer modified epoxy resins for structural applications, *SAMPE Quarterly*, *6*, 4, 1975.
177. Brown, H. P., Crosslinking reactions of carboxylic elastomers, *Rubber Chem. Technol.*, *36*, 25, 1963.
178. Siebert, A. R., and Riew, C. K., The chemistry of rubber-toughened epoxy resins, *ACS Meeting, Organic Coating and Plastic Division*, Los Angeles, 1971.
179. Butta, E., Levita, G., Marchetti, A., and Lazzeri, A., *Polym. Engng. Sci.*, *26*, 63, 1986.
180. Okamoto, Y., *Polym. Engng. Sci.*, *23*, 222, 1983.
181. Bucknall, C. B., and Partridge, I. K., *Polymer*, *24*, 639, 1983.

182. Raghara, R. S., *J. Polym. Sci., Part B, Polym. Phys.*, *25*, 1017, 1987; *ibid.*, *26*, 65, 1988.
183. Bucknall, C. B., and Partridge, I. K., *Br. Polym. J.*, *15*, 71, 1983.
184. Diamant, J., and Moulton, R., *29th National SAMPE Symposium*, 1984, 422.
185. Hedrick, J. H., Yilgor, I., Wilkens, G. L., and McGrath, J. E., *Polym. Bull.*, *13*, 201, 1985.
186. Hedrick, J. H., Yilgor, I., Jerek, M., Hedrick, J. C., Wilkens, G. L., and McGrath, J. E., *Polymer*, *32*, 2020, 1991.
187. Yoon, T. H., Liptak, S. C., Priddy, D., and McGrath, J. E., *ANTEC'93*, New Orleans, 1993, 3011.
188. Martuscelli, E., Musto, P., Ragosta, G., and Scarinzi, G., *Fourth European Symposium on Polymer Blends*, Capri, Italy, 1993, 335.
189. American Cyanamid, U.S. Patent 3,472,730, 1979.
190. American Cyanamid, U.S. Patent 4,604,319, 1986.
191. Matheswaran, M., and Padmanabhan, K., Static and impact behaviour of thermoplastic modified glass fabric/epoxy composites, *J. Polym. Sci., Polym. Phys. Ed.*, *33*, 981, 1995.
192. Bucknall, C. B., Partridge, I. K., Jayle, L., Nozue, I., Fernybough, A., and Hay, J. N., *Polym. Prepr.*, *33*, 378, 1992.
193. Raghava, R. S., *J. Polym. Sci., Phys. Ed.*, *26*, 65, 1988.
194. Jijima, T., Suzuki, N., Fukuda, W., and Tomoi, M., *Polym. Int.*, *38*, 343, 1995.
195. Jijima, T., Miura, S., Fukuda, W., and Tomoi, M., *J. Appl. Polym. Sci.*, *57*, 819, 1995.
196. Jijima, T., Suzuki, N., Fukuda, W., and Tomoi, M., *Eur. Polym. J.*, *31*, 775, 1995.
197. Jijima, T., Takemoto, T., and Tomoi, M., *J. Appl. Polym. Sci.*, *43*, 1683, 1991.
198. Albert, P., Langer, J., Kressler, J., and Mulhaupt, R., *Acta Polymerica*, *46*, 68, 1995.
199. Langer, J., Albert, P., Gronski, W., and Mulhaupt, R., *Acta Polymerica*, *46*, 74, 1995.
200. Ewans, A. G., *J. Am. Ceram. Soc.*, *73*, 187, 1990.
201. Claussen, N., *J. Am. Ceram. Soc.*, *59*, 49, 1976.
202. Low, J. M., Moi, Y. W., Bandyopadhyay, S., and Silva, V. M., *Mater. Forum*, *10*, 241, 1987.
203. Kim, J. K., and Robertson, R. E., *J. Mater. Sci.*, *27*, 161, 1992.
204. Kubotera, K., and Yee, A. F., *ANTEC'93*, New Orleans, 1993, 3290.
205. Banthia, A. K., Chaturvedy, P. N., Jha, V., and Pendyala, V. N. S., in *Rubber Toughened Plastics*, Riew, C. K., Ed., ACS Series, 222, 1989.
206. Wang, H. B., Li, S. J., and Ye, J. Y., *J. Appl. Polym. Sci.*, *44*, 789, 1992.
207. Lee, Y. D., *J. Appl. Polym. Sci.*, *32*, 6317, 1986.
208. Touhsaent, R. E., Thomas, D. H., and Sperling, L. H., in *Toughness and Brittleness of Plastics*, ACS Series, 154, 1976, 206.
209. Sperling, L. H., and Sarge, H. D., *J. Appl. Polym. Sci.*, *16*, 3041, 1972.
210. Takahashi, T., Nakajima, N., and Saito, N., in *Rubber Toughened Plastics*, Riew, C. K., Ed., ACS Series, 222, 1989, 243.
211. Warrik, E. L., Pierce, O. R., Polmanteer, K. E., and Saam, J. C., *Rubber Chem. Technol.*, *52*, 437, 1979.
212. Riffle, J. S., Vilgor, I., Banthia, A. K., Tran, C., Wilkes, G. L., and McGrath, J. E., *ACS Symposium Series*, 221, 1982, 425.
213. Yorkgitis, E. M., *Adv. Chem. Series*, *208*, 111, 1984.
214. Yorkgitis, E. M., Eiss, N. S., Tran, C., and McGrath, J. E., *Adv. Polym. Sci.*, *72*, 79, 1985.
215. Cecere, J. A., Hedrick, J. L., and McGrath, J. E., *Polym. Prepr.*, *27*, 298, 1986.
216. Kasemura, T., Kawamoto, K., and Kashima, Y., *J. Adhes.*, *33*, 19, 1990.
217. Lanzetta, N., Laurienzo, P., Matincorico, M., Martuscelli, E., Ragosta, G., and Volpe, M. G., *J. Mater. Sci.*, *27*, 786, 1992.

218. Twardowski, T. E., and Geil, P. H., *J. Appl. Polym. Sci.*, *42*, 69, 1991.
219. Mijovic, J., Pearce, E. M., and Foun, C. C., in *Rubber Toughened Plastics*, Riew, C. K., Ed., ACS Series, 222, 1984, 93.
220. Rosser, R. W., Chen, T. S., and Tylor, M., *Polym. Comp.*, *5*, 198, 1984.
221. Mascia, L., Zitouni, F., and Tonelli, C., *J. Appl. Polym. Sci.*, *51*, 995, 1994.
222. Mascia, L., Zitouni, F., and Tonelli, C., *Polym. Engng. Sci.*, *35*, 1069, 1995.
223. Pascal, T., Bonneau, J. L., Biolley, N., Mercier, N., and Sillion, B., Approach to improving the toughness of TGMDA/DDS epoxy resin by blending with thermoplastic polymer powder, *Polym. Adv. Technol.*, *6*, 219, 1995.
224. Boyd, J. D., U.S. Patent 5,037,689, 1991.
225. Folda, T., Eur. Pat. Appl. 377,294, 1990.
226. Frigione, M. E., Mascia, L., and Acierno, D., Oligomeric and polymeric modifiers for toughening of epoxy resins, *Eur. Polym. J.*, *31*, 1021, 1995.
227. *Khim. Prom. za Rubejom*, 1975, No. 10, p. 35 (Russian).
228. *Epoxy Resin Based Materials*, *Catalog*, NIITEHIM, Moscow, 1977 (Russian).
229. Moshinski, L. Ia., *Solidifiers of Epoxy Resins*, *Catalog*, NIITEHIM, Moscow, 1976 (Russian).
230. Rosenberg, B. A., and Enicolopov, M. C., Curing materials, *Zh. Vses. Khim. D. I. Mendeleev*, 23, 272, 1978 (Russian).
231. Rosenberg, B. A., and Kardasova, D. A., Polymeric adhesives and their properties, *Zhur. Vses. Khim. D. I. Mendeleev*, 23, 298, 1978.
232. Evans, U. R., and Taylor, C. A., *Trans. Inst. Met. Finishing*, *39*, 188, 1962.
233. Wiggle, R. R., Smith, A. G., and Petrocelli, J. V., *J. Paint. Tech.*, *40*, 174, 1968.
234. Anderson, W. A., *Off. Dig.*, *36*, 1210, 1964.
235. Jain, P. C., Rosato, J. J., and Agarwala, V. S., *Corrosion J.*, *42*, 700, 1986.
236. Leidheiser, H., and Wang, W., *J. Coat. Technol.*, *53*, 77, 1981.
237. Leidheiser, H., *Ind. Eng. Chem. Prod. Res. Dev.*, *20*, 547, 1981.
238. Tinh Nguyen and Byrd, E., *Polym. Prepr.*, *56*, 585, 1987.
239. Nielsen, P. O., *Adhesive Age*, 1982, April, p. 82.
240. Feldman, D., Banu, D., Nathanson, A., and Wang, J., Structure-properties relations of thermally cured epoxy-lignin polyblends, *J. Appl. Polym. Sci.*, 42, 1537, 1991.

8

Sulfur-Containing Oligomers

I. INTRODUCTION

Synthesized as early as 1943 [1], polysulfide oligomers (PSOs) soon became rather important raw materials for producing sealant compositions. Ten years later, Thiokol Chemical Corp. produced them industrially, and in the Soviet Union, a team of research workers led by Apukhtina elaborated a technology of their own [2,3]. At present, developed countries produce important quantities of PSOs. There are also data on production capacities in other countries, such as Poland and Romania; the latter in fact has its own technology [4–6].

Most industrial PSOs have a low branched structure with final SH groups in keeping with the following formula:

$$\begin{array}{l} HS{\sim}R{-}S{-}S{-}R'{-}S{-}S{\sim}R{-}SH \\ \qquad\qquad\qquad\quad | \\ \qquad\qquad\qquad\quad S{-}S{\sim}R{\sim}SH \end{array} \tag{1}$$

Polysulfide sealants have found wide acceptance in construction, aircraft, automotive, marine, and insulating glass. In fact, polysulfide sealants were the first elastomeric sealants used in the construction industry and continue today, to a large extent, to be the standard with which other sealants are compared.

Polysulfide sealants, although facing strong challenges from silicones and polyurethanes, continue to be the dominant segment of the elastomeric sealant market (Table 1) [2].

II. SYNTHESIS

Several methods of synthesizing oligosulfides are known: degradation of polysulfides with high molecular mass by means of "Bunte Salts," or radical polymerization of olefins or dienes in the presence of sulfur. Oligosulfides can also be obtained by polymerization of thiiranes and by oxidation of low molecular weight dimercaptanes.

A. Degradation of Polysulfides

The polysulfide polymers are prepared by condensation of organic polyhalides with inorganic sulfides in aqueous suspension. The principal monomer is bis-2-chloroethyl formal. Ethylene dichloride is used in special polymers. Generally, most

Table I The Use of Various Sealants in the United States (in tons)

Polymer	1969	1975	1980
Polysulfides	7,200	10,000	12,800
Polyurethanes	4,800	7,600	11,200
Polysiloxanes	2,400	4,400	6,800
Polyacrylates	1,200	3,200	5,600
Poly(vinyl chloride)	8,000	11,200	13,600
Polybutene	2,000	2,000	2,400

Source: Ref. 2.

products contain some crosslink precursor, for example, 1,2,3-trichloropropane. Bis-4-chlorobutyl formal and bis-4-chlorobutyl ether are used in small amounts where improvement in low-temperature performance is required. The polymerization reaction producing high molecular weight polymer is

$$nClC_2H_4OCH_2OC_2H_4Cl + nNa_2S_{2.25} \rightarrow$$
$$—(C_2H_4OCH_2OC_2H_4S_{2.25})_n— + 2nNaCl \quad (2)$$

The sulfur is present as a mixture of disulfide and trisulfide.

Liquid polysulfide polymers, which are commercially more important than the solid elastomers, range from a molecular weight of 1000 to 8000. They are prepared from the high molecular weight polymers described earlier. The high molecular weight polymer is split into segments that are simultaneously terminated by mercaptan groups:

$$R—S—S—R + NaSH + NaHSO_3 \rightarrow 2R—SH + Na_2S_2O_3 \quad (3)$$

The concentration of the splitting salts regulates the average molecular weight of the liquid polysulfides. For example, Thiokol liquid polymer LP-2 has the following average formula:

$$HS—(C_2H_4OCH_2OC_2H_4—S—S)_{23}—C_2H_4OCH_2OC_2H_4—SH \quad (4)$$

The main characteristic of reaction (2) is that, in equimolar proportions of reactants, polymer average molecular weight does not exceed 5000 Da. Final groups are of the form $—S_xNa$ but also —R—Cl. The presence of the latter is undesirable. The utilization of polysulfide in excess leads to the formation of polysulfide or hydroxyl end groups. The latter ones are determined by the presence of OH groups in the reaction medium, in keeping with the reaction

$$\sim—R—Cl + HO^- \rightarrow —R—OH + Cl^- \quad (5)$$

The excess of polysulfide in the reaction medium plays the role of chain extending agent:

$$\sim S_x\text{—}R\text{—}S_x\text{—}Na + Na\text{—}S_x\text{—}R\text{—} \rightarrow \sim S_x\text{—}R\text{—}S_x\text{—}R\text{—}\sim + Na_2S_x \qquad (6)$$

or of initiator of the degradation reaction for chains with final groups:

$$\sim S_x\text{—}R\text{—}S_x\text{—}R\text{—}OH + Na_2S_x \rightarrow \sim S_x\text{—}R\text{—}S_x\text{—}Na + HO\text{—}R\text{—}S_x\text{—}Na \qquad (7)$$

In industrial practice, polysulfide in excess of 10–20% is used (expressed in moles).

The synthesis of linear polymer is accompanied by the formation of cyclics. The amount and structure of cyclics depends on the amount and structure of dihalide monomer Cl—R—Cl in the following order [7]:

$$R = \text{—}(CH_2)_6\text{—} < \text{—}(CH_2)_5\text{—} < \text{—}C_2H_4\text{—}O\text{—}C_2H_4\text{—} < \text{—}(CH_2)_4\text{—}$$

The occurrence of cyclic structures also depends on the sulfur content of initial polysulfides: the lower the sulfur content is, the more cyclic structures are formed. For this reason, to yield polysulfide polymers, the use of sodium tetrasulfide is preferred to sodium disulfide. More than 100 different monomers have been used [8–15], and based on experimental data, an attempt was made to establish reactivity–structure correlation. For halide derivatives, the most reactive are allylic and benzylic halide derivatives; the primary derivatives are more active than the secondary or tertiary derivatives. As far as industrial monomers are concerned, 1,2-dichloroethane is more active than 2,2-dichloroethane or 2,2-dichlorodiethyl formal.

To induce formation of branches, branching agents such as CCl_4, hexachlorohexane, and 1,2,3-trichloropropane are used [16,17].

The polysulfide polymers are prepared as a suspension by condensation of the dihalide monomer with polysulfide in aqueous solution [18]. Bis-2-dichloroethyl formal is the monomer usually used, and 0.1–0.4% 1,2,3-trichloropropane is added as a crosslinking monomer [19]. In conducting polymerization, the standard practice is to feed the dihalide monomer into the aqueous polysulfide solution containing specific suspending and nucleating agents. A combination of an alkyl naphthalene sulfonate with magnesium hydroxide sol prepared *in situ* is commonly used.

The synthesis of sulfide-containing oligomers can be also achieved by means of Bunte salts—a high molecular weight compound, obtained by the following reaction [20]:

$$\sim\text{-R-Cl} + Na_2S_2O_3 \longrightarrow \sim \text{R-S-}\overset{\overset{\displaystyle O}{\|}}{\underset{\underset{\displaystyle O}{\|}}{S}}\text{-ONa} + NaCl \qquad (8)$$

The Bunte salt reacts with sodium sulfide, according to the following reaction:

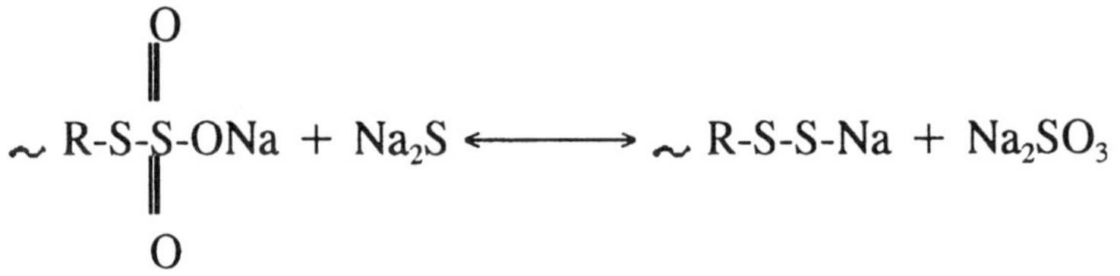

(9)

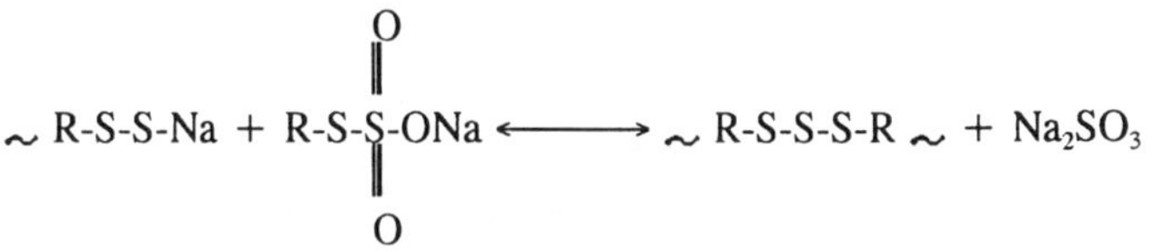

(10)

The resulting polymer is then desulfured and degraded with sodium hydrosulfide [21]. During the degradation process, the following cyclization take place [7]:

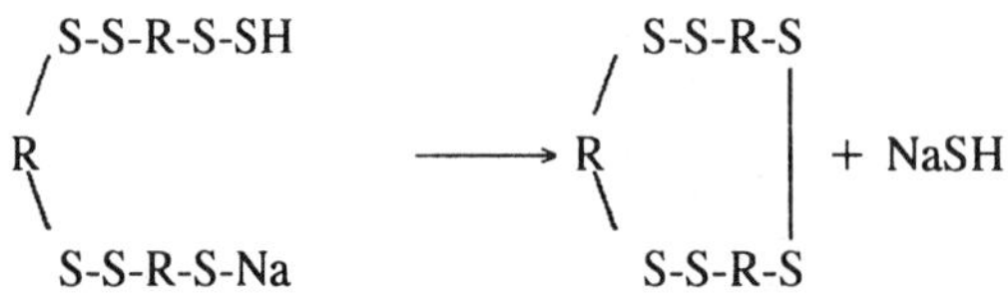

(11)

Another method of obtaining sulfur-containing oligomers and polymers is thiirane polymerization. In the presence of proton donors (e.g., H_2S and ditiols), thiiranes polymerize with the formation of polythyolesterditiols.

Polymerization of propylene sulfide in the presence of carbonates and zinc, cadmium, or alkaline thiolates produces oligosulfides with —HS end groups [22]. It was found that at anionic polymerization of propylene sulfide, chains with regular structure of the type $—(CH_2—CH(CH_3)—S—S)_n—$ are obtained. In the catalytic BuLi–LiOR system, propylene sulfide polymerization take place according to the following mechanism [2]:

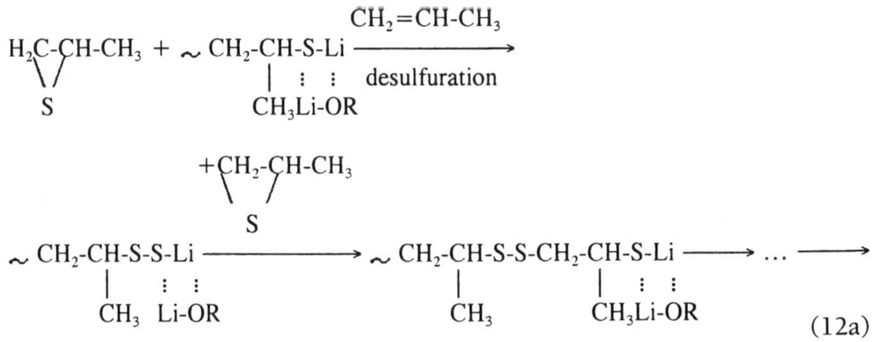

(12a)

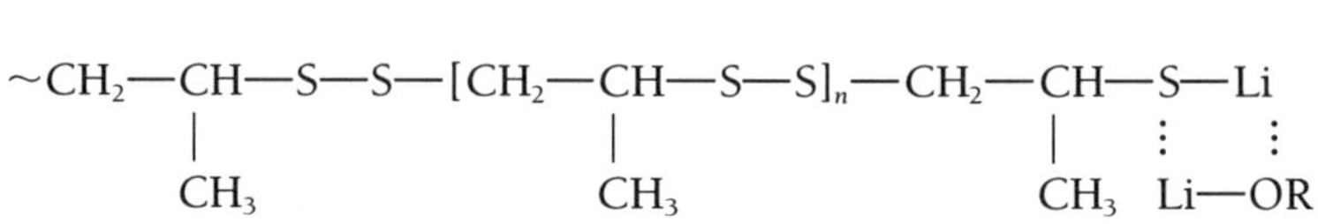

(12b)

There are experimental data attesting to the synthesis of polysulfide oligomers by organic dimercaptan oxidation. By this method, polyalkylene di- and polysulfide silicium containing oligomers were produced, characterized by increased stability at lower temperatures and in organic solvents [23].

Polysulfidic oligomers with final —OH groups can also be obtained from ditiodiglycols according to the following reactions [24]:

$$nHO—CH_2—CH_2—S—S—CH_2—CH_2—OH + nCH_2O \xrightarrow{+H^+}$$

$$\sim—[CH_2—CH_2—O—CH_2—O—CH_2—CH_2—S—S]_n—\sim + H_2O \qquad (13)$$

We can also mention the production of polysulfide oligomers using dienic monomers as a raw material. The following reaction takes place [25]:

$$CH_2{=}CH—R—CH{=}CH_2 + (n + 1)H_2S_x \rightarrow$$

$$H—S_x—[CH_2—CH_2—R—CH_2—CH_2—S_x]_n—H \qquad (14)$$

The production of sulfur-containing oligomers with —SH end groups is based on the analogous polymer reactions. Called oligomercaptanes, these products can be obtained by radicalic telomerization of unsaturated compounds in the presence of thiols and xantogenate disulfides [26]. The amount of telogen usually ranges between 2 and 20% (by weight), and polymerization is achieved in emulsion or solution. As monomers, dienic hydrocarbons are used in a mixture with styrene or various acrylic monomers.

The production of oligoesterdithiols by radicalic polymerization of nonsaturated monomers in the presence of hydrogen sulfide or ditiols has been suggested [27]. The end group and molecular weight of products is controlled by reactant ratio: The excess in tiol determines the formation of an oligomer with M_n = 250–400 and mercapto type end groups. Low amounts of thiol lead to an increase of oligomer molecular weight and formation of nonsaturated end groups.

Quite of advantage is synthesis of oligomers with —SH end groups by processing Li-containing living polymers, according to the following reactions [28]:

$$\sim CH_2Li + S \longrightarrow \sim CH_2\text{-}S\text{-}Li \qquad (15)$$

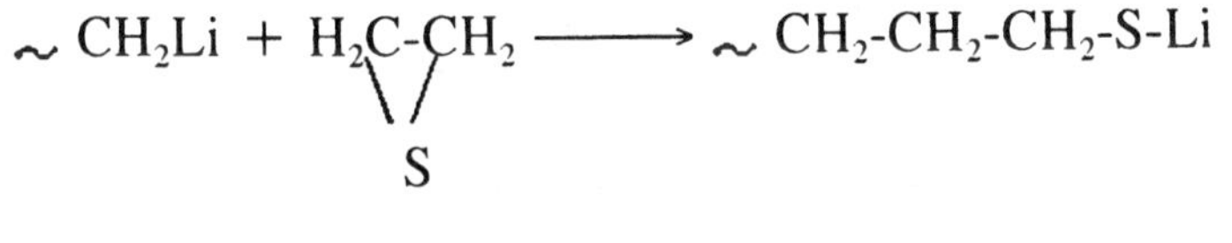

(16)

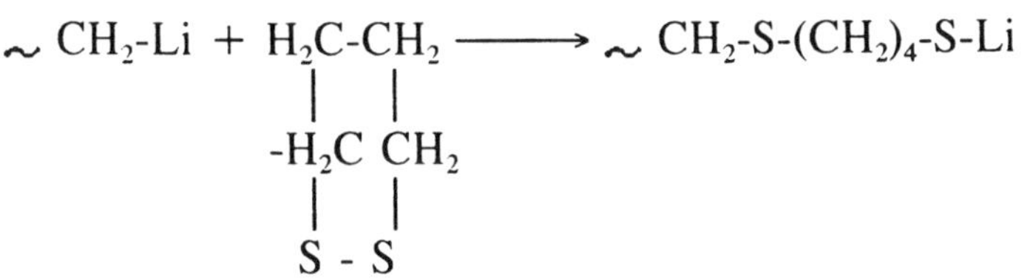

(17)

Oligomercaptane synthesis methods by chemical transformation of oligomers with halide end groups, or of epoxidized oligodienes are also known [29].

Use of a mixed dihalide monomer feed in a conventional polymerization will produce random copolymers. A block copolymer cannot be prepared by a stepwise addition of the monomers. During copolymerization interchange takes place, resulting in randomization. To prepare block copolymers, it is best to prepare individual mercaptan-terminated polymers, blend them in the desired proportion, and then process the blended system by conventional techniques.

Liquid polysulfide polymers readily react with epoxy resins under basic conditions to form a block copolymer:

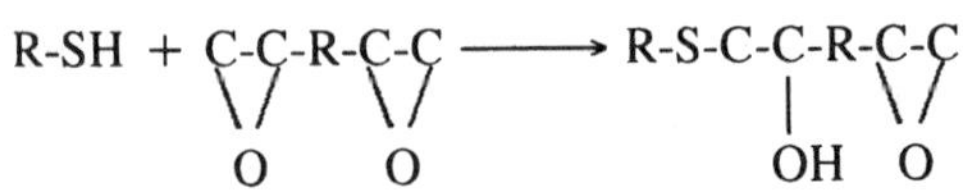

(18)

The polysulfide–epoxy block copolymer has significantly higher impact resistance and more flexibility than epoxy [30,31].

Generally, the lower molecular weight liquid polymers, such as Thiokol LP-3 and LP-33, are useful for epoxy modification. For most applications, the diglycidyl ethers of bisphenol A–derived liquid epoxies are most suitable. The most widely used epoxies are in the viscosity range of 80 to 200 P and have an epoxy equivalent of 175 to 210.

The reaction between liquid polysulfide polymers and epoxies is catalyzed by organic amines, for example, dimethylamine methyl phenol, tri-dimethylaminomethyl phenol, diethylenetriamine, and benzyldimethylamine. The reaction of liquid polysulfide polymer with an epoxy resin and a primary amine is

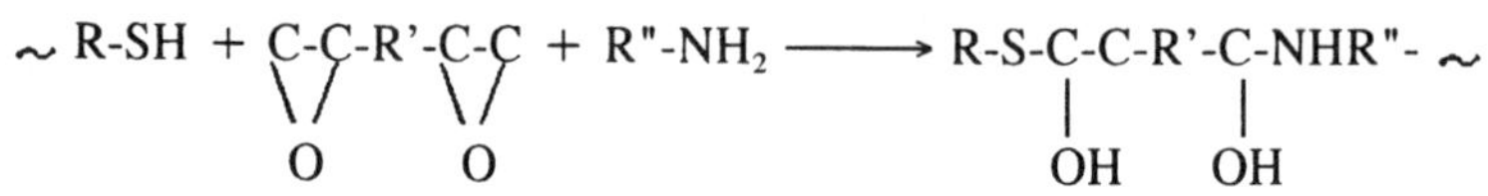

(19)

Villa [32–34] has modified liquid polysulfide polymers by reacting them with either vinyl cyclohexane diepoxide or abietic acid. The modified products were used as an effective additive. While the reaction between liquid polysulfide polymer and abietic acid is not fully known, Villa has theorized that the reaction proceeds with inversion of a mercaptan terminal to hydroxyl terminal followed by esterification [32]. The following reaction takes place:

$$HS\text{-}(C_2H_4\text{-}O\text{-}CH_2\text{-}O\text{-}C_2H_4\text{-}S\text{-}S)_n\text{-}C_2H_4\text{-}O\text{-}CH_2\text{-}O\text{-}C_2H_4\text{-}SH$$

Liquid polysulfide

$$\downarrow \text{heat}$$

$$HS\text{-}(C_2H_4\text{-}O\text{-}CH_2\text{-}O\text{-}C_2H_4\text{-}S\text{-}S)_n\text{-}C_2H_4\text{-}S\text{-}CH_2\text{-}CH_2\text{-}CH_2\text{-}OH \quad (20)$$

The reaction product of the liquid polythiol polymer and abietic acid is believed to be predominantly of the general type shown in

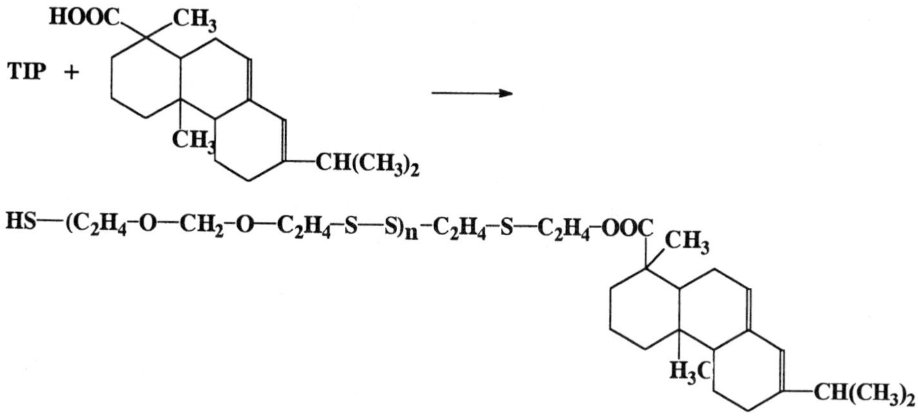

(21)

However, it should be observed here that the reaction product may in fact contain a mixture of chemical structures. Thus, it is likely that in addition to the foregoing structure, the reaction product may contain structures wherein there are abietate terminals at both ends of the polysulfide oligomer chain, and in addition it may contain unreacted liquid polythiol oligomer with —SH terminals at both ends of the oligomer chain. The predominant type of structure obtained is influenced in large measure by the reaction conditions employed.

The epoxy-modified polysulfide oligomer, useful as additive, was prepared by Villa [32–34] by reacting low molecular weight polysulfide with a diepoxide in a solvent medium in the presence of an acid catalyst. The chemical reactions proceeds as in

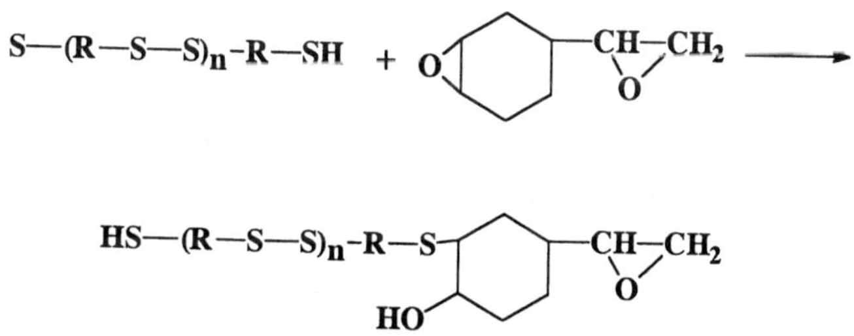

(22)

where R is the ethyl formal radical $—(C_2H_4—O—CH_2—O—C_2H_4)—$.

In reaction (22), the epoxy cyclohexane group is reactive under acidic conditions, whereas the epoxy ethyl group is reactive under basic conditions. The chemical structure shown is believed to be the predominant form of the reaction product. However, it should be observed that the reaction product very possibly may consist of a mixture of materials with varying structures. Thus, the reaction product may, in addition to the structure, may contain a product wherein the polysulfide oligomer chain or a portion of the reaction product has —SH terminals at both ends of the oligomer chain.

III. PHYSICAL PROPERTIES

The tensile properties of unfilled polysulfide oligomers are rather poor. However, suitable pigment reinforcement can lead to products with adequate tensile and elongation properties. The molecular weight of liquid oligomer before cure also influences the physical properties of the cured oligomer. It is difficult, however, to make a generalization about the effect of molecular weight on physical properties, since oxidative curing also enters into this aspect. The liquid oligomer can be cured with a wide variety of oxidizing agents.

Higher tensile strength is obtained with the higher molecular weight materials. The effect of branching on the physical properties of polysulfide liquid oligomers is shown in Table 2 [35].

Cured liquid polysulfide compositions have excellent resistance to a wide variety of oils and solvents, for example, aliphatic and aromatic hydrocarbons, esters, ketones, and dilute acids and alkalis. Table 3 shows the properties of two polysulfide oligomers cured with conventional filled formulations [36]. The data presented are only trends and not absolute, since the results depend on the efficiency of cure. Systems that are not properly compounded have poorer solvent resistance.

The glass transition temperatures (T_g) of polysulfides depend on the hydrocarbon moiety and the length of the polysulfide chain. The amount of crosslinking monomer used is small and therefore does not influence T_g. Generally, the greater the hydrocarbon content, the lower the T_g; the higher the rank of the polysulfide, the higher the T_g. In Table 4 the glass transition temperatures of the elastomeric polysulfides are given [36]. Polysulfides based on bischloroethyl formal show no evidence of crystallization at low temperatures.

The thermal stability of polysulfide polymers depends on the curing agent used to vulcanize the polymer. Commercially available polysulfide polymers are based on the ethyl formal disulfide backbone; therefore, the characteristics of the ethyl formal structure will regulate the upper temperature limits. The initial attack on the ethyl formal structure is an acid-catalyzed hydrolytic attack on the formal group by trace amounts of water, releasing free formaldehyde. Formaldehyde reduced the disulfide bonds to mercaptan. The formic acid so generated catalyzes the hydrolysis of the formal group. The terminal mercaptan group can react with the hydroxyl group to give a monosulfide bond and release water. The sequence of reactions is as follows:

$$\sim S—CH_2—CH_2O—CH_2—CH_2—S—\sim + H_2O \rightarrow$$

$$\sim S—CH_2—CH_2—OH + HO—CH_2—CH_2—S—\sim \qquad (23)$$

$$\sim CH_2—CH_2—S—S—CH_2—CH_2\sim + CH_2O \rightarrow$$

$$2\sim CH_2—CH_2—SH + HCOOH \qquad (24)$$

$$\sim CH_2—CH_2—SH + HO—CH_2—CH_2—S—\sim \rightarrow$$

$$\sim CH_2—CH_2—S—CH_2—CH_2—S—\sim + H_2O \qquad (25)$$

Degradation results in weight loss and loss of flexibility due to formation of monosulfide structure, since disulfide and formal groups offer a flexibilizing effect

Table 2 Effect of Crosslinking on Physical Properties of Polysulfide Oligomers

Formulation						
Polymer	100	100	100	100	100	100
Crosslinking (%)[a]	4.0	2.0	1.5	1.0	0.5	0.1
SRF black (%)	30	30	30	30	30	30
Stearic acid (%)	1	1	1	1	1	1
C-5 accelerator[b] (%)	15	15	15	15	15	15
Set time (h)	2	4	5	5	6	6
Physical Properties of Molded Sheets after Curing 10 min at 160°C						
Tensile (psi)	550	500	550	675	575	500
Modulus 300% (psi)	500	375	350	275	250	200
Elongation (%)	310	420	700	850	930	1000
Hardness, Shore A	55	53	53	53	45	40
Physical Properties of Sheets after Heat Aging 70 h at 100°C						
Tensile (psi)	550	525	675	650	650	650
Modulus 300% (psi)	—	424	450	375	375	300
Elongation (%)	250	370	500	600	600	650
Hardness, Shore A	62	58	54	53	53	50

[a]Mol percent of trifunctional monomer in polymer feedstock.
[b]A 50% suspension of lead oxide in a plasticzer, e.g., dibutyl phthalate.
Source: Ref. 35.

through free rotation. Calcium oxide is an effective stabilizer since it can both neutralize formic acid and absorb water. However, practical cure rates cannot be achieved in anhydrous systems by metal dioxide curing agents normally used with liquid polysulfides.

Another source of thermal instability arises from metal incorporation in the polymer via reaction of mercaptan with metal oxides. Formation of mercaptide groups can be minimized by incorporation of small amounts of sulfur.

The viscoelastic properties of polysulfide polymers has been extensively studied by Tobolsky [37].

Polysulfide polymers are unique in their ability to internally relieve stress in a cured state by interchange reactions between mercaptan and disulfide linkages. The stress decay of crosslink elastomer follows the equation

$$F_{(t)} = F_{(0)} + e^{-t/\tau} \tag{26}$$

where $F_{(t)}$ = final stress; $F_{(0)}$ = initial stress; t = time, and τ = relaxation constant. This ability of polysulfide to relieve stress is extremely valuable in maintaining adhesion in joints that are subject to movement.

The relationship between viscosity and molecular weight is regulated by the following equation:

$$\eta = KM^a \tag{27}$$

where η = viscosity in poise, $K = 5.0 \pm 1.0 \times 10^{-8}$, $a = 2.75 \pm 0.03$, and M =

Table 3 Solvent Resistance of Cured Polysulfide Oligomers

Solvent	Volume Increase (%) after 30-Day Immersion at 80°F	
	LP-2[a]	LP-32
Toluene	95	140
Xylene	40	60
Motor oil	5	5
Diesel oil	5	5
Ethyl alcohol	5	5
Butyl alcohol	5	5
Ethylene glycol	5	5
Ethyl cellosolve	15	25
Acetone	40	50
Methyl ethyl ketone	55	90
Ethylene dichloride	440	600
Trichloroethylene	275	400
Perchloroethylene	30	45
Chlorobenzene	270	475

[a]Manufactured by Thiokol Speciality Chemical, San Diego, CA.
Source: Ref. 36.

molecular weight. Yet relation (27) is also influenced by the amount of the cross-linking agent used in oligomer synthesis. Thus, for linear oligosulfides,

$$\lg \eta = -1.05 + 1.68 \times \lg M_n \tag{28}$$

and for crosslinking oligosulfides,

Table 4 Glass Transition Temperature (T_g) of Elastomeric Polysulfides

Polymer	T_g (°C)
Poly(ethylene disulfide)	−27
Poly(ethylene tetrasulfide)	−24
Poly(ethyl ether disulfide)	−53
Poly(ethyl ether tetrasulfide)	−40
Poly(ethyl formal disulfide)	−59
Poly(pentamethylene disulfide)	−72
Poly(hexamethylene disulfide)	−74
Poly(butyl formal disulfide)	−76
Poly(butyl ether disulfide)	−76

Source: Ref. 36.

$$\lg \eta = -2.94 + 2.41 \times \lg M_n \tag{29}$$

and

$$\lg \eta = -5.48 + 3.241 \times \lg M_n \qquad \text{(with 2 mol\% 1,2,3-trichloropropane)} \tag{30}$$

Determination of molecular weight distribution (MWD) in oligosulfides leads to contradictory results. The existence of some intermolecular reactions of the diol-disulfide type leads in the case of linear oligosulfides to the assumption of an equilibrium MWD similar to the most probable distribution [38–40]. Experimental data show oligosulfides with broad MWDs ($M_w/M_n = 2.8$–4.5). These contradictory findings can be explained by the history of the material, by the presence of impurities, or by structural differences in polysulfidic sequences of oligomers.

IV. CHEMICAL PROPERTIES

Oligosulfides, which contain as much as 40% chemically bonded sulfur, participate in two important categories of reactions: modifying —SH end groups, and modifying polysulfide sequences inside the chain. The former category of reactions differs very little from classical reactions of low molecular weight mercaptanes. As far as the latter category is concerned, we should first mention that sulfur atom reactivity depends on its position in the chain. Thus, it is already known that in polythionate type ions, interatomic distances (in nm) —S—S— vary in terms of sequence length, viz.

$$\begin{array}{lll} SO_3^- & SO_3^- & SO_3^- \\ |\ 0.212 & |\ 0.212 & |\ 0.210 \\ S & S & S \\ |\ 0.202 & |\ 0.204 & |\ 0.204 \\ S & S & S \\ |\ 0.212 & |\ 0.204 & |\ 0.204 \\ SO_3^- & S & S \\ & |\ 0.212 & |\ 0.204 \\ & SO_3^- & S \\ & & |\ 0.212 \\ & & SO_3^- \end{array} \tag{31}$$

The —S—S— link-breaking process can take place under the action of nucleophilic or electrophilic agents [41].

For example, under the action of nucleophilic agents, desulfuration can occur according to the following reactions [41]:

$$\sim\!R{-}S{-}S{-}S{-}S{-}R\!\sim + HO^- \rightleftarrows \sim\!R{-}S{-}S{-}S{-}OH + \sim\!R{-}S^- \tag{32}$$

$$\sim\!R{-}S{-}S{-}S{-}OH + HO^- \rightleftarrows \sim\!R{-}S{-}S{-}OH + \sim\!HO{-}S^- \tag{33}$$

$$\sim\!R{-}S{-}S{-}OH + \sim\!RS^- \rightleftarrows \sim\!R{-}S{-}S{-}R\!\sim + HO{-}S^- \tag{34}$$

These reactions are based on the exchange of tioanion with HO^- ion, the affinity of which toward sulfur is larger.

More frequent are nucleophilic decompositions of —S—S— links. These can be achieved with the participation of sulfides, disulfides, and inorganic hydrosulfide, used individually or in mixture with Na_2SO_3. The following reactions take place [42]:

$$\sim R—S—S—R\sim + 2NaSH \rightleftarrows 2\sim R—SNa + Na—S—S—Na \quad (35)$$

$$\sim R—S—S—R\sim + Na—S—S—Na \rightleftarrows \sim R—S—S—Na + Na—S—S—R \quad (36)$$

Sodium hydrosulfide is usually employed together with sodium sulfite, and in this case the following chemical reactions take place:

$$A\begin{cases}\sim R—S—S—R\sim + NaSH \rightleftarrows R—SH + R—SNa + S \\ S + Na_2SO_3 \rightarrow Na_2S_2O_3\end{cases} \quad (37)$$

$$B\begin{cases}\sim R—S—S—R\sim + Na_2SO_3 \rightleftarrows R—SNa + R—S—SO_3Na \\ \sim R—S—SO_3Na + NaSH \rightleftarrows RSH + R—SNa\end{cases} \quad (38)$$

Variants A and B can also be expressed by the following general reaction:

$$\sim R—S—S—R\sim + NaSH + Na_2SO_3 \rightarrow \sim RSH + \sim RSNa + Na_2S_2O_3 \quad (39)$$

Experimental checking of the mechanism showed that variant B does not work in the absence of NaSH; Na_2SO_3 does not decompose the polymer. So the conclusion was that NaSH determines the decomposition of the oligomer and Na_2SO_3 ensures its irreversibility. Apukhtina and coworkers [43] studied reaction (39) using labeled NaSH. Data obtained emphasized the following steps of the process:

$$\sim R—S—S—R\sim + Na\overset{*}{S}H \rightarrow \sim R—SH + R—S—\overset{*}{S}—Na \quad (40)$$

$$\sim R—S—\overset{*}{S}—Na \rightarrow \sim R—SNa + \overset{*}{S} \quad (41)$$

$$\overset{*}{S} + Na_2SO_3 \rightarrow O_2S(ONa)SNa \rightleftarrows OS(ONa)_2 \quad (42)$$

Breaking of —S—S— links can also occur under the action of *in situ* formed hydrogen, due to the reaction between metals and acids or by means of $LiAlH_4$. The initially formed complex,

$$2\sim R—S—S—R\sim + LiAlH_4 \rightarrow \sim(RS)_4—Li—Al + 2H_2 \quad (43)$$

hydrolyzes in acid medium, forming polymers with mercaptanic end groups [44].

Di- and polysulfide links also break under the action of other reducing agents. Thus,

$$\sim R—S—S_n—S—R'\sim \xrightarrow{NaBH_4} \sim R—SH + nH_2S + \sim R'—SH \quad (44)$$

Similar results are also obtained by means of $LiBH_4$ or with $Na_2S_2O_4$ and NaOH, according to the reaction [45]

$$\sim R—S—S—R\sim + Na_2S_2O_4 + 2NaOH \rightarrow 2\sim R—SH + 2Na_2SO_3 \qquad (45)$$

Decomposition of polysulfide oligomers by means of thiols was also investigated [46]. On this occasion, it transpired that the process is accelerated by the presence of nucleophilic agents. The mechanism of the process is determined by the working conditions: in nonpolar media and at a higher temperature, breaking is homolytic; whereas in polar media, breaking has an ionic character [47]. The reactions that take place are the following:

In nonpolar medium

$$R—S—S—R \rightarrow 2RS'$$

$$RS^{\bullet} + HS—R' \rightarrow R—SH + R'S^{\bullet}$$

$$R'S^{\bullet} + R'S^{\bullet} \rightarrow R—S—S—R' \qquad (46)$$

In polar medium

$$R\text{-}S\text{-}S\text{-}R' + HO^- \longleftrightarrow R\text{-}S\text{-}Oh + RS^-$$

$$R\text{-}S\text{-}OH + R'S^- \longleftrightarrow R\text{-}S\text{-}S\text{-}R' + HO^-$$

$$R'SH + HO^- \longleftrightarrow R'S^- + H_2O$$

$$R\text{-}S\text{-}S\text{-}R' + R'S^- \longleftrightarrow R\text{-}S\text{-}S\text{-}R' + RS^- \qquad (47)$$

or

$$R\text{-}S\text{-}S\text{-}R \xrightarrow{H^+} \underset{\displaystyle \mid \atop H}{\overset{+}{[R\text{-}S\text{-}S\text{-}R]}} \longrightarrow RS^+ + RSH \qquad (48)$$

$$RS^+ + R'\text{-}S\text{-}S\text{-}R' \longrightarrow R\text{-}S\text{-}S\text{-}R' + R'S^+ \qquad (49)$$

$$R'S^+ + RSH \longrightarrow R'\text{-}S\text{-}S\text{-}R + H^+ \qquad (50)$$

It is obvious that reactions (48)–(50) have an electrophilic character.

V. NOMENCLATURE AND INDUSTRIAL PRODUCTION

Polysulfide oligomer nomenclature is specific for each country. Thus, in the United States polysulfide oligomers are produced with names prefixed by LP (liquid polymer), in Japan they are given the prefix S, and in Germany the prefix Th or G. Table 5 lists several examples of polysulfide oligomers produced in the United States, Japan, and Germany [2].

The polysulfide oligomers ZL-560 and ZL-616 are also produced in United States. The chemical structure of these oligomers is given in the following formulas: ZL-560 has the structure

$$HO\text{-}CH_2\text{-}\underset{\underset{S}{\|}}{S}\text{-}(CH_2\text{-}CH_2\text{-}O\text{-}C_2H_6\text{-}S_{3.5})_n\text{-}C_2H_4\text{-}O\text{-}CH_2\text{-}O\text{-}C_2H_4\text{-}\underset{\underset{O}{\|}}{S}\text{-}CH_2OH \qquad (51)$$

where n = 3–30; and ZL-616 has the structure

$$HS\text{-}X\text{-}(OCH_2\text{-}\underset{\underset{CH_3}{|}}{C}H\text{-})_n\text{-}OX\text{-}SH \qquad (52a)$$

where n = 5–50 and

X=

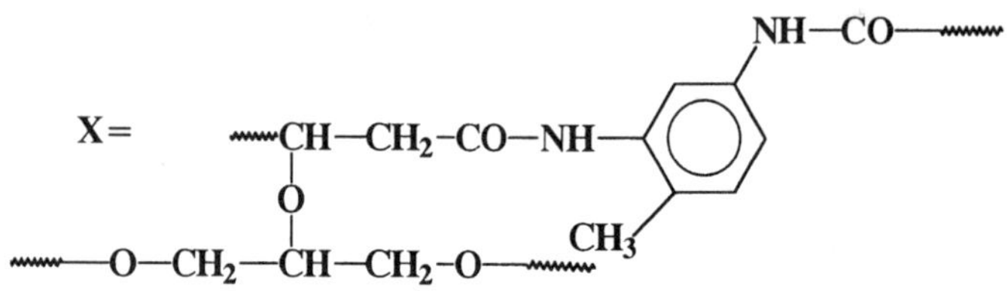

(52b)

The characteristics of polysulfide oligomers produced in Russia are listed in Table 6 [2].

VI. CURING OF POLYSULFIDE OLIGOMERS

Solidification of polysulfide oligomers and oligomercaptans is based essentially on oxidation reactions via organic or inorganic oxidizing agents. The rate and depth of oxidation reactions depend on numerous factors, among which the most important are oligomer structure, nature of oxidizing agent, the presence of activating or inhibiting additives, temperature, and medium humidity.

Subsequent to the oxidation process, both molecule expansion and, more seldom, their reticulation occur in keeping with the following reaction:

Table 5 Physicochemical Characteristics of Industrial Polysulfide Oligomers

Country	Trade name	$M_n \times 10^{-3}$	Viscosity at 25°C (Pa·s)
United States	LP-2	4	37.5–42.5
	LP-3	1	0.7–1.2
	LP-5	2.9	7.5–12.5
	LP-8	0.6	0.29–0.44
	LP-12	4	35–45
	LP-31	7.5	80–140
	LP-32	4	37.5–42.5
	LP-33	1	1.4–1.65
	LP-205	1.2	1.3–1.81
	LP-370	1.2	1.3–1.80
Germany	G-1	3.3–6.6	—
	G-2	1.9–3.3	—
	G-3	1.2–1.9	—
	Th-91	2.79	10.0
	Th-92	2.26	10.0
	Th-101	2.42	7.0
Japan	S-300	—	50–90
	S-340	—	50–90
	S-380	—	50–90
	S-840	—	200–300

Source: Ref. 2.

$$\sim \text{R-SH} + \text{Oxidizing agent} + \text{HS-R} \sim \longrightarrow$$

$$\sim \text{R-S-S-R} \sim + H_2O + \text{products resulted from reduction reaction of oxidizing agents} \quad (53)$$

The oxidizing agents most commonly used are oxygen donor materials, for example, lead dioxide, activated manganese dioxide, calcium peroxide, cumene hy-

Table 6 The Characteristics of Polysulfide Oligomers Produced in Russia

Trade name	$M_n \times 10^{-3}$	Viscosity at 25°C (Pa·s)
NVB-2	1.7–2.5	7.5–11.0
NVB-I	2.2–3.7	15–30
NVB-II	2.8–5.5	30–50
FH-1.0	2.7–3.5	15–25
FH-0.5	2.3–3.3	15–25

Source: Ref. 2.

droperoxide, the various alkaline dichromates, and *p*-quinonedioxime. Among metal oxides, lead or manganese ones are preferred, with manganese having the advantage of low toxicity.

Manganese dioxide is the classic oxidizing agent used in organic chemistry. For alcohols, the mechanism of oxidation reaction is [48]

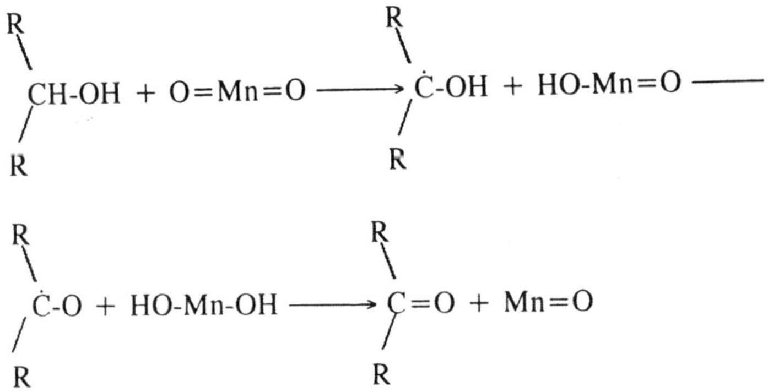

(54)

Such a mechanism can be extended to the oxidation of polysulfide oligomers, although the literature does not mention it.

Lead dioxide includes among its impurities lead monoxide, which is also formed in the process of PbO_2 reduction:

$$2 \sim R\text{-}SH + PbO_2 \longrightarrow \sim R\text{-}S\text{-}S\text{-}R \sim + PbO + H_2O \tag{55}$$

The presence of PbO can be the cause of lead dimercaptide formation, in keeping with the reaction

$$2 \sim R\text{-}SH + PbO \longrightarrow \sim R\text{-}S\text{-}Pb\text{-}S\text{-}R \sim + H_2O \tag{56}$$

after which their partial oxidation with PbO_2 takes place:

$$\sim R\text{-}S\text{-}Pb\text{-}S\text{-}R\text{-} \sim + PbO_2 \longrightarrow \sim R\text{-}S\text{-}S\text{-}R \sim + 2PbO \tag{57}$$

To avoid lead dimercaptide formation in PbO_2 oxidation of polysulfide oligomers, an additional oxidizer (sulfur) is added or more intensive thermal conditions are applied:

$$R\text{-}S\text{-}Pb\text{-}S\text{-}R\text{-} + S \longrightarrow R\text{-}S\text{-}S\text{-}R + PbS \tag{58}$$

$$R\text{-}S\text{-}Pb\text{-}S\text{-}R\text{-} \xrightarrow{\text{heat}} R\text{-}S\text{-}R + PbS \tag{59}$$

Another point of view regarding the mechanism of curing polysulfide oligomers with metal dioxides has been elaborated. According to it, the breaking of

—S—C— bonds requires the extraction of metal atoms from the crystalline network. These two phenomena, the breaking of the bond and the extraction from the crystalline network, require large energy consumption, and these large amounts of energy do not develop during the oxidation process. As a result, the oxidation process can develop in keeping with the following reactions [1]:

$$2 \sim \text{R-SH} + \text{MnO}_2[\text{MnO}_2]_{\text{cryst}} \longrightarrow \sim \text{R-S-S-R} \sim + \text{MnO}[\text{MnO}_2]_{\text{cryst}} + \text{H}_2\text{O} \qquad (60)$$

$$2 \sim \text{R-SH} + \text{MnO}[\text{MnO}_2]_{\text{cryst}} \longrightarrow \begin{array}{l} \sim \text{R-S} \diagdown \\ \qquad\quad \text{Mn}[\text{MnO}_2]_{\text{cryst}} + \text{H}_2\text{O} \\ \sim \text{R-S} \diagup \end{array} \qquad (61)$$

The strong oxidative agents of polysulfide oligomers also are peracids. In this latter case, reaction mechanism foresees initial formation of ion radicals:

$$\begin{array}{l} \text{S}_2\text{O}_8^{2-} \rightleftharpoons 2 \cdot\text{SO}_4^- \\ \cdot\text{SO}_4^- + \sim \text{R-SH} \longrightarrow \sim \text{RS}\cdot + \text{H}^+ + \text{SO}_4^{2-} \\ 2 \sim \text{RS}\cdot \longrightarrow \sim \text{R-S-S-R} \sim \end{array} \qquad (62)$$

The use of peroxides or hydroperoxides as oxidizers makes oligomercaptanes change into high molecular weight disulfides [1]:

$$2 \sim \text{R-SH} + \text{C}_6\text{H}_5\text{-C(CH}_3)_2\text{OOH} \longrightarrow \sim \text{R-S-S-R} \sim + \text{C}_6\text{H}_5\text{C(CH}_3)_2\text{OH} + \text{H}_2\text{O} \qquad (63)$$

Products obtained under these conditions have reduced thermal stability due to the occurrence of a degradation process generated by acid compounds formed in the reaction medium by free radicals generated by the excess of peroxides or hydroperoxides:

$$\begin{array}{l} \text{ROOH} \rightarrow \sim\text{RO}^\bullet + \text{HO}^\bullet \\ \sim\text{R}'\text{—S—S—R}''\sim + \sim\text{RO}^\bullet \rightarrow \sim\text{R}'\text{S}^\bullet + \sim\text{R}''\text{—S—OH} \\ \sim\text{R}''\text{—SH} + \text{HO}^\bullet \rightarrow \sim\text{R}''\text{S}^\bullet + \text{H}_2\text{O} \end{array} \qquad (64)$$

Increased thermal stability of the cured product is achieved by means of activating agents such as amines, 2,4,6-tris(dimethylaminoethyl)phenol, benzyliden methyl amine, Sb trivalent compounds, maleic anhydride and SiO_2, and iodine or iodine derivatives. In case of iodine addition (0.01–0.2% in mass), the following reaction takes place:

$$\begin{array}{l} \sim\text{R—SH} + \text{I}_2 + \text{HS—R}\sim \rightarrow \sim\text{R—S—S—R}\sim + 2\text{HI} \\ 2\text{HI} + \text{C}_6\text{H}_5\text{—C(CH}_3)_2\text{OOH} \rightarrow \sim\text{H}_2\text{O} + \text{I}_2 + \text{C}_6\text{H}_5\text{—C(CH}_3)_2\text{OH} \end{array} \qquad (65)$$

Amine activating agents added in the process of peroxidic curing of polysulfidic oligomers are introduced directly into the system, before peroxide or hydroperoxide introduction. The mechanism of aminic compounds action is based on the formation of ion radicals, in keeping with the following reactions:

$$C_6H_5{-}C(CH_3)_2OOH + RNH_2 \rightarrow [R{-}NH_2]^+ + C_6H_5{-}C(CH_3)_2O^- + HO^{\bullet}$$

$$C_6H_5{-}C(CH_3)_2O^- + [R{-}NH_2]^+ \rightarrow R{-}NH + C_6H_5{-}C(CH_3)_2OH \qquad (66)$$

The radicals formed do not attack —S—S— links but participate in recombination reactions with radicals formed through peroxide decomposition. These processes prevent oligomer degradation and provide adequate thermostability of the cured product.

Aliphatic dimercaptane oxidation with $R \leq 12C$ is achieved with diacylsulfides, for example, tiuramidisulfides, dixantogenates, and benztiazoldisulfides. Numerous secondary reactions also take place in this process, and especially formation of cyclics. Thus, from 1,4-butandimercaptan does not result a linear product but a cycloditian one [49,50].

By oxidation of aliphatic or aromatic mercaptanes with tetrahydrazine, disulfides emerge according to the following reactions [51]:

$$(C_6H_5)_2\text{-N-N}(C_6H_5)_2 \longrightarrow 2(C_6H_5)_2N\cdot$$

$$(C_6H_5)N\cdot + \sim\text{R-SH} \longrightarrow (C_6H_5)_2NH + \sim RS\cdot$$

$$2 \sim RS\cdot \longrightarrow \sim\text{R-S-S-R}\sim \qquad (67)$$

To obtain solid polysulfide oligomers, other organic compounds are used, for example, lead tetraacetate, dichlorodiphenol, organic compounds of Sn, hydrazine, and piperidine. In the latter case, the solidification process envisages the formation of an intermediate with the following structure [52]:

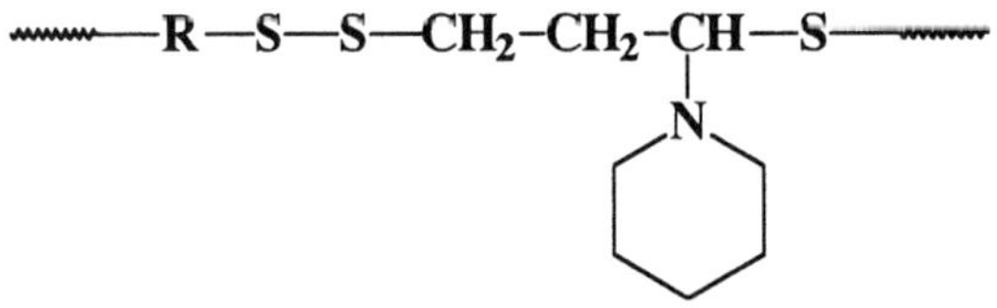

(68)

Polysulfide oligomers can also be solidified with furfurol together with aminic activating agents [2]:

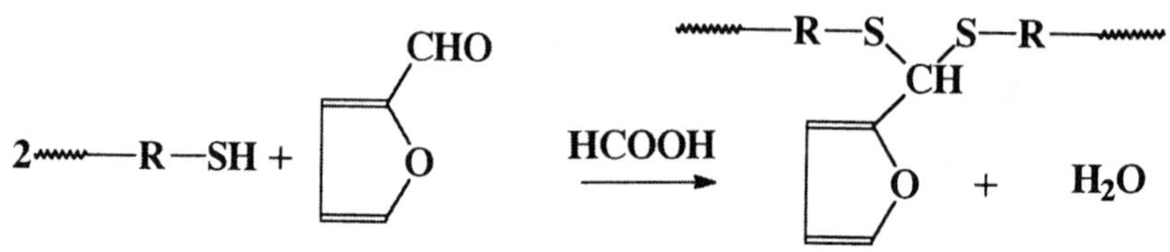

(69)

The advantage of cured products obtained in keeping with the reaction (69) are reduced viscosity and adequate thermal stability.

The process of polysulfide oligomer curing can also be performed with thalium alcoholates, magnesium oxide, as well as with sulfoxides [53,54]:

$$2\sim R—SH + R'_2SO \rightarrow \sim R—S—S—R\sim + R'_2S + H_2O \quad (70)$$

The use of *p*-quinonedioxime or of di- or trinitrobenzene as curing agents of polysulfide oligomers is known. In this case, the following reactions take place [2]:

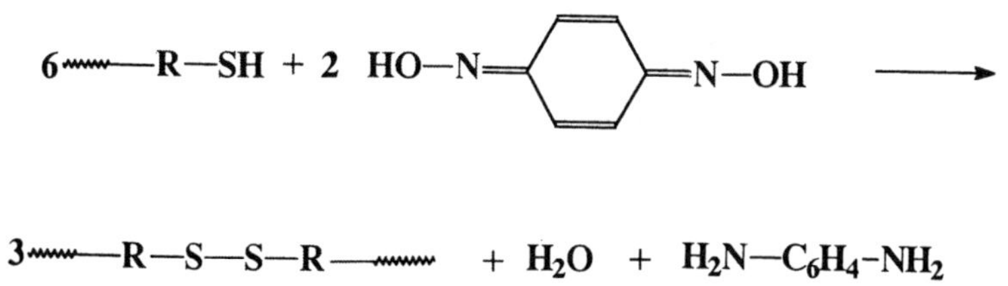

(71)

and

$$12\sim R—SH + C_6H_4(NO_2)_2 \rightarrow C_6H_4(NH_2)_2 + 6\sim R—S—S—R\sim + 4H_2O \quad (72)$$

A simple procedure for strengthening oligomercaptanes uses oxygen. The mechanism of the process is based on formation of thiolanons (RS—), which subsequently yield electrons and transform into the radical $\sim RS^{\bullet}$, which following recombination forms disulfide oligomers [55].

Polysulfide oligomers can solidify as a result of their copolymerization with various monomers. The process has in general a radicalic character and develops in the presence of radicalic initiators. Many times, it is the mercapto group that plays the role of the initiator. Among monomers, acrylates [56] and dienes [57] are used.

The interaction between polysulfide oligomer and methyl methacrylate or other acrylates is accelerated by an increase of temperature. The process determines a diminution of —SH groups into the polysulfide oligomer [58]. In this case, the homolytic breaking of —S—S— links in the oligomer chain is possible. Thus, the appearance of $\sim R—S—S—R^{\bullet}$ type radicals enables attack of the monomer double bond. Less possible is the formation of $\sim R—S—S^{\bullet}$ radicals and, implicitly, their participation in the polymerization reaction. In its general form, the polysulfide oligomer–methyl methacrylate interaction can be rendered by the following reaction [59]:

$$\sim R\text{-}S\text{-}S\text{-}R\sim \longrightarrow 2\sim RS\cdot \xrightarrow{+CH_2=CH(CH_3)\text{-}COOCH_3}$$

$$\sim R\text{-}S\text{-}CH_2\text{-}\dot{C}(O{=}C\text{-}OCH_3)\text{-}CH_3 \xrightarrow{+RS\cdot} R\text{-}S\text{-}CH_2\text{-}C(CH_3)(O{=}C\text{-}OCH_3)\text{-}S\text{-}R \quad (73)$$

At the same time, chain transfer reactions are to be expected:

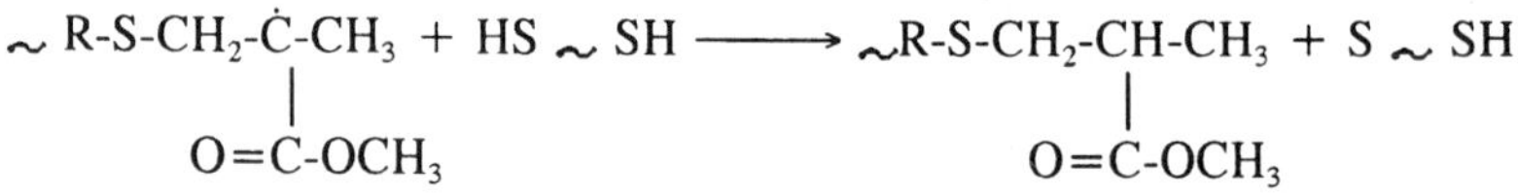

(74)

Curing of polysulfide oligomers is also possible by means of UV radiation, in the presence of photosensitizers such as benzophenone, anthraquinone, or mixtures of these substances [60].

With a view to diversify structure and, implicitly, polysulfide oligomer properties, chemical modification of these oligomers is performed by end group reaction with various organic substances. The best-known reactions are the modifications of polysulfide oligomers with epoxy resins. The admixture of significant amounts of epoxy resin with the polysulfide oligomer can generate two processes, formation of a block copolymer:

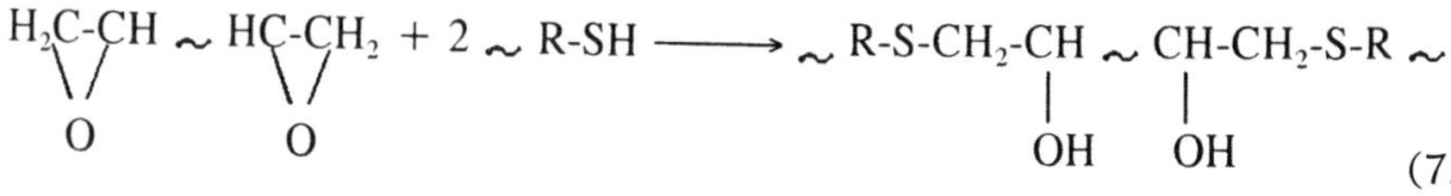

(75)

or the production of a physical blend [60].

If the mixing is performed in the presence of diamines, crosslinking is favored via some linear products [60]:

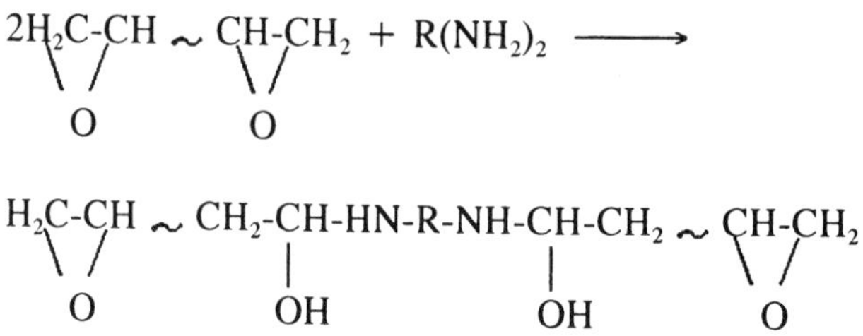

(76)

Further on, secondary amine groups react with epoxy groups, the crosslinked product being thus formed:

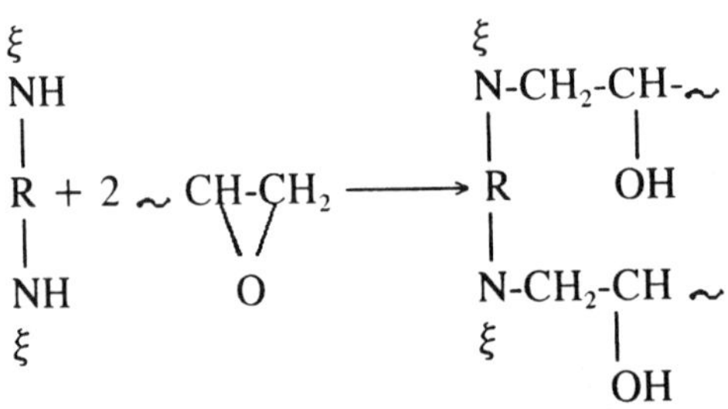

(77)

Reaction (76) can also have an ionic pathway:

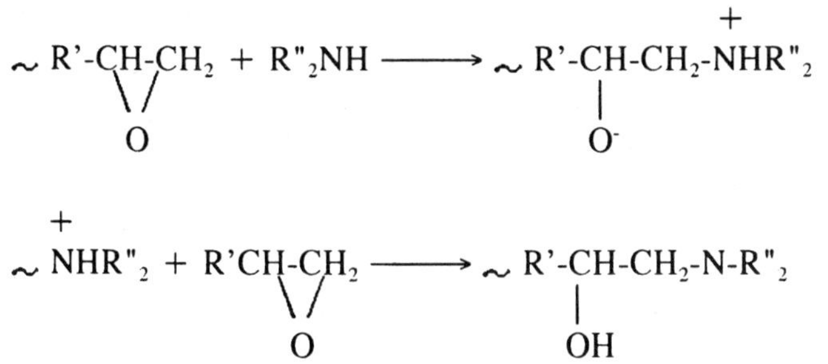

(78)

In the presence of tertiary amines, the process has only an ionic pathway:

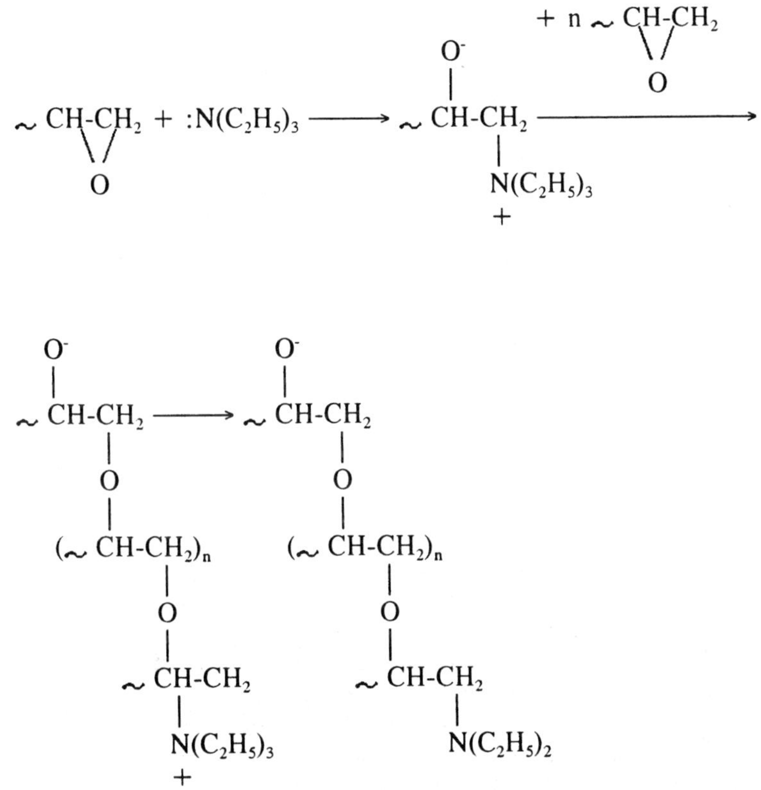

(79)

Just the same, another mechanism of polysulfide oligomer–epoxy resin interaction similar to that suggested by Teverovskaia and coworkers for polychloroprene curing is possible [61]. According to this mechanism, the essential fact of the process lies in the breaking of the —S—S— link and interaction of $—RS_n^{\bullet}$ macroradicals with epoxy resin, in keeping with the following reactions:

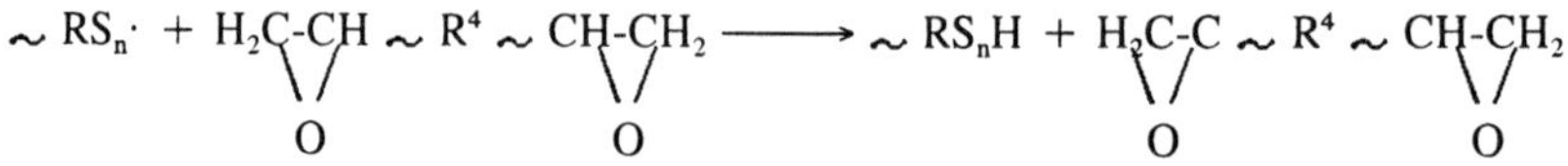

(80)

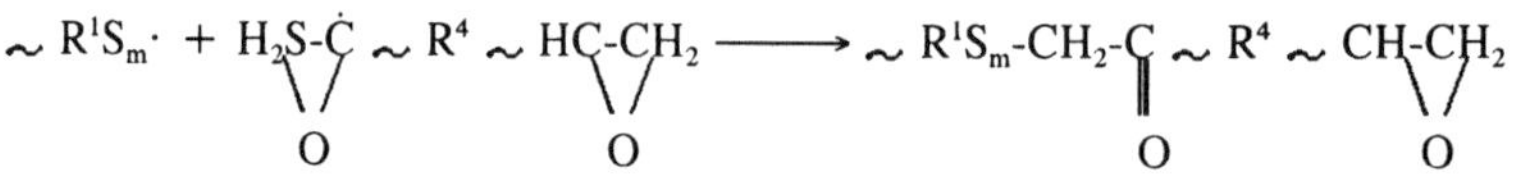

(81)

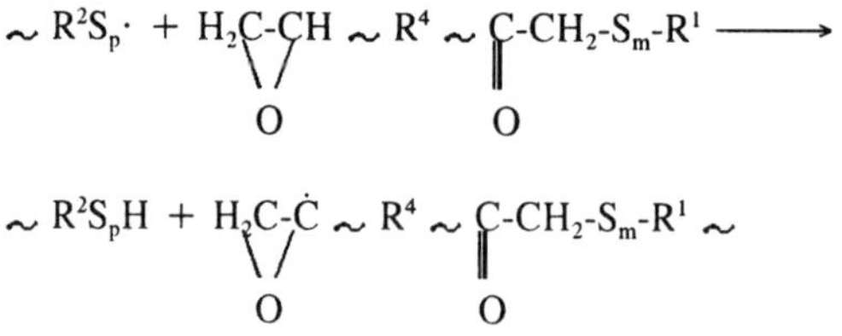

(82)

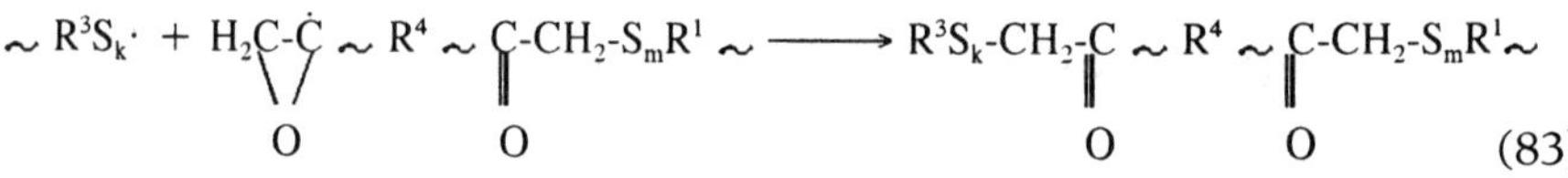

(83)

The only argument in favor of this mechanism is the presence of radicals in these systems [62].

A special place among products obtained by modifying polysulfide oligomers is occupied by thiokolurethanes obtained according to the reaction [63]

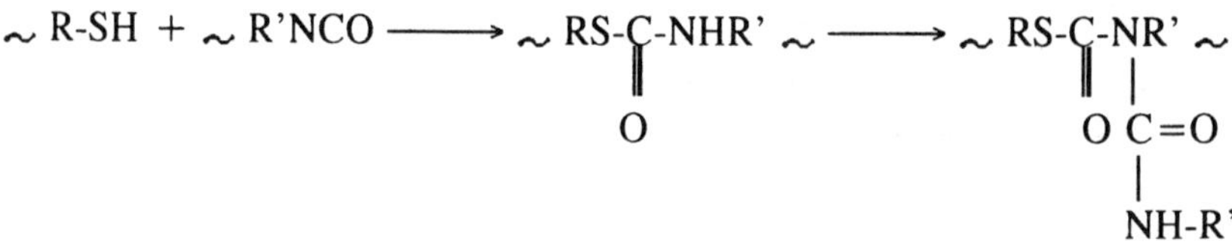

(84)

If for this reaction a diisocyanate of low molecular weight is used, the process develops at high temperature and in the presence of catalysts (sodium acetate, stanium dibutyllaureate, etc.) [64]. Solid products obtained from polysulfide oligomers modified with urethanes have inadequate physical and chemical properties. With a view to obtaining products with adequate properties, first adducts are achieved from polysulfide oligomers and an excess of diisocyanate, and later these adducts are modified with compounds containing hydroxylic groups.

Sometimes, for modification with diisocyanates, polysulfide oligomers with —OH end groups are used [65]. For the same purpose, adducts between diisocyanates and polypropylene glycol with polyesters, bismaleiimides, or mixtures of furfurol and ketones are also used [66,67].

A broad application involves composites made up of polysulfide oligomers and products of carbochemistry origin—coke plant, pitcoal tar, anthracenic oil, pitch, bitum, and tar. These composites are economical and are used in the building industry [68].

To obtain optimum properties, a polysulfide oligomer must be specifically formulated to meet the desired requirements. Working life should be adjusted in the desired range—the catalyst and compound adjusted for storage suitability. Suitable fillers should be incorporated. The various aspects of compounding such as curing agent, curing modifier (retarder, accelerator), filler, plasticizer, and adhesion additive are discussed in the following.

Sealants are generally produced in the form of products with two or three components. The main component (component A) is made from polysulfide oligomer, filler, adhesive, and other additives. The second component (component B) is a mixture of curing agent, plasticizer, stabilizer, and thixotropic agent. The third component (component C) is made up of curing accelerator. Mixing of the three components is done before use.

The curing agents most commonly used in industry are oxygen-donating materials, for example, lead dioxide, activated manganese dioxide, calcium peroxide, amine hydroperoxide, the various alkaline dichromates, and *p*-quinonedioxine [69].

Industry makes prevalent use of lead and manganese oxides in an admixture with a determined composition: MnO_2 = 1/3.2 [70]. In the case of monocomponent materials, the product should be well dried (humidity $\leq$ 0.1%) and processed by microencapsulation. At 100 parts by weight of polysulfide oligomers, 20 parts by weight of microspheres with ϕ = 0.1–0.2 mm are introduced, as obtained by microencapsulation of PbO_2 with gelatin.

For sealants with two components, this mixture is prepared separately.

The mixture of bichromates is used in Russia in the form of aqueous solutions, pastes, or solutions in organic solvents. To adjust the solidification rate, acid additives to these composites are used, for example, stearic acid mixed with butyl ester of oleic acid. Aqueous solutions of sodium or potassium bichromate are more active than PbO_2/MnO_2 mixtures and have some economic advantages [71].

In polysulfide oligomer curing, cumene hydroperoxide is quite frequently used. This process is applied when the manufacture of colorless products is sought [72].

The use of potassium permanganate as curing agent is also reported, as it has the advantage of a high curing rate. Materials thus obtained are used in dental surgery [73].

The composition of monocomponent sealants includes curing agents covered by a film made up of polymer aqueous solutions. Materials obtained through polysulfide oligomer curing with alcoholates or glycolates of alkaline or alkaline-earth metals (4 parts alcoholates at 100 parts oligomer) or with anhydrous alkaline peroxides are known [74].

Acceleration of the polysulfide oligomer curing process with a view to obtaining monocomponent sealants is achieved by means of some hygroscopic substances, for example, Al_2O_3, $Ca(OH)_2$, $Ba(OH)_2$ combined with aminic compounds or with acid catalysts (acetic, propionic, oleic, or stearic acids, etc.) [75].

Water activation is applied in case of silanized sealants. In these composites, the main component is polysulfide oligomer to which silicoorganic monomer is added (methyltriacetoxysilave, dimethyldiacetoxysilane, etc.) [76]. The silicoorganic compound should contain easily hydrolyzable groups.

For sealants obtained by polysulfide oligomer copolymerization with unsaturated monomers (1,2-butadiene, acrylates) as accelerators, aluminum, manganese, or calcium chlorides are used.

A product of high stability is obtained from polysulfide oligomer curing in the presence of fatty acid salts combined with zeolite-supported alkyl amines [77,78].

Bicomponent sealants contain the accelerator in the curing paste. In tricomponent sealants, the accelerators are an independent component.

Generally, a wide range of substances are used to accelerate the polysulfide oligomer curing process. Thus, it is reported that diphenyl guanidine activates —SH groups of polysulfide oligomers and in this way accelerates the curing process [79]. The same action is shown by o-toluidine, aliphatic acid salts with at most five carbon atoms [80], and mono- or diethanol amine. In this last case, if the curing agent is of peroxidic nature, some free radicals should also occur [81]:

$$C_6H_5\text{—}C(CH_3)OOH + HN(CH_2CH_2OH)_2 \xrightarrow{-H_2O}$$

$$C_6H_5\text{—}C(CH_3)_2O\text{—}N(CH_2CH_2OH)_2 \rightarrow$$

$$C_6H_5\text{—}C(CH_3)_2O^{\bullet} + {}^{\bullet}N(CH_2CH_2OH)_2 \qquad (85)$$

If the rate of polysulfide oligomer curing is to be lowered, numerous procedures may be applied. One of these requires the substitution for a part of PbO_2 with PbO. The same effect is produced by the introduction into the system of oleic or stearic acids. The efficiency of these agents decreases with medium humidity and during use. Yet some curing agent–accelerator–inhibitor combinations are preferred (Table 7).

Fillers increase the strength, affect rheological properties, and reduce the cost of sealants. Filler reinforcement of polysulfide sealants significantly increases the tensile properties. The increase is related to the type of filler, its particle size, and the type of cure. Indiscriminate selection of the filler can ruin the performance of a polysulfide sealant. Therefore, filler selection and filler loading should be carefuly established. Factors such as pH, particle size, surface area, and surface treatment should be taken into consideration.

The classes of fillers used in formulating polysulfide oligomers are calcium carbonate (wet or dry ground limestone, chalks, or precipitated carbonates), carbon blacks (furnace and thermal), clays (calcined), silica and silicate fillers, and titanium dioxide (rutile is chalk resistant and preferred over anatase). In general, sealants are prepared using a combination of fillers to attain desired properties.

New filler materials are used today. They consist of mixtures of classical filler with modifiers, and the blend obtained is processed in order to improve resistance and elasticity of sealants destined for hydraulic construction. In this category, modified aerosil with $\text{—}OC_2H_5$, $\text{—}O(CH_2)_2\text{—}NH\text{—}C_6H_5$, or $\text{—}OCH_2CH_2OH$ groups are known [82]. The use as filler of sand, cement, Al_2O_3, or asbestos is also reported [83].

Table 7 Curing Agent–Accelerator–Retarder Systems

Curing agent	Accelerator	Retarder	Observations
PbO_2	Sulfur, diphenylguanidine, 2,4,6-tris(dimethylaminomethyl)phenol	Stearic acid, stearates	Widely used formulation
MnO_2	Diphenylguanidine, *o*-toluolguanidine, amines, sulfur	Stearic acid, stearates, sulfur	Ensures thermal resistance of products; resistance to hydrolysis and lack of toxicity during processing
$Na_2Cr_2O_7$	Water, dimethylformamide, ethanolamine, diphenylguanidine	Stearic and oleic acids, esters of fatty acids	Used in thermoresistant sealants
Hydroperoxide of isopropyl benzene	Amines, 2,4,6-tris-(dimethylaminoethyl), phenol, cuprous abietate	—	Used in cast composites
n-Quinone dioxime	Diphenylguanidine, sulfur, manganese dioxide, or lead dioxide	Not necessary	For composite with long storage stability
CuO, ZnO_2, BaO_2	Water, amines, sulfur, tetramethyl-thiouramidesulfide	Molecular sieve	For one-component sealants

Source: Ref. 2.

The filler role is not confined to inorganic compounds. Polysulfide oligomer processing also employs as fillers numerous organic substances: polypropylene powder ($M_n = 10 \times 10^4$, particle size 1–5 μm), polyethylene, or copolymers of ethylene with olefins having three to six carbon atoms [84]. These materials provide stress dissipation on the contact surface between support and sealant, eliminate sealant stratification, and prevent crack forming.

Plasticizers improve the working properties of the sealants while lowering their modulus. Plasticizers must be compatible with the cured sealant and should have low volatility. Plasticizers fulfill, among other roles, that of dispersion agent contributing to homogeneization of curing agents in the polysulfide oligomer mass. From a chemical point of view, plasticizers should be neutral, liquid, and hydrophobic. Most frequently, dibutyl- and dioctylphthalates are used as plasticizers; less commonly used are hydrogenated terpenes, terphenyls, and thioesters. With a view to reducing sealant cost, especially for those applied in road coverings, indenecoumarone and pitcoal tar as well as anthracenic oils are used as plasticizers. In composites used in the building industry, polyisobutylene or copolymers of butadiene with an average molecular weight of 30,000 are used as plasticizers for polysulfide oligomers [85,86].

In the case of MnO_2 curing of polysulfide oligomers, the plasticizers not only improve flow properties, but also contribute to the occurrence of optimal conditions for more rapid solidification.

The incorporation of *adhesion additives* is known to improve adhesion. Typical examples are phenolic resin additives, for example, Methylon AP-108, Durez 10674, Bakelite Resin BRL 2741, and Resinox 568. Epoxies also perform well as adhesion promoters. In some instances, silanes (e.g., A-187 and A-189) are known to improve the polysulfide oligomer adhesion.

As a rule, silanes are introduced 10 parts by weight of epoxy resin. At room temperature, polysulfide oligomer–epoxy resin mixtures solidify slowly; this fact demands addition of activating agents. In this case, activating agents of the aminic type (ethylenediamine, dimethylurea, urotropine, polyalkylenepolyamine, aminophenols, and phenols) are applied [87].

Polysulfide oligomer solidification can be also accelerated by means of NaOH, alkaline chromates, and acid or neutral esters of polybasic acids or of anhydrides of polybasic acids. Significant improvement in physicochemical properties of cured products is obtained when polysulfide oligomer composites are mixed with polyesteric resins, polyacrylates, pitcoal tars, or halogenated poly(vinyl chloride) [88].

Curing of polysulfide oligomer–epoxy resin mixtures with MnO_2 implies partial binding by chemical links of the resin.

The nonbonded part can be chemically extracted with organic solvents. The viscosity of polysulfide oligomer–epoxy resin mixtures are modified during preservation. This phenomenon results from the interaction of —OH groups with the epoxy resins.

The mobility of the hydrogen atom in the —SH group conditions polysulfide oligomer modification with phenol–formaldehyde resins, so that the following chemical reaction occurs [89]:

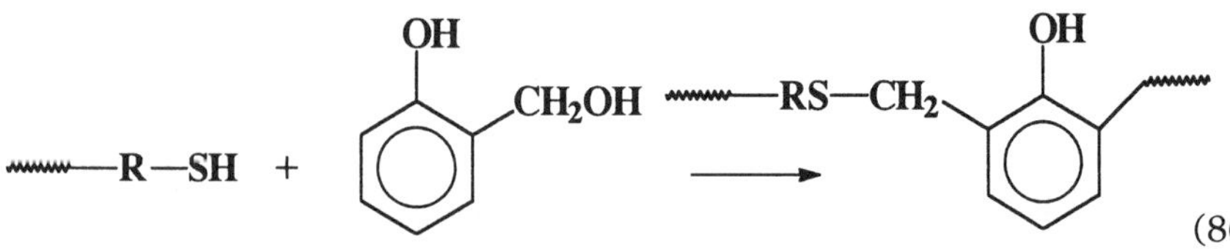

(86)

Introduction of phenol–formaldehyde resins improves polysulfide oligomer adhesivity. In these composites, the preferred curing agent is PbO_2.

Efficient adhesion additives are silanes and siloxanes with a content in vinyl, aminoalkyl, and ethoxy groups [90]. A wide application as additives is known for the *p*-glycidoxypropylbicyclo-2,2,1-heptane, vinyltriethoxysilane, methacryloxypropyltrimethoxysilane, and γ-mercaptopropyltrimethoxysilane [91,92].

Composites using as adhesion additives maleic anhydride, diepoxide, and condensation products of dicarboxylic acids with mercapto compounds are known and used [93].

To prevent leakage of sealants from vertical surfaces, thixotropic additives are used, such as SiO_2, bentonites, and aqueous solutions of polyvinylic alcohol, which provide for an increase in viscosity from 500–600 to 3000 Pa·s.

By the use of additives in formation, the density of composites based on polysulfide oligomers can be modified. An improvement of the thermophysical and damping properties can thereby be attempted.

The process of pore formation is carried out by the decomposition of some substances, either under the action of temperature or due to the water formed within the system during the process of curing [94].

The foaming agent is chosen so that its decomposition temperature is lower than that of the polysulfide oligomer. Here, alkaline metal carbonates and hydrides, azodicarboxylic acid derivatives, dicarboxylic acid derivatives or dicarboxylic acid azides, hydrazine or its salts and derivatives (*N,N′*-dimethylhydrazine, hydrazine formate, hydroxyalkylhydrazines, benzensulfonohydrazine) and *n*-toluenesulfo-semicarbazide are used. Quantitatively, foaming agents form 0.1–10 parts by weight of composite mass [95]. The action mechanism of the sodium salt of azo-dicarboxylic acid is the following:

$$2NaO{-}CO{-}N{=}N{-}CO{-}ONa + 4H_2O \rightarrow$$

$$N_2 + H_2N{-}NH_2 + 4NaHCO_3 \quad (87)$$

Nitrogen can also be formed through the following reactions:

$$H_2N{-}NH_2 + 2PbO_2 \rightarrow N_2 + 2H_2O + 2PbO \quad (88)$$

$$NaO{-}CO{-}N{=}N{-}CO{-}ONa + PbO_2 + H_2O \rightarrow$$

$$N_2 + 2NaHCO_3 + PbO \quad (89)$$

The foregoing reactions also show that the foaming agent action is determined by the chemical processes in which water (resulting from the solidification process) and the curing agent are involved.

The foaming process can also be determined by introducing into the system hydrogen evolved from hydrides according to the following reaction:

$$LiAlH_4 + 2H_2O \rightarrow 4H_2 + LiAlO_2 \quad (90)$$

Practically, the porous form of a sealant can be obtained by introducing the foaming agent into any of the common industrial products.

Polysulfide oligomer solidification is conventionally divided in three steps: The first one represents the time in which the product retains its plastic properties, so that the product can be applied on various surfaces (2 to 10 h). In the second period, a gradual loss of plastic properties (12 to 72 h) is found, followed by a third period in which the product achieves its optimal physicomechanical properties (120 to 160 h). The duration of polysulfide oligomer curing depends on a series of factors. This dependence can be also represented by the following relation:

$$T = f(t, h, g_1, g_2, g_3, A) \quad (91)$$

where T represents the viability of the product (the strengthening process duration); t the temperature of the environment (in °C); h the air relative humidity (%); g_1 the amount of curing agent (g); g_2 the amount of curing accelerator (g); g_3 the quantity of impurities in the polysulfide oligomer (free sulfur, free salts of metals with various valences); and A the activity of curing agent depending on quantity, degree of dispersion, and impurities. As a rule, g_1, g_2, g_3, and A are established by producer recipe. Only t and h act as variables in use. As a result, the relation (91) becomes

$$T = f(t, h) \tag{92}$$

It was found that an increase of 10°C in temperature (at a constant humidity of 50%) determines a decrease in viability of a product by 25%, and an increase of humidity by 10% (at a temperature of 26.5°C) determines a decrease of viability by 10 to 15% [96]. Knowing these data, the following empirical relation can be established:

$$T = K \exp(-at - bh) \tag{93}$$

where the values of constants a and b can be experimentally established. Thus, polysulfide oligomer viability is estimated in the lab by gelation time. In this respect, determination of viscosity [97] or of stirrer power consumtion [98,99] or special devices are used [100,101].

Sealant viability is also influenced by product mass, the mixing process, and the shape of the equipment wherein the processing takes place. In industrial and laboratory practice, polysulfide oligomer viability can also be determined by the following relation [93]:

$$T = \frac{T_L}{1 + KV} \tag{94}$$

where T_L represents viability as determined under laboratory conditions, V is the sealant mass volume, and K is a safety constant.

In polysulfide oligomer curing, as a rule, an excess of curing agent is used. In practice, the quantity of curing agent q is calculated by means of

$$q = \frac{KabM}{33 \times 100C} \tag{95}$$

where a represents the content in —SH groups of polysulfide oligomer (%); b is the amount of polysulfide oligomer (g); C is the stoichiometric coefficient of —SH groups oxidized by a molecule of oxidizing agent; M is the molecular weight of oxidizing agent; 33 is the molecular weight of the —SH group; and K is a safety constant given by

$$K = K_1K_2K_3 \tag{96}$$

where K_1 represents the excess of oxidizing agent (including additives) and has the value 1.1–1.3; K_2 is the excess of pure oxidizing agent and has the value 1.1–1.35; and K_3 is a constant specific for the case in which oxidizing agent is used is paste form (includes dispersion agent and inhibitors) and has the value 1.8–2.0. Taking into account the values of constants K_1, K_2 and K_3, the general constant K can take values ranging between 2.2 and 3.5.

The curing rate is significantly influenced by polysulfide oligomer structure and branching degree. The number average functionality of a polysulfide oligomer is set by the constant of the branching agent (1,2,3-trichloropropane). Neglecting the presence of cyclic (ca. 3–4% in industrial products), a relation between the content m of 1,2,3-trichloropropane in the initial monomer mixture (in %) and M_n of the oligomer can be established:

$$M_n = \frac{166(100 - m) + 137m}{m} = \frac{16{,}600}{m} - 29 \qquad (97)$$

Here, 166 and 137 represent molecular weights of the sequences $—S—(CH_2CH_2O)_2—CH_2CH_2—S—$ and $—S—CH(—S)—CH_2—CH_2—S—$, respectively.

VII. INDUSTRIAL PRODUCTION OF POLYSULFIDE OLIGOMERS

The largest production capacities of thiokolic sealants are held by United States through such firms as Products Research Corp., (Dallas, TX), Minnesota Mining Manufacturing Corp. (Indianapolis, IN), Boston Blacking Comp. (Boston, MA), and Thiokol Corp. (San Diego, CA). In Europe, these products are manufactured in France (Le Joint Francaise) (La Rochelle), England (Expandite Butyl Putty, Bostik Ltd.) (Manchester), Germany (Lechler Diring Dichtungwerke KG, Ewald Dörken AG) (Frankfurt), Austria (Determann & Krasq Corp.) (Vienna), the former Yu-

Table 8 Manufacturers and Main Characteristics of Thiokolic Sealants

Trade name	Country	Viability (h)	Density (kg/m^3)	Working temperature (°C)	Application
PR-1422A2	United States	0.5–2.0	1500	−55–135	Aerospace technology
PR-1422B2	United States	0.5–2.0	1450	−55–135	Aerospace technology
EC-801 A	United States	0.5–2.0	1450	−55–135	Aerospace technology
EC-801 B	United States	2–3	1550	−69–93	Aerospace technology
PR-380M	United States	2	1500	−70–105	Aerospace technology
PR-1221B	United States	0.25–12	1530	−70–105	Aerospace technology
PR-1221BT	United States	0.25–8	1510	−70–105	Aerospace technology
Pro-seal 918,947 917,915 930,907 980,940	United States	2	—	−50–100	Building technology
PRS-150	France	3–36	—	−40–95	Building technology
PR-391HT	France	3–36	1690	−40–105	Building technology
Thioflex-600	England	—	1580	−40–100	Building technology
Thiochrom 202 C	Yugoslavia	2	1350	−50–120	Building technology
Thiochrom 211 C	Yugoslavia	2	1350	−50–120	Building technology
Thiochrom C	Yugoslavia	2	1400	−50–120	Building technology
Thiochrom T	Yugoslavia	2	1400	−50–120	Building technology
Thiochrom	Romania	2	1700	−50–120	Building technology

Source: Ref. 102.

Table 9 Physicochemical Characteristics of Thiokol Sealants Produced in Russia

Trade name	Viability (h)	Density (kg/m^3)
U-30M	2–9	1400
U-30-mes-5	2–10	1400
U-30-mes-10	2–10	1400
UT-31	2–9	1750
TU-32	2–8	1750
UT-34	2–8	1550
51-UT-36A	1–3	1450
51-UT-36B	1–5	1450
51-UT-37	1–4	1400
51-UT-38A	2–6	1350
51-UT-38B	2–10	1300
51-UT-38G	4–12	1300
VITEF-1	2–8	1500
VITEF-2	0	1470

Source: Ref. 102.

goslavia (Kemijki Combinat Hromos) (Belgrad), and Romania (Oltchim) (Rîmnicu Vâlcea). Table 8 lists the main characteristics of the polysulfide oligomers produced at industrial scale.

In Russia, thiokolic sealants are produced in a series of composites, the characteristics of which are listed in Table 9 [102]. Several details on sealants produced in Russia need mention. The products U-30-mes-5, UT-32, UT-34, and 51-UT-37 are used either without substrates, or, when used in an organic solvent–containing

Table 10 Thiokolic Sealants Produced in Russia for Building Technology

Trade name	Viability (h)	Density (kg/m^3)
51-UTO-40	unlimited	1550
51-UTO-42	unlimited	1550
51-UTO-43	unlimited	1550
51-UTO-44	unlimited	1550
AM-0.5	≥2	—
KB-0.5	≥2	—
PA	1.5–8	—
PL	1.5–8	—
Ps-B	≥40 min	—
Hydrom-1	1–10	—
Hydrom-2	1–10	—

Source: Ref. 102.

Table 11 The Characteristics of Two-Component Thiokol Sealants Produced in Europe

Trade name	Country	Density (kg/m^3)	Working temperature (°C)	Application range
Terostat 2K15	Germany	1600	−40–80	Dwelling building (window sealing)
Terostat 2K70	Germany	1680	−30–80	Dwelling building, industrial building (the link between semifabricated plates and metal structure)
Terostat 75	Germany	1400	−40–80	
Widolastik KD-402	Germany	1500	—	Industrial and dwelling buildings (joint of prefabricated plate sealing)
Duro-Deltal	Germany		−40–80	Sealing of prefabricated plate joints in dwelling and industrial buildings, airfield, highway joint
M 70959		1500		
M 72047		1500		
M 72079		1550		
M 72747		1520		
Duro-Deltal	Germany	1280		Sealing of tank joints and in industrial buildings
X-72038			—	
C-72395			−60–80	
C-72251			−60–80	
Duro-Deltal K-72041	Germany	1330	—	Industrial and dwelling buildings, underground building, highway
Isavit-170	Germany	—	—	Sealing of prefabricated plates
Usekol 947	Belgium	—	−40–100	Putties for window glass sealing
PRS-250	France	—	−40–100	Highway joints, water tank, pools, airfield sealing
Thiochrom H	Yugoslavia	1400	—	Airfields
Bostik 2135	England	1480	—	Dwelling building
Bostik 2137		1620	−40–90	
Bostik Yulk	England	1450		Dwelling building
Seal-108		—	−40–90	
Bostik 3222				
Plastikol-15	Germany	1300	−45–90	In dwellings; not recommended for water tank sealing
Bostik 1138	England	1500	−40–90	Sealing of vertical joints of prefabricated concrete plates

medium, in combination with epoxy resins. Products 51-UT-38 (types A and B) are two-component sealants. The former is made up of polysulfide oligomer and carbon black, whereas the latter is made in aqueous solution of $K_2Cr_2O_7$ and tar. Sealants 51-UT-36, 51-UT-37, VITEF-1, VITEF-2, U-30-mes-5, U-30-mes-10, UT-31, UT-32, and UT-34 are tricomponent products: a curing paste, an epoxy resin, and diphenylguanidine.

Russia produces a large assortment of polysulfide oligomers (Table 10). For example, Hydrom-1 and Hydrom-2 are meant for road and airplane runway building, AM-0.5 and KB-0.5 for dwelling and industrial buildings. These products are solidified with $K_2Cr_2O_7$-based paste. PA, PB, and PS-B products are used in the automotive industry, on pleasure boats, and for household applications, respectively.

Among sealants produced in the United States, 80% are used in construction. Due to their excellent resistance to hydraulic and aircraft fuels, weather, water, water chemical treatment, UV light, the fields of polysulfide oligomer applications and their end uses are

1. Aircraft (sealing integral fuel tanks, sealed pressurized cabins, potting electrical connectors, sealing bolted steel tanks)
2. Automotive (glazing of windshields, glazing of rear lights, recreational vehicles, vibration damping, gas tank liner)
3. Construction (curtain wall, building exterior joint, airfield, insulated glass, canal, swimming pool)
4. Marine (flight deck, wooden and fiberglass decks of pleasure craft)
5. Miscellaneous (solid rocket fuel binder, relief maps, printing roll, dental impression, hose, gasket)

Table 11 lists bicomponent thiokolic sealants used in construction. The essential characteristic of these products is practical deformation, which allows estimation of the strength of these materials under a permanent tensile and compressive stress at an elasticity modulus of 0.2–0.5 MPa. Practical strain shows in joint thickness and its strain at permanent stress.

In conclusion, the polysulfide sealants have unique properties that can be put to very demanding uses. These are excellent resistance to various physical and chemical agents, excellent flexibility over a wide temperature range, adhesion, and long-term performance.

REFERENCES

1. Bertozzi, E. R., Chemistry and technology of elastomeric polysulfide polymers. *Rubber Chem. Technol.*, *41*, 114, 1968.
2. Averko-Antonovich, L. A., Kirpichnikov, P. A., and Smislov, R. A., *Polysulfide Oligomers and Polysulfide Based Sealants*, Khimia, Leningrad, 1983 (Russian).
3. Shlyachter, R. A., Nasonova, T. P., Apukhtina, N. P., and Sokolov, V. N., Study of the functionality and branching of liquid polysulfide polymers, *Visokomol. Soedin.*, *B*, *14*, 32, 1972 (Russian).
4. Corciovei, M., RO Patent 56,965, 1970.
5. Corciovei, M., RO Patent 56,966, 1970.
6. Corciovei, M., RO Patent 56,967, 1970.

7. Nasonova, T. P., Shlyachter, R. A., and Apukhtina, N. P., The influence of monomer structure upon the formation of cyclics during the synthesis of polysulfide oligomers, *Visokomol. Soedin., B, 13*, 635, 1971 (Russian).
8. Ansoupulos, P. A., Berbino, C. E., and Heilman, W. J., U.S. Patent 4,218,555, 1980.
9. Scott, W. B., and Simpson, P. L., GER Patent 2,159,392, 1972.
10. Bostik, S. A., FR Patent 2,261,300, 1974.
11. Bostik, S. A., U.S. Patent 3,813,368, 1974.
12. Bertozzi, E. R., U.S. Patent 3,817,947, 1974.
13. Leon, N. H., GB Patent 1,376,901, 1974.
14. Gar, L. W., U.S. Patent 3,892,686, 1975.
15. Collins, G. L., Costanza, R., and Conciatori, B., U.S. Patent 4,207,156, 1980.
16. Wysnewski, T., POL Patent 89,171, 1976.
17. Hermann, H., GER Patent 2,531,107, 1976.
18. Patrick, J. C., and Ferguson, H. R., U.S. Patent 2,466,963, 1946.
19. Dumitriu, P., and Uglea, C. V., RO Patent 75,850, 1981.
20. Milligan, B., Saville, B., and Swan, J. M., Trisulphides and tetrasulphides from Bunte salts, *J. Chem. Soc.*, 1963, 3608.
21. Milligan, B., Saville, B., and Swan, J. M., New syntheses of trisulphides, *J. Chem. Soc.*, 1961, 4850.
22. McKillop, D. A., Morgan, D., and Woodhaus, R. T., Reactive liquid polymers of propylene sulfide, *Polym. Prepr., 10*, 779, 1969.
23. Bonsignore, P. V., Mervel, C. S., and Bonerjee, S., Synthesis of polysulfide compounds. I. Organic dimercaptanes oxidation, *J. Org. Chem., 25*, 237, 1960.
24. Thiokol Corp., U.S. Patent 2,728,748, 1956.
25. Thiokol Corp., U.S. Patent 2,919,262, 1959.
26. Fokina, T. A., Apukhtina, N. P., Klebanskii, A. L., Pavlova, L. V., and Fikhtengalts, V. S., Radical telomerization of unsaturated hydrocarbons with the use of diisopropylxanthic disulfide as a telogen, *Vysokomol. Soedin., A, 13*, 1972, 1971 (Russian).
27. Crivello, J. V., Polyimidothioethers, *J. Polym. Sci., Chem. Ed., 14*, 159, 1976.
28. Boscato, J. F., Catala, J. M., Franţa, S., and Brossas, J., Living polymers as raw materials for synthesis of polysulfide oligomers, *Makromol. Chem., 180*, 1571, 1979.
29. Umbach, W., Meheren, R., and Stern, W., Transformation of halogenated oligomers and epoxidized oligodiens in polysulfide oligomers with —SH end groups, *Fette, Seifen, Anstrichmittel, 71*, 199, 1969.
30. Fettes, E. M., and Cannon, J. A., U.S. Patent 2,789,958, 1957.
31. Berenbaum, M. B., and Panek, J. R., in *Polyethers*, Part III, Gaylord, W. G., Ed., Interscience, New York, 43.
32. Villa, J. L., U.S. Patent 4,017,554, 1977.
33. Villa, J. L., U.S. Patent 3,882,091, 1975.
34. Villa, J. L., U.S. Patent 3,919,147, 1975.
35. Larson, R., Influence of branching upon the physical properties of polysulfide oligomers, *Polym. Paint and Color J., 162*, 555, 1972.
36. Usmani, A. M., Liquid polysulfide oligomers and sealants, *Polym. News, 10*, 231, 1985.
37. Tobolsky, A. V., and McKnight, W. J., *Polymeric Sulfur and Related Polymers*, Wiley-Interscience, New York, 1965.
38. Genkin, A. N., Nasonova, T. P., Podubnii, T. P., and Shlyachter, R. A., Study of molecular weight distribution in polysulfide oligomer by chromatographic fractionation, *Vysokomol. Soedin., A, 4*, 1088, 1962 (Russian).
39. Shlyahter, R. A., and Nasonova, T. P., *Synthetic Rubber*, Garmanova, I. V., Ed., Khimia, Leningrad, 1976, 552 (Russian).

40. Bertozzi, E. R., Davis, F. O., and Fettes, E. M., Disulfide interchange in polysulfide polymers, *J. Polymer Sci.*, *19*, 17, 1956.
41. Fettes, E. M., and Mark, H., Molecular weight distribution in polysulfide polymers, *J. Appl. Polymer Sci.*, *7*, 2239, 1963.
42. Parker, A. J., and Kharasch, N., Sulfur reactivity, *Chem. Rev.*, *59*, 583, 1959.
43. Novoselok, F. B., Sokolov, V. N., Apukhtina, N. P., Shlyahter, R. A., and Pablov, V. N., Mechanism of —S—S— bond fission in polysulfide polymers, *Vysokomol. Soedin.*, *A*, *7*, 1726, 1965 (Russian).
44. Eliel, E. L., Rao, V. S., Smith, S., and Hutchins, R. O., A convenient and stereoselective dithiol synthesis, *J. Org. Chem.*, *40*, 524, 1975.
45. Recalic, V. I., Fission of di- or polysulfide bonds upon the influence of reducing agents, *J. Polymer Sci.*, *Polymer Chem. Ed.*, *18*, 2033, 1980.
46. Sugiama, K., Nakaya, T., and Imoto, M., Vinyl polymerization. 275. Polymerization of vinyl monomers photosensitized by tetramethyltetrazene, *J. Polym. Sci.*, *A-1*, *10*, 205, 1972.
47. Fava, A., and Iliceto, A., Kinetics of displacement reactions at the sulfur atom. II. Stereochemistry, *J. Am. Chem. Soc.*, *80*, 3478, 1958.
48. Pratt, E. F., and McGovern, T. P., Oxidation by solids. I. Oxidation of selected alcohols by manganese dioxide, *J. Org. Chem.*, *26*, 2973, 1961; Pratt, E. F., and Van de Castle, J. F., Oxidation by solids. II. The preparation of eithertetraarylethanes or diarylketones by oxidation of diarylmethanes with manganese dioxide, *J. Org. Chem.*, *28*, 638, 1963; Pratt, E. F., and McGovern, T. P., Oxidation by solids. III. Benzanilines from N-benzylanilines and related oxidations by manganese dioxide. *J. Org. Chem.*, *29*, 1540, 1964.
49. Fettes, E. M., and Jorczak, J. S., Polysulfide polymers, *Ind. Eng. Chem.*, *42*, 2217, 1960.
50. Kobayashi, N., Osawa, A., and Fujisawa, T. J., Cyclics in polysulfide polymers, *J. Polymer Sci.*, *Polym. Chem. Ed.*, *13*, 2863, 1975.
51. Wallace, T. J., Mahon, L. J., and Kelliher, J. M., Synthesis of disulfides in the presence of tetraphenylhidrazine, *Nature*, *206*, 709, 1965.
52. Apukhtina, N. P., Solidification mechanism of polysulfide oligomers, *Cauciuk i Rezina*, 1967, No. 6, p. 7 (Russian).
53. Yiannios, C. N., and Karabinos, J. V., Oxidation of thiols by dimethyl sulfoxide, *J. Org. Chem.*, *28*, 3246, 1963.
54. Goethals, E. J., and Silis, C., Oxidation of dithiols to polydisulfides by means of dimethylsulfoxides, *Makromol. Chem.*, *119*, 249, 1968.
55. Oaki, S., *Organic Chemistry of Sulfur Compounds*, Khimia, Moscow, 1975, 512.
56. Thiokol Corp., U.S. Patent 3,640,923, 1972.
57. Zapp, R. L., and Oswald, A. A., Polysulfide polymers, *Rubber Chem. Technol.*, *48*, 860, 1975.
58. Minkin, V. S., Averko-Antonovioch, L. A., and Kirpichnikov, P. A., Study of the vulcanization mechanism of liquid thiokols by sodium bichromate, *Vysckomol. Soedin.*, *B*, *17*, 26, 1975; Minkin, V. S., Yastribov, V. N., Mikhutdinov, A. A., and Kirpichnikov, P. A., Study of the crystallization kinetics of poly(ethylene adipates) by pulse NMR methods, *Vysokomol. Soedin.*, *B*, *17*, 101, 1975; Minkin, V. S., Averko-Antonovich, L. A., Romanova, G. V., and Kirpichnikov, P. A., NMR study of molecular mobility in polysulfide oligomers and their modification products, *Vysokomol. Soedin.*, *B*, *17*, 394, 1975; Minkin, V. S., Averko-Antonovich, L. A., Kachalkina, I. N., and Kirpichnikov, P. A., NHR-Study of the molecular mobility of polysulfide oligomers, *Vysokomol. Soedin.*, *B*, *17*, 782, 1975; Minkin, V. S., Romanova, G. V., Averko-Antonovich, L. A., Svortsova, O. V., and Kirpichnikov, P. A., NMR impulse

study of the vulcanization kinetics of liquid thiokols modified by ethylene glycol dimethacrylate, *Vysokomol. Socdin., B, 17*, 831, 1975 (Russian).
59. Minkin, V. S., Romanova, G. V., Averko-Antonovich, L. A., Vorotnikova, G. P., and Kirpichnikov, P. A., Study of the interaction mechanism of polysulfide oligomer with methyl methacrylate, *Vysokomol. Soedin., A, 17*, 1009, 1975 (Russian).
60. Thipkol Corp., U.S. Patent 3,853,727, 1974.
61. Teverovskaia, E. N., Zorina, V. B., Zbuch, N. R., and Bloh, G. A., Vulcanization mechanism of polysulfide oligomers, *Cauciuk i Rezina*, 1979, No. 9, p. 13 (Russian).
62. Ulberg, Z. R., Kompaniets, V. A., and Ilina, Z. T., Radicalic characteristics of vulcanization process of polysulfide oligomers solidification, *Koll. Zhur., 32*, 278, 1970 (Russian).
63. Apukhtina, N. P., Novoselok, F. B., Kurovskaia, L. S., and Ternovskaia, G. K., *Synthesis and Physical-Chemistry of Polymers*, Naukova Donka, Kiev, 1970, 141 (Russian).
64. Hastings, G. W., and Joleston, D., Polysulfide oligomer modification, *Br. Polymer J., 3*, 83, 1971.
65. Thiokol Corp., U.S. Patent 4,045,472, 1977.
66. Thiokol Corp., U.S. Patent 3,951,898, 1976.
67. Crivello, J. V., and Juliano, D. C., Polyimidothioether-polysulfide block polymers, *Polymer Prepr., 14*, 1220, 1973.
68. Gudev, N., and Stoianova, V., Petroleum derivatives-polysulfide oligomer composites. Potential applications in construction, *Stroitelsvo (Bulgaria)*, 1974, No. 1, p. 11 (Bulgarian).
69. Thiokol Corp., U.S. Patent 3,207,696, 1965.
70. Higashi, S., Metallic oxides as curing agents pf polysulfide oligomers, *J. Nippon Univ. Sch. Dent., 13*, 93, 1971.
71. Averko-Antonovich, L. A., Minkin, V. S., Nefedov, E. S., and Rubanov, V. E., Sodium bichromates as curing agent of polysulfide oligomers, *Cauciuk i Rezina*, 1965, No. 9, p. 23 (Russian).
72. Higashi, S., Sodium permanganate as curing agent of sulfur containing polymers, *J. Nippon Univ. Sch. Dent., 14*, 22, 1972.
73. Averko-Antonovici, L. A., Cumenhydroperoxide as curing agent of polysulfide oligomers, *Cauciuk i Rezina*, 1965, No. 9, p. 23 (Russian).
74. Thiokol Corp., U.S. Patent 4,110,295, 1978.
75. Thiokol Corp., U.S. Patent 4,082,693, 1978.
76. Milkin, R., Brit., GB Patent 1,547,171, 1979.
77. Norris, V. K., U.S. Patent 3,681,301, 1972.
78. Thiokol Corp., U.S. Patent 3,402,151, 1968.
79. Kirchof, F., Accelerators in oligosulfides vulcanization, *Gummi u. Asbest, 9*, 384, 1956.
80. Villa, J. L., U.S. Patent 3,748,314, 1973.
81. Averko-Antonovich, L. A., Kirpichnikov, P. A., and Prohorov, F. S., Mechanism of vulcanization in polysulfide oligomers processing, *Cauciuk i Rezina*, 1968, No. 5, p. 18 (Russian).
82. Minkin, V. S., SU Patent 655,753, 1979.
83. Apukhtina, N. P., SU Patent 566,858, 1976.
84. Thiokol Corp., GER Patent 2,720,534, 1978.
85. Danev, I. V., BG Patent 19,689, 1978.
86. Nuralov, A. R., Application of petroleum derivatives in formulation of polysulfide oligomers, *Stroit. Materialov*, 1977, No. 9, p. 8 (Russian).
87. Charlesworth, J. W., Mechanical relaxation in episulfide network polymers, *J. Polymer Sci., Polymer Phys. Ed., 17*, 329, 1979.

88. Freidin, A. S., and Bu Ba Kim, Polysulfide oligomers in metalurgy, *Vestn. Maschinostroen.*, *50*, 51, 1970 (Russian).
89. Averko-Antonovich, L. A., Muhudnikova, T. E., and Kirpichnikov, P. A., Polysulfide oligomers modification using phenol-formaldehydic resins, *Cauciuk i Rezina*, 1975, No. 4, p. 18 (Russian).
90. Thiokol Corp., U.S. Patent 3,714,132, 1973.
91. Thiokol Corp., U.S. Patent 4,070,328, 1978.
92. Thiokol Corp., U.S. Patent 4,070,329, 1978.
93. Dachselt, E., *Thioplaste*, VEB, Leipzig, 1971, 164.
94. Tons, E., Thixotropic properties of polysulfide oligomers, *Adhesive Age*, *8*, 29, 1965.
95. Hunter, V. A., and Kleinfeld, M., Foaming agents of polysulfide oligomers vulcanization, *J. Rubb. World*, *153*, 84, 1965.
96. Smislova, B. A., Thiokol based selants, TNIT, Moscow, *Neftekhimia Ser.*, 1974, p. 83 (Russian).
97. *Elastomerics*, *109*, 15, 1977.
98. Engeldinger, H. K., The behaviour of polysulfide oligomers during the aering process, *Adhäsion*, *19*, 312, 1975.
99. Ivanov, I. A., SU Patent 329,474, 1972.
100. Sviridov, V. I., and Sokolov, A. D., Curing of polysulfide oligomers, *Plasmasî*, 1976, No. 6, p. 66 (Russian).
101. Averko-Antonovich, L. A., Minkin, V. S., and Iasterov, V. N., Chromatographic characterization of polysulfide oligomers, *Vysokomol. Soedin.*, *B*, *15*, 24, 1973 (Russian).
102. Smislova, R. A., and Kotliarova, S. V., *Sealants Handbook*, Khimia, Moscow, 1976, 71 (Russian).

9

Formaldehyde-Based Oligomers

I. PHENOL–FORMALDEHYDE OLIGOMERS

Synthesis of phenol–formaldehyde oligomers uses as the main raw materials phenol and formic aldehyde. Both are used in the form of aqueous solutions of 90% and 40–45% concentration, respectively [1]. Since formaldehyde contains small amounts of formic acid, the condensation process can take place in the absence of catalysts, so that phenol–formaldehyde oligomers can be obtained. The reaction rate is, however, quite low when the process takes place at atmospheric pressure and a temperature around 100°C. The use of acid or basic catalysts is compulsory in industry to produce phenol–formaldehyde oligomers of desired characteristics. Novolac type oligomers are obtained in acid catalysis; in basic catalysis, resol type oligomers are produced.

Generally, the process can be represented by the following two reactions:

$$C_6H_5OH + CH_2O \xrightarrow{k_1} HO{-}C_6H_4{-}CH_2OH \qquad (1)$$

$$HO{-}C_6H_4{-}CH_2OH + C_6H_5OH \xrightarrow{k_2} HO{-}C_6H_4{-}CH_2{-}C_6H_4{-}OH \qquad (2)$$

According to reaction (1), mono-, di-, and trihydroxymethylene derivatives of phenol can be formed. Both in acid and in alkaline media, the first stage of the process should be the formation of a monohydroxymethylene derivative of phenol. Subsequent development of the process is determined by the value of the constants k_1 and k_2. If $k_1 > k_2$, the process leads to the accumulation of hydroxymethylene derivatives of low molecular weight. If $k_1 < k_2$, monohydroxymethylene derivatives of phenol react with each other so that linear oligomers are obtained.

To obtain novolac type oligomers, the most active catalyst is hydrochloric acid. Its concentration is determined by two factors: formaldehyde acidity and the pH value of the reaction medium (1.8–2.2). As a rule, for HCl a concentration of 0.1 to 0.3% is used vs. the phenol concentration introduced.

Novolac oligomers obtained in the presence of HCl develop with violent heat of reaction. This is why the introduction of HCl (37% solution) is performed step by step. In the acid catalysis achieved with sufuric acid, the reaction product is brown-colored, and ultimately H_2SO_4 neutralization with $Ba(OH)_2$ or $Ca(OH)_2$ is required.

Because of corrosion, the synthesis of novolac type oligomers catalyzed by mineral acids has been abandoned. To this, the remark may be added that in the situation in which HCl and CH_2O concentration in the gaseous phase exceeds

0.01%, chlorurated derivatives of phenol can be formed, which are carcinogenic products [1].

The use of oxalic acid for the synthesis of novolac type oligomers allows the realization of the process under accessible experimental conditions, without violent reaction heat. The end product is colorless and stable. Oxalic acid is separated by washing or thermal decomposition at 180°C in vacuum. Simultaneously, it can be eliminated by distillation of phenol excess [2].

Trichloroacetic acid can also catalyze the process of condensation. In this case, polycondensation takes place at 60 to 85°C, and subsequent heating of the system at 100 to 200°C determines catalyst decomposition in $CHCl_3$ and CO_2 [2].

The reaction between phenol and formic aldehyde takes place in the homogeneous medium only in the first stages (up to 60–65% conversion of initial monomers). The initial stage of the polycondensation can be kinetically described by

$$\frac{vn^2b[w]}{CC_\Phi} = 3(k_1 - k_2) + k_2 \frac{3C_\Phi^\circ - 2C^\circ + 2C}{C_\Phi} \tag{3}$$

in which v is the consumption rate of formaldehyde, C_Φ, C°, and C are the initial concentrations of phenol, formic aldehyde, and of HCl, respectively, and b is a parameter that is related to the methanol and water content of the system.

In the final stages of the process, the system separates into layers, ultimately forming a biphasic oligomer–water system. The dependence between concentrations (in mass percents) at equilibrium of phenol, water $(C_\Phi)_e$, and oligomer $(C_\Phi)_e$ is nonlinear [3]. The distribution coefficient of phenol, K_d, in the oligomer–water system is determined by the Nernst equation:

$$K_d = \frac{(C_\Phi)_e^m}{(C_\Phi)_e} \tag{4}$$

where m is an index representing the number of modified molecules distributed between the two phases as a result of the phenomenon of association. This coefficient is a function of hydrochloric acid concentration in the aqueous phase (in mass %). The character of K_p dependence on concentration C of hydrochloric acid in aqueous phase may be evaluated by [4]

$$K_p = 1075 - 0.005t - 0.3 \exp\left(-\frac{C}{2}\right) \tag{5}$$

Relation (4) allows the correlation of the distribution coefficient of phenol, K_d, with the content of HCl in the aqueous phase.

It was also seen that the HCl concentration in the aqueous phase is four to nine times greater than that in organic phase (oligomer) [5]. On the other hand, distribution of formaldehyde between the two phases does not depend on the HCl concentration in the aqueous phase or on working temperature.

The conditions of system transition from the homogeneous to the nonhomogeneous state are represented by the following empirical relation:

$$10^{-3} \log(C_\Phi)_e = 355 - 75(C_M)_o + [395 + 360(C_M)_o]\log \frac{C_\Phi}{C_o} \tag{6}$$

in which $(C_\phi)_e$ is phenol concentration at the interface in mol/L and $(C_M)_o$ is methanol concentration in the initial mixture.

Polycondensation of phenol with formaldehyde in acid medium forms novolac type oligomers, the composition of which depends on the pH value. If pH < 0.5, the main species formed are *p*-, *p'*-; if pH = 0.5–1.0, besides *p*-, *p'*-species, *o*-, *p'*-species are also formed; and if pH = 1.05–5.0, *o*-, *o'*-species are formed in the main [2]. Table 1 lists the compositions obtained under various experimental conditions.

If novolac is produced with lanthanum acetate or Zn oxide as catalyst, orthonovolacs are obtained in the main, in which the content of *o*-, *o'*-methylene bridges is as high as 45–50% while *m*-, *m'*-methylene bridges do not exceed 5%. Also interesting is that the condensation process without catalyst at pH = 4–5 yields a significant amount of *o*-, *o'*- and *o*-, *p*-dihydroxydiphenylmethane; in these products, *p*-, *p'*-isomer is lacking.

Performing synthesis in two stages (where in the former stage a catalyst that orients substitution in the ortho position is used, and the latter stage uses an acid as catalyst) provides the formation of a reaction product the isomeric composition of which is similar to novolac obtained by acid catalysis (*o*-, *p*-isomer 80%).

Some metal cation (Zn^{+2}, Mg^{+2}, Ca^{+2}, Sr^{+2}, Ba^{+3}, Mn^{+2}, Ni^{+3}, Fe^{+3}, Pb^{+2}) catalysts provide ortho orientation in the condensation reaction of phenol with formaldehyde at pH values between 4 and 7 [6].

As a rule, to obtain novolac type oligomers, metal salts of weak carboxylic acids are preferred.

Orthonovolac resins are obtained in two stages. The former takes place at 100°C when the addition reaction takes place; after separation of water from the reaction mixture, the second stage is performed at 140 to 160°C when the so-called condensation reaction takes place. The orientation effect of zinc acetate is found in both stages of the process.

Table 1 Isomeric Content of Phenol–Formaldehyde Oligomers

	Working conditions			Composition (%)		
Oligomer	*T* (°C)	pH	Catalyst	*o*-, *o'*-	*o*-, *p*-	*p*-, *p'*-
Obtained at room temperature	23	0.5	HCl	2	64	34
Industrial product (Resin 18)	100	1.5–2.0	HCl	4	67	29
Industrial product (Resin 104 continuous procedure)	100	0.5	HCl	6	66	28
Industrial product (Resin 18 continuous procedure)	100	0.5	HCl	10	65	25
Industrial product obtained by two-step procedure	100	4.0	HCl	2	80	18
Orthonovolac	100	0.9	HCl	—	—	—
	100	4.0–5.0	HCl	45	53	2
	120–140	4.0–5.0	no catalyst	52	48	—

Source: Ref. 5.

Novolac isomeric composition is also influenced by the nature of the solvent. When the reaction occurs in organic solvents (i.e., ethanol, toluene, tetrachloroethylene), ortho isomers are diminished. Still, this diminution is influenced by the nature of the acid used in the following order: acetic > oxalic > benzenesulfonic > hydrochloric.

In anhydrous conditions and in the presence of aliphatic carboxylic acids, a reaction product with a high ortho isomer content is obtained. Such oligomers show an increased capacity for crosslinking.

Alkaline catalysis of the polycondensation process takes place in the presence of various bases, for example, NaOH, $Ba(OH)_2$, $Ca(OH)_2$, $Mg(OH)_2$, Na_2CO_3, and hexamethylenetetramine and other tertiary amines.

Industrial synthesis prefers NaOH (0.1% concentration), $T = 70-100°C$, pH > 9, and a phenol:formaldehyde molar ratio of 1.2 to 2.5. Resolic resins obtained under these conditions contain free phenol (1.0 to 20%), a water-soluble fraction with low molecular weight (hydroxymethylene derivatives of phenol) amounting to 5 to 15%, and oligomers (80–95%). Unless reaction temperature does not exceed 60°C, hydroxymethylenephenols are the only reaction product. Under actual conditions ($T = 100°C$) hydroxymethylenephenols react with phenol, forming oligomers.

Sodium hydroxide provides for the formation of products with good solubility in the reaction medium, hinders clotting processes, and allows one to obtain "dry" resols. In resol drying phase, sodium hydroxide is neutralized with weak carboxylic acids (i.e., lactic, benzoic, or oxalic acids) [7].

Another basic catalyst is barium hydroxide, which provides the advantage of employing accessible working conditions; Catalyst elimination is made by CO_2 barbotage into the reaction mixture [7].

The most commonly used basic catalyst is ammonium hydroxide employed in a proportion of 0.5 to 3% vs. phenol. Because ammonium reacts with formaldehyde, the actual catalyst is hexamethylenetetraamine (urotropine). Under these conditions, it is not necessary to eliminate catalyst from the reaction product, and oligomers formed have high molecular weight and are yellow-colored (because of —CH=N— bonds).

Isomeric composition of resolic oligomers is determined by the nature of the catalyst, pH value, and the composition of the reaction medium. Thus, substitution for NaOH by hydroxides or oxides of alkaline-earth metals increases the ortho derivative content of the reaction product. The same modification is obtained by reducing pH value to as low as 6–8 or carrying out the process in a medium with a low water content. As a rule, resolic oligomers have an extremely heterogeneous composition, which represents one of the main causes of structural faults found in resite. These faults determine the alteration of the physicomechanical performance of resite [8].

The rate of substitution reaction in ortho and para positions is determined by catalyst in the following order: K < Na < Li < Ba < Sr < Ca < Mg.

The use of ammonium as catalyst yields 2- and 4-hydroxymethylphenol, 2-butoxbenzylamine, and 2,2-dihydroxybenzylamine. An oligomer of increased oligomer chain length occurs through mutual reactions between hydroxymethylphenol with phenol and nitrogen-containing compounds.

Subsequent properties of resol depend in the main on the initial CH_2O:NH_3 molar ratio; the lower this ratio, the richer the reaction product will be in species with aminic groups, which in their turn will also determine a reduced rate of resol solidification in the product.

Phenolic resins have found widespread use in numerous areas, in particular, molded and cast plastic articles, adhesives for plywood as laminating resins, and thermal insulation. Because these resins are versatile and inexpensive, numerous reactants have been used to modify the resins in order to improve certain properties, for example, dispersing ability, or to inhibit specific properties that are undesirable for certain uses.

By way of illustration of modifying phenolic–formaldehyde resins, there are patents describing sulfomethylating a phenol–formaldehyde resin and its use as a synthetic tanning agent [9] and a process for breaking emulsions with phenol resins that have been reacted with alkylene oxide [10]. D'Alelio [11] describes soluble copolymers of epoxyalkoxy hydrocarbon substituted phenol–aldehyde resin in cements, impregnates, coatings, and the like. Walz [12] explains a synthesis process of dyestuff composition consisting of the reaction product of an alkylene oxide and the condensation product of a phenolic compound, formaldehyde, and an amine. Falkehay and Bailey [13] produce a sulfomethylated phenolic–formaldehyde resin in which some or all of the free phenolic hydroxyl groups are blocked; it has use as a surfactant, a dispersing agent, or in dyestuff compositions.

Hindered phenol–formaldehyde resins, that is, substituted compounds having at least the ortho positions of the phenyl blocked by alkyl groups, such as bulky branched alkyl groups, have been commercially employed as stabilizers and antioxidant additives in a wide variety of polymers [14–16].

The cyclic phenol–formaldehyde resins were naturally produced in low yield as by-products of linear resins [17]. In the cyclic phenol–aldehyde resin, the end valences are joined to form a cyclic structure, for example,

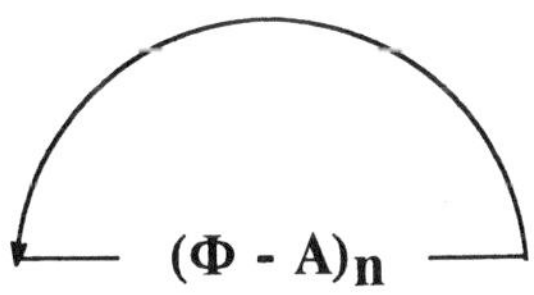

(7)

where the circular line indicates a cyclic structure where n is, for example, 4–16 or greater;

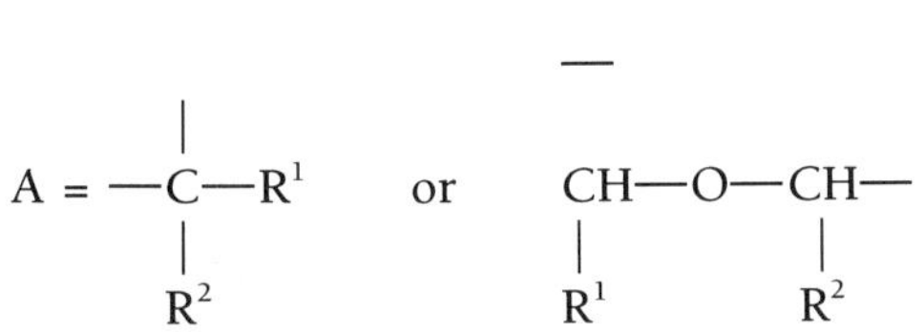

ϕ = phenyl, and R^1, R^2 = alkyl. The preferred structure is the cyclic tetramer where $n = 4$:

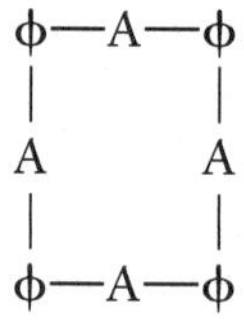

(8)

The cyclic resins are in general infusible. These materials show considerably less solubility in common organic solvents than their linear counterparts. The cyclic resins derived from a substituted phenol and an aldehyde are, for example, insoluble or slightly soluble in paraffinic hydrocarbons, ketones, esters, alcohols, water, dimethylformamide, and so forth. They show greater solubility in aromatic solvents and certain halogenated solvents, for example, chloroform, and fair to good solubility in certain basic nitrogen solvents, such as pyridine and piperazine. These solubility characteristics may be affected by changes in the bridge between the phenolic nucleus [18]. Thus, for example, a long-chain substitution will make the cyclic resins more hydrocarbon soluble than the corresponding short-chain-substituted material.

The particular effect of structure on melting point can be clearly appreciated from the experimental data given in the Table 2, which lists the melting points for the crystalline form of the compound A:

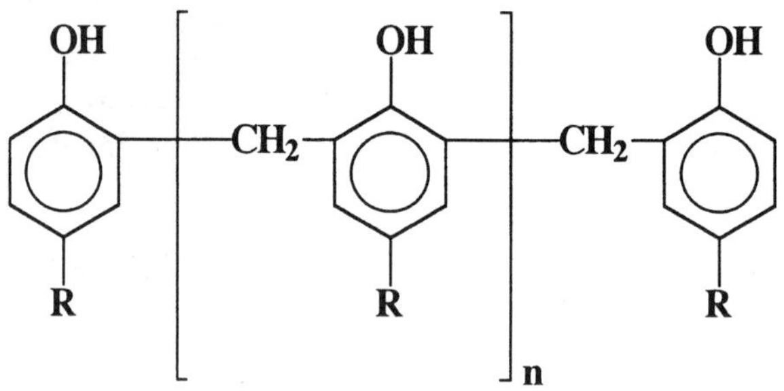

(9)

Table 2 Melting Points of the Phenol–Formaldehyde Oligomers with General Formula A

n	Melting point (°C)
0	160
1	217
2	206
3	200
4	250
5	247
6	255

where R = *tert*-butyl. In comparison, the cyclic tetramer of general structure

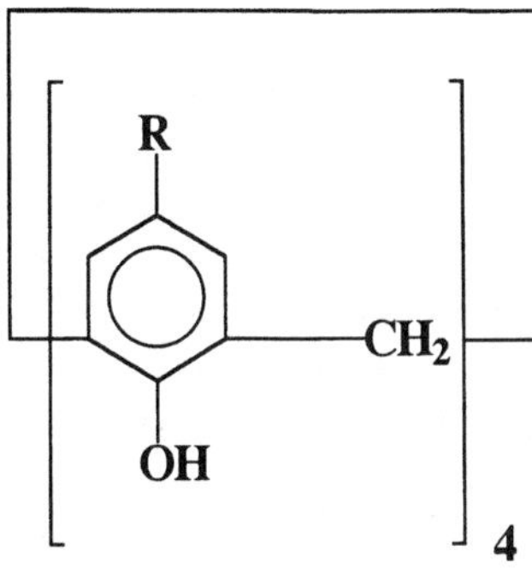

(10)

has a melting point above 360°C.

The nature of the products formed from a phenolic compound and a carbonyl compound varies greatly depending on the method and conditions of reaction. Thus, if a phenol of type R—C_6H_4—OH is reacted with an aldehyde, for example, paraformaldehyde, under acidic conditions, the linear resin is formed almost exclusively. Under neutral conditions, little or no reaction takes place. Under strongly basic conditions, some cyclic tetramer is formed but the formation of a linear product is strongly favored. Under mildly basic conditions, substantial amounts of the cyclic tetramer can be found in the resinous products.

To achieve higher yields of the cyclic tetramer, in addition to having mild basicity, it is desirable not only to carry out the condensation in a relatively nonpolar solvent, but also to carry out the reaction at a slower rate than employed in forming the linear resin.

By the use of preformed phenolic compounds of the formula

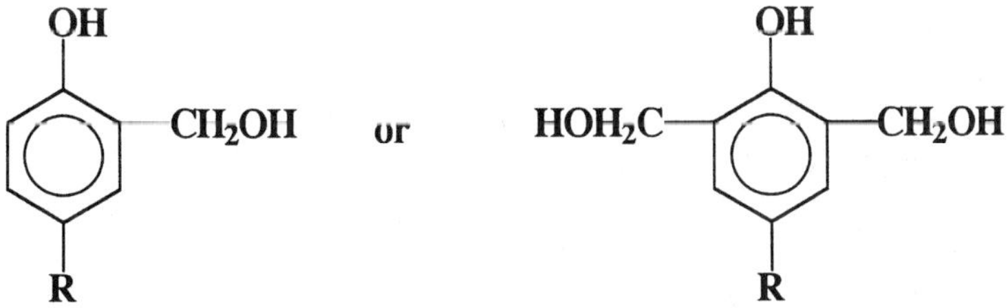

(11)

it is possible to form cyclic compounds with —CH_2— and/or —CH_2OCH_2— bridges, depending on the reaction conditions. When the methylol phenol compound is condensed under essentially neutral conditions using sufficiently high temperature to eliminate water, but not so high as to cause elimination of formaldehyde, a cyclic resin is obtained having 4–8 phenolic units (almost exclusively an even number of units rather than odd) bridged by ether-containing linkages, —CH_2OCH_2—. However, since it is practically impossible to entirely exclude the elimination of formaldehyde, some of the bridging will be means of methylene groups —CH_2—.

II. UREO–FORMALDEHYDE RESINS

The main raw materials used for the synthesis of ureo–formaldehyde resins are urea and formaldehyde. Urea is a solid, crystalline, water-soluble substance. It has a weak basic character. Formaldehyde is used in solution form at 30–55% concentration. It is of advantage to use either high-concentration formaldehyde solutions or precondensates. Both processes provide a satisfactory economic efficiency and diminish residual water content. Precondensates can be obtained through the reaction between urea and formaldehyde (molar ratio 1:1.1–6), at $T = 45°C$ and pH = 6.8–8.0 [19].

An advantage of precondensate application is a diminished content of free CH_2O in the reaction product.

The technology of ureo–formaldehyde oligomers yield is based mainly on guiding pH value via the nature of the catalyst and the molar ratio of the reactants. The technological process of obtaining these oligomers is made up of several stages, namely [20].

1. Formation of urea and formaldehyde mixture in molar ratio 1:2 at 80–85°C
2. Achieving pH = 7.0–7.4 and condensation of the mixture for 30 min at the same temperature
3. Reducing pH to 4.8–5.8 and continuing condensation in acidic medium until the viscosity of reaction mixture reaches 15–50 MPa·s
4. Establishing the pH within the range of 5.8–6.5, urea addition until a urea:CH_2O molar ratio of 1:1 to 1:1.3 is reached, and reducing temperature to 50°C
5. Final value of pH must be raised to 7–8

Industrial production of ureo–formaldehyde oligomers can be achieved by continuous or discontinuous processes. The latter processes have some economical advantages.

A certain interest is aroused by the continuous process that envisages the barbotage of gaseous formaldehyde in the urea solution in the presence of a catalyst (cationite based on transitional metal salts). In this way, ureo–formaldehyde oligomers could be produced, and they can crosslink in a rather short time [21].

The defining peculiarity of ureo–formaldehyde oligomer production processes is the essential effect of reaction mass acid/basic character on the composition of the reaction mixture: in an alkaline medium, monohydroxymethylcarbamide is the main product; in neutral alkaline medium (pH = 7–8), mono- and dihydroxymethylcarbamide are obtained (the possibility of forming tri- and tetrahydroxymethylcarbamide is small because a small excess of formaldehyde is being employed); and in acidic medium (pH = 1 to 4), only methylcarbamide is obtained.

To avoid formation of methylcarbamide, industrial production of ureo–formaldehyde oligomers is achieved by a two-stage process. In the first stage, in neutral or weakly alkaline medium (pH = 7–8) ureo–hydroxymethylene derivatives are formed, since hydroxymethylenecarbamide formation rate increases with increasing pH value of reaction mixture. In the second stage, in a weakly acidic medium (pH = 4.6–6) polycondensation of the formed hydroxymethylenecarbamide takes place.

Under these conditions, it becomes necessary to carefully supervise medium pH. This can be modified via the following reactions:

$$2CH_2O + O_2 = 2HCOOH \tag{12}$$

and

$$2CH_2O + NaOH = CH_3OH + HCOONa \tag{13}$$

Formation of some unstable salts, through the reaction between urea and some of the initial formaldehyde, determines from the beginning an increased medium pH; in the subsequent stages of the reaction, due to urea consumption, a reduction of medium pH takes place. This pH variation necessitates permanent control of reaction medium pH by the addition of buffers (sodium acetate, $(H_4N)_2CO_3$, citric acid, and sodium acetate mixture) or urotropine. The latter forms, with formic acid, medium-buffering compounds [21].

The catalyst and medium pH have a decisive effect on the ureo–formaldehyde oligomer structure [22].

It is possible that linear oligomers are formed with amidic groups in the main chain,

$$\text{—NH—CONH—CH}_2\text{—NH—CONH—CH}_2\text{O—} \tag{14}$$

as well as those with tertiary amine groups,

$$\text{—CO—}\underset{|}{\overset{|}{\text{N}}}\text{—CH}_2\text{—}\overset{|}{\text{N}}\text{—CO—NH—CH}_2\text{—} \tag{15}$$

The use of substantial excess amine can lead to the production of oligomers with triazine cycles in the main chain. These oligomers are more stable and show good water solubility [23,24].

Occurrence of the reaction in acidic media has considerable impact on process development in the polycondensation stage [25]. Thus, with pH values of 5.0, 4.9, 4.8, and 4.7, the duration of the polycondensation stage is 1800, 100, 50, and 20 min, respectively.

Production technology of ureo–formaldehyde oligomers currently employs formaldehyde introduced into a reactor heated to 40°C; after this, urotropine is introduced in powder form, until pH = 7.0–8.0 is reached. Then urea is introduced (urea:CH_2O molar ratio = 1.5–1.6). After the latter has been dissolved, an aqueous solution of oxalic acid is introduced [26]. Occasionally, higher temperatures may be required. Thus, in Russia, to obtain water-soluble ureo–formaldehyde oligomers, the first stage of the reaction is carried out at 102°C [27]. Similarly, Nobel Hoechst Chemie in France (Dieppe) recommends that acid condensation be done at the boiling temperature of the reaction medium [25]. A substantial shortening of reaction time can be thus achieved [28].

The molar ratio between reactants plays an important role in ureo–formaldehyde oligomer production. An increased molar fraction of urea results in a lower content of dimethylenetric linkages in the resins. This structural characteristic allows the production of better-quality products through increased stability and diminished content of low molecular weight species [25,28].

In some cases [25,28], a modification of reactant molar ratio is applied during the process. Thus, after the acid stage of condensation is over, the amount of urea introduced is supplemented.

Ureo–formaldehyde oligomer production technology has introduced some catalytic systems that imply the formation of some complex combinations [29]. This has been possible because urea and formaldehyde contain electron donor sites. As acceptors, aprotic acids can be used.

Open pore urea–formaldehyde structures have been prepared by adding phosphoric or oxalic acid to an aqueous solution of a nonetherified urea–formaldehyde resin [30]. Open pore polymers are not foams in the usual sense. Foams are colloidal systems containing a gas as the dispersed phase. The continuous phase may be liquid or solid. Foams possess a very large interfacial area. Conventional foams consist of a porous structure having more or less regular unit cells, either open or closed, which are usually formed by generating a gas in a polymerizing resin matrix. Foam structure can be classified into the following six categories, based on the type of cell structure: closed cell dodecahedron, open cell dodecahedron, interstices between spherical particles, irregular pores formed by filler removal, syntactic filled foams, and small pores between randomly packed chains of spherical particles bonded to one another.

Open pore polymers fall into the last category, because small interconnecting pores exist between randomly packed chains or spheres. Open pore structures are not like conventional foams that have unit cells, because they are composed of agglomerated spherical particles (1–10 μm diameter) bonded to one another in a rigid, highly permeable structure.

The conversion of ureo–formaldehyde resin that is soluble in water to an insoluble open pore urea–formaldehyde material involves the following reactions [31,32]:

Formation of methylene bridges between methylol and amino groups of two neighboring molecules:

$$R{-}NH{-}CH_2OH + H_2NR \rightarrow R{-}NH{-}CH_2{-}NHR + H_2O \tag{16}$$

Formation of ether bridges between methylol groups of two neighboring molecules; both inter- and intramolecular reactions are possible:

$$\begin{gathered} R{-}NH{-}CH_2OH + HOCH_2{-}NH{-}R \rightarrow \\ R{-}NH{-}CH_2{-}O{-}CH_2{-}NH{-}R \end{gathered} \tag{17}$$

Formation of methylene bridges between two methylol groups; here, the reaction is mostly intermolecular, but intramolecular reaction is also possible:

$$\begin{gathered} R{-}NH{-}CH_2OH + HOCH_2{-}NH{-}R \rightarrow \\ R{-}NH{-}CH_2{-}NH{-}R + H_2O + CH_2O \end{gathered} \tag{18}$$

The formation of ether bridges between methylol groups of two neighboring molecules involves the following mechanism, where CH_2OH represents the water-soluble urea–formaldehyde resin [32,33]:

$$\sim CH_2OH + H^+ \rightleftharpoons \sim CH_2{-}\overset{\overset{H}{|}}{\underset{+}{O}}{-}H \rightleftharpoons \sim \overset{+}{C}H_2 + H_2O \tag{19}$$

$$\sim \overset{+}{C}H_2 + \sim CH_2OH \rightleftharpoons CH_2{-}\overset{+}{\underset{\underset{\underset{\xi}{CH_2}}{|}}{O}}{-}H \xrightleftharpoons{-H^+} \sim CH_2{-}O{-}CH_2\sim \tag{20}$$

Foamed polymers based on reactive oligomers have been reviewed by Shutov [34].

Depending on the formulation and reaction conditions, open pore urea–formaldehyde oligomers can be prepared in block, thin-sheet, rod, or powder form. They can be made into soft, resilient, or hard products. The resiliency of open pore urea–formaldehyde oligomers can be improved by incorporating up to 20% of a phenol–formaldehyde comonomer into the composition [35].

Open pore urea–formaldehyde structures have unique properties, and their spherical and pore sizes can be controlled to make them suitable for many applications. Filtration structures, chromatographic columns, porous urea–formaldehyde pigmented polystyrene, smog dispersal agents, moisture retentive fertilizers, fruit coatings, and porous polymer-bound multicomponent corrosion inhibitors have been prepared. Development of technologies based on open pore urea–formaldehyde structures is a distinct possibility.

III. MELAMINE–FORMALDEHYDE OLIGOMERS

The production of melamine–formaldehyde oligomers uses as raw materials formaldehyde and melamine (2,4,6-triamino-1,3,5-triazine). The latter is a crystalline substance, heavily soluble in water (5% at 100°C), and it shows a stronger alkaline character than urea or formaldehyde. From the very beginning, one should note that melamine–formaldehyde oligomer structure depends strongly on working conditions.

Water-soluble melamine–formaldehyde oligomers are produced in two stages: The former takes place in alkaline medium (pH = 7.5–9.0) and yields melamine hydroxymethylene derivatives. At the same time, oligomer formation is avoided. The latter stage achieves hydroxymethylene derivative condensation in acidic medium. A large excess of aldehyde (formaldehyde:melamine molar ratio = 8–9) is used.

Melamine–formaldehyde colloidal solutions are produced by melamine condensation with formaldehyde in aqueous suspension at pH = 4–9 and CH_2O: melamine molar ratio = 3–5. The precondensate thus formed is mixed with an aqueous solution of formaldehyde (5 to 20 mol per mol of melamine). The duration of the process is much more reduced if a Männich base is used as a catalyst [36].

Melamine–formaldehyde oligomers have found numerous applications in the production of pressed and stratified materials for paper, lacquers, and dyestuffs.

Thus, melamine–formaldehyde oligomers for the production of powders required to obtain melaminated plates can be achieved by means of a 30% formaldehyde solution, leading to pH = 8.0–8.5 with a 10% concentration Na_2CO_3 solution. The melamine:CH_2O molar ratio = 1:2.2. The mixture is heated to 75°C. Subsequently, the catalyst necessary for the polycondensation reaction is introduced (phthalic acid monoureide) [36].

Highly dispersed crosslinked melamine–formaldehyde structures have been prepared by Renner [37]. In a typical open pore melamine–formaldehyde preparation, a melamine resin that has a low memelamine:formaldehyde molar ratio (1: 1.6) and that contains 15% sucrose as an additive was converted into an open pore melamine–formaldehyde oligomer by acidification with 7% phosphoric acid (based on resin solid) at 50–60°C.

The reaction of melamine and formaldehyde in nearly neutral solution produces methylol melamine or its low molecular weight polycondensates. Applying the melamine solution, then removing water by drying at elevated temperature, and finally reacting the condensate with heat produces hardened and crosslinked melamine–formaldehyde oligomers that are nonmelting and insoluble.

In a British patent [38], highly dispersed crosslinked melamine–formaldehyde oligomers having good reinforcing properties in natural rubber and synthetic elastomers are described. The reinforcing properties are due to uniform structure and the surface reactivity. In yet another patent [39], open pore melamine–formaldehyde powders made from melamine–formaldehyde oligomers are claimed to be useful as a filler in rubber and as vehicles for pesticides used in the form of a dust. Their use as a thixotropizing agent is also claimed [40]. Processes for preparing cured particulate melamine–formaldehyde cleaning agents using a melamine–formaldehyde resin are described in a U.S. patent [41]. The particles obtained are elipsoidal.

REFERENCES

1. Knop, A., and Scheib, V., *Phenolic Resins and Related Materials*, Khimia, Moscow, 1983, 280 (Russian).
2. Bahman, A., and Miller, K., *Phenolic Resins*, Khimia, Moscow, 1978, 288 (Russian).
3. Jubanov, B.A., Boico, G. I., and Zainulina, A. S., Kinetics of phenol-formaldehyde condensation, *Izv. Akad. Nauk Kaz. SSSR, Ser. Khim.*, 1981, No. 1, C46 (Russian).
4. Demkin, V. M., Ivanov, P. S., and Kuzmina, L. A., Characterization of condensation reactions. I. The reaction of formaldehyde with phenol, *Zur. Prikl. Khim.*, *44*, 1142, 1971 (Russian).
5. Jubanov, B. A., Boico, G. I., and Umerzakova, M. B., Kinetics of condensation processes. I. Condensation of phenol with formaldehyde, *Izv. Akad. Nauk Kaz. SSSR*, 1981, No. 1, 42 (Russian).
6. Virspcha, Z., and Bschezinskii, Ia., *Aminoplastics*, Khimia, Moscow, 1973, 344 (Russian).
7. Barg, E. I., *Technology of Plastic Materials*, GHI, Leningrad, 1951, 656 (Russian).
8. Siling, M. I., *Progress in Science and Technology*, VINITI, Moscow, 11, 119, 1977 (Russian).
9. Komarek, E., U.S. Patent 3,065,039, 1962.
10. De Groote, E., U.S. Patent 3,165,039, 1962.

11. D'Alelio, G. F., U.S. Patent 2,658,885, 1956.
12. Walz, K., U.S. Patent 3,606,988, 1964.
13. Falkehay, S. I., and Bailey, C. W., U.S. Patent 3,870,681, 1975.
14. Osberg, E. V., and Beck, W., U.S. Patent 3,699,173, 1972.
15. Moss, E. K., U.S. Patent 3,953,645, 1976.
16. Moss, E. K., U.S. Patent 3,876,620, 1975.
17. Mange, F. E., U.S. Patent 3,320,208, 1967.
18. Buriks, R. S., and Mange, F. E., U.S. Patent 4,032,514, 1977.
19. Hoechst, A. B., GER Patent 2,351,799, 1977.
20. Miller, H., GER Patent 3,444,203, 1985.
21. Korshak, V. V., *Technology of Plastic Materials*, Khimia, Moscow, 1985, 607 (Russian).
22. Psenchina, V. P., and Molotcova, N. N., *Processing and Technology of Synthetic Resins*, NIIPM, Moscow, 1976, 73 (Russian).
23. Korshak, V. V., Cyclic urea–formaldehyde oligomers, *Plast. Mass.*, 1982, No. 6. p. 47 (Russian).
24. Prokofiev, V. S., *Technology of Plastic Materials*, Publications of Moscow Inst. of Wood Res., 1984, 22.
25. Nobel Chemie Hoechst A.G., GER Patent 2,347,401, 1975.
26. Matvelaschvili, G. C., Romanov, N. M., and Mambisch, E. I., *Progress in Science and Technology*, VINITI, 14, 79, 1981 (Russian).
27. Popova, T. A., *Synthesis of Urea Formaldehyde Resins*, NIITEHIM, 10, 54, 1980 (Russian).
28. Nobel Hoechst Chemie A.G., GER Patent 2,402,202, 1977.
29. Serenkov, V. I., *News in the Chemistry and Technology of Polycondensates*, Nalchik, 1979, 39 (Russian).
30. Salyer, I. O., and Usmani, A. M., Open pore urea formaldehyde resins, *J. Appl. Polymer Sci.*, *23*, 381, 1979.
31. Usmani, A. M., and Salyer, I. O., Open pore polymer structures: A review, *Polym. Plast. Technol. Eng.*, *12*, 61, 1979.
32. Usmani, A. M., and Salyer, I. O., Porous urea/formaldehyde polymers. *J. Sci. Ind. Res.*, *39*, 555, 1980.
33. Patrick, I. G., U.S. Patent 4,203,779, 1982.
34. Shutov, F. A., Foamed polymers, *Adv. Polym. Sci.*, *39*, 1, 1981.
35. Usmani, A. M., Applications of porous urea/formaldehyde polymers, *J. Macromol. Sci.-Chem.*, *A19*, 1237, 1983
36. Rauner, H., GER Patent 2,241,713, 1979.
37. Renner, A., Highly dispersed, crosslinked melamine-formaldehyde polymers, *Makromol. Chem.*, *120*, 68, 1968.
38. Ciba, Ltd., GB Patent 1,029,441, 1966.
39. Ciba, Ltd., GB Patent 1,043,437, 1966.
40. Ciba, Ltd., GB Patent 1,071,307, 1966.
41. Cooley, W. E., and Vanden, P. L., U.S. Patent 3,251,800, 1966.

10

Oligomer Science: Between Tradition and Innovation

Two decisive aspects in the field of science are tradition and innovation. Tradition is the basis, for it is the cumulation of wisdom in the body of knowledge. To know what a subject is all about and to control it creates self-confidence, thus paving the way for innovations. Innovation is the adventure, since the challenge can call into question (or even cause one to lose) one's own scientific identity, gained through tradition.

Persisting in tradition without innovation, however, soon leads to tiresome routine, to the science of yesterday; the longing for new adventures withers and dies. On the other hand, pure innovation harbors the danger of superficiality. The sum of knowledge is immense and growing! Tradition and solid, successful work are honored and admired. Nevertheless, science can only be justified by a challenge, and it demands the willingness to give up long-held classical or traditional views in the attempt to discover new horizons.

Right from its very beginnings, polymer science (including oligomers) was a field on the border between chemistry and physics—that is, between two classical disciplines. Thus, scientists in this field had to dare to innovate. In the meantime, macromolecular chemistry, too, has become a classical discipline, a mature science with all the advantages and handicaps of maturity. Harvest is plentiful, the results are abundant—but one has to ask, where is the future, where are the adventures?

In 1981 Herman Mark asked the question of what direction polymer science was about to take [1]. Is it going in the direction of life science? Is it possible to reduce polymer science, cell biology, and medicine to a common denominator? So from that point of view, polymer science finds itself not only on the border between chemistry and physics, but also between material science and life science. Confronting such questions with the present reality allows us to examine some perspectives on the contribution of oligomer chemistry to the development of future human knowledge.

It is my belief that the future of the chemistry of oligomers will include a deep penetration into life science, including the synthesis of new biomaterials, utilization of oligomers as informational carriers, and their involvement in the living cell's dynamics. The research initiated by Tirell [2] and the discoveries made by Pedersen [3] and Lehn [4] support these assertions. It is, however, necessary to add that the science of oligomers in the near term will make essential contributions to the existing technologies of preparation and processing of high molecular weight polymers. In the following, detailed comments on the aforementioned perspectives will be made.

We consider that the polymer and oligomer industry together with electronics, atomic energy, and the use of planetary ocean resources have represented determining forces of technical progress. The participation of polymer science in this "technical war" calls, however, for numerous and significant changes in the concept and methodology of research in this domain.

Classical technology of polymeric material production has a multistage character. Any product obtained from a polymer (yarn, sheets, materials for aeronautics, parts of a car body) have to pass through three significant stages: monomer, semiproduct, and finite product. These stages represent, in real-world applications, synthesis, purification, and polymerization (or polycondensation) of monomer followed by polymeric material processing (by casting, pressing, extrusion, or injection) with a view to obtaining an object with an adequate shape and appearance.

Classical technology presents some obvious advantages. It allows control of shrinkage in the process of transition from monomer to polymer, and it permits setting the intensity of caloric energy release during polymerization. Yet classical technology has quite large economical drawbacks: huge investments, the need for a large number of qualified workers, and additional transport costs. Another problem raised by classical technology is quality improvement of products. This technology works within the contradiction between quality and productivity on the one hand, and that between quality and technology progress on the other.

Indeed, the overwhelming majority of polymers are solid substances. As a result, the solid semiproduct (powder, grains) that cannot be applied as such must be turned into an object with a determined shape and adequate appearance. This change is performed by melting followed by casting in certain forms. The process is also accompanied by other secondary operations to meet commercial requirements (outer appearance, color, etc.).

It is also interesting to follow the intimate changes of the polymer during the process that yields the finite product. In the solid state, the macromolecule has a given conformation, variable to a low degree, due to rotation, free or hampered, around the links forming the main chain. The whole system is in a state of equilibrium governed by the well-known laws of thermodynamics.

During the melting process, the macromolecules gain freedom of movement. They become more flexible, and intermolecular interactions diminish. The system tends to a new state of equilibrium. To achieve this, a certain period of time is necessary, determined by experimental conditions and the nature of the polymer. From a physical point of view, the system passes from the solid into a liquid state, from which it can adopt desired geometrical shapes. The necessary time to pass from the solid into the liquid state (relaxation time) is of the order of seconds or minutes for small molecules, and grows quite impressively for macromolecules. We add, however, that for polymers the relaxation time is an average quantity; in other words, the entire system does not simultaneously reach the equilibrium state. Some zones in the system remain in a state of quasi-equilibrium, making it impossible to achieve some desired physicomechanical or physicochemical parameters in the material. To reach a state of ideal equilibrium, a slow cooling process is required. This is not possible, because the production of polymer materials is based on technologies the duration of the processes of which is limited. This situation hardly compromises the range of application of polymeric materials. Indeed, not all of these materials need the highest performance characteristics in application. Decorative polystyrene plates do not require a high shock resistance; a car body does not need to be elastic. On the other hand, due to the polymer diversity, manufactured materials can benefit from desired properties achieved via polymer chemical structure. This solution, however, is a palliative with confined application. It is well known that in many cases, optimal properties are found in poorly soluble and infusible polymers. Clearly, performance, under classical processing operations, is heading toward a limit as established by the unwritten law of performance: a good performance is achieved through laborious processing.

The alternative, with great economic promise, is achieving polymer synthesis and processing in one step. Thus, the application of oligomers opened a new stage in the technology of polymers, based on the direct transition of liquid compounds (i.e., monomers or oligomers) into articles (finite products without synthesis, separation, and processing of high molecular weight compounds). In other words, here polymer formation and article production are combined in the same process as in, for example, *reaction injection molding* (RIM) for the production of urethane foams and elastomers (RIM is also referred to in many cases as high-pressure impingement mixing, HPIM, or liquid reaction molding, LRM).

Reaction injection molding processes utilize rapid chemical reaction, allowing the production of solid parts from liquid monomers with reaction times on the order of seconds to minutes.

Work is under way to evaluate potential RIM processes that do not involve urethane or urea products. Epoxy, polyester, polyacrylate, and polystyrene systems all show required reactivity without the evolution of volatiles. Additionally, the anionic ring-opening of caprolactam has been investigated as an avenue for potential polyamide RIM systems [5,6].

Oligomers able to participate in polycondensation reactions (which we shall call polycondensation oligomers, PCOs), especially phenyl–formaldehyde resins, played an important role in the industrial applications of polymers. About 1904 to 1910, due to Bakeland's research work in Belgium and Petrov's in Russia, the first synthetic polymers were obtained. Later, phenol–formaldehyde, carboxylic, and glyptalic resins played an important part in producing materials used in the electronics, varnish, and plastics industries. Due to the liberation of side products, which caused on the one hand intense pollution of the surrounding environment and, on the other hand, numerous difficulties in finite product casting, PCOs could no longer face the competition of products made by means of oligomers able to participate in polymerization reactions (which we shall call polymerization oligomers, PMOs).

Over the last two decades, the yield of PMO-based polymeric materials knew a real boom, which saw a 200% increase of their production facilities (as compared with the 5% increase for PCOs).

What are the advantages of using PMOs in the technology of polymeric material production? Let us recall only few. First, oligomers are easily fusible liquid or solid substances. It is already known that liquids are more "technological" than solids. Let us remember that in metallurgy, in order to obtain profiled products, casting the molten metal into form is preferred to the processing of solid semifinished products. Second, conversion of oligomers into polymers requires a considerably lower number of exothermal stages as compared with direct polymerization of monomers. This also allows for the lighter "organization" of liberated heat elimination. Also related to this is the significant diminution of volume contraction. Third, the use of PMOs eliminates the need to use high pressures and temperatures. This determines an important reduction of costs, opens new vistas for the introduction of automation, reduces risks, and finally, increases productivity. Last but not least, the use of PCOs and PMOs for the production of polymeric materials allows a good guiding of finite product physical and chemical properties. *Thus, oligomers become carriers of properties—they are better carriers of information to the finite products.*

The oligomers' quality of information carrying is not to be found only in technological domains. Several interdisciplinary investigations focus attention on the determining role synthetic oligomers will play in the future, as information carriers in the chemical and physical processes specific to living organisms. In our opinion, the interdisciplinary research represents the key to modern chemistry's successful penetration into both medicine and biology, as a means of stimulating imagination and intuition.

The fascination evoked in human beings by biological processes is ancient. The fascination of simulating these processes and transferring them into technical dimensions has just begun. The field is open, and modern chemistry can offer essential contributions to it. Where to go? At present, material science certainly can profit more from life science than vice versa. A technical analogue for biological membranes mimicking the specificity of their recognition has not yet been found—despite the countless varieties of synthetic polymer membranes already known. However, the liquid membranes may be considered as an example of a synthetic model able to reproduce the transmembrane transport through biological membranes. In these synthetic systems, the linear or cyclic oligomers, with different chemical structures, play the role of "carriers," in the same way that valinomycin and other natural antibiotics facilitate the transport of certain molecular species through the cell membranes of the living organisms [7].

There have been many attempts to mimic the natural prototypes. The breakthroughs are still missing: the knowledge is available, but innovation is a long time coming.

Physicists tend to like "too simplified" models, while the biologists tend to reject all approaches that are not in accord with the "too complicated" prototypes of nature. Thus, organic chemistry and macromolecular chemistry (including oligomer chemistry) can play the important role of mediator. One hint that this might work is given in the contribution by J. M. Lehn [8]. Molecular recognition processes for organic and inorganic compounds are the first step toward simulating highly specific recognition reactions common in biology [9].

The art of synthesizing exceedingly complex oligomers (linear or cyclic) and characterizing them perfectly has reached dizzying heights and opens fascinating opportunities in the area of molecular engineering [10]. But what is it good for? This art certainly will remarkably increase our chemical knowledge—but is it enough? It is not unlikely that, independent of the enormously increasing number of new synthetic compounds, the number of true innovations or innovative compounds has steadily decreased during the last few decades. In his aggressively and optimistically written contribution on "Chemical Needs and Possibilities at the

End of the Century," Frejaques [11] points out the importance of chemical systems. He stresses that, above and beyond the chemistry of covalent bonds, a chemistry of supermolecular systems is now emerging.

The synthesis of new oligomers may lead to functional units that can contribute to the understanding of biological processes, and to the development of new biomaterials. Both tradition and innovation are needed for that. On the border between the different disciplines of science, adventures are waiting. Neither uncritical optimism nor obstructive pessimism are justified. The excitement and the courage to set out for new frontiers and the willingness for close comparison are basic prerequisites for the adventure of science. All the knowledge is available—we only have to learn to use it.

Such a survey of future possibilities for oligomer chemistry cannot ignore the realization that some systems possess the capacity for self-organization. Molecular biology today can understand the central role of polysaccharides or nucleic acids in the life processes. But it has to be pointed out that their functions are based in all cases on molecular mobility and high ordered structure. This is achieved by their incorporation into membranes, by self-organization in solution, or by orientation at cell surfaces. These are properties, the combination of which are typical of liquid-crystalline behavior. Thus, synthetic oligomers with liquid-crystalline properties are self-organizing systems, which are important for material science as well for life science.

Medicine and pharmacology are viewed nowadays as the most fascinating and promising domains for polymer and oligomer applications, from both current and long-term perspectives. Targeting drugs that would ideally ignore the normal part of the body and home in on diseased areas in need of treatment has been for the most part of this century only a little more credible than the philosopher's stone. This situation remained at the beginning of the second half of the twentieth century, even if Paul Ehrlich, a brilliant chemist, foresaw "bodies which possessed a particular affinity for certain organ . . . as carriers by which to bring therapeutic groups to the organ in question" [12].

If we consider the effect of drugs on living system as an interference based on a mutual exchange of information, it is quite obvious that the communication potential of classical drugs is attenuated and primitive. The advent of macromolecular drugs widened the adaptation and pharmacological guidance capacities of drugs.

The polymeric systems that are being explored for pharmacological application include (1) polymer drugs (polymer or copolymers that are physiologically active themselves), (2) drug-carrying polymers (polymers that have active drugs bound to a parent polymer backbone), (3) time-release drug polymers (polymers used for controlled release of the drug; the drug may be encapsulated with water-soluble polymer coatings that dissolve at different rates and release the drug at various times, or it may be embedded into a polymer matrix from which it diffuses at specific rates), and (4) site-specific drugs (polymers that have special chemical groups attached to the polymer carrying the drug; these groups can combine with specific receptor sites on a protein, on the cell surface, or in the lipid areas to achieve specific binding and drug delivery [13].

Binding of drugs to oligomeric matrices, that is, matrices having average molecular weights from a few hundred to a few thousand, is now receiving increasing attention. The results obtainable with oligomeric carriers are not exactly the same as those with high molecular weight carriers. In particular, oligomeric matrices can be absorbed through the gastrointestinal tract and may act as transport agents to facilitate absorption of drugs; the same may be true as far as skin permeability is concerned. Nevertheless, prolonged activity may also be obtained, and toxic side reactions may be reduced. Thus, oligomeric matrices are, in this respect, alternatives to high molecular weight ones.

Many parts of this book are certainly incomplete. This is due partly to a lack of sufficient insight, partly to insufficient data. The reader should, therefore, not look at this contribution as a crystal clear definition of facts and problems, but instead as an impetus to possible developments.

REFERENCES

1. Mark, H., *Angew. Chem. Int. Ed. Engl.*, *20*, 303, 1981.
2. Tirell, D. A., The interface of the biological sciences with polymer science, *ACS Meeting*, Washington, D.C., 1992.
3. Pedesen, C. J., Cyclic polyethers and their complexes with metal salts. *J. Am. Chem. Soc.*, *89*, 7017, 1967.
4. Lehn, J. M., Chemistry of transport processes—design of synthetic carrier molecules, in *Physical Chemistry of Transport Membrane Ion Motion*, Spach, G., Ed., Elsevier, Amsterdam, 1983, 181.
5. Kubiak, R. S., and Harper, R. C., New procedures in processing of polymers, *SPE Antec*, November, 1979, p. 12.
6. Becker, W. E., *Reaction Injection Molding*, Van Nostrand Reinhold Co., New York, 1979.
7. Uglea, C. V., and Zănoagă, C. V., Transport of amino acids through bulk liquid membranes, *J. Membr. Sci.*, *47*, 285, 1989; *ibid.*, *65*, 47, 1992.
8. Lehn, J. M., *Angew. Chem. Int. Ed. Engl.*, *27*, 89, 1988.
9. Lehn, J. M., *Science*, *227*, 849, 1985.
10. Uglea, C. V., and Negulescu, I. I., *Synthesis and Characterization of Oligomers*, CRC Press, Boca Raton, Florida, 1991.
11. Frejaques, C., *Chem. Ind. (London)*, 1985, 780.
12. Elrlich, P., *Coll. Stud. Immunol.*, 2, 42, 1906.
13. Ottenbrite, R. M., *Polymers in Biotechnology*, ACS Symposium Series, Washington, D.C., vol. *362*, 1988, 122.

Index